Prestressed Concrete Designer's Handbook

Prestressed Concrete Designer's Handbook

THIRD EDITION

by

P. W. Abeles
DSc, CEng, FIStructE, FAm Soc CE, MConsE

and

B. K. Bardhan-Roy
BE, CEng, FICE, FIStructE, MConsE

A Viewpoint Publication

Viewpoint Publications

Books published in the Viewpoint Publications series deal with all practical aspects of concrete, concrete technology and allied subjects in relation to civil and structural engineering, building and architecture.

Contributors to Viewpoint Publications include authors from within the Cement and Concrete Association itself and from the construction industry in general. While the views and opinions expressed in these publications may be in agreement with those of the Association they should be regarded as being independent of Association policy.

First edition 1962
12.054 Second edition 1976

12.084 (limp)
12.086 (hardback) } Third edition 1981

ISBN 0 7210 1227 2 (limp)
0 7210 1232 9 (hardback)

Viewpoint Publications are designed and published by the Cement and Concrete Association, Wexham Springs, Slough SL3 6PL

Printed by H. Charlesworth & Co Ltd.

Any recommendations made and opinions expressed in this book are the authors' based on their own personal experience. No liability or responsibility of any kind (including liability for negligence) is accepted by the Cement and Concrete Association, its servants or agents.

PREFACE TO THE THIRD EDITION

Dr Paul W. Abeles, the first author of the book, died soon after the publication of the second edition, and the undersigned has become the sole author for this, and any subsequent editions. Apart from being a co-author for the earlier edition he also worked in close association with Dr Abeles for a number of years and shares his design philosophy. The continuity of the original concept of the work has thus been retained. Dr Abeles was a pioneer in the field of prestressed concrete and indeed the originator of the concept of 'partial prestressing'. The publication of the third edition, in fact, is an appreciation of his invaluable contribution towards the development of prestressed concrete.

On the technical side, a significant factor which has emerged since publication of the second edition is the CEB/FIP Model Code for Concrete Structures. Virtually all the chapters have been updated to accommodate the Model Code recommendations parallel with those of the British and American Codes.

Sections on lightweight aggregate concrete have been substantially revised in the light of further research and experience. Specific design values have been suggested for lightweight aggregate concrete for all aspects of serviceability design criteria, similar to those for dense concrete.

Chapter 16 which deals with fire resistance has been re-written to incorporate the current thoughts on analytical approaches for predicting fire resistance of structural members, consistent with the principles of limit state design. Simplified design curves and formulae in line with the guidance given in the joint publication of the Institution of Structural Engineers and The Concrete Society, *Design and Detailing of Concrete Structures for Fire Resistance*, have also been included.

Finally the author wishes to express his thanks to Mr Jan Bobrowski, Dr Andrew Short, Dr T. Barta and colleagues in Messrs Jan Bobrowski and Partners for their continued assistance and help.

B. K. Bardhan-Roy

(March 1981)

PREFACE TO THE SECOND EDITION

The first edition of this book was already sold out in 1969 but the authors postponed preparing a new edition until the British Unified Code, which has become CP 110, and the new American Building Code 1971 had appeared.

However, Mr Turner, who works both as a consulting engineer and as a senior member of a contracting firm, was occupied in these capacities to such an extent that he was unable to continue in carrying out the work. Mr Bardhan-Roy joined as the second author and he and the first author completed the revision and amplified also those parts already previously revised. For future editions, Mr Bardhan-Roy will become the sole co-author. His co-operation made it possible for a great number of new examples to be incorporated since he is a partner of the firm Jan Bobrowski and Partners, consulting engineers in Great Britain and Canada (Calgary). Quite a number of new examples of precast members (double-T beams) and other examples such as the precast roof structure at Doncaster Grandstand have been included. Special problems as the use of lightweight concrete and the fire resistance of prestressed concrete with which the first and second authors have been concerned for some time have been investigated.

As a new international notation has been accepted, almost all symbols have been changed to correspond with these international signs with the few exceptions pointed out. However, specific stresses have been signified in the same manner within the framework, as was done in the first edition.

The Chapters on materials (2) and on structures (3) were greatly extended based on research results which have been established since the first edition was written some time ago, with special reference to the first author's experience during his activity as visiting professor at Duke University, North Carolina and at the University of Kentucky.

A new Chapter 5 on criteria for analysis and design has been incorporated based on the new conditions of limit state design which have been introduced in CEB and CP 110. In all examples the new SI units, introduced in these two specifications, have been given in addition to the imperial units which are still generally used in the United States and by the older engineers in Great Britain. Furthermore, the equivalent values in the metric system have been added.

Chapters 6, 7 and 8 deal with losses, analysis and design respectively, more or less based on the first edition. A very comprehensive Chapter 9 contains examples of simply supported beams; this has been greatly expanded and contains constructions in normal weight and lightweight concrete and includes an unsymmetrical section. Chapter 10 deals with end blocks and Chapter 11 with composite members, the importance of the latter being emphasized and extended by new examples. It is also shown in one example that with shallow parts of sections different stresses may be obtained when certain tolerances are considered. Chapter 12 is devoted to 'deflection' and Chapter 13 to design considerations which particularly deal with bending up of cables and the provision of shear reinforcement based on the limit state of collapse; also the question of lateral stability is considered in this chapter.

Chapter 14 is devoted to statically indeterminate structures, more or less on the basis of the first edition except that a new example is shown in which precast members are combined to statically indeterminate structures. Chapter 15 deals with other applications of prestressing such as bridge structures, masts, piles, shells and piles and tanks.

A completely new Chapter 16 covers design for satisfactory fire resistance. This problem is particularly important because it lies in the hands of the designer to increase the fire resistance considerably by the provision of satisfactory details.

The second part of the book contains design charts and additional data. The examples are given in Imperial as well as in the new SI and kilogrametric units. Many examples were computed by calculating machines and the results comprise many digits sometimes two decimals. Obviously such numerical refinement is unnecessary and slide rule results would suffice in general. Designers should not assume that this apparent accuracy reflects the exactitude of the stresses computed which in fact depends on the correctness of the assumptions made. On the other hand, it should be borne in mind that good detailing and construction are as important as good analysis.

The authors would like to refer to the senior author's two-volume work *Introduction to Prestressed Concrete* which deals with many questions in greater detail. Ten simple principles are also set out on page 481.

The modern designer should appreciate that the time has passed when it was a 'must' for a prestressed concrete structure to be under permanent nominal compression. This claim advanced 35 years ago by the then leaders in prestressed concrete construction is considered essential by some authorities even today. However, within the FIP, discussion is proceeding with a view to the acceptance of limited cracking under service load provided that there is no danger of corrosion. In fact it has been suggested that reinforced concrete and fully prestressed concrete are the extreme conditions, with partially prestressed concrete as the general solution. As in life, so in design, the middle of the road solution may be more satisfactory than the extreme ones.

Finally, the authors wish to express their gratitude to Dr T. Barta, Mr Jan Bobrowski and to the staff of Jan Bobrowski & Partners, Consulting Engineers, for their continuous assistance during the preparation of the manuscript.

P.W.A – B.K.B (November 1975)

CONTENTS

Part 1

Part 2

Acknowledgements

Thankful acknowledgements are due to:

(1) The British Standard Institution (BSI) for kind permission to reproduce extracts from BS 4757, 4486, 4625, 5896 and CP 110.
Complete copies of the Standards can be obtained from the BSI, 2 Park Street, London W1A 2BS.

(2) Comité Euro International du Béton (CEB) for permission to use some particulars from the CEB/FIP Model Code for Concrete Structures and the Manual of Lightweight Aggregate Concrete. The French edition of the Model Code can be obtained from CEB Office, 6 rue Lauriston. F 75116 Paris, and the English edition from Cement and Concrete Association, Wexham Springs, Slough SL3 6PL.

(3) The Prestressed Concrete Institute (PCI) Committee on Fire Resistance Rating, especially to Mr Gustaferro for allowing inclusion of tabulated data from unpublished code provision of 1971.

(4) Institution of Structural Engineers (England) for permitting reproduction for some design graphs and tables from the joint Institution of Structural Engineers/Concrete Society publication 'Design and Detailing of Concrete Structures for Fire Resistance'.

(5) McGraw Hill Book Co. for permission to use some formulae and tabular data from the book '*Theory of Elastic Stability*' by S. P. Timoshenko.

NOTATION

1. Strength*, Stress and Strain

f_k	Characteristic strength generally
f_{sk}	Characteristic strength of steel
f_{ck}	Characteristic strength of concrete
f_{su}	Ultimate strength of steel
f_y	Yield stress of steel
f_{sp}	Proof stress
f_{cu}	Cube strength of concrete
f_{cyl}	Cylinder strength of concrete
f_{prism}	Prism strength of concrete
f_{ct}	Direct tensile strength of concrete
f_{cr}	Flexural tensile strength of concrete (rupture)
γ_m	Partial safety factor of the strength of material, e.g. $\frac{f_k}{\gamma_m}$
f	Stress generally
f_s	Steel stress
f_c	Concrete stress
$f_{s\,max}$	Maximum steel stress
$f_{c\,max}$	Maximum concrete stress
f_{sh}	Shear stress in concrete (homogeneous section)
v	Shear stress in concrete (cracked section)
f_{mn}	Concrete stress in a homogeneous section at position 'm' and loading 'n' (e.g. f_{bt} = stress at bottom face at transfer) 'm' will be 'b' or 't' depending on the position being bottom face or top face. Similarly 'n' will be 't' at transfer, 'w' at working load**
E_s	Elastic modulus of steel
E_c	Elastic modulus of concrete
α_e	E_s/E_c = Modular ratio
ϵ	Strain
ϵ_e	Elastic strain
ϵ_c	Creep strain
ϵ_s	Shrinkage strain
η	Relaxation

*According to International Notation strength is to be denoted by σ and stress by f. In the notation both strength and stress are denoted by f.
**Compressive stresses (+), tensile stresses (−).

$\bar{f}_{xy}$	Permissible stress in concrete for 'x'; 'c' = compressive 't' = tensile for 'y'; 't' at transfer 'w' at working load ('t' = tensile and 't' at transfer: Permissible tensile stresses are (+))

Note: Thus concrete tensile stresses are to be denoted as $-\bar{f}_{tt}$ or $-\bar{f}_{tw}$.

2. Loading and Bending Moment

G_k	Characteristic dead load
g_k	Characteristic dead load per unit length
Q_k	Characteristic superimposed load
q_k	Characteristic superimposed load per unit length
w_k	Characteristic wind pressure per unit length
γ_g	Partial load factor for dead load
γ_q	Partial load factor for superimposed load
γ_w	Partial load factor for wind load
Total Load	$\gamma_g G_k + \gamma_q Q_k + \gamma_w W_k$
l	Span
M_g	Bending moment due to dead load
M_w	Bending moment due to working load
M_u	Ultimate bending moment
M_R	Minimum resistance moment
V	Shear force generally
V_u	Ultimate shear force

3. Cross-Section

h	Entire height
d	Effective depth to tensile steel
d'	Depth to compressive steel
d_{ft}	Depth of top flange
d_{fb}	Depth of bottom flange
b	Breadth
b_w	Breadth of web

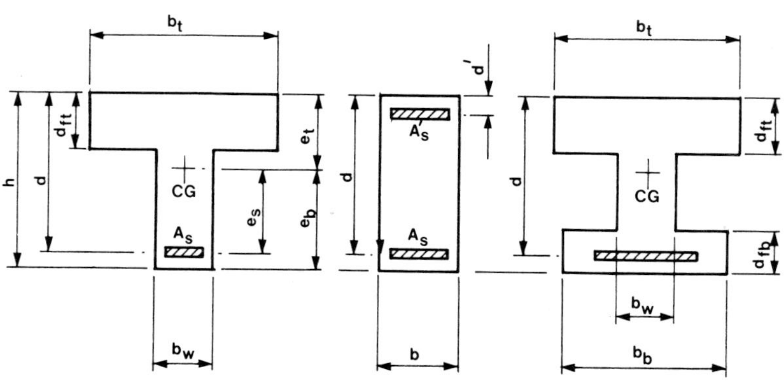

b_t	Breadth of top flange
b_b	Breadth of bottom flange
e_s	Eccentricity of tendons from CG
e_b	Eccentricity of bottom fibre from CG
e_t	Eccentricity of top fibre from CG
A	Area generally
A_c	Area of concrete
A_s	Area of tensile steel
A'_s	Area of compression steel
A_{sn}	Area of non-tensioned steel
A_{ps} or A_p	Area of prestressing tendons
A_{sv} or A_v	Cross-sectional area of shear reinforcement (2 legs of a stirrup)
I	Second moment of area generally (Moment of inertia)
I_x	Second moment of area about x-axis
I_y	Second moment of area about y-axis
I_{xy}	Second moment of area about any other axis
i	$\sqrt{I/A}$ = Radius of gyration
J	Polar moment of inertia
S	First moment of area related to y (general)
Z_t	First moment of area (Section modulus) of top flange
Z_b	First moment of area (Section modulus) of bottom flange
ρ	$\frac{A_s}{bd}$ = reinforcement ratio

4. Prestressing

P	Prestressing force
P_t	Prestressing force at transfer
P_e	Effective prestressing force
$P_{e\,min}$	Effective prestressing force after maximum possible losses
also P_{et}	Effective prestressing force at definite time t
f_{pi}	Initial prestress in steel
f_{pt}	Prestress in steel at transfer
f_{pe}	Effective prestress in steel
Δf_{pt}	Losses of prestress at transfer
Δf_{pe}	Loss due to elastic shortening
Δf_{pc}	Loss due to creep
Δf_{pr}	Loss due to relaxation of steel
Δf_{ps}	Loss due to shrinkage
Δf_{pf}	Loss due to friction
$\Delta f_{p\,total}$	Total losses

5. Ultimate Load Conditions

F_{cu}	Ultimate force in concrete
$F_{c\,max}$	Maximum force in concrete
F_{su}	Ultimate force in steel
$F_{s\,max}$	Maximum force in steel
z	Lever arm

6. Various terms

T	Temperature
t	Time
K, k	Constants having dimensions
ν	Poisson's ratio
μ	Coefficient of friction
ϕ	Creep factor
ϕ_0	Shear strength reduction factor for lightweight concrete
ϕ_1	Capacity reduction factor
γ	Weight density

PART 1

CHAPTER 1

INTRODUCTION

1.1 Principles of prestressed concrete

Prestressing may be defined as the purposeful and controlled creation of permanent stresses in a structural member, before the full dead and live loads are applied, so as to counteract all or part of these loads. It serves two main purposes: to improve the resistance of the member to the dead and live loads (service load) and to modify the behaviour of the member or structure in such a way as to make it more suitable for its intended purpose.

There are significant differences in principle between reinforced concrete and prestressed concrete. In the design of reinforced concrete beams it is assumed that the tensile strength of the concrete is negligible, and the tensile forces created by the bending moments are resisted by reinforcement, to which the forces are transferred by bond. Cracking and, to a large extent, deflections are virtually irrecoverable in ordinary reinforced concrete, with relatively poor bond between the steel and concrete, though with high-strength concrete and good bond a substantial degree of recovery may take place. The reinforcement usually exerts no forces on the member on its own account. In prestressed concrete, on the other hand, the primary purpose of the prestressing steel is to apply a force to the concrete, either by bond or by means of special anchoring devices; hence the whole of the concrete can be made to act structurally. The steel required to produce the prestressing force is thus used actively to preload the member and cracking and deflections are recoverable to a higher degree.

However, it should be noted that under overload conditions, as soon as the flexural tensile strength of the concrete has been exceeded, prestressed concrete behaves in a manner similar to reinforced concrete, and at the ultimate load or collapse condition of a flexural member, the tensile and compressive resistances required to withstand these conditions are the same for both reinforced and prestressed members. Moreover, as the prestress becomes zero, the behaviour of a prestressed member becomes almost the same as that of a reinforced concrete member; thus, there are basic similarities, as well as basic differences, between reinforced and prestressed concrete.

The idea of prestressing is by no means new; but it became practicable only when it was recognized that the initial prestressing force was reduced appreciably by losses, and hence high-strength steel (and therefore high initial tensioning stresses) were essential. Prestressing is applied mainly to flexural members, but it is also used with advantage in direct tensile members, ties, tanks, pipes, and steel structures.

1.2 Advantages of prestressed concrete

The possibility of avoiding permanent cracks is one of the major advantages of prestressed structures. The improved durability, compared with reinforced concrete, is also useful particularly in members exposed to corrosive atmospheres or aggressive ground conditions, and in marine structures. The advantages for liquid-containing structures, pipes, and tanks are obvious. In addition, much less steel is required, since the weight of the high-strength prestressing steel is only a fraction of the weight of reinforcement which it replaces; the cross-section of the member can be smaller, since the whole of the concrete is put to structural use; and the resistance of beams to shearing and flexural cracks is considerably increased. It is therefore possible to design longer spans or cantilevers using comparatively shallow and slender sections. However, this may tend to enhance the susceptibility of the member to flutter because of reduced natural frequency of vibration and in certain cases lead to dynamic instability either due to wind excited oscillation or vibration of some other origin unless adequate damping is introduced or the stiffness is increased.

By controlling the amount of prestress a structure could be made either rigid or flexible without affecting its ultimate resistance. A flexible structure is obviously more resilient and can absorb considerable energy before failure due to impact and thus show improved performance under seismic and dynamic conditions. Examples of such structures include fender-piles of wharfs or jetties. On the other hand, a very rigid structure is more suitable to resist heavy vibration (for example turbo-foundations).

1.3 Economics of prestressed concrete

A prestressed member, when compared with an equivalent reinforced concrete member, requires less concrete, and about $\frac{1}{5}$ to $\frac{1}{3}$ of the amount of steel; but the difference in initial cost is not proportional to the difference in weights of materials. Steel as well as concrete in a prestressed concrete member, needs to have high strength and high quality. The unit costs of such steel and/or concrete are higher than those of the materials required for reinforced concrete construction. Formwork or moulds may be more expensive, and the additional cost of the prestressing operation itself must be considered. However, in general, there is little difference between the initial costs of reinforced and prestressed members, provided that large numbers of precast prestressed units are required. On the other hand, the indirect savings which accrue from prestressing are often substantial and should be taken fully into account. They include the reduction or total avoidance of maintenance and the longer working life due to the greater durability of the material (arising from the absence of permanently open cracks), the reduction in the dead weight imposed on the supporting members and foundations, the achievement of longer spans and fewer supports, and reductions in the structural depths of members (or longer spans with the same depth).

1.4 Principles of design

The permanent stresses which are created by prestressing can be produced in many ways. The most suitable methods for concrete structures subjected to bending or tension are those in which tensioned steel wires, strands, cables, or bars (commonly called 'tendons') are used to apply a compressive force to the concrete. It is necessary to provide sufficient steel in the tensile zone to ensure suitable resistance at ultimate load (see Chapter 3), either all or part of the steel is tensioned; the eccentric prestressing force so obtained produces stresses of the required nature – that is, the tensile zone is subjected to a prior compression. The initial tensioning stress in the steel is reduced, in the course of time, by various 'losses' (see Chapter 6) and the residual prestressing force which remains after all of these losses have occurred is termed the 'effective' prestressing force. At the time when the prestress is applied the critical factor which limits its magnitude is the compressive strength of the relatively young concrete; hence it is important to assume that the *losses which occur at transfer are the minimum*, thereby ensuring that, at this stage, the greatest possible prestressing force has been allowed for in the design. On the other hand, the critical factor at working load is usually the tensile stress which can be permitted in the concrete; for this purpose the smallest possible 'effective' prestressing force should be considered, and it should therefore be assumed that the *maximum losses occur after transfer.* The magnitude of the effective prestressing force required under working load conditions can be calculated for any given load and working stress. In other words, a particular set of conditions calls for a particular force at a particular eccentricity.

It is also important to note that the losses which occur are largely independent of the magnitude of the initial stress in the prestressing steel, being generally in the order of 20 000 to 40 000 lbf/in^2 (1400–2800 kgf/cm^2; 138–276 N/mm^2*) in the steel and 300 to 600 lbf/in^2 (21–42 kgf/cm^2; 2.07–4.14 N/mm^2) in the concrete. Hence, in order that the remaining prestress should still be a substantial proportion of the initial prestress, steel and concrete of high strength and quality are essential; typical initial stresses are 150 000 to 200 000 lbf/in^2 (10 500–14 000 kgf/cm^2; 1050–1380 N/mm^2) in the steel, and 1800 to 3000 lbf/in^2 (126–210 kgf/cm^2; 12.4–20 N/mm^2) in the concrete. The latter value may be exceeded provided the concrete strength is high enough.

1.5 General methods of prestressing

(i) Pre-tensioning. When the prestressing steel is tensioned against independent anchorages before the concrete is placed around it, the process is termed 'pre-tensioning'. This term applies to the steel only, and it is therefore incorrect in the strict sense to speak of 'pre-tensioned beams' or 'pre-tensioned concrete', though these terms are often used in this book in connection with examples. In this case the concrete is bonded to the tensioned steel; when the concrete has attained the necessary strength the tension at the anchorages is gradually released and the steel is severed at the ends of the member. This application of the prestressing force to the concrete is termed the 'transfer' of the prestressing force. The full prestressing force is transferred to the concrete over a certain 'transmission length', the extent of which depends on the surface condition and cross-sectional profile of the steel, its diameter, and the strength of the concrete. The transmission length is also influenced by the 'wedging' effect at the severed end of the tensioned steel, which is due to the tendency of the steel to revert to its original diameter from the reduced diameter which it possesses when tensioned (Figure 1.1). These considerations are discussed in Chapter 4. It is only necessary

*According to ISO it has been agreed to replace 1 N/m^2 by Pascal. Thus 1 N/mm^2 = 1×10^{-6} N/m^2 = 1×10^6 Pascal

1 kp (kilopond) = 1 kgf is used in Germany

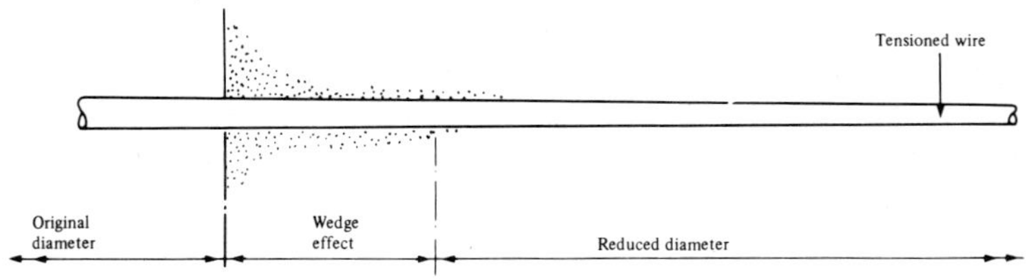

Figure 1.1 Wedge-effect at de-tensioning

to note here that the essential advantages of pre-tensioning over other methods of prestressing are:

(1) Excellent and reliable bond can be obtained between the tensioned steel and the concrete along the entire length.
(2) As pre-tensioning is commonly carried out in a permanent or temporary factory, satisfactory supervision is more easily obtained than on the site. It is also easier to ensure that the concrete is properly cured.

When many similar members are required pre-tensioning is best applied by means of 'long-line' prestressing beds on which the members are cast end-to-end around continuous tendons tensioned against anchorages at each end of the beds. The tendons are severed between the members when the concrete has hardened. The tensioning force should be gradually released at the anchorage and the members should be able to slide along the bed, to avoid the development of internal forces; a sudden application of the prestress may cause an undesirable shock load. It is also possible to produce single members, either on shorter beds or by anchoring the steel to the ends of the moulds. If the condition of the surface of the steel is such that the concrete is not properly bonded to it, slipping will occur, and adequate transfer of the prestressing force to the concrete will not be obtained. In this event the member cannot be considered to be prestressed, and its resistance to bending will be much less than that of an ordinary reinforced concrete member. Consequently, only approved steel, and concrete of a certain minimum strength, should be employed in the pre-tensioning process. The concrete strength should preferably not be less than 5000 lbf/in^2 (350 kgf/cm^2; 34.5 N/mm^2) before transfer as generally considered to be necessary.

As the prestressing force should not be transferred until the concrete has obtained this minimum strength it is necessary either to allow time for hardening or to provide special facilities, such as steam curing, for the rapid attainment of strength. Special care should also be taken to ensure that the concrete is kept moist during the hardening period in order that no shrinkage cracks may develop before the prestress is transferred.

Prestressing may be applied to monolithic concrete members or to members composed of numbers of separate elements. In the latter case it is preferable to place the steel in grooves in the elements. The tensioned steel is embedded in concrete or mortar and the joints between the separate elements are filled with concrete or mortar at the same time; the prestressing force is transferred when this has hardened sufficiently. The prestressed concrete member thus obtained has monolithic properties to some degree, though not usually to the same extent as a member in which the whole of the concrete is cast around the tensioned steel.

(ii) Post-tensioning. When the steel is tensioned after the concrete has hardened the process is termed 'post-tensioning'. The prestressing force is transferred to the concrete by means of end anchorages at the time when the steel is tensioned (see Chapter 4). There are many systems for post-tensioning, a number of which are described in Part 2. The steel must not be bonded to the concrete before tensioning or the attainment of the necessary elongation would be impossible, and contact between the steel and the concrete should be reduced as much as possible. The tensioned steel may be individual wires or strands, cables composed of separate wires or strands, or alloy bars, and may be placed in ducts, sheaths, open grooves or recesses in the concrete, or outside the concrete altogether. If, however, unbonded tendons (i.e. tendons covered with grease and plastic coating) are used, they can be cast into the concrete. Some reduction of the total tensioning force occurs at sections distant from the tensioning end due to friction between the steel and the concrete or sheath even if the steel is straight throughout its length. The magnitude of this reduction depends on the type and profile of the steel and the condition of the surface of the duct, groove, or sheath, and also on the total angle through which the steel is bent.

The term 'post-tensioning' is frequently mis-applied in the same manner as the term 'pre-tensioning'. In the correct sense this term should be applied only to the steel, and not to the member or the concrete which, of course, is compressed.

As with pre-tensioning, members prestressed with post-tensioned steel may either be monolithic or composed of separate elements. In the latter case, they may be assembled without mortar joints, if the end faces of each block fit perfectly together; otherwise thin caulked mortar joints of sufficient strength, or concrete joints of adequate thickness (preferably in the form of stiffners with ducts for threading tendons) must be provided before the prestressing force is applied. Whether or not such members act monolithically depends entirely on the details of design and the workmanship.

BONDED TENDONS

After the steel has been anchored at both ends it is possible to obtain some degree of bond between the steel and the concrete by forcing grout into the space between them. Unless unbonded tendons are used, the efficiency of the bond obtained in this way depends on the spacing of the wires, bars, or cables and, when grouted, on the effective grouting; the arrangement of the spaces between the wires or strands comprising the cables when cables are used, and that between the concrete and the steel when single strands or bars are used. If the steel is surrounded by a sheath, the condition of the surface of the sheath will also affect the efficiency of the bond. In some instances when sufficient space is available and the surface of the sheath, if any, is sufficiently rough, it is possible to obtain good bond. In all cases the attainment of satisfactory bond is dependent on the quality of the workmanship. The bond of post-tensioned steel is of great importance when conditions at ultimate load and the factor of safety against failure are considered (see Chapter 3), and also with regard to cracking, as described in the next paragraph. If satisfactory bond is obtained it is also possible with some systems to remove the anchorage and transfer the prestress solely by bond. Either grout or mortar may be used.

NON-BONDED TENDONS

If the spaces between the post-tensioned steel and the concrete or sheath are intentionally left without grout, or if the grout is insufficient or unreliable, or if unbonded tendons are used the behaviour of the member under excessive load will be entirely different from the behaviour of a member with efficiently grouted steel, and different conditions will apply at failure. A member with unbonded steel develops a small number of wide cracks when it is overloaded, in the same way as a member reinforced with mild steel bars of large diameter. Moreover, although the elongation of the steel caused by the tensioning force is nearly uniform along its whole length in both cases, with bonded steel the further elongation due to superimposed load occurs only at the cracks; with unbonded steel it remains uniformly distributed.

The provision of non-tensioned bonded reinforcement improves the distribution of cracks in this case, and reduces their widths; it also improves the behaviour at ultimate load. If the prestressing steel is not grouted, precautions must be taken to prevent corrosion; inert gases, such as nitrogen, vapour phase inhibitors and certain types of grease have been successfully used.

Particular problems occur in the application of prestressing to floor slabs and flat slabs when these are cast in place; the relatively small continuous spans necessitate the use of undulating tendons and the extra operations involved in grouting such tendons add to the construction time and cost of the structure. To overcome these drawbacks, a type of tendon has been introduced in which the wires forming each cable are coated with grease and the complete cable is wrapped with a paper or plastic sheath; the tendons, complete with their sheaths, are preformed and are cast into the concrete. The grease acts both as a lubricant, to reduce friction, and as a protection against corrosion. Care must be taken to ensure that no grease is present at the anchors; that the anchors are fully embedded within the concrete; and that sufficient non-tensioned bonded reinforcement is added to control cracking under overload conditions.

(iii) Other methods of prestressing. The structures discussed in this book are subjected mainly to bending and are generally prestressed by means of tensioned steel. It is possible, however, to prestress a structure by the use of expanding cement. This was suggested by M. Henri Lossier, but the practical application of the method has advanced to some extent only in Russia. An expanding cement has been developed and its properties investigated at the University of California; the results of this work indicate that there are some possibilities for future development, although both the magnitude of the prestress obtained in the concrete, and the strength of the concrete itself, appear to be still rather limited.

Another method is the jacking-up or anchoring-down of the supports of continuous or framed structures, and the jacking apart of compressive members such as arches. Flat jacks are preferably employed for this method. The losses in these cases will differ from those occurring in structures prestressed with tensioned steel, and re-stressing is often necessary.

The tensioning of steel bars by means of heat is another possibility. The bars are preferably heated by means of an electric current which is passed through them, the elongated steel being secured by independent anchorages and then cooled to develop the required tension. The force in this pre-tensioned steel is then transferred to the concrete in the usual way. This method was introduced on a large scale in Russia. In general, however, the repeated passage of electric currents through prestressing wires or strands can cause rapid electrolytic corrosion, and this practice should normally be avoided.

The use of dead-weight or gravity loading to obtain the required prestress is also a possibility which has been used for many years. The vertical prestress in the walls of cylindrical tanks can be partly obtained in this way.

The possible use of glass fibre instead of steel has also received some consideration. This method is used for wrapping pipes, but the practical application of this material to the prestressing of beams seems to be even more remote than the use of expanding cement. Research has shown that the resistance of glass fibre reinforcement in a concrete beam after cracking occurs is rather limited. In addition, it has a limited resistance to fire and is open to damage by sabotage when it is accessible; thus no use can be made of its resistance to corrosion.

Lastly, mention should be made of a system in which the bottom flanges of high-strength rolled steel joists are encased in concrete at a time when the joists have been artificially loaded (preflexed), the prestress being applied to the concrete when the loads are removed. The development of visible cracks is delayed, and a better distribution of cracks is obtained, if secondary reinforcement is provided close to the outer edge of the concrete. This system offers a member of smaller depth and limited deflection, although of doubtful cost advantage.

1.6 Types of prestressed concrete

The behaviour of prestressed concrete is outlined in Chapter 3; it is sufficient to note here that there is a wide difference between the conditions which apply at working load and those at failure, as indicated in the simplified load-deflection diagram in Figure 1.2. In an under-reinforced section (that is a section in which the steel is the weaker part at failure, and yields before the concrete crushes in the compressive zone) there are three stages. In Stage 1, the resistance of the whole section is available; the member deflects in an elastically rigid manner until cracking occurs (in uncracked members) or until joints or previous cracks open or re-open (in non-homogeneous or cracked members). During Stage 2, in which the resistance of the tensile zone is not available, the deflection remains elastic but the slope of the load-deflection curve changes, and the member becomes more flexible. Within both these stages, the deflection is almost completely recoverable. In Stage 3, the member gradually yields; this stage does not occur in over-reinforced sections, which undergo a sudden compressive failure within Stage 2, before the steel begins to yield. During Stage 1, prestressed concrete behaves as a completely homogeneous material, with fully recoverable deflections; in Stage 2 the deflections are still reversible, but the behaviour is otherwise similar to that of reinforced concrete; in Stage 3 the behaviour is entirely like that of reinforced concrete.

When prestressed concrete was first introduced, the presence of a permanent compression at the 'tensile' face was considered to be essential. For the last few decades, however, designs have been based on two different conditions; namely, that within the first range the behaviour of the member, with regard to cracking and deflection, should be satisfactory under working load or service load conditions, and that the ultimate resistance of the member, at failure or collapse, should provide a known and definite safety factor over the working load. From these conditions the concept of 'limit-state' or 'limit-stage' design has been

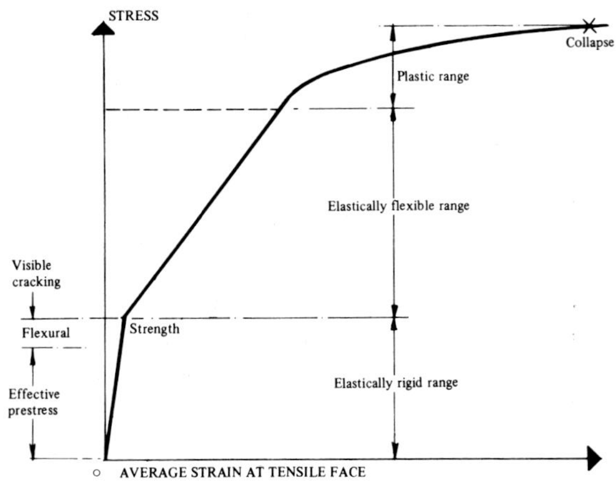

Figure 1.2 Simplified load-deflection diagram of prestressed concrete

developed by the Comité Euro International du Béton, and is incorporated in CP 110(1)* and also in CEB/FIP Model Code(2). In CP 110 three classes of prestressed concrete are defined. These agree very closely with the proposal put forward by the first-named author in 1950(3) and incorporated in the First Report on Prestressed Concrete issued by the Institution of Structural Engineers in 1951(4).

'Classes' of prestressed concrete are sometimes mistakenly identified with the classes or standards of work. Perhaps for that reason in CEB/FIP Model Code and the ACI Building Code(5) the term class has been omitted and the division is confined to 'uncracked' and 'cracked' categories.

Figure 1.3 shows comparison of classification in the First Report (1951) with that in CP 110, Model Code and ACI Building Code. As can be seen from this figure the classification in the First Report is far more comprehensive than in the others. Of special interest is the subdivision of class 3 (or cracked category) into 'type (ii) and type (iii)'. While type (iii) allows *permanent* cracks albeit of limited width, the type (ii) allows cracks only under the maximum service load which can occur rarely and be of short duration. No crack is allowed under permanent (or frequently occurring) service load.

The property of almost complete recovery of deflection which prestressed concrete possesses ensures that any such cracks close again completely under normal service loading; hence this type of design represents, in the authors' view, the most suitable and rational solution, as the factor of safety against collapse is related to the abnormal service load. The design for the normal service loading can therefore be limited to that required for Class 1 structures, and for the abnormal service loading to Class 3. For all classes, however, the conditions required at ultimate load are the same; thus the required resistance of the steel at failure is the same, but the magnitude required for the effective prestressing force depends on the loading at which complete counteraction of the tensile bending stress is required. The computation of the prestressing force required

*References are given at the end of each chapter.

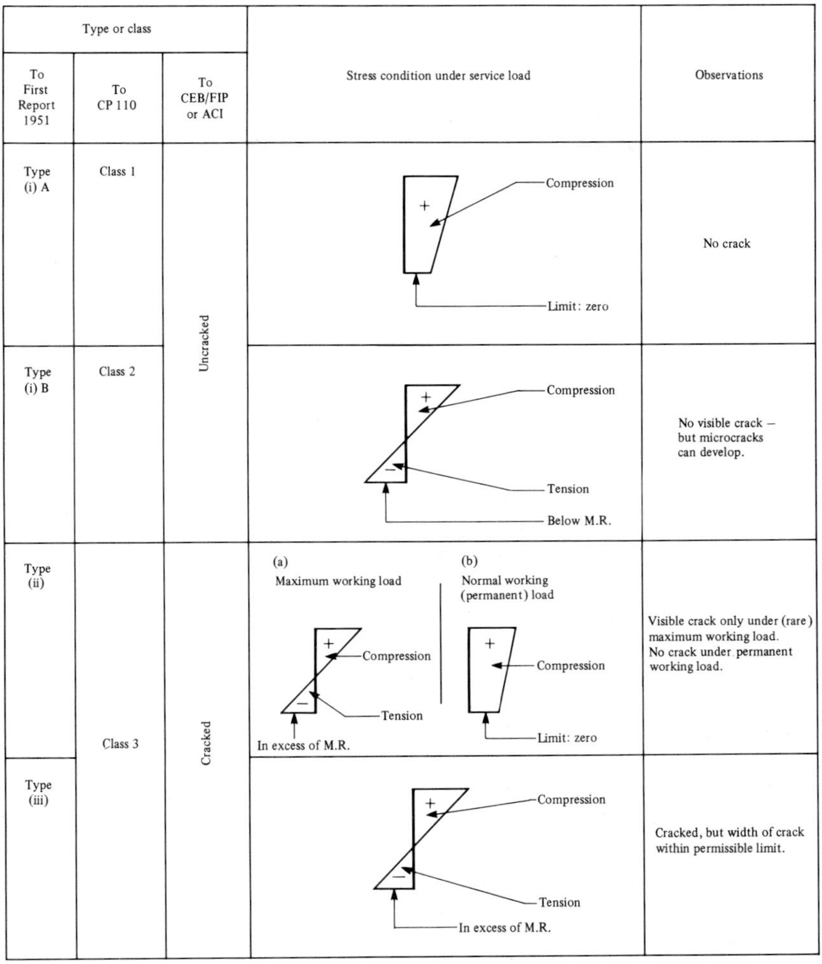

M.R. = Modulus of Rupture

Figure 1.3 Classification of prestressed concrete

for members of Class 3 or cracked category is based on hypothetical tensile stresses in a homogeneous section, which have been shown by experience to produce cracks of permissible widths and deflections of permissible magnitudes, and these values are specified by the ACI Building Code[5] and in CP 110 (See also Figure 1.4).

This method is rather simple to apply and has a great attraction for the designer. It should, however, be borne in mind that such limiting stresses depend on so many various factors that in each individual case these may have to be determined by the designer. Siriaksorn and Naaman[6] showed that the fictitious tensile stresses may range from 435 lbf/in^2 (30.58 kgf/cm^2, 3 N/mm^2) to 3.335 lbf/in^2 (234.50 kgf/cm^2, 23 N/mm^2) to cause the same crack width (0.008 in, 0.2 mm).

Another approach is based on calculations of 'crack width' and 'deflection' of

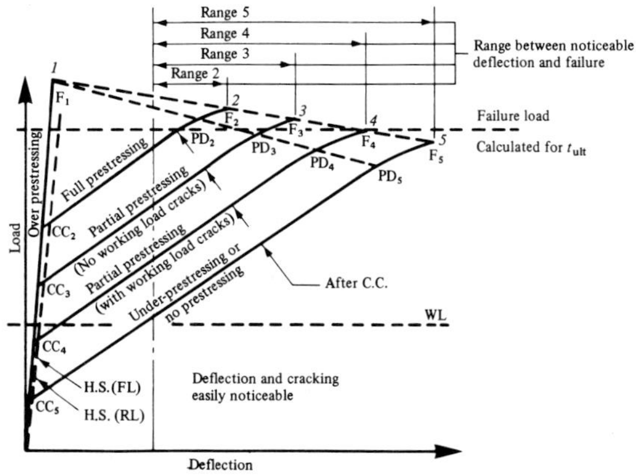

C.C.	Commencement of cracking	P.D.	Permanent deflection
F.	Failure	W.L.	Working load
H.S.	Homogeneous section	F.L.	First loading
M.O.R.:	Modulus of rupture	R.L.	After repeated loading
P.C.	Prestressed concrete		

Figure 1.4

a member and limiting these to specified maximum values. In this case great difficulties arise in the design as the formulae suggested to date for predicting 'crack width' or 'deflection' are not only very complicated, but also appear to be rather unreliable on most occasions.

There are some schools of thought who are of the opinion that the definition of Class 1 structures (i.e. no tensile stress under service load) could be widened to allow limited tensile stresses up to the formation of microcracks provided the maximum possible losses have been taken into account and shrinkage cracks before prestressing have been avoided.

It is interesting to note that even Freyssinet, who strongly believed that Class 1 structure was the only solution for prestressed concrete, later said that a tensile stress as high as 50 kgf/cm^2 (710 lbf/in^2; 4.9 N/mm^2) could be permitted in a bridge if this loading occurred only rarely[7].

REFERENCES

1. BRITISH STANDARDS INSTITUTION. CP 110. 1972 Parts 1–3. *Code of Practice for the structural use of concrete.* London.
2. CEB/FIP. *Model code for concrete structures.* Comité Euro International du Béton, 1978, p. 348.
3. ABELES P. W. Further notes on the principles and design of prestressed concrete. *Civil Engineering and Public Works Review.* Vol. 45, No. 529, July 1950, pp. 443–445.
4. INSTITUTION OF STRUCTURAL ENGINEERS. *First report on prestressed concrete.* Institution of Structural Engineers. September 1951, p. 285.
5. ACI COMMITTEE 318. *Building Code Requirements for Reinforced Concrete.* (ACI 318–77). American Concrete Institute, Detroit. 1977.
6. SIRIAKSORN, A., and NAAMAN, A. E. *Serviceability based design of partially prestressed beams.* PCI Journal, Vol. 24, No. 3. May/June 1979. Part 2: Computerised design and evaluation of major parameters.
7. FREYSSINET, E. *The birth of prestressing.* London, Cement and Concrete Association, 1956. Translation No. 59, pp. 1–38.

CHAPTER 2

MATERIALS

A – STEEL

2.1 Types of steel

Steel for prestressed concrete must have high tensile strength and adequate ductility. These qualities are found in (a) carbon or alloy steel, hot rolled, but otherwise untreated; (b) cold worked steel, which is drawn or deformed, and preferably tempered; and (c) hot rolled and tempered steel.

Carbon or alloy steel has a carbon content not greater than 1 per cent and this is mainly responsible for its high tensile strength. Alloying elements such as manganese, nickel and chromium may also be added to improve the mechanical properties of the steel, and various heat treatments have a beneficial effect; these treatments make use of the fact that if steel is heated to a temperature higher than about 850°C (1550°F) (termed the 'transformation temperature'), its final structure and the extent to which its properties are improved depend on the rate of cooling. If the steel is cooled slowly from its transformation temperature, the treatment is termed 'annealing'; if it is allowed to cool from the transformation temperature at its normal rate, the treatment is termed 'normalizing'. If, on the other hand, the steel is suddenly cooled from above the transformation temperature to room temperature by immersion or 'quenching' in oil, the hardness and brittleness of the material are appreciably increased. Quenching is usually followed by a tempering process in which the steel is reheated to about 400°C (750°F) and allowed to cool in air; this reduces the brittleness of the steel. If the steel is rapidly cooled from above the transformation temperature to about 450°C (850°F) and then allowed to cool slowly to room temperature the process is termed 'patenting', and has an effect similar to that of quenching and tempering. The term 'stress relieving' is used to describe heat treatment for a prolonged period at about 260°C (500°F) or a short period at about 500°C (950°F). The term 'stabilizing' denotes heat treatment at about 400°C (750°F) combined with a tensile stress of about 65 per cent of the ultimate strength of the steel.

Cold working of steel increases its strength, and is mainly carried out by drawing wire through a series of dies, with progressive reductions in the diameter of each die, and consequently of the wire. Rolling may also be used to produce the same result. Rolling, whether hot or cold, enables the steel to be deformed or indented, if required.

It is usual to apply heat treatment to all prestressing steels except those of natural hardness, and secret or proprietary processes are often used. Ordinary oil quenching is generally considered to be unsatisfactory. Other methods, termed 'mar tempering', are used and cooling is often carried out in lead, salt or oil baths.

Although the effects of the foregoing processes are known qualitatively, the actual properties of any steel can be determined only by tests. It is essential that sufficient satisfactory data should be available before any type of steel is used for prestressing.

2.2 Strength of prestressing steel

The practice of specifying a minimum strength for prestressing steel has been superseded, in recent British Standards and Codes of Practice by the concept of a 'characteristic strength', defined as that value below which not more than 5 per cent of test results fall. Typical values are 210 000–240 000 lbf/in^2 (14 800–16 900 kgf/cm^2; 1450–1660 N/mm^2) for wire: 240 000–260 000 lbf/in^2 (26 900–18 300 kgf/cm^2; 1660–1800 N/mm^2) for strand; and 150 000 lbf/in^2 (10 550 kgf/cm^2; 1040 N/mm^2) for alloy bar.

2.2.1 Stress–strain relationship

An ideal stress–strain diagram for prestressing steel is shown in Figure 2.1, which meets the following requirements.

(1) It is imperative to have a high tensile stress which must be accompanied by only a small amount of creep. This is achieved if the permanent elongation at the working stress is small, and the type of steel for which the stress–strain diagram is linear for a large proportion of the ultimate load is used. This property is measured by the 'proof stress', which is defined as the stress which produces a certain permanent deformation (usually 0.2 per cent but sometimes 0.1 per cent) on first loading, and a steel which is suitable for prestressing should have a high proof stress (Figure 2.2).

(2) It is also most desirable that an ultimate elongation of appreciable magnitude should be obtained in order to reduce as much as possible the chance of sudden fracture; this may occur, for example, with piano wire, which has a very small elongation at failure. Prestressing wire and strand have a minimum elongation of between 3 per cent and 5 per cent, which is quite sufficient with satisfactory bond; the value for alloy bars is about 10 per cent.

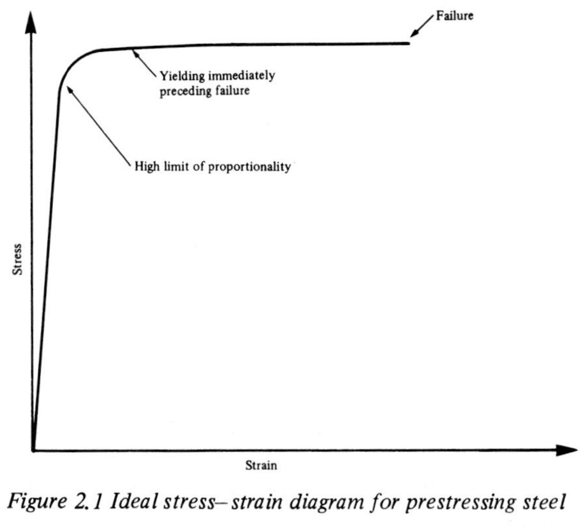

Figure 2.1 Ideal stress–strain diagram for prestressing steel

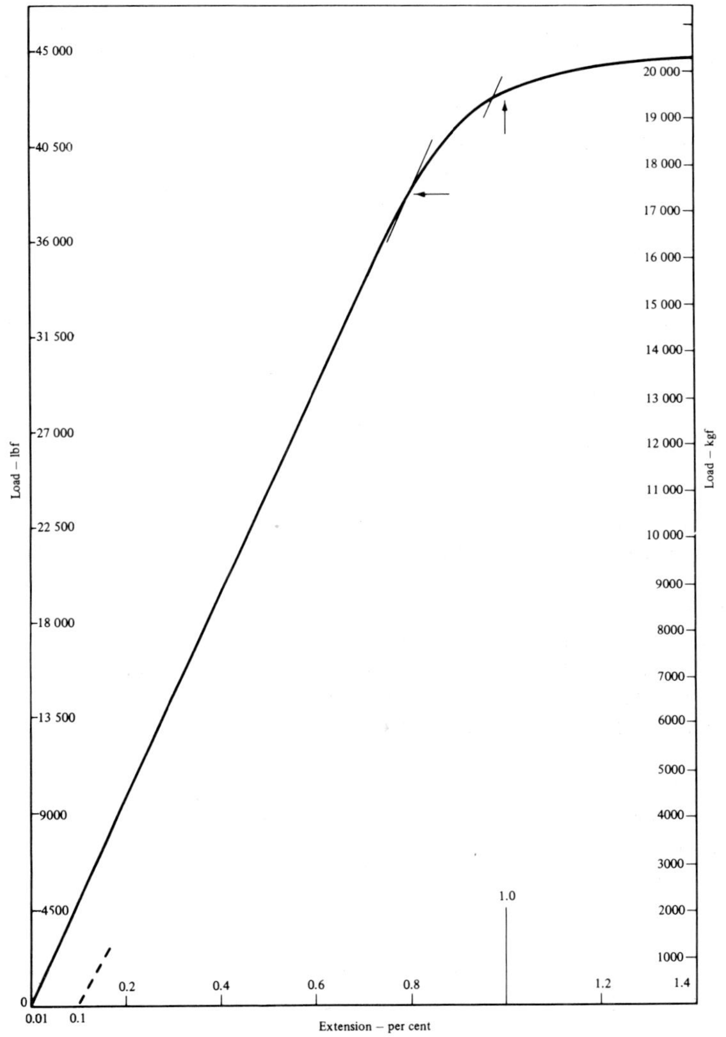

0.5 in. (12.70 mm) diameter Dyform strand
0.01% off-set 38.250 lbf (17.350 kgf)
0.1% off-set 42.800 lbf (19.414 kgf)
Load at 1% ext. 43.100 lbf (19.550 kgf)
Breaking load 48.000 lbf (21.772 kgf)
Modulus of elasticity
27.88×10^6 lbf/in^2 (19.602 kgf/mm^2)
Area = 0.174 in^2 (112.25 mm^2)

Figure 2.2 Load-extension diagram (proof load)

The stress–strain diagrams for various types of steel in Figure 2.3 indicate that the ultimate elongation tends to decrease as the ultimate strength increases. It is clear, therefore, that piano wire is not entirely suitable for prestressing, despite its high proof stress, as its ultimate elongation is very limited. On the other hand, mild steel and deformed bars, which have a large ultimate elongation, are unsuitable because of their low yield point or proof stress. Figure 2.3 also shows that a distinct yield point occurs in low-alloy bars, and this influences the ultimate strength of structures with bonded steel in which the steel is the weaker part and failure is initiated by its excessive deformation in some cases by its fracture.

The modulus of elasticity for prestressing steel depends on the type of steel employed, and values should be obtained from the supplier of the steel. Typical values are 25×10^6 lbf/in^2 (1.76×10^6 kgf/cm^2; 0.173×10^6 N/mm^2) for low-

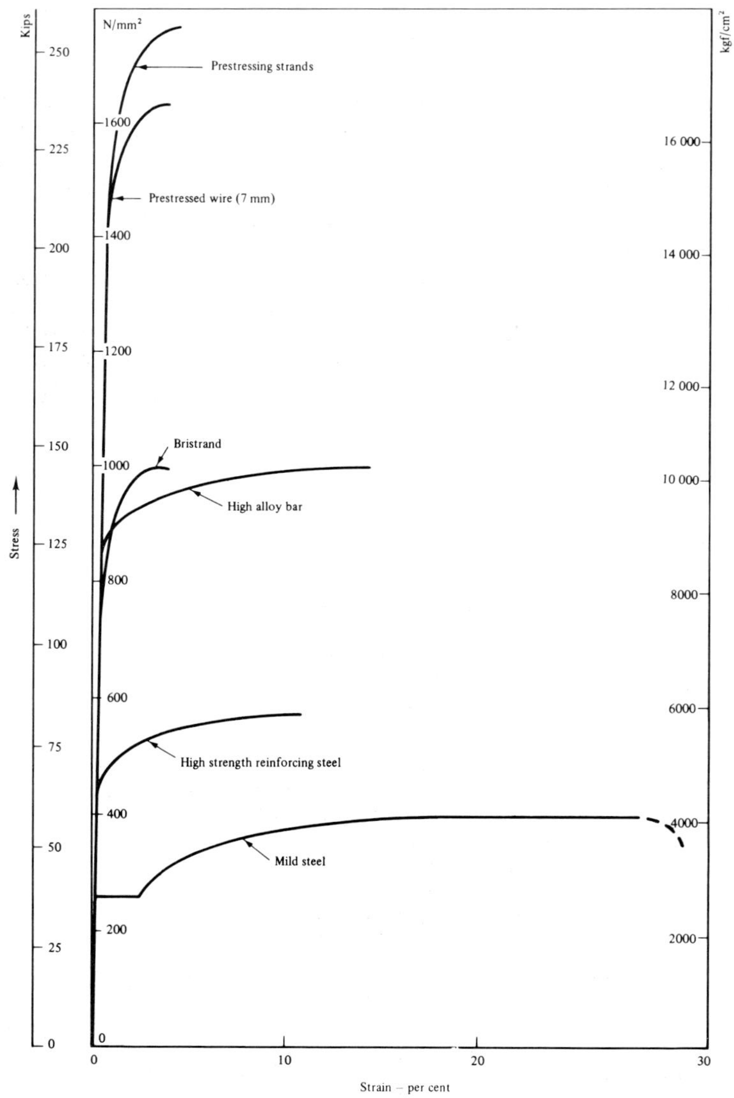

Figure 2.3 Stress – strain diagram for various steel

alloy bars, 28×10^6 lbf/in^2 (1.97×10^6 kgf/cm^2; 0.194×10^6 N/mm^2) for carbon steel wires, and between 23.5×10^6 and 29×10^6 lbf/in^2 (1.65×10^6 kgf/cm^2; $0.163 \times 10^6 - 0.2 \times 10^6$ N/mm^2) for strands. A typical load–extension curve for strands remains linear for only about a half of its length, and a typical 0.2 per cent proof stress is between 85 per cent and 95 per cent of the breaking load.

2.2.2 Resistance to fatigue

In a prestressed concrete structure the tensioned steel is subjected to sustained loads of considerable magnitude, and to a further repeated loading whenever the live load is applied at short intervals of time. The effect of this repeated loading depends on the number of times the load is applied, and is much less severe in structures with well-bonded steel than in structures with non-bonded or poorly-bonded steel.

The fact that repeated loading influences the ultimate strength of a material is not a new discovery; it was known to Wöhler in 1870[(1)]. Figure 2.4 indicates the relationship between the number of cycles of applied load and the maximum stress sustained for three materials (after Moore)[(2)]. The maximum stress is termed the 'endurance limit', or 'fatigue strength', and is defined as the maximum stress that can be attained without failure when the material is subjected to a given number of repetitions of loading. It is dependent on the number of cycles of loading and on the range of the applied cyclic stress. At one time an endurance limit based on one million repetitions of load was considered to be sufficient, but to-day two or more million repetitions are frequently applied. For any particular number of load cycles, for example 10^6, the fatigue resistance of steel

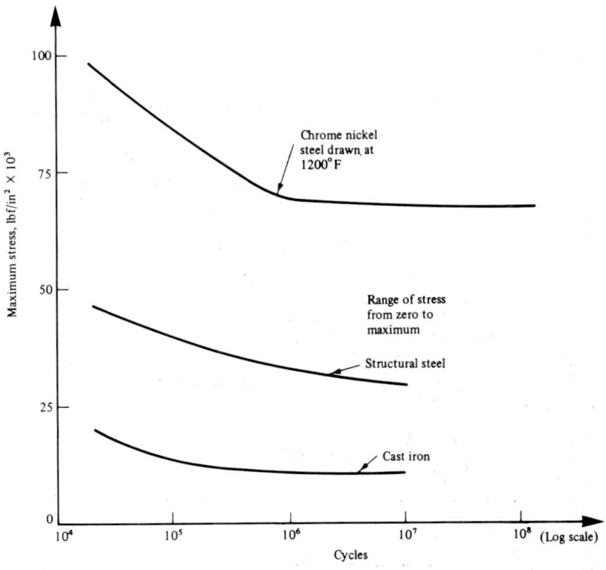

10 000 lbf/in^2 = 700 kgf/cm^2
= 70 N/mm^2

Figure 2.4 Endurance limit for iron and steel

is generally indicated by a 'Goodman' diagram[3] (Figure 2.5), which shows maximum, minimum and mean stresses; in this case, obtained from tests on axle steel[1]. If the lower stress is zero and only tensile or compressive stresses are applied, the range of stress sufficient to cause failure after 10^6 cycles is usually about 50 per cent of the stress at which static failure occurs. When the mean stress is zero (that is, equal tensile and compressive stresses are alternately applied), the range increases to two thirds of the static failure stress, each separate stress being one third of this value. When the lower limit of stress is half the upper limit, the stress range reduces to one third of the static failure stress. This is indicated in Figure 2.5 – Goodman diagram. With prestressing wires and strands, the ranges shown in Figure 2.6 are obtained[4]. In this case, the ratio between the upper limit of the range and a lower limit of zero stress becomes 0.3 instead of 0.5. Clearly, for prestressing tendons only tensile stresses need be considered, but for non-tensioned steel the effects of some compressive stress must also be taken into account as indicated in the figure.

Recent investigations for 10^7 load cycles on strands have produced even lower stress ranges, of about 10 per cent of the static failure stress[5].

The endurance limit may be exceeded occasionally without causing adverse effects. In tests carried out by French[6] on a material with an endurance limit for ten million repetitions of 36 000 lbf/in² (2550 kgf/cm²; 250 N/mm²), this limit was still reached despite occasional stresses of 55 000 lbf/in² (3875 kgf/cm², 380 N/mm²) applied 40 000 times, 45 000 lbf/in² (3180 kgf/cm²; 310 N/mm²) applied 100 000 times, and 40 000 lbf/in² (2880 kgf/cm²; 280 N/mm²) applied 800 000 times. Figure 2.7 presents these results so that a 'zone of damage' is shown between the endurance-limit line and the 'damage' line derived from the foregoing results.

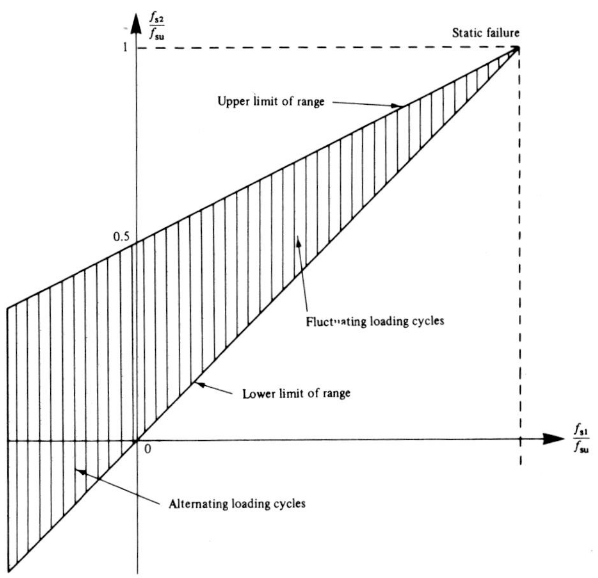

Figure 2.5 Goodman diagram for steel

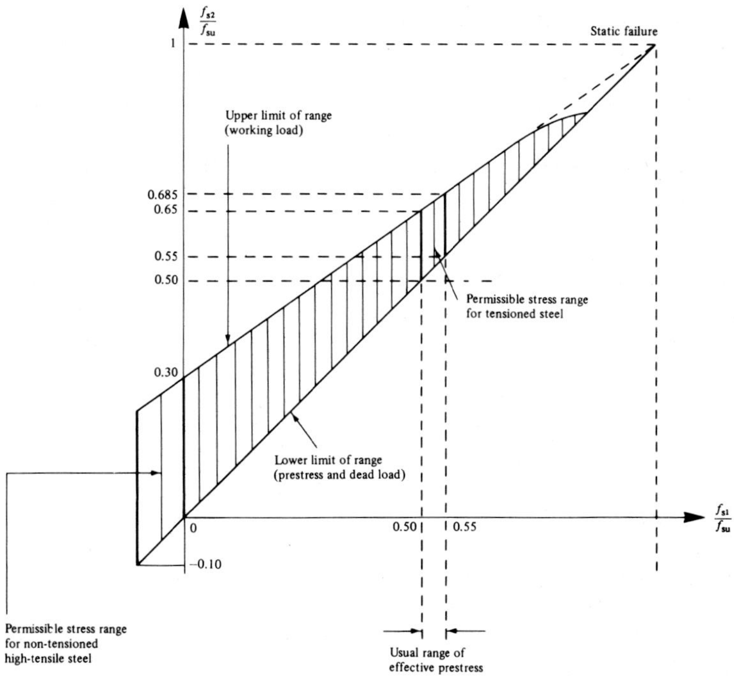

Figure 2.6 Goodman diagram for prestressing steel

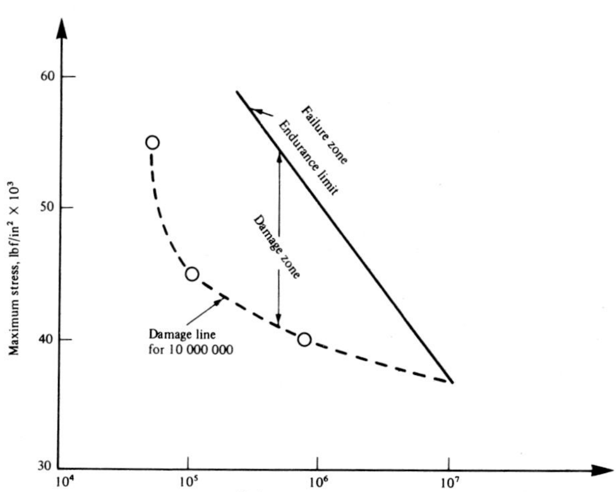

$$10\,000\ \text{lbf/in}^2 = 700\ \text{kgf/cm}^2 = 70\ \text{N/mm}^2$$

Figure 2.7 Zones of damage and failure

The preceding notes assume that the cyclic loading is applied continuously. It has been established that rest periods have a beneficial effect on the subsequent elastic behaviour of the material, and Wilson[7] has shown that the fatigue strength of mild steel is also improved if the rest periods are not shorter than thirty minutes.

It can normally be assumed that the endurance limit of cold-drawn wire for two million repetitions lies between 55 per cent and 70 per cent of the ultimate strength when the range of cyclic stress lies between zero and the endurance limit, and that the limit may be between 66 per cent and 75 per cent of the ultimate load for a much smaller range of fatigue stress. With strands or indented wires the resistance to fatigue is reduced. Certain alloy steels have a lower endurance limit, and the permissible range of dynamic stress which is available at the effective prestress is therefore rather small to withstand fatigue.

Other effects, termed 'stress corrosion' and 'corrosion fatigue', may occur when the tensioned steel is exposed to air. These may lead to failure at a lower load than the ultimate load, and Figure 2.8 (due to Judge[3]) illustrates this. Only tensioned steel which is entirely free and exposed to corrosion is subject to these types of attack. Cases of stress corrosion have occurred with hot-rolled wire, before the present types of prestressing steel were developed. In one instance, however, a failure which was initially attributed to stress corrosion was found on investigation to have been due to the addition of calcium chloride to the concrete when hot-water curing was applied[8]. Tests by Evans[9] have since shown that the use of calcium chloride combined with steam curing may lead to rapid corrosion of steel in concrete, whether it is tensioned or not. It should be pointed out that there is generally no danger of stress corrosion if there is no corroding medium.

2.2.3 Creep and relaxation

Creep may be defined with sufficient accuracy as the change in elongation which occurs over a period of time as a result of sustained loading. When considering prestressing steel it is necessary to determine the property known as 'relaxation',

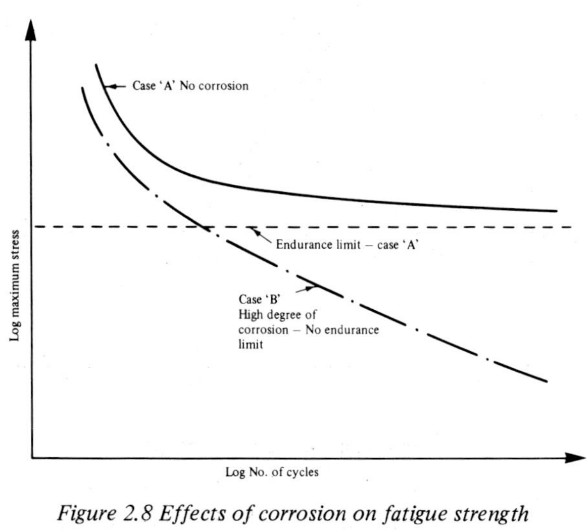

Figure 2.8 Effects of corrosion on fatigue strength

which is the reduction in tensile stress that occurs as a consequence of creep throughout the length of the wire or bar when the length is kept constant. (Secondary effects due to the small change in length caused by creep of the concrete are ignored in the design of prestressed concrete.)

It was previously noted that creep is relatively small when the stress–strain relationship of the steel is suitable. The magnitude of the creep, however, depends entirely on the treatment the steel undergoes; two types of steel with very similar stress–strain diagrams may differ appreciably in their extent of creep (or relaxation). It is therefore important to obtain accurate data, relating the magnitude of the creep to the applied stress, for any steel used in prestressed concrete. In general, the creep increases as the tensile stress increases. The relaxation depends on the quality of the steel, and may be between 4 per cent and 10 per cent of the original stress when the stress due to tensioning is 70 per cent of the ultimate stress; with some modern steel the relaxation may be smaller even than 4 per cent. It also depends on the form in which the steel is used; strands typically have larger relaxation than wires or bars. For stabilized strands (including Dyform strands) the manufacturers specify low relaxation, between 1 and 3 per cent of the tensile stress, when stressed to 70 per cent of the strands. The values now given in various British Standards are noted in Chapter 4. There is no objection to the use of a steel with a greater creep provided that it is taken into account.

2.2.4 Effect of high temperatures

An alloy steel which is hot rolled but otherwise untreated (and consequently of limited strength) undergoes a reduction in strength when its temperature is raised above 350°C (650°F) but regains its original strength when it cools. Cold-worked steels and heat-treated steels suffer a permanent reduction in strength under similar conditions, and if the temperature exceeds 400°C (750°F) this reduction is considerable. When considering the resistance of a prestressed concrete structure to fire, it is therefore important that the cover of concrete should provide sufficient insulation to delay the heating of the embedded steel (See Chapter 16). It is also necessary to ensure that no accidental heating of cold-worked steels takes place before stressing. Welding of prestressing steels is not permissible in any form – except that occasional welds in the wires comprising stranded cables are sometimes permissible.

2.2.5 Effect of low temperatures

Prestressed concrete has been used for the storage of cryogenic liquids at temperatures down to –183°C (–297°F). Tests on wires and alloy bars have shown that the strength increases at these low temperatures by about 5 to 10 per cent and the yield point increases by up to 20 per cent. The impact strength, from notch tests, is much reduced. The coefficient of thermal movement is 10.8×10^{-6} per °C (6×10^{-6} per °F) between 24°C (75°F) and –100°C (–150°F); and 10.2×10^{-6} per °C (5.7×10^{-6} per °F) between 24°C (75°F) and –157°C (–250°F). Tests on strands appear to give similar values.

2.2.6 Corrosion of prestressing steel

In reinforced concrete the steel must be bonded to the concrete in order to ensure a proper transfer of stress, and since cement-based materials provide good protection against corrosion, it is necessary to ensure only that the concrete is properly compacted and that the cover is sufficient. The same is true of pre-tensioned prestressing steel.

With post-tensioning, it is usual to fill the tendon ducts with grout after completion of tensioning; this provides a bond, and also inhibits corrosion. While care should be taken to ensure that no excess water remains in the duct, it has been found that the presence of small pockets of water is not a corrosion hazard since the chemical activity soon exhausts itself. Non-bonded post-tensioning steel should always be protected even if the non-bonded condition is only temporary. Galvanizing has been employed; in this case there is the possibility that a breach of the coating may cause electrochemical conditions leading to rapid corrosion, so that careful handling is advisable. Greases, surface coatings, and gases such as nitrogen, have been tried, with limited success. A combination of greasing and plastic coatings (usually PVC or polyethylene) is commercially available, and this has proved entirely successful.

Apart from atmospheric attack, corrosion has sometimes been caused by water pockets containing high proportions of calcium chloride, and by the earthing of electric currents through tendons, during construction. There is evidence to suggest that when corrosion occurs, it may take place at a more rapid rate in a highly-stressed tendon, an effect which is sometimes termed 'stress corrosion'.

Non-corroding steels (such as stainless or high-nickel steels) are available; in general, their high cost makes their use economically unattractive for prestressing, but in some special cases (particularly with external cables) this relative disadvantage may still be acceptable for the achievement of an overall economy. However, even with stainless steel there is a possibility of stress corrosion at elevated temperatures (about 90°C, i.e. 194°F)[10,11].

There have been a number of instances of failure due to corrosion. They have occurred in only a very small proportion of the total number of prestressed structures constructed, and the cause has almost always been found to lie in malpractice (such as insufficient cover, bad grouting, omission of grout) during design and/or construction. It should be borne in mind that prestressing steel is less 'forgiving' in this respect than ordinary reinforcement and appropriate precautions should be taken throughout all stages of the work.

B – CONCRETE

2.3 Concrete

Concrete is a heterogeneous material of a most complicated nature, the properties and qualities of which vary widely; no attempt is made here to describe the specific properties or the problems involved in the production of high-strength concrete and reference is made to other publications[12,13]. However, the designer has to specify the strength required and has also to take into account the variations in strength under various types of loading, their relationship with the modulus of elasticity, and such physical properties as shrinkage and creep, all of which affect the deformation of concrete. Obviously the effects of ageing of the concrete on these properties should also be considered. These aspects are briefly discussed in the following.

2.3.1 Type of cement and mix

Ordinary and rapid-hardening Portland cements are generally used for prestressed concrete, the latter being called 'early strength cement' in the United States. The very early achievement of high strength is usually important, and the use of high alumina cement (which has this property) is sometimes advantageous. However, it should be adopted only in cases where the concrete manufacturer has direct experience in the production of this type of concrete and is reliable. A very low water/cement ratio (i.e. less than 0.35) and a very good compaction ensuring a minimum strength of 9000 lbf/in^2 (630 kgf/cm^2; 62 N/mm^2) at 24 h should be employed and the concrete should not be exposed to humid and warm conditions; chemical and physical changes leading to a marked decrease in strength may occur as a result of the conversion of the hydrated cement to a more porous form. For the same reason there is a reduction in the strength at working load, compared with that at transfer (which is the opposite of the case with Portland cement) and allowances should be made for this at ultimate load conditions. Any contact with alkalis which may result in alkaline hydrolysis should be avoided. In normal circumstances Portland cement concrete can be bonded to high alumina cement concrete and vice versa; but under continuous exposure to water, alkali can be transferred from Portland cement concrete to high alumina cement concrete with deleterious results.

In order to obtain a concrete of suitably high strength and quality, a low water/cement ratio should be adopted; this in turn requires a high degree of compaction. Admixtures are sometimes used to increase the workability with a relatively low water/cement ratio; the properties of the aggregates themselves also affect the workability. The mix usually consists of graded aggregates, sand and cement. Gap-graded aggregates, as suggested by Stewart(14), offer significant advantages, though compaction becomes more difficult. A mix comprising single-sized aggregates with mortar result in greater uniformity and strength, so that the modulus of elasticity is also more uniform and other long-term characteristics are more consistent. This is important in connection with the mass production of repetitive units, which are intended to have uniform deformation(15).

With single-sized aggregates the surface area is greatly reduced as compared with graded aggregates and consequently under equal conditions a lower water content is required resulting in higher strength and lower shrinkage and creep.

The highest concrete strength could be obtained by a mix using single-sized aggregates and low water content combined with phased vibration (i.e. low frequency vibration at the beginning to move the concrete to fill the mould and high frequency vibration afterwards to consolidate the matrix (i.e. mortar) in the voids of the aggregates).

2.4 Strength of concrete

Most of the codes of practice currently in use in the United Kingdom have specified minimum strengths for concrete, but in CP 110 the concept of a 'characteristic strength' is employed. This is defined as that value below which not more than a certain proportion (5 per cent in this case) of the test results fall and not any individual result by more than 15%. CEB/FIP Model Code also accepted this concept.

2.4.1 Compressive strength

The compressive strength of concrete increases with age as indicated in Figure 2.9. The shape of the curve depends partly on whether ordinary or rapid-hardening (high-early-strength) cement is used, and is also influenced by the magnitude of the strength. In most countries it is usual to accept the crushing strength of cubes as a measure of the compressive strength of concrete, but a closer agreement between the test results and the actual strength of concrete in place is obtained if the tests are made on cylindrical or prismoidal specimens, the most reliable being those obtained from cylinders. The relationship between the strength of cubes and cylinders or prisms is complex. It is sometimes assumed that the relationship is linear, the strength of the cylinder or prism being assumed to be 80 per cent of the strength of a cube of equal area, but if this were correct there would clearly be no necessity to use cylinders or prisms. In fact the ratio of the strength of cylinders or prisms to the strength of cubes depends on the composition of the concrete and the dimensions of the cylinder or prism, and may vary between 0.6 and 0.9; for concretes made with the same aggregate the ratio increases with strength. In most cases it is between 0.67 and 0.80.

It should be noted that the results obtained from static tests on cubes, cylinders, or prisms may require to be modified if a structure is subjected to sustained or cyclic loading. For example, tests in Switzerland[(16)] showed that the strength of prisms subjected to two million cycles of repeated compressive loading were reduced as indicated in Figure 2.10. If the normal strength of a prism is denoted by f_{prism} and the 'endurance limit' is defined as the maximum stress that can be resisted without failure at two million repetitions of loading, the results obtained varied, according to the mixture of concrete used, between $0.58 f_{prism}$ and $0.72 f_{prism}$ when the range of variations extended from zero to the endurance limit, and between $0.75 f_{prism}$ and $0.85 f_{prism}$ when the lower limit of the loading was half the endurance limit. The limit for sustained loading varied between $0.85 f_{prism}$ and $0.95 f_{prism}$. Figure 2.10 is based on average values of the foregoing tests and clearly indicates that the endurance limit is reduced when the range of cyclic loading increases.

2.4.2 Tensile strength

The curve showing the increase in tensile strength of concrete with time is quite different from that for compressive strength (Figure 2.9). In this case a reduction in strength due to differential dehydration takes place when the concrete

Figure 2.9 Compressive and tensile strength of concrete

hardens, and only a small increase occurs as the concrete ages. If the concrete is properly cured or stored in water, however, an increase in strength occurs immediately, and a decrease occurs only when the concrete dries. The tensile strength mainly depends on the properties of the mortar between the coarse aggregate, whereas the compressive strength depends, generally, to a certain extent on the compressive strength of the coarse aggregate. The tensile strength of a material is normally determined by applying a force to a specimen such that a uniform direct tensile stress is obtained and increasing this force until failure occurs. Practical difficulties arise when this procedure is applied to concrete specimens; if briquettes are used the stress distribution is usually non-uniform, a higher apparent strength being obtained as a consequence of concentration of stress, and with prisms or cylinders it is difficult to ensure that the ends of the specimens are properly gripped and that failure does not occur at the ends. A method whereby cylinders are loaded along a generator is now widely used, and appears to overcome these disadvantages. The loading arrangement for this test is shown in Figure 2.11; failure occurs by splitting along the vertical diameter[16]. Tests carried out in Holland have shown that cubes may also be used for splitting-tests; the tensile stress f_{ct} is then given by $0.64\,F_{cu}/a^2$ in which a is the length of side of the cube and F_{cu} is the failure load. It is also usual to determine the tensile strength from bending tests on prisms, from which the tensile strength in

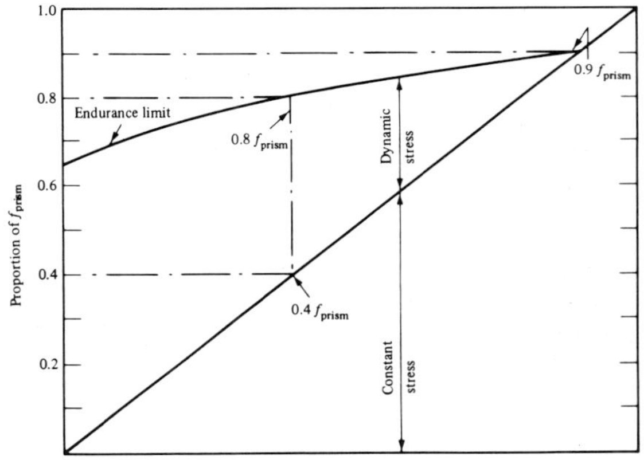

Figure 2.10 Goodman diagram for concrete

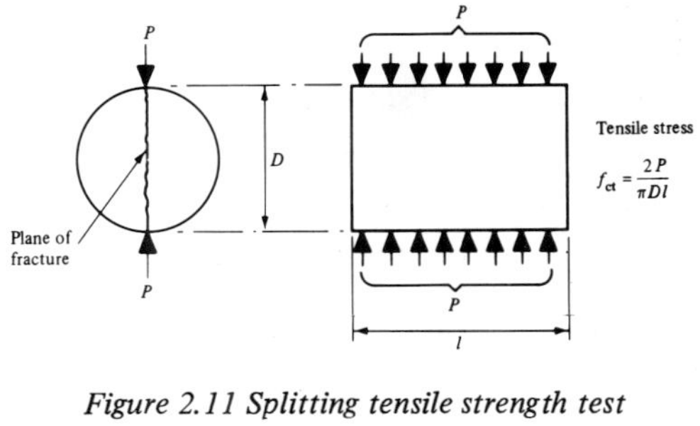

Figure 2.11 Splitting tensile strength test

bending, or modulus of rupture, is obtained. This is a nominal value based on the assumption that the specimen is homogeneous, and its magnitude varies to a great extent with the dimensions of the prism and the manner in which it is loaded. The modulus of rupture is in most cases between 1.75 and 2.25 times the direct tensile strength, and may, with sufficient accuracy, be assumed to be twice the direct tensile strength. The tensile strength of a small specimen should be taken as no more than a guide to the potential strength of the concrete in a large member, and not necessarily as indicating the actual strength. Recent research has shown that the heat of hydration generated before the concrete sets leads to cooling and consequent thermal shrinkage at the time of hardening. There is also an additional irreversible shrinkage when the concrete dries out, and either or both of these effects can cause cracking very early in the life of the member. This is particularly important in members designed to Class 2 of CP 110.

No linear relationship exists between the compressive and tensile strengths of concrete, but the tensile strength does vary to some extent with the compressive strength. The relationship $f_{ct} = 0.5 f_{cu}^{0.75}$ in which f_{ct} is the direct tensile strength and f_{cu} is the compressive cube strength (both in lbf/in^2) has been proposed by the *Department of Road Research*, and values for the expression are given in Figure 2.12. Table 2.1 gives tensile strength and moduli of rupture, taken from the *Journal American Concrete Institute*, the compressive strength f_{cyl} in this case being obtained from tests on cylinders.

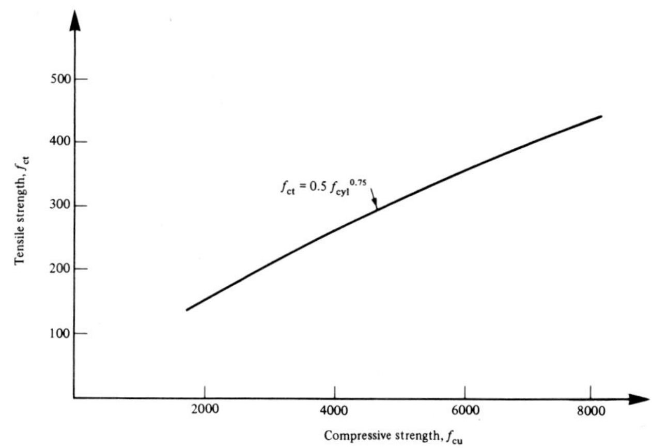

f_{cu}	2000	3000	4000	5000	6000	7000	8000
f_{ct}	150	205	254	300	341	348	723

All values given in lb/in^2
10 000 lbf/in^2 = 700 kgf/cm^2 = 70 N/mm^2

Figure 2.12 Relationship between compressive and tensile strength of concrete

Table 2.1 Modulus of rupture and tensile strength of concrete

f_{cyl} (Cylinder strength)	2 000	3 000	4 000	5 000	6 000	7 000	8 000	lbf/in²
	140.6	210.9	281.2	351.5	421.8	492.1	562.5	kgf/cm²
	13.79	20.68	27.58	34.47	41.37	48.26	55.16	N/mm²
Approximate f_{cu} (Cube strength)	2 670 3 000	4 000 4 500	5 340 6 000	6 670 7 500	8 000 9 000	9 340 10 500	10 670 12 000	lbf/in²
	187.7 187.7	281.2 316.4	375.4 421.8	468.9 527.3	562.5 632.8	656.7 738.2	750.2 843.7	kgf/cm²
	18.41 20.68	27.58 31.03	36.82 41.37	45.99 51.71	55.16 62.05	64.40 72.40	73.57 82.74	N/mm²
(Modulus of rupture) f_{cr}	375	485	580	675	765	855	930	lbf/in²
	26.37	34.09	40.78	47.46	53.78	60.11	65.39	kgf/cm²
	2.58	3.34	4.00	4.65	5.27	5.89	6.41	N/mm²
(Tensile strength) f_{ct}	200	275	340	400	460	520	580	lbf/in²
	14.06	19.33	23.90	28.12	32.34	36.56	40.78	kgf/cm²
	1.37	1.89	2.34	2.76	3.17	3.59	4.00	N/mm²

2.4.3. Relation between strength and modulus of elasticity

The magnitude of modulus of elasticity has an important effect on the behaviour of prestressed concrete and it is essential to make a realistic assessment of its value. It is usually considered to be related to the compressive strength of concrete.

In the British Codes of Practice (CP 110 and CP 115), the German Code of Practice for prestressed concrete (DIN 4227) and in the CEB/FIP Model Code, numerical values have been suggested for different grades of concrete. ACI Building Code suggests the following empirical formula for determination of modulus of elasticity of normal weight concrete:

$E_c = 57\,000\sqrt{f_{cyl}}$ where E_c is the elastic modulus and f_{cyl}, the cylinder strength (both in lbf/in²).

or

$E_c = 4733.59\sqrt{f_{cyl}}$ when both E_c and f_{cyl} are in N/mm².

For the density range of 90lb/ft³ to 155lb/ft³ (i.e. 1440 kg/m³ to 2483 kg/m³),

ACI Building Code also permits assessing modulus of elasticity of concrete in accordance with Pauw's[17] formula:

$$E_c = 33\,(\gamma)^{1.5}\,\sqrt{f_{cyl}}$$

when E_c is in lbf/in^2, γ in lb/ft^3 and f_{cyl} in lbf/in^2

or

$$E_c = 0.0427\,(\gamma)^{1.5}\,\sqrt{f_{cyl}}$$

when E_c and f_{cyl} are in N/mm^2 and, γ in kg/m^3.

Two more widely used empirical formulae relating compressive strength of concrete with modulus of elasticity are given below:

(a) M. Roš (Switzerland)[18]

$$E_c = 7\,810\,000 \times \frac{f_{prism}}{(f_{prism} + 2130)}$$

when E_c and f_{prism} in lbf/in^2

or

$$E_c = 53\,862 \times \frac{f_{prism}}{(f_{prism} + 14.69)}$$

when E_c and f_{prism} in N/mm^2

(b) Inge Lyse (Norway)[19]

$$E_c = 1\,800\,000 + 460\,f_{cyl}$$

If E_c and f_{cyl} are in lbf/in^2

or

$$E_c = 12\,413.79 + 460\,f_{cyl}$$

If E_c and f_{cyl} are in N/mm^2

Figure 2.13 shows the results obtained in accordance with the above formulae and also compares them with the numerical values given in the British, German and the model codes.

The above formulae and values are quite realistic in general; but it is important to note that much lower values may be obtained when certain types of aggregates such as chalk, some sandstones and shales are used.

As discussed in section 2.7, the modulus of elasticity of structural lightweight aggregate concrete is appreciably lower than that of dense concrete of equal compressive strength. Pauw's formula may be used to assess the modulus of elasticity of lightweight aggregate concrete, as this takes variations of density into account.

According to CP 110 and CEB/FIP Model Codes the modulus of elasticity of structural lightweight concrete may be obtained by multiplying the corresponding values given for normal weight concrete by a factor.

This factor is $\left(\frac{\gamma}{2300}\right)^2$ for CP 110 and $\left(\frac{\gamma}{2400}\right)^2$ for the Model Code where γ is the density of lightweight concrete in kg/m^3.

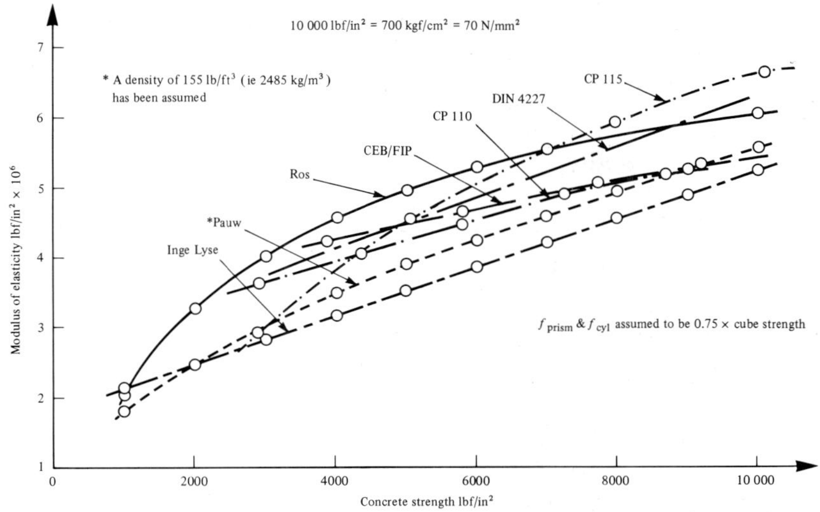

Figure 2.13 Relationship between modulus of elasticity and compressive strength of concrete (for dense concrete)

Non-destructive methods of testing, using supersonic waves or gamma rays, have been introduced in recent years by means of which the dynamic modulus of elasticity may be determined. The dynamic modulus is generally slightly greater (about 15–20%)[20] than the static modulus, ascertained by a static test on a prism or cylinder, and it has been suggested that the dynamic modulus should supersede the static modulus for design purposes. Results obtained so far by these methods, however, show certain variations in the relationship between the dynamic modulus and the cube strength, and further research appears to be necessary before such a change is made.

2.4.4 Relationship between stress and strain

Figure 2.14 shows the relationship between stress and strain for the first loading of concrete in compression. Curve 'A' illustrates the behaviour of low-strength concrete. The relationship is linear over a large extent of deformation; curve 'B' represents high-strength concrete and shows that the relationship in both cases is linear over a certain range of stresses which is greater than with curve 'A'. Figure 2.15 shows the variation of strain with time for concretes A and B of Figure 2.14 when subjected to a constant stress f. When a load producing a stress f is applied at the time t_0 an immediate elastic shortening $\epsilon_{e0} = f/E_{c0}$ occurs where E_{c0} is the modulus of elasticity at time t_0. The magnitude of this elastic deformation will clearly be much greater with concrete A than with concrete B. As time passes a continuous increase in the deformation takes place. This increase is also greater with A than with B, and reaches its maximum after several years at time t_∞. The effect of the removal of the load is as shown in Figure 2.16, which indicates the deformation of concrete under a sustained load applied at time t_0 and after its removal at time t_1 when the stress is reduced to zero. If the material were elastic no permanent deformation would remain, and with a perfectly elastic material all deformation would disappear immediately the load

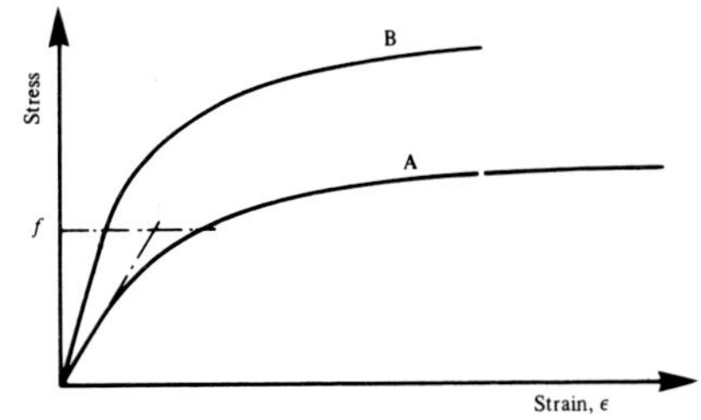

Figure 2.14 Stress – strain curve for concrete

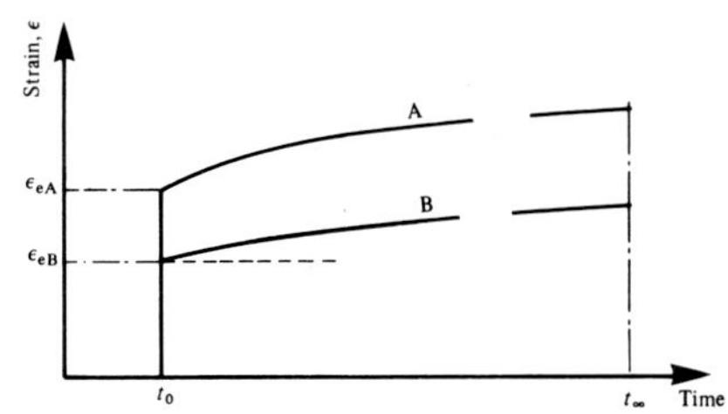

Figure 2.15 Deformation under sustained loading

was removed. Figure 2.16, however, shows that the deformation is instantaneously reduced by an elastic recovery ϵ_{e1}, and is further reduced by a gradual recovery ϵ_{cr}, termed 'creep recovery', but that a permanent deformation ϵ_p remains at time t_∞.

It should be noted that the same elastic modulus occurs for differing stresses, if the strength of the concrete has reached a constant value before the load is applied. In Figure 2.16, however, the value of E is greater at time t_1 than at time t_0, with the result that the elastic recovery ϵ_{e1} is less than ϵ_{e0}.

Another aspect of the behaviour of concrete is shown in Figure 2.17, which was published by the first named author some years ago[(21)]. The curve shown with a broken line was obtained during a continuous loading test, and the full line was obtained when the loading was interrupted and reduced to zero several times during the test. This latter curve shows the so-called 'hysteresis' effect – that is different deformations occur when the load is removed and when it is re-applied – and it should also be noted that the straight lines which can be drawn

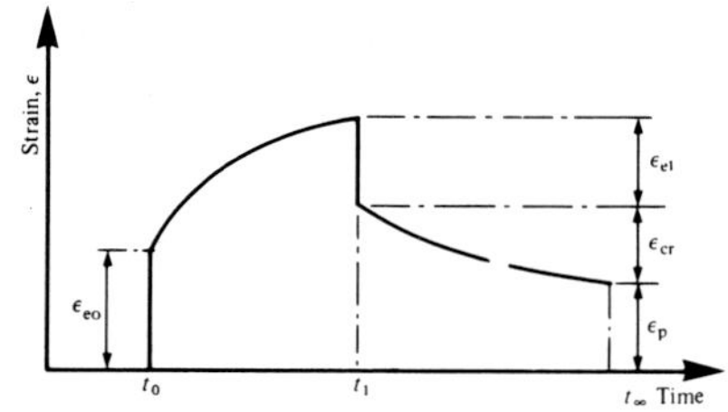

Figure 2.16 Deformation during and after loading

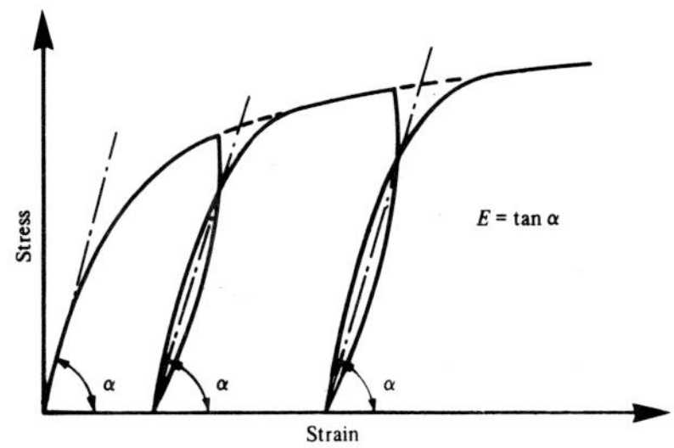

Figure 2.17 Stress – strain curve for interrupted loading

through each end of the hysteresis loop are parallel to the tangent at the origin of the curve, which represents the modulus of elasticity. These considerations apply when the load is removed and re-applied slowly at any stage below the lower plastic limit.

A further characteristic of the behaviour of concrete is that after a previous relatively high loading the stress–strain curve remains straight over the range from zero stress to a stress of at least 25 per cent of the ultimate strength, and departs very little from this line up to a stress which is 40 per cent to 50 per cent of the ultimate strength (Figure 2.18). This is also true for concrete that gives a curved deformation graph on first loading, such as concrete A in Figure 2.14.

2.4.5 Shrinkage and creep

In Figure 2.15 in addition to elastic deformation, shrinkage and creep also appear. Many divergent opinions and theories have been advanced to account for these phenomena, and separate chapters would be required to discuss them in full. The notes which follow deal briefly with the aspects that are important to the designer.

It is generally assumed that shrinkage is entirely independent of the applied loading, and that creep is largely dependent on the magnitude and duration of the applied loading.

A distinction must be made between the initial irreversible shrinkage which occurs when the concrete sets and dries and the reversible shrinkage and swelling that occur whenever the moisture content of the hardened concrete varies or its temperature changes. Many factors affect the magnitude of the shrinkage. Tests

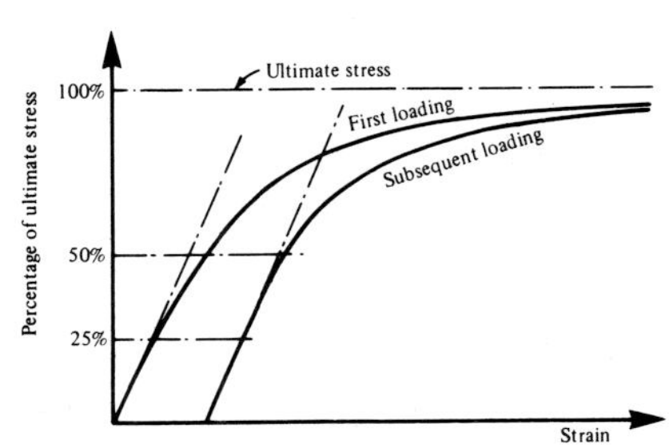

Figure 2.18 Stress – strain curves for the first and subsequent loadings

have shown, for example, that the shrinkage of concrete made with sandstone aggregate may be twice that of concrete made with limestone aggregate[22]. Shrinkage increases as the water/cement ratio increases, the water being more influential than the cement. The magnitude and development of shrinkage are also affected by the size of the member; with large elements the shrinkage takes place over a number of years, and only 40 per cent or so of the total shrinkage may occur during the first year. With smaller elements almost all of the shrinkage takes place during the first six months, and a greater total shrinkage occurs than with larger elements[23]. The magnitude of the shrinkage may vary between 0.03 per cent and 0.06 per cent. In high-strength concrete, as used for prestressing, a maximum shrinkage of 0.03 per cent is normal, but higher values may occur in regions where there are great heat and little humidity.

A very important factor influencing the magnitudes of both shrinkage and creep is the humidity of the environment, both being much greater in dry air than in a humid atmosphere. When a concrete is stored in water, swelling occurs instead of shrinkage. If the concrete is later removed from the water shrinkage will eventually take place, though it may take several years to develop fully.

The magnitude of creep is also greatly influenced by the strength of concrete at the time of loading (Figure 2.19). If prestress is applied where the concrete has attained a high strength, the creep will be much less than if the prestress is applied at a low strength. It follows that, if the same prestress is applied to two concretes of equal early age, one being made with ordinary Portland cement A and the other with rapid-hardening cement B much less creep will occur in concrete B than in A; the shrinkage of B will, however, be greater. Research has also shown that the magnitude of creep is directly proportional to the stress (provided that the stress does not exceed 40 per cent of the ultimate strength of the concrete), that it increases as the water/cement ratio increases, and that it decreases as the cement content increases. It is also influenced by the properties of the aggregates; concrete made with basalt, granite, or sandstone has greater creep than concrete made with limestone[24].

Figure 2.20 indicates the effect of different compressive stresses on the creep. For example when a part of the prestress is transferred at time t_0, a creep strain is developed; this may be increased at a later time t_1, when a second part of the prestress is transferred (when the resistance to creep has increased, so that the relative creep strain is less). At a time t_2 the additional dead load is applied; this reduces the compressive stress, and thus prevents any further substantial increase in the creep.

Figure 2.21 illustrates another feature of creep – namely, its dependence upon the humidity. The two curves (a) and (b) may indicate two creep curves for

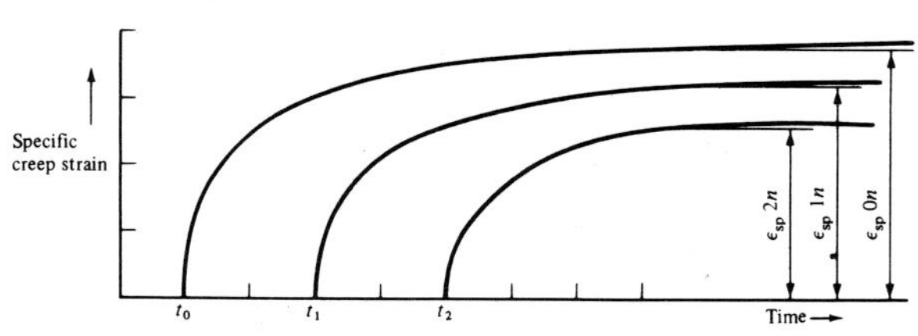

Figure 2.19 Relationship between creep strain strength of concrete and time

different humidities and temperatures. If a concrete is now subjected to compression under conditions corresponding to the larger creep 'a' (lower humidity) for a time between t_0 and t_1 and afterwards to conditions corresponding to a lower creep 'b', a resultant creep deformation of the type indicated will take place. It should be noted that in the two preceding examples the modulus of elasticity E_c has been assumed to be constant. In fact, an increase of the strength and in the value of E_c in the course of time causes a reduction of the creep strain.

The total strain in the concrete, due to direct load, creep and shrinkage, may be expressed (according to Neville[13]) by the equation:

$$\epsilon_\infty = \frac{f_0}{E_0}(1 + \phi_\infty) + \frac{f_\infty - f_0}{E_0}(1 + \eta) + \epsilon_s$$

in which

f_0/E_0	is the elastic strain in the concrete
ϕ_∞	is the creep factor

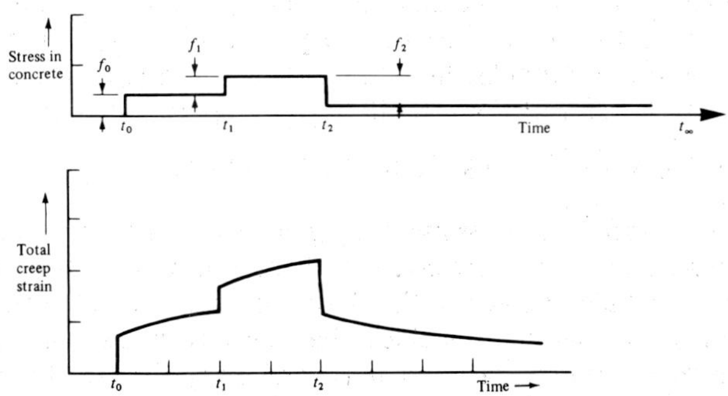

Figure 2.20 Strain in concrete (elastic and creep) for different stresses

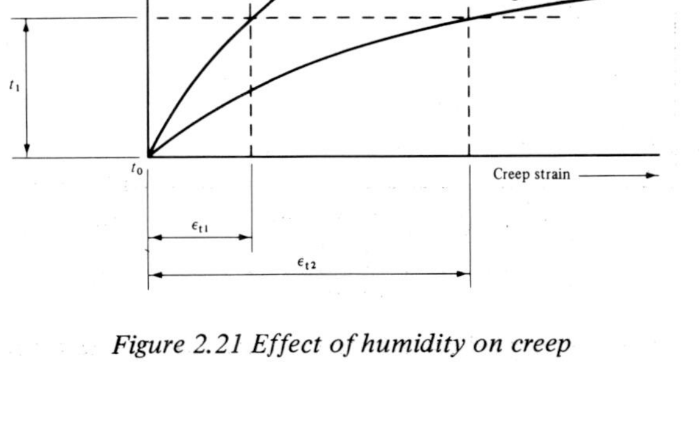

Figure 2.21 Effect of humidity on creep

$(f_\infty - f_0)/E_0$	is the increase in strain between the initial stage and infinite time
η	is a relaxation factor
ϵ_s	is the shrinkage strain after infinite time

The relaxation factor η is introduced to allow for the fact that the direct load may be reduced by creep. Obviously, if the load were constant, this factor would be 1. For many cases a value of 0.9 may be appropriate; η usually ranges between 0.5 and 1.

Both the CEB/FIP and CP 110 recommend, for design purposes, a stress–strain curve for concrete as shown in Figure 2.22.

In view of all these causes of variation, it is not easy to forecast the magnitude of creep accurately. The relationship between creep and time is usually expressed as an exponential curve, as first derived by Clerk Maxwell in 1867[(25)]. Provided that the prestress is not greater than 40 per cent of the ultimate strength of the concrete the limiting creep may be considered to be proportional to the prestress, and with a prestress of 1000 lbf/in^2 (70.3 kgf/cm^2; 6.93 N/mm^2) the limiting creep may, in climates similar to that in Great Britain, where the humidity is relatively high, approximate the limiting shrinkage. In such climates and under normal conditions the limiting specific creep for concrete with a minimum strength at transfer of 6000 lbf/in^2 (422 kgf/cm^2; 41.6 N/mm^2) may be assumed to be 0.33×10^{-6} per lbf/in^2 stress (48×10^{-6} per N/mm^2 stress). For lower strengths a value of $\dfrac{6000}{f_{cu}} \times 0.33 \times 10^{-6}$ (f_{cu} in lbf/in^2) may be assumed (Figure 2.23), and for climates with less humidity greater values should be allowed. Details of losses due to creep are given in Chapter 6. An alternative method, relating the magnitude of creep to the elastic compression has been employed for some time in Germany; it is included here because it is simple, and because it provides a better approximation than is possible by the use of a single factor. In this method the curve shown in Figure 2.24 is adopted in conjunction with the limiting values of shrinkage and creep shown in Table 2.2. In this table the values are related to the environmental conditions, and a coefficient k is intro-

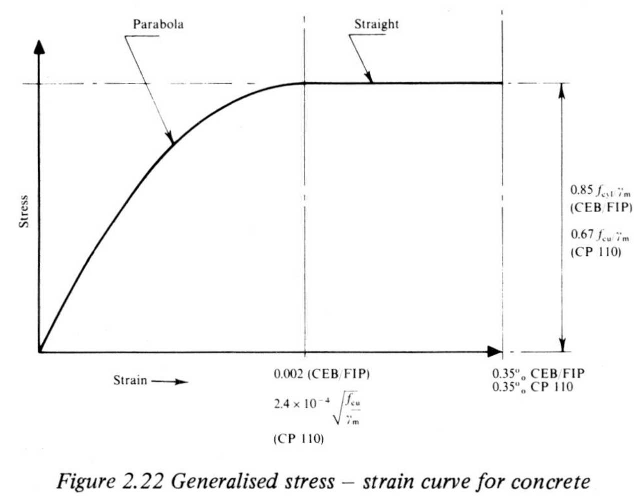

Figure 2.22 Generalised stress – strain curve for concrete

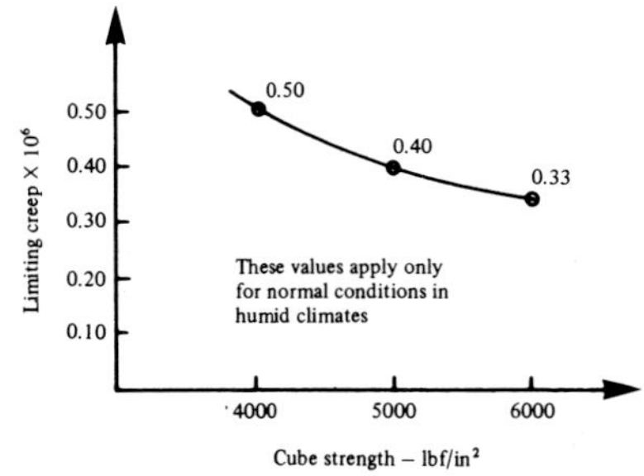

Figure 2.23 Relationship between creep and concrete strength (according to CP 115 and CP 110)

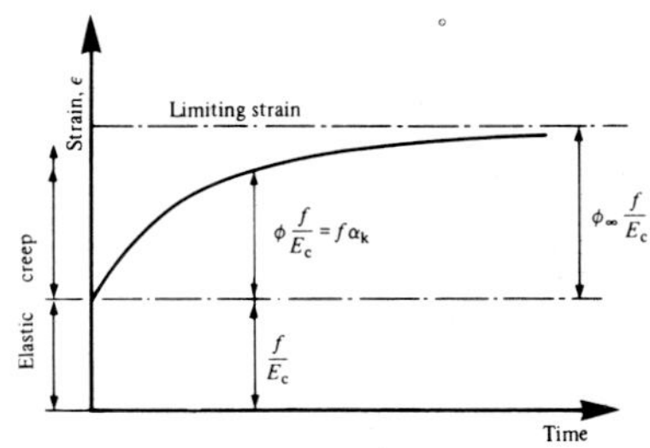

$$\epsilon = \frac{f}{E_c}(1 + \phi) = f\left(\frac{1}{E_c} + \alpha_k\right)$$

$$\phi = \frac{\text{Creep strain}}{\text{Elastic strain}} = E_c\alpha_k$$

Figure 2.24 Determination of creep (DIN 4227)

duced (Figure 2.25) whose magnitude depends on the ratio of strength of the concrete at the time of stressing to the limiting strength of the concrete. The limiting strength is taken as 1.3 times the cube strength after 28 days for ordinary Portland cement concrete, and 1.15 times the cube strength after 28 days for concrete made with rapid-hardening Portland cement. For example, if the 28-day cube strength is 6000 lbf/in² (422 kgf/cm²; 41.6 N/mm²) at the time the prestress is applied, then

$$\frac{f_{cu}}{f_{cu\infty}} = 1/1.30 = 0.77$$

for ordinary Portland cement or 1/1.15 = 0.87 for rapid-hardening cement, giving values for k of 0.95 and 0.72 respectively (Figure 2.25). Hence, for a member located in the open air, the creep factor ϕ is between $2k$ and $3k$ – that is, between 1.90 and 2.85 for ordinary Portland cement or between 1.44 and 2.16 for rapid-hardening cement (Table 2.2).

Assuming the modulus of elasticity is 5×10^6 lbf/in² (352 000 kgf/cm²; 34 650 N/mm²) (from Figure 2.13), the creep strain per unit stress = ϕ/E_c (Figure 2.24), which gives values of 0.29×10^{-6} to 0.43×10^{-6} for ordinary

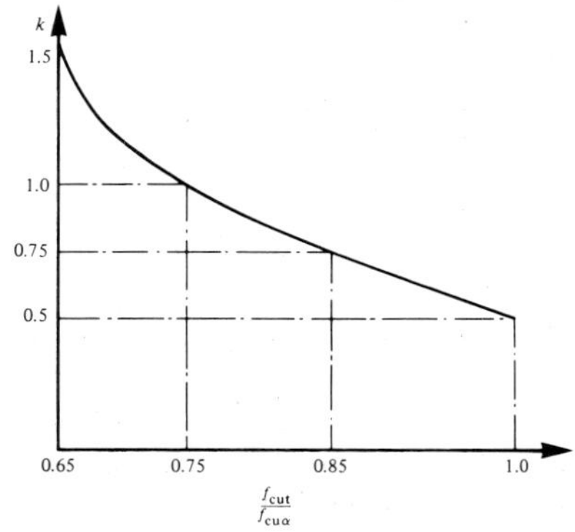

f_{cut} = Compressive strength at time, t
$f_{cu\,28}$ = Compressive strength at 28 days
$f_{cu\alpha}$ = Limiting compressive strength
= 1.30 $f_{cu\,28}$ for ordinary Portland cement concrete
= 1.15 $f_{cu\,28}$ for rapid-hardening cement concrete

Figure 2.25 Determination of K (DIN 4227)

Table 2.2 Effect of environment on creep and shrinkage

Environment	Creep ϕ	Shrinkage $\epsilon_s \times 10^{-5}$
In water	$0.5k$–$1.0k$	0
Very humid air – e.g. Directly above water	$1.5k$–$2.0k$	10
Open air	$2.0k$–$3.0k$	20
Dry air – e.g. in dry inner rooms	$2.5k$–$4.0k$	30

Portland cement, and 0.38×10^{-6} to 0.57×10^{-6} for rapid-hardening cement. The elastic strain per unit stress, by contrast, is 0.2×10^{-6} lbf/in^2.

A third method, proposed by Ross, relates the total deformation to the modulus of elasticity of the concrete. In this method, the secant modulus of elasticity is considered, which takes into account the entire deformation at any given stress(26).

The magnitude of creep recovery (Figure 2.16) also depends on the age and strength of the concrete. Figure 2.26 indicates a method of assessment, based on the law of superposition, proposed by McHenry(27).

2.4.6 Prediction of shrinkage and creep

(1) Early proposals

As previously noted, many formulae for creep and shrinkage have been devised. A hyperbolic formula for the creep strain ϵ_{ct} after time t was proposed by Ross as early as 1936(26), in the form:

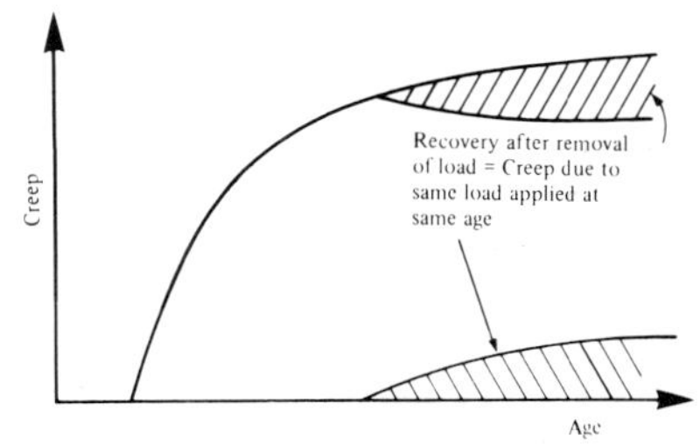

Figure 2.26 Recovery of creep according to McHenry

$$\epsilon_{ct} = \frac{t}{k_1 + k_2 t}$$

in which k_1 and k_2 are empirical constants, determined experimentally. This expression may be transformed into:

$$\epsilon_{ct} = \frac{k_3 t}{k_4 + t} \times f_c$$

in which f_c is the stress in the concrete and k_3 and k_4 are constants. The same expression was obtained independently by Lorman[28]; it may be used to compute the limiting creep strain after infinite time. In this case a value of $\epsilon_{c_\infty} = k_3 f_c = \frac{1}{k_2} f_c$ is obtained, in which $\epsilon_{c\infty}$ represents the specific creep (that is, limiting creep per unit stress). When the time t is equal to $k_4 = \frac{k_1}{k_2}$, the creep strain is $0.5\ k_3 f_c$; that is, half of the total creep takes place within the time $t = k_4$.

(2) ACI Proposals, 1970
Branson, Meyers and Kripanarayanan[29] have obtained the following modified equations from various test results:

$$\epsilon_{ct} = \frac{t^c}{d + t^c} \times \epsilon_{c_\infty} \text{ and } \epsilon_{st} = \frac{t^e}{f + t^e} \times \epsilon_{s\infty}$$

in which ϵ_{ct} and ϵ_{st} are the creep and shrinkage strains respectively after infinite time c, d, e and f are constants. Neville and Meyers[30] have investigated the accuracy of any method which uses creep values ϵ_{ct} measured after t weeks under load to predict the creep strain ϵ_{cl} after one year under load, based on n specimens loaded for t weeks; they have obtained an error coefficient

$$K = \frac{\sqrt{(\epsilon_{ct} - \epsilon_{cl})^2 / n}}{\epsilon_{cl}}$$

In order that the error should be less than 10 per cent, 20 weeks of loading are required.

Meyers, Branson and Schumann[31] have proposed prediction formulae using 28-day test data; they suggest the following values for the constants: $c = 0.6$, $d = 10$, $e = 1$, and $f = 35$ (for moist-cured concrete) or 55 (for steam-cured concrete). The limiting creep and shrinkage strains obtained in this way are therefore:

$$\epsilon_{c\infty} = \frac{\epsilon_{ct}}{\left(\dfrac{t^{0.6}}{10 + t^{0.6}}\right)}$$

$$\epsilon_{s\infty} = \frac{\epsilon_{st}}{\left(\dfrac{t}{35 + t}\right)} \quad \text{for moist-cured concrete}$$

$$\epsilon_{s\infty} = \frac{\epsilon_{st}}{\left(\dfrac{t}{55 + t}\right)} \quad \text{for steam-cured concrete}$$

(3) CEB/FIP Model Code Recommendations (1978):
The foregoing expressions relate only to uniform conditions. Allowances for variations in the conditions are included in the CEB/FIP Recommendations.

Provided that the stress in the concrete is within the range of service stresses, the creep strain $\epsilon_c\ (t \,.\, t_0)$ after time 't' under a constant stress f_c applied at time 't_0' is given by

$$\epsilon_c(t - t_0) = \phi(t - t_0) \times \frac{f_c}{E_{c28}}$$

in which $\phi\,(t - t_0)$ is the creep coefficient.

Creep coefficient can be determined from the following relationship:

$$\phi\,(t - t_0) = \beta_a(t_0) + \phi_d\, \beta_d\,(t - t_0) + \phi_f\ [\beta_f(t) - \beta_f(t_0)]$$

$$\beta_a(t_0) = 0.8\left[1 - \frac{f_{cyl}(t_0)}{f_{cyl\ \alpha}}\right]$$

ϕ_d = delayed modulus of elasticity taken as 0.4
ϕ_f = flow coefficient and consists of
ϕ_{f1} which depends on ambient environment and given in Table 2.3 column 3.
and ϕ_{f2} which depends on notional thickness and is given in Figure 2.27.
β_d = Function corresponding to the development with time of delayed elastic strain and may be obtained from Figure 2.28.
β_f = Function corresponding to the development with time of delayed plasticity (Figure 2.29) depending on notional thickness.

Table 2.3 Basic coefficients of creep and shrinkage

Ambient environment	Relative humidity	coefficients of		Coefficient λ
		creep ϕ_{f1}	shrinkage ϵ_{s1}	
1	2	3	4	5
Water		0.8	+0.00010	30
Very damp atmosphere	90%	1.0	–0.00013	5
Outside in general	70%	2.0	–0.00032	1.5
Very dry atmosphere	40%	3.0	–0.00052	1

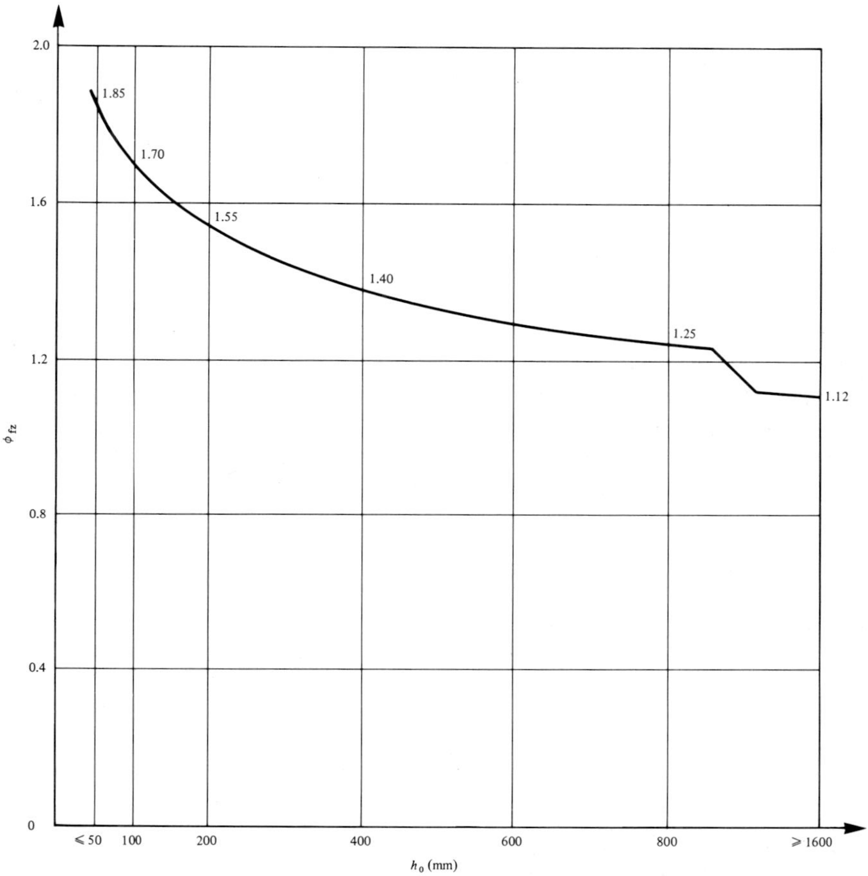

Figure 2.27 Influence of the notional thickness on creep

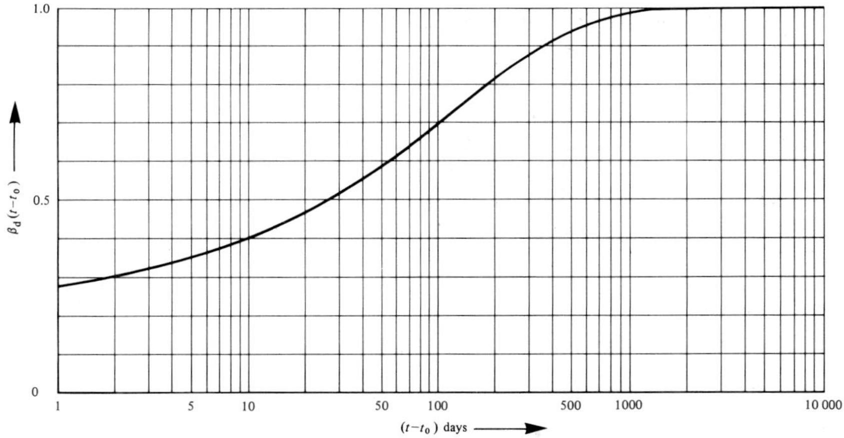

Figure 2.28 Development with time of the delayed elastic strain

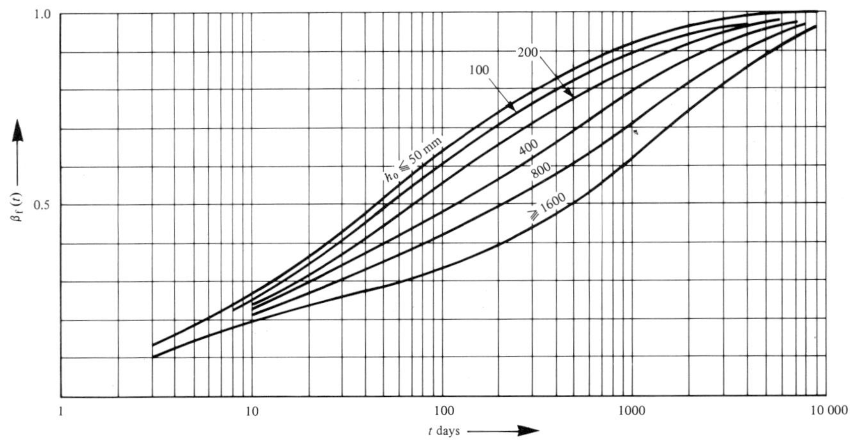

Figure 2.29 Development with time of the delayed plastic strain

In the above equation the first term on the right hand side represents the irreversible part of the deformation which occurs soon after the load is imposed. The second term represents the recoverable part of delayed deformation (delayed elasticity) while the third part represents the irreversible delayed deformation. This term is affected by the age at which load is applied, but the second term is independent of age of loading.

The strain due to shrinkage which develops in an interval of time $(t - t_0)$ is given by

$$E_s(t - t_0) = E_{so}\ [\beta_s(t) - \beta_s(t_0)]$$

where E_{so} is basic shrinkage coefficient and consists of

E_{s1} which depends on the environment (some values are given in Table 2.3) and E_{s2} which depends on the notional thickness and shown in Figure 2.30

β_s = Function corresponding to the change of shrinkage with time depending on notional thickness as shown in Figure 2.31.

In prediction of creep and shrinkage the concept of notional thickness has been introduced. This is defined as $\eta_0 = \lambda \dfrac{2A_c}{u}$

where λ = a constant depending on environment and tabulated in column 5 of Table 2.3

A_c = area of concrete section

and u = perimeter in contact with atmosphere.

2.4.7 Effect of high temperatures

When concrete is heated to temperatures above 100°C (212°F), it loses first the free water; the remaining water (that is, other than free water) is lost at temperatures above 200°C (392°F), and phase changes in the cement also start to take place at about this temperature. Cracking may occur at much lower temperatures,

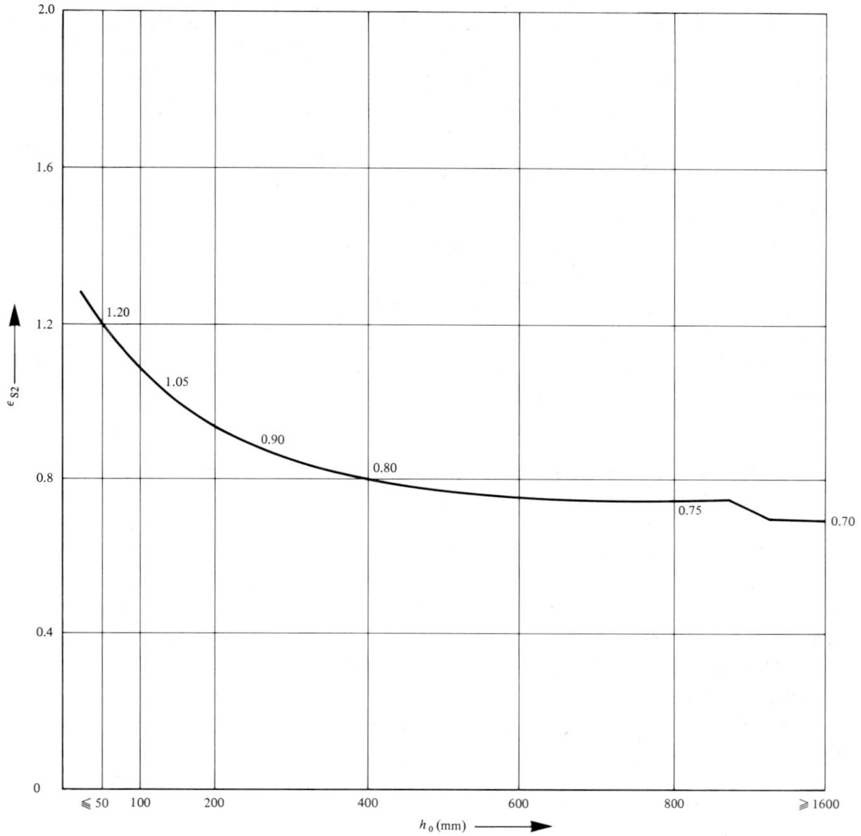

Figure 2.30 Development of the shrinkage with time

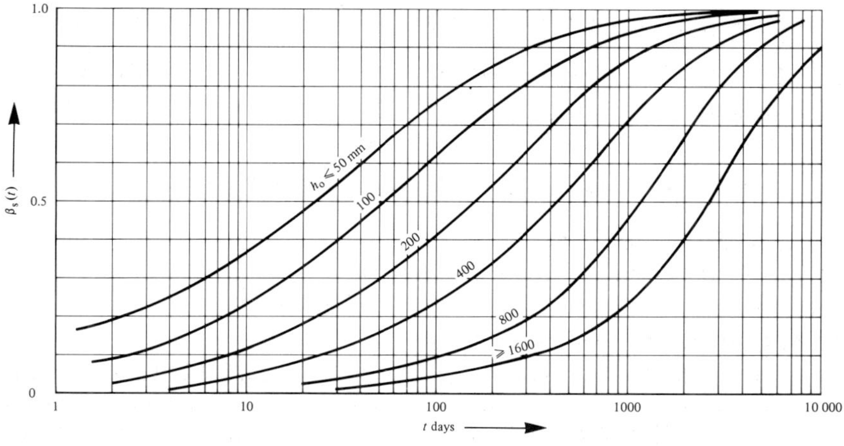

Figure 2.31 Influence of the notional thickness on shrinkage

particularly if the temperatures are not uniform throughout the member, but no significant chemical or physical change takes place below 200°C (392°F).

The fire resistance of concrete depends as well on the nature of the aggregate and the water content as on the properties of the cement. The least satisfactory aggregates from this point of view are flint, gravel, granite and all crushed stone (except limestone) (See also Chapter 16).

2.4.8 Effect of low temperatures

At very low temperatures, the behaviour of concrete depends markedly on its water content. Oven-dry specimens, under test, show an increase in compressive strength, tensile strength and elastic modulus of about 5 to 10 per cent at −150°C (−238°F); but moist specimens show an increase of between 200 and 300 per cent in compressive and tensile strengths, together with an increase of about 50 per cent in the modulus of elasticity.

2.4.9 Movements due to temperature changes

The magnitude of the expansion or contraction of concrete due to temperature is affected by many factors, and the coefficient of expansion may lie between 8.5×10^{-6} and 14×10^{-6} per °C (3.85×10^{-6} and 6.4×10^{-6} per °F). The coefficient is low when limestone, basalt or granite aggregate is used (8.6×10^{-6} to 9.6×10^{-6} per °C; 3.8×10^{-6} to 4.35×10^{-6} per °F) and is higher for river aggregates, gravel, pumice or slag (10.1×10^{-6} to 11.9×10^{-6} per °C; 4.6×10^{-6} to 5.4×10^{-6} per °F). It increases as the cement content increases, as the water content increases, and as the sand content decreases.

At low temperatures, average values of 8×10^{-6} per °C (4.5×10^{-6} per °F) have been obtained for concrete with a strength of 5000 lbf/in^2 (350 kgf/cm^2; 35 N/mm^2) throughout the temperature range from 24°C (75°F) to −157°C (−250°F).

2.5 Durability of concrete

Durability may be defined as the ability of the concrete to retain both its strength and its appearance over a long period of time. It is improved when chemically-inert aggregates are employed, and when dense, thoroughly-compacted concrete is ensured. It is also affected by the occurrence and recurrence of freeze-thaw cycles; the repetition of the expansion of internal water when it is frozen exercises a cumulative effect which can lead to the disruption of the concrete. This effect can be mitigated or avoided by ensuring full compaction of the concrete; alternatively the use of air entrainment, in which the mix contains an additive which creates a large number of separated small air bubbles, is commonly adopted.

Intense radiation – particularly in the form of gamma rays – has a long-term weakening effect on concrete.

2.6 Admixtures

When classified according to the effects they are intended to produce, four main groups of admixtures can be distinguished.

Plasticisers (Water-reducing admixtures): these permit the water content of the mix to be reduced; their action arises by virtue of their lubrication properties, or by air entrainment (see later). The lower water content allows a higher strength to be attained, but the entrained air reduces this again to a value about equal to, or slightly less than, that which would otherwise be obtained. However, compaction is more easily achieved, and durability is often improved.

Air-entraining agents: these create a mass of small air bubbles within the mix, and improve its resistance to freeze-thaw conditions. They also serve to lubricate the mix during placing and compaction, reducing segregation and allowing the water content to be reduced.

Quick-setting agents: the setting of Portland cement can be accelerated by the addition of a relatively small percentage of calcium chloride. This is the active ingredient of many quick-setting agents; it has been found to cause corrosion of reinforcement and prestressing steel, particularly when in solution or in the presence of water. The rate of corrosion also increases with the temperature. Therefore such agents should not be used.

Expanding agents: these are useful and important admixtures when used in the grout for filling ducts. They often contain aluminium powder, and also a very small proportion of calcium chloride; its use under these circumstances is harmless.

2.7 Structural lightweight concrete

Structural lightweight concrete has been widely used in the USA and to a rather lesser extent in Great Britain. It usually has a density of 100 to 125 lbf/ft^3 (1600 to 2000 kgf/m^3) together with a cube strength of at least 4500 lbf/in^2 (316 kgf/cm^2; 31.2 N/mm^2) or a cylinder strength of 3000 lbf/in^2 (210 kgf/cm^2; 20.8 N/mm^2); higher strengths are obtained without difficulty. Lightweight aggregates may be of mineral origin, occurring naturally; or as by-products of industrial processes; or they may be especially manufactured. The natural lightweight aggregates are usually of volcanic origin, such as pumice; expanded slag is obtained by rapidly cooling the molten slag from a blast furnace. Clay, shale and slate aggregates may be obtained by expanding, but can also be produced by sintering after applying heat.

Short and Kinniburgh[32] describe three lightweight aggregates which have been introduced in Great Britain: Aglite (similar to the corresponding USA product and consisting of expanded and sintered clay obtained from the waste of coal washing plants); Leca (introduced from Denmark, and comprising a round smooth clay aggregate, expanded in a rotary kiln); and Lytag (introduced in Great Britain; a sintered pulverized fuel ash). According to a recent publication[33] five kinds of lightweight aggregates suitable for structural concrete are available in Great Britain in 1980:

(a) Leca (see above), which produces concrete of a very low density (80 to 100 lb/ft^3; 1280 to 1600 kg/m^3) with a 28-day cube strength of 2500 lbf/in^2 (176 kgf/cm^2; 17.3 N/mm^2) although strengths up to 4000 lbf/in^2 (280 kgf/cm^2; 27.8 N/mm^2) may be obtained.
(b) Foamed slag made from molten blast-furnace slag giving concrete with densities in the range from 100 to 135 lb/ft^3 (1600 to 2165 kg/m^3) and strengths between 2000 and 8000 lbf/in^2 (140 to 560 kgf/cm^2; 14 to 56 N/mm^2).
(c) Pelletised foamed slag is a further development of the foamed slag manufacturing process. Molten slag is run on to a revolving, fluted drum to which

jets of water are directed. The slag is flung into the air and expands rapidly, thus forming smooth-surface aggregate. Density and strength range is similar to foamed slag.

(d) Aglite (see above), of irregular shape, giving concrete with a density in the range from 100 to 115 lb/ft^3 (1600 to 1840 kg/m^3; 15.7 to 18.1 kN/m^3) and strengths of 6000 lbf/in^2 (420 kgf/cm^2; 41.6 N/mm^2) or more.

(e) Lytag (see above), giving concrete with a density of 110 to 120 lb/ft^3 (1760 to 1920 kg/m^3; 17.3 to 18.8 kN/m^3) and strengths up to 8000 to 10 000 lbf/in^2 (560 to 700 kgf/cm^2; 56 to 70 N/mm^2).

Solite (slate expanded in rotary kiln) ceased to be manufactured in Great Britain in the early 1970's. Another lightweight aggregate called Sintag (sintered colliery shale) is being produced in this country, but it is used mainly in 'block' manufacture.

The strength of ordinary concrete depends on the strength of the coarse aggregate, the voids being filled up with a uniform matrix of lower strength. With lightweight concrete the strength of the matrix is mostly greater than that of the coarse aggregate; hence the strength of the matrix and the extent of its arching action over the aggregate[34] govern the strength of the concrete and this is obtained by rational mix design and good compaction[34, 35]. The coarse aggregate is usually porous and this is a factor which should be remembered when determining the water/cement ratio; the different absorption characteristics of various lightweight aggregates must also be considered when designing the mix. Both coarse and fine aggregates may be lightweight, or lightweight coarse aggregates may be used in conjunction with normal weight fine aggregates. In the latter case it is easier to achieve higher strengths. Bobrowski and Bardhan-Roy[35] give details of mixes using Lytag and Solite employed in precast prestressed concrete roof beams when a minimum strength of 7500 lbf/in^2 (526 kgf/cm^2; 52 N/mm^2) was required; densities of 115 to 120 lb/ft^3 (1845 to 1920 kg/m^3) and 113 to 119 lb/ft^3 (1810 to 1900 kg/m^3; 177 to 186 kN/m^3) gave 28-day cube strengths of 7400 to 9250 lbf/in^2 (520 to 650 kgf/cm^2; 51 to 64 N/mm^2) and 7200 to 8850 lbf/in^2 (505 to 622 kgf/cm^2; 50 to 61 N/mm^2) respectively.

An advantage of the use of lightweight aggregates is that definite single-sized aggregates are obtainable (and need not be sieved off as with graded aggregates obtained from a quarry). Moreover, they are often round and thus provide the most suitable shape. Owen of C & CA[36] has established that no cracks occur when curing is carried out under elevated temperatures because of the better thermal properties of lightweight aggregate. Whereas in normal aggregate owing to the differential rate of expansion between the aggregate and the softer paste cracking may occur at the interspace. It is also claimed that in the case of concrete made of gravel aggregates the cube strength over-estimates the strength in the structure whereas with lightweight concrete the reverse is the case.

It is a feature of lightweight concrete that the modulus of elasticity is between 50 per cent and 70 per cent of that for concrete of equal strength made with ordinary aggregates (with the exception of chalk, sandstone, and shale, as mentioned earlier). The shrinkage and creep of lightweight concrete are usually greater than those of ordinary concrete, but these depend mainly on the properties of the aggregate. American and German tests have shown that for certain types of lightweight aggregates shrinkage and creep are no greater than those of ordinary concrete, though the elastic deformation is almost always about twice that of ordinary concrete (that is, the modulus of elasticity is halved). Information regarding shrinkage and specific creep should be obtained from the manufacturers of the particular lightweight aggregate. In absence of any information the following values are suggested for design purpose[25]: Shrinkage strain (ultimate) $400 - 600 \times 10^{-6}$ for ambient condition of temperature (20°C) and relative humidity (65–70%) and for round, regular size aggregate specific creep: $0.7 - 0.9 \times 10^{-4}$ per N/mm^2 ($0.45 - 0.60 \times 10^{-6}$ per lbf/in^2) strain.

REFERENCES

1. TIMOSHENKO, S. P. *History of the strength of materials.* New York, McGraw-Hill, 1953. pp. 452.
2. MOORE, H. F. *Textbook of the materials of engineering.* New York, McGraw-Hill, 1947. pp. 500.
3. JUDGE, A. W. *The testing of engineering materials.* Vol. 3. London, Pitman, 1947. pp. 624.
4. EKBERG, C. G., WALTHER, R. E. and SLUTHER, R. G. *Fatigue resistance of prestressed concrete beams in bending.* Structural Division of the American Society of Civil Engineers. July 1957. Paper No. 1304. pp. 1–17.
5. TIDE, R. H. R. and VAN HORNE, D. A. *A statistical study of the static and fatigue properties of high strength prestressing strands.* Progress Report No. 2, Bond in Prestressed Concrete. Bethleheim, Pa., Lehigh University, 1966. Fritz Engineering Laboratory Report No. 309.
6. FRENCH, H. G. *Transactions of the American Society for Steel Treating.* Vol. 21. 1933.
7. WILSON, W. M. *Engineering experimental station.* University of Illinois. Bulletin 327. pp. 58–62.
8. LEGGET, R. F. Failure of prestressed concrete pipe at Regina. Saskatchewan. *Proceedings of the Institution of Civil Engineers.* May 1962. pp. 11–20.
9. EVANS, R. H. Application of prestressed concrete to water supply and drainage. *Proceedings of the Institution of Civil Engineers.* Part III, December 1955. pp. 725–783.
10. *Journal for Stainless Steel Industry*, Vol. 1, No. 3, September 1973.
11. BARDHAN-ROY, B. K. *New grandstand for the Sandown Park Racecourse at Esher.* Paper presented to the Seventh FIP Congress, New York, May 1974.
12. ABELES, P. W. *An introduction to prestressed concrete.* London, Cement and Concrete Association 1964. Vol. 1. pp. 118–199. Publication 12.013.
13. NEVILLE, A. M. *Properties of concrete.* London, Pitman and Sons Ltd, 1963. pp. 532.
14. STEWART, D. A. *The design and placing of high-quality concrete.* Second edition, London, Spon 1962. pp. 162.
15. BOBROWSKI, J. Personal communication.
16. *Concrete and Constructional engineering* Vol. 51 No. 9, September 1956.
17. PAUW, A. Static modulus of elasticity of concrete as affected by density. *Journal of the American Concrete Institute, Proceedings.* Vol. 57 Pt 1, December 1960. pp. 679–688.
18. ROŠ, M. and SARRASIN, A. Die material-technischen Grundlagen und Probleme des Eisenbetons im Hinblick auf die zukünflige Gestaltung der Stahlbetonbauweise. Zurich. Eidgenoesiche Materialprefungs und Versuchanstalt fur Industrie Bauwesen und Gewerbe, 1950. Bericht Nr 162.
19. HOGNESTAD, E., HANSON, N. W. and McHENRY, D. Concrete stress distribution in ultimate strength design. *Journal of the American Concrete Institute, Proceedings.* Vol. 52 December 1955. pp. 447–454.
20. SWAMY, R. N and BANDYOPADHYAY, A. K. The Elastic Properties of Structural Lightweight Concrete. *Proceedings of the Institution of Civil Engineers* 59 (1975). pp. 381–94.
21. ABELES, P. W. Elastizität des Betons. *Zement* No. 37, 38 and 40. 1937.

22. LEA, F. M. and LEE, C. R. Shrinkage and creep in concrete. *Symposium on the shrinkage and cracking of cementive materials.* London, Society of Chemical Industry, 1947. pp. 7–17.
23. DAVIS, R. E. and TROXELL, G. E. Volumetric changes in Portland cement mortars and concrete. *Journal of the American Concrete Institute, Proceedings.* Vol. 25. 1929. pp. 210–260.
24. DAVIS, R. E. and DAVIS, H. E. Flow of concrete under the action of sustained loads. *Journal of the American Concrete Institute, Proceedings.* Vol. 2, March 1931. pp. 837–901.
25. MAXWELL, J. C. On the dynamical theory of gases. Report of the British Association for the Advancement of Science, Part 2, 1859. pp. 9.
26. ROSS, A. D. Concrete creep data. *The Structural Engineer.* Vol. 15. 1937. No. 8. p. 314.
27. McHENRY, D. New aspect of creep in concrete and its application to design. *Proceedings of the American Society for Testing Materials.* Vol. 43. 1943. pp. 1069–1084.
28. LORMAN, W. R. The theory of concrete creep. *Proceedings of the American Society for Testing Materials.* Vol. 40. 1940. pp. 1082–1102.
29. BRANSON, D. E., MEYERS, B. L. and KRIPANARAYANAN, K. M. *Time-dependent deformation on non-composite and composite sand-lightweight prestressed concrete structures.* Highway Research Board, Washington DC. January 1970. Report 70-1.
30. NEVILLE, A. M. and MEYERS, B. L. Creep of concrete influencing factors and prediction. *Symposium on Creep of Concrete.* Detroit American Concrete Institute, 1964. pp. 1–31. SP-9.
31. MEYERS, B. L., BRANSON, D. E. and SCHUMANN, C. G. *Prediction of creep and shrinkage behaviour from 28-day data.* Department of Civil Engineering, University of Iowa. October 1970. Report No. 70.
32. SHORT, A. and KINNIBURGH, W. *Lightweight concrete.* 2nd edition. London, Applied Science Publishers, 1968. pp. 368.
33. HORLER, D. B. An Update of Lightweight Aggregate Production in UK. *Proceedings of Concrete International CI '80.* April 1980. The Construction Press. pp. 16–19.
34. BOBROWSKI, J. and BARDHAN-ROY, B. K. Structural assessment of lightweight concrete. *Concrete.* Vol. 5. July 1971. No. 7. pp. 229–235.
35. BOBROWSKI, J., ABELES, P. W. and BARDHAN-ROY, B. K. The impact of research on the use of lightweight concrete in major structural work. Paper presented to the Sixth CIB Congress, The impact of research on the built environment, Budapest, October 1974. Vol. 1. pp. 576–584.
36. OWEN, P. G. A technique for the production of high early strength lightweight concrete. *Advances in concrete.* Proceedings of the symposium of the University of Birmingham, September 1971. London, Cement and Concrete Association, 1972. pp. 5.

CHAPTER 3

BEHAVIOUR OF PRESTRESSED CONCRETE BEAMS

Prestressed concrete is a compound material which can be made to possess properties ranging from those of a brittle elastic material susceptible to sudden failure to those of a yielding or ductile material. It is possible to design prestressed concrete members in such a way that they exhibit either great flexibility with little permanent deformation, or great rigidity. Flexural cracks may close completely when the load is reduced (provided that it does not exceed 70 per cent of the load causing failure), and the outstanding property of prestressed concrete is this recovery of deformation. The versatility of the material is well illustrated by prestressed concrete beams, the behaviour of which depends on the physical properties of the concrete and the steel, the size and shape of the section, the amount of steel, and its distribution in tensile zone of concrete, the magnitude of the prestressing force, and the efficiency of the bond. As such large differences in behaviour are possible, it is important that the load causing cracks, the distribution and size of cracks, and the ultimate static, fatigue and impact loads should be carefully considered and the corresponding factors of safety assessed.

3.1 Deformation of reinforced and prestressed concrete beams

Reinforced concrete. In reinforced concrete construction the design is based on a cracked section, and cracks are normally unavoidable since cracks due to shrinkage develop in all constructions of appreciable length. Consider, however, the case of a short beam in which cracks need not occur before the tensile resistance of the concrete has been overcome by loading. The relation between load and deflection for the first loading of two reinforced concrete beams of equal cross-section but with different amounts of steel is shown in Figure 3.1. A small part of the diagram, representing a beam before cracking occurs, corresponds to the deformation of a homogeneous section, and the amount of reinforcement has little influence on the slope of the curve. After cracking has occurred the deflection is similar to that of a section composed only of a concrete compressive zone and separate tensile reinforcement, and concrete in tension is neglected in the usual way. Consequently the amount of reinforcement has a much greater influence on this part of the diagram (Figure 3.1). Immediately the cracks develop noticeable permanent deformation takes place if the bond is poor, and the cracks remain permanently open. Hence in this case Stage I is available only for the first loading of the beam, and for second and subsequent loadings the deflection is shown by the full line in Figure 3.1. Stage 2 then applies for all loads until plastic yielding takes place in the third stage and the diagram becomes curved.

Thus with ordinary reinforced concrete in bending there are three distinct states. The first is here termed the 'uncracked elastic' or 'rigid elastic' state, when

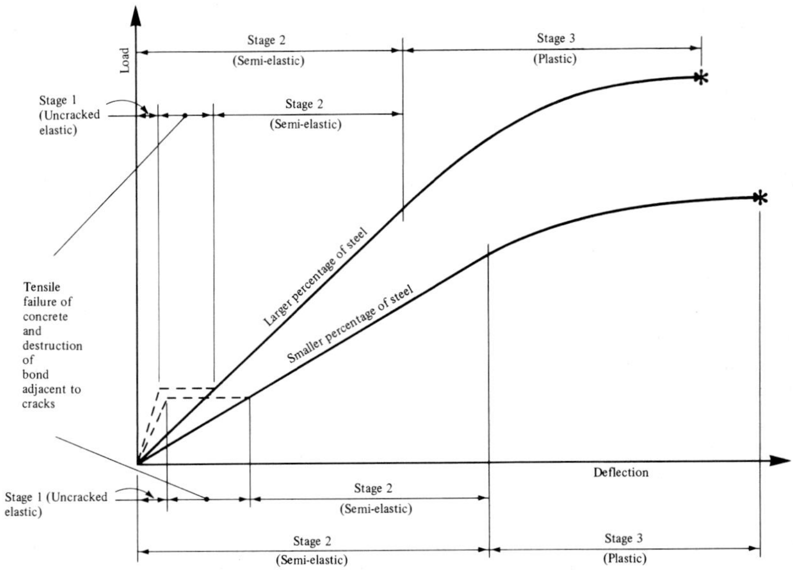

Figure 3.1 Deformation of under-reinforced, reinforced concrete beams

the deformation is linear and the permanent deflection is negligible, and which will occur only in members which are free from cracks due to shrinkage; the second is the 'semi-elastic' or 'flexible elastic' state, in which the immediate deformation is linear but there is an appreciable permanent deflection; the third is the 'plastic' state, when the deformation is no longer linear.

Prestressed concrete. The behaviour of prestressed concrete in bending is similar in principle to that described for ordinary reinforced concrete, but the following major differences should be noted.

(1) The transition from the uncracked elastic state to the second state, as evidenced by the change in slope of the deflection diagram, occurs at a much greater load, since it is necessary to overcome the effective pre-compression as well as the tensile resistance (where tensile resistance is available) before cracks occur.

(2) In the case of prestressed concrete the second state is more accurately termed the 'cracked elastic' state, since the permanent deformation is very small and cracks close completely when the load is removed. Under subsequent loadings the member remains in the first state until the pre-compression has been overcome and appreciable permanent deformation takes place only in the plastic stage, after a large proportion of the load causing failure has previously been applied. The assumption that cracks reappear when the stress at the tensile face is zero, which is often used to ascertain the load causing zero stress and hence the effective prestress, is accurate only if substantial cracks are present; otherwise the apparent effective prestress may be greater than that actually available. If the steel is well bonded, and if the previous loading has not greatly exceeded the load caus-

ing cracking, the curve of deflection will be intermediate between lines 1 and 2 (Figure 3.2) and cracks will become visible at zero stress only if the previous loading was sufficient to cause cracks of substantial depth and or was repeated several times.

These points are illustrated by Figure 3.2 which is a diagram of the deflection of a member subjected to static loading of reasonably short duration. Figures 3.1 and 3.2 relate to so-called 'under-reinforced' members, in which the prestressing steel or reinforcement is the weaker part, and failure occurs primarily as a result of yielding of the prestressing steel or reinforcement in the cracks, with consequent excessive deformation of the beam. In an 'over-reinforced' section failure occurs by crushing of the concrete before the steel starts to yield. The failure is sudden, and takes place during Stage 2; Stage 3 (the plastic stage) does not occur.

Sustained loading reduces the load-bearing capacity of the concrete (see page 60). This has less effect in the case of an under-reinforced member, in which the primary cause of failure is the yielding of the steel, but the load-bearing capacity of an over-reinforced member may be noticeably reduced when the loading is sustained.

3.2 Cracking

The tensile resistance of a reinforced concrete or prestressed concrete beam depends on the modulus of rupture. This is the nominal tensile stress due to bending at which the unreinforced concrete fails, and is determined on the assumption of a linear distribution of stress in a homogeneous material (see Chapter 2). The strength of the concrete in direct tension is appreciably less than the modulus of rupture, and may be only half its value. As soon as this tensile strength (direct) is exceeded, very fine and short cracks develop which are visible only under a microscope or are indicated by the readings of strain-gauges, the total

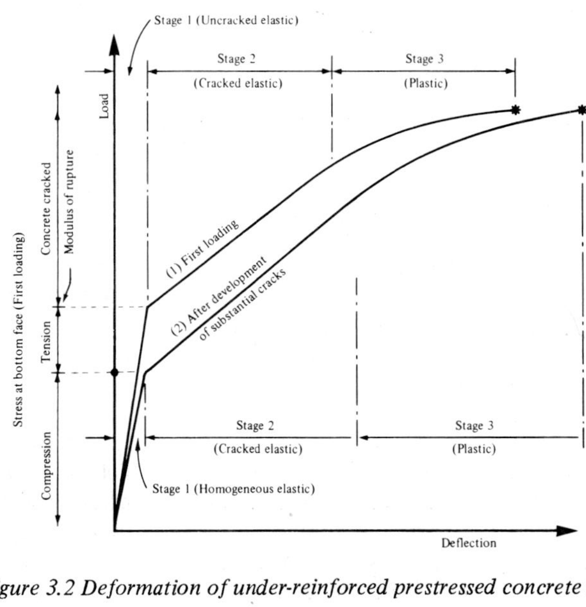

Figure 3.2 Deformation of under-reinforced prestressed concrete beam

strain at the edge exceeding the extensibility of the concrete. The cracks become visible when they are deeper at a greater load. Close examination of the relation between stress and deformation before and during cracking shows that the sharp change of slope between Stages 1 and 2 shown in Figures 3.1 and 3.2 is not correct. In fact the linear diagram of the first stage becomes curved as soon as the tensile strength of the concrete is reached when microscopic cracks develop (see Figure 3.3).

The stage at which cracks become visible depends mainly on the amount of reinforcement or prestressing steel and on its distribution. If the amount is small the increase in deformation which occurs during cracking will be greater than that which occurs with a larger quantity, and consequently cracks will be visible under a smaller stress in the first case than in the second. Similarly, if a small number of reinforcing bars or prestressing cables, each of large diameter, are used cracking will become visible at a smaller load than would be the case if an equal area of bars or cables, each of smaller diameter, were provided. Another major influence is the thickness of concrete covering the steel; the greater the cover the sooner cracks will become visible. The experience and eyesight of the observer are also of importance, since it is only to a skilled observer that the stage at which cracks become visible is quite definite.

The stress at which microscopic cracks occur depends solely on the tensile strength of the concrete; such cracks therefore develop when the tensile stress is between 300 and 600 lbf/in^2 (21 and 42 kgf/cm^2; 2.1 and 4.1 N/mm^2), depending on the strength. On the other hand the stress at which cracks become visible depends on many factors and can be predicted only within wide limits. These

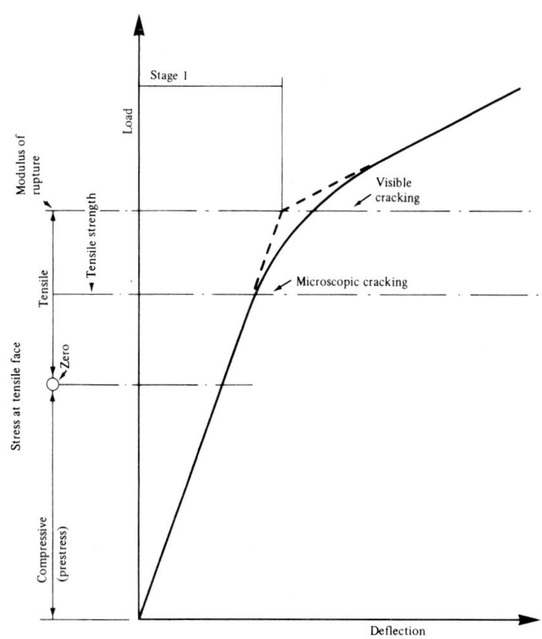

Figure 3.3 Visible and microcracking

may be 600 and 1200 lbf/in^2 (42 and 84 kgf/cm^2; 4.1 and 8.3 N/mm^2) for cracking visible to a trained observer, whereas an unskilled observer may not notice cracking in a beam with a high percentage of steel until the nominal tensile stress is well above 1000 lbf/in^2 (70 kgf/cm^2; 6.9 N/mm^2). The limits of stress given in the foregoing correspond to concretes of medium and high strength respectively. With pre-tensioned steel which is well distributed and concrete with a strength of about 7000 lbf/in^2 (500 kgf/cm^2; 45 N/mm^2), visible cracks may occur at a nominal tensile stress of 1000 lbf/in^2 (70 kgf/cm^2; 6.9 N/mm^2).

The shape of the section also has an effect on the cracking. In a rectangular or I-shaped section the cracks are restrained by the steel, to a greater extent (depending on the amount of steel); their greatest width occurs at the tensile face of the member and they extend gradually to the neutral axis. In a T-shaped section the neutral axis is usually either within the flange or close to it. As soon as they appear, the cracks penetrate almost to the neutral axis. The steel in the web does little to restrain them, and they are usually wider in the web than they are at the bottom face, except when longitudinal steel is provided on both sides of the web.

3.2.1 Development of cracks

While the nominal tensile stress at which visible cracks first occur depends mainly on the quality of the concrete and the quantity and distribution of the steel, the further development of cracking is greatly influenced by the efficiency of the bond between the steel and the surrounding concrete. If the steel is well distributed and the bond is satisfactory, more cracks develop and the width of each separate crack remains small. Full elongation of the steel takes place only close to each crack, consequently the total deformation is limited and Stage 2 is comparatively long. In the case of a prestressed beam with non-bonded or inefficiently bonded steel, however, the steel elongates uniformly along its entire length and the deformation of the beam is therefore much greater. The pattern of cracking is also much less satisfactory, as fewer but wider cracks develop, which fork as they approach the neutral axis and the range of Stage 2 is much curtailed. Such a crack formation is generally a sign of insufficient bond; it can be avoided by providing well-bonded non-tensioned steel near the tensile face.

3.3 Failure

The behaviour of a prestressed concrete beam at the load causing failure depends basically on whether the section is under-reinforced or over-reinforced, but may be modified by the degree of bond between the steel and the concrete.

3.3.1 Sections with well-bonded steel

In a section with well-bonded steel the only important factors affecting the strength are the ultimate resistance of the concrete to compression, the ultimate tensile resistance of the steel, and the lever arm between these forces.

In an under-reinforced section the steel is the weaker part and failure is primarily due to the steel reaching its ultimate resistance, although with really good bond there may be an apparent increase in the ultimate strength of the steel. This is due to the fact that the local reduction in cross-sectional area which normally occurs in the steel as the stress approaches the ultimate is not so severe when the efficiency of the bond is unimpaired (see page 55); however this property is unreliable and is ignored in design. If the steel has a distinct yield point failure will occur when this yield point is exceeded. The magnitude of the prestressing force influences the magnitude of the deformation, but has hardly any effect upon the ultimate load. The neutral plane rises as failure is approached and large deformations take place. The steel may either rupture (if the concrete has a very high strength) or reach a stress close to the proof or yield stress, in which case failure occurs by crushing of the concrete after the limit of deformation is reached. It is also possible in sections with bonded steel for failure to occur simultaneously with cracking, if the section is either so highly prestressed that cracks cannot develop before the ultimate compressive strength of the concrete is reached, or the section is so under-reinforced that the tensile resistance of the steel is less than the tensile resistance of the uncracked concrete. Such cases should be avoided.

In an over-reinforced member the concrete is the weaker part, and failure will occur when the resistance of the concrete in the compressive zone is reached. The magnitude of the prestressing force also has an effect upon the ultimate load, since this depends on the greatest deformation or strain which can occur in the concrete. This deformation is caused partly by the prestressing force and partly by the load, and therefore an increase in the prestressing force would reduce the ultimate load.

For purposes of practical design the approximation is usually made that the distribution of stress in the compressive zone at ultimate load is rectangular.

3.3.2 Sections with non-bonded steel

Since the steel is free to elongate over its entire length it will rarely reach its ultimate resistance before the concrete fails in compression. Experiments have shown that in this case also the distribution of strain across the section is linear. As the load is increased, the depth of the compressive zone is reduced and the neutral axis, or axis of zero strain, moves upwards.

The relation between the greatest contraction of the concrete and the corresponding elongation of the steel depends on the percentage of steel and the properties of the two materials. It therefore follows that the ultimate load depends solely on the greatest strain reached in tension and compression, since the rate of strain of the steel determines the extent of cracking and hence the depth of the compressive zone and the intensity of the compressive stress. The rate of strain in the steel under additional loading, however, is dictated by the magnitude of the effective tensile stress in the steel (Figure 3.4) and therefore the ultimate strain in the concrete (and hence the ultimate resistance of the section) depends on the magnitude of the prestress also.

It should be noted that this reasoning cannot be applied to sections with bonded steel, since test results appear to indicate that the elongation of the steel under excessive load is not uniform, as the strain near a crack must be greater than the strain between cracks. Tests have also shown that the strength of non-bonded tendons can be utilized as fully as that of pre-tensioned steel, provided that additional bonded non-tensioned steel is placed close to the tensile face. This co-operates fully with the concrete up to the point of failure; it also restrains the width of the cracks and improves their distribution.

Summary. In general, then with bonded steel the critical factors for under-reinforced sections are the total ultimate force developed in the steel and also to

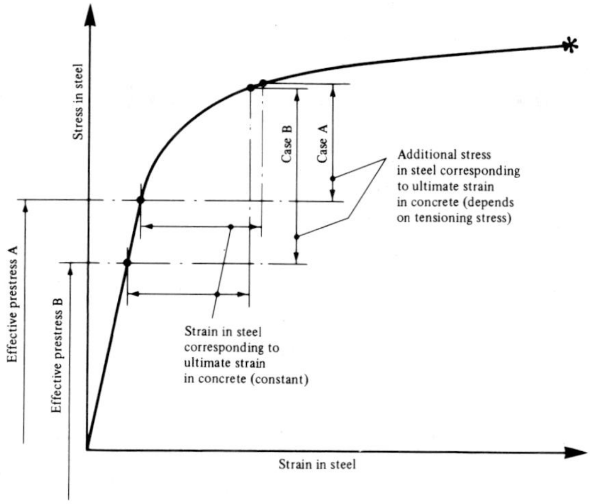

Figure 3.4 Effect of prestress on ultimate strain (unbonded steel)

a certain extent, the compressive strength of the concrete, as this affects the depth of the compressive zone and hence the lever arm. For over-reinforced sections the critical factor is the ultimate compressive force developed in the concrete, which depends on the maximum strain in the concrete. With non-bonded steel the magnitude of the prestress and the maximum tensile and compressive strains determine the ultimate load.

3.4 Resistance to shear

Tests show that as in flexure so in shear performance characteristics of a reinforced concrete member and a prestressed (fully or partially) concrete member are very similar at ultimate condition. Yet in all codes of practice without exception, the method of assessment of resistance to shear is different for reinforced concrete from that of prestressed concrete even when the limit state of collapse is considered.

In both cases (reinforced or prestressed), however, the method of assessment is primarily based on 'nominal shear stress' across the full section regardless of cracks which almost invariably develop before failure. Such approach is in fact a factored extension of working stress design principle and does not appear to be fully in line with the limit state philosophy.

From observations of tests on flexural members it is also apparent that there are basically three modes of failure as outlined hereunder:

(1) Flexural failure mode characterised by vertical cracks: This mode has already been discussed and it appears to be the governing consideration when the shear span/effective depth ratio exceeds 6. (Shear span is defined as the ratio of bending moment to shear at the section considered).

(2) Failure mode characterised by diagonal splitting under biaxial stress field in the web: This is predominant normally within the shear span/effective depth ratio of 2 and mostly in sections having thin web in that zone.

(3) Failure mode characterised by bond imposed inclined cracks initiated by flexural cracks: The combined effect of shear and bending moment is the

critical factor for such failure mode and this usually occurs within the shear span/effective depth range of 2 to 7.

A realistic treatment of shear is possible only when the characteristics of each mode are recognised and due account is taken in the design procedure.

In addition of course, failure may be initiated by slip at anchorage either due to insufficient development length at bearing or close spacing of tendons. Good detailing practice can take care of such failure.

In the design of reinforced concrete it is customary to consider the resistance to shear and bending as though they are independent without any interconnection. This practice may be adequate when one of the actions (i.e. shear or bending) is predominant relative to the other, (like mode 1 or 2 above), but it could be dangerous to ignore the combined effect in case of mode 3 where large bending moment is accompanied by a large shearing force at a section (for example under point loads or at supports of continuous beams).

The existence of interconnection was shown by Whitney(4) and Abeles(5). More recently Bobrowski and Bardhan-Roy(6) have devised a suitable method of design based on test results which is described in Chapter 7. This has been incorporated in CEB/FIP Manual of Lightweight Aggregate Concrete(7) and in the reference (3).

Figure 3.5 taken from reference (6) shows the moment shear interconnection and the safe zone for design. It also shows that the full flexural resistance is available only when the shearing force is small. Thereafter it reduces significantly as the shear increases.

As stated earlier the treatment of shear in the codes of practice is based on permissible nominal shear stress.

In CP 110 these stresses are related to the concrete grade and percentage of steel while prestressing effect has been separately allowed for. In the Model Code allowable stresses depend only on the concrete quality although the percentage of steel as well as prestressing effect has been taken care of in a different way. In the ACI Building Code the limiting shear stress is related to the concrete strength, the moment and shear at the section concerned and the effective prestress.

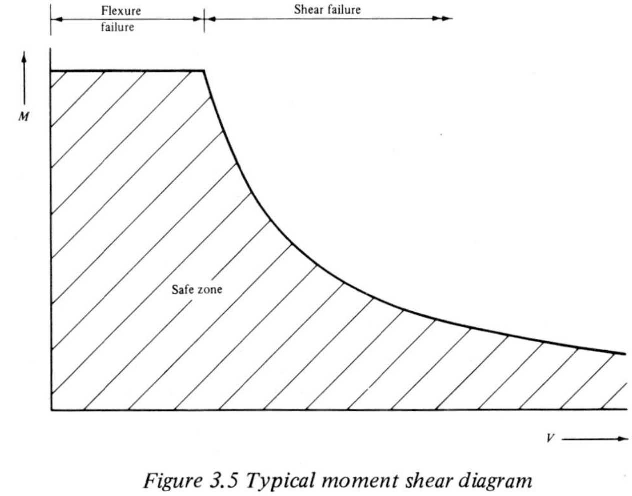

Figure 3.5 Typical moment shear diagram

Values of nominal shear stresses in CP 110 are based on the report of the Shear Study Group[8] Figure 3.6 obtained from the Shear Study Group and Regan[9] is of interest in as much as the solid line shown in the diagram refers to a safe limit up to ν_u/ν ratio of 4. The dotted line inserted by the authors shows that higher values are possible for lower shear span/depth ratio without endangering the safety of the structure.

Prestressing reduces the principal tensile stresses which may cause crack due to shear. Bent-up cables also reduce the principal tensile stresses under working load, and consequently there is sometimes no need to provide stirrups to resist shearing forces. However this may sometimes lead to reduced margin of safety at ultimate condition. Wide cracks caused by shearing forces remain open when the load is removed, and should at all times be avoided. Furthermore, failure due to shear occurs suddenly, with little warning, and it is therefore essential that the factor of safety against failure due to shear should be at least equal to that against failure due to bending. Failure near the ends of a member due to shear alone can occur only when the prestress is small, although failures due to horizontal splitting at the ends may occur during or immediately after prestressing if stirrups are not provided. In the great majority of cases failure would be caused primarily by diagonal cracking due to combined bending and shearing stresses; this may lead directly to crushing of the concrete, or to bond-slip or yielding of the steel with consequent crushing of the concrete.

As cracks due to bending become diagonal only above the level of the main steel, the danger of failure due to diagonal cracking is increased if the steel is remote from the tensile face, as the depth of concrete available to resist the diagonal tension is thereby reduced. It is therefore undesirable for all, or a large part of the steel to be remote from the tensile face at positions of point loads; this should be particularly considered when designing cable profiles for continuous beams or framed structures.

It should also be noted that the width of inclined cracks can be greatly reduced by the provision of sufficient closely-spaced stirrups; in this way a stage can be reached at which failure occurs due to excessive principal compressive stresses, rather than principal tensile stresses.

3.5 Resistance to fatigue

Prestressed concrete is not usually subjected to a large range of repeated loading, except in the case of special constructions such as railway sleepers. For such structures it is, of course, important that the resistance to fatigue should be investigated over the full range of load, but in normal structures in which the dead weight is considerable the range is limited to that portion of the live load which recurs regularly. Occasional loads have little long-term effect if they occur at intervals of thirty minutes or longer, and may be considered as static loads. It is usually sufficient to ensure that the factor of safety against failure under static loading is obtained after one or two millions of cycles of the range of fatigue loading have been applied, and it is not generally necessary to relate the required factor to the limit of endurance.

Prestressed concrete structures should preferably be under-reinforced (when the ultimate load depends mainly on the steel) or the design may be balanced (when the compressive resistance of the concrete is also important).

It has been established that failure of concrete due to fatigue is highly unlikely

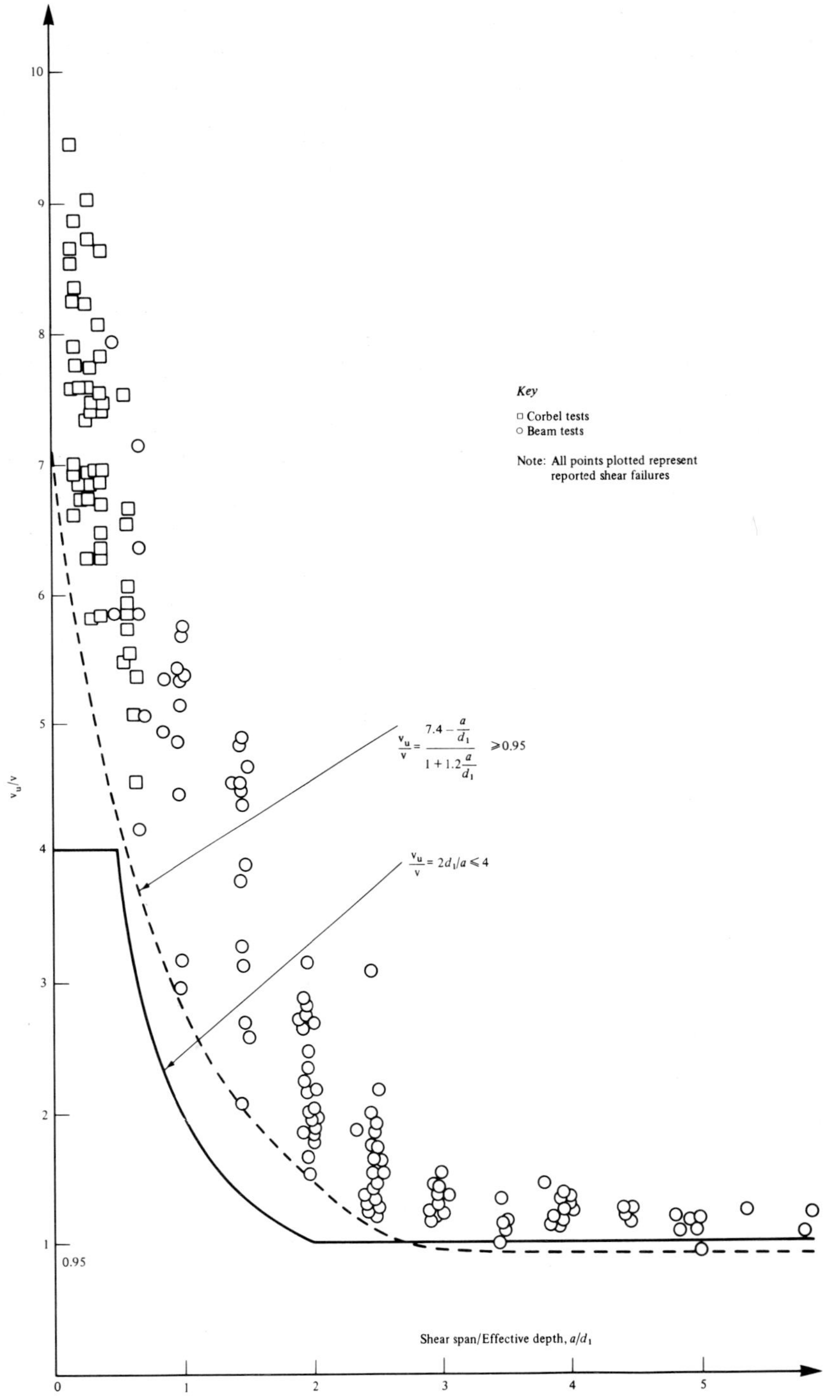

Figure 3.6 Effect of shear span on shear stress

within the range of normal permissible compressive stresses. The steel, however, is subjected to high stresses at all times, and the question of resistance to fatigue requires careful consideration.

It is shown in Chapter 2 that different considerations apply to structures with well-bonded steel and structures with poorly-bonded or non-bonded steel. If the structure remains uncracked little increase of the stress in the steel takes place due to live load and the range of repeated stress in the steel is small, being the dynamic stress in the concrete multiplied by the modular ratio. There is therefore no danger of failure occurring due to fatigue, even if the initial tensile stress in the steel is high, so long as the concrete adjacent to the steel remains in compression. When the steel is well bonded its resistance to fatigue is improved by the adjacent concrete. This is particularly desirable if tension occurs in the concrete provided no visible cracks develop. The development of these cracks depends on the range of fatigue stress (Figure 3.7); tests carried out by British Railways (Eastern Region) showed that specimens remained free from cracks when the repeated stresses ranged from 650 lbf/in^2 (46 kgf/cm^2; 4.5 N/mm^2) in tension to 100 lbf/in^2 (7 kgf/cm^2; 0.7 N/mm^2) in compression for 1×10^6 loading cycles[10]. Although microscopic cracks developed under a tensile stress of 500 to 550 lbf/in^2 (35 to 39 kgf/cm^2; 3.5 to 3.9 N/mm^2), no visible cracks occurred after a million cycles of the same range of stress were applied. If the range is large, commencing with a compressive stress of the order of 1600 lbf/in^2 (113 kgf/cm^2; 11.1 N/mm^2), the permissible limit of the tensile stress, based on the results of similar tests, may be assumed to be about 400 lbf/in^2 (28 kgf/cm^2; 2.8 N/mm^2). These tests have shown that this fatigue loading has no effect on the load at which visible cracks occur in a subsequent static test.

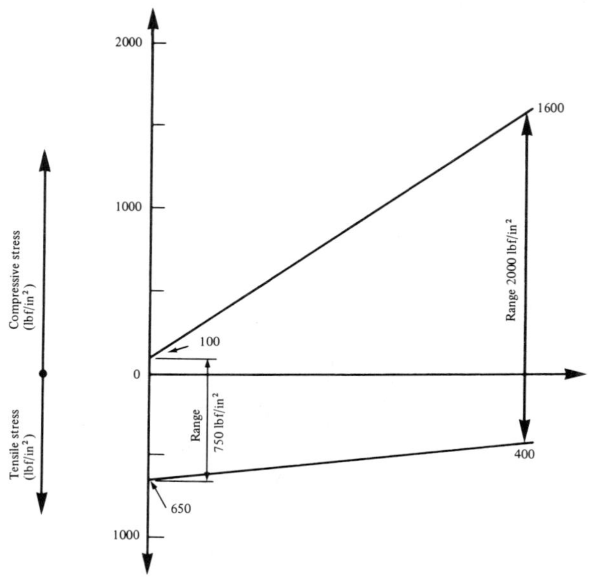

10 000 lbf/in^2 = 700 kgf/cm^2 = 70 N/mm^2

Figure 3.7 Influence of range of fatigue on cracking

The appearance of hair cracks has no effect on the resistance to fatigue so long as the cracks remain fine. Tests have shown that the resistance to fatigue is not reduced even after many millions of loading cycles are applied. As soon as the cracks become wider, however, the bond is destroyed near the cracks, and the resistance to fatigue then depends only on the magnitude of the prestress (which governs the range of stress) and the properties of the steel; for example, the resistance to fatigue of indented wire is reduced by the 'notch' effect.

In the case of structures with non-bonded steel, the range of stress in the steel due to live load is much greater, and the resistance to fatigue of the structure depends solely on the resistance of the steel (see Figure 2.4). This is because the steel is free to deform as though it were acting alone, and its properties are unaffected by the concrete. When such structures are subjected to static loading a relatively high initial pre-stress in the steel is usually advantageous, since it tends to increase the resistance of the structure. Where fatigue loading is likely, however, the effective stress in the steel due to dead load only, and the range of repeated stress, must conform to the resistance of the steel to fatigue, and this may require the avoidance of high tensioning stresses.

The modulus of elasticity of an uncracked structure, or a structure with hair cracks, is little affected by cyclic loading, but a small permanent deformation may occur which tends to increase slightly with the number of cycles, the extent of cracking, and the range of load. In Figure 3.8 curve A shows the relation between load and deflection for the first static loading of a beam up to cracking(11); curve B shows the relation after the application of a million cycles of a range of fatigue stress from 100 lbf/in^2 (7 kgf/cm^2; 0.7 N/mm^2) in compres-

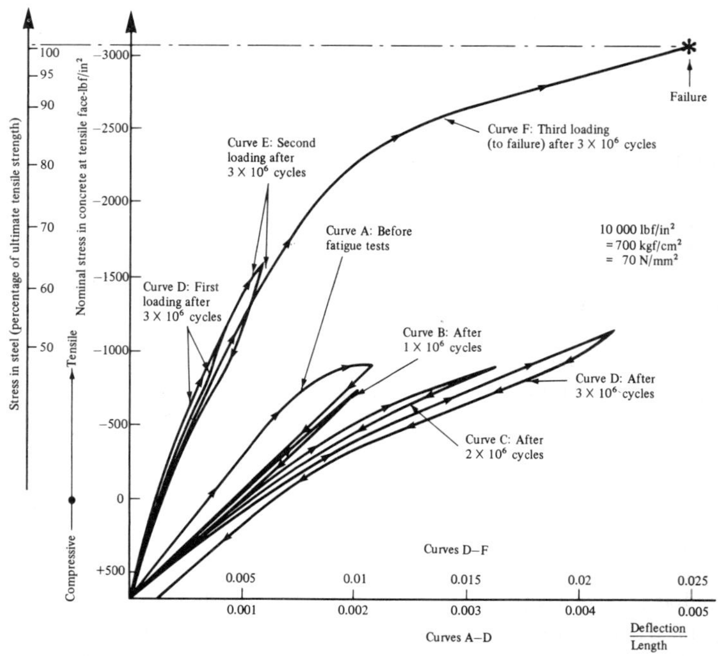

Figure 3.8 Fatigue loading (Liege, 1951)

sion to 600 lbf/in^2 (42 kgf/cm^2; 4.2 N/mm^2) (nominal) in tension; curve C shows the relation after the application of another million cycles of stress ranging from 100 lbf/in^2 (7 kgf/cm^2; 0.7 N/mm^2) in compression to 800 lbf/in^2 (56 kgf/cm^2; 5.6 N/mm^2) (nominal) in tension; and curve D shows the relation after a third million cycles of stress ranging from 400 lbf/in^2 (28 kgf/cm^2; 2.8 N/mm^2) (nominal) in tension to 1000 lbf/in^2 (70 kgf/cm^2; 6.9 N/mm^2) (nominal) in tension. (The tensile stresses given are based on the uncracked section and are therefore nominal only, as cracking occurred during the first static loading.) Curves D, E and F are to a different scale from curves A, B and C; curve E is for a higher static load and curve F is for static loading to failure. The magnitude of the static load causing failure was not reduced by the preceding three million cycles of loading in the cracked state, as compared with a companion beam, statically loaded to failure.

Tests at Duke University(12, 13) demonstrated that prestressed concrete can withstand a relatively large number of loading cycles over a large loading range. For example, 200 000 cycles of loading ranging from 30 per cent to 70 per cent of the static failure load (corresponding to the dead load and to 1.82 times the maximum design load respectively) were withstood, as were 30 000 cycles ranging from 30 percent to 80 per cent of the static failure load (2.27 times the maximum design load) and even a few thousand cycles between 30 per cent and 90 per cent of the static failure load (2.78 times the maximum design load). The results were compared with tests obtained from non-embedded prestressing strands. The cumulative effect was found on average to agree with Miner's hypothesis(14); that is

$$\sum \frac{n_i}{N_i} \leqslant 1$$

in which n_i is the number of cycles over a particular loading range actually applied and N_i is the number of cycles over the same loading range which would cause failure. This results in a damage line giving a curve of smaller capacity than the $S - N$ curve (see Chapter 2). In practical application of Miner's hypothesis some safety margins should be introduced. In other words the left hand term of the equation should be multiplied by a suitable safety margin, such as 1.25 for prestressing steel, 1.50 for prestressed concrete with excellent bond and 2.00 for prestressed concrete with moderate bond(15).

3.6 Resistance to impact

The two most important properties which a structure subjected to impact must possess are the ability to absorb the impact and a high resistance to local damage. It is therefore essential to provide a structure which may safely deform to a considerable extent (that is, the load causing cracking should be only a small proportion of the load causing failure) and local brittleness must be overcome by the provision of stirrups and secondary reinforcement, such as additional non-tensioned reinforcement in the tensile and compressive zones. Prestressed concrete has the great advantage that it can be designed and constructed to ensure that deformations due to impact disappear almost completely, whereas a reinforced concrete structure with an equal capacity to absorb impact would sustain a substantial permanent deformation. On the other hand, structures which are of highly-prestressed concrete, without non-tensioned steel and possessing little flexibility, will absorb much less impact than a corresponding reinforced concrete structure and would therefore collapse under impact, such as that due to an earthquake or explosion.

It is vitally important that the tendons should be securely anchored, especially when they are non-bonded; otherwise they may be dislodged by a sudden impact, and their stored energy is such as to cause them to leave the member like a missile, leading to the complete collapse of the members. Earthquakes are known to have caused failures of this type.

3.7 Effects of vibration

Vibration is of little importance in normal reinforced concrete structures, as their section and mass are usually such as to provide adequate stiffness and damping with little chance of resonance, and it is considered only in the case of heavy machinery. With prestressed concrete, however, the tendency is to use thinner and shallower members which could bring the natural frequency of vibration of the structure close to the frequency of the normally applied superload and thus cause near resonance conditions. Moreover very little inherent damping is available in prestressed concrete. This also applies to test specimens subjected to cyclic loading; the natural frequency of the specimen must not be resonant with the frequency of the applied loading.

It is therefore necessary to investigate the natural frequency of thin prestressed concrete members employed in such structures as bridges, and long span floors where vibration would be important (e.g. dance halls) and in structures which may be exposed to wind-excited oscillations(16). The approximate method which follows is often sufficient, but reference should be made to a standard text book(17, 18) whenever more accurate information is required.

The fundamental natural frequency of a prismatic beam is $k/\sqrt{y_d}$ where y_d is the maximum static deflection of the member and k is the coefficient, depending on the end condition and loading. For uniformly distributed load it is 3.89 for cantilevers, 3.52 for simply supported beams and 3.51 for built-in members, where y_d is in inches and the frequency is in Hertz (cycles/sec). Higher frequencies can occur if the mode of vibration differs from the fundamental, but these are of less importance. The natural frequency should be compared with that due to the imposed loading. Resonance occurs when these are equal, and when the ratio of the imposed frequency is less than about 1.5 times the natural frequency(17, 18) the resultant amplitude of the vibration virtually equals the static deflection, upon which they are superimposed (see Figure 3.9).

The dissipation of energy by internal friction exercises a damping effect on the amplitude of any vibration. This is due to the property of hysteresis which concrete possesses (page 57). The loss of energy increases when cracks develop and the deflection of the cracked sections is also greater; the frequency is therefore different from that of the uncracked member, and a structure which is resonant under normal conditions may no longer be so after cracking takes place.

Some observed frequencies of various types of loading are given in Building Research Station Digest No. 78. In addition to the avoidance of resonance, it is necessary to limit the amplitude of any vibration to a value which will not unpleasantly affect human sensitivity. Little information is available on this subject, but Figure 3.10 (from BRS Digest No. 78) gives some indication of maximum values.

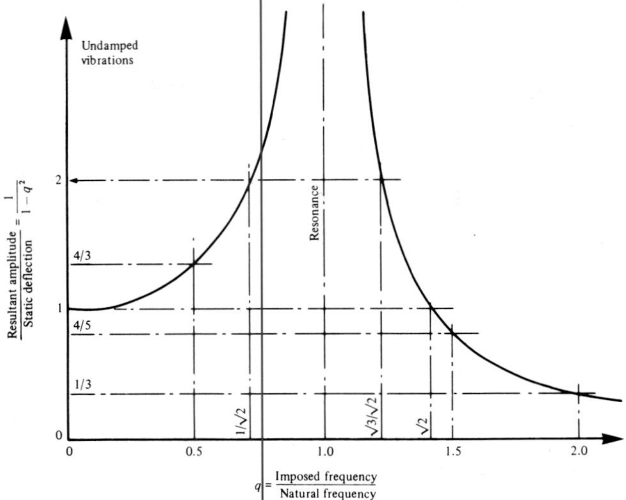

Figure 3.9 Resonance

In case of vibration, special problems may occur with excessive oscillations, as the natural damping of prestressed concrete is relatively low. Aerodynamic wind tunnel tests on models may be necessary in the case of such excitations due to vortex shedding and external damping may be necessary to control the amplitude of oscillations(16)

3.8 Resistance to earthquakes

Experience has shown prestressed concrete to be a very suitable form of construction for earthquake-resistant structures, provided that it is designed to be sufficiently flexible to absorb the impact, and that the steel is well bonded. With non-bonded steel, secure anchors are essential, as stated in the previous section.

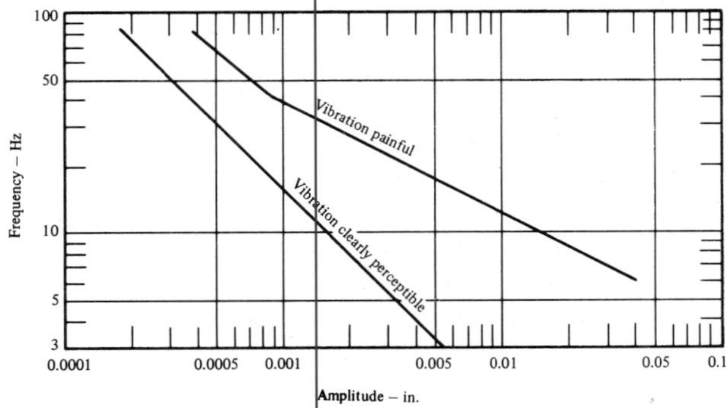

Figure 3.10 Human reactions to vibration

3.9 Sustained loading

Little information is available concerning the long-term increase in deformation which occurs under sustained loading, and it is difficult to design for a definite maximum deformation. Chapters 2 and 6 give details of the numerous variables which affect the magnitude of creep; the maximum possible deformation may be obtained by assuming safe limiting values for these variables, but this will rarely coincide with the actual deformation.

The resistance of plain concrete to sustained compression is between 70 per cent and 80 per cent of its ultimate resistance (Chapter 2), but the resistance of an under-reinforced prestressed concrete member in which the ultimate load is dependent on the steel is usually higher. Experiments have shown that such beams are able to sustain for a considerable time a load equal to 80 per cent to 90 per cent of the ultimate load, without showing any sign of approaching failure, although the deflections may increase considerably. It may therefore be concluded that there is no need to allow for a smaller ultimate load in consequence of sustained loading when the steel is the weaker part, but a reduction should be allowed when the concrete is the weaker part.

3.10 Resistance to fire

The resistance to fire of a load-bearing element is defined as the length of time during which the element may be exposed to fire without failure occurring. In general, resistances of half an hour, one hour, two hours and four hours are considered.

It is stated in Chapter 2 that the strength of prestressing steel decreases when its temperature is increased. The resistance to fire of prestressed concrete is governed by the rate at which the temperature of the steel increases and sufficient concrete cover to protect the steel should therefore be provided in simply supported beams. Tests at Fire Research Station England[19] show that a period of resistance of one hour is obtained for a freely supported member when the cover is 38 mm (1½ in.). If the required period is two hours a cover of 64 mm (2½ in.) is suitable, and a cover of 100 mm (4 in.) is required for a resistance of four hours. When the thickness of the cover is 64 mm (2½ in.) or more, mesh reinforcement with a cover of 25 mm (1 in.) should be provided to obviate spalling of the concrete. Tests in Holland have demonstrated that the resistance time is affected by the shape and size of the cross-section as well as the cover. This is why the tabulated data for fire rating suggests minimum cover in relation to minimum width and thickness. The rate at which the temperature of the steel rises may be reduced by the provision of insulating layers to protect the steel. A layer of vermiculite concrete 25 mm (1 in.) thick, for example, increases the period of resistance by about two hours, as also does a layer of vermiculite-gypsum plaster 22 mm ($\frac{7}{8}$in.) thick or a layer of sprayed asbestos 19 mm ($\frac{3}{4}$in.) thick; lightweight concrete or burnt clay also provide efficient insulating covers.

The tests at FRS[19] relate to beams with post-tensioned cables. In American tests a fire resistance of 2 hours was obtained on double-T slabs and channels with pre-tensioned strands, the cover being 41 mm ($1\frac{5}{8}$ in.) at the bottom (compared with 64 mm (2½ in.) given earlier).

Failure of prestressed concrete due to fire is unlikely to occur suddenly. As the strength of the steel decreases the prestressing force is gradually reduced. A progressively increasing deflection takes place, with a consequent increase in cracking, which together give a clear visual warning of approaching collapse. Prestressed beams which have survived fire without collapsing may still have some permanent residual deflection and loss of prestress. If water is present in the ducts, however, this turns to steam when it is heated and may cause local explosive damage, probably followed by sudden failure.

Tests have been made in the USA[20, 21] to compare the resistance to fire of prestressed beams made with ordinary and lightweight concrete; indicating superiority of lightweight. They show that an increase in thickness of the cover does not lead to a proportionate increase in the resistance time. Other tests showed that the resistance time is greatly increased when the ends of the beams are restrained or continuous. In members with ends restrained or continuous, the cover is relatively unimportant.

The results of a test made on restrained T-beams[22] are summarized in Figure 3.11. These show that the beams remained fully serviceable, with small deflections, for one hour, and collapse was still avoided after four hours. The determination of the two durations relating to the limit states of serviceability and collapse, by means of a single test, is a most useful development.

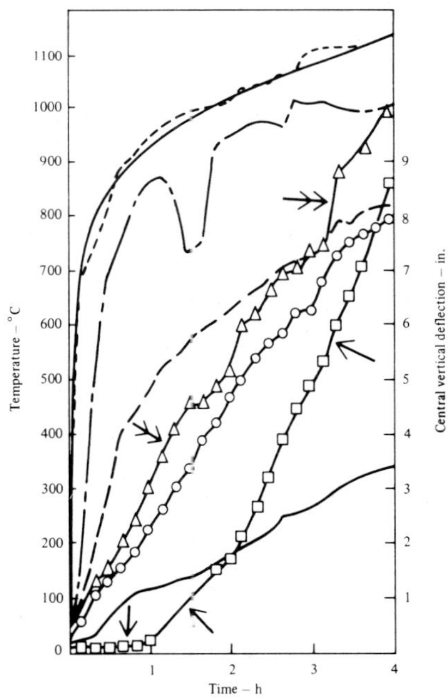

——— Standard time-temperature curve
- - - - Actual mean furnace temperature
△——△ Maximum temperature of Dyform prestressing strands in webs
○——○ Mean temperature of Dyform prestressing strands in webs
—.—.—. Maximum temperature of 5 mm (0.2 in.) wires in flanges
– – – – Mean temperature of 5 mm (0.2 in.) wires in flanges
□——□ Central vertical deflection
——— Mean temperature of reinforcing mesh in structural topping

Figure 3.11 Results of fire test of restrained T-beam

3.11 Corrosion and durability

Reference should be made to the appropriate sections in Chapter 2. The durability of prestressed concrete is generally better than that of reinforced concrete, provided that the aggregates are satisfactory and the concrete is well compacted; with pre-tensioning this is essential in any case, to obtain good bonding. With post-tensioning, it is necessary to ensure that the tendons are properly encased, either by good grouting or by other and suitable forms of protection, to avoid corrosion.

REFERENCES

1. COMITÉ EUROPÉEN DU BÉTON. Manual technologie des structures en betons legeos. Final draft, May 1973. Paris C.E.B. 1973.
2. THE CONCRETE SOCIETY/INSTITUTION OF STRUCTURAL ENGINEERS. *Design and Detailing of Concrete Structures for Fire Resistance.* Interim Guidance by a Joint Committee of the Institution of Structural Engineers and the Concrete Society, May 1978.
3. BOBROWSKI, J. and BARDHAN-ROY, B. K. Partial Prestressing – Development and Design Recommendations. Paper presented at an FIP Symposium, Bucharest, September 1980.
4. WHITNEY, Charles S. Ultimate shear strength of reinforced concrete flatslab, footings, beams and frame members without shear reinforcement. *Journal of ACI*, October 1957. Vol 29, No. 4.
5. ABELES, P. W. *Diagonal shear failure in prestressed concrete beams with well-bonded steel.* Proceedings of the Third FIP Congress, Berlin 1958. pp. 133–139.
6. BOBROWSKI, J. and BARDHAN-ROY, B. K. A method of calculating the ultimate strength of reinforced and prestressed concrete beams in combined flexure and shear. *The Structural Engineer*, Vol. 47 No. 5, May 1969. pp. 197–209.
7. CEB/FIP. CEB/FIP Manual of Lightweight Aggregate Concrete – Design and Technology. The Construction Press, 1977.
8. INSTITUTION OF STRUCTURAL ENGINEERS. Shear Study Group. *The shear strength of reinforced concrete beams.* London 1969. p. 2, 47, 84.
9. BAKER, A. L. L., YU, C. W. and REGAN, P. E. Explanatory note on the proposed Unified Code clause on shear in reinforced concrete beams with special reference to the Report of the Shear Study Group. *The Structural Engineer*, Vol. 47, No. 7, July, 1969. pp. 285–293.
10 ABELES, P. W. Prestressed concrete bridges. Cumulative effect and range of fatigue loading. Paper presented at the IABSE Sixth Congress. Stockholm, 1960. 27 June–1 July 1960. Zurich, The Association, 1961, Final Report. pp. 377–384.
11. ABELES, P. W. Fatigue tests on partially prestressed concrete members. Paper presented at the IABSE Fourth Congress. Cambridge, 25 August–5 September 1952, Zurich, The Association, 1953. Final Report. pp. 463–467.
12. ABELES, P. W., BROWN, E. I. and SLEPETZ, J. M. Fatigue resistance of partially prestressed concrete beams to large range loading. Paper presented at the IABSE Eighth Congress. New York, 9–14 September 1968. Zurich, The Association, 1969. Final Report. pp. 925–936.
13. ABELES, P. W., BROWN, E. I. and HU, C. H. Fatigue resistance of under-reinforced prestressed beams, subjected to different stress ranges: Miner's Hypothesis. Abeles Symposium. Fatigue of Concrete, Detroit, American Concrete Institute, 1974. pp. 277, SP 41.
14. MINER, M. A. Cumulative damage in fatigue. *J. Appl. Mech. Transactions of the American Society of Mechanical Engineers*, Series E. Vol. 67. September 1945. pp. A159–A164.
15. ABELES, P. W. and BOBROWSKI, J. The resistance of prestressed concrete members to dynamic loading. Paper presented to the FIP Seventh International Congress, New York, 26 May–1 June 1974. London, Concrete Society, 1974. pp. 11.

16. BOBROWSKI, J., ABELES, P. W. and BARDHAN-ROY, B. K. The design of cantilever roofs to control dynamic effects of wind by external damping. Paper presented to the FIP Seventh International Congress. New York. 26 May–1 June 1974. London, Concrete Society, 1974. pp. 17.
17. TIMOSHENKO, S. P. and YOUNG, D. H. *Vibration problems in engineering.* Third edition. London, Van Nostrand, 1955. pp. 468.
18. STEFFENS, R. I. Some aspects of structural vibration. Skip, B. O. ed. *Proceedings of Symposium on Vibration in Civil Engineering*, London, April 1965. London, Butterworth 1966. pp. 1–30.
19. ASHTON, L. A. and BATE, S. C. C. The fire resistance of prestressed concrete beams. *Proceedings of the Institution of Civil Engineers.* Paper No. 6444. Vol. 17, October 1960. pp. 15–38.
20. CARLSON, C. C. Fire resistance of prestressed concrete beams. Study A – Influence of thickness of concrete covering over prestressing steel strand. Skokie Portland Cement Association, 1962. pp. 40. Research and Department Bulletin.
21. GUSTAFERRO, A. H. and CARLSON, C. C. An interpretation of results of fire tests on prestressed concrete building components. *Journal of the Prestressed Concrete Institute.* Vol. 1, No. 5, October 1962. pp. 14–22.
22. ABELES, P. W. and BOBROWSKI, J. Fire resistance and limit state design. *Concrete.* Vol. 6, April 1972. pp. 33–35.

CHAPTER 4

TYPES OF PRESTRESSING STEEL

4.1 Prestressing tendons

Prestressing tendons normally take the form of separate wires, wires spun together helically to form strands, or bars. For pre-tensioned steel, wires, strands, and occasionally bars are used singly, to permit the concrete to bond directly to them; when post-tensioning is used, it is common practice to group the separate tendons together, so as to reduce the number of anchorages and ducts required to accommodate them. When grouped in this way, the tendons in each duct are usually termed a 'cable'.

British Standards for Tendons. The requirements for prestressing tendons are set out in the following British Standards:

BS 5896:1980: Steel wire for prestressed concrete – High tensile steel wire and strand for the prestressing of concrete
BS 4757:1971: Nineteen – wire steel strand for prestressed concrete
BS 4486:1969: Cold worked high tensile alloy steel bars for prestressed concrete

Wires and Strands. The wire is required to be cold drawn from plain carbon steel (BS 5896) or patented plain carbon steel (BS 4486). The usual chemical composition is shown in Table 4.1.

Table 4.1 Chemical composition of alloying elements of prestressing wires and strands

Element	Minimum %	Maximum %
Carbon	0.60	0.90
Silicon	0.10	0.35
Manganese	0.50	0.90
Sulphur	–	0.04
Phosphorus	–	0.04

The drawn wire is to be free from surface or other defects, and the finished wire or strand must be free from oil and grease unless otherwise specified by the purchaser. Superficial rusting is allowed, provided that there is no visible pitting of the surface. In the case of *wire* (BS 5896), the coils supplied to the purchaser must not include welds; except that, by agreement between the buyer and the supplier, special lengths may be drawn from rods welded before the patenting process is applied. For *strands*, no length of strand may be joined to another by any method, though separate wires within the strand may be welded together prior to patenting. No welding is allowed after patenting or during or after wire drawing. If special lengths of seven-wire strand (BS 5896) are required, and provided the user is made fully aware of the reduced mechanical properties involved, not more than one wire in any 40 m (130 ft) may be welded after patenting or drawing; this relaxation applies to seven-wire strand only.

The tolerance on the nominal diameter of *prestressing wire* (BS 5896) is ±0.04 mm (0.0016 in.) for wires under 4.5 mm (0.177 in.) in diameter, and ±0.050 mm (±0.002 in.) for wires of 4.5 mm (0.177 in.) or more in diameter. If the wire is to pay out straight from the coil, the internal diameter of the coil shall not be less than 1.8 m (6 ft) for wires of 7 mm (0.276 in.) diameter or greater, or 1.2 m (4 ft) for wires less than 7 mm (0.276 in.) diameter. In *seven-wire strands* (BS 5896) the nominal diameter of the centre wire is to be at least 2 per cent greater than that of the surrounding wires; after heat treatment, it is to be wound onto coils of such a size [and not less than 600 mm (2 ft) in any case] that it pays off 'reasonably straight'. The tolerances on the nominal diameter of the finished strand are +0.4 mm (+0.016 in.) and –0.2 mm (–0.008 in.) for 12.5 and 15.2 strands. For *nineteen-wire strand* (BS 4757), different requirements are laid down for treated and 'as-spun' strands. Treated strand has a nominal diameter of 18 mm (0.725 in.) with tolerances on diameter of +0.5 mm (+0.02 in.) and –0.25 mm (–0.01 in.). The treatment comprises low-temperature heating as a continuous linear process, after which it is to be wound on to coils with a minimum diameter of 900 mm (3 ft), from which it pays off 'substantially straight'. 'As-spun' strands, with nominal diameters of 25.4 mm (1 in.), 28.6 mm (1.125 in.) and 31.8 mm (1.25 in.) have tolerances on diameter of +0.6 mm (+0.024 in.) and –0.25 mm (–0.01 in.) in all cases; no heat treatment is required, and the minimum coil diameter is 1.5 m (5 ft).

Testing. In the case of *wire* the manufacturer should provide one load/extension curve for each parcel of wire, a parcel being defined as any quantity of finished wire presented for testing at any one time. Tests are to be made on samples taken from the end of one coil in every five within the parcel, but the results of these are only required to be kept available for inspection by the purchaser. Specimens are tested for characteristic strength, proof stress, and reverse bend tests; relaxation tests may also be called for. The specified values are shown in Table 4.2; the wire is deemed to comply with the requirements for specified characteristic strength provided that not more than two of any 40 consecutive results fall below the specified value, no results are less than 95 per cent of the specified value, and none are more than 230 N/mm^2 (24 kgf/mm^2; 33 600 lbf/in^2) above it.

For *strands*, the manufacturer is required to provide dated test certificates prepared from the relevant test results. Tests are to be made on samples cut from each coil; they comprise a tensile test, an elongation test, a proof-load test (for seven-wire strand only), and – if required – relaxation test results. For seven-wire strand, proof-load tests and load-extension curves are called for only for one test piece in every five; for nineteen-wire strands a proof-load test is required only for treated strands; for these, the test, and the plotting of a load-extension curve, are specified for one test piece in every three. For 'as-spun' strand, load-extension curves are to be plotted for every test piece. The values specified are

Table 4.2 Mechanical properties of wires

Nominal wire diameter		Specified characteristic strength			Reverse bend radius		Conditions in which wire supplied (see below)
(mm)	(in.)	(N/mm²)	(kgf/mm²)	(lbf/in²)	(mm)	(in.)	
7	0.276	1570	160	227 650	20	0.8	1, 2
7	0.276	1670	170	242 150	20	0.8	1, 2
6	0.236	1670	170	242 150	15	0.6	1, 2
6	0.236	1770	180	256 650	15	0.6	1, 2
5	0.197	1670	170	242 150	15	0.6	1, 2, 3
5	0.197	1770	180	256 650	15	0.6	1, 2, 3
4.5	0.177	1620	165	235 000	15	0.6	1, 2, 3
4	0.1575	1670	170	242 150	10	0.4	1, 2, 3
4	0.1575	1770	180	256 650	10	0.4	1, 2, 3
3	0.118	1770	180	256 650	7.5	0.3	3
3	0.118	1860	190	269 700	7.5	0.3	3

Conditions in which wire is supplied

Number	1	2	3
Description	Cold drawn, pre-straightened, normal relaxation	Cold drawn, pre-straightened, low relaxation	Cold drawn (Mill Coil)
0.1% proof stress, as percentage of specified characteristic strength	83	83	80
Maximum in % relaxation after 1000 hours from:			
70% Initial stress	8	2.5	10
80% Initial stress	12	4.5	–

summarized in Table 4.3; the specified characteristic strength is defined in the same way as that for wire, except that no upper limiting value is imposed. The minimum elongation at failure is specified as 3.5 per cent, except for 'as-spun' nineteen-wire strand; no value is specified for this.

All three specifications included provisions for re-testing, in the event of failure of a sample to meet the requirements.

Bars. No chemical composition is given for the steel, except that sulphur and phosphorus must not exceed 0.04 per cent, but the manufacturer is required to provide the chemical analysis on request. Threads, if provided, are to be cold-rolled; no welds are permitted, and the bars are to be protected at all times from the effects of local heat. Tolerances are specified only on the mass; on the basis that the density of the steel is 7850 kg/m^3 (0.283 lb/in^3) the variations permitted are +4 per cent and −2 per cent for a batch (defined as a number of lengths of one size from one cast) and +6 per cent and −2 per cent for any one bar.

From the purchaser's viewpoint, the requirements for testing are less satisfactory than those included in the standards for wires and strands. The manufacturer is not required to provide any documentary evidence of the test results obtained, though the records of the tests must be 'available for inspection by the purchaser or his representative'. Further, unlike the standards for wires and strands, no option of independent testing before delivery is available to the purchaser. The routine tests comprise a tensile test, a proof-load test and a minimum elongation (of 6 per cent); one sample is to be taken from each 5 metric tonnes within a

Table 4.3 Mechanical properties of strands* (BS 5896, 4757)

Nominal diameter of strand		Nominal area of steel		Specified characteristic load			Conditions in which strand supplied	BS No.
(mm)	(in.)	(mm^2)	(in^2)	(kN)	(kgf)	(lbf)		
9.3	0.366	52.3	0.083	92	9382	20 683	1, 2	5896
11	0.430	71.0	0.110	125	12 747	28 102	1, 2	(7-wire
12.5	0.492	94.2	0.146	164	16 724	36 870	1, 2	standard
15.2	0.600	138.7	0.216	232	23 659	52 158	1, 2	strand)
18	0.7	210	0.325	370	37 730	83 180	1, 2	4757
25.4	1.0	423	0.656	659	67 200	148 150	3	(19-wire
28.6	$1\frac{1}{8}$	535	0.830	823	83 920	185 000	3	strand)
31.8	$1\frac{1}{4}$	660	1.020	979	99 830	220 100	3	

Condition in which strand is supplied

Number	1	2	3
Description	Normal relaxation heat treated	Low relaxation heat treated	As spun
0.1% proof stress as percentage of specific characteristic strength	85	8.5	–
Maximum in % relaxation after 1000 hours from 70% initial stress	8	2.5	9
80% initial stress	12	4.5	14

*For super strand and drawn strand see BS 5896.

batch. The samples may be cut from the ends of processed tendons, or from off-cuts produced during processing. The specified characteristic load and the 0.2 per cent proof load (defined as the load at 0.7 per cent total strain) are given in Table 4.4; the breaking load for the thread is required to be at least 95 per cent of that for the bar. The modulus of elasticity is to be determined from the test readings and recorded. The breaking load is required to be not less than 95 per cent of the specified characteristic load, and not more than two out of the last 40 test results may be less than the specified values; no upper limiting value is specified.

Provision is made in the standard whereby a purchaser may, if he wishes, check that a batch attains the specified characteristic load. After delivery, ten bars are selected at random, and test pieces are cut from one end of each bar; if one should fail at less than 95 per cent of the specified value, that bar is rejected. If two fall below this value, the whole batch is deemed not to comply with the requirements of the standard.

The purchaser may require the maker to provide evidence of the relaxation properties of the tendon. The standard also includes provisions for re-testing, if samples tested by the maker should fail to meet the requirements.

4.2 Wires, strands and bars for pre-tensioning

It was thought at one time that a satisfactory bond between tensioned steel and concrete could be obtained only by the use of wires of small diameter, and piano wire of 2 mm (0.08 in.) diameter is still used occasionally. This type of wire has a smooth hard surface which for large wires prevents the development of good bond, and its unsatisfactory behaviour at ultimate load has been described in Chapter 2. These disadvantages led to the use of indented wires abroad. Single smooth wires of 0.2 in.and 0.276 in.(5 and 7 mm) diameters were introduced in Great Britain in 1939 and 1952 respectively and have proved satisfactory for pre-tensioning since their surface conditions are such as to ensure good bond. This is due to a very slight corrosion of the surface of the wire, such that no peeling of the surface is likely. Indented wires also provide good bond, but the indentations

Table 4.4 Properties of cold-worked high-tensile alloy steel bars (BS 4486)

Nominal size		Specified characteristic load			Minimum 0.2% proof load		
(mm)	(in.)	(kN)	(kgf)	(lbf)	(kN)	(kgf)	(lbf)
*20	0.78	325	32 850	73 100	275	27 750	61 900
22	0.87	375	37 900	84 400	325	32 850	73 100
*25	0.985	500	50 600	112 500	425	42 900	95 800
28	1.11	625	63 000	140 600	525	53 000	118 100
*32	1.26	800	80 900	180 200	675	68 050	152 500
*40	1.57	1250	126 200	281 500	1050	106 000	236 500

**Preferred sizes*

must not be so large as to reduce appreciably the cross-sectional area of the wire or cause fatigue failure at the notches. Seven-wire strands are also widely used for pre-tensioning; in addition to the normal surface bond they provide a mechanical bond with the concrete, because of the configuration of the wires comprising the strand.

With pre-tensioned steel, a certain minimum embedded length, termed the 'transmission length', is necessary, along which the force is gradually developed in the concrete by bond. The transmission length required increases when the diameter of the wire increases and also to some extent when the strength of the concrete is reduced. With small wires the prestress in the concrete is developed over a very short length, but with larger sizes the required length may be 2 to 3 ft (0.67 to 0.9 m). It should be noted that the rate of transmission is not uniform (Figure 4.1). More than half of the prestressing force is transferred to the concrete in the first quarter of the transmission length and up to 85 per cent may be transferred in the first half. In CP 110, it is noted that the transmission length for wire may vary between 50 and 160 diameters, and the following general recommendations are given:

Plain or lightly crimped wire: 100 diameters; 80 per cent transfer in first 70 diameters
Heavily-crimped wire: 65 diameters; 80 per cent transfer in first 54 diameters
Strand, 9.3 mm (0.366 in.) diameter: 200 mm ± 25 mm (8 in. ± 1 in.)
Strand, 12.5 mm (0.492 in.) diameter: 330 mm ± 25 mm (13 in. ± 1 in.)
Strand, 17.8 mm (0.7 in.) diameter: 500 mm ± 50 mm (20 in. ± 2 in.)

Occasionally the length available is less than the transmission length required for smooth wires, and in such cases it is advisable to use well-deformed wires, or strands, both of which have a shorter bond length. Figure 4.2 shows three types of wire, two of German and one of British manufacture, which possess this property. The first is the oval Sigma steel, made by Stahlwerke Rheinhausen; it is a hot-rolled patented steel on which special ribs are rolled. The second is a cold-drawn heat-treated twisted oblong wire, named Neptun, produced by Stahlwerke Felten & Guilleaume. The third is a cold-drawn crimped round wire produced by the Somerset Wire Company. In addition to deformed wires, round wires with indentations are produced by most British and European manufacturers. These may have transmission length slightly shorter than that required for plain wire, but by no means as small as those for deformed wires.

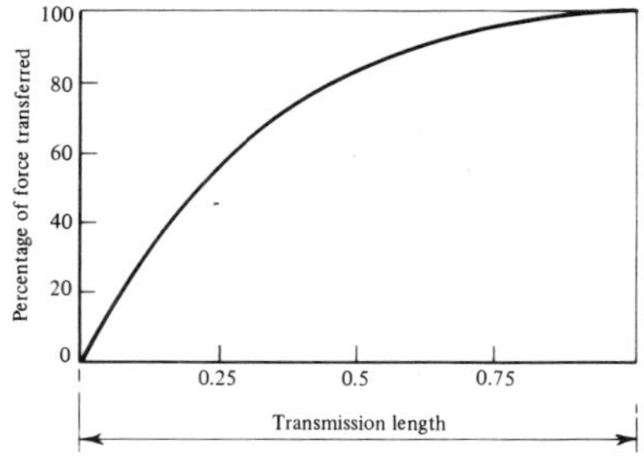

Figure 4.1 Transmission length

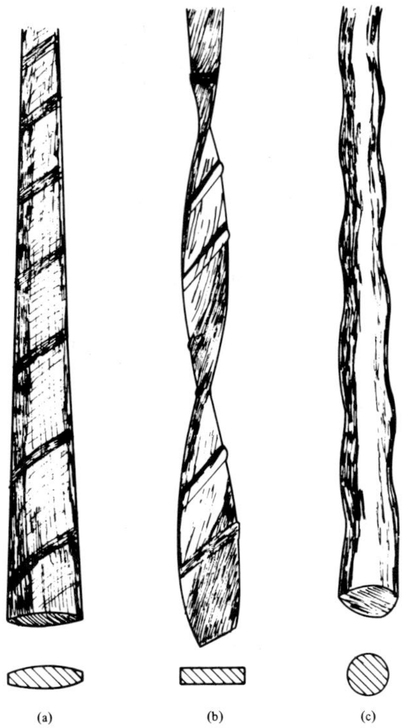

(a) 'Sigma' hot-rolled oval ribbed wire
(b) 'Neptun' cold-drawn oblong twisted wire
(also supplied as hot-rolled wire).
(c) Cold-drawn crimped wire (Somerset Wire Co. Ltd.)

Figure 4.2 Deformed bars for pre-tensioning

A special strand, known as Dyform, is made by British Ropes Ltd. The strand is first formed in the normal way and is then compacted to form the cross-section shown in Figure 4.3. In this case, the objective is primarily to increase the load which a strand of given diameter can apply; there is no gain in transmission properties.

High-alloy steel bars with special indentations have also been used for pre-tensioning. In this case the bar is tensioned in a manner similar to that described later for post-tensioned alloy bars, but after the grout hardens the end anchors are removed and the prestressing force is transmitted solely by bond.

When selecting a suitable size of wire it is desirable to ensure that the number of wires is sufficient to distribute the compressive stress uniformly over the concrete but not so great as to impede the placing of the concrete. In general it appears desirable to provide at least six wires in the tensile zone, but to avoid the adoption of a multitude of wires. If the number of wires is less than six, the failure of one would greatly reduce the factor of safety of the member. For this reason the use of only two or three wires in a member should be avoided where possible.

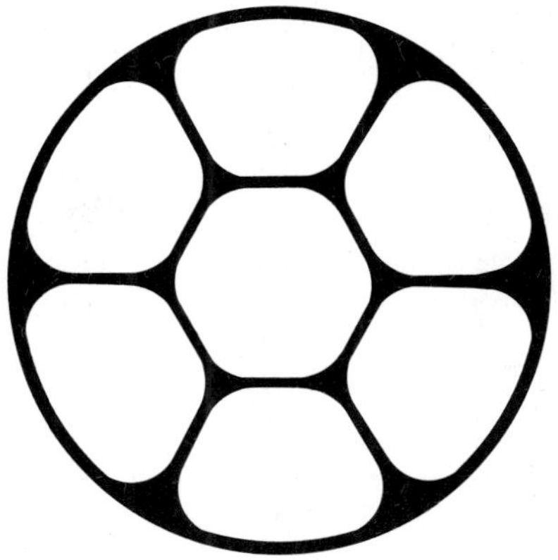

Figure 4.3 Cross-section of Dyform strand

Details of prestressing wires, strands and bars are given in Charts 2 and 3 in Part 2.

4.3 Tendons for post-tensioning

Many cables with different arrangements of wires and strands and different methods of anchorage are available for post-tensioning. The main types are briefly described here; more complete data are given in Charts 16 to 49 in Part 2.

Historically, the two basic types are represented first by the Freyssinet cable and later by the Magnel cable. In the Freyssinet cable the wires, which usually number twelve, are closely spaced around a central spring, or core helix, thus forming a cable of annular cross-section. The cables may be very small and special care is then necessary to ensure satisfactory grouting, neat cement and water being used for the purpose. The spacing of cables should also be considered. In the Magnel cable multiples of four or eight wires are provided in horizontal layers and the wires are well separated by spacers which allow easy grouting with cement mortar. With this type of cable a considerable prestressing force may be concentrated in a single cable. The Magnel system itself is no longer commercially available in Great Britain.

Many other types are available, including CCL Systems and PSC (Great Britain) and Leoba, Holzmann, and Beton-und-Monier Bau Gesellschaft (Germany), which more or less follow the Magnel principle with regard to the use of spacers; Systems Franki-Smet (Belgium), Morandi (Italy), Hochtief, and Grün & Bilfinger (Germany), which more or less follow the Freyssinet principle. Other early types like Gifford-Udall and Gifford-Burrow systems are no longer commercially available. BBRV and Losinger VSL (Switzerland), PI and Prescon (USA), which represent a type intermediate between Magnel and Freyssinet. In some systems of this type, the wires may be distanced by spacers, but are not necessarily separated by them. In the latter case, if the wires are bent up they touch each other, forming a group into which the grout cannot be inserted.

In some types of cable spacing may be obtained automatically if Neptun or Sigma wires with diagonal cross ribs are used instead of round wires. The cross ribs or adjacent wires touch each other at points only, since the ribs on opposite sides of a wire slope in opposite directions, and sufficient space is available between the wires to allow the easy admission of grout.

The cables may be inserted into ducts formed in the concrete or placed in tubes or sheaths. The Baur-Leonhardt cable (Germany) may also be placed around the outside of the concrete, forming closed loops. This cable consists of closely spaced stranded wires, and is therefore of the Magnel type.

Cables comprising single or multiple strands, which can be inserted in ducts or placed in tubes, have been introduced by Anderson and Roebling (USA), Rheinhausen (Germany), Freyssinet and SEEE (France), PSC, Stress Block and CCL Systems (Great Britain), and most suppliers of post-tensioning systems now provide components for anchoring strands. In order to obtain the greatest possible strength of a large stranded cable it is necessary that the outer wires should be stranded in the same direction as those forming the inner core; such strands are described as 'parallel lay'. As a consequence an untwisting torque may occur during tensioning in systems in which the jack is restrained from rotating (though most strand jacks now permit rotation during stressing). At transfer, this torque may be transmitted to the prestressed unit, and it may occasionally be large enough to warrant consideration in the design.

In addition to the medium and large cables already described there are several types of cables with two, three, four, six and eight wires which are used to provide smaller prestressing forces. In all these cables, wires of 0.2 in. and 0.276 in. (5 and 7 mm) diameter are generally used. The two-wire and four-wire ducts of the PSC system are the smallest in size as no spacers are used, the arrangement of the wires being such that the space available for grout is not less than that obtained with spacers. Particulars of cables are given in Charts in Part 2.

4.4 Bars for post-tensioning

Two types of steel bar have been developed for use in post-tensioning. They are used in the Macalloy system (Britain), which is known as 'Stressteel' in the USA and the Dywidag system (Germany). In the Macalloy system, high-alloy steel bars from 18 mm (¾ in.) up to 40 mm (1⅝ in.) diameter are used. In the Dywidag system the bars are of low-alloy steel of natural hardness, but with a definite yield point and are usually 25 mm (1 in.) in diameter although bars of lesser diameter are also available. In the Dywidag system high-alloy bars with greater strength have also been introduced.

The bars may be inserted into ducts or placed within tubes or sheaths in the concrete, in the same way as cables; the Macalloy system also includes a four-bar tendon. It is possible to obtain good bond if the grouting is carefully done with neat cement and water. The bars may be placed relatively close to one another, in the same way as Freyssinet cables. Full details of these bars are given in Part 2.

4.5 Anchoring prestressing steel

There are four basic methods of anchoring the steel after it has been tensioned. Three of these are represented by the Freyssinet, Magnel and Macalloy methods respectively; the fourth is represented by CCL, PSC and BBRV systems.

In the Freyssinet system all the wires or strands of the cable are wedged between a cylinder which is embedded in the concrete and a cone which is inserted therein. In the Magnel system pairs of wires are anchored by flat wedges to plates, termed 'sandwich plates', which in turn transfer the prestressing force to the concrete through an anchor-plate; as previously mentioned, this system is no longer available. In the Macalloy system the prestressing force is also transferred to the concrete through an anchor-plate, by means of a nut tightened on a thread, rolled on to the end of the bar. With SEEE system, soft steel cylinders, containing the strands are pushed through a die; threads are rolled onto the swaging cylinders, and nuts are tightened to anchor the tendons. In the Dywidag system, bars with threads throughout their length are available.

In the fourth method single wires or strands are secured to cylindrical grips by means of one or more wedges, or alternatively by button heads formed on the wire (BBRV and Prescon). The wedge system has also been adopted for bars (Stressteel and Macalloy alternative anchorage) and for cables of single or multiple strands (CCL Systems, PSC, Stress Block, and Anderson).

When the wires are secured by wedges, whether they be concrete cones or steel wedges, some slipping is unavoidable when the pull on the prestressing steel is relaxed. This may affect the tensioning stress substantially if the prestressing tendon is short. When the steel is secured by a nut the process is simple and no slip occurs during transfer. Moreover, no difficulty is experienced in regulating the prestressing force at any time. Because of the many advantages of this method of anchoring, it is employed in several of the cable systems previously mentioned, including some systems which use strand. In several systems, the separate wires of the cable are secured to a threaded anchor-head before tensioning. The wires are then tensioned simultaneously and anchored by means of a nut. The wires are connected to the anchor-head by upsetting and enlarging the ends of the wires (BBRV, Prescon), by concreting the wires into the head (Beton-und-Monier Bau, Holzmann A.G.), or by looping them around a cross-bar with a threaded hole (Leoba). In some of these systems it is also possible to employ a temporary anchor-plate and dispense with it as soon as cement mortar or concrete, which is inserted round the anchor-head, hardens and secures the head to the concrete. The anchor-plate is then removed. The anchor-head is usually conical. The advantage of a positive anchorage which will not slip is thereby retained without the cost of a permanent anchor-plate.

In the original Holzmann large-cable system (Germany) the cable comprises layers of four oval-shaped well-spaced wires, and is secured by means of a wedge to a large prefabricated member (corresponding to a female cone or large grip) which also forms the anchor-plate. In the improved Holzmann K.A. system (Klem Anker i.e. clamp anchor system), now being used up to forty oval-shaped or rectangular wires with diagonal ribs are clamped by means of transverse bolts and nuts which are tightened against outer plates.

In the Losinger VSL system (Switzerland), up to 36 wires are secured to an anchor head by means of a single conical wedge with circumferential grooves in which the wires are housed.

A continuous cable is used in the Baur-Leonhardt system, in which the tensioning is done by jacking apart two parts of the structure around which the cables are looped. Alternatively, separate cables may be used, one end being anchored in the structure and the other in a movable anchor-block to which the jacks are applied.

Another post-tensioning system, which is in fact the oldest, is that developed by Coyne, and used mainly for retaining walls, dams, and barrages. It comprises a straight cable of 600 to 800 wires of 0.2 in.(5 mm) diameter, strapped together to form a bundle. One end is embedded in a bulkhead of concrete and held by bond, and the other end is fixed to a large steel drum filled with cement mortar, thereby forming an anchor head to which the jacking force is applied. This system was used to prestress the first prestressed concrete pressure vessels for nuclear reactors, at Marcoule (France).

Cables may have the same type of anchorage at both ends, or the wires may be embedded in the concrete at one end before they are tensioned; loops or other shapes which ensure a satisfactory anchorage are then formed at the end of the cable.

The systems described in the foregoing can also be used for prestressing circular containers or pipes by arranging the cables in overlapping arcs; special anchors are available with some systems to simplify the work. Circular structures can also be prestressed by means of wire under tension wound around them in the form of a continuous helix (Preload, BBRV and Dywidag systems). The British contractors Taylor Woodrow Ltd., have also developed a wire-winding system for large pressure vessels.

4.6 Efficiency of anchors

A common basis for measuring the behaviour of anchors is given in BS 4447 (1969): *Methods of establishing the performance of prestressing anchorages for post-tensioned construction.* Two tests, for gripping efficiency and anchor efficiency, are described.

In the grip test, suitable lengths of wire, strand or bar may be held by grips at both ends, or by grips at one end and a device at the other which is designed to ensure that failure occurs at a point remote from that end. The test load is applied either by jacks or by a tensile testing machine. Failure is deemed to occur when the specimen no longer supports any increase in loading. The gripping efficiency of the anchor is the ratio of this load to the ultimate tensile strength of the tendon itself. For unbonded tendons, additional requirements for cyclic loads are given.

The anchor efficiency test simulates an end block with a ratio of width of loaded area to width of prism (see Chapter 10) of 0.6. Again, the efficiency is given by the ratio of the load at which the specimen can no longer support an increase to the ultimate tensile strength of the tendon, and requirements for cyclic loading are also given in the case of unbonded tendons.

'Characteristic values' are derived in both tests, from results for at least ten specimens. No minimum values are specified.

CHAPTER 5

CRITERIA FOR ANALYSIS AND DESIGN

5.1 The essential loading stages

In materials such as steel and timber there is a similarity between the distribution of stress at working load and at failure; consequently there is a direct relationship between the stresses at failure and at working load under elastic conditions, and a definite factor of safety is therefore available. This is not the case with reinforced concrete, and is even less true in the case of prestressed concrete when the design is based only on elastic conditions; with these materials the conditions of stress at working load and ultimate load are completely different except in the case of brittle failure in prestressed concrete, when cracking and failure occur simultaneously.

Various opinions have been expressed on this question. It has been stated, for example, that if the stresses at the working load are within those permitted a sufficient factor of safety against failure is automatically obtained. This is certainly true in some cases (for example, under-reinforced members to which large prestressing forces are applied so that the area of steel necessary to provide the prestressing force exceeds that needed for the conditions at ultimate load) but the limited deformation of such a structure may render the design unsuitable. However, when the prestressing force is smaller, additional non-tensioned steel might be required, and in the case of over-reinforced structures the required factor of safety against failure may not be available if elastic conditions only are considered.

On the other hand, it has been suggested that if a sufficient factor of safety against failure is obtained, based solely on conditions at ultimate load, the conditions at working load need not be considered. In this case the factor of safety against failure is undoubtedly sufficient but the possibility of excessive deformations and cracking at the working load may be insufficiently guarded against.

It is therefore essential to investigate the conditions at both working load and failure. This is in conformity with the recommendations given in the *First Report on Prestressed Concrete* published by the Institution of Structural Engineers in 1951, the British Code CP 115: 1959, and CP 110: 1972 and most of the national codes; it ensures that definite margins of safety are provided against both cracking and failure. A uniform factor of safety against failure for all forms of construction is recommended in CP 115; namely, 1½ for the dead load and 2½ for the imposed or live load, or 2 for the combined load, whichever gives the smaller value. The corresponding factors adopted in the *American Concrete Institute Building Code* ACI 318 of 1977 are 1.4 for the dead load and

1.7 for the imposed load, or an overall factor of 0.75 × (1.4 × dead load + 1.7 × live load + 1.7 × wind load) for the combined load. In Germany the factor is 1¾ for the combined load. It is preferable for the factor to be varied to take into account such matters as the probability of overloading, the danger to life in the event of failure, and the economic consequences of failure. For example, in a structure containing liquid increasing the live load is only possible by storing a liquid heavier than that assumed in the design and a thus smaller factor of safety with respect to the live load might therefore be employed. The possibility of inaccuracies in design and construction however makes it essential to use a factor of at least $1\frac{1}{3}$ for the combined load.

CP 110 and CEB/FIP Model Code recommend the use of two 'partial safety factors', one of which is applied to the loading and the other to the strengths of the materials. The values for the load factor range up to 1.6 in CP 110 and 1.5 in Model Code; the (reduction) factor for the strength is 0.67 for concrete and 0.87 for steel. These values are for the limit state of collapse; other values are given for other limit states.

The resilience of the structure should also be considered. For example, the deformation permissible at working load should not be such that the serviceability of the structure is impaired, but the structure should not be so rigid that little deformation occurs as failure is approached. The ratio of the factor of safety against failure to that against cracking should therefore, in the authors' opinion, be at least 1½. If the ratio is less than, say, 1¼, little warning of impending failure would be given; furthermore, the structure would offer little resistance to impact and therefore its ability to absorb shock (due to an explosion, for example) would be small.

Three types of structure are considered in the *First Report:* (1) Structures in which cracking is not permissible at working load; (2) Structures in which temporary cracking is permitted at abnormal loads provided tensile stresses do not occur at working load; (3) Structures of minor importance in which cracking is permitted at ordinary working load. A structure of Type (1) may be either (a) a structure in which no tensile stress is permitted at working load, or (b) a structure in which the tensile stress is less than that likely to cause visible cracks; in this case the factor of safety against cracking is less than for a structure of Type (1a). If it is composed of separate units, or if cracks develop before the prestress is applied, there is no safety against cracking in structures of Type (1a) unless there are residual compressive stresses at full load. A monolithic structure of Type (1b) therefore has a greater factor of safety against cracking than a non-monolithic structure of Type (1a). Structures of Types (1b) and (2) are suitable when a considerable degree of elasticity is desired before failure occurs, and the ratio of the factor of safety against failure to that against cracking is large. For example, if the factor of safety against failure is 2 and that against cracking is 1.2, the ratio is $1\frac{2}{3}$ and there would be noticeable deformation at a load equal to 70 per cent of the load causing failure. If the factor of safety against failure is 2 and that against cracking is $1\frac{2}{3}$, the ratio of the two factors is 1.2; cracking would not occur until the load is 83 per cent of the load causing failure, and there would be no visible warning that failure was impending. These conditions must be borne in mind when selecting suitable stresses for the conditions at working load, the only purpose of which is to ensure the serviceability of the structure.

CP 110 also divides prestressed concrete into three classes: (1) those with no tensile stresses; (2) those with no visible cracking; and (3) those with cracks not exceeding 0.1 mm (in aggressive environments) or 0.2 mm (in normal environments).

The factor of safety against failure, the serviceability at working load, and the elasticity of the structure should therefore be considered as equally important. If the factor of safety against cracking is required to be large, that against failure should also be increased to ensure that visible warning of overloading is given.

The conditions described in the foregoing relate to statically-determinate structures. In the case of statically-indeterminate structures there may be further complications, but the permissible stresses are the same although some redistribution of stress may occur. Secondary bending moments, which may occur due to the profiles of cables, may reduce the counteraction due to the prestressing steel under conditions of working load and positions of cables which are suitable for the working load may not be suitable for conditions at failure. It has been suggested that the conditions at failure are not affected by the position of the prestressing steel and that plastic hinges form in positions that ensure a satisfactory factor of safety against failure. This behaviour has not yet been satisfactorily proved. Even if it does such a solution is undesirable since the position of the prestressing steel remains far from the tensile face and large cracks develop in the event of overloading. Moreover, early failure due to shearing forces may occur in members subject to rolling loads, since the resistance to diagonal tensile stresses is reduced when the distance of the steel from the compression edge is reduced. A member designed for conditions at failure only may suffer from the same disadvantages as a statically-determinate structure, namely, excessive deformation and cracking at working load. It is therefore necessary to consider conditions at working load and at failure. The prestressing steel should be near the tensile face, where it is most effective at failure and for the prevention of wide cracks if overloading occurs. The conditions at working load should be investigated, the permissible stresses being such that excessive deformation and cracking are avoided.

The distribution of bending stress at working load (Figure 5.1) is totally different from that at failure (Figure 5.2), although the strains distribution remains virtually linear in both cases. Under service conditions, the distribution of stress is linear, as in a homogeneous material; this assumption is entirely correct for sections which are wholly in compression (Figure 5.1a) and is also applicable when small tensile stresses (less than the tensile strength of the concrete) occur in an uncracked section (Figure 5.1b). It may also be used as a

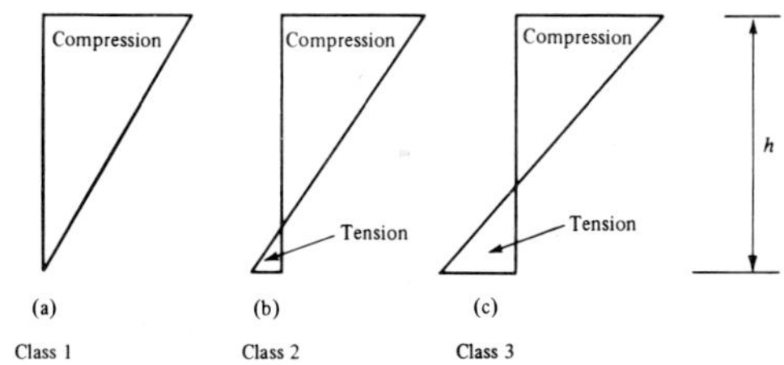

Figure 5.1 Stress distribution at service load

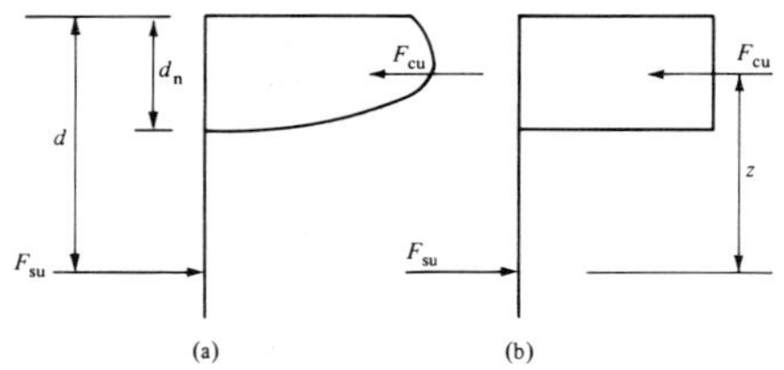

Figure 5.2 Stress distribution at failure

reasonable approximation when larger tensile stresses (up to the modulus of rupture) are present, and it is possible to consider nominal tensile stresses even in a cracked section (Figure 5.1c). In each of these cases, the stress pattern is the resultant of those due to the prestressing force (usually eccentric) after all losses and to the bending moments caused by the loading. In the case of a cracked section, the distribution shown in Figure 5.1c may be replaced by that due to the combination of bending moments and eccentric prestressing force acting on a cracked section.

The stress conditions at ultimate load may be derived solely from considerations of the equilibrium of the ultimate forces in a cracked section (Figure 5.2a and 5.2b; the latter is a simplified equivalent version), or may be obtained by considering the distribution of strain throughout the depth of the member. The latter is discussed in more detail in Chapter 7.

5.2 The 'limit-state' approach

The general object of all Codes of Practice and Standards for structural design irrespective of the materials used for construction, has always been to ensure that structures and members designed in accordance with them would be serviceable at working loads and would have an adequate factor of safety against failure due to accidental overloading or variations in the quality of the materials of which they are composed. Until very recently, however, the method by which this objective was achieved (with quite satisfactory results in the great majority of cases) was to specify in detail parameters such as permissible loading, permissible stresses, load factors, limiting deflections, and so on; it was assumed that compliance with these requirements would produce structures which behaved satisfactorily in practice, and in the great majority of cases this assumption is justified.

On the other hand, there are a number of objections to such a method of approach. It is illogical, and may occasionally be dangerous, to assume that compliance with a set of requirements of this type will ensure a safe and serviceable structure, and it is clearly more sensible and suitable to place the sequence in the reverse order; as stated in CP 110, 'The purpose of design is the achievement of acceptable probabilities that the structure being designed will not become unfit for the use for which it is required, i.e., that it will not reach a limit state'.

Thus a 'limit-state' is defined as the stage at which the structure becomes unfit for its envisaged use; CP 110 lists seven such states (collapse, deflection, local damage, vibration, fatigue, durability, and fire resistance) which should be considered in all cases. The statement in the previous paragraph also serves as a reminder that all design is based on probability, and the determination or assessment of 'acceptable probabilities' that a structure will not reach a limit-state (whether through overload, fatigue, fire or some other cause) is a specific responsibility for the designer (and often, also, his client).

In the past, the design of reinforced concrete has been based either on the elastic theory – that is, on the equilibrium of the internal stress resultants and external forces and moments acting on a cracked section at working load, using a constant or variable modular ratio – or on the conditions at ultimate load. For prestressed concrete on the other hand, the design has for many years been based on a combination of both of these widely-differing methods. So while the recommendations of CP 110 involve a radical change in the method of design for reinforced concrete, forcing the designer to consider more carefully the probable service and collapse loadings, the changes affecting the design of prestressed concrete are of a more minor nature.

The relative importance of the various limit states may vary widely from one structure to another. For example, in a floor slab supporting vibrating machinery, the limit states of vibration, fatigue and deflection may be especially important, whereas the design of a slab supporting static loading may be determined almost entirely by the limit-states of local damage and collapse. In the following, for ease of presentation, these limit states are grouped under two broad headings: 'serviceability' (which includes deflection, local damage, vibration and durability) and 'collapse' (which includes also fatigue, durability, and fire resistance).

Serviceability. To satisfy the limit-state of serviceability, it is necessary to ensure that no excessive deformation, major cracking, and spalling takes place under the expected service conditions; permissible stresses at working load are still specified, as a means of ensuring that the limit state of local damage is avoided. In the design of reinforced concrete, endeavours are being made to predict the deflections and widths of cracks from the percentage of reinforcement; if the predicted values are excessive, the cross-section would be increased appropriately, in accordance with procedures based on test results. A similar procedure could well be applied to prestressed concrete at service conditions, particularly when cracks of specific widths form a major part of the limit-state requirements.

However, with prestressed concrete it is essential to consider the conditions at transfer as well as those arising in service. For this purpose it is necessary to know the stresses in the concrete (based on a linear stress distribution across a homogeneous section) at the time of application of the prestress in order to determine the magnitude of the elastic and creep losses (see Chapter 6). It is also useful to know the loading which reduces the pre-compression in the concrete tensile zone to zero, and also to know the stress conditions in a homogeneous section at the service load; these are required for members of Classes 1 and 2, in which either no tensile stress or only a limited tensile stress is permitted under service load.

The design should also guard against the structure being rendered unserviceable by corrosion or fire; this is achieved partly by the provision of sufficient cover and good detailing. Damage and excessive deformation must be avoided at the time of prestressing: this requires (a) sufficient protection against cracking, including sufficient steel in the anchorage zone; (b) checking of the cross-section and the prestressing force to limit the initial camber of the member due to prestressing; and (c) the avoidance of local crushing and/or high creep losses, by ensuring adequate concrete strength at the time of prestressing. This may involve a clear definition of the minimum strength required which in turn may mean a delay in manufacturing the members. While these conditions have little direct connection with permissible stresses it seems sensible to use them as suitable criteria to determine acceptable conditions at transfer; this procedure was adopted in CP 115 and is now included in CP 110, which limits, for example, the initial tensioning stress to not more than 70 per cent of the characteristic strength of the tendon and the concrete stress to not more than 50 per cent of the characteristic compressive strength available at the time. Further requirements are related to the transmission length (for pre-tensioned steel) and to the losses which occur between transfer and service conditions when maximum losses are assumed to have taken place. ACI Building Code has also similar limitation in the initial tensioning stress in tendons and compressive stress in concrete. In the CEB/FIP Model Code the initial stress in tendons is limited to 75% of characteristic strength or 85% of 0.1% proof stress. For concrete the limiting compressive stress is 60% of characteristic cylinder strength.

CP 110 includes specific limitations on the minimum dimensions of the section, compared with the length, to safeguard the stability of the member; these restrictions may often be severe. However, the transport and placing of long slender members is often a matter of experience, care, and forethought; when desired, it is often possible to use more slender constructions than those permitted by CP 110 on the responsibility of the engineer, as shown in the examples in Chapters 9 and 11. General limits on the maximum camber and maximum deflection are also given; the actual magnitudes will obviously depend largely on the quality of the concrete, and hence on its modulus of elasticity.

Collapse. For the limit state of collapse, a sufficient factor of safety against collapse under abnormal loading must be ensured, and it is essential to consider all possibilities of overloading, including in some cases the cumulative effect of loading. It is thus necessary for the designer to know the expected lifetime of the structure, since the cumulative effect of live load (such as a moving load, or wind pressure) depends largely on this. Similar considerations apply when earthquakes need to be considered; in earthquake zones it is possible to make an assessment of the probability of their occurrence and magnitude, based on previous records, during the required lifetime of the structure.

In some cases, it appears that the introduction of a compound or multiple factor of safety – one as a result of the use of 'characteristic strengths' and 'characteristic loads', and the other as a separate load factor – can lead to wasteful designs. It is clearly essential to ensure that there is some margin between the notional 'collapse' conditions upon which the design is based and those causing actual failure; but test results reported at a symposium at the University of Salford in 1972 showed that this margin might be as high as 33 per cent.

For fire resistance, it is necessary to know the times for which a structure is expected to resist a fire (a) without becoming unserviceable and (b) without collapsing. These are two different conditions which should be investigated separately*; they affect the magnitude of the cover to steel (see Chapter 3) and the detail design.

*See Abeles and Bobrowski: Fire resistance and limit state design. *Concrete*, April 1972.

The introduction of limit-state methods involves a major change in the philosophy of design, but its practical effect on the mechanics of the procedures for the analysis and design of prestressed concrete is quite small; for service load conditions the design is still based on the effective prestressing force and the linear stress distribution in a homogeneous section, and for collapse conditions the ultimate load resisted by the cracked section is considered.

5.3 The magnitude of the prestress and the prestressing force

When computing stresses on the basis of a linear stress distribution in a homogeneous material (as in Figure 5.1) it is obviously essential to assess the magnitude of the prestressing force acting at the loading stage considered.

The requirements for design and analysis are different. In the case of analysis the conditions at transfer (such as the strength of the concrete at the time of prestressing, the magnitude of the initial prestressing force and so on) are known, and it is possible to compute the magnitude of the prestressing force available at the stage considered. Several such stages must usually be investigated. In the *initial* stage, during tensioning, the steel stress f_{pi} is applied to the area A_{st} of the tensioned steel, producing the initial prestressing force P_i. The value f_{pi} and P_i vary along the beam, when post-tensioned steel is used, as a result of losses due to friction. At *transfer*, when this force is applied to the concrete, elastic shortening takes place with pre-tensioned steel. (With post-tensioned steel some elastic shortening of the concrete also occurs, except when the whole of the steel is tensioned in a single operation.)

The effective stresses are obtained after the prestress f_{pt} (and prestressing force P_t) which occur at transfer are reduced by further losses until the maximum losses have taken place and the minimum effective prestress f_{pe} in the steel (corresponding to the prestressing force P_e) has been reached. Details of these losses, which occur mainly as a result of shrinkage and creep of the concrete and relaxation of the steel, are given in Chapter 6.

The process of design is usually the reverse of that for analysis, since it works back from the minimum effective prestress required to meet the design conditions to determine the magnitudes of the initial prestressing force which will ensure that the required effective prestress is obtained. This is shown in Chapter 8; analysis is discussed in Chapter 7 and the design of members subjected to bending is described in Chapter 8.

5.4 Notation

The notation adopted in the second edition of this book has been retained in principle because it has proved satisfactory over many years of practical use. A new international notation has recently been proposed by a joint committee of the CEB/FIP in which the type of letter adopted is related to the property described. For example, Roman capital letters designate a force, a product of force and length, an area to a power, or a temperature (that is, bending moments, normal forces, shearing forces, concentrated loads, overloads, first and second moments of area, and so on). A Roman small letter indicates a length, a length per time, a force per unit length, or an area (that is unit moments, unit shears, linear dimensions, strengths, and stresses). Greek lower case letters denote dimensionless values, such as strains or angles and ratios.

In the following, the concrete strength is denoted by f with appropriate suffixes: 'cu' for cube, 'cyl' for cylinder, 'prism', 'r' for rupture i.e. flexural strength, and 't' for tensile strength. Additional suffixes usually denote the age of the concrete and a star superscript denotes characteristic strength; for example $f^*_{cu\,28}$ denotes the characteristic cube strength of the concrete at an age of 28 days, whereas $f_{cyl\,t}$ signifies an actual cylinder strength (from a test on a specimen) at the time of transfer.

The concrete stresses are also denoted by f; two suffixes are added for a homogeneous section. The first suffix denotes the position at which the stress occurs: e.g. 'b' for bottom, 't' for top edge of the section, while 's' signifies the plane of the centroid of the tensioned steel. The second suffix indicates the loading causing the stress, and by implication the time at which it occurs; typical examples are 'T' for prestress at transfer, 'E' for effective prestress, 'D' for full dead load, 'L' for full working-life load and 'W' for total working load. Small letters are used instead of capitals to indicate that the combined effect of loading and prestress is considered. For instance, f_{tW} is the stress at the top edge of the beam which would be caused by the full working load acting alone, and f_{tw} is the stress due to the combined effects of the effective prestress and full working load. Hence,

$$f_{tw} = f_{tW} + f_{tE}$$

The only exception to this rule occurs in the case of the effective prestress. This relates to the action of the prestress only, and as the combined effect must include some additional load (such as the full working load, the full dead load or any other specified load) the symbols f_{be} and f_{te} are defined as including the stresses due to weight of the member at prestress. It will be noted that each symbol represents the initial letter of the expression or word referred to. Compressive stresses are taken as positive.

The prestress in the steel is denoted by the symbol f_p with a second suffix 'i', 't' and 'e', which are used to denote 'initial', 'transfer', and 'effective' respectively. The prestressing force is denoted by P with suffixes as with the prestress of 'i', 't' and 'e'.

Permissible stresses are generally denoted by $\bar{f}$.

5.5 Permissible stresses in concrete

5.5.1 Permissible compressive stress at transfer

In CP 115 it is recommended that the compressive stress in the concrete at transfer should not exceed 50 per cent of the crushing strength of works cubes at transfer, or 40 per cent when the prestress is distributed more or less uniformly over the cross-section of the member, and in no case more than 3000 lbf/in^2 (210 kgf/cm^2; 21 N/mm^2). This recommendation is unchanged in CP 110 except that no limiting value has been specified. However, since creep increases when the ratio of stress to strength exceeds one-third (see Chapter 2), and sometimes even when it is less (see Chapter 3), it is advisable to restrict the stress to 40 per cent of the crushing strength of cubes irrespective of the distribution of stress, as was recommended in the First Report. In the USA, the ACI Building Code (No. 318, 1977) allows 60 per cent of the crushing strength of cylinders which is equivalent to about 40 per cent of the strength of cubes if the strength of a cylinder is two-thirds that of a cube. In the German Specification (DIN 4227) the crushing strength at transfer is assumed to be γ of the specified strength of the concrete, of which there are three types, having strengths varying from 4260 to 8520 lbf/in^2 (300 to 600 kgf/cm^2; 30 to 60 N/mm^2). The stresses permissible at transfer are given as a percentage of the strength namely: for T-beams and hollow beams bending in one plane, 42 per cent to 54 per cent; for rectangular beams bending in one plane; 44 per cent to 58 per cent; and for members in direct compression 35.5 per cent to 46 per cent.

5.5.2 Compressive stresses permissible at working load

It is recommended in CP 115 and in CP 110 that the compressive stress in bending due to the working load should not exceed $\frac{1}{3}$ of the crushing strength of cubes at 28 days except at the supports of continuous beams and other statically-indeterminate structures in which a stress of 40 per cent of the cube strength is permitted. The compressive stress recommended in direct compression is a quarter of the cube strength. The permissible stresses due to the greatest working load may be exceeded by up to 25 per cent if the excess is due solely to the effect of wind. In CP 116, however, the maximum allowable compressive stress in bending due to working load is $f_{cu}/2.73$ and in compression $f_{cu}/3.65$.

The permissible stress in the ACI Building Code is 45 per cent of the crushing strengths of cylinders and is equivalent to 30 per cent of the cube strength on the assumption that the ratio of the cylinder to cube strength is $\frac{2}{3}$. The stresses permitted at working load by the German Specification are relatively small and are 1560–2270 lbf/in^2 (110–160 kgf/cm^2; 11–16 N/mm^2) for rectangular beams, 1705–2415 lbf/in^2 (120–170 kgf/cm^2; 12–17 N/mm^2) at the corners of rectangular beams subjected to bending in two planes, 1420–2130 lbf/in^2 (100–150 kgf/cm^2; 10–15 N/mm^2) for T-beams, and 1130–1850 lbf/in^2 (80–130 kgf/cm^2; 8–13 N/mm^2) for members subjected to direct compression. In each case the lower stress applies to concrete having the lower strength as given for the stresses permitted at transfer.

In the CEB/FIP Model Code there is no specific mention about the permissible compressive stress in concrete under service load; but a value of 0.6 times the characteristic cylinder strength has been implied under rare combinations of service load.

5.5.3 Tensile stresses permissible at working load

Some designers prefer that there should be no tensile stress in the concrete at working load, and this condition is satisfactory if economy is not important, or if it is not essential to reduce the member to the least possible depth. There should be no tensile stress in members, such as turbo-generator bases which are subjected to a large number of cycles of dynamic load of high frequency, or in similar structures. In other cases there is no objection structurally to the existence of tensile stresses within certain limits or to temporary hair-cracks or to permanent microcracks.

The serviceability of a structure at working load may be impaired if cracks are produced. This may happen at a high tensile stress with well-bonded steel in a monolithic member or at a low stress with badly-bonded steel. It is possible to prevent visible cracks even when appreciable tensile stresses are permitted at working load; however, this will be achieved only when provision is made to ensure that no cracks occur before the prestress is applied. In addition to careful design, close supervision of the work and effective moist curing are essential if cracks are to be avoided. The permissible tensile stresses in bending as recommended in CP 115 for members with pre-tensioned and post-tensioned steel are given in Table 5.1 in which Case A applies when the greatest working load occurs frequently or is of long duration or both, and Case B applies when the greatest working load occurs rarely and is of short duration. The pressure of the wind and the weight of snow on roofs are mentioned in the Code as examples of the latter case. These stresses may be increased by not more than 250 lbf/in^2 (17.5 kgf/cm^2; 1.7 N/mm^2) provided that the increased stress is not greater than

Table 5.1 Permissible tensile stresses (CP 115)

Cube strength			3000	4500	6000	7500	lbf/in^2
			210	316	420	525	kgf/cm^2
			21	31	42	52	N/mm^2
Prestressed Concrete	Pre-tensioned steel	Case A			300	325	lbf/in^2
					21	23	kgf/cm^2
					2.1	2.24	N/mm^2
		Case B			450	500	lbf/in^2
					32	35	kgf/in^2
					3.1	3.5	N/mm^2
	Post-tensioned steel	Case A		175	200	225	lbf/in^2
				12	14	16	kgf/cm^2
				1.2	1.4	1.6	N/mm^2
		Case B		275	300	325	lbf/in^2
				19	21	23	kgf/cm^2
				1.9	2.1	2.24	N/mm^2
	Added concrete	Case A	200	250	300		lbf/in^2
			14	17	21		kgf/cm^2
			1.4	1.7	2.1		N/mm^2
		Case B	300	375	450		lbf/in^2
			21	26	32		kgf/cm^2
			2.1	2.6	3.1		N/mm^2

Case A: The service load occurs frequently
Case B: The service load occurs rarely and is of short duration
The table is for normal weight concrete (for lightweight concrete – no recommendation)

three-quarters of the stress at which the first crack appears as determined by test. In such a case, therefore, for concrete with a cube strength of 7500 lbf/in^2 (530 kgf/cm^2; 53 N/mm^2) at 28 days the greatest tensile stress permissible is 750 lbf/in^2 (53 kgf/cm^2; 5.2 N/mm^2) with pre-tensioned steel or 575 lbf/in^2 (40 kgf/cm^2; 4 N/mm^2) with post-tensioned steel. When the increased stresses are adopted the minimum prestress in the concrete should be at least 1500 lbf/in^2 (105 kgf/cm^2; 10 N/mm^2) and the pre-tensioned steel should be well distributed; post-tensioned steel (which is usually less well distributed) should, if necessary, be supplemented by non-tensioned steel provided near the tensile face of the concrete. Test results indicate that a minimum prestress of 1500 lbf/in^2 (105 kgf/cm^2; 10 N/mm^2) is unnecessarily high; in the authors' opinion it is sometimes prohibitive to a satisfactory design.

Experience shows that it is possible to avoid cracks becoming visible even when high tensile stresses occur. Tensile stresses of 750 lbf/in^2 (53 kgf/cm^2; 5.2 N/mm^2) in roofs, and 650 lbf/in^2 (46 kgf/cm^2; 4.5 N/mm^2) in road bridges have been adopted with satisfactory results, the adequacy of the design in these cases being controlled by performance tests. When such tests are made, the factor of safety against cracking, despite higher stresses, may exceed that of a member with post-tensioned steel designed to resist only compressive stress but made under poor supervision. If the concrete is inadequately cured, cracks due to shrinkage of the concrete may occur before the prestress is applied.

A method of design, whereby there is no limit to the tensile stresses caused in the concrete by occasional heavy loads provided that no tensile stresses are caused by the normal load, is economical and complies with the recommendations of CP 115. In this case the factor of safety required against failure must be provided for the total load including the abnormal load. CP 110 allows only a limited increase of permissible tensile stresses due to loading of temporary duration. CP 115 method is particularly useful in the design of continuous beams or beams with cantilevered ends in which the load on some spans may cause reversals of bending moments. The envelopes of the diagrams of the greatest possible bending moment and greatest normal bending moment on a beam with a cantilever at one end are shown in Figure 5.3a. If the beam is designed to resist the entire range of bending moments with little or no tensile stress in the concrete a very large member is required. If, however, the load causing the greatest possible bending moments is likely to occur rarely, if at all, the method described in the foregoing would lead to a more economical beam. A cantilever will generally be subjected to the full working load and it is therefore necessary to design it to resist the maximum bending moment at the support. The occurrence of the distribution of loading causing the greatest possible bending moments at the supports of a continuous beam (Figure 5.3b), however, may be rare and generally the bending moment actually occurring at a support would be less than the maximum possible bending moment.

The requirements of CP 110 for the limit state of local damage are broadly similar to those of CP 115; the permissible flexural tensile stresses are zero for structures of Class 1 and as given in Table 5.2 for Class 2 structures. The distinction between Case A and B has been omitted. These stresses may be increased by 245 lbf/in^2 (17 kgf/cm^2; 1.7 N/mm^2) provided that the increased value is not more than three-quarters of the stress causing cracking (as in CP 115). For Class 2 structures the ratio of the tensile stress to the flexural tensile strength of

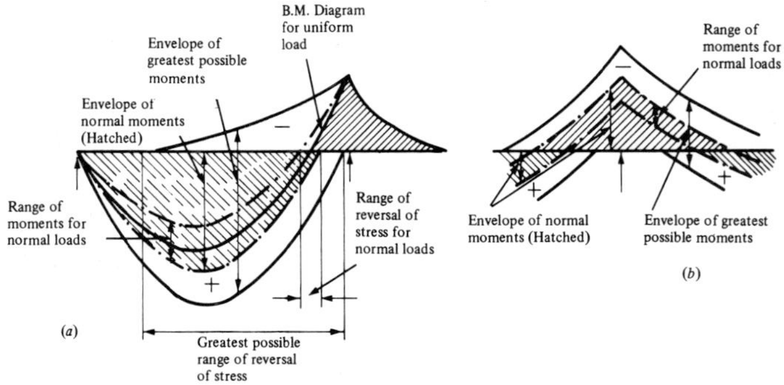

Figure 5.3 Bending moment envelope for cantilevers

Table 5.2 Permissible tensile stresses (CP 110) for normal weight concrete

Concrete grade (characteristic cube strength)	4350	5800	7250	8700	lbf/in²
	305	410	510	610	kgf/cm²
	30	40	50	60	N/mm²
Pre-tensioned member	–	420	465	510	lbf/in²
		29	32	35	kgf/cm²
		2.9	3.2	3.5	N/mm²
Post-tensioned member	305	335	360	410	lbf/in²
	21	23	26	28	kgf/cm²
	2.1	2.3	2.55	2.8	N/mm²

the concrete should not exceed 1 (with pre-tensioned steel) or 0.8 (with post-tensioned steel). Permissible flexural tensile stresses – in the CEB/FIP Model Code for uncracked category (equivalent to class 2 of CP 110) are given in Table 5.3. For Class 3 structures the requirements are based primarily on crack widths; for permissible crack widths of 0.1 mm and 0.2 mm, values of nominal tensile stresses ranging between 460 lbf/in² (32 kgf/cm²; 3.2 N/mm²) [for 0.1 mm cracks in concrete with a cube strength of 4260 lbf/in² (300 kgf/cm²; 30 N/mm²)] and 1080 lbf/in² (75 kgf/cm²; 7.5 N/mm²) [for 0.2 mm cracks in concrete with a cube strength of 7500 lbf/in² (525 kgf/cm²; 52.5 N/mm²)]. These stresses may be increased by 570 lbf/in² (40 kgf/cm²; 4 N/mm²) (for 0.2 mm cracks) or 430 lbf/in² (30 kgf/cm²; 3 N/mm²) for every 1 per cent of non-tensioned reinforcement up to a limiting value of one quarter of the cube strength. There is no specific recommendation for permissible tensile stresses in lightweight aggregate concrete. In absence of any other information

80% of permissible stresses for the respective grade of normal weight concrete may be assumed. Reference may be made to CEB/FIP Manual on lightweight aggregate concrete.

In a member which is not exposed to the weather or to a corrosive atmosphere, the permissible tensile (flexural) stresses in accordance with the ACI Building Code are:

(a) for a member with pre-tensioned steel, $12\sqrt{f_{cyl}}$ lbf/in² ($0.997\sqrt{f_{cyl}}$ N/mm²)

(b) for a member with post-tensioned steel, $6\sqrt{f_{cyl}}$ lbf/in² ($0.498\sqrt{f_{cyl}}$ N/mm²)

The German regulations permit 'limited' or 'partial' prestressing. The permissible tensile stress varies from 570 lbf/in² (40 kgf/cm²; 4 N/mm²) for concrete with a strength of 4260 lbf/in² (300 kgf/cm²; 30 N/mm²) to 850 lbf/in² (60 kgf/cm²; 6 N/mm²) for concrete with a strength of 8520 lbf/in² (600 kgf/cm²; 60 N/mm²). These values may be increased to 710 lbf/in² and 990 lbf/in² (50 and 70 kgf/cm²; 5 and 7 N/mm²) respectively at the corners of members subjected to bending in two planes. However, when partial prestressing is employed for road bridges, the permissible stresses are from 426 lbf/in² to 639 lbf/in² (30 kgf/cm²; 3 N/mm² to 45 kgf/cm²; 4.5 N/mm²). When the tensile zone is at the top of the member, these stresses apply only if the member is well protected. The above relates to cube strength.

Table 5.3 Permissible direct tensile stresses $\bar{f}_{ct}$ (to CEB/FIP Model Code)

Concrete grade (characteristic cylinder strength)	4350	5800	6525	7250	lbf/in²
	305	410	459	510	kgf/cm²
	30	40	45	50	N/mm²
$\bar{f}_{ct}$	406	493	536.50	580	lbf/in²
	28.50	34.67	37.73	40.79	kgf/cm²
	2.80	3.40	3.70	4.00	N/mm²

Flexural tensile stress $\bar{f}_{cr} = \left[0.6 + \frac{0.4}{4\sqrt{h}}\right] \cdot \bar{f}_{ct}$ (in N/mm²)

where h = height of element (in metres)

$4\sqrt{h} \not> 1$

Note: For lightweight aggregate structural concrete

permissible fluxural tensile stress = $\bar{f}_{cr}$ for dense concrete × $\left[0.30 + 0.70 \frac{\gamma}{2400}\right]$

where γ = density of lightweight concrete in kg/m³.

5.5.4 Temporary tensile stresses at transfer and during transport

It is recommended in CP 115 that temporary tensile stresses should not exceed 175 lbf/in² to 225 lbf/in² (12–16 kgf/cm²; 1.2–1.5 N/mm²) for concrete having a cube strength of 4500 lb/in² to 7500 lbf/in² (320–530 kgf/cm²; 32–53 N/mm²) at transfer. However, greater stresses may be permitted for periods not exceeding 48 hours. Tensile stresses in the ends of beams with pretensioned steel should preferably be less than the foregoing stresses, as imperfect compaction of the concrete in these regions may affect the development of the prestress and may cause undesirable cracking at the ends.

The recommendations of CP 110 allow a stress of 140 lbf/in^2 (10 kgf/cm^2; 1 N/mm^2) in Class 1 structures. For Class 2, the values are as given in Table 5.2. Class 3 structures should also be designed for the same stresses; if these are exceeded the section should be considered as cracked and the second moment of area is then based on the cracked section.

When calculating the stresses likely to occur during transport and erection of a member, allowance should be made for the probability of impact; in the authors' opinion, it is desirable to allow an impact factor of 1½ to 2, and to limit the permissible tensile stress to not more than 500 lbf/in^2 (35 kgf/cm^2; 3.5 N/mm^2). In the USA Building Code the permissible stresses are as follows:

monolithic member without non-tensioned steel $3\sqrt{f_{\text{cyl t}}}$ lbf/in^2

monolithic member with non-tensioned steel $6\sqrt{f_{\text{cyl t}}}$ lbf/in^2

member composed of separate blocks and with non-tensioned steel $3\sqrt{f_{\text{cyl t}}}$ lbf/in^2

in which $f_{\text{cyl t}}$ is the crushing strength in lbf/in^2 of cylinders at the age at which transfer takes place. Relating the permissible stress to $\sqrt{f_{\text{cyl t}}}$ is not entirely satisfactory.

The permissible tensile stresses specified in the German regulations vary from 425 lbf/in^2 (30 kgf/cm^2; 3 N/mm^2) for concrete with a strength of 4260 lbf/in^2 (300 kgf/cm^2; 30 N/mm^2) to 640 lbf/in^2 (45 kgf/cm^2; 4.5 N/mm^2) for a strength of 8520 lbf/in^2 (600 kgf/cm^2; 60 N/mm^2).

No specific limiting stresses at transfer have been recommended in the Model Code.

5.5.5 Principal tensile stresses

In some regulations, permissible principal tensile stresses are specified for working-load conditions only. In CP 115, however, maximum principal tensile stresses have been recommended both for ultimate and working load conditions. The principal tensile stresses at ultimate load shall not exceed 300, 350 and 400 lbf/in^2 (21, 24.5 and 28 kgf/cm^2; 2, 2.4 and 2.8 N/mm^2) concrete having cube strengths of 4500, 6000 and 7500 lbf/in^2 (315, 420 and 525 kgf/cm^2; 31, 42 and 52 N/mm^2) at 28 days. The corresponding stresses at working load are 125, 150 and 175 lbf/in^2 (9, 10.5 and 12 kgf/cm^2; 0.9, 1.0 and 1.2 N/mm^2). In all cases, the principal tensile stress should be computed for the effective prestress and the stresses due to the specified loading. When the principal tensile stresses at working load exceed the permissible values, reinforcement to resist shearing should be provided. If the principal tensile stress is more than 1½ times the tensile stress permissible at working load, the reinforcement should be designed to resist the entire shearing force; for stresses between 1 and 1½ times the permissible stress the proportion of the shearing force to be resisted by the reinforcement should be interpolated linearly from 0 to 100 per cent. If the principal tensile stress at the ultimate load exceeds the permissible stress, the entire shearing force in excess of that resisted by any inclined prestressing steel should be resisted by reinforcement. It is generally advisable to provide stirrups unless the calculations show that they are not needed to resist shearing and complete freedom from cracks due to shrinking before the prestress is applied can be assured. (See Chart 1 in Part 2).

In CP 110 the limit state of collapse is taken as the basis of design for shear; in this case the appropriate principal tensile stress is given as $0.24\sqrt{f_{cu}}$ in which f_{cu} is in N/mm^2. Similar principle has been adopted in the Model Code.

The permissible principal tensile stresses need not be considered to apply at sections where the effects of bending predominate, but only at those sections where shearing is more important. The suddenness of failure due to shearing makes such a case potentially more dangerous than the gradual failure due to bending, and while cracks caused by bending due to occasional overloading disappear when the overload is removed, diagonal cracks due to shearing do not close, unless inclined cables are provided, and therefore constitute a permanent weakness. The authors have reservations however about this approach based on principal tensile stresses in containing shear especially at ultimate conditions.

The formulae which are given in the USA Building Code are intended to ensure that failure due to shearing cannot occur before failure in bending, and are based on the yield stress of the steel and the strength of the concrete. It is also doubtful whether the formulae fulfil their purpose.

In the German Regulations permissible principal tensile stresses are specified for shearing only, torsion only, and combined shearing and torsion, and are given in Table 5.3. The smaller stresses are for concrete having a strength of 4260 lbf/in^2 (300 kgf/cm^2; 30 N/mm^2) and the larger for a strength of 8520 lbf/in^2 (600 kgf/cm^2; 60 N/mm^2). No calculations for the shearing reinforcement are required if the permissible stress at ultimate load is not greater than the stress in line (2) of Table 5.4. Sufficient well-distributed reinforcement must be provided to resist the entire tensile force due to the principal stress wherever the stress exceeds ¾ of the stress in line (3). If the stress exceeds the appropriate stress in line (3), the cross-section must be increased or the stress reduced by increasing the prestressing force. Stirrups must always be provided in beams, even if the principal stress is less than those given in lines (1) and (2). It is surprising that the principal stresses permitted in partially prestressed members exceed those in fully prestressed members. This requirement is probably based on the assumption that with partial prestressing a smaller factor of safety against cracking may be allowed. This assumption may be true when cracking due to bending is being considered, but it seems better to ensure that a larger factor of safety against cracking due to shearing is obtained in view of the suddenness with which failure due to shearing occurs.

5.5.6 Permissible stress in bending in concrete at ultimate load

In CP 115 it is recommended that, when the steel is initially stressed to between 60 and 70 per cent of its strength, the mean compressive stress in the concrete at ultimate load may be assumed to be 40 per cent of the cube strength and the depth of the centroid of the stress block at the load causing failure may be assumed to be 40 per cent of the depth to the neutral plane. In CP 110 a choice is allowed between compressive stress of 40 per cent of the cube strength and the compression derived from stress–strain curve with $\gamma_m = 1.5$. In both cases the limiting strain is restricted to 0.0035. The stress given in the USA Building Code is $0.85\,f_{cyl}$ which corresponds to 64 per cent of the strength of cubes when the ratio of the strength of a cylinder to the strength of a cube is ¾, and 57 per cent when this ratio is ⅔. The stress specified in the German regulations is two-thirds of the strength of cubes; the distribution of stress assumed, however, is such that the results obtained are about the same as those given in the foregoing.

Table 5.4 Principal tensile stresses (German Regulations)

Line	Condition	Permissible stress (lbf/in^2)			
		(a) Shear only	(b) Torsion only	(c) Shear + torsion	
(1)	When design permits no tensile stress due to bending at working load	8–12	6–10	10–15	kgf/cm^2
		114–172	85–142	142–213	lbf/in^2
		0.8–1.2	0.6–1	1–1.47	N/mm^2
(2)	When design permits tensile stress due to bending at working load	16–24	12–20	20–30	kgf/cm^2
		227–341	170–284	284–426	lbf/in^2
		1.57–2.35	1.2–1.95	1.95–2.95	N/mm^2
(3)	At load causing failure	31–47	24–39	39–59	kgf/cm^2
		454–682	341–568	568–852	lbf/in^2
		3.1–4.7	2.35–3.9	3.9–5.9	N/mm^2

High values correspond to strength of 8520 lbf/in^2 (600 kgf/cm^2; 60 N/mm^2)
Low values correspond to strength of 4260 lbf/in^2 (300 kgf/cm^2; 30 N/mm^2)
Intermediate values obtained by linear interpolation.

5.6 Stress in steel

5.6.1 Stress permissible in steel at transfer and working load

Regulations concerning the permissible stress in the steel at transfer differ considerably. In France, tensioning stresses of up to 90 per cent of the tensile strength or the yield-point stress are permitted. The German regulations specify that the stress after all losses have taken place shall not exceed 55 per cent of the tensile strength or 75 per cent of the yield stress or proof stress, whichever is less. Although a stress of 55 per cent is low, it is preferable to an initial stress of 90 per cent of the ultimate since at this high stress a very large part of the elongation of the steel (and consequently of its ductility) are no longer available. The German regulations allow the initial stress in pre-tensioned steel to be increased to a maximum of 80 per cent of the yield stress or proof stress, and if friction occurs the stress at working load in post-tensioned steel may be increased by 5 per cent.

It is recommended in CP 115 that the initial tensioning stress should not exceed 70 per cent of the ultimate strength of cold-drawn wire or strands. For alloy-steel bars, 70 per cent of the strength or 85 per cent of the 0.2 per cent proof stress (whichever is less) is recommended, the proof stress being based on the smallest cross-section of the bar. When alloy-steel bars are used in the pre-tensioning method, the recommended stress is increased to 75 per cent of the strength or 90 per cent of the 0.2 per cent proof stress, whichever is less. An increase of 10 per cent may be allowed during tensioning, to overcome the effects of friction. In CP 110 a maximum initial prestress of 80 per cent of the characteristic strength is specified.

The maximum tensioning stress given in the *USA Building Code* is 70 per cent of the strength; a stress of 80 per cent is permitted for a short period provided that this stress is reduced to 70 per cent when the tensioning operation is completed. The effective stress in the steel after losses have been deducted should in no case be assumed to exceed 60 per cent of the strength or 80 per cent of the yield stress or proof stress.

In the Model Code the maximum tensioning stress is limited to 75% of characteristic strength or 85% of 0.1% proof stress of steel.

5.6.2 Stress due to bending permissible in steel at ultimate load

The greatest stress in pre-tensioned steel recommended in CP 115 is equal to the tensile strength, except when the ratio of the tensile strength of the steel to the strength of the concrete multiplied by the percentage of reinforcement is very high. This is termed the 'percentage/strength ratio' or 'reinforcement index'; permissible stresses corresponding to particular values are given in Table 8.1, Chapter 8. If the depth of the compression zone exceeds 60 per cent of the effective depth to the steel, the permissible maximum stress is slightly less than the tensile strength. The maximum permissible stress for well-bonded post-tensioned steel is between 75 per cent and 100 per cent of the tensile strength, the lower stresses corresponding to higher values of the percentage/strength ratio and to greater depths of the compressive zone. With non-bonded post-tensioned steel it is advisable for the greatest stress to be the same as the initial tensioning stress. The greatest stress in non-tensioned steel should be equal to the yield stress or, in the case of high-strength steel, equal to the 0.2 per cent proof stress. The values given in CP 110 for the limit state of collapse are derived from the strain at the level of the steel. They are comparable in magnitude with those given in CP 115.

In the USA Building Code formulae are given in which a nominal tensile stress is used, which is equal to $\left(1 - \dfrac{\rho f_{su}}{2f_{cyl}}\right) f_{su}$, in which f_{su} is the tensile strength of the steel and f_{cyl} is the crushing strength of concrete cylinders, both being in lbf/in^2. For example, if f_{su} = 200 000 lbf/in^2 (14 000 kgf/cm^2; 1400 N/mm^2) and ρ = 1 per cent, the permissible stresses corresponding to f_{cyl} = 5000 and 10 000 lbf/in^2 (350 and 700 kgf/cm^2; 35 and 70 N/mm^2) are 0.8 f_{su} and 0.9 f_{su} respectively. Only mild steel may be used as non-tensioned reinforcement, and the greatest permissible stress is therefore the yield-point stress.

In the German regulations the greatest permissible stress is specified as the yield stress or proof stress for high-tensile steel because definite yielding occurs in most German wire when the proof stress is exceeded. If non-tensioned steel is used, it must be of a type normally used in reinforced concrete; the yield points or proof stresses for the three types so used are 31 000 lbf/in^2 (2200 kgf/

cm^2; 220 N/mm^2) (mild steel), 48 000 lbf/in^2 (3350 kgf/cm^2; 335 N/mm^2) and 57 000 lbf/in^2 (4000 kgf/cm^2; 400 N/mm^2). This sometimes leads to the provision of large areas of non-tensioned steel; in such a case the effect on the losses should be considered (see Chapter 6).

The use of high-tensile steel for non-tensioned reinforcement is therefore in accordance only with the recommendations of CP 115 and CP 110 and the greatest stress permissible in such reinforcement may be up to 90 per cent of the tensile strength, if the 0.2 per cent proof stress is 90 per cent of the strength.

5.6.3 Permissible stresses in reinforcement resisting shearing

No specific recommendations concerning tensile stresses permissible in shearing reinforcement at working load are given in CP 115 or the USA Building Code, but it is recommended in CP 115 that under ultimate conditions the tensile stress of shear reinforcement should be 80 per cent of the yield point or the 0.2 per cent proof stress. This recommendation implies that non-tensioned high-tensile steel may be used as shearing reinforcement.

The requirements of CP 110 are based on collapse criteria; they imply that the limiting stress in shear reinforcement should be taken as 0.87 times its characteristic strength, which should not be considered as more than 60 000 lbf/in^2 (4250 kgf/cm^2; 425 N/mm^2). In the Model Code the limiting stress in shear reinforcement is 0.87 × 500, i.e., 63 075 lbf/in^2 (4350 kgf/cm^2; 435 N/mm^2).

The German regulations specify that the maximum stresses shall be the same as those due to bending, that is, 31 000 lbf/in^2, 48 000 lbf/in^2, or 57 000 lbf/in^2 (2200, 3350, 4000 kgf/cm^2; 220, 335, 400 N/mm^2).

5.7 Stresses permissible in composite members

No recommendations or requirements for composite members are given in the USA Building Code or the German regulations. In CP 115 it is stated that prestressed beams may be considered to act in conjunction with concrete cast in place if provision is made to resist the horizontal shearing forces at the surface of contact by the use of 'shear connectors', or by means of roughening or irregularities of the surfacing of the concrete. Composite members may therefore be considered to behave as homogeneous members. CP 115 is the first document to recognize this behaviour, which has been proved by numerous tests, and to permit tensile stresses in the concrete cast in place. The recommended stresses are only small, but in practice the concrete cast in place is prevented from cracking (except for harmless microscopic cracks) by the restraining effect of the adjacent compressed concrete.

In accordance with CP 115, if the greatest working load is of frequent occurrence and of long duration (Case A), then the permissible tensile stresses in the added concrete vary from 200 lbf/in^2 (14 kgf/cm^2; 1.4 N/mm^2) for concrete having a strength of 3000 lbf/in^2 (210 kgf/cm^2; 21 N/mm^2), to 300 lbf/in^2 (21 kgf/cm^2; 2.1 N/mm^2) when the cube strength is 6000 lbf/in^2 (420 kgf/cm^2; 42 N/mm^2). The corresponding stresses, when the maximum load occurs rarely and is of short duration, as in the case of wind and snow on roofs (Case B), are 300 lbf/in^2 and 450 lbf/in^2 (21 and 31 kgf/cm^2; 2.1 and 3.1 N/mm^2). These stresses are given in Table 5.1. The stresses in the added concrete may be increased by 50 per cent if the stresses $\overline{f}_{tw}$ are reduced by the same numerical amount. For example, for concrete having a strength of 6000 lbf/in^2 (420 kgf/cm^2;

42 N/mm^2), prestressed by means of pre-tensioned steel, and subjected to sustained maximum loading (Case A), the tensile stresses recommended are 300 lbf/in^2 (21 kgf/cm^2; 2.1 N/mm^2) in the prestressed concrete and in the added concrete. If the stress in the added concrete is increased by 50 per cent to 450 lbf/in^2 (31 kgf/cm^2; 3.1 N/mm^2) the stress permissible in the prestressed concrete must be reduced by the same amount, that is to 150 lbf/in^2 (10.5 kgf/cm^2; 1 N/mm^2). If tests show that the increased tensile stress that is 300 + 250 = 550 lbf/in^2 (21 + 17.5 = 38.5 kgf/cm^2; 2.1 + 1.75 = 3.85 N/mm^2) does not exceed 75 per cent of the stress at which the first crack appears, the stress permissible in the prestressed concrete would then become 550 – 150 = 400 lbf/in^2 (38.5 – 10.5 = 28 kgf/cm^2; 3.85 – 1.0 = 2.85 N/mm^2).

Many tests have shown that there are no harmful effects if much greater tensile stresses occur in the concrete, and several structures designed on the basis of these higher stresses are now in satisfactory service. As an example mention may be made of a road bridge designed by British Rail (Eastern Region), which has been in use since 1950. The limiting stress for loading of this type (Case A) is given in CP 115 as 450 lbf/in^2 (31 kgf/cm^2; 3.1 N/mm^2) yet the design stress of 720 lbf/in^2 (50 kgf/cm^2; 5 N/mm^2) has proved to be satisfactory for this structure.

In CP 110 as also in the Model Code the main basis of design is shifted to conditions at ultimate load, and this has led to a simplification of the requirements for composite structures, these are designed on the same basis as non-composite structures. For the limit state of local damage, additional requirements relating to horizontal shear, tensile stresses in the added concrete, differential shrinkage, and continuity are given.

5.8 Other criteria for local damage

5.8.1 Deflection

In CP 115, the limiting upward deflection (camber) is specified as 1/300 of the span, where finishes are to be applied. The CP 110 gives a wider range of values; at working load, a maximum of 1/250 of the span is specified.

Where brittle finishes or partitions are used, guidance values of 1/350 of the span or 20 mm are given. A maximum upward deflection of 1/300 of the span is also suggested.

In the USA, ACI Building Code gives maximum values for the immediate deflection of 1/180 of the span (for flat roofs) and 1/360 of the span (for floors): these values are reduced to 1/240 and 1/480 of the span, respectively, in cases where damage might otherwise be caused.

In the roof construction if the resultant deformation is downwards, so-called ponding may occur which should be avoided.

No specific limitation of the magnitude of deflection has been mentioned in the Model Code except stating that 'Deflection should be limited in order to satisfy the functional requirements better'.

5.8.2 Cracking

No recommendations relating specifically to crack widths are given in CP 115. As stated before CP 110 includes values of nominal tensile stresses which are intended to ensure that crack widths are limited to 0.1 mm or 0.2 mm in structures of Class 3.

In the USA, the *ACI* Building Code also gives a nominal tensile stress ($12\sqrt{f_{cyl}}$, in which f_{cyl} is in lbf/in^2) as a limiting value based on maximum crack widths. This value may be exceeded if experiments or analysis show that the performance of the structure is not impaired. The adoption of a single limiting value of this type is rather too general. If f_{cyl} = 6400 lbf/in^2 (450 kgf/cm^2; 45 N/mm^2), for example, the limiting value is 960 lbf/in^2 (67.5 kgf/cm^2; 6.75 N/mm^2); this may cause either appreciable cracking (with badly distributed steel) or no visible cracking (if the reinforcement is well distributed).

The German Code does not allow Class 3 structures.

5.9 Safety factors

The basis of the limit state design is to ensure an adequate margin against the appropriate limiting conditions of the design such that in general

$$\text{Effect } [\gamma_c(\gamma_{f1}G_k + \gamma_{f2}Q_k + \gamma_{f3}W_k)] \leqslant \text{function}\left[\frac{f_k}{\gamma_m \text{ conc } \gamma_m \text{ steel}}\right]$$

G_k, Q_k and W_k = characteristic dead load, super load and wind load respectively
γ_f = partial safety factor for load
γ_m = partial safety factor for materials
γ_c = partial safety factor to account for the mode and consequence of failure
f_k = characteristic strength

The above factors are determined on the basis of statistical probabilities and given in national and international Codes of Practice. Obviously two distinctly different sets of values of the factor are required for the ultimate and serviceability limit. Table 5.5 compares the respective values of these factors in the British Codes (CP 110 and CP 115), the ACI Building Code and the CEB/FIP Model Code for ultimate conditions. It should be noted that although the importance of the factor γ_c (for mode and consequence of failure) has been fully recognised, as yet no national or international code actually suggests any definite acceptable value for this – currently it appears that a part of it is indirectly incorporated in other partial safety factors.

In the CEB/FIP Code an additional factor ψ has been introduced to take account of the reduced probabilities of a combination of loading. For the limit state of serviceability γ_m has been replaced in all Codes by definite allowable stresses and γ_f has been broadly taken as 1 both for dead load (G_k) and superload (Q_k). When wind load (W_k) acts together with dead load and superload CP 110 allows a factor of 0.8 both to W_k and Q_k. The approach in the CEB/FIP Model Code is to divide the working load under three headings, viz. (a) service load (Q_{ser}), (b) frequent load ($G_k + \psi_1 Q_k$) and (c) semi-permanent load ($G_k + \psi_2 Q_k$). The coefficients ψ_1 and ψ_2 are reduction factors, some approximate values for which have been recommended on the basis of the occupation characteristics of the structure.

CEB approach is very rational, and with realistic values of coefficients (ψ_1 and ψ_2) it will undoubtedly contribute to more economic yet perfectly satisfactory design, particularly in partially prestressed concrete.

Table 5.5 Safety factor for ultimate limit state

		CP 110	CP 115	CEB/FIP Model Code	ACI Building Code
Load	*Load case 1* Dead load + superload	$1.4\ G_k + 1.6\ Q_k$	$1.5\ G_k + 2.5\ Q_k$ or $2(G_k + Q_k)$ whichever is less	$1.35\ G_k + 1.5\ Q_k$	$1.4\ G_k + 1.7\ Q_k$
Load	*Load case 2* Dead load + 2 variable superloads			(a) $1.35\ G_k + 1.5\ Q_{k1} + 1.5\ \psi\ Q_{k2}$ (b) $1.35\ G_k + 1.5\ Q_{k2} + 1.5\ \psi\ Q_{k1}$	
Load	*Load case 3* Dead load + wind	$0.9\ G_k + 1.4\ W_k$		$G_k + 1.5\ W_k$	$0.9\ G_k + 1.3\ W_k$
Load	*Load case 4* Dead load + superload + wind	$1.2 \times (G_k + Q_k + W_k)$		$1.2 \times (G_k + Q_k + W_k)$	$0.75 \times (1.4\ G_k + 1.7\ Q_k + 1.7\ W_k)$
Material	Concrete	1.5 1.3 for accidental load	No material safety factor. Design based on minimum strength	1.5 1.3	No material safety factor but capacity reduction factor ϕ_1 $\phi_1 = 0.90$ for bending $\phi_1 = 0.85$ for shear $\phi_1 = 0.70$–0.75 for axial compression
Material	Steel	1.15 1 for accidental local damage		1.15 1.00	

5.10 General recommendations

Unless the designer is sure that the assumptions made in the design can be fulfilled, it may be better to use reinforced concrete (preferably with a small prestress to reduce cracking) instead of prestressed concrete. Unless the maximum possible losses are assessed properly, the calculated stresses may bear little relation to those which may occur in practice and the adoption of conservative stresses is not in itself sufficient to ensure a satisfactory structure.

Different considerations govern the selection of permissible stresses at transfer, working load, and ultimate load. It is particularly desirable that conditions at transfer should be safely determined since the adoption of excessive compressive stresses may cause immediate failure, either local or complete, if the specified strength of the concrete is not obtained; if the difference between the strength required and that obtained is not so great, such stresses may cause losses greatly in excess of those expected with a consequent reduction in the factors of safety both against cracking and against failure. Such a reduction will also adversely affect conditions at working load. A reduction of the initial tensioning stress in the steel can also reduce the factor of safety against cracking if the reduction factor for the losses is not amended at the same time. If the initial tensioning stress is 150 000 lbf/in^2 (10 500 kgf/cm^2; 1050 N/mm^2), for example, and losses are 25 per cent this corresponds to a total loss of 37 500 lbf/in^2 (2600 kgf/cm^2; 260 N/mm^2). If the initial stress is reduced to 120 000 lbf/in^2 (8400 kgf/cm^2; 840 N/mm^2) the loss corresponding to the same proportions would be only 30 000 lbf/in^2 (2100 kgf/cm^2; 210 N/mm^2) and this underestimation would result in an effective prestress of 90 per cent of the predicted value and a consequent reduction in the factor of safety against cracking. No advantage is gained, however, by providing a greater prestress than is necessary, since this leads to an increase in the rigidity of the structure, and affects adversely its resilience. It is therefore important to ensure that the ratio of the ultimate bending moment to that causing cracking is at least 1.5. The bending moment M_{cr} causing visible cracking is $Z(f_{ce} + f_{ct})$, in which f_{ce} is the effective prestress and f_{ct} the stress at which cracks become visible. The stress f_{ct} will usually vary between 900 and 1100 lbf/in^2 (64 and 78 kgf/cm^2; 6.4 and 7.8 N/mm^2) if pretensioned steel is used, and between 600 and 1000 lbf/in^2 (42 and 70 kgf/cm^2; 4.2 and 7 N/mm^2) if post-tensioned steel is used, depending on the distribution of the steel, the strength of the concrete, and the efficiency of the bond. If a prestressed structure is composed of separate precast blocks, no tensile resistance should usually be taken into account, but to obtain the greatest possible value of M_{cr} a stress of 300 lbf/in^2 (21 kgf/cm^2; 2.1 N/mm^2) may be allowed if there are satisfactory mortar joints. The foregoing points must be considered when deciding on suitable permissible stresses. It is also necessary to ensure that the risk of sudden failure due to shearing, principal stress, or compression is eliminated.

CHAPTER 6

LOSSES OF PRESTRESSING FORCE

6.1 Introduction

The methods of analysis and design of prestressed concrete generally used in Great Britain are based on the assumption that the prestressing force applied to the concrete (which is also termed the prestressing force 'at transfer'), and the minimum effective prestressing force acting on a section, are both known. In a simply-supported beam, for example, an effective eccentric compressive force of definite magnitude is taken into account, and the stresses due to this force are directly superimposed on those due to the load, the whole system being assumed to remain statically determinate. Although this method is not entirely exact, numerous tests have proved that actual behaviour conforms safely and with sufficient accuracy to these assumptions.

A more precise method of calculation is often used, for example in Germany, which allows for the fact that a simply-supported beam becomes statically indeterminate when it is prestressed, the initial effect of the prestressing force and the effect of changes occurring with the passage of time being therefore considered independently of each other. Although this may be theoretically correct, its application to practical design produces complications. The designer should know in advance the conditions under which the concrete will be cast and cured, and should also allow for the effect of weather conditions on creep. It is rarely possible to anticipate these conditions accurately, and it is essential that the most unfavourable assumptions be made in order to allow for the maximum possible losses of the initial prestress. The accuracy of the method is therefore unlikely to exceed that of the simpler method.

If the design is based on the minimum effective prestressing force, it is essential to determine the maximum possible losses of prestress which the various influences can cause. When these are known the calculation is much simpler than, and quite as accurate as, a detailed computation in which each influence is investigated separately. A more detailed calculation may sometimes be worthwhile when the actual deformation of a structure is required to be known at a definite time or when the conditions at a certain time are of special interest (for example in the case of the loading of a test specimen), but even in this case a simple design with an additional eccentric force included to allow for the reduction in losses is usually adequate.

The minimum effective prestress is obtained when maximum losses are assumed to occur after the prestressing force is applied to the concrete, and it follows that the worst conditions for the factors governing the losses should be assumed unless any particular factor is definitely known to be more favourable.

For example, the magnitude of creep increases as the stress in the concrete increases. The loss of prestress as a result of creep therefore varies considerably with different stresses in the concrete, and it is clearly incorrect to adopt a constant overall value for losses (for example 15 per cent) which includes no allowance for the magnitude of the compressive stress in the concrete adjacent to the prestressing steel and the ambient conditions. Values which are even lower than 15 per cent may sometimes occur, in especially favourable instances; but it is completely incorrect to use such values generally, as is sometimes done. On the other hand, the concrete may be moist cured for the whole of the time until the member is placed in position, and in such an unusual case little creep need be considered for the first stage when the stress in the concrete adjacent to the steel is high. It is also possible that in some special cases the construction when in use will always be exposed to moist surroundings, and in this case shrinkage and creep will be very small. In general, however, more unfavourable conditions need be considered. Figure 6.1 adapted from CEB/FIP Model Code shows diagrammatically the losses and effective prestress in a section with pre-tensioned steel. Sections stressed with post-tensioned steel will have additional losses due to friction, but the magnitude of losses due to elastic shortening as well as creep and shrinkage will be reduced.

The factors that lead to losses of prestress are discussed in the following, and suitable values are suggested.

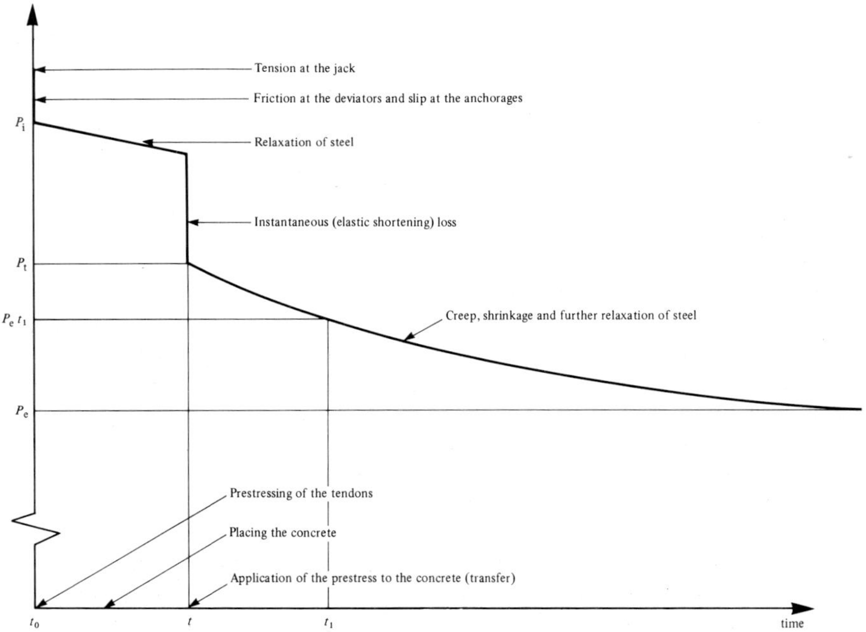

P_i = Initial prestressing force
P_t = Prestressing force at transfer
$P_e t_1$ = Effective prestressing force after time t_1
P_e = Effective prestress after full losses

Figure 6.1 Losses in a given section (pre-tensioning)

6.2 Loss of stress between the jack and the anchorage

Post-tensioned steel may be anchored in a number of ways. When wedges are used, some slipping between the wires and wedges is unavoidable when the tensile force applied to the steel by means of the jack is transferred to the wedges. When anchor-plates are used in conjunction with nuts and threads, or the steel forms loops around the concrete, a small shortening of the steel may occur due to the anchor-plate or loop being pressed into the concrete. All these factors should be allowed for at the time when the tension is applied to the steel; similarly, any errors in calibration, and any losses due to friction in the connection between the pump and the dial-gauge, should be determined and allowed for at the time of tensioning by increasing the force applied by the jack. It is essential that there be approximate agreement between the measured elongation of the steel and the stress corresponding to the force indicated by the gauge.

The designer need consider these losses only when wedges are used to anchor short wires. A slip of the order of $\frac{1}{8}$ in. (3 mm) usually occurs at a wedge; values for particular systems may be obtained from the suppliers. The extension of steel with a modulus of elasticity of 28.8×10^6 lbf/in^2 (2×10^6 kgf/cm^2; 200 kN/mm^2) due to a stress of 144 000 lbf/in^2 (10 000 kgf/cm^2; 1 kN/mm^2) is 6 in. in 100 ft (150 mm in 30 m) and a reduction of $\frac{1}{8}$ in. (3 mm) in this extension corresponds to a loss of $\frac{1}{48}$, or 3000 lbf/in^2 (210 kgf/cm^2; 21 N/mm^2) of the initial stress in the steel. This is easily allowed for by adding 3000 lbf/in^2 (210 kgf/cm^2; 21 N/mm^2) to the initial tensioning stress. Such losses may occur with both pre-tensioned and post-tensioned steel. If the steel were only 10 ft (3 m) long, however, a shortening of $\frac{1}{8}$ in. (3 mm) would correspond to a loss of prestress of 30 000 lbf/in^2 (2100 kgf/cm^2; 210 N/mm^2) and this is far too much to add to the specified initial tensioning stress. This loss should be allowed for in the design, when the length is less than 30 ft (9 m).

When preparing the specification and drawings, it should be made clear which losses should be allowed for at the site by increasing the initial tensioning stress, and that the specified prestressing force is that which is to be applied when allowance has been made for these additional losses.

6.3 Elastic shortening

6.3.1 Pre-tensioned steel (Figure 6.2)

When the steel is pre-tensioned a reduction of the initial stress in the steel occurs at transfer since, as soon as the concrete is compressed, an elastic shortening of the concrete takes place. This is accompanied by an equal reduction in the length of the steel, with a consequent reduction in the initial prestress, and is given by

$$\frac{f_{sT}}{E_c} = \frac{\Delta f_{pe}}{E_s} \qquad (6.1)$$

in which Δf_{pe} is the loss of the initial tensioning stress in the steel due to the elastic shortening of the concrete, f_{sT} is the stress in the concrete adjacent to the steel and E_c and E_s are the moduli of elasticity of the concrete and steel; the ratio $E_s/E_c = \alpha_e$. If the compressive stress in the concrete adjacent to the steel is known equation 6.1 becomes

$$\Delta f_{pe} = \alpha_e f_{sT} \qquad (6.2)$$

In general the compressive stress f_{sT} is unknown, since it depends on the prestressing force at transfer (P_t), which in turn depends on the loss due to elastic shortening[1]. In this case

$$f_{sT} = \frac{P_t}{A} + \frac{P_t e_s}{\frac{I}{e_s}} = \frac{P_t}{A}\left(1 + \frac{e_s^2}{i^2}\right) \qquad (6.3)$$

in which A is the area of the concrete section and $i^2 = \frac{I}{A}$. Also

$$P_t = A_{st}(f_{pi} - \Delta f_{pe}) \qquad (6.4)$$

in which A_{st} is the area of tensioned steel and f_{pi} the initial tensioning stress in the steel. Therefore, combining equations 6.1, 6.3, and 6.4,

$$\frac{\Delta f_{pe}}{E_s} = \frac{(f_{pi} - \Delta f_{pe})A_{st}}{E_c A}\left(1 + \frac{e_s^2}{i^2}\right) \qquad (6.5)$$

from which

$$\Delta f_{pe} = \frac{\alpha_e K_e}{1 + \alpha_e K_e} f_{pi} = \lambda f_{pi} \qquad (6.6)$$

in which

$$\lambda = \frac{\alpha_e K_e}{1 + \alpha_e K_e} \qquad (6.6a)$$

$$K_e = \frac{A_{st}}{A}\left(1 + \frac{e_s^2}{i^2}\right) \qquad (6.6b)$$

Thus, equations 6.6, 6.6a and 6.6b give exact values. Expression 6.2, however, can generally be used with sufficient accuracy even when the stresses in the concrete at the time when the prestress is applied are only assumed, as the stress f_{sT} in the concrete adjacent to the steel may then be estimated with sufficient accuracy from the maximum compressive stress. If full agreement between the assumed and actual values of f_{sT} is not obtained at the first attempt, the improved value arising from the first attempt should be inserted in equation 6.2 and the calculation repeated. The ratio between f_{sT} and the maximum compressive stress in the concrete depends on the arrangement of the steel in the cross-section, and may be determined exactly when the position of the steel is known. Values of the ratio $\frac{100\Delta f_{pe}}{f_{pi}}$ (that is, $\frac{\Delta f_{pe}}{f_{pi}}$ in percentage form) occurring in equation 6.6 are given in Table 6.1, for values of α_e between 4 and 12 (to allow for lightweight concrete), when the percentages of steel are 0.2 and 0.6 and the ratio e_s^2/i^2 has values of 1.3 and 2.6.

As a check over the suitability of the values considered in Table 6.1, reference may be made to the four sections given in Table 8.2 – namely, a rectangular beam, two T-beams, and an I-beam. The corresponding values of e_s^2/i^2 for these sections are 1.92, 1.18, 2.50 and 2.78 respectively. Hence the values of 1.3 and 2.6 used in Table 6.1 are quite appropriate.

When an eccentric prestressing force produces a bending moment which opposes that due to a downward load, a concrete member supported at its lower

Table 6.1 Values of $\frac{100\Delta f_{pe}}{f_{pi}}$

α_e	4				8				12			
$\rho\%$	0.2		0.6		0.2		0.6		0.2		0.6	
$\frac{e_s^2}{i^2}$	1.3	2.6	1.3	2.6	1.3	2.6	1.3	2.6	1.3	2.6	1.3	2.6
$\frac{100\,\Delta f_{pe}}{f_{pi}}$	1.8	2.8	5.2	7.9	3.6	5.5	10.0	14.6	5.2	8.0	14.0	20.5

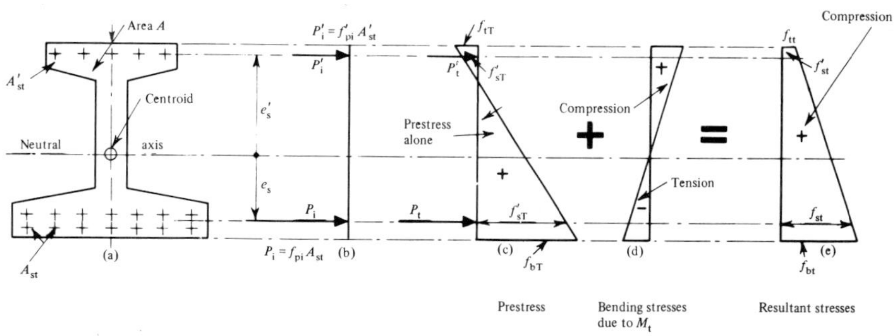

Figure 6.2 Elastic losses with pre-tensioned steel

surface will deflect upwards and the dead weight at the time of prestressing (that is at transfer) will then produce a bending moment M_t of opposite sign to that due to the prestressing force. The stresses then developed will be the resultant of those due to the prestressing force at transfer (denoted by the subscript T) and those due to the bending moment M_t (the resultant being denoted by the subscript t). Consequently, this stress will include the tensile bending stress; therefore

$$f_{bt} = f_{bT} - \frac{M_t}{\frac{I}{e_b}}$$

It is sometimes stated, incorrectly, that loss at transfer due to elastic shortening is offset by elastic elongation when the member is loaded. The elongation of both steel and concrete which occurs in the tensile zone when a member is loaded is accompanied by tensile bending stresses which reduce the pre-compression of the concrete. The effective elongation of the steel (related to zero extension of the concrete) is unaffected by any combined elongation of steel and concrete, and the loss of stress can be caused only by the shortening of the steel due to the elastic shortening of the concrete at transfer. This is given by equation 6.2, which is based on the stress f_{sT} in the concrete adjacent to the steel that is produced by the prestress alone; the resultant stress f_{st} shall not be used. It is therefore immaterial whether lifting of the member occurs at the instant of prestressing or whether it is delayed, as the combined elongations due to bending cannot affect the magnitude of the loss due to elastic shortening.

6.3.2 Post-tensioned steel (Figure 6.3)

In the case of post-tensioned steel there is no loss of stress due to elastic shortening when the whole of the prestress is applied to the concrete in one operation, as the steel is not anchored until the complete prestressing force is obtained, by which time the elastic deformation of the concrete has taken place. If all the steel is not tensioned at the same time, however, tensioning of each steel member or group of members after the first produces an elastic deformation which modifies the length of, and the stress in, every anchored steel member, and losses due to elastic shortening at each operation are obviated only in the steel member or group of members tensioned during the operation. Consequently, if equal tensile forces are required in each steel member, the initial forces in each must be adjusted to allow for subsequent losses due to elastic shortening.

When the order of tensioning each tendon or group of tendons is known, it is not difficult to evaluate the loss of stress in each due to elastic shortening. The calculation is frequently tedious, however, and a simplified method is available. This is based on the general condition that the average elastic shortening of every tendon will never exceed one-half of the elastic shortening which would occur if all the tendons were pre-tensioned; equation 6.7 therefore gives a safe approximate value, although the stress in each tendon differs from the mean value so obtained. Strictly speaking, the effect only of the prestress at transfer should be considered, but the stress f_{st}, which includes the stress due to the bending moment M_t, may appropriately be used in this case as the counter-action of M_t is gradually offset as each tendon is tensioned.

$$\Delta f_{pe} \text{ (post-tensioning)} = \tfrac{1}{2}\, \alpha_e f_{st} \qquad (6.7)$$

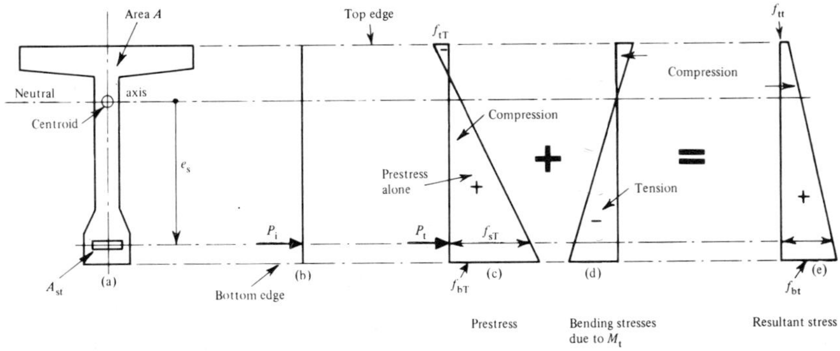

$$P_i = A_{st} f_{pi} \qquad P_t = (f_{pi} - \Delta f_{pe}) A_{st}$$

$$f_{sT} = \frac{P_t}{A}\left(1 + \frac{e_s^2}{i^2}\right)$$

Figure 6.3 Elastic losses with post-tensioned steel

6.3.3 General considerations (see Figure 6.3)

Suitable values for the modular ratio α_e are discussed in Chapter 2; the values given in Table 6.2 are based on the relationship between the cube strength at transfer and the modulus of elasticity for concrete as given in CP 115 and CP 110 and on a modulus of elasticity for steel of 29×10^6 lbf/in² (2×10^6 kgf/cm²; 200 kN/mm²). If tensioned steel is provided only in the tensile zone the losses due to elastic shortening may be calculated with sufficient accuracy by adopting a value for f_{sT} equal to the stress in the concrete at the centroid of the tensioned steel, and assuming that the loss is the same for each wire or bar. The resultant stress in the concrete adjacent to the centroid of the steel will then be

$$f_{st} = f_{sT} - \frac{M_t}{\frac{I}{e_s}}$$

If the steel is distributed throughout the section, however, or separate tensioned steel is provided at the top and bottom, this assumption is not justified.

In the case shown in Figure 6.4a, for example, $e_s = \frac{h}{6}$ and for a rectangular section $i^2 = \frac{h^2}{12}$; assuming that $\alpha_e = 5$ and $A_{st} = 0.01A$, then from equation 6.6b,

$$K_e = 0.01 \left\{ 1 + \frac{\left(\frac{h}{6}\right)^2}{\frac{h^2}{12}} \right\} = 0.0133$$

and from equation 6.6,

$$\Delta f_{pe} = \left(\frac{5 \times 0.0133}{1 + [5 \times 0.0133]} \right) \times f_{pi} = 0.0625\, f_{pi}, \text{ or } 6.25\%.$$

In the case shown in Figure 6.4b, however, the prestressing steel is placed at a distance of $0.1h$ from the outer edges.

From the two conditions,

$$A_{st} + A'_{st} = A_s \text{ and } (A_{st} + A'_{st}) \times \frac{h}{6} = (A_{st} - A'_{st}) \times 0.4h$$

$$A'_{st} = 0.411A_{st}, \text{ and therefore } A_{st} = \frac{A_s}{1.411} = 0.71A_s, \text{ and } A'_{st} = 0.29A_s.$$

An exact solution may be obtained by considering separately the influences of $A_{st} = 0.71A_s$ and $A'_{st} = 0.29A_s$, the combined value being obtained by superposition. It is simpler however, to assume a value for the loss and obtain an improved value by repeating the calculation. Again assuming that $\alpha_e = 5$ and $A_{st}/A = 0.01$, the calculation is as follows:

Assume f_{pi} = 150 000 lbf/in² (10 000 kgf/cm²) (1000 N/mm²)
Assume loss = 5 per cent
Then

$$f_{bT} = 0.95 \times 0.01 \times 2 \times f_{pi} = 2850 \text{ lbf/in}^2 \quad (190 \text{ kgf/cm}^2) \quad (19 \text{ N/mm}^2)$$

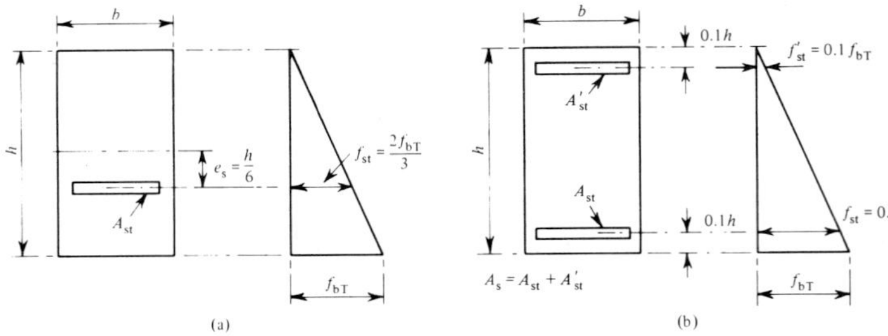

Figure 6.4 Effect of position of steel on losses

Hence

$f_{sT} = 0.9 f_{bT} =$ 2565 lbf/in^2 (171 kgf/cm^2) (17.1 N/mm^2)

and

$f'_{sT} = 0.1 f_{bT} =$ 285 lbf/in^2 (19 kgf/cm^2) (1.9 N/mm^2)

Therefore,

$\Delta f_{pe} = \alpha_e f_{sT} =$ 12 825 lbf/in^2 (855 kgf/cm^2) (85.5 N/mm^2)

and

$f_{pt} = f_{pi} - \Delta f_{pe} =$ 137 175 lbf/in^2 (9145 kgf/cm^2) (914.5 N/mm^2)

Also

$\Delta f'_{pe} = \alpha_e f'_{sT} =$ 1425 lbf/in^2 (95 kgf/cm^2) (9.5 N/mm^2)

and

$f'_{pt} = f_{pi} - \Delta f'_{pe} =$ 148 575 lbf/in^2 (9905 kgf/cm^2) (990.5 N/mm^2)

For sections such as that shown in Figure 6.4b the effects of A_{st} and A'_{st} on the stress f_{bT} at the bottom edge are expressed by the coefficients $k_b = 1 + \frac{e_b e_s}{i^2}$ and $k'_b = 1 - \frac{e_t e_s}{i^2}$. In this case $i^2 = \frac{h^2}{12}$, $e_s = e'_s = 0.4h$, and $e_b = e_t = 0.5h$.

Hence

$$\frac{e_b e_s}{i^2} = \frac{e_t e_s}{i^2} = \frac{0.5h \times 0.4h \times 12}{h^2} = 2.4$$

and

$$k_b = 1 + 2.4 = 3.4; \; k'_b = 1 - 2.4 = -1.4$$

Therefore

$$f_{bT} = (3.4 \times 0.71 \times f_{pt} - 1.4 \times 0.29 \times f'_{pt}) \times 0.01$$
$$= 2711 \text{ lbf/in}^2 \; (189.8 \text{ kgf/cm}^2) \; (18.85 \text{ N/mm}^2)$$

Repeating the calculation with the improved value of f_{bT},

$f'_{sT} =$ 271 lbf/in^2 (19 kgf/cm^2) (1.9 N/mm^2)

f_{sT} = 2437 lbf/in^2 (171 kgf/cm^2) (17 N/mm^2)

$\Delta f_{pe} = \alpha_e f_{sT}$ = 12 185 lbf/in^2 (850 kgf/cm^2) (89 N/mm^2)

f_{pT} = 137 815 lbf/in^2 (9600 kgf/cm^2) (955 N/mm^2)

$\Delta f'_{pe} = \alpha_e f'_{sT}$ = 1355 lbf/in^2 (95 kgf/cm^2) (9.5 N/mm^2)

f'_{pt} = 148 650 lbf/in^2 (10 400 kgf/cm^2) (1040 N/mm^2)

$f_{bT} = (3.4 \times 0.71 \times f_{pT} - 1.4 \times 0.29 \times f'_{pT}) \times 0.01$

= 2718 lbf/in^2 (190 kgf/cm^2) (19 N/mm^2)

The stress f_{bT} in the concrete corresponding to f_{pi}

= 150 000 lbf/in^2 (10 500 kgf/cm^2) (1000 N/mm^2)

is

$f_{pi} \times 2 \times 0.01$ = 3000 lbf/in^2 (210 kgf/cm^2) (20 N/mm^2)

Hence the loss is

$$\frac{f_{bT} - f_{bt}}{f_{bT}} \times 100 = 9.4\%$$

This is 50 per cent more than the value of 6.25 per cent obtained on the assumption that the steel was concentrated at its centroid. Similar results (but with slightly smaller differences) are obtained in the case of well-distributed wires.

It is obvious that the tensioned steel which prestresses the concrete cannot take part in resisting the prestress. Consequently, the values of A, I, and e_s in equations 6.3, 6.5, and 6.6a should be related to the net cross-sectional area, that is the area of the concrete less the area of the tensioned steel (in the case of post-tensioning the area of the ducts) plus $(\alpha_e - 1)$ times the cross-sectional area of any non-tensioned steel. On the other hand, when calculating $M_t/(I/e_s)$ the entire area should be considered by adding $\alpha_e A_{st}$ to the area of the cross-section. Generally this distinction is unnecessary in design, particularly with pre-tensioned steel, as the difference between the area of concrete A_c and the net area

$$A_0 = A_c - A_{st} + (\alpha_e - 1)A_{sn}$$

is usually negligible. The area A_{st} is very small, but the magnitude of $(\alpha_e - 1)A_{sn}$ may be large enough to warrant consideration. Similarly, the difference between A_c and

$$A_e = A_0 + \alpha_e A_{st}$$

is usually very small. With post-tensioned steel, however, the area A should be reduced by the area of the ducts and this should be taken into account by reducing the area A_c by the sum of the areas of all the ducts. The same is true for A_e, unless it is ensured that the ducts are completely grouted. Obviously, if test results have to be analysed it is essential to consider in equations 6.3, 6.5, and 6.6a the correct cross-sectional area A_0 and not A_c.

Structural lightweight concrete: As mentioned in Chapter 2, the modulus of elasticity for structural lightweight concrete is much less than that for ordinary concrete, and hence the losses due to elastic shortening are usually about twice as much as those for ordinary concrete.

6.4 Losses due to shrinkage and creep of concrete

The nature and effect of shrinkage and creep are considered in Chapter 2, and the following notes concern the maximum losses arising from these phenomena that should be allowed for in design.

With pre-tensioning little or no shrinkage takes place before transfer when moist curing is carefully employed, and it should therefore be assumed that the entire shrinkage occurs after transfer. A smaller loss is frequently allowed for when post-tensioning is employed, on the assumption that an appreciable amount of shrinkage occurs before the prestress is applied. While this may be so, however, it is preferable to ensure that satisfactory curing so delays the shrinkage that the full amount may be considered to occur after transfer even if this results in a slight reduction of the effective prestressing force. If this is achieved the reduction is offset by the fact that the entire tensile resistance of the concrete remains available when preliminary shrinkage cracks are prevented by careful curing, although this may be difficult to ensure for members more than 100 ft (30 m) long.

For high-strength concrete and pre-tensioned steel the ultimate shrinkage to be allowed for may be taken as 0.03 per cent for normal exposure (at 70 per cent relative humidity), which corresponds to a reduction of $0.3 \times 10^{-3} \times E_s$ in the tensioning stress.

When E_s is taken as 30×10^6 lbf/in^2 (2.1×10^6 kgf/cm^2; 0.2 MN/mm^2) this amounts to a reduction of 9000 lbf/in^2 (630 kgf/cm^2; 60 N/mm^2) in the tensioning stress. For constructions which are permanently exposed to very humid conditions (90 per cent relative humidity) the reduction in the tensioning stress is only about 3000 lbf/in^2 (210 kgf/cm^2; 20 N/mm^2). These values are given in CP 110, based on the assumption that transfer takes place between three and five days after casting; if moist curing is employed, the same values can be taken for shorter or longer periods between casting and transfer. In the earlier CP 115, only the value for normal exposure was given.

In the authors' opinion a third case should be considered, corresponding to permanent exposure in dry conditions (such as a heated room with a relative humidity of 60 per cent); for this case a shrinkage of 0.04 per cent should be allowed.

For post-tensioned steel, tensioned between seven and fourteen days after casting, CP 110 gives values which are 33 per cent less than those for pre-tensioned steel; namely 0.02 per cent and 0.007 per cent respectively. The authors recommend the use of a higher value of 0.027 per cent in dry conditions. When post-tensioning is applied to a member, consisting of precast segments, usually most of the shrinkage of the precast concrete has already taken place. In this case only the shrinkage of the mortar at the joints need be considered.

The amount of shortening, and therefore the losses due to creep, depend on the ratio of the compressive stress to the strength, and this ratio will vary with time. Attempts have been made to solve this problem mathematically, the exponential formula for the total losses due to creep being obtained by integration. However, such an analysis ignores the fact that the effective prestress is decreased not only by creep but also, and at the same time, by shrinkage and relaxation of the steel. Consequently, the analytical formulae based on a definite relationship between the creep and the elastic deformation are not correct; as shown later, they are based on an idealized deformation curve. It seems sufficiently accurate to assume that the mean compressive stress from which the total creep may be calculated is 80 per cent to 90 per cent of the stress adjacent to the centroid of the tensioned steel at transfer (f_{st}).

In addition to the effect of the stress/strength ratio, the magnitude of the creep also depends on the ambient conditions. Hence, in relatively dry surroundings, and with a relatively high compressive stress applied at an early age, the loss due to creep will be much greater than if the surroundings are humid and for the compressive stress is reduced immediately after prestressing to a relatively small value (due to dead load counteracting the prestress, for example). For members with pre-tensioned steel both CP 110 and CP 115 recommend a maximum allowance for creep of 0.33×10^{-6} for a stress of 1 lbf/in^2 (4.7×10^{-6} per kgf/cm^2); (48×10^{-6} per N/mm^2) in the concrete adjacent to the steel. For an average stress of 1000 lbf/in^2 (70 kgf/cm^2; 7 N/mm^2) applied at the centroid, the losses due to creep would exceed those due to shrinkage by 10 per cent (i.e. 0.033 per cent as compared with 0.03 per cent with shrinkage); an allowance should, however, be made for the reduction in the concrete stress at the centroid of the steel which occurs due to the other losses, and it is suggested that the losses due to shrinkage and creep be assumed to be equal when the stress in the concrete adjacent to the steel is 1000 lbf/in^2 (70 kgf/cm^2; 7 N/mm^2).

If the ratio of the strength of the concrete at transfer to its final strength is known, its effect may be allowed for by introducing the coefficient k given in the German Specification (see Chapter 2). In general, however, this ratio will not be known, and the maximum values given in the foregoing should be adopted.

For members in which the steel is post-tensioned between two and three weeks after the concrete is placed, both CP 110 and CP 115 specify a maximum creep of 0.25×10^{-6} for a stress of 1 lbf/in^2 (3.5×10^{-6} per kgf/cm^2; 36×10^{6} per N/mm^2) in the concrete adjacent to the steel, and the losses due to creep are consequently 75 per cent of those arising when the steel is pre-tensioned.

The losses given in the foregoing in the case of pre-tensioned and post-tensioned steel are based on the assumption that a minimum cube strength $f_{ct\ min}$ of 6000 lbf/in^2 (420 kgf/cm^2; 41.5 N/mm^2) is obtained at the time the concrete is prestressed. When this strength is not obtained, CP 110 and CP 115 both recommend that the losses should be increased in the ratio $\dfrac{f_{cu\ min}}{f_{cu\ t}}$ ($f_{cu\ t}$ is the cube strength when the prestress is applied).

The maximum stress in the concrete when the prestress is applied and the specified cube strengths are both known to the designer, who can therefore decide whether the ratio $\dfrac{6000}{f_{cu\ t}}$ given in the British Codes or the factor k given in German Standard is the more appropriate. It should be borne in mind that losses due to creep are proportional to the stress in the concrete only when this stress does not exceed 33 per cent of the cube strength(2), and it is therefore preferable to allow for a stress at transfer of $0.4 f_{cu\ t}$ and not $0.5 f_{cu\ t}$ (with a maximum of 3000 lbf/in^2 (210 kgf/cm^2; 21 N/mm^2) as recommended in CP 115. It is of interest that the German Standard permits maximum stresses at transfer of 2960 lbf/in^2 (210 kgf/cm^2; 21 N/mm^2) for rectangular sections in simple bending, 3130 lbf/in^2 (220 kgf/cm^2; 22 N/mm^2) at the corners of rectangular sections subjected to bending in two directions, and only 2840 lbf/in^2 (200 kgf/cm^2; 20 N/mm^2) in I or T sections, compared with the single maximum value of 3000 lbf/in^2 (210 kgf/cm^2; 21 N/mm^2) recommended in CP 115. On the other hand, the German Standard requires the use of concrete with a compressive strength of about 8500 lbf/in^2 (600 kgf/cm^2; 60 N/mm^2) at 28 days (corresponding to about 6400 lbf/in^2 (450 kgf/cm^2; 45 N/mm^2) when the prestress is applied) while the British Codes recommend a strength of 6000 lbf/in^2 (420 kgf/cm^2; 41.5 N/mm^2) when the prestress is applied. It appears desirable therefore to specify that the concrete shall have a compressive strength of 2½ times the maximum stress in the concrete at the time when it is prestressed; the stress f_{st} in the concrete adjacent to the steel will then not exceed about $\frac{1}{3}$ of the strength of the concrete (since the stress in the concrete adjacent to the steel is less than the maximum) and the relationship between f_{st} and the losses due to creep will be linear. In assessing this stress the effect of bending stresses due to dead load at the time of prestressing may be taken into account if provision is made to ensure their presence; it may be necessary during prestressing to separate the concrete from its bottom form. A requirement to this effect is included in CP 110; there a higher creep is specified when the stress adjacent to the centroid of the steel at transfer is more than 33 per cent of the strength at transfer. The increase is linear, with a maximum of 25 per cent when the stress is half the strength.

If it is known when the design is made that the concrete will have a high percentage of its final strength when the prestress is applied, a smaller loss due to creep may be allowed. This is the case, for example, when prestressing is carried out in two stages with a known time interval between them. Reduction factors of 0.75 for an interval of three months or 0.65 for an interval of six months appear to be safe.

The ratio of the losses due to maximum creep [on the assumption that E_s = 29 X 10^6 lbf/in^2 (2 X 10^6 kgf/cm^2; 0.2 MN/mm^2) for wires and small diameter strands, and 25 X 10^6 lbf/in^2 (1.75 X 10^6 kgf/cm^2; 0.175 MN/mm^2) for large strands and alloy bars] and those due to elastic shortening for the same stress in the concrete is 9.6/α_e for pre-tensioned steel and 7.2/α_e for post-tensioned steel. If the limits of α_e are 4.7 and 8 (see Table 6.2), the ratio varies between 0.9 and 2.05. In the USA, where there are greater variations of temperature and humidity

Table 6.2 Relation between cube strength and modular ratio

Approx. cube strength		lbf/in^2	4270	5350	6400	7450	8550	9600
		kgf/cm^2	300	375	450	525	600	675
		N/mm^2	30	37.5	45	52.5	60	67.5
Modular Ratio α_e	Wires and small strands		8	7.14	6.66	6.06	5.71	5.3
	Large strands and alloy bars		7	6.25	5.83	5.3	5	4.72

than in Great Britain, limiting ratios of 1 for very humid conditions and 3 for very dry conditions are recommended[3].

If the steel is distributed throughout the section, or if separate tensioned steel is provided at the top and bottom, the calculation of the losses due to creep should be modified in the same way as those due to elastic shortening (see page 104/5). CEB/FIP Model Code recommendations for prediction of ultimate creep and shrinkage has been discussed in Chapter 2. Where a high degree of accuracy is not required the values for the final creep co-efficient and shrinkage strain may be taken from Table 6.3. These values which depend on (i) notional size of the section, (ii) age of concrete at loading and (iii) relative humidity are based on suggestions in the Model Code. They are valid up to a concrete stress level of 40% of characteristic cylinder strength.

Structural lightweight concrete. The magnitude of the shrinkage and creep in structural lightweight concrete, and therefore of the corresponding losses, depends to some extent on the nature of the lightweight coarse aggregate and also on whether ordinary or lightweight fine aggregate is used. With some lightweight aggregates, the losses due to shrinkage and creep may even be of the same order as for ordinary concrete; but in general they are greater. They should be computed from data relating to the actual materials employed, since the range of variation in the properties of these materials is significant. For high strength lightweight concrete using British aggregate (e.g. Lytag) of mostly rounded edges and regular shape a specific creep value of $0.45 - 0.60 \times 10^{-6}$ per lbf/in^2 stress ($70 - 90 \times 10^{-6}$ per N/mm^2) and a shrinkage strain of $400 - 600 \times 10^{-6}$ may be assumed.

Table 6.3 Final values for the creep coefficient $\phi(t_\infty, t_0)$ and of the basic values shrinkage $\epsilon_s(t_\infty, t_0)$ (to CEB/FIP Model Code)

	Humid atmospheric conditions outside (RH ~ 75%)		Dry atmospheric conditions inside (RH ~ 55%)	
Notional size: $\frac{2Ac}{u}$ u = Perimeter in contact with atmosphere	small ⩽ 200 mm	large ⩾ 600 mm	small ⩽ 200 mm	large ⩾ 600 mm
Creep: $\phi(t_\infty, t_0)$ Age at commencement of loading to				
fresh (3–7 days)	2.7	2.1	3.8	2.9
medium (7–60 days)	2.2	1.9	3.0	2.5
mature (> 60 days)	1.4	1.7	1.7	2.0
Shrinkage: $\epsilon_s(t_\infty, t_0)\ 10^3$ Age of concrete at the instant from which the shrinkage effect is being considered.				
fresh (3–7 days)	0.26	0.21	0.43	0.31
medium (7–60 days)	0.23	0.21	0.32	0.30
mature (> 60 days)	0.16	0.20	0.19	0.28

6.5 Losses due to relaxation of the steel

Relaxation varies considerably according to the type of steel, and low losses due to relaxation should be assumed only when the steel is known to possess such properties. If the type of steel to be used is not known when the design is made safe values should be allowed. In the case of German wire it is claimed that no relaxation at all occurs below the so-called 'creep limit'. In the case of USA wire or strand a loss of 5000 lbf/in^2 (350 kgf/cm^2; 35 N/mm^2) is usually stated to be sufficient when the initial tensioning stress does not exceed $\frac{2}{3}$ of its ultimate tensile strength. CP 115 tentatively recommends that a loss of 15 000 lbf/in^2 (1050 kgf/cm^2; 105 N/mm^2) be allowed for wire in the 'as drawn' condition, which may be reduced to 10 000 lbf/in^2 (700 kgf/cm^2; 70 N/mm^2) when it is straightened and subsequently heat-treated, or when an overstress of 10 per cent of the initial tensioning stress is applied for two minutes during tensioning. It is also left to the discretion of the designer to reduce this last-mentioned allowance if he is satisfied that the properties of the steel warrant it. The use of any wire in the 'as drawn' condition is not usually desirable. Losses of 15 000, 10 000 and 5000 lbf/in^2 (1050, 700 and 350 kgf/cm^2; 105, 70 and 35 N/mm^2) are considered in this Chapter.

In CP 110, the loss of prestress due to relaxation is related to the jacking force in the steel. Unless better information is available from the suppliers, it is recommended that a relaxation loss of 8 per cent should be allowed when the jacking force is 70 per cent of the characteristic strength, and 10 per cent when the jacking force is 80 per cent of the characteristic strength. When a smaller jacking force is used, a relaxation loss varying linearly from 0 per cent (for a jacking force of 50 per cent of the characteristic strength) to 8 per cent (for a jacking force of 70 per cent of the characteristic strength) is recommended. The recommendations apply to wires, strands and bars. There are certain specially treated low-relaxation wires and strands for which the manufacturers guarantee 1½ per cent to 2½ per cent relaxation losses when tensioned to 70 per cent of the ultimate.

6.6 Losses due to friction

Friction may be caused by variations of the profile of a duct intended to be straight, with consequent contact between the steel and the sides of the duct, and by curvature of the steel. The prestressing force is therefore reduced when the length of the wire or bar is increased and its slope changes. The loss in a straight wire or bar depends mainly on the type of duct or sheath employed; when a flexible sheath is used the loss may also be affected by the extent to which vibration is used in placing the concrete. The loss due to curvature depends on the total angle through which the steel is curved and on the coefficient of friction between the steel and the surface along which it moves.

The loss due to accidental variations in the profile of the duct may be evaluated from the expression

$$P_{x1} = P_0 \,(1 - e^{-Kx}) \quad \ldots \quad (6.8)$$

in which P_0 is the prestressing force in the steel at the jacking end, 'e' is the base of Napierian logarithms, K is a constant, and x is the distance from the jack to the point at which losses are required to be known (see Figure 6.5).

The value of K depends on the type of duct or sheath, the type of steel, and the vibration employed in placing the concrete, and CP 115 recommends a normal value of not less than 1×10^{-3} per foot of length (3.3×10^{-3} per metre length). When rigid ducts are used, or when special care is taken to support them firmly, K may be reduced to 0.5×10^{-3} per foot of length (1.7×10^{-3} per metre). Similar values are given in CP 110.

Losses due to curvature of the steel are given by

$$P_{x2} = P_0\,(1 - e^{-\mu x/R}) = P_0\,(1 - e^{-\mu\theta}) \quad \ldots \quad (6.9)$$

in which P_0 and 'e' are as previously defined, x is the arc length measured along the curve, μ is the coefficient of friction, R is the radius of curvature of the steel, and $\frac{x}{R} = \theta$ is the angle turned through (see Figure 6.6). Values of μ recommended in CP 115 are given in Table 6.4, together with additional data which reduce the work of evaluating expressions 6.8 and 6.9. When Kx, $\frac{\mu x}{R}$, or $\left(Kx + \frac{\mu x}{R}\right)$ are not more than 0.2, the exponential equation may be assumed to equal $(1 - Kx)$, $\left(1 - \frac{\mu x}{R}\right)$ or $1 - \left(Kx + \frac{\mu x}{R}\right)$ as appropriate. Hence for small values

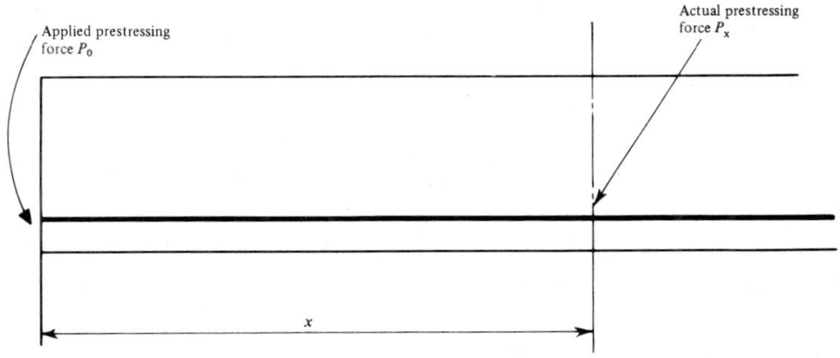

Figure 6.5 Friction loss of the prestressing force in straight tendons

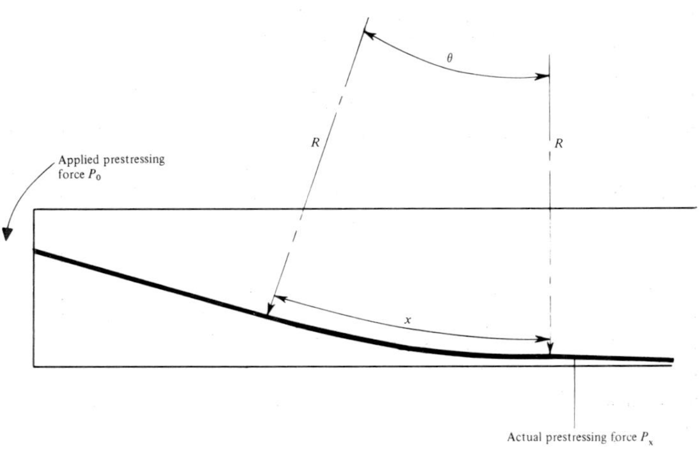

Figure 6.6 Friction loss of the prestressing force due to curvature

of Kx or $\frac{\mu x}{R}$, the losses are given by

loss of force $\Delta P_0 = P_0\,Kx$ or $P_0\,\mu x/R$ . . . . (6.9a)

loss of stress $\Delta f_{pf} = f_{pf}\,Kx$ or $f_{pf}\,\mu x/R$ . . . . (6.9b)

Losses due to friction may be reduced either by tensioning from both ends (when the maximum value of x is halved) or by a temporary increase in the prestressing force (Figure 6.7) or by combination of both. When the prestressing force is reduced to its correct value, the consequent reduction of prestress is a minimum at the point farthest from the jacks. Other methods have been suggested by Leonhardt, Zerna and Hahn. In Leonhardt's method, which has been used with success for the long stranded cables of the Leonhardt-Baur system, additional jacks are placed at openings formed along the cable (Figure 6.8). In Zerna's method an auxiliary cable, which is not anchored, is placed between the anchored main cable and the upper side of the duct. After the main cable has been tensioned, the tension being maintained constant, the auxiliary cable is tensioned and pulled out from the duct; as a result the original frictional losses in the main cable are offset by the friction developed by the extraction of the auxiliary cable (Figure 6.9). Dr Hahn suggests that the tensioned steel should be heated; the expansion of the steel on heating and its contraction on cooling reduce the frictional losses in a manner similar to that shown in Figure 6.7.

When losses due to friction are allowed for in the calculations, the required prestressing force and the expected elongation should both be specified. The loss should not be computed for the greatest value of x, but for the length at the end of which the maximum effective prestress is still required. If the steel is tensioned from both ends, this is at midspan for symmetrical loading; in this case the elongation of the steel at prestressing should be computed for the stress $f_{pi\ av} = f_{pi} - \frac{1}{2}\Delta p_f$. If the steel is tensioned from one end the same expression applies but the loss $\Delta f_{p\ max}$ corresponding to $x = l_e$ is used, in place of Δfp corresponding to $x = \frac{1}{2}\,l_e$; l_e is the entire length of the member, and not the clear span.

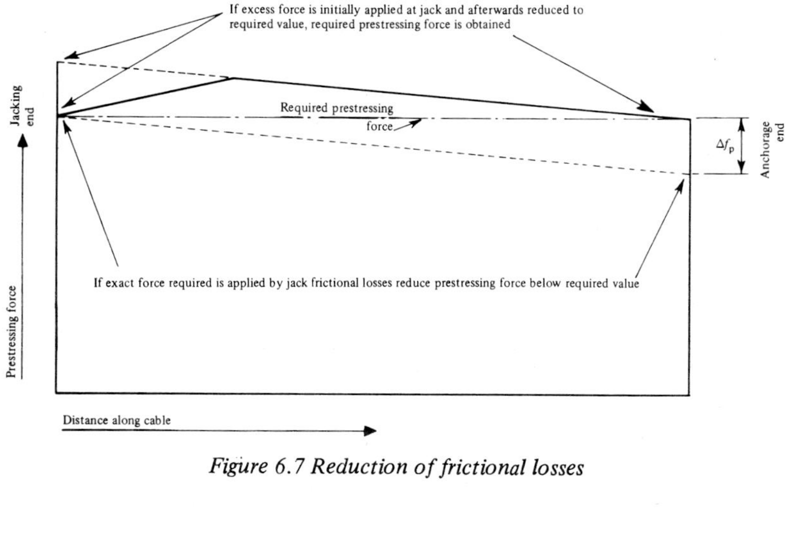

Figure 6.7 Reduction of frictional losses

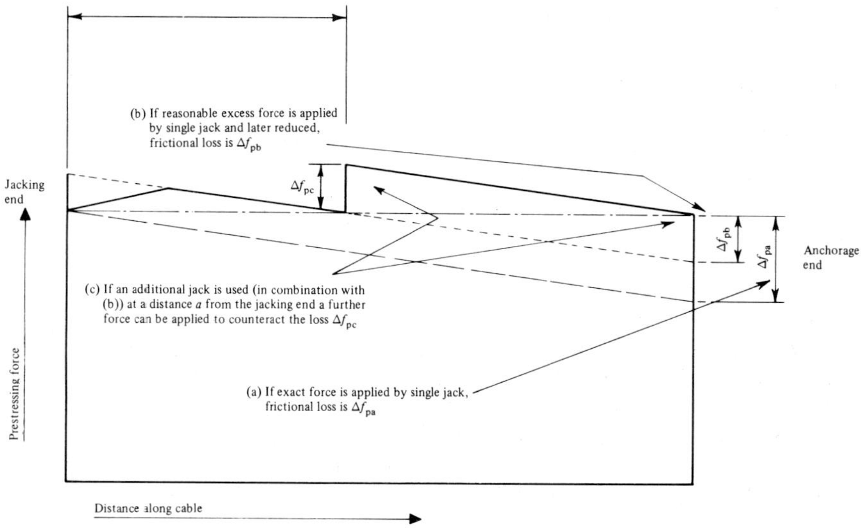

Figure 6.8 Dr. Leonhardt's method

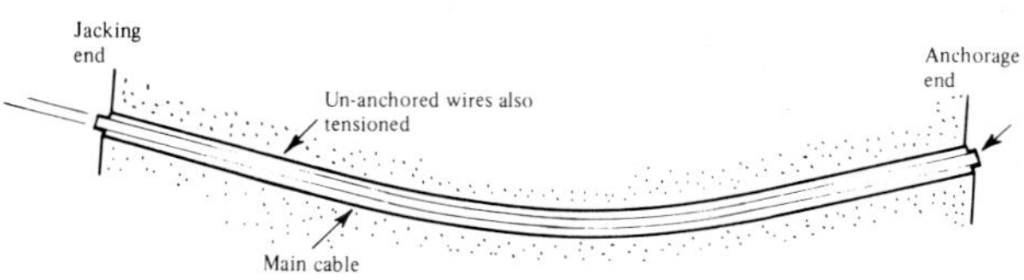

Figure 6.9 Professor Zerna's method

The losses due to accidental variation (wobble) and to curvature are additive, since variations can occur on both the straight and curved positions of a tendon. Further, it is necessary to make a reasonable allowance for friction at an early stage in the design, before the tendon profiles are known. An example of a preliminary calculation of this type is given in the following; the numerical values for the three different systems of units have been rounded off. The beam is assumed to be simply supported, with uniform loading; the tendons are jacked from both ends.

Span of beam	60 ft (18 m)
Depth of beam	3 ft (0.9 m)

Prestress applied at jacking end 150 000 lbf/in^2 (10 000 kgf/cm^2; 1000 N/mm^2)

Assume that 50 per cent of the tendons are deflected through 75 per cent of the beam (Figure 6.10). Then the angle θ is given approximately by:

$$\theta \text{ (radians)} = \frac{3h}{4} \bigg/ \left(\frac{l}{4}\right) = \frac{2.25}{15}$$

$$= 0.15$$

Table 6.4 Factors for frictional losses

Values of μ recommended in CP 115 and CP 110

Steel moving on smooth concrete:	0.55
Steel moving on steel (fixed to duct) (CP 115):	0.35
Steel moving on steel (fixed to cable) (CP 115):	0.25
Steel moving on steel (CP 110):	0.30
Steel moving on lead:	0.25

Values for circular construction

Steel moving on smooth concrete:	0.45
Steel moving on steel (fixed to concrete):	0.25
Steel moving on steel rollers:	0.10

Values of Kx or $\mu x/R$	Values of e^{-Kx} or $e^{-\mu x/R}$
0.01	0.990
0.02	0.980
0.03	0.970
0.04	0.961
0.05	0.951
0.06	0.942
0.07	0.932
0.08	0.923
0.09	0.914
0.10	0.905
0.11	0.896
0.12	0.887
0.13	0.878
0.14	0.869
0.15	0.861
0.16	0.852
0.17	0.844
0.18	0.835
0.19	0.827
0.20	0.819

Assume $\mu = 0.30$ and $k = 1 \times 10^{-3}$

Then

$$\mu\theta = 0.3 \times 0.15 = 0.045 \text{ and } kx = 1 \times 10^{-3} \times 30 = 3 \times 10^{-2}$$

Hence

$$\Delta f_{\text{pf}_1} \text{ (due to wobbling)} = f_{\text{po}} kx = 4500 \text{ lbf/in}^2 \text{ (330 kgf/cm}^2\text{; 33 N/mm}^2\text{)}$$

and

$$\Delta f_{pf2} \text{ (due to curvature)} = f_{po}\, e^{\mu\theta} = 6750 \text{ lbf/in}^2 \text{ (450 kgf/cm}^2\text{; 45 N/mm}^2\text{)}.$$

Hence the loss in half of the tendons is 11 250 lbf/in^2 (780 kgf/cm^2; 78 N/mm^2) and the average loss in all the tendons is 7875 lbf/in^2 (555 kgf/cm^2; 58.8 N/mm^2).

6.6.1 Reverse friction

It is good practice to compare the direct reading obtained from the pressure gauge on the jack with the prestressing force computed from the measured elongation of the tendon; these should be in reasonably close agreement. It is sometimes overlooked that if the tension in the tendon is reduced for some reason, before anchoring, the final elongation obtained will not be proportional to the reduction in the applied force.

Consider, for example, the case shown in Figure 6.11, in which a tendon is jacked from one end with an initial force P_i. It is assumed that the frictional loss along the tendon is 0.2 f_{pi}, so that at this stage the elongation is given by

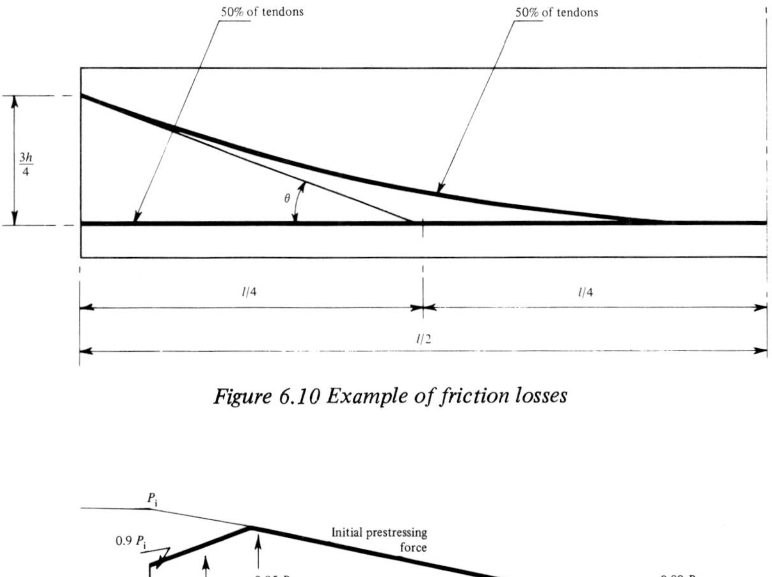

Figure 6.10 Example of friction losses

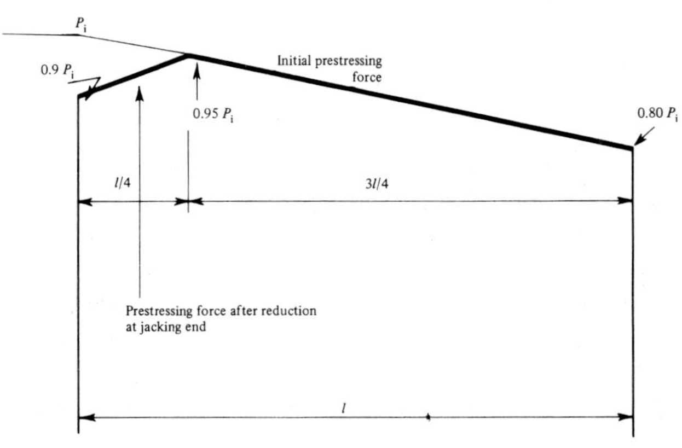

Figure 6.11 Further example of friction losses

$$\frac{P_i l}{A_s E_s}\left(\frac{0.8 + 1.0}{2}\right) = \frac{0.9 P_i l}{A_s E_s}$$

If now the initial force P_i is reduced to 90 per cent of its previous value, the effect of reverse friction near to the jacking end is to give a residual elongation of

$$\frac{P_i l}{A_s E_s}\left(\frac{0.9 + 0.95}{2} \times \tfrac{1}{4} + \frac{0.95 + 0.8}{2} \times \tfrac{3}{4}\right) = 0.887 \frac{P_i l}{A_s E_s},$$

$$\text{and not } 0.9 \times 0.9 = \frac{0.81 P_i l}{A_s E_s}$$

6.6.2 Pre-tensioned steel

Frictional losses may also occur with pre-tensioned steel, due to friction between the tendons and the spacing diaphragms which are often employed. These losses may be reduced by tapping the diaphragms or by vibration. Frictional losses may also occur in deflected or draped tendons, when their profile is obtained by placing strong transverse reinforcement across the mould.

6.7 Losses due to non-tensioned steel

The presence of longitudinal non-tensioned steel in a prestressed section affects primarily the losses due to elastic compression, shrinkage, and creep. The effect on the elastic losses may be allowed for with sufficient accuracy by adding $(\alpha_e - 1)$ times the area of the non-tensioned steel to the area of the cross-section.

The presence of non-tensioned steel helps to reduce the magnitude of the shrinkage and creep strains, to an extent which depends on the percentage of non-tensioned steel provided; but the strains which do occur reduce a compressive force in the non-tensioned steel. Recent tests[4] have shown that these compressive forces increase in almost direct proportion to the percentage of non-tensioned steel, and are also proportional to the magnitude of the shrinkage and creep strains. The magnitude of this loss may be taken approximately as

$$\Delta f_{pn} = (\bar{\epsilon}_s + \bar{\epsilon}_c) A_{sn} E_s$$

in which ϵ_s and ϵ_c represent the maximum shrinkage and creep strains ϵ_s and ϵ_c multiplied by a factor $\dfrac{1}{1 + \alpha_n}$; $\alpha_n = \dfrac{\alpha_e A_{sn}}{A_c}$; A_{sn} is the area of the non-tensioned steel; and A_c is the area of the concrete. Clearly, if the compressive stress adjacent to the tensioned steel is small, these losses will also be small.

6.8 Losses due to other causes

Losses due to changes of temperature and also, in certain cases, to steam curing, may occur when pre-tensioning is employed. Losses due to changes of temperature after the concrete has hardened and bonded with the steel are negligible, since the coefficients of temperature of concrete and steel are about equal, but changes of temperature before the concrete has sufficiently hardened tend to cause movement of the steel relative to the concrete, thereby reducing or even destroying the bond between them. This must be avoided in any circumstances since, after slipping occurs, part of the prestress is retained by friction and lost when the member is loaded, so that such a structure cannot be considered as being prestressed.

Losses due to steam curing may possibly arise in a similar manner if it is applied before a certain amount of bond has developed, which usually takes place after about three or four hours. The rise in temperature of the steel may result in relative movement between the concrete and steel which would be detrimental to the development of bond, and contraction of the concrete when the temperature is lowered may reduce the stress in the steel. If steam curing is applied after this degree of bond has developed, as is usually the case, the losses are negligible.

In view of these possibilities, it is important when the long-line process is used to specify that, when varying temperatures are expected to occur, the tensioning process should be carried out when the temperature is high, and that the concrete should be quickly bonded to the tensioned steel whenever large changes of temperature may occur. This ensures that bonding does not occur at a time when the stress in the steel is reduced. Similarly, steam curing should be delayed until after the bond has developed.

6.9 Combined losses

An expression for the combined effect of elastic deformation, creep, shrinkage, and relaxation, derived by Neville and Trost, is given in Chapter 2. Trost[5] has also derived an expression for the corresponding losses in tendons located at their centroidal position with an eccentricity e_s, in the following terms:

$$\Delta f_{pesc} = \alpha_e \, \delta f_s (2 + \alpha_e \eta) + E_s \, \epsilon_{st} \quad \text{(6.10)}$$

in which Δf_{pesc} is the loss of prestress due to elastic shortening, shrinkage and creep after time 't', α_e is the modular ratio, δf_s is the change of stress in the concrete at the level of the steel, after a time 't', η is the relaxation factor at time 't', and ϵ_{st} is the shrinkage strain after time 't'.

6.10 The total losses

Losses of prestress can be considered to depend on seven factors, which may be denoted by *a, b, c, d, x, y, z*. The factors *a, b*, and *c* depend on the conditions of casting, curing and use; *d* depends on the type of duct and the curvature of the steel, if any; *x* represents the stress at transfer; *y* is the average stress in the concrete adjacent to the centroid of the steel in the tensile zone; and *z* represents the prestress applied by the jack.

The total loss is given by the expression

$$\Delta f_{p\ total} = a + bx + cy + dz = (\Delta f_{ps} + \Delta f_{pr}) + b\, f_{sT} + c\, f_{s\,av} + \Delta f_{pf} \quad \text{(6.11)}$$

upon which Charts 4 and 5 in Part 2 are based. The losses Δf_{ps} (due to shrinkage) and Δf_{pr} (due to relaxation) are represented by the constant *a*; for example, the maximum loss due to shrinkage is 8700 lbf/in^2 (610 kgf/cm^2; 60 N/mm^2) when $E_s = 29 \times 10^6$ lbf/in^2 (2×10^6 kgf/cm^2; 200 kN/mm^2), and a safe value for the loss due to relaxation is 10 000 lbf/in^2 (700 kgf/cm^2; 70 N/mm^2). The factor *b*, on which the loss due to elastic shortening depends, is equal to α_e for pre-tensioned steel and may be 0.5 α_e for post-tensioned steel; the creep factor *c*, in *cy* of equation 6.11, has maximum values of 9.6 for pre-tensioned steel and 7.2 for post-tensioned steel, assuming $E_s = 29 \times 10^6$ lbf/in^2.

The stress $f_{s\,av}$ is the average stress in the concrete adjacent to the steel, upon which the extent of the creep depends. It should be remembered that the stress during the first month after prestressing has the same importance as the average stress during the remainder of the life of the structure. The reduction-factor of 0.9 should not be used in this connection, as $f_{s\,av}$ is assessed on this basis.

6.11 Losses with pre-tensioning

It is normal practice to specify the pre-tensioning force required, from which the initial stress f_{pi} is determined. The stress f_{pt} in the steel when the prestress is applied to the concrete is obtained from $f_{pt} = f_{pi} - \Delta f_{pt}$, in which Δf_{pt} represents the losses occurring at transfer.

In order to determine the maximum possible stress in the concrete and the minimum effective prestress, it is necessary to allow for minimum losses before prestressing and maximum losses after. Consequently it is advisable to assume that no losses due to shrinkage occur before prestressing, the only loss considered being that due to elastic shortening Δf_{pe}, and therefore for maximum stress in the concrete $\Delta f_{pt} = \Delta f_{pe}$ and $f_{pt} = f_{pi} - \Delta f_{pe}$. Clearly, any 'pull-in' on anchoring should also be allowed for. No loss due to relaxation of the steel has been allowed for; however, the relaxation depends to some extent on the time which elapses between tensioning and transfer, and if the rate at which losses due to relaxation occur is known a suitable allowance may be made. The minimum effective prestress f_{pe} is given by

$$f_{pe} = f_{pt} - \Delta f_{ps} - \Delta f_{pc} - \Delta f_{pr} = R_0 f_{pt} = R_e f_{pi},$$

in which Δf_{ps} is the loss due to shrinkage, Δf_{pc} the loss due to creep, and Δf_{pr} the loss due to relaxation of the steel.

The reduction factors R_0 and R_e vary considerably, being dependent mainly on the stress in the concrete adjacent to the steel. It should be noted that this stress may be much less after the loading is applied than at transfer, and allowance may be made for this when calculating the losses due to creep if the length of time is known which will elapse, before the full dead load is applied. Table 6.5 gives the equivalent stress which may be used for this purpose, for various values of initial stresses and stresses due to dead load, and for various periods, when at transfer the stress in the concrete adjacent to the steel is $0.9 f_{bt}$.

Table 6.5 Ratios of losses due to creep

$\frac{f_{s\,av}}{0.9 f_{bt}}$	Time after transfer during which M_t acts alone							
	1 week	2 weeks	3 weeks	1 month	2 months	3 months	6 months	1 year
0	0.25	0.35	0.43	0.50	0.58	0.64	0.75	0.90
0.1	0.325	0.415	0.487	0.55	0.622	0.676	0.775	0.91
0.2	0.40	0.48	0.544	0.60	0.664	0.712	0.80	0.92
0.3	0.475	0.545	0.601	0.65	0.706	0.748	0.825	0.93
0.4	0.55	0.61	0.658	0.70	0.748	0.784	0.85	0.94
0.5	0.675	0.675	0.715	0.75	0.790	0.820	0.875	0.95
0.6	0.70	0.74	0.772	0.80	0.832	0.856	0.90	0.96
0.7	0.775	0.805	0.829	0.85	0.874	0.892	0.925	0.97
0.8	0.85	0.87	0.886	0.90	0.916	0.928	0.95	0.98
0.9	0.925	0.935	0.943	0.95	0.958	0.964	0.975	0.99
1	1	1	1	1	1	1	1	1

The evaluation of the reduction factors R_0 and R_e is facilitated by Chart 4. The chart is entered on the scale marked f_{pt} and, after losses due to elastic shortening are taken into account, f_{pt} may be determined as shown by the chain-dotted line. Losses due to creep, relaxation, and shrinkage (for normal conditions of 70 per cent relative humidity) are next allowed for, the value of f_{pe} being then determined. A straight-edge laid between f_{pe} and f_{pi} and f_{pt} and f_{pe} enables R_0 and R_e to be determined on the nomograms. In the example shown on the chart, f_{pi} is assumed to be 145 000 lbf/in^2 (10 200 kgf/cm^2; 1 kN/mm^2) and f_{bT} is assumed to be 1000 lbf/in^2 (70 kgf/cm^2; 6.9 N/mm^2), the value of f_{pt} is found to be 136 000 lbf/in^2 (9550 kgf/cm^2; 935 N/mm^2). Losses due to creep are determined assuming $f_{s\ av}$ to be 600 lbf/in^2 (42 kgf/cm^2; 4.1 N/mm^2). Losses due to shrinkage and relaxation are assumed in this example to be 150 000 lbf/in^2 (1050 kgf/cm^2; 103 N/mm^2); f_{pe} is found to be 116 000 lbf/in^2 (7950 kgf/cm^2; 780 N/mm^2) and R_e and R_0 are found to be 0.795 and 0.84 respectively.

If a long period is likely to elapse before the full dead load is applied to the member, the maximum losses due to creep should be taken into account, as shown by the full lines on the chart. If, however, it is known that the full dead load will be applied soon after transfer, and that $f_{s\ av}$ will be appreciably less than $0.9f_{bt}$, the value of $f_{s\ av}$ may be taken from Table 6.5, and the losses determined from the lines shown in chain-dot on the chart. Losses indicated by the chart are those recommended in CP 115 and CP 110.

6.12 Losses with post-tensioning

The conditions assumed in computing the losses with post-tensioned steel are identical in principle with those for pre-tensioning, but different values are employed. Losses occurring at transfer include those due to friction (Δf_{pf}) in addition to those due to elastic shortening (Δf_{pe}) where applicable: $f_{pt} = f_{pi} - \Delta f_{pe} - \Delta f_{pf}$. The calculated elongation will agree with that actually observed only when losses due to friction are allowed for accurately. The common practice of increasing the initial prestressing force, regardless of its magnitude, until the calculated elongation is obtained may cause dangerous local compressive stresses, and the designer should clearly specify whether a definite elongation is required or whether losses due to friction have been allowed for; in the latter case a definite prestressing force and an expected elongation should be specified. The first method may cause an appreciable increase in the initial tensile stress in the steel, and it is necessary to ensure that the permissible value is not exceeded. If the full elongation is obtained when losses due to friction have been allowed for, then a higher effective prestress than that assumed in the calculations may occur, the extent of which will depend entirely on the variation of stress along the steel due to frictional losses where the slope changes.

The effective prestress f_{pe} is given, as before, by

$$f_{pe} = f_{pt} - \Delta f_{ps} - \Delta f_{pc} - \Delta f_{pr} = R_0 f_{pt}$$

and the same considerations apply. Shrinkage and creep may be less in view of the greater strength of the concrete when the prestress is applied, and f_{pt} may be less when losses due to friction are large. Chart 5 reduces the work involved in evaluating the expression. The chart is similar to that for pre-tensioned steel but also includes curves for frictional losses. In the example shown on the chart, the value of f_{pi} is 145 000 lbf/in^2 (10 200 kgf/cm^2; 1 kN/mm^2), and losses due to friction (assuming e^{-kx} to be 0.95) and elastic shortening [assuming f_{bt} = 1000 lbf/in² (70 kgf/cm²; 6.9 N/mm²)] are obtained. The value of f_{pt} is found to be 133 000 lbf/in² (9350 kgf/cm²; 0.915 kN/mm²). Losses due to creep are determined, the value of $f_{s\ av}$ being assumed to be 600 lbf/in² (42 kgf/cm²; 4.1 N/mm²), and the losses due to shrinkage and relaxation are assumed to be 15 000 lbf/in² (1050 kgf/cm²; 103 N/mm²) f_{pe} is found to be 112 500 lbf/in² (7900 kgf/cm²; 0.775 kN/mm²), R_e is 0.775 and R_0 is 0.84.

The assessment of the losses that arise when prestressing is carried out in two or more stages is slightly more complicated, and is not considered here.

6.13 Examples

The following examples illustrate the method of computing losses occurring with pre-tensioned and post-tensioned steel for various values of prestress in the concrete, making certain assumptions concerning the age of the concrete when the prestress is applied and the stresses due to the dead load at the time of pre-stressing. Similar calculations can be made for any other value of prestress, and the corresponding reduction factor R_0 obtained.

6.13.1 Example 1. Pre-tensioned steel

A symmetrical precast beam with pre-tensioned wires is considered, in which the application of the prestressing force gives rise to a triangular distribution of stress, and the maximum bending stresses due to weight of the beam are ±500 lbf/in² (35 kgf/cm²; 3.5 N/mm²). It is also assumed that the centroid of the tensile steel is at a distance $0.1h$ from the bottom, the stress in the concrete in the plane of the centroid amounting therefore to $0.9f_{bt}$ (Figure 6.12). If the modular ratio α_e corresponding to the strength of the concrete at the time of transfer is 5, then the losses due to elastic shortening are $4.5f_{bt}$.

Young's modulus for the wire is assumed to be 29 × 10⁶ lbf/in² (2 × 10⁶ kgf/cm²; 200 kN/mm²), and all the shrinkage of the concrete is assumed to occur after transfer, the concrete being moist cured until it is prestressed. Losses due to shrinkage are therefore 0.03 per cent, or 8700 lbf/in² (610 kgf/cm²; 60 N/mm²). At midspan the stresses due to weight of the beam oppose those due to the prestressing force. The initial stress is therefore reduced from f_{bT} to f_{bT} − 500 lbf/in² (f_{bT} − 35 kgf/cm²; f_{bT} − 3.5 N/mm²) and the stress in the plane of the centroid of the tensile steel is between 85 per cent and 95 per cent

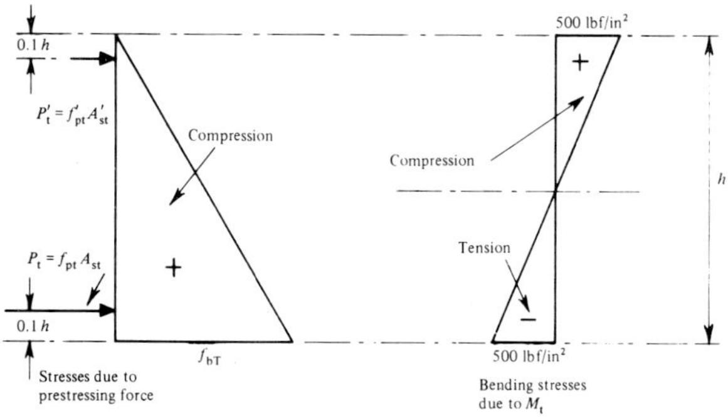

Figure 6.12 Stress diagram for example 6.13.1

of the maximum stress at the bottom face. It is assumed that the beam may not be loaded further for a considerable time, and that most of the creep will therefore take place with this distribution of stress; however, since shrinkage of the concrete and relaxation of the steel will also occur during this period, and the creep will occur gradually, it is sufficient to allow for an average creep corresponding to 90 per cent of the stress at the centroid of the steel. If this stress is assumed to be f_{st}, the losses due to creep are $0.9\ \epsilon_c E_s f_{st}$, in which ϵ_c is the creep strain per unit stress, and are evaluated in Table 6.6 for various stresses at transfer. In every case a maximum relaxation of 10 000 lbf/in^2 (700 kgf/cm^2; 70 N/mm^2) in the steel has been taken into account. A graph showing the reduction factors R_0 and R_e (the latter indicating the entire losses between the initial prestress and the minimum possible effective prestress) is also given in Figure 6.13 based on an initial tensioning stress of 157 000 lbf/in^2 (11 000 kgf/cm^2; 1.1 kN/mm^2). Losses and reduction factors for the steel near the upper face of the beam may be calculated in the same way and values of the reduction-factors R_0' and R_e' for this steel are given in the table and on the graph. In certain circumstances – for example, for acceptance tests – it may be required to calculate the losses after a certain period; in this case the full elastic loss should be allowed for, and the combined values of the total losses due to shrinkage, creep, and relaxation may be multiplied by the factor given in Figure 6.14, although the time at which the losses due to relaxation occur is not known with any accuracy. Figure 6.14 is in accordance with Table 6.5 (relating only to creep of concrete) CP 115, and CP 110. It is based on the assumption that losses due to shrinkage and relaxation occur at the same rate as those due to creep, and R_0 is assumed to represent the losses after two years. Generally R_e and R_0 indicate the reduction of the initial prestress and that occurring at transfer respectively, and both give the same value for the minimum effective prestress that may occur after a long period of time: $f_{pe} = R_{0\infty} f_{pt} = R_e f_{pi}$. When additional reduction-

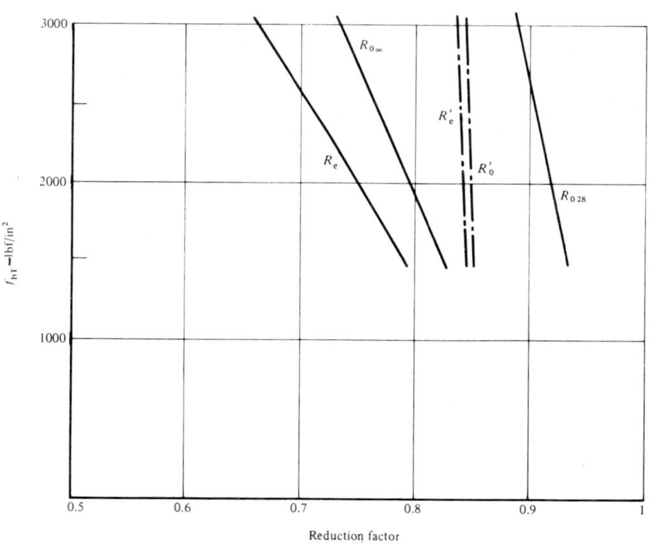

Figure 6.13 Reduction factors for example 6.13.1

Table 6.6 Calculations for Example 1

$$f_{pi} = 157\ 000\ \text{lbf/in}^2\ (11\ 000\ \text{kp/cm}^2;\ 1.1\ \text{kN/mm}^2),\ \Delta f_{pe} = 4.5 f_{sT}$$

$$\Delta f_{pc} = 0.33 \times 10^{-6} \times 29 \times 10^6 \times 0.9 f_{st} = 8.62 f_{st}\ (\text{lbf/in}^2);\ 4.7 \times 10^{-6} \times \Delta f_{pc} = 2 \times 10^6 \times 0.9 f_{st} = 8.64 f_{st}\ (\text{kgf/cm}^2);$$

$$48 \times 10^{-6} \times 0.2 \times 10^6 \times 0.9 f_{st} = 8.64 f_{st}\ (\text{N/mm}^2)$$

$$\Delta f_{ps} = 8700\ \text{lbf/in}^2\ (610\ \text{kp/cm}^2;\ 60\ \text{N/mm}^2).\ \Delta f_{pr} = 10\ 000\ \text{lbf/in}^2\ (700\ \text{kgf/cm}^2;\ 70\ \text{N/mm}^2)$$

$$f_{bt} = f_{bT} - 500\ \text{lbf/in}^2\ (f_{bT} - 35\ \text{kgf/cm}^2;\ f_{bT} - 3.5\ \text{N/mm}^2)$$

$$\bar{f}_{tt} = 500\ \text{lbf/in}^2\ (35\ \text{kgf/cm}^2;\ 3.5\ \text{N/mm}^2)$$

	lbf/in²				kgf/cm²				N/mm²			
Δf_{bT}	1500	2000	2500	3000	105	140	175	210	10.5	14	17.5	21
f_{pe}	6750	9000	11 250	13 500	472	630	788	945	47.2	63.0	78.8	94.5
f_{pt}	150 250	148 000	145 750	143 500	10 528	10 370	10 212	10 055	1052.8	1037.0	1021.2	1005.5
f_{bT}	1000	1500	2000	2500	70	105	140	175	7.0	10.5	14.0	17.5
f_{st}	950	1400	1850	2300	66.5	98.0	129.5	161.0	6.65	9.80	12.95	16.10
Δf_{pc}	8200	12 100	15 900	19 800	562	829	1095	1362	57.5	84.7	112.0	139.1

Δf_{ps} + Δf_{pc} + Δf_{pr}	26 900	30 800	34 600	38 500	1872	2139	2405	2672	187.5	214.7	248.0	269.1
f_{pe}	123 350	117 200	111 150	105 000	8656	8231	7807	7383	865.3	822.3	779.2	736.4
$R_{o\infty} = \dfrac{f_{pe}}{f_{pt}}$	0.821	0.792	0.763	0.732	0.821	0.792	0.763	0.732	0.821	0.792	0.763	0.732
$R_e = \dfrac{f_{pe}}{f_{pi}}$	0.786	0.747	0.710	0.670	0.786	0.747	0.710	0.670	0.786	0.747	0.710	0.670
$R'_{o\infty} = f'_{pe}/f'_{pt}$	0.850	0.846	0.844	0.841	0.850	0.846	0.844	0.841	0.850	0.846	0.844	0.841
$R'_e = f'_{pe}/f'_{pi}$	0.846	0.841	0.837	0.833	0.846	0.841	0.837	0.833	0.846	0.841	0.837	0.833
$R_{o28} = f_{pe28}/f_{pt}$	0.934	0.920	0.906	0.891	0.934	0.920	0.906	0.891	0.934	0.920	0.906	0.891

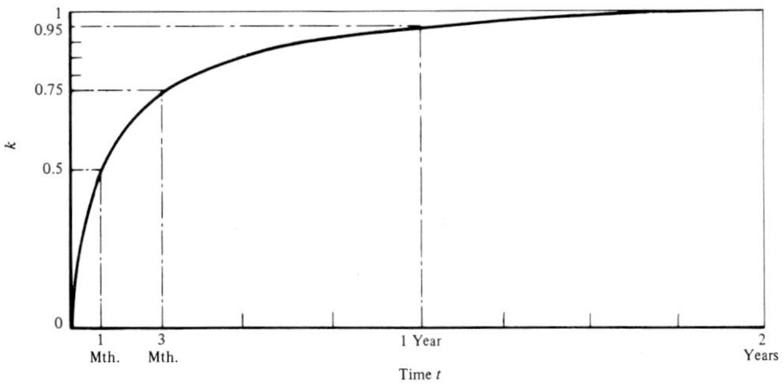

Figure 6.14 Relation between losses and time

factors are required R_0 is termed $R_{0\infty}$ to distinguish it from R_{0t} or $R_{0\,28}$; similarly $R_{e\infty}$, R_{et}, and R_{e28} should be introduced when specific times are considered. When $R_{e\infty}$ and $R_{0\infty}$ are used without further subscripts, they represent R_e and R_0.

Many variations are, of course, possible in the value of the initial tensioning stress, the distribution of the prestress, the stresses due to weight of the beam, and the age of the beam when it is first loaded. If the beam be loaded at an early age, for example, losses due to creep will be smaller, and it may perhaps be assumed that 50 per cent of these losses occur before loading and the remainder after loading when the stress causing the creep is much less; alternatively, the values given in Table 6.7 may be used. The data in Table 6.7 are calculated on the assumption that the creep which occurs when the stress in the concrete adjacent to the steel is high is one-third of the total creep, and that the stress after the member is placed in position is 750 lbf/in² (53 kgf/cm²; 5.2 N/mm²). The equivalent stress upon which the calculation of the loss due to creep is based is then

$$f_{s\,av} = \tfrac{1}{3}f_{st} + \tfrac{2}{3} \times 750 = \tfrac{1}{3}f_{st} + 500 \text{ lbf/in}^2 \; (\tfrac{1}{3}f_{st} + 35 \text{ kgf/cm}^2;$$
$$\tfrac{1}{3}f_{st} + 3.5 \text{ N/mm}^2)$$

In the method described in the foregoing an attempt is made to ascertain the maximum possible losses in order to determine the minimum effective prestress that can reasonably be expected to occur, and to ensure that the beam will behave as predicted by the calculations. These considerations also require that the strength of the concrete should be at least 2.5 times the maximum stress in the concrete at transfer, although CP 115 and CP 110 recommend a minimum strength of only twice the stress at transfer when the distribution of stress is triangular.

A method sometimes adopted is to ignore the stress in the concrete at transfer, which may be greater than that permitted by the British Code – for example, even two-thirds of the strength – and to simplify the calculation in such a way as to allow for as little loss as possible. Losses due to elasticity and creep, for example, are calculated for the stress in the concrete adjacent to the centroid of the entire steel, instead of the centroid of the steel in the tensile zone. Losses

Table 6.7 Reduction of losses due to creep

$$f_{sav} = \frac{f_{st}}{3} + 500\ \text{lbf/in}^2 ; \frac{f_{st}}{3} + 35\ \text{kgf/cm}^2 ; \frac{f_{st}}{3} + 3.5\ \text{N/mm}^2$$

$$\Delta f_{pc} = 8.6\, f_{sav}$$

	lbf/in²				kgf/cm²				N/mm²			
f_{bT}	1 500	2 000	2 500	3 000	105	140	175	210	10.5	14.0	17.5	21.0
$\frac{f_{st}}{3}$	317	467	617	767	22.2	32.7	43.2	53.7	2.22	3.27	4.32	5.37
f_{sav}	817	967	1117	1267	57.2	67.7	78.2	88.7	5.72	6.77	7.82	8.87
Δf_{pc}	7030	8320	9620	10 910	492	582	673	762	49.2	58.2	67.3	76.2
f_{pe}	124 420	120 980	117 530	113 890	8726	8478	8229	7983	873.6	848.8	823.9	799.3
R_o	0.827	0.817	0.806	0.793	0.827	0.817	0.806	0.793	0.827	0.817	0.806	0.793
R_e	0.792	0.770	0.748	0.726	0.792	0.770	0.748	0.726	0.792	0.770	0.748	0.726

due to creep are calculated on the assumption that the member is immediately loaded, although in fact it may be previously exposed for some time in the open air without curing. Further, if the margin between the compressive stress and the strength at the time of application of the prestress is small, the actual creep is much greater than that assumed, since creep is proportional to stress only when the stress is less than one-third of the strength. This value is based on tests by Davis, as described on page 30. Non-linear creep has been shown to occur at stress/strength ratios in excess of 0.25 for a cylinder strength of about 6500 lbf/in^2 (460 kgf/cm^2; 45 N/mm^2). This means that proportionality between creep and stress may be limited to a stress as small as $0.75 \times 0.25 = 0.19$ of the cube strength, assuming a ratio of cylinder strength to cube strength of 0.75. It is therefore reasonable to conclude that the effective prestress remaining in the concrete may be much less than that assumed in the calculations, and the factor of safety against cracking may be greatly reduced or even non-existent. If the steel is well distributed throughout the tensile zone this will not be particularly disadvantageous, since high tensile stresses in the concrete or even fine hair cracks that may occur under working load will usually be harmless, but it is the authors' opinion that the purpose of a calculation is to obtain results which agree as closely as possible with the actual behaviour; moreover, the application of such a method of calculation to members in which the steel is not so well distributed and the bond is less efficient – particularly with post-tensioned steel – may have undesirable consequences.

6.13.2 Example 2. Post-tensioned steel

Consider a symmetrical member with post-tensioned steel, part of which is bent up so that the stresses due to the weight of the beam immediately counteract those due to the prestressing force, if measures are taken to ensure the destruction of any adhesion between the soffit of the beam and the form. As in Example 1, it is assumed that the distribution of stress due to the prestress alone is triangular, the prestress at the top face being zero and that at the bottom being f_{bT} at the time of prestressing, and that the maximum bending stresses due to the weight of the beam are ±500 lbf/in^2 (35 kgf/cm^2; 3.5 N/mm^2). The resultant stresses at transfer are therefore $f_{bt} = f_{bT} - 500$ lbf/in^2 ($f_{bT} - 35$ kgf/cm^2; $f_{bT} - 3.5$ N/mm^2). If the tensioning force is applied simultaneously to all the steel then there is no loss due to elastic shortening, but if the tendons are tensioned one after another some elastic shortening will occur. It is assumed that the net initial tensioning stress is the same for all the steel members, except for the differences in losses due to friction; the average loss due to elastic compression will then not exceed half the loss that occurs with pre-tensioning. Thus for $\alpha_e = 5$ the maximum loss due to elastic shortening will be $2.5f_{st}$, in which f_{st} is the resultant stress in the concrete in the plane of the centroid of the steel. Table 6.8 and Figure 6.15 show losses for various values of prestress for an initial net tensioning stress of 145 600 lbf/in^2 (10 200 kgf/cm^2; 1000 N/mm^2). This reduced stress allows a margin for frictional losses.

Losses due to shrinkage are assumed to be the same as those in Example 1, although two-thirds of this value would be permissible. However, such a reduction is based on the assumption that some shrinkage has occurred before prestressing, in which case cracks may have occurred, and an assumed factor of safety against cracking does not apply; it is therefore wiser, as already stated, to

Table 6.8 Calculations for Example 2

f_{pi} = 145 600 lbf/in² (10 200 kgf/cm²; 1 kN/mm²). $\Delta f_{pe} = 2.5 f_{st}$. $\Delta f_{pc} = 4 f_{st}$

f_{ps} = 8700 lbf/in² (600 kgf/cm²; 60 N/mm²). Δf_{pr} = 10 000 lbf/in² (700 kp/cm²; 70 N/mm²)

$f_{bt} = f_{bT}$ − 500 lbf/in² (f_{bT} − 35 kgf/cm²; f_{bT} − 3.5 N/mm²)

$\bar{f}_{tt}$ = 500 lbf/in² (35 kgf/cm²; 3.5 N/mm²)

	lbf/in²				kgf/cm²				N/mm²			
f_{it}	1500	2000	2500	3000	105	140	175	210	10.5	14	17.5	21
f_{st}	1400	1850	2300	2750	98	129.5	161	192.5	9.8	12.95	16.1	19.25
Δf_{pe}	3500	4600	5800	6900	245	324	402	481	24.5	32.4	40.2	48.1
f_{pt}	142 100	141 000	139 800	138 700	9955	9876	9798	9719	975.5	967.6	959.8	951.9
Δf_{pc}	5600	7400	9200	11 000	392	518	644	770	39.2	51.8	64.4	77
$\Delta f_{pc} + \Delta f_{ps} + \Delta f_{pr}$	24 300	26 100	27 900	29 700	1702	1828	1954	2080	169.2	181.8	194.4	207
f_{pe}	117 800	114 900	111 900	109 000	8253	8048	7844	7639	806.3	785.8	765.4	744.9
$R_{0\infty}$	0.829	0.815	0.800	0.786	0.829	0.815	0.800	0.786	0.829	0.815	0.800	0.786
R_e	0.809	0.790	0.769	0.750	0.809	0.790	0.769	0.750	0.809	0.790	0.769	0.750
R_{028}	0.915	0.900	0.883	0.868	0.915	0.900	0.883	0.868	0.915	0.900	0.883	0.868

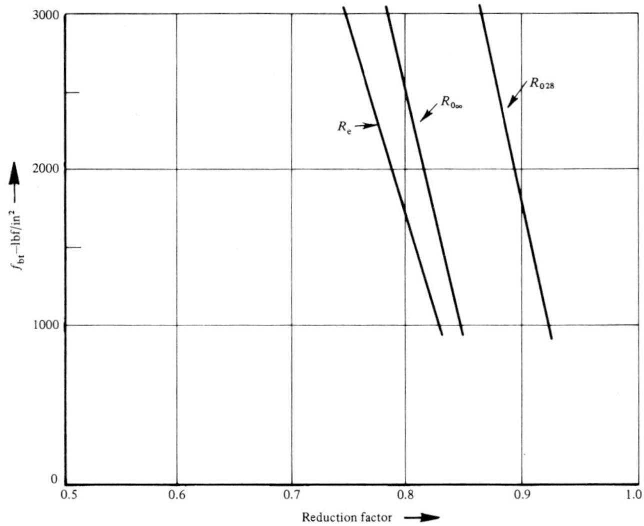

Figure 6.15 Reduction factors of example 6.13.2

consider the entire loss due to shrinkage but to insist on reliable moist curing of the concrete until it is prestressed.

The strain due to creep is reduced from 0.33×10^6 to 0.25×10^6 per lbf/in² (3.5×10^{-6} per kgf/cm²; 36×10^{-6} per N/mm²), on the assumption that the age of the concrete when it is compressed is considerably greater than is the case in the pre-tensioning process, and also because members with post-tensioned steel are usually loaded shortly after prestressing, and only a small part of the creep will take place under the stress f_{st}, the remainder occurring with a much smaller stress $f_{s\,av}$. The actual loss due to creep may therefore be between 0.5 and 0.75 of that considered in Example 1; that is between 0.5×0.75 and 0.75×0.75 of $8.62f_{st}$, which amounts to $3.23f_{st}$ to $4.83f_{st}$. In Table 6.8 a value of $4f_{st}$ is allowed, together with a loss of 10 000 lbf/in² (700 kgf/cm²; 70 N/mm²) due to relaxation of the steel.

Here again many variations are possible, but Table 6.8 and the graphs indicate the manner in which the reduction factors R_0 and R_e may be represented. Values of $R_{0\,28}$ are also included, calculated on the same basis as for Example 1. Numerous similar graphs and tables can easily be prepared so that the values of R_0 for any given requirements are immediately obtained.

Smaller losses due to relaxation may also be considered, provided that it is certain that such losses will occur with the steel used. Examples of the calculation of losses for particular members are given in Chapter 9.

If the section contains an area A_{sn} of non-tensioned steel the prestress of the concrete is reduced, as noted previously, since the effective area of the section is increased by $(\alpha_e - 1)A_{sn}$. Moreover, additional losses occur due to the redistribution of stress caused by shortening of the non-tensioned steel as a consequence of shrinkage and creep. This shortening is of the same order as that occurring in the tensioned steel, and produces a compressive force in the non-tensioned steel; this in turn induces an equal and opposite force in the concrete which reduces the prestress of the concrete. Tests have proved that if A_{sn} is relatively small, there is little reduction of the effective prestress; however, with a large area of non-tensioned steel an appreciable reduction of the effective prestress may take place and the prestress may even be cancelled out entirely if the stress in the concrete is low. It seems preferable, therefore, to use high-tensile steel instead of mild steel as non-tensioned reinforcement when this is required for conditions at ultimate load, as the effect of this re-distribution of stress will then be negligible. Only the British Codes permit this; ACI Building Code and the German Specification require non-tensioned reinforcement to be of the same type as that permitted for reinforced concrete (see page 92).

In the case of high-alloy bars, the factors R_0 are different. On the one hand the initial stress in the steel is much less, and on the other hand, so far as is known, no losses due to pull-in need be considered when threaded anchors are used. Despite the latter advantage, however, the comparative reduction factors may be smaller in view of the lower initial stress in the steel, of which the losses obviously represent a greater proportion.

REFERENCES

1. ABELES, P. W. Fully and partly prestressed reinforced concrete, *Journal of the American Concrete Institute,* January 1945. Vol. 16 No. 3, pp. 181–214.
2. FREUDENTHAL, A. M. *Creep effect in the analysis of reinforced concrete structures.* Papers presented to the IABSE Fifth Congress. Lisbon. 25 June–2 July 1956. Zurich. The Association 1956, Preliminary Report, pp. 85–99.
3. Tentative Recommendations for Prestressed concrete, A.C.I.-A.S.C.E. Joint Committee 323., *Journal of the American Concrete Institute,* February 1962.
4. ABELES, P. W. and KUNG, R. Prestress losses due to the effect of shrinkage and creep on non-tensioned steel, *Journal of the American Concrete Institute,* January 1973 Vol. 70, No. 1 pp. 19–27.
5. TROST, H. Auswirkungen des Superpositionsprinzipes auf Kriech- und Relaxationsprobleme bei Beton- und Spannbeton (The Principle of Superposition and its effect on problems of creep and relaxation in concrete and Prestressed Concrete), *Beton und Stahlbetonbau* (Berlin-Wilmersdorf), October and November 1967. V. 62, No. 10 and 11.

CHAPTER 7

ANALYSIS OF STRESSES

7.1 General

This chapter relates only to analysis; that is, to the determination of stresses in sections with known dimensions and properties. Matters relating to design (that is, the determination of the dimensions and properties of sections to suit given conditions of loading and permissible stresses) are discussed in Chapter 8. The system of notation used in this and succeeding chapters is described in section 5.4. It should be noted that compressive stresses are considered positive and tensile stresses negative (this conforms with normal design practice but is the opposite of that convention used in research). A summary of most of the symbols used in this book is given at the beginning of the book.

7.2 Determination of elastic stresses in a homogeneous member

In this section, only a linear distribution of stress is considered. The stresses are assumed to be within the elastic range, and the member is assumed to be homogeneous; these assumptions also give good results after an excess load has caused microscopic or even visible cracking, so long as the steel is well distributed and properly bonded to the concrete. The equations which follow are directly applicable to members with pre-tensioned and post-tensioned steel.

Sufficiently accurate results are usually obtained by considering the gross cross-sectional area A of the concrete, the co-operation of the steel being ignored when there are no large cable holes. Formulae in which the effective area of the section is taken into account are required for more accurate analysis of prestressed members (when an allowance is made for the co-operation of the steel), and for members with post-tensioned steel and large cable holes (when the area of the holes is deducted from the area of the section, Figure 7.1). In the case of members without large cable holes, the initial prestressing force is applied to a cross-section the area of which A_0 excludes that of the prestressing steel, and is given by

$$A_0 = A_c + (\alpha_e - 1)A_{sn} - (A_{st} + A'_{st})$$

Where α_e = modular ratio

Any subsequent loading is resisted by an area

$$A_e = A_c + (\alpha_e - 1)A_{sn} + (\alpha_e - 1)(A_{st} + A'_{st})$$

The nett area A_0 should also be used for evaluating losses due to elastic shortening of the concrete and relaxation of the steel, as the steel cannot cause such effects and at the same time resist them.

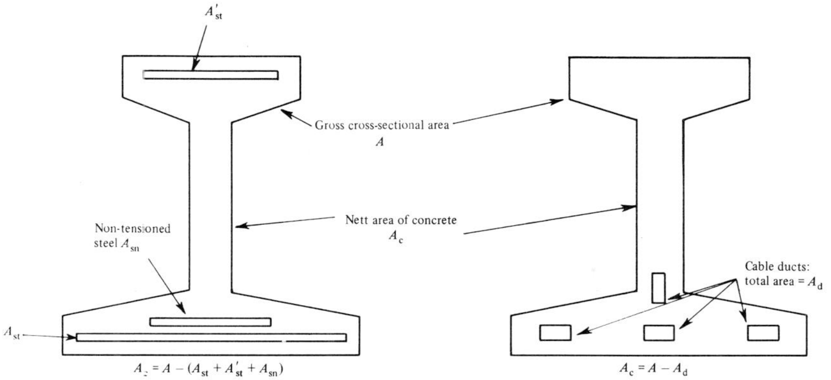

Figure 7.1 Effect of cable ducts on area of section

Such 'exact' formulae are necessary for the analysis of test results; however, they are rather involved, and the improvement in accuracy which results is usually small, except in the case of members with large ducts, when the actual stresses in the concrete at transfer are appreciably greater than those given by approximate formulae. Similarly, with relatively large areas A_{sn} of non-tensioned steel the prestress in the concrete will be much less, as already stated and therefore A_{sn} should be taken into account. The area A_c above may be adopted if neither the area of the cable ducts nor $(\alpha_e - 1)A_{sn}$ exceeds, say 3 per cent of the gross cross-sectional area. There are some basic formulae and they apply to all methods of prestressing; that is, to pre-tensioning, post-tensioning, full and partial prestressing, members in which the whole of the steel or only a part of it is tensioned, and members which are monolithic or composed of separate parts. The formulae can also be applied to composite construction.

7.3 Stresses due to prestressing force and weight of member

7.3.1 Stresses due to prestressing force only

At transfer: Figure 7.2 illustrates the stresses in the concrete of a section with one axis of symmetry due to the prestressing force alone. The maximum stresses are:

At the bottom

$$f_{bT} = \frac{P_t}{A} + \frac{P_t e_s}{Z_b} = \frac{P_t}{A}\left(1 + \frac{e_b e_s}{i^2}\right) = \frac{k_b P_t}{A} \qquad (7.1)$$

At the top

$$f_{tT} = \frac{P_t}{A} - \frac{P_t e_s}{Z_t} = \frac{P_t}{A}\left(1 - \frac{e_t e_s}{i^2}\right) = \frac{k_t P_t}{A} \qquad (7.2)$$

in which

$k_b = 1 + \dfrac{e_b e_s}{i^2}$ and $k_t = 1 - \dfrac{e_t e_s}{i^2}$. The constants k_b and k_t are of importance as

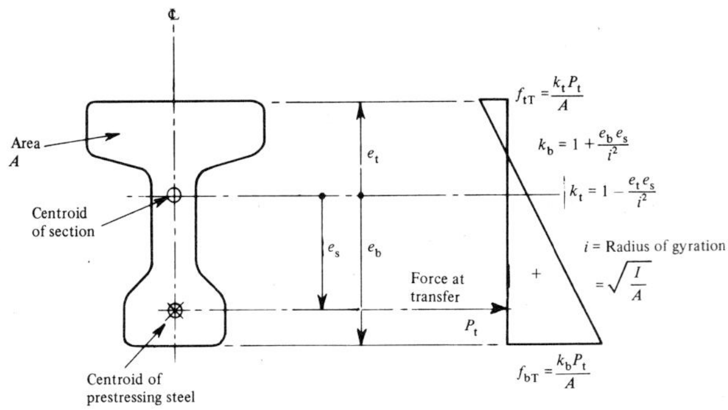

Figure 7.2 Stresses at transfer due to prestress only

they indicate the shape of section required (see Chapter 8). The prestressing force at transfer (P_t) is the initial force P_i less the losses up to transfer as described in Chapter 6.

Effective stresses:
The prestressing force at transfer P_t is reduced by the losses to an effective force $P_e = R_0 P_t$; R_0 is a reduction factor which is assessed as described in Chapter 6. The effective stresses due to P_e are therefore:

At the bottom

$$f_{bE} = \frac{k_b P_e}{A} = \frac{k_b R_0 P_t}{A} \quad \ldots\ldots\ldots\ldots \quad (7.3)$$

At the top

$$f_{tE} = \frac{k_t P_e}{A} = \frac{k_t R_0 P_t}{A} \quad \ldots\ldots\ldots\ldots \quad (7.4)$$

UNSYMMETRICAL SECTION

The general expression for the stress f at a point x, y due to an eccentric longitudinal force P acting on an unsymmetric member (Figure 7.3) is given by:

$$f = \frac{P}{A} + \frac{M_y - M_x\,(I_{xy}/I_{xx})}{I_{yy} - (I_{xy}^2/I_{xx})}\,x + \frac{M_x - M_y\,(I_{xy}/I_{yy})}{I_{xx} - (I_{xy}^2/I_{yy})}\,y \quad \ldots\ldots \quad (7.5)$$

in which x and y are any two perpendicular axes through the centroid (the positive directions being such that e_x and e_y are positive);

e_x and e_y are the eccentricities of the force P
$M_x = Pe_x$; $M_y = Pe_y$
$I_{xx} = \int y^2\,dA$; $I_{yy} = \int x^2\,dA$; $I_{xy} = \int xy\,dA$
dA = elemental area at point where stress is required
A = total area of section.

For analysis, the axes x and y should be the principal axes, which are inclined at an angle θ given by $\tan 2\theta = \dfrac{2\,I_{xy}}{I_{xx} - I_{yy}}$. Further particulars relating to the design of unsymmetrical prestressed sections are given in Chapter 9, Example 9.2.9.

7.3.2 Combined stresses due to prestressing force and weight of member

Figure 7.4 illustrates the combined stresses due to the prestressing force P_t at transfer and to the bending moment M_t caused by the weight of the beam. If this moment operates immediately the prestressing force is applied to the concrete, the stresses in the concrete at transfer are:

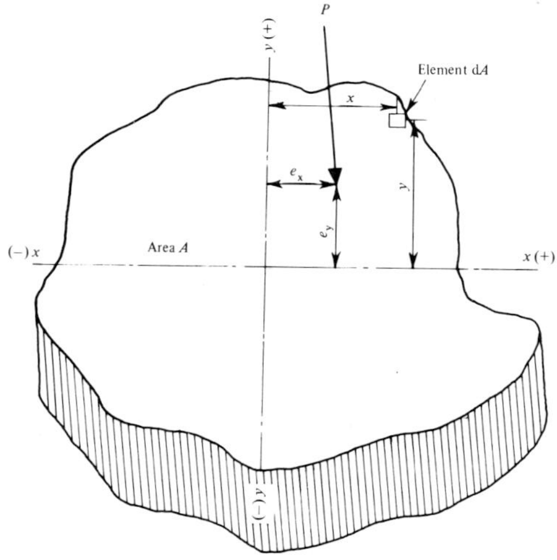

Figure 7.3 Unsymmetrical section

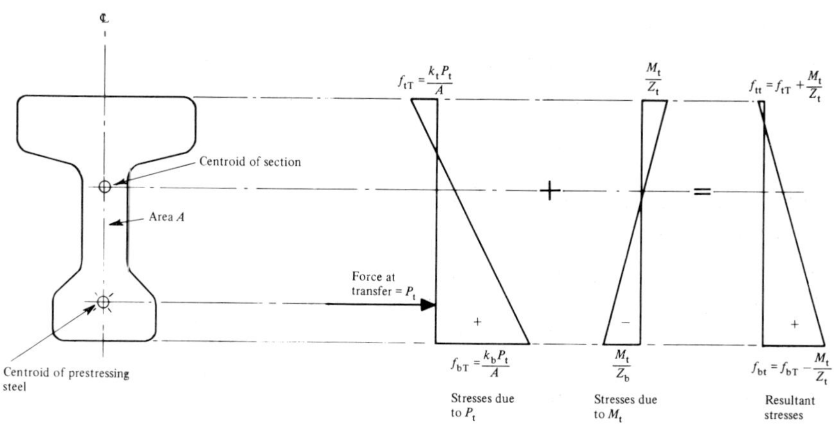

Figure 7.4 Combined stresses at transfer

At the bottom

$$f_{bt} = \frac{k_b P_t}{A} - \frac{M_t}{Z_b} \quad \ldots \quad (7.1a)$$

At the top

$$f_{tt} = \frac{k_t P_t}{A} + \frac{M_t}{Z_t} \quad \ldots \quad (7.2a)$$

7.4 Stresses due to working load

Figure 7.5 illustrates the stresses which occur when the full working bending moment M_w is applied. All possible losses are assumed to have taken place, and M_w is assumed to include the moment M_t due to the weight of the member. The combined stresses are:

At the bottom

$$f_{bw} = f_{bE} - \frac{M_w}{Z_b} = f_{bE} - f_{bW} \quad \ldots \quad (7.3a)$$

At the top

$$f_{tw} = f_{tE} + \frac{M_w}{Z_t} = f_{tE} + f_{tW} \quad \ldots \quad (7.4a)$$

Similarly, when the bending moment M_d due to the complete dead load is being considered, the stresses are:

At the bottom

$$f_{bd} = f_{bE} - \frac{M_d}{Z_b} = f_{bE} - f_{bD} \quad \ldots \quad (7.3b)$$

At the top

$$f_{td} = f_{tE} + \frac{M_d}{Z_t} = f_{tE} + f_{tD} \quad \ldots \quad (7.4b)$$

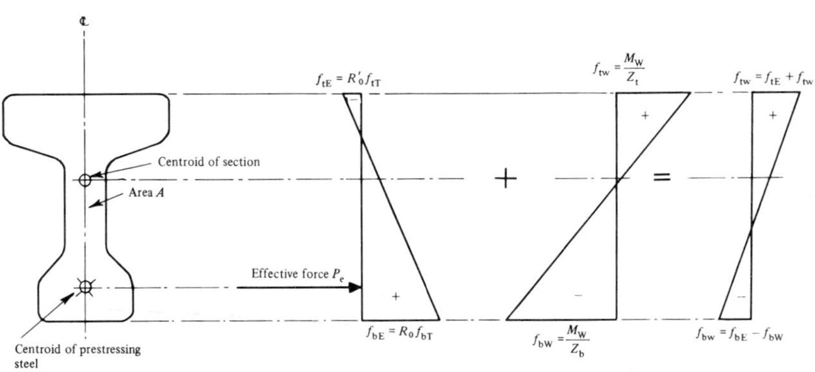

Figure 7.5 Resultant stresses under working load

It is sometimes necessary to calculate the minimum combined stresses which may occur immediately after the application of the dead weight or the working load. In this case the minimum possible losses should be allowed for, and instead of f_{bD} and f_{bE} (indicating the lowest possible effective prestress) the maximum possible effective prestress (denoted by $f_{bE\,max}$ and $f_{tE\,max}$) is required.

7.5 Stresses due to direct load

7.5.1 Concentric load

Figure 7.6 shows a uniformly prestressed member subjected to a concentric tensile force P_w. The stresses are given by:

Prestress at transfer, $f_T = \dfrac{P_t}{A}$ (assuming no moment due to weight of member).

Effective prestress, $f_E = \dfrac{P_e}{A}$

Imposed tensile stress, $f_W = -\dfrac{P_W}{A}$

Resultant working stress, $f_w = f_E + f_W = \dfrac{P_e - P_W}{A}$

If the area of steel forms a large percentage of the total area, then the use of A_0 in determining the values of f_W and f_w may prove wasteful and the more accurate expressions should be used. These are:

Prestress at transfer, $f_T = \dfrac{P_t}{A_0}$

Effective prestress, $f_E = \dfrac{P_e}{A_0}$

Imposed tensile stress, $f_W = -\dfrac{P_W}{A_e}$

Resultant working stress, $f_w = f_E + f_W = \dfrac{P_e}{A_0} - \dfrac{P_W}{A_e}$

7.5.2 Eccentric load

Figure 7.7 shows a prestressed member subjected to an eccentric tensile force P_W. The effective stresses f_{bE} and f_{tE} are given by equations 7.3 and 7.4

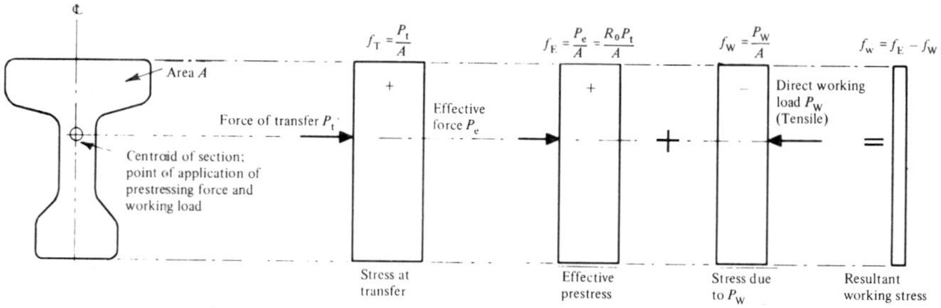

Figure 7.6 Combined effect of concentric prestress and tensile force

(assuming no bending moments due to the weight of the member), and the stresses imposed by the load P_W are:

At the bottom

$$f_{bW} = -\frac{P_W}{A} - \frac{P_W e_W}{Z_b} = -\frac{P_W}{A}\left(1 + \frac{e_b e_W}{i^2}\right) = -\frac{k_{bW} P_W}{A}$$

At the top

$$f_{tW} = -\frac{P_W}{A} + \frac{P_W e_W}{Z_t} = -\frac{P_W}{A}\left(1 - \frac{e_t e_W}{i^2}\right) = -\frac{k_{tW} P_W}{A}$$

in which

$$k_{bW} = 1 + \frac{e_b e_W}{i^2} \quad \text{and } k_{tW} = 1 - \frac{e_t e_W}{i^2}$$

The resultant working stresses are therefore:

At the bottom

$$f_{bw} = f_{bE} + f_{bW} = f_{bE} - \frac{k_{bW} P_W}{A} \qquad (7.6)$$

At the top

$$f_{tw} = f_{tE} + f_{tW} = f_{tE} - \frac{k_{tW} P_W}{A} \qquad (7.7)$$

If A_0 and A_e greatly differ, areas A_0 and A_e should be used in place of A. For a concentric force,

$$k_{bW} = k_{tW} = 1$$

7.6 Combined stresses due to bending and direct compressive load

Figure 7.8 shows a prestressed member subjected to a direct load P_W and a

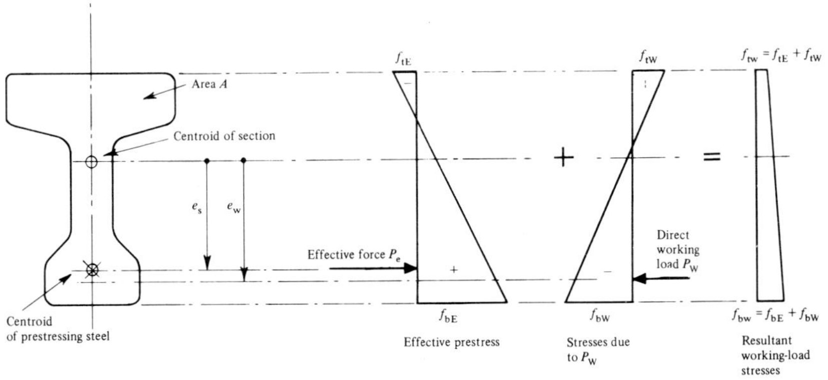

Figure 7.7 Combined effect of prestress and eccentric tensile force

bending moment M_W. Expressions 7.1, 7.2, 7.3, 7.4, 7.1a, and 7.2a still apply, but the values of the working stresses f_{bw} and f_{tw} are now given by:

At the bottom

$$f_{bw} = \frac{k_b P_e}{A} + \frac{k_{bW} P_W}{A} - \frac{M_W}{Z_b} \quad \ldots \quad (7.8)$$

At the top

$$f_{tw} = \frac{k_t P_e}{A} + \frac{k_{tW} P_W}{A} + \frac{M_W}{Z_t} \quad \ldots \quad (7.9)$$

If the force P_W is applied at the centroid of the section $k_{bW} = k_{tW} = 1$. Again, if the prestressing steel forms a large percentage of the total area of the section, areas A_0 and A_e should be used in place of A.

7.7 Stresses due to shearing only

In a homogeneous isotropic material the shearing stress at any level in the section is given by the general formula $f_{sh} = VAy/bI$ in which b is the breadth of the section at the level where the shearing stress is required. The remaining symbols are defined in Figure 7.9. Chart 6a in Part 2 gives values of Ay/bI which correspond to the maximum shearing stress $f_{sh\ max}$ occurring in commonly used sections.

If the prestressing steel is bent up (Figure 7.10), the effective shearing force at the section is reduced from V to $V - P_{ev}$ and

$$f_{sh} = \frac{(V - P_{ev})Ay}{bI} \quad \ldots \quad (7.10)$$

where P_{ev} is the vertical component of P_e.

The foregoing formulae, and those in the next three sections, relate to homogeneous, isotropic members where sections which are initially plane remain plane after bending. In prestressed concrete, stirrups are usually provided, and the equations, though they give a reasonable approximation in many cases, cannot be expected to represent the true conditions.

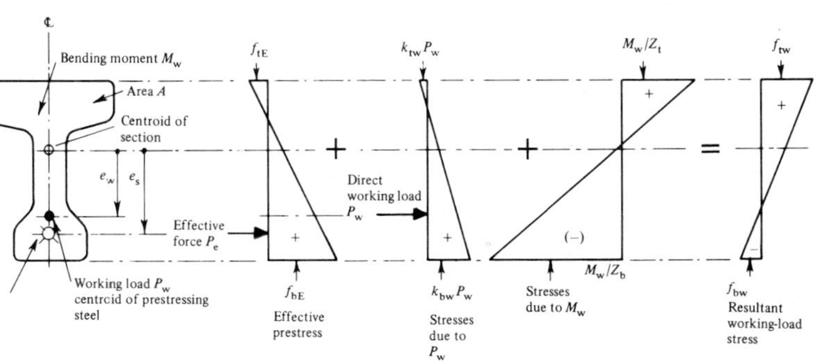

Figure 7.8 Combined stresses due to prestress, direct load and bending moment

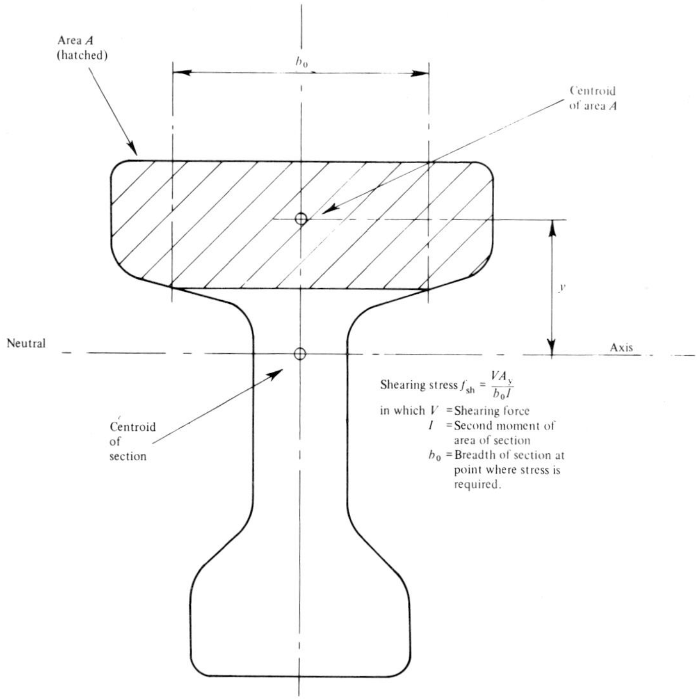

Figure 7.9 Shearing stress in a homogeneous section

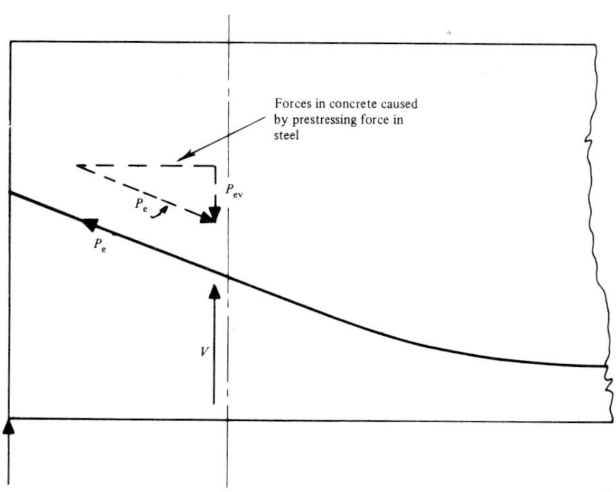

V Shearing force
P_e Effective prestressing force
P_{ev} Vertical component of P_e

Figure 7.10 Effect of bent-up tendon on the shearing force

The design methods recommended in CP 115 for conditions at working load are based on the principal stress theory outlined in the foregoing.

7.8 Stresses due to shearing and bending

The principal tensile and compressive stresses in a homogeneous isotropic material are given by

$$f_{pt} = \frac{f_x + f_y}{2} - \sqrt{\frac{(f_x - f_y)^2}{4} + f_{sh}^2} \quad \text{. . . . . (7.11)}$$

and

$$f_{pc} = \frac{f_x + f_y}{2} + \sqrt{\frac{(f_x - f_y)^2}{4} + f_{sh}^2} \quad \text{. . . . . (7.12)}$$

(The symbols are defined in Figure 7.11; a graphical method for obtaining f_{pt} and f_{pc} is given in Chart 1 in Part 2). When no transverse moment M_y occurs the expressions are simplified to

$$f_{pt} = \frac{f_x}{2} - \sqrt{\frac{f_x^2}{4} + f_{sh}^2} \text{ and } f_{pc} = \frac{f_x}{2} + \sqrt{\frac{f_x^2}{4} + f_{sh}^2} \quad \text{. . . . . . (7.13)}$$

At the neutral axis, $f_x = 0$ and $f_{sh} = f_{sh\ max}$

Hence

$$f_{pt} = -f_{sh\ max} \text{ and } f_{pc} = +f_{sh\ max} \quad \text{. . . . . (7.14)}$$

Similarly, at the top and bottom of the section, $f_{sh} = 0$, $f_x = f_{x\ max}$, and the principal stress $= f_{x\ max}$. For uniformly distributed load the greatest bending and shearing stresses in a rectangular beam are normally the principal stresses at extreme fibres and centroid respectively of sections at critical points. In an I-beam however, the maximum principal stresses often occur at the junction of the flange and the web.

In prestressed concrete, f_x and f_y are the stresses due to the combined prestress and bending moment at the section considered.

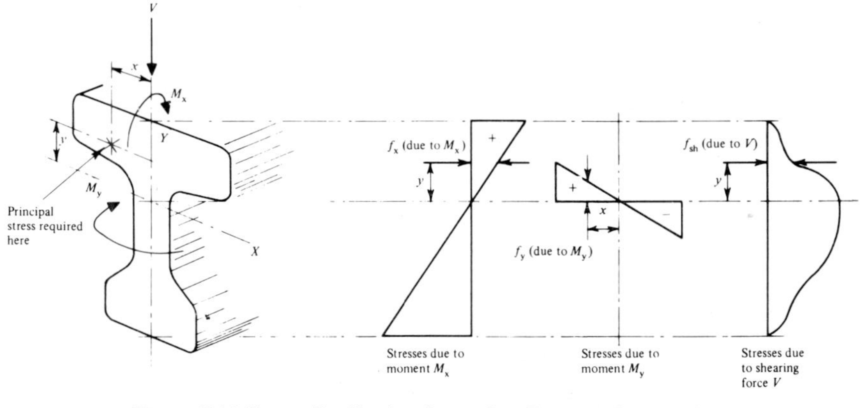

Figure 7.11 Stress distribution due to bending and shearing force

7.9 Stresses due to torsion

7.9.1 Torsion: St Venant (elastic)

The classical theory of torsion in a homogeneous isotropic material, derived by St Venant[1], assumes that sections initially plane will remain plane after twisting, and that shearing stresses only are produced by twisting. This is true only for shafts of annular or circular cross-section, but is also very nearly true for thick members of almost any section. The general formula, derived for circular shafts, is

$$\tau_{max} = \frac{T_M r}{J} \quad (7.15)$$

in which τ_{max} is the maximum shearing stress due to the twisting moment T_M; r is the radius of the section; and J is its polar moment of inertia. Solutions have been obtained for members with cross-sections other than circular; a number of these are summarized in Chart 7 in Part 2. Sections such as I- or T- beams, composed of a number of rectangles, can be analyzed by means of the expression

$$\tau = \sum_{n=1}^{n=N} \frac{T_M d}{\frac{b_n d_n^3}{3}} \quad (7.16)$$

in which d is the thickness of the section at the point considered, b_n is the width and N is the number of rectangles comprising the section (Figure 7.12).

In dealing with more irregular sections, analogue methods are sometimes helpful. The liquid continues to move as a result of its inertia, and the stream-lines of the moving liquid give a qualitative representation of the lines of equal torsional stress in the section. The application of the analogy to a triangular section is shown in Figure 7.13; it is apparent that the section is approximately equivalent to the inscribed circle.

7.9.2 Torsion in plastic condition

Numerical values may be also obtained by means of Nadai's sand-heap analogy[2]. If completely cohesionless sand is heaped at a plastic state on the cross-section the volume of the sand-heap is then equivalent to half of the torsional moment at the section, and the slope of the sand (which is constant) corresponds to the magnitude of the shearing stress. For a triangular section (Figure 7.14) for example, the slope is equal to h/r, where r is the radius of the inscribed circle.

Thus $h = \dfrac{3V_0}{A} = \dfrac{3T_M}{2A}$ and $r = \dfrac{A}{s}$, then torsional stress = slope $= \dfrac{h}{r} = \dfrac{3T_M s}{2A^2}$ in which V_0 is the volume of the sand heap, A is the area of the triangle, T_M is the torsional moment, and s is half the periphery of the triangle. An identical formula can be obtained for a circular section.

The British Code of Practice CP 110 accepts Nadai's theory for evaluation of shear stress in concrete section due to torsion. For some typical simple sections the shear stress due to torsion can be calculated as shown in Figure 7.15.

The design procedure of a member subjected to combined shear stress due to torsion and vertical shear has been outlined on pages 155/6. When reinforcement is provided for torsion the transverse reinforcement must be closed links and the longitudinal reinforcement should be adequately anchored. The reinforcement is

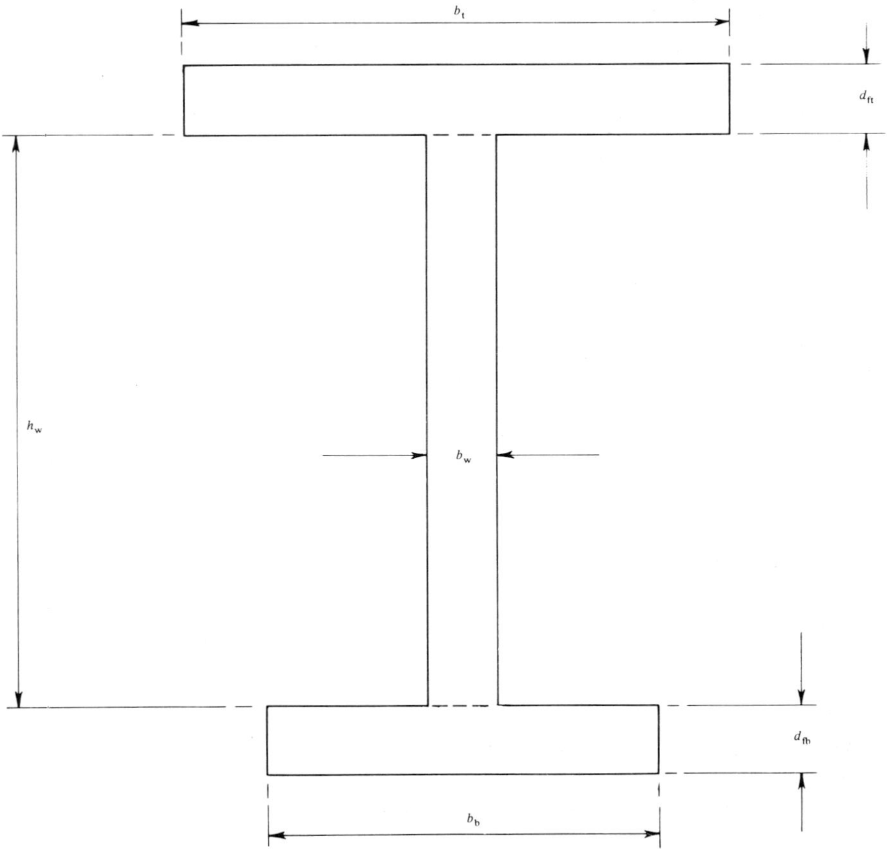

Figure 7.12 I-beams divided into rectangles to obtain torsional shear stress

determined on the basis of plastic analysis of the space truss model in which reinforcements take the tension.

7.9.3 Warping

This is based on the condition that the cross-section remains undistorted but can warp. Maillart was the first to propose (in 1921) the idea of warping and this was later further developed by Vlassov and others. In this case in addition to the torsional shear stresses (St Venant) direct stresses due to warping have to be taken into account. Their effect is more pronounced when the cross-section is rather thin. The theory is too complex to be summarized here. Reference may be made to the text book by Kollbrunner and Basler(3).

7.10 Combined stresses due to torsion and bending (uncracked section)

For a circular member subjected to a torsional moment T_M and bending moment M, the maximum and minimum principal stresses are given by

$$f_{max} = \frac{16}{\pi d^3}\left[M + \sqrt{M^2 + T_M{}^2}\right] \text{ and } f_{min} = \frac{16}{\pi d^3}\left[M - \sqrt{M^2 + T_M{}^2}\right]. \quad (7.17)$$

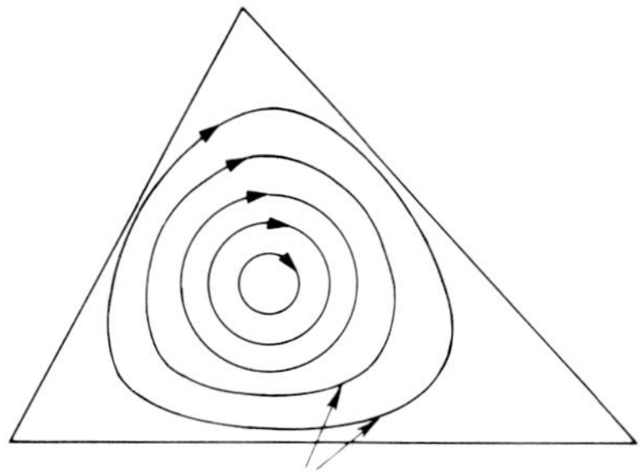

Figure 7.13 Hydrodynamic analogy for a triangular section

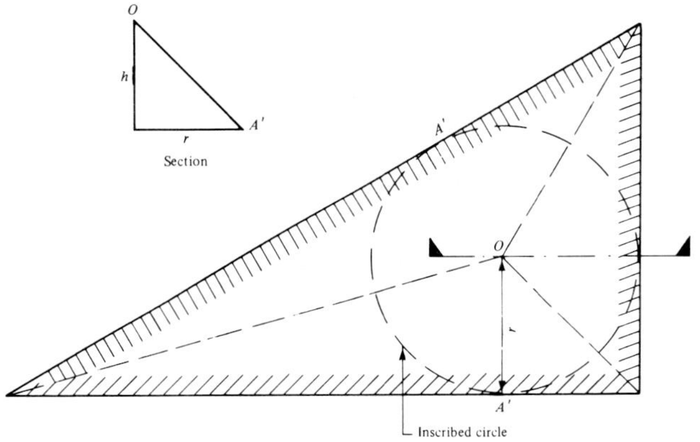

Figure 7.14 Nadai's sand-heap analogy for a triangular section

The maximum torsional shearing stress is given by

$$\tau = \frac{16}{\pi d^3} \sqrt{M^2 + T_M{}^2} \qquad (7.18)$$

When torsional stresses occur in combination with bending stresses in sections other than circles, the inscribed circle should be assumed to be the equivalent section.

7.11 The determination of flexural stresses at failure

In this section, the analysis of stresses at failure in bending is considered. The two methods discussed are based respectively on the distribution and balancing of ultimate forces only, and on the balancing of ultimate forces calculated from the ultimate strains. The effect of shear and torsion is considered in Section 7.12. In the following the suffix 'u test' is used to indicate measured test results, and 'max' to indicate predicted or calculated values.

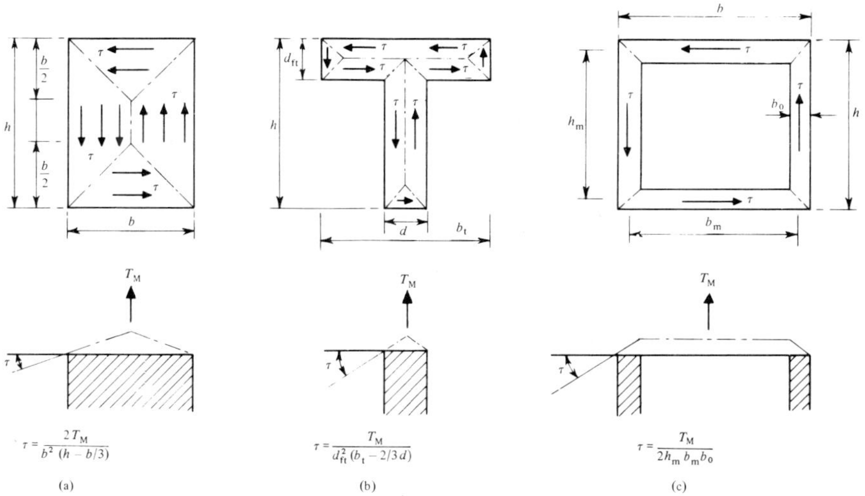

Figure 7.15 Torsional shear stress based on sand heap analogy

7.11.1 Equilibrium of ultimate forces (strains ignored)

If the compressive zone of the member is of constant breadth, the actual section shown in Figure 7.16(a) may be replaced by the equivalent section shown in Figure 7.16(b). The actual distribution of the compressive stress at failure may be as in Figure 7.16(c) to (e), but experience confirms that if the distribution is assumed to be rectangular [Figure 7.16(f)], satisfactory results are obtained. As discussed in Chapter 3, with under-reinforced sections the neutral axis moves upwards with increasing load until the stresses at the compressive face equals the compressive strength of the concrete. Consequently, the assumed depth of the neutral axis ($d_{n\ max}$) corresponds only approximately with the actual depth ($d_{n\ test}$).

For analysis, it is necessary to know the cube strength f_{cu} or, preferably, the cylinder strength f_{cyl} or prism strength f_{prism} of the concrete, and the tensile strength f_{su} or yield point f_y of the steel. If f_{prism} is unknown it may be taken as being between 0.6 f_{cu} and 0.9 f_{cu} (see Chapter 2). The stress $f_{c\ max}$ in the concrete in the equivalent stress diagram (Figure 7.17) is assumed to be equal to f_{prism}. From the equivalent stress diagram (Figure 7.17) the maximum compression is given by

$$F_{c\ max} = d_n\, b f_{c\ max} = d'_n\, b d f_{c\ max} \qquad (7.19)$$

where $d'_n = \dfrac{d_n}{d}$

The maximum tensile force $F_{s\ max} = k_s A_s f_{su}$, in which k_s is a factor that is introduced to allow for a possible difference between the maximum stress actually obtained and the true ultimate strength or yield point of steel[4].

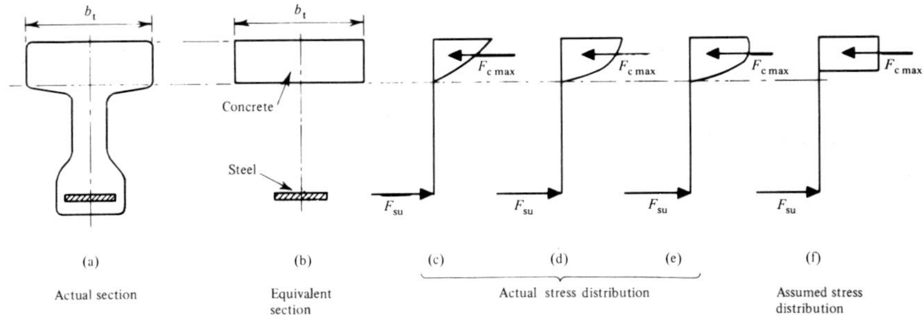

Figure 7.16 Equilibrium condition at failure

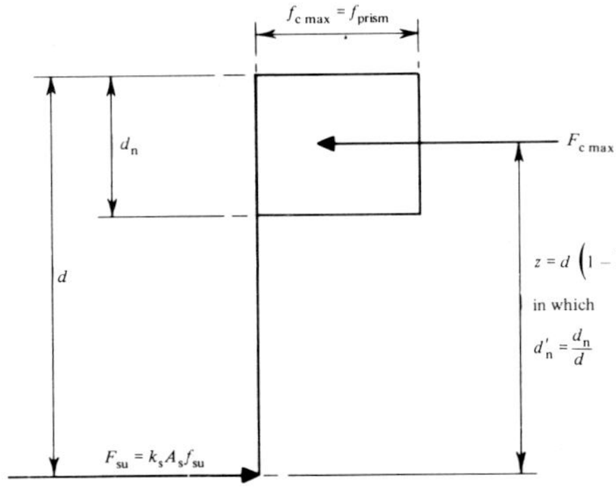

Figure 7.17 Stress distribution at failure

When $k_s = 1, f_{s\,max} = f_{su}$

$$F_{c\,max} = F_{s\,max} \text{ and } M_{max} = F_{c\,max}\, z = F_{s\,max}\, z \quad \text{. . . . . (7.20)}$$

Replacing $F_{c\,max}$ and $F_{s\,max}$ by their equivalents and substituting

$$\rho = \frac{100\,A_s}{bd}$$

$$k_s A_s f_{su} = d'_n\, bd f_{c\,max}$$

$$\frac{d'_n}{k_s} = \frac{A_s f_{su}}{b d f_{c\,max}} = \rho \frac{f_{su}}{f_{c\,max} \times 100}$$

or, assuming

$$f_{c\,max} = f_{cyl} = f_{prism} = \tfrac{3}{4} f_{cu}$$

$$\frac{d'_n}{k_s} = \frac{4}{3} \cdot \rho \cdot \frac{f_{su}}{100 f_{cu}}$$

The expression $\frac{\rho f_{su}}{100 f_{cu}}$ is often termed the 'reinforcement index' or 'percentage/strength ratio' and is here denoted by k_r. The expression $\frac{\rho f_{su}}{100 f_{c\,max}}$ in which $f_{c\,max}$ (corresponding to the cylinder strength) replaces f_{cu} is denoted by $k_{r\,cyl}$.

Hence

$$\frac{d'_n}{k_s} = \tfrac{4}{3}\,k_r = k_{r\,cyl} \text{ and therefore}$$

$$M_{max} = F_{s\,max}\,z = k_s A_s f_{su}\left(d - \frac{d_n}{2}\right)$$

$$= k_s A_s f_{su}\, d\left(1 - \frac{d'_n}{2}\right)$$

$$= k_s A_s f_{su}\, d\left(1 - \frac{k_s k_{r\,cyl}}{2}\right) \quad . \; . \; . \; . \; . \; . \; . \; . \; (7.21)$$

or,

$$M_{max} = k_s\left(\frac{\rho b d^2}{100}\right) \times \left(1 - \frac{k_s k_{r\,cyl}}{2}\right) f_{su}$$

$$= k_s k_{r\,cyl} \times \left(1 - \frac{k_s k_{r\,cyl}}{2}\right) b d^2 f_{c\,max} \quad . \; . \; . \; . \; . \; . \; . \; . \; . \; (7.21a)$$

If $f_{c\,max} = f_{prism}$ (an assumption which tests have shown to be reasonable), the only unknown in these equations is k_s, which can therefore be evaluated. If $k_s = 1$ the nominal ultimate tensile stress in the steel is equal to f_{su}; if k_s exceeds 1 the nominal stress exceeds f_{su}. This may occur with well-bonded steel, when the cracks in the concrete are numerous but narrow. In consequence, the strain of the steel as it approaches failure is less than that of non-bonded or poorly-bonded steel, and as f_{su} is determined in effect by testing non-bonded steel the smaller strain in a narrow crack (and consequently the greater cross-sectional area) of well-bonded steel gives a nominal ultimate stress $f_{s\,max}$ in excess of f_{su}. The value of k_s is therefore a measure of the quality of the bond; with non-bonded steel k_s will be less than 1, since the large deformation of the beam would lead to failure of the concrete before f_{su} was reached.

A safe limiting value of d_n for a section with well-bonded steel is about $0.5\,d$; $F_{c\,max}$ then becomes $0.5\,b\,d\,f_{prism}$ and M_{max} is $\frac{3}{8}\,b\,d^2\,f_{prism}$ or $9/32\,b\,d^2\,f_{cu}$ (assuming $f_{prism} = \frac{3}{4} f_{cu}$). However, although these approximations are safe, the maximum force theory does not give accurate results when applied to the analysis of over-reinforced sections, since the neutral axis, after rising when cracking occurs, moves downwards again as failure approaches, the greatest depth being reached at failure. With over reinforced sections $f_{c\,max}$ becomes f_{cu}.

7.11.2 Equilibrium of ultimate strains

This method, proposed by Baker[5] is based on the observation that the distribution of ultimate compressive and tensile strains remains linear. It is assumed that the maximum compressive strain in the concrete, the depth of the neutral axis, and the strength of the steel and concrete are known. The strain in the steel is deduced from measurements made on the concrete in the tensile zone, and as these include cracks as well as true strain the effect of the cracks is allowed for by introducing a factor R_B, and defining the strain in the steel as R_B times the apparent strain in the concrete adjacent to the steel. This is illustrated in Figure 7.18, from which

$$\epsilon_s' = R_B\,\epsilon_c'\,\frac{d - d_{nu}}{d_{nu}} \qquad (7.22)$$

in which ϵ_c' is the ultimate strain in the concrete and ϵ_s' is the change in strain in the steel, from that which produces zero stress in the concrete adjacent to the steel to the ultimate strain. The strain producing zero stress in the concrete is assumed to be equal to the strain in the steel before transfer, the effects of creep being ignored. If the stress in the steel before transfer is $\beta f_{s\,max}$, the strain equation may be transformed as follows. (See Figure 7.19)

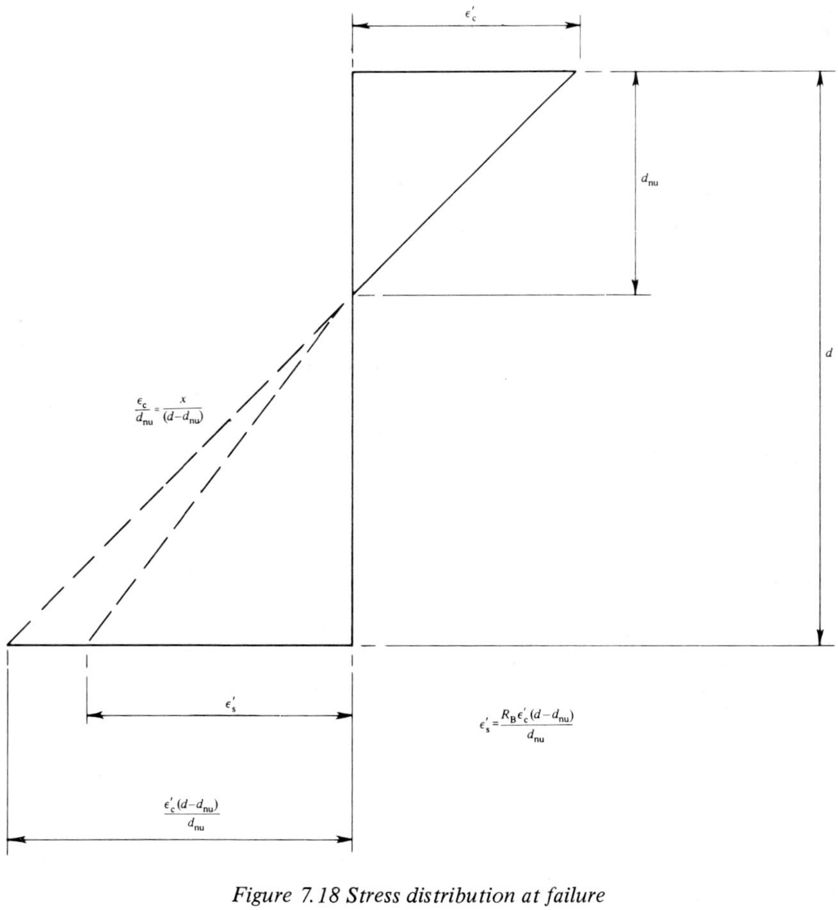

Figure 7.18 Stress distribution at failure

$$\frac{f_{s\,max}\,(1-\beta)}{E'_s} = \frac{f_{prism}}{E'_c}\left(\frac{d-d_{nu}}{d_{nu}}\right) R_B \qquad (7.23)$$

Therefore

$$f_{s\,max} = \frac{E'_s}{E'_c}\left(\frac{d-d_{nu}}{d_{nu}}\right)\frac{R_B}{(1-\beta)}\,f_{prism} \qquad (7.24)$$

Also

$$F_{s\,max} = k_s f_{s\,max} A_s$$

And

$$M_{max} = F_{s\,max}\,z_{max} = z_{max}\frac{E'_s}{E'_c}\left(\frac{d-d_{nu}}{d_{nu}}\right)\frac{R_B}{(1-\beta)}\,k_s A_s f_{prism} \qquad (7.25)$$

in which E'_s and E'_c are the secant moduli of elasticity of the steel and concrete respectively (that for steel being defined as $\frac{f_{su} - \beta f_{s\,max}}{E'_s}$, the symbols being found in Figure 7.19). The unknown quantities in this equation comprise z_{max} and R_B.

Baker recommends that $F_{s\,max}$ should be determined by direct measurement in the case of beams with non-bonded bars. In this case the lever arm z_{max} is calculated from the formula

$$z_{max} = \frac{M_{max}}{F_{s\,max}} \qquad (7.26)$$

and R_B is obtained from the expression

$$R_B = (1-\beta)\frac{E'_c}{E'_s}\,\frac{d_{nu}}{d-d_{nu}}\,\frac{M_{max}}{z_{max} A_s k_s f_{prism}} \qquad (7.27)$$

For sections with non-bonded steel this method gives satisfactory results. For sections with bonded steel, Baker recommends that $F_{s\,max}$ should

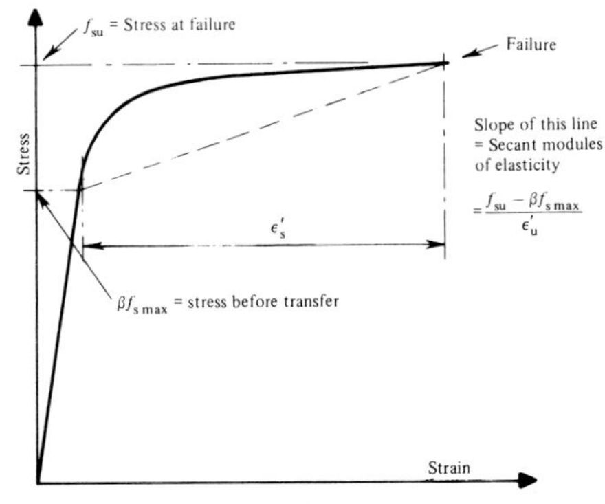

Figure 7.19 Steel stress at failure according to Baker

be determined by tests on similar sections prestressed by means of an equivalent area of bonded bars of a size sufficient to enable the stress to be measured directly. However, the replacement of bonded wires by large bars would cause appreciable differences in the extent and pattern of cracking, and much lower values of $F_{s\,max}$ would be expected. Consequently the value of z_{max} determined in this way would be excessive (perhaps even appearing to exceed the total depth of the section) since the actual strains in the cracks obtained with bonded wires would be greater than the average strains actually measured with non-bonded wires.

It seems to be better, therefore, to assess z_{max} from the measured depth of the neutral axis, by assuming a suitable distribution of stress in the concrete. This may be taken to be parabolic (Figure 7.20) giving $z_{max} = d - \frac{3}{8}\ d_{nu}$. Alternatively, Jensen suggests(6) that the distribution of stress in the compressive zones at failure may be assumed to be given by the stress–strain diagrams obtained from tests on cylinders (Figure 7.21). The centroids of those diagrams can be found, and as the depth of the neutral axis is known the lever arm is determinate. If the average of several such tests is taken the accuracy of the result is probably improved.

Another method by which the lever arm is assessed is based on extensive tests

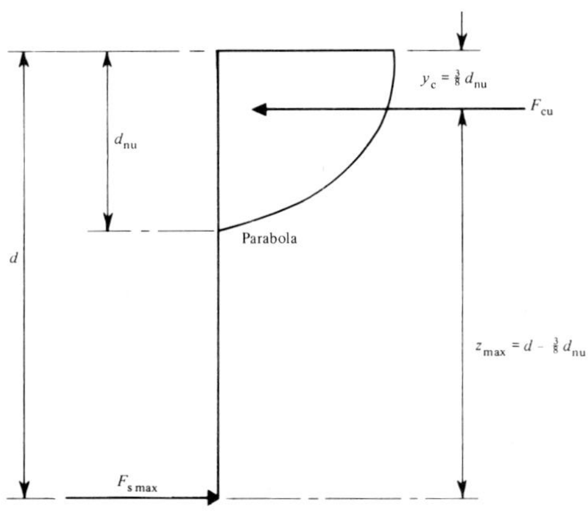

Figure 7.20 Stress distribution at failure

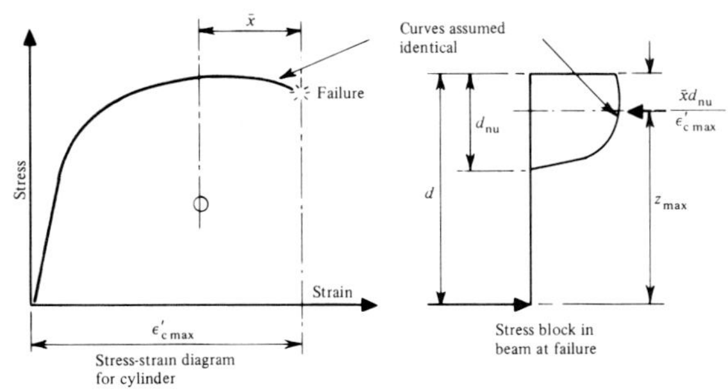

Figure 7.21 Jensen's method

made by Hognestad[7] (in which the distribution of compressive stress was measured). This primarily is a method of design, and is discussed further in Chapter 8; only the application to analysis is discussed here. With the notation given in Figure 7.22, the compressive force at failure $F_{c\ max}$ is assumed to be

$$F_{c\ max} = k_1\, k_3\, k'_u\, b\, d\, f_{cyl} \quad (7.28)$$

in which f_{cyl} is the cylinder strength of the concrete. Hence, when k_2 (and thus z_{max}) is known d_n is determined by direct measurement, Baker's formulae may be used for analysis if the strain data are known. Data for k_2 obtained by Hognestad are given in Table 7.1. However, these are based on the cylinder strength, and as noted in Chapter 2, the ratio of cylinder strength to cube strength varies with the proportions and properties of the constituents, independently of the strength, between 0.6 and 0.9. Values of 0.67, 0.75 and 0.85 are often used, and cube strengths corresponding to these are included in the table, together with the limiting values corresponding to 0.6 and 0.9. The variation in k_2 is, fortunately, small, so that an accurate estimate of the ratio is rarely required. With a cube strength of 6000 lbf/in^2 (420 kgf/cm^2; 42 N/mm^2), for example, k_2 would lie between 0.44 and 0.45; for a cube strength of 7500 lbf/in^2 (525 kgf/cm^2; 52 N/mm^2) the corresponding values would be 0.43 and 0.44 and this degree of accuracy is normally adequate for analysis. When $M_{u\ test}$ only is known, the factors k_1, k_2, and k_3 given in Table 7.1 may be used for analysis. Kaar et al[8] have extended the data for concrete up to f_{cyl} = 14 000 lbf/in^2 (1000 kgf/cm^2, 100 N/mm^2).

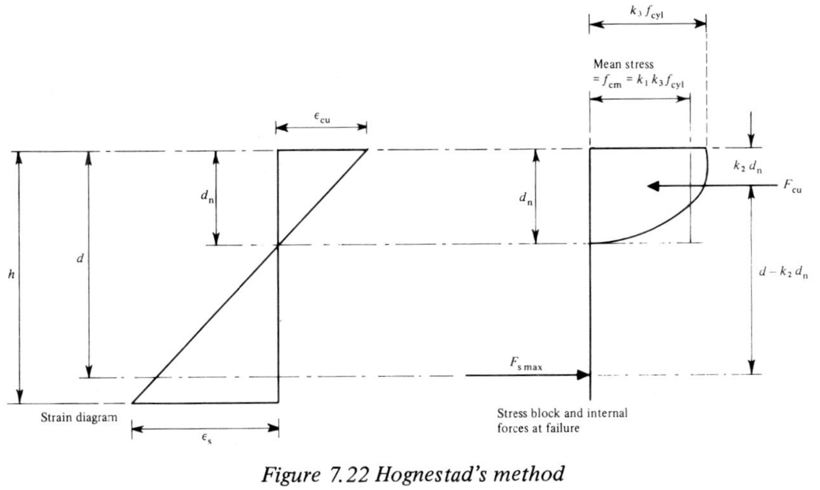

Figure 7.22 Hognestad's method

7.12 The determination of shearing strength

7.12.1 CP 110 Method

The ultimate shear resistance (V_{cr}) of a prestressed section cracked in flexure is given as:

$$V_{cr} = \left(1 - \frac{0.55\, f_{pe}}{f_{sk}}\right) \cdot v_c \cdot b\,.\,d + M_0\,\frac{V}{M} \quad (7.29)$$

Table 7.1 Data for Hognestad's method

Cylinder strength f_{cyl}	lbf/in²	3000	4000	5000	6000	7000	8000
	kgf/cm²	210	280	350	420	490	560
	N/mm²	21	28	35	42	49	56
k_1		0.82	0.79	0.75	0.71	0.67	0.63
k_2		0.46	0.45	0.44	0.42	0.41	0.40
k_3		0.97	0.94	0.92	0.92	0.93	0.94
Approximate equivalent limits of cube strength	lbf/in²	3300–5000	4400–6670	5600–8330	6700–10 000	7800–11 670	9000–13 330
	kgf/cm²	230–350	310–470	390–590	470–700	550–820	630–940
	N/mm²	23–35	31–47	39–59	47–70	55–80	63–94
Cube strength for $f_{cyl} = 0.85 f_{cu}$	lbf/in²	3500	4700	5900	7050	8250	9400
	kgf/cm²	245	330	415	495	580	660
	N/mm²	24.5	33	41.5	49.5	58	66
Cube strength for $f_{cyl} = 0.75 f_{cu}$	lbf/in²	4000	5330	6670	8000	9330	10 670
	kgf/cm²	280	375	470	565	655	750
	N/mm²	28	37.5	47	56.5	65.5	75
Cube strength for $f_{cyl} = 0.67 f_{cu}$	lbf/in²	4500	6000	7500	9000	10 500	12 000
	kgf/cm²	320	420	530	630	740	840
	N/mm²	32	42	53	63	74	84

where d is the distance from the extreme compression fibre to the centroid of the tendons at the section considered (effective depth).

M_0 is the moment necessary to produce zero stress in the concrete at the depth d

$M_0 = 0.8 f_{yE} \frac{I}{y}$ where f_{yE} is the stress due to prestress only at depth d and distance y from the centroid of the concrete section which has second moment of area I.

f_{pe} is the effective prestress in steel after all losses have occurred. For the purposes of this equation f_{pe} shall not be put greater than $0.6 f_{sk}$.

v_c = allowable ultimate shear stress in beam as given for different reinforcement ratio and concrete strength. (γ_m for concrete already taken into account)

V & M are the shear force and bending moment respectively at the section considered due to ultimate loads.

V_{cr} should be taken as not less than $0.1 \sqrt{f_{cu}} . b . d$

b = width of web. For flanged beam $b = b_w$

f_{sk} = characteristic strength of steel. As this is accepted as the manufacturers' guaranteed ultimate strength, f_{su} and f_{sk} are virtually same. If $V > V_{cr}$ reinforcement is required for the amount $(V - V_{cr})$ such that

$$\frac{A_{sv}}{s} \geqslant \frac{V - V_{cr}}{0.87 f_y . d} \quad \ldots \quad (7.30)$$

where A_{sv} is the cross sectional area of two legs of a link and s spacing (pitch) of links.

Equation 7.29 is also valid for lightweight aggregate concrete with appropriate values of ν_c specified for lightweight concrete.

7.12.2 CEB/FIP Model Code Method (S.I. units)

$$V_{cr} = \nu_c \times k\,(1 + 50\,\rho_L) . b . d \times \beta_1 \quad \ldots \quad (7.31)$$

where ν_c=allowable shear stress as given for different concrete (grades slightly different from CP 110 value).

$k = 1.6 - d$ (in M) but not less than 1.

$\rho_L = \dfrac{A_{st}}{b.d.} \not> 0.02$ b and d as in CP 110.

A_{st} = area of tensile reinforcement properly anchored.

$\beta_1 = \left(1 + \dfrac{M_0'}{M} \not> 2\right)$

M_0' = decompression moment related to extreme tensile fibre (in CP 110 this is related to centroid of steel).

This equation is applicable only to sections without shear reinforcements. Equation 7.31 is applicable to lightweight concrete with suitable values for ν_c.

7.12.3 ACI Method (units are Imperial in this case)

$$V_{cr} = \phi_1\, b . d . \nu_c \quad \ldots \quad (7.32)$$

ϕ_1 = capacity reduction factor given in Table 5.5.

For lightweight aggregate concrete V_{cr} in equation 7.32 should be reduced by a factor ϕ_0 which varies between 0.75 to 0.85

$$\nu_c = 0.6\sqrt{f_{cyl}} + \frac{V_d + \dfrac{(V . M_{cr})}{M}}{b . d}$$

or $3.5\sqrt{f_{cyl}} + 0.3 f_{bE} + \dfrac{V}{bd}$ whichever is less, but need not be taken less than $1.7\sqrt{f_{cyl}}$

f_{cyl} = cylinder strength of concrete in Imperial Units

ν_d = shear due to design dead load

$M_{cr} = (I/e_t) . (6\sqrt{f_{cyl}} + f_{bE} - f_{bd})$

f_{bE} = compressive stress in concrete due to prestress only after all losses at extreme fibre of a section at which tensile stresses are caused by applied load.

f_{yE} = as f_{bE} but at the centroid of section or junction of flange and web in case of T beam with centroid lying in flange.

f_{bd} = stress due to dead load at the point where f_{bE} is considered.

ν_{ev} = vertical component of effective prestressing force at the section.

b = width of beam and equal to b_w for flanged beam.

Reinforcement is needed if $V > V_{cr}$ for the amount $(V - V_{cr})$ such that

$$\frac{A_{sv}}{s} = \frac{V - V_{cr}}{f_y . d} \quad \ldots \quad (7.33)$$

7.12.4 Authors' recommendation

While the above equations (CP 110, Model Code and ACI) may perhaps give a perfectly safe result, they cannot, however, be relied upon to predict reliably the ultimate shear capacity.

The authors are of the opinion that a more realistic treatment of shear is possible when the following two distinct modes of shear failure are recognised and due account is taken of the characteristics of each mode in the design procedure. This has been discussed in 3.4. Recommendations given below apply equally to reinforced as well as fully or partially prestressed concrete.

(a) failure mode characterised by diagonal splitting under biaxial stress field in the web.

(b) failure mode characterised by bond imposed inclined cracks initiated by flexural tension.

In addition, of course, failure may be initiated by slip at anchorage either due to insufficient development length at bearing or close spacing of tendons.

7.12.4.1 The diagonal splitting phenomenon is normally predominant within the shear span/effective depth ratio of 2 and mostly in sections having thin web in that zone. A design method for this mode of failure can be adapted from *Design and Detailing of Concrete Structures for Fire Resistance*(9). If the reduction coefficient due to elevated temperature is deleted from the equations given in these 'guidelines' they can be used to determine the ultimate strength against diagonal splitting failure at normal temperature.

Thus, referring to Figure 7.23 (which has also been taken from these 'guidelines') the following relationship can be established:

$$V_{cr} = B.\frac{f_{cyl}}{\gamma_{mc}}.\sin\phi \qquad (7.34)$$

where B = effective bearing area to be taken as lesser of:

(i) $\frac{b.h}{3}$

(ii) $b \times$ width of bearing

b = minimum width in the cross-section within the shear span

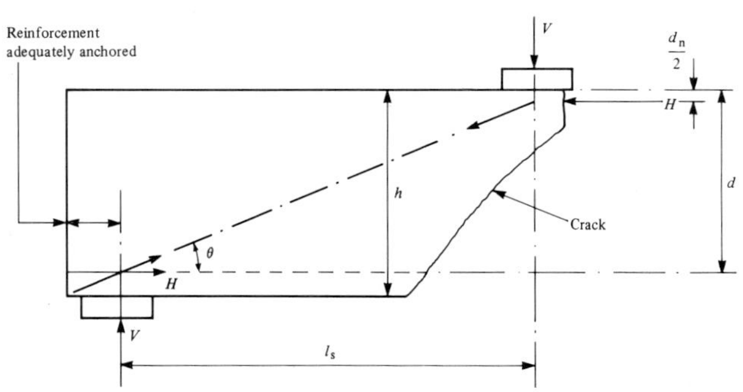

Figure 7.23 Force diagram for end blocks and brackets ($l_s/d \leqslant 2$)

f_{cyl} = characteristic cylinder strength of concrete and may be taken as 0.7 × characteristic cube strength (i.e. 0.7 f_{cu})

ϕ = angle as shown in Figure 7.23.

When V exceeds V_{cr} the cross-section needs to be increased. In addition sufficient horizontal and vertical steel should be provided such that:

$$A_{sv} \cdot \frac{f_y}{\gamma_{ms}} . \text{Cos} . \phi + A_{st} \cdot \frac{f_y}{\gamma_{ms}} . \text{Sin}\ \phi \geqslant V \qquad (7.35)$$

where A_{sv} = total area of vertical steel within the shear span.

A_{st} = area of horizontal steel in excess of what is required to resist H.

A minimum percentage of transverse steel, however, is essential within the shear span.

Also from the truss action, H = tensile force = $\dfrac{V}{\tan \phi}$ should be determined and adequately anchored reinforcement should be provided to ensure that:

$$A_{st} \cdot \frac{f_y}{\gamma_{ms}} \geqslant H \qquad (7.36)$$

All the above equations are also valid for lightweight concrete.

7.12.4.2 The range of shear span/depth ratio where combined effect of flexure and shear becomes critical is generally over 2. An empirical formula based on tests on reinforced, fully prestressed and partially prestressed beams with different types of shapes and sizes has been given in references 9 and 10. The formula is also applicable to lightweight aggregate concrete members and included in the CEB/FIP Manual on Lightweight Aggregate Concrete[11]. However for some lightweight concrete V_{cr} obtained by equation 7.37 may preferably be reduced by 10%. The formula, which can be applied to a wide variety of cross-sections and reinforcement ratio is reproduced below:

$$M = 0.875\, d\, l_s \left(0.342\, b_1 + 0.3 \frac{M_R}{d^2} \sqrt{\frac{z}{l_s}} \right) \sqrt[4]{\frac{16.66}{\phi f_{su}}} \not> M_R \qquad (7.37a)$$

$$V_{cr} = \frac{M}{l_s}$$

M = moment capacity in N mm

M_R = ultimate moment of resistance in pure flexure at the section (in N mm).

l_s = shear span (in mm)

f_{su} = steel strength (may be taken as f_{sk})

b_1 = effective width (mm) as shown in Figure 7.24

$$\phi = \frac{\text{area of tensile steel}}{\text{concrete area to effective depth}}$$

When M and M_R in lbf in. and z, l_s, b and d are in inch units and f_{su} in lbf/in^2

$$M = 0.875\, d . l_s \left(50\, b_1 + 0.3 \frac{M_R}{d^2} \sqrt{\frac{z}{l_s}} \right) K_1 \not> M_R \qquad (7.37b)$$

$$K_1 = \sqrt[4]{\frac{2419.20}{\phi f_{su}}}$$

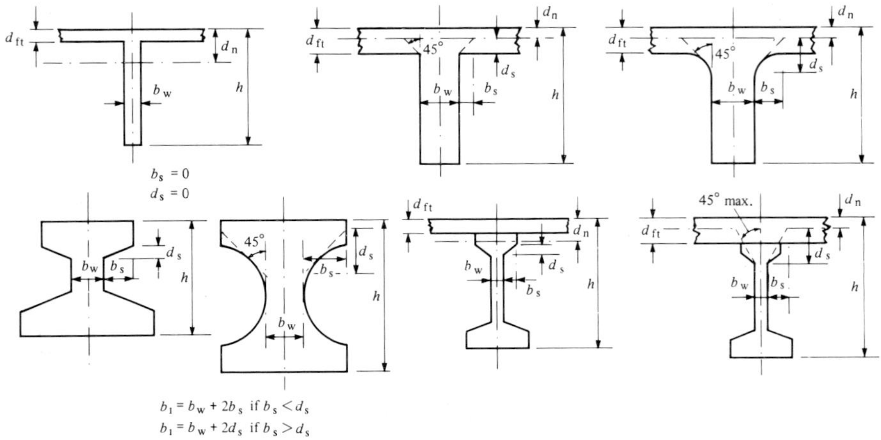

Figure 7.24 Effective width b_1 for different cross sections

When M and M_R in kgf/cm units and z, l_s, b and d are in cm and f_{su} in kgf/cm^2

$$M = 0.875\, d.\, l_s\left(3.5\, b_1 + 0.3\frac{M_R}{d^2}\sqrt{\frac{z}{l_s}}\right)K_1 \not> M_R \quad . \quad . \quad . \quad . \quad (7.37c)$$

$$K_1 = \sqrt[4]{\frac{169.34}{\phi f_{su}}}$$

Typical curves given by this expression are shown in Figure 7.25 taken from reference 10.

The effective width b_1 as shown on Figure 7.24 depends on width of the web geometry at the junction of flange and web, and on the depth of compression block.

When V_{cr} is less than V, reinforcement needs to be provided in accordance with the formula

$$\frac{A_{sv}}{s} = \frac{V - V_{cr}}{0.87 f_y\, .d}$$

The formulae 7.29 to 7.37 are intended primarily for design purposes; but since these are based on test results, they are included in this chapter.

7.12.5 CP 115 and 116 Method

In CP 115 and 116 the effect of shear is considered by calculating principal tensile stresses at critical sections due to prestress, shear and bending for working load as well as ultimate load and ensuring that the principal stresses are within the specified sets of limits. If the stresses calculated on the assumption of uncracked section exceed the permissible limit reinforcement is to be provided.

7.12.6 Torsion

7.12.6.1 CP 110 Method

As mentioned in section 7.9.2 the current approach in CP 110 for treatment of torsion is to assess the torsional shear stress at any section assuming plastic stress distribution in accordance with Nadai's Theory. If the calculated shear

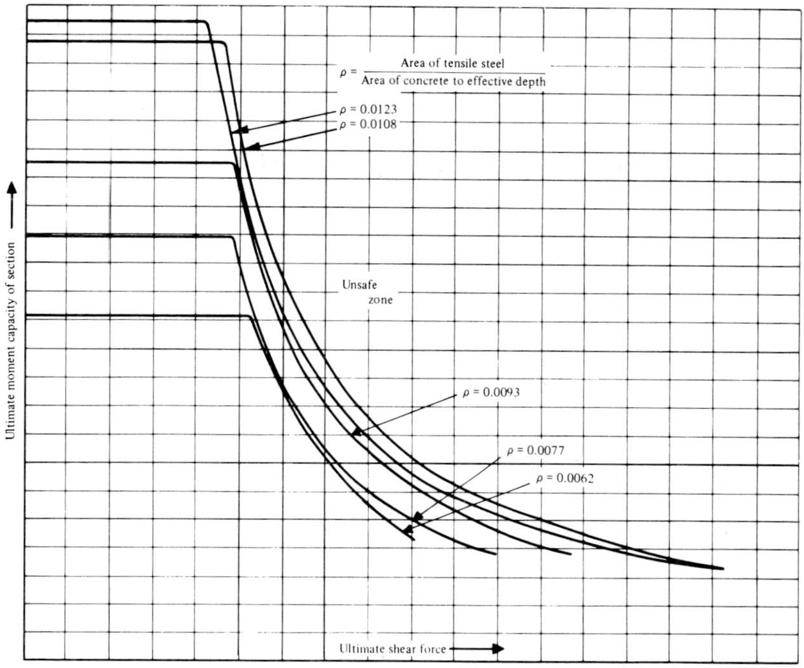

Figure 7.25 Typical moment capacity and shear force relationship of a particular section with different areas of steel (formula 7.37)

stress exceeds specified limiting values then reinforcement will be required. The reinforcement for torsion is additional to any requirements for shear or bending and consists of

(a) longitudinal reinforcement to be distributed round the inside perimeter of the links, at least four bars being placed at four corners of the link
(b) transverse reinforcement in the shape of closed links or stirrups.

The area of transverse steel can be obtained from the equation

$$\frac{A_{sv}}{s} \geqslant \frac{T_M}{0.8\,A_0 \times 0.87\,f_y} \qquad (7.38)$$

where T_M = Torsional moment
$A_0 = b_0 \times d_0$ (b_0 and d_0 as shown in Figure 7.26)
A_{sv} = area of legs of closed links.

The amount of longitudinal steel is obtained from the equation

$$A_L = (A_s + A_s') = \frac{A_{sv}}{s} \times (b_0 + d_0) \qquad (7.39)$$

7.12.6.2 CEB/FIP Model Code Method

The approach in the Model Code is to calculate the torsional resistance of the section on the basis of model space truss (see Figure 7.27 taken from reference 12) in which the forces are balanced by the reinforcement in tension and by

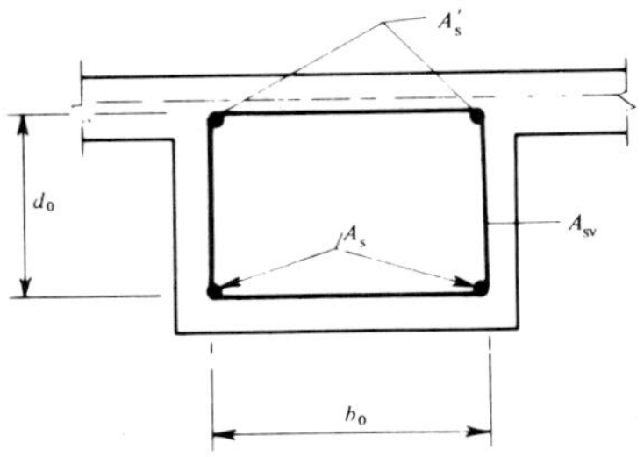

Figure 7.26 Parameters for obtaining reinforcement for torsion

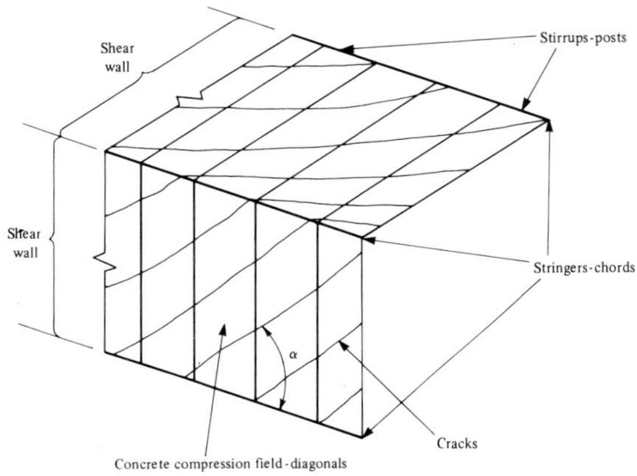

Figure 7.27 Space truss model

concrete struts in compression. The torsional strength of the section is governed by the following:

(1) T_{RC} – the resistance as a concrete strut in a space truss and is given by the equation

$$T_{RC} = 0.50\,\frac{f_{cyl}}{\gamma_m}\,.\,A_{ef}\,.\,h_{ef}\,.\,\text{Sin}\,2\alpha \quad \ldots\ldots\ldots\ldots \quad (7.40)$$

A_{ef} and h_{ef} are the area and wall thickness respectively of a hollow section real or notional as shown in Figure 7.28. The angle α is the inclination of the strut to centre line of the element and should normally be between 30° and 60°. To a certain extent the inclination can be controlled by choice of transverse and longitudinal steel.

(2) $T_{RS\nu}$ i.e. resistance of transverse steel and is given by

$$T_{RS\nu} = \frac{A_{sv}}{s}\,.\,2\,.\,A_{ef}\,.\,\frac{f_y}{\gamma_m}\,.\,\text{Cot}\,\alpha \quad \ldots\ldots\ldots\ldots \quad (7.41)$$

A_{sv} = cross sectional area of links in the effective wall.

(3) T'_{RSL} or resistance of longitudinal steel given by

$$T_{RSL} = \frac{A_L}{p_{ef}}\,.\,2\,.\,A_{ef}\,.\,\frac{f_y}{\gamma_m}\,.\,\tan\alpha \quad \ldots\ldots\ldots\ldots \quad (7.42)$$

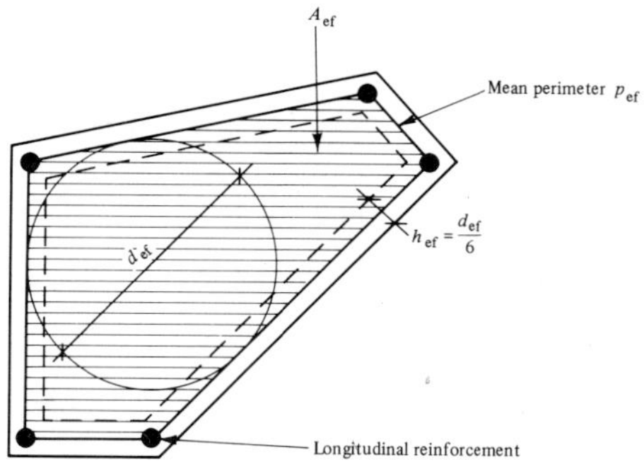

Figure 7.28 Equivalent hollow section

where A_L = sum of cross sectional areas of longitudinal steel. For tendons having an area of A_{st}, the area A_L to be taken in equation 7.42 is the lesser of the following values:

$$A_L = \frac{f_p}{f_y} . A_{st} \quad (7.43)$$

$$\text{or } A_L = \frac{f_{pe} + f_y}{f_y} . A_{st} \quad (7.44)$$

REFERENCES

1. ST. VENANT B. *Memoires des Savants Etranges.* Vol. 14, 1855.
2. NADAI, A. *Theory of flow and fracture of solids.* Vol. 1, New York, McGraw Hill, 1950. pp. 572.
3. KOLLBRUNNER, C. F. und BASLER, K. *Torsion in structures – An engineering approach,* Berlin, Springer, 1969. pp. 280.
4. ABELES, P. W. *The use of high-strength steel in ordinary reinforced and prestressed concrete beams.* I.A.B.S.E. 4th Congress, Cambridge, 25 August-5 September 1952. Preliminary Publication pp. 871–891.
5. BAKER, A. L. L. *The Ultimate-load theory applied to the design of reinforced and prestressed concrete frames.* London Concrete Publications Ltd., 1956. pp. 91.
6. JENSEN, V. P. *Ultimate strength of reinforced concrete beams as related to the plasticity ratio of concrete.* Urbana University of Illinois, June 1943. pp. 62. Bulletin 345.
7. HOGNESTAD, E. HANSON, N. W. and McHENRY, D. Concrete stress distribution in ultimate strength design. *Journal of the American Concrete Institute.* Vol. 27. December, 1955. No. 4. pp. 447–454.
8. KAAR, P. H., HANSON, N. W. and CAPELL, H. T. Stress-strain characteristics of high strength concrete. *Research and Development Bulletin*, Portland Cement Association.
9. THE CONCRETE SOCIETY/INSTITUTION OF STRUCTURAL ENGINEERS. *Design and detailing of concrete structures for fire resistance.* Interim guidance by a Joint Committee of the Institution of Structural Engineers and the Concrete Society, May 1978. pp. 60.
10. BOBROWSKI, J. und BARDHAN-ROY, B. K. A method of calculating the ultimate strength of reinforced and prestressed concrete beams in combined flexure and shear. *The Structural Engineer,* Vol. 47 May, 1969, No. 5. pp. 197–209.
11. CEB/FIP. *CEB/FIP Manual of Lightweight Aggregate Concrete – Design and Technology.* The Construction Press, 1977.
12. COMITÉ EURO INTERNATIONAL DU BETON. *Bulletin D'Information No. 126 – Shear and torsion.* June 1978.

CHAPTER 8

FLEXURAL DESIGN OF PRESTRESSED MEMBERS

8.1 General design considerations

8.1.1 General remarks

Much time can be saved and repetition of work avoided when the design is prepared in the following sequence.

It is obvious that the accuracy and quality of a design are determined by the accuracy of the basic data; care should be taken to see that these conform as closely as possible to the actual conditions of manufacture and construction, and that any reasonable divergence between the actual and assumed conditions can lead only to a safer structure (that is, the design assumptions should always be on the safe side). This accuracy is in no way improved by excessive arithmetical precision. The loading and bending moments used in design are often conventional; in assessing the losses it is therefore desirable to consider whether a more probable distribution of load would lead to increased losses. The permissible stresses should be selected in accordance with the principles described in Chapter 5; where any doubt exists as to actual conditions conservative values should be chosen. The same principle should be applied when estimating the losses; they should give the maximum probable prestressing force at transfer and the minimum probable at working load. It is also necessary on occasion to investigate the stresses when minimum losses are assumed to occur.

8.1.2 Design concept

As stated in Chapter 5, there are three main considerations for design, namely:

(1) The member must have sufficient strength to support, without collapse, an 'ultimate load', which should represent the greatest load that might occur under the most unfavourable of conditions – possible, though highly improbable, loading. All possible types of load action should also be considered, including fatigue (i.e. the cumulative effect of all repetitive loadings throughout the expected life of the structure); earthquakes (when allowances for the probable occurrence and probable degree of an earthquake should be made); impact (when sufficient flexibility should be provided); and continuous vibration (when sufficient stiffness and damping shall be provided and resonance must be avoided).

It is thus apparent that in many cases the expected lifetime of the structure has an important effect on the design, and must therefore be known with some accuracy.

(2) The deformation under service loading should not exceed some definite and pre-determined value.

(3) If visible cracking is permitted at all, the width of the cracks should be

limited to such magnitudes as not to impair the performance and durability of the structure under service conditions. Condition (1) relates to the behaviour at ultimate load, but conditions (2) and (3) involve an investigation of the elastic behaviour of the member. It is essential to base the computation of the prestressing force on these two latter conditions, in order to determine the load causing zero compression at the tensile face, at which a change of stress from compressive to tensile occurs.

In the early days, the design of prestressed concrete was based solely on elastic conditions; the consideration of ultimate-load conditions was first introduced in about 1940. In the first edition of this book, the design method advocated was based on elastic conditions, with a check on ultimate-load conditions, and this combined method of design was shown to be quite suitable. In the following, it is now advocated that the designs should be based primarily on the collapse load, with some of the features determined from elastic conditions, and a check on some of the permissible stresses to ensure that particular limit-states are not exceeded.

It should, however, be noted that the relative importance of these separate aspects of the design may vary considerably according to the type of the member; this has been discussed in Chapter 3. The three classes of prestressed concrete in CP 110 have already been discussed in Chapter 5. Class 1 is characterized by the requirement that no tensile stresses should occur at service load; in this case condition (3) does not apply, and excessive deformations would not normally be expected, so that the investigation of condition (2) can often be dispensed with except perhaps for camber. In the case of shallow members, however the deflection may be an important consideration (even if the section is uncracked), particularly when the loading is sustained. Although avoidance of collapse as a result of an overload should be the main basis of design, the possibility of collapse at transfer need not be investigated provided the elastic conditions at this stage are satisfactory. Neither is it necessary to compute the compressive stresses at service loading, when the compressive zone complies with ultimate-load requirements.

Class 2 is defined by the condition that no visible cracks are permitted, though microcracks may occur at service load; hence limited tensile stresses (not more than 75 per cent of the modulus of rupture of the reinforced member) should be permitted at working load but only if the development of shrinkage cracks before prestressing has been avoided. In principle the method of design is much the same as for Class 1, except that the deflection may sometimes become a limiting factor; this is partly because of the non-linear nature of the load-deflection relationship, which becomes curved after microcracks occur.

Class 3 relates to structures in which visible cracks are permitted at service load, and the importance of conditions (2) and (3) then increases markedly. While some research has been made, the data available are not yet sufficient to enable reliable design rules to be formulated. In CP 110 the simplified design principle for this class of structure is based on permissible 'fictitious tensile stresses'. It is recommended in this book that Class 3 should be replaced where possible by a combination of Classes 1 and 3, in which a member is designed as of Class 1 at frequent or normal service load and Class 3 at occasional or abnormal service load, the latter being of rare occurrence. In this case the design basis for conditions at 'normal' service load is similar to that described for

Class 1 members, while the ultimate load is obtained by applying a sufficient factor of safety to the 'total' service load with abnormal loading. It is not usually necessary to investigate the deformations and maximum crack widths due to the 'abnormal' service load since this should be of short duration and would occur only temporarily. A realistic assessment of the proportion of the total service load which will be permanent or frequently occurring is, of course, essential. In the CEB/FIP Model Code some factors have been introduced to assess the permanent, semipermanent or quasipermanent service loading.

In every case, it is important to investigate the camber, or maximum deformation after transfer, before the added loading counteracts the effect of the prestress. This is particularly so for members of Class 1, especially when the stress in the concrete at transfer is high and the strength is relatively low. The camber may increase considerably with time, when the period between transfer and the application of the load is long, and a large part of the creep should be allowed for. With the smaller prestressing forces which are typical of Classes 2 and 3, the camber at transfer is less likely to be critical.

8.2 Various methods of design

As noted previously, the original concept of prestressing was limited to constructions in which only compressive stresses were present; Freyssinet once said[(1)] that any intermediate solution between 'full' prestressing and reinforced concrete was worse than either, though he later modified this opinion[(2)] with the statement that limited tensile stresses of 50 kgf/cm^2 (720 lbf/in^2; 5 N/mm^2) due to occasional rare overloads would be harmless. Nevertheless, most text books have been based on the concepts of 'full' prestressing (Class 1) and elastic design.

The loading at which the resultant stress at the tensile face of the concrete is zero was widely termed the 'transformation load', as a result of this philosophy; it is also, and more aptly, termed the 'decompression load' (or zero stress load). The loading causing zero stress is undoubtedly important; but so also are the conditions at ultimate load, and in all cases it is necessary to know both of these conditions.

The introduction of the concepts of limit-states, and of the three classes of members, appears to involve the investigation of many more conditions than before particularly for reinforced concrete, though not so far for prestressed concrete. The design procedure advocated in this book gives priority to ultimate-load requirements; the design should start with this investigation, and the additional calculations needed to satisfy all the requirements relevant to a particular Class are described in the next section. It is sometimes suggested that all limit-states should always be investigated; the authors believe that this is rarely necessary, but it is essential to determine as accurately as possible, the decompression (zero-stress) load. This involves a careful assessment of the maximum and minimum losses, to determine either the loading at which no cracks will occur (as with Class 1) or that at which they close completely.

Lin has presented[(3)] a design method based on the concept of 'balanced loading'; that is, that at some definite loading the combined deflection due to this and the equally definite prestressing force is nearly zero. This selected loading stage may be either the dead weight alone, or a specified percentage of the added service loading may be included. Strictly speaking, the case considered is that in which uniform compressive stresses are present at the critical section of the member; it does not automatically lead to zero deflection, since the tendon profile will not usually correspond exactly with that of the bending moment diagram, even when the tendons are bent up (also termed 'deflected' or 'draped'), and there will also be variations in the elastic modulus of the concrete.

This method has been widely used in the USA; in the hands of experienced and knowledgeable designers, it leads to partially-prestressed members with quite satisfactory deformation and cracking characteristics. Ultimate-load conditions should however be checked, although the computation of the elastic stresses at various loadings is not necessary.

Very rarely can all four of the permissible stresses in the concrete be reached exactly at transfer and service load and if a fifth condition – that of Lin's 'balanced load' – is introduced it becomes impossible. This is illustrated by Nilson(4); he specified the balanced load as 25 per cent of the live load of 500 lbf/ft (745 kgf/m; 7320 N/m) on a simply-supported beam spanning 30 ft (9.15 m). He also specified permissible tensile stresses of 165 and 380 lbf/in^2 (11.55 and 26.6 kgf/cm^2; 1.14 and 2.62 N/mm^2) at transfer and service load respectively, and a permissible compressive stress of 1800 lbf/in^2 (126 kgf/cm^2; 12.4 N/mm^2) for both conditions together with a loss ratio of 0.20 to obtain the effective prestress. For a beam of rectangular cross-section, as assumed by Nilson, the stress distributions are given in Figure 8.1; it is apparent that only three of the five conditions (permissible stresses at working load, and the balanced load) are fulfilled. This example was re-designed by the first-named author(5), starting with the ultimate-load requirements, and employing the combination of Classes 1 and 3 recommended in the foregoing, in that the beam conforms to Class 1 for a 'normal' load of 50 per cent of the total superimposed loading and to Class 3 for the entire 'abnormal' superimposed load (Figure 8.2). The permissible compressive stress is the same as that used by Nilson, but the tensile stress at transfer is larger, and obviously the nominal tensile stress due to the 'abnormal' service load has the much higher value of

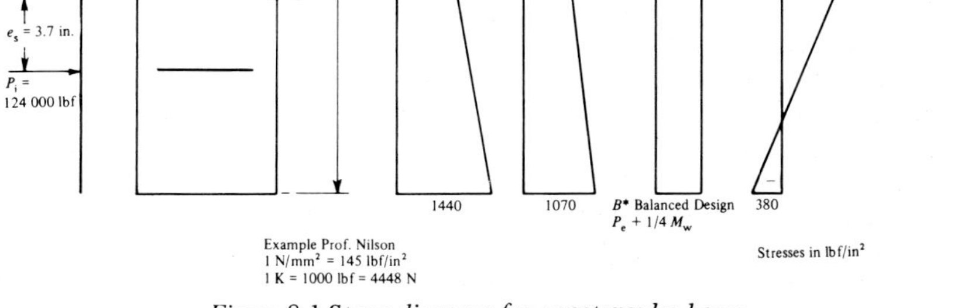

Figure 8.1 Stress diagrams for a rectangular beam

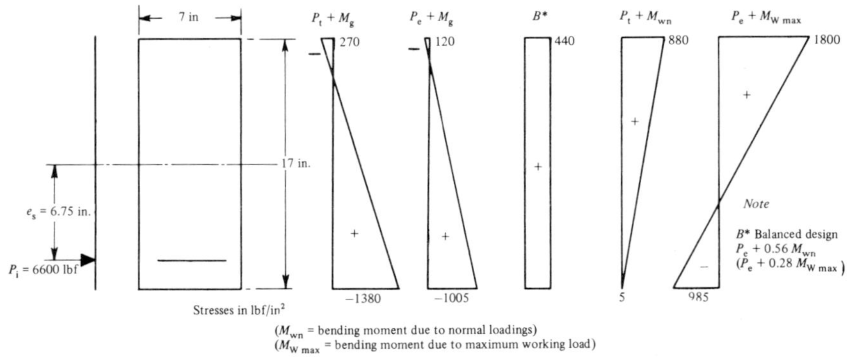

Figure 8.2 Stress diagrams for redesign of the section in Figure 8.1

996 lbf/in² (70 kgf/cm²; 6.9 N/mm²). Fine visible cracks would be expected at this level of stress, but these would close completely at the 'normal' superimposed loading, after 'decompression', when there is a residual compressive stress of 5 lbf/in² (0.35 kgf/cm²; 0.03 N/mm²). The ultimate-load conditions are complied with, even though the areas of concrete and reinforcement are smaller, and the initial prestressing force is reduced to 53 per cent of Nilson's value. The tendons, 3 strands of ½ in. (12.7 mm) diameter with a cover of 1½ in. (38 mm) must be bent up or 'debonded' near the ends, to obtain suitable stress distributions at the ends of the beam ('de-bonding', or 'sleeving', is a technique by which the bond of a pre-tensioned tendon is broken). The 'balanced loading' condition occurs at 29 per cent of the full service load.

A flow diagram for use with this particular design method is given in Figure 8.3[(6)]; in this case, again, the primary emphasis is placed on meeting the conditions at ultimate load; the cross-section (compressive zone and lever arm) and area of steel being selected on this basis. The effective prestressing force is then computed to meet the condition that no tensile stress is to occur at the 'normal' service load.

However, since it is based on the method used in Figure 8.2, the flow diagram in Figure 8.3 includes no provision for a computation of the tensile stress, maximum crack width, and deflection, under 'abnormal' loading, since all of these as explained earlier are of only short duration.

8.3 Outline of design procedure

From the previous section it is apparent that a logical design procedure can be evolved, in which the major or primary requirements are first satisfied, and those which are less likely to be critical are considered subsequently. The preparation of a complete design for a prestressed member often involves a considerable amount of work, and it is obviously sensible to adopt methods which minimize the need for repetitive calculations. The recommended sequence is as follows.

8.3.1 Selection of trial section

Since the selection of an inappropriate trial section can result in the waste of much time and work, it should be chosen with proper care and forethought. The depth may be restricted in advance to some minimum value, or selected

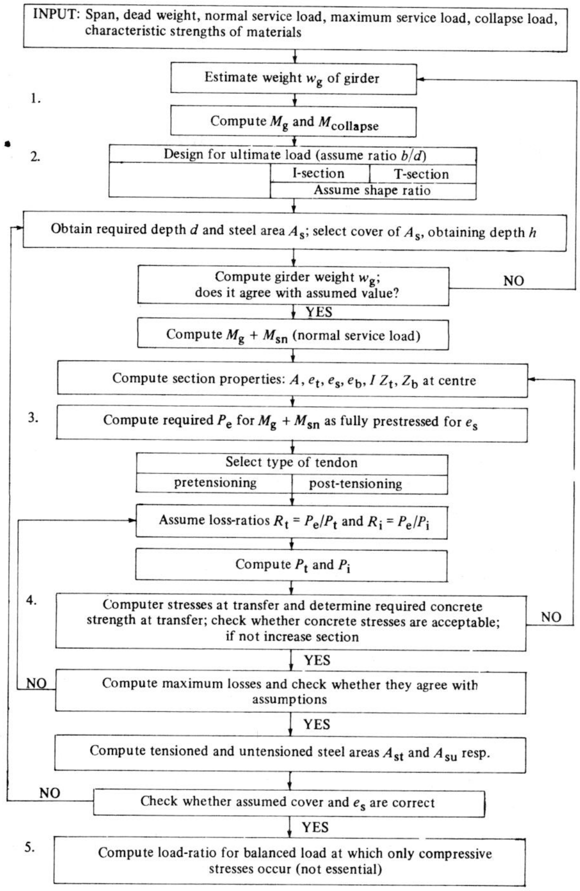

Figure 8.3 Flow chart for design shown in Figure 8.2

empirically from typical span/depth ratios, or from considerations of deflection and stress, and an approximate estimate of the area is needed to obtain the bending moment due to the weight of the beam.

The shape of the section is best determined in two parts. The dimensions of the compression flange are based to suit the conditions at ultimate load, while those of the bottom flange are derived from the appropriate conditions at service load; these depend on the Class of the member (see section 8.1) and the permissible compressive stress (which is related to the strength) at transfer.

8.3.2 Positions and area of steel

The total area of steel required and its position is determined to suit the conditions at ultimate load, in the manner which is usual for reinforced concrete, and using a similar arrangement of steel and lever arm. The effective prestressing force required, is derived from the conditions at service load depending on the class of the structure. This can be obtained either by tensioning part of the steel to the permissible limit or all steel to a lesser stress.

8.3.3 Secondary checks

The remaining parts of the design are secondary only in that they are less likely to prove critical; many of them are not to be considered as unimportant. These include the computation of stresses and their comparison with permissible values; checks on deflections and camber; the selection of a suitable cable profile; a consideration of the anchorage and end-block requirements; the investigation of the resistance to shearing; and, if necessary, a check on the lateral stability of the member.

8.3.4 Comments

The general summary of the procedure given in the foregoing is considered in greater detail in the following sections and in subsequent chapters; a number of illustrated examples appear in Chapter 9. It is restated that though some of the section dimensions are determined by ultimate-load criteria, others depend on conditions at service load; hence, while the limit states are considered separately in the text which follows, the design of a member should be based, not on one or the other, but on a combination of both.

When the cross-section is fixed in advance (in order to use an existing mould, for example), a check is made to see whether it is satisfactory. If the ultimate compressive zone is smaller than that required, an increase in size by means of concrete cast in place may be possible. Similarly, if the compressive stresses in the tensile zone at transfer are excessive, the prestressing may need to be carried out in two or more stages; and if the tensile stresses at service loading are too high, a larger prestressing force will be needed.

8.4 Ultimate (collapse) - load design

The object of ultimate-load design is to ensure that the member will not fail before the bending moment exceeds some predetermined multiple of that at working load; that is to say, the moment of resistance M_R at some predetermined multiple of the service load (which may be 1 if the maximum possible load is known) is given by

$$M_R \geqslant \gamma_{(f,m)} M_w = M_{ult}$$

in which γ_m is the required factor of safety (see Chapter 5). The greatest moment of resistance is also given by $M_R = F_{s\,max}\, z_{max}$ or $F_{c\,max}\, z_{max}$ depending on whether the steel or the concrete is the weaker part of the member; the smaller value must be taken. As stated in Chapter 3, the design should be such that the steel is the weaker part of a prestressed member; the large increase in strain which occurs in the steel as the stress in the steel approaches the failure stress causes the neutral axis to rise, until the area of concrete in compression becomes so small that the stress in the concrete reaches its ultimate value. Consequently, for an under-reinforced section the maximum force of the steel based on the maximum primary cause of the compressive failure is the extension of the steel; the value of $F_{s\,max}$ is therefore the critical value and hence $F_{c\,max} = F_{s\,max}$. If $F_{s\,max}$ is made greater (over reinforced section), by the provision of a larger area of steel, then the depth of the compressive zone must increase, until a limit is reached at which failure of the concrete occurs before the strain in the steel becomes large. Hence, in order to obtain $F_{c\,max}$ it is necessary to calculate the values of $F_{s\,max}$ and z_{max}.

Lever arm – As z_{max} depends on the shape and depth of the compressive stress block at failure, the values of z_{max} and $F_{c\,max}$ are considered together.

Maximum compression – The exact distribution of compressive stress which occurs at failure is difficult to predict, and several simplified methods of determining the area and centroid of the stress block are considered here.

In the first, the approximation is made that the distribution of stress at ultimate load is as shown in Figure 8.4. A uniform compressive stress f_{cm} is assumed to exist throughout a certain depth d_n of the beam; tests have shown that f_{cm} may be taken as equal to the prism strength f_{prism} (that is, between 60 and 80 per cent of the cube strength). Hence

$$M_R = f_{prism}\, b d_n \left(d - \frac{d_n}{2}\right) \qquad (8.1)$$

For an under-reinforced beam, it is safe to assume that d_n will not exceed $0.5\,d$; z_{max} will therefore have a least value of $0.75\,d$, and for a rectangular section the greatest moment of resistance is given by

$$M_R = f_{prism}\, b\, 0.5\, d\, 0.75\, d$$

and if $f_{prism} = 0.6\, f_{cu}$

$$\boxed{M_R = 0.225\, bd^2 f_{cu}} \qquad (8.2)$$

This can be compared with the required moment M_{ult} and the values of b and d adjusted to make $M_R = M_{ult}$

In the case of a flanged section M_R is given by

$$\boxed{M_R = 0.6 f_{cu}\, b\, d_n\, (d - \tfrac{1}{2} d_n)} \qquad (8.3)$$

The maximum d_n when the contribution of web to the compressive strength is ignored is d_f, i.e. the depth of the flange.

The second method, given in CP 115, applies to rectangular sections only. In this case the mean compressive stress f_{cm} is assumed to be $0.4\, f_{cu}$ (Figure 8.5)

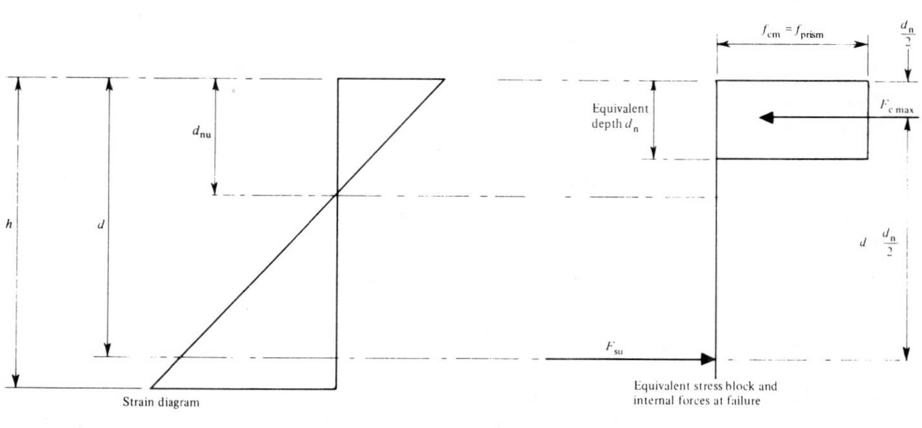

Figure 8.4 Conditions at ultimate load (simplified stress-block)

and the lever arm z_{max} is given by $z_{max} = d - 0.4\,d_{nu}$. Since the strain in the steel is the primary cause of failure the value of d_n is related to the percentage/strength ratio, as given in Table 8.1. It is therefore necessary first to calculate or estimate the percentage/strength ratio, and the appropriate value of d_n is then selected from Table 8.1 and M_R obtained from the expression:

$$M_R = 0.4\,f_{cu}\,bd_{nu}\,(d - 0.4\,d_{nu}) \quad (8.4)$$

It is shown in the following that these two methods give similar results when the safe mean compressive stress in the first diagram is assumed to be $0.5\,f_{cu}$. If $F_{s\,max}$ is the same in cases (a) and (b) (Figure 8.6), then $F_{c\,max}$ must also be the same, and for equal moments of resistance it follows that

$$F_{c\,max} = f_{cm\,1}\,d_n\,b = f_{cm\,2}\,d_{nu}\,b = 0.4\,f_{cu}\,d_{nu}\,b \quad (8.5)$$

(where $f_{cm\,2} = 0.4\,f_{cu}$)

and

$$M_R = F_{c\,max}\left(d - \frac{d_n}{2}\right) = F_{c\,max}\,(d - 0.4\,d_{nu}) \quad (8.6)$$

From equations 8.5 and 8.6 it can be solved that $f_{cm\,1} = 0.5\,f_{cu}$

Table 8.1 Coefficients for ultimate-load design (CP 115 and CP 110)

$\frac{A_{st}\,f_{su}}{bd\,f_{cu}}$	Pre-tensioned steel: $k_u = f_{s\,max}/f_{su}^*$	Pre-tensioned steel: d_n/d CP 115	Pre-tensioned steel: d_n/d CP 110	Post-tensioned steel with effective bond: $k_u = f_{s\,max}/f_{su}^*$	Post-tensioned steel with effective bond: d_n/d CP 115	Post-tensioned steel with effective bond: d_n/d CP 110	
0.025	1.0	0.06	0.054	1.0	0.06	0.054	*In CP 110 f_{su} is the characteristic strength divided by the material safety factor (1.15)
0.05	1.0	0.125	0.109	1.0	0.125	0.109	
0.10	1.0	0.25	0.217	1.0	0.25	0.217	
0.15	1.0	0.375	0.326	1.0	0.375	0.326	
0.20	1.0	0.50	0.435	0.95	0.475	0.414	
0.25	1.0	0.625	0.542	0.90	0.56	0.488	
0.30	1.0	0.75	0.655	0.85	0.64	0.558	
0.40	0.9	0.90	0.783	0.75	0.75	0.653	

Post-tensioned steel without bond

$\frac{A_{st}\,f_{pe}}{bd\,f_{cu}}$	$k_u = f_{s\,max}/f_{pe}$ CP 115	CP 110 $l/d = 30$	CP 110 $l/d = 20$	CP 110 $l/d = 10$	d_{nu}/d CP 115	CP 110 $l/d = 30$	CP 110 $l/d = 20$	CP 110 $l/d = 10$
0.025	1.7	1.23	1.34	1.45	0.11	0.10	0.10	0.10
0.05	1.6	1.21	1.32	1.45	0.20	0.16	0.16	0.18
0.10	1.4	1.18	1.26	1.45	0.35	0.30	0.32	0.36
0.15	1.3	1.14	1.20	1.36	0.49	0.44	0.46	0.52
0.20	1.2	1.11	1.16	1.27	0.60	0.56	0.58	0.64

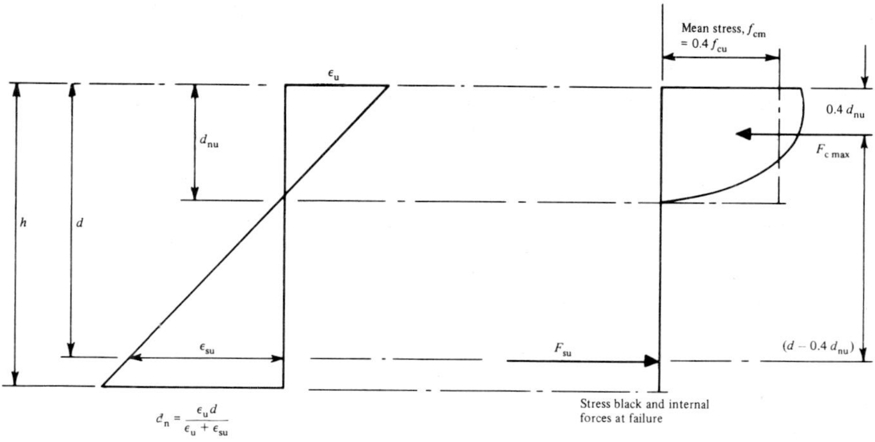

Figure 8.5 Conditions at ultimate load (CP 115)

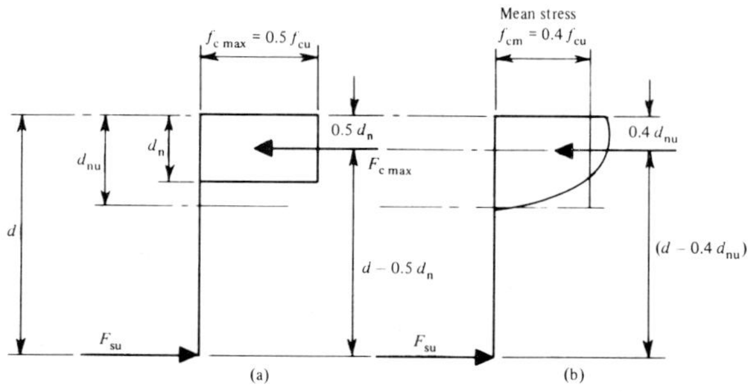

Figure 8.6 Comparison of stress blocks in Figures 8.4 and 8.5

Hence from equation 8.2: $d - 0.5\, d_n = d - 0.4\, d_{nu}$. Therefore $d_n = 0.8\, d_{nu}$. Inserting this value in equation 8.1 it follows that $0.8\, f_{cm1}\, d_{nu} = 0.4\, f_{cu}\, d_{nu}$ and $f_{cm1} = 0.5\, f_{cu}$. Since d_{nu} approximates to the true position of the neutral axis, as determined from tests, d_n may be considered as an 'equivalent' depth which is four-fifths of the true depth.

As the recommendations given in CP 115 relate only to rectangular sections, this equivalence of the two methods can be used to derive a logical extension of the recommendations of CP 115 to include the case of a flanged section. The percentage/strength ratio is determined as for a rectangular section with b equal to b_t and the corresponding depth of the neutral axis obtained from Table 8.1. If the contribution of the web is ignored the flange thickness should be four-fifths of this depth; the mean stress is assumed to be $0.5\, f_{cu}$ and the lever arm is $d - \frac{d_f}{2}$.

For flanged beams Bobrowski and Bardhan-Roy(7) suggest a limiting uniform compressive stress of $0.625 f_{cu}$ based on test results. M_R for under-reinforced flanged beam is then given by:

$$M_R = 0.625 f_{cu}\, b_t d_n \left(d - \frac{d_n}{2}\right) \qquad (8.7)$$

when d_n is within the flange thickness.
If d_n exceeds d_f and the web contribution to compressive stress is not ignored then

$$M_R = 0.625 f_{cu}\, b_t d_f \left(d - \frac{d_f}{2}\right) + 0.625 f_{cu}\, b_w (d_n - d_f) \left(d - d_f - \frac{d_n - d_f}{2}\right) \qquad (8.8)$$

d_n is obtained by balancing the compressive force ($F_{c\,max}$) with $F_{s\,max}$ the limiting value of d_n, i.e. d_{nu} is $d/2$. Partial safety factor of materials (γ_m) has not, however, been taken into account in equations 8.7 and 8.8.

The ACI Building Code (ACI 318) permits a rectangular stress block with a concrete stress of 0.85 f_{cyl} assumed to be uniformly distributed over a depth of $\beta_1 d_{nu}$ (Figure 8.13) where d_{nu} is the depth of actual neutral axis corresponding to strain distribution with a maximum compressive strain of 0.003. The value of β_1 is 0.85 for f_{cyl} up to 4000 lbf/in^2 (280 kgf/cm^2; 28 N/mm^2) reducing continuously at a rate of 0.05 for each 1000 lbf/in^2. Assuming f_{cyl} as $0.75 \times f_{cu}$ the ACI value of limiting uniform compressive stress would amount to $0.85 \times 0.75\, f_{cu}$, i.e. 0.63 f_{cu}. This compares very well with the value suggested above.

Two methods are given in CP 110; they allow either a mean compressive stress of $0.4 f_{cu}$ to be considered over the full depth of the compressive zone (as in CP 115), or the use of stress block with the profile shown in Figure 8.7. In both cases the strain at the compressive face, at failure, is assumed to be 0.0035. Similarly the stress in the steel is obtained from the appropriate stress-strain diagram (see Figures 8.8 to 8.10). Design charts based on these diagrams are given in CP 110 for the case of rectangular beams.

For rectangular beams, the value of M_R may be obtained from the expression

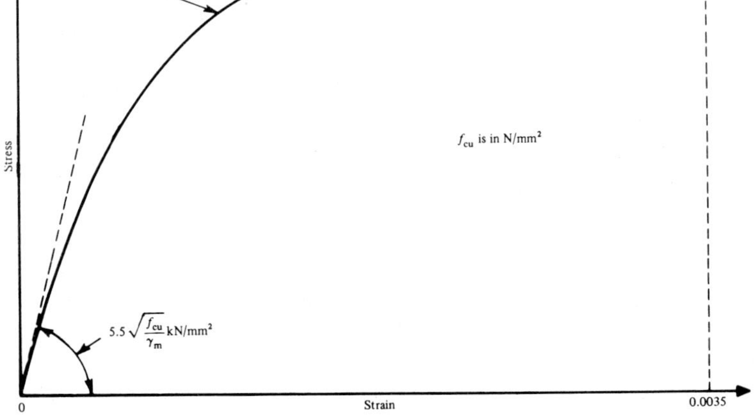

Figure 8.7 Short-term design stress–strain curve for normal weight concrete (CP 110)

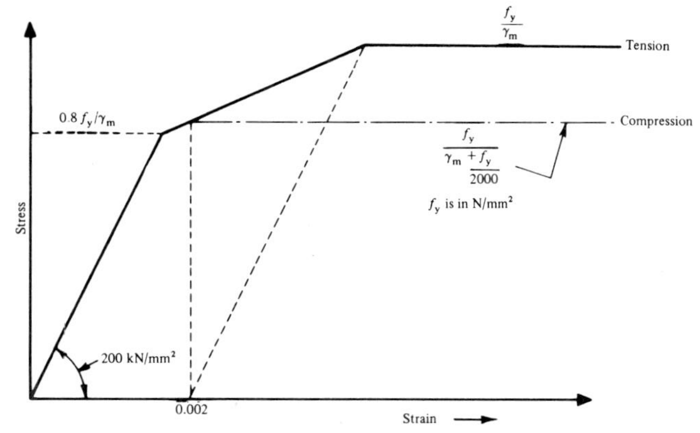

Figure 8.8 Short-term design stress–strain curve for reinforcement (CP 110)

$$M_R = f_s A_{st} (d - 0.5\, d_n) \quad (8.9)$$

or

$$0.15\, bd^2 f_{cu} \text{ when } d_{nu} > d/2 \quad (8.9a)$$

in which

- f_s is the stress in the tendons at failure (obtained from Table 8.1) (Maximum = the characteristic strength divided by material safety factor)
- d_n is the neutral axis depth (obtained from Table 8.1)
- d is the effective depth, and
- A_{st} is the area of the prestressing tendons.
- f_{cu} is the characteristic concrete strength (cube)

In the CEB/FIP Model Code computation of moment of resistance (M_R) is based on the idealised stress-strain diagrams for concrete and steel as shown in Figures 8.11 and 8.12 respectively. The strain distribution is linear with the maximum for concrete (in compression) being 0.0035 and for steel (in tension) 0.010. A simplified assumption of a rectangular distribution of the compressive stress equal to $0.85\, f_{cyl}/\gamma_m$ can be made for 0.8 × depth of the neutral axis obtained by strain distribution in accordance with above. (If the width in compression zone reduces towards the extreme fibre in compression then $0.8\, f_{cyl}/\gamma_m$ should be taken).

Figure 8.13 (a), (b), and (c) illustrate the simplified method of calculation of moment of resistance (M_R) in accordance with CP 110, ACI Building Code and the CEB/FIP Model Code respectively.

The limiting strain of 0.010 for steel in the Model Code is in the opinion of the authors, unduly restrictive particularly in case of T beams and flanged beams. It should also be noted that in the Model Code the characteristic strength of tendons is defined as 5% fractile of the proof-strength and not the ultimate strength as in CP 110.

For lightweight aggregate concrete no difference has been made in CP 110 and ACI Building Code in the assessment of M_R (the moment of resistance).

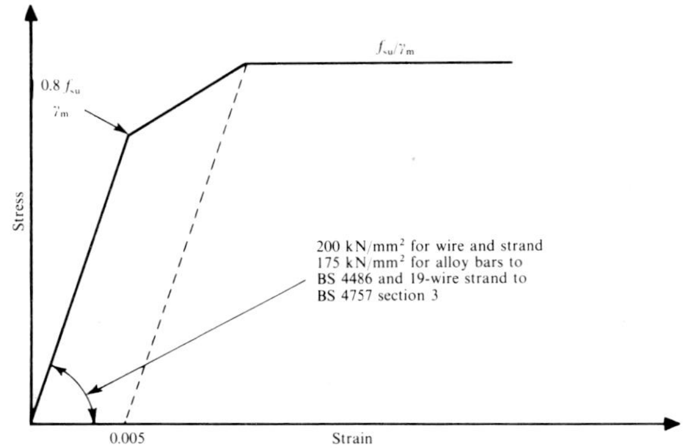

Figure 8.9 Short-term design stress–strain curve for normal and low relaxation steel (CP 110)

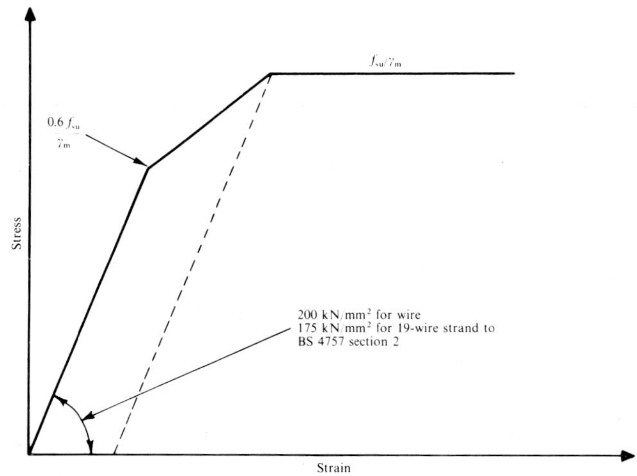

Figure 8.10 Design stress–strain curve for 'as drawn' wire and 'as spun' strand (CP 110)

Kaar et al[8] however recommend that in ACI code β_1 should be 0.65 for all grades of lightweight aggregate concrete. In the Model Code on the other hand the suggestion is to reduce the maximum value of the ordinate of the stress-strain diagram in 8.11 to 95% of what has been shown for normal weight concrete. For rectangular distribution of compressive stress, the maximum value of $0.80\, f_{cyl}/\gamma_m$ (for a section with constant width or width increasing towards extreme compression fibre only – otherwise $0.75\, f_{cyl}/\gamma_m$) is to be taken and for a depth of $0.75\, d_n$ only. Tests with lightweight aggregate like 'Lytag' 'solite' and similar show that the Model Code method of calculation for moment of resistance is very much on the conservative side.

The method, proposed by Hognestad, Hanson and McHenry as described in Chapter 7, may be applied as follows. With the notation given in Figure 8.14

$$M_R = k_1\, k_3\, f_{cyl}\, b\, d_n\, (d - k_2\, d_n) \qquad (8.10)$$

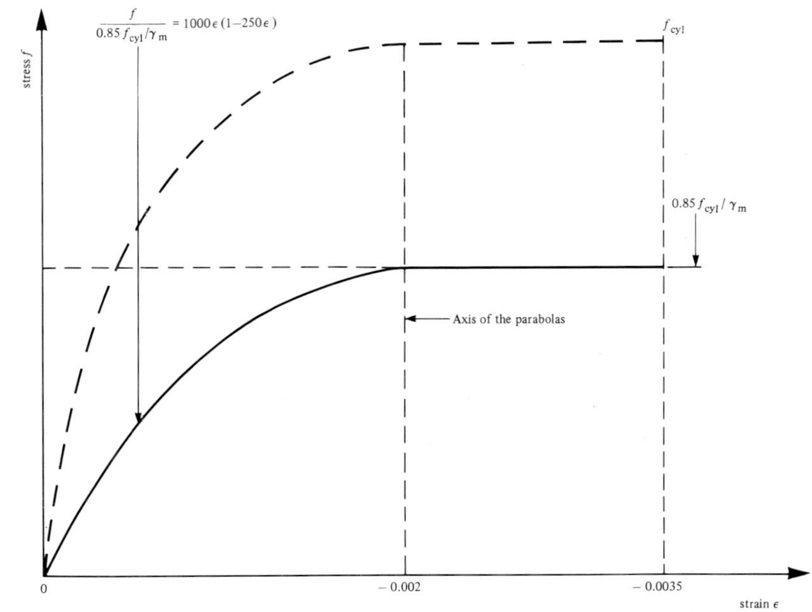

Figure 8.11 Design stress–strain diagram for normal weight concrete (to CEB/FIP Model Code)

in which f_{cyl} is the cylinder strength. The value of d_n is obtained from the ratio $\frac{\epsilon_u}{\epsilon_{su} + \epsilon_u}$; ϵ_{su} in turn is obtained from a stress–strain diagram for the steel and k_2, and $k_1 k_3$ from Table 7.1. The cylinder strength may be related to the cube strength by means of Table 2.1. Instead of a constant depth of centroid as $0.4\, d_{nu}$ or $0.5\, d_{nu}$ the values of k_2 given in the table vary in accordance with the test results; they are therefore based on practice rather than theory. For light-weight aggregate concrete, the value of k_2, according to Kaar et al[(8)], varies usually between 0.34 and 0.37 for all grades above 4000 lbf/in^2 (281 kgf/cm^2; 27.6 N/mm^2) cylinder strength. For normal weight concrete such values of k_2 are obtained for grades above 8000 lbf/in^2 cylinder strength.

Maximum tension – The tensile force $F_{s\,max}$ required is calculated from

$$F_{s\,max} = \frac{M_{ult}}{z_{max}} \qquad (8.11)$$

The maximum resistance which the steel can actually provide is given, in principle, by the expression $M_R = F_{s\,max}\, z_{max} = k_u A_s f_{su}\, z_{max}$, in which A_s is the area of the entire tensile reinforcement, both prestressed and non-stressed; f_{su} is the ultimate tensile strength of the steel and k_u is a factor indicating the effectiveness of the bond. When the required value of $F_{s\,max}$ is known, A_s can be determined. If the area A_{st} of tensioned steel is less than A_s, then an area A_{sn} of non-tensioned reinforcement must be provided such that

$$k_u A_s = k_{ut} A_{st} + k_{un} A_{sn}$$

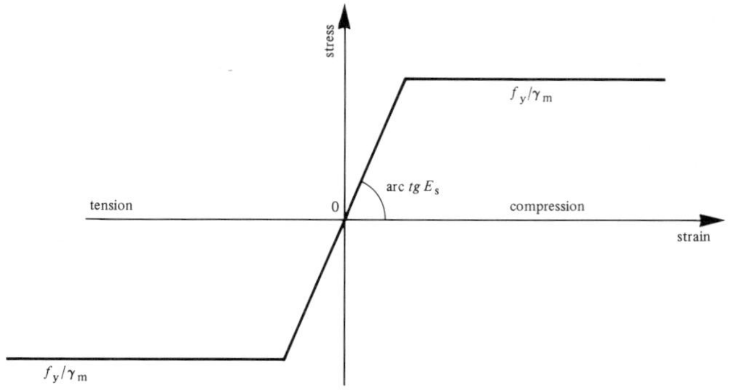

Figure 8.12 Simplified stress–strain diagram for steel (to CEB/FIP Model code)

in which k_{ut} and k_{un} are the respective values indicating the efficiency of the bond of A_{st} and A_{sn} respectively.

In both CP 115 and CP 110 the values of k_u are related to the percentage/strength ratio, as given in Table 8.1. It is therefore necessary first to calculate the area of tensioned steel, from which the percentage/strength ratio and k_u are then obtained. The values of k_u are from 0.9 to 1.0 for well-bonded pre-tensioned wires, bars, or cables. For non-bonded cables the effective prestress f_{pe} is used in place of f_{su}, and the corresponding values of k_u range from 1.2 to 1.7 in CP 115; the values in CP 110 are also related to the span/depth ratio.

In under-reinforced rectangular sections the steel is the weaker part, and it is therefore necessary to investigate only whether the area of the tensioned steel A_{st} required at working load is adequate also at ultimate load; that is, to compare A_s with A_{st}. Similarly, in over-reinforced sections, the resistance only of the concrete need be investigated but in under-reinforced I- or T-sections the resistance of both the concrete (that is, the size of the compressive flange) and the steel should be checked. If A_s required is greater than A_{st}, additional non-tensioned steel A_{sn} is needed, such that $A_s = A_{st} + A_{sn}$ when the prestressing steel of the same ultimate strength is used for A_{sn}. Otherwise A_{sn} obtained by the equation has to be multiplied by the factor f_{su}/f_y. With I- and T-sections it is desirable to ensure from the beginning that the area of the top flange is sufficient to resist the ultimate compressive force without exceeding the specified ultimate compressive stress.

In the case of rectangular slabs, for example, the minimum depth required may be influenced by conditions at ultimate load. Using the first method

$$M_R = b\,(0.5\,d)(0.75\,d)f_{prism} = 0.375\,b\,d^2\,f_{prism} \quad \text{(8.12)}$$

and hence

$$\boxed{d_{min} = \sqrt{\frac{M_R}{0.375\,b f_{prism}}}} \quad \text{(8.13)}$$

Alternatively, from CP 115 the limiting percentage/strength ratio (also called reinforcing index)

$$\frac{A_s}{bd}\frac{f_{su}}{f_{cu}} = 0.40; \text{ therefore } d_{min} = \frac{2.5\,A_s}{b}\frac{f_{su}}{f_{cu}}$$

8.5 Design for working load (elastic conditions)

As mentioned in section 8.2, the primary requirement in the design of prestressed members is to ensure that the limit state of failure is avoided, by basing the design on ultimate-load conditions. Such a procedure establishes the size of the compressive zone and the total area of steel required. On the other hand, the dimensions of the tensile zone and the magnitude of the prestressing force can be determined only from a consideration of the elastic stresses in a homogeneous section.

It is also important to investigate some of the stress conditions for a homogeneous section at service load. At one time this was widely and incorrectly taken as the sole basis for design; however, it is equally incorrect to ignore elastic conditions entirely. There is certainly no need to compute the maximum compressive stress at working load, since a compressive zone designed with adequate factors of safety for conditions at ultimate load is large enough to ensure that the permissible stress at service load is not exceeded. Similarly the total area of steel required for ultimate-load conditions will normally exceed that needed for service conditions.

One very important criterion is the permissible compressive stress $\bar{f}_{ct}$ in the concrete at transfer (which depends on the specified strength). Another is the permissible tensile stress $\bar{f}_{tw}$ under service conditions; in structures of Class 1, this value is zero, while in Classes 2 and 3 a limited tensile stress is permitted at service loading. If Classes 1 and 3 are combined, as the authors advocate, the value is zero at the 'normal' service load.

These two criteria can be combined to give the stress range:

$$f_{br} = R_0\bar{f}_{ct} + \bar{f}_{tw} \qquad \text{(see equation 8.18)}$$

The section modulus for the bottom flange is then given by the expression

$$z_{b\,min} = \frac{M_w - R_0 M_t}{f_{br}} \qquad \text{(see equation 8.18)}$$

and the effective prestressing force required is given by:

$$P_e = \frac{A_0}{R_0 k_b}\left(\bar{f}_{ct} + \frac{M_t}{z_{b\,min}}\right) \qquad \text{(see equation 8.22)}$$

The full conditions for the elastic design of a homogeneous section are set out in the following mainly because there may be some unusual conditions in which a full elastic check may be necessary. In general, however, only the conditions relating to the bottom flange will require investigation.

If $\bar{f}_{ct}$ and $-\bar{f}_{tt}$ are the permissible compressive and tensile stresses at transfer, and $\bar{f}_{cw}$ and $-\bar{f}_{tw}$ are those at the time the working load is applied, equations 7.1 to 7.2 may be rewritten as follows.

$$f_{bt} = \frac{k_b P_t}{A} - \frac{M_t}{Z_b} \leqslant \bar{f}_{ct} \qquad (8.14)$$

$$f_{tt} = \frac{k_t P_t}{A} + \frac{M_t}{Z_t} \geqslant -\bar{f}_{tt} \qquad (8.15)$$

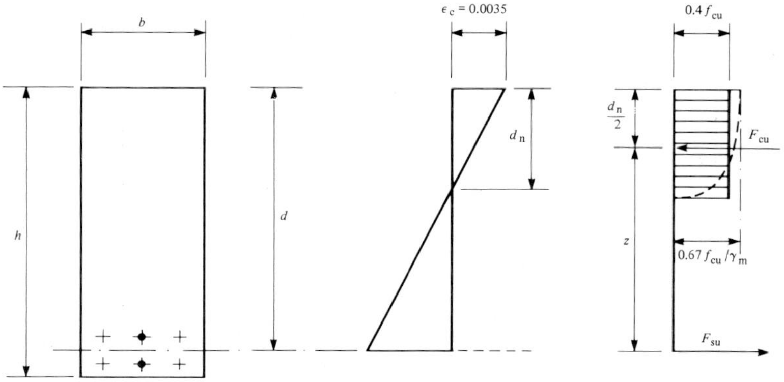

f_{cu} = characteristic cube strength of concrete

$F_{cu} = F_{su}$

$d_n \leqslant d/2$

$$F_{su} = \frac{1}{1.15} \Sigma \left[A_{st} f_{su} + A_{sn} f_y \right]$$

or $(0.2\, b\, d\, f_{cu})$ whichever is less

$M_R = F_{su} \cdot z$

Prestressing tendons = A_{st} shown thus +
Non-tensioned steel = A_{sn} shown thus ✦

(a) to CP 110

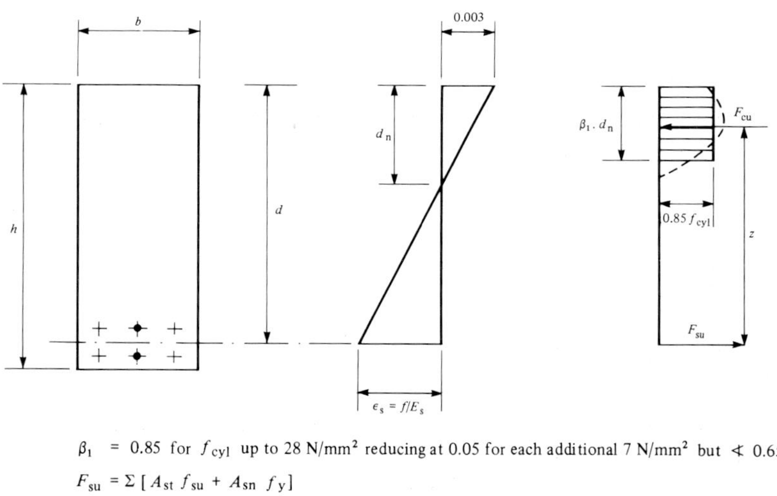

β_1 = 0.85 for f_{cyl} up to 28 N/mm² reducing at 0.05 for each additional 7 N/mm² but $\nless$ 0.65

$F_{su} = \Sigma \left[A_{st} f_{su} + A_{sn} f_y \right]$

$= (b. \beta_1 . d_n\ 0.85 f_{cyl})$

f_{cyl} = cylinder strength of concrete

$M_R = \phi . F_{su} . z$

Prestressing tendons = A_{st} shown thus +
Non-tensioned steel = A_{sn} shown thus ✦

(b) to ACI building code

Figure 8.13 Simplified method of calculating moment of resistance

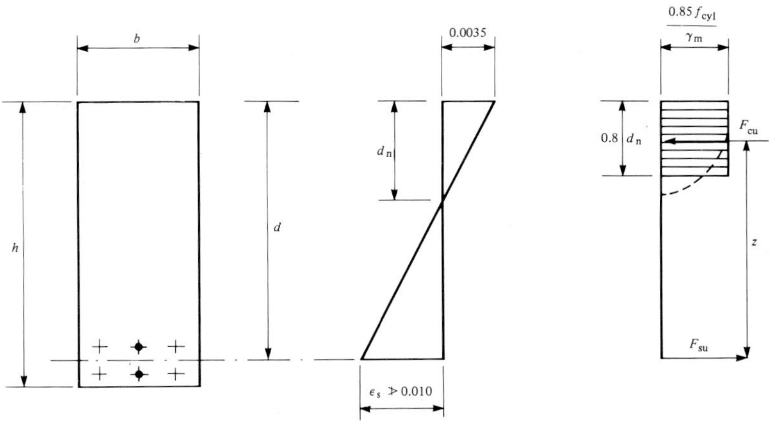

$$F_{su} = \frac{1}{1.15} \Sigma [A_{st} f_y + A_{sn} f_y]$$
$$= (b \,.\, 0.8\, d_n \times \frac{0.85 f_{cyl}}{1.5})$$

f_{cyl} = characteristic cylinder strength of concrete

f_y = characteristic strength defined as 5 % fractile of proof stress or 0.2 % yield strength

$M_R = F_{su} \,.\, z$

Prestressing tendons = A_{st} shown thus +
Non-tensioned steel = A_{sn} shown thus ◆

(c) to CEB/FIP Model Code

Figure 8.13 Simplified method of calculating moment of resistance

(provided that $\frac{k_t P_t}{A} \geqslant -\bar{f}_{tt}$ at the end of the member). This means that the stress is either positive (that is, compressive) or negative (that is, tensile), in which case it is numerically smaller than $-f_{tt}$.

$$f_{bw} = \frac{R_0 k_b P_t}{A} - \frac{M_w}{Z_b} \geqslant -\bar{f}_{tw} \quad \text{. . . . . (8.16)}$$

$$f_{tw} = \frac{R_0 k_t P_t}{A} + \frac{M_w}{Z_t} \leqslant \bar{f}_{cw} \quad \text{. . . . . (8.17)}$$

As $-\bar{f}_{tt}$ and $-\bar{f}_{tw}$ represent permissible tensile stresses they are preceded by a negative sign; k_t may be positive or negative, depending on the position of the steel. Even when k_t is negative $\frac{k_t P_t}{A}$ may be numerically less than $\frac{M_t}{Z_t}$ leaving f_{tt} positive.

For elastic conditions the greatest economy is often achieved when the actual stresses equal the permissible stresses, as in quations 8.14a to 8.17a. In equations 8.16a and 8.17a the appropriate values of $\frac{kP_t}{A}$ obtained from equations 8.14a and 8.15a are inserted.

$$\boxed{\bar{f}_{ct} = \frac{k_b P_t}{A} - \frac{M_t}{Z_{b\,min}}} \quad \text{. . . . . (8.14a)}$$

$$-\bar{f}_{tt} = \frac{k_t P_t}{A} + \frac{M_t}{Z_{t\,min}} \qquad (8.15a)$$

$$-\bar{f}_{tw} = \frac{R_0 k_b P_t}{A} - \frac{M_w}{Z_{b\,min}} = R_0 \left(+\bar{f}_{ct} + \frac{M_t}{Z_{b\,min}} \right) - \frac{M_w}{Z_{b\,min}} \qquad (8.16a)$$

$$\bar{f}_{cw} = \frac{R_0 k_t P_t}{A} + \frac{M_w}{Z_{t\,min}} = R_0 \left(-\bar{f}_{tt} - \frac{M_t}{Z_{t\,min}} \right) + \frac{M_w}{Z_{t\,min}} \qquad (8.17a)$$

The last two equations determine the least moment of resistance which will suffice for conditions at working load; when the terms containing P are eliminated these equations may be restated in the following form.

$$M_w - R_0 M_t = Z_{b\,min} (R_0 \bar{f}_{ct} + \bar{f}_{tw}) = Z_{b\,min} f_{br} \qquad (8.18)$$

$$M_w - R_0 M_t = Z_{t\,min} (\bar{f}_{cw} + R_0 \bar{f}_{tt}) = Z_{t\,min} f_{tr} \qquad (8.19)$$

in which $f_{br} = R_0 \bar{f}_{ct} + \bar{f}_{tw}$ is the range of stress at the bottom of the beam corresponding to the bending moment $M_w - R_0 M_t$, and $f_{tr} = \bar{f}_{cw} + R_0 \bar{f}_{tt}$ is the range of stress at the top of the beam.

These equations may be solved for $Z_{b\,min}$ and $Z_{t\,min}$. The value of $Z_{t\,min}$ so obtained is almost always less than that derived from ultimate-load considerations; in such a case it is not possible to reach the maximum compressive and tensile stresses at both transfer and working load, since the section will be larger than that required for this purpose. Thus the range of stress actually occurring at the top of the beam will be less than the maximum permissible, and is given by $f_{tr} = \bar{f}_{cw} - R_0 f_{tt\,lim}$ in which $f_{tt\,lim}$ is the greatest tensile (or the least compressive) stress which may occur at the top flange. The sign of $f_{tt\,lim}$ depends on the eccentricity of the steel and on the magnitude of M_t. This condition is shown in Figure 8.15 from which

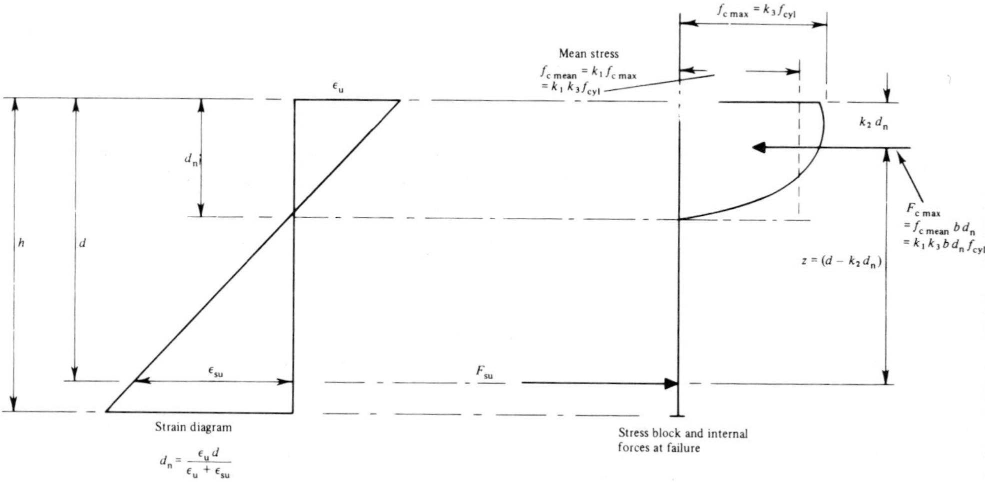

Figure 8.14 Conditions at ultimate load (Hognestad)

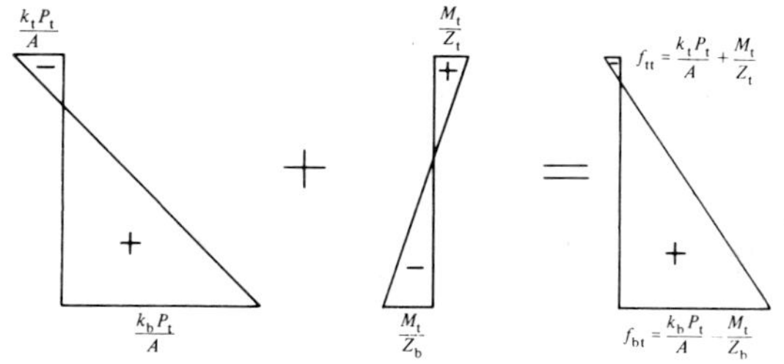

Figure 8.15 Stress distribution at transfer

$$f_{tt\ lim} = \frac{k_t P_t}{A} + \frac{M_t}{Z_t} \qquad (8.20)$$

The term $R_0 M_t$ should be omitted if the weight of the beam does not counteract the effect of the prestressing force, for example if the beam is prevented by friction from separating from the mould at transfer, or if a section near the end (where M_t is small) is considered. If there is a possibility that the beam will be transported upside down while supported near the ends, or that it will be suspended at the centre while it is the right way up, the bending moment $-M_t$ should be inserted in equations 8.18 and 8.19, the first term then will be $M_w + R_0 M_t$. If the bending moment M_t always counteracts the bending moment due to the prestressing force its effect is beneficial, as in general no tensioned wires or bars are required in the top of the beam; the depth of the beam can also be reduced in this case but a greater tensioning force is then required.

If the prestressing force P acts at a distance e_s from the centroid of a section (Figure 8.16), the stresses at the bottom and top of the beam are $k_b \frac{P}{A}$ and $k_t \frac{P}{A}$. The ratio $\frac{k_t}{k_b}$ is $\frac{i^2 - e_s e_t}{i^2 + e_s e_b}$; this is also the ratio of the stresses at the top and bottom and depends on the eccentricity of the prestressing force and the shape of the section. In Table 8.2 four sections are shown, together with their properties; in each section the prestressing steel is assumed to be at a distance of 0.1 h from the bottom face. This is, of course an approximation; in practice, the distance has to comply with the requirements for cover. The ratio k_t/k_b is then a measure of the efficiency with which the prestressing force is utilized, on the assumption that the distribution of the prestress should be approximately triangular and also that the greatest possible amount of the steel should be near to the tensile edge (for ultimate-load conditions). The smaller the value of k_t/k_b the more closely is this condition approached.

The economy of the sections in resisting a given bending moment is compared in Table 8.3 for the simple case when no tensile stresses are allowed at transfer and working load. It is clear that the rectangular section requires the greatest amount of concrete and prestressing steel and the I-section the least. It should be noted that these results take no account of conditions at ultimate load. These require that as much of the steel as possible should be concentrated near the outer tensile edge, and the dimensions of the compressive zone are usually

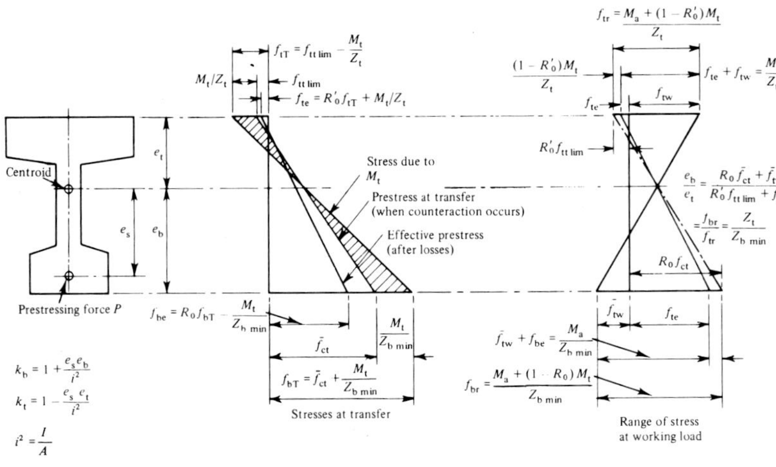

Figure 8.16 Ranges of stresses at transfer and working load

Table 8.2 Distribution of stress in different sections

	A	B	C	D
SECTION	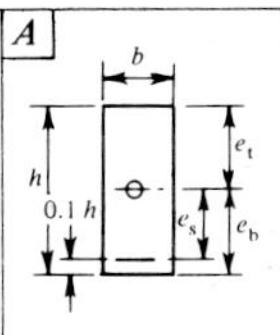	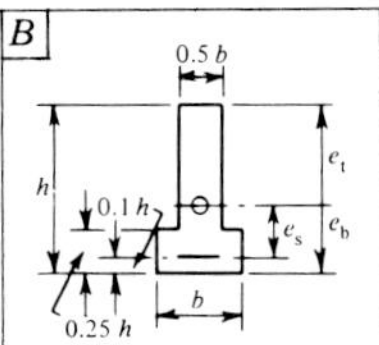	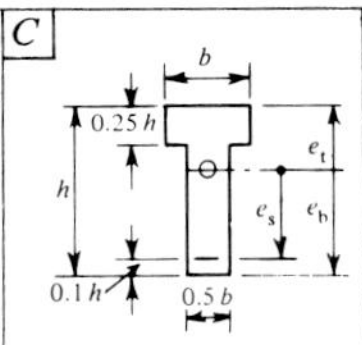	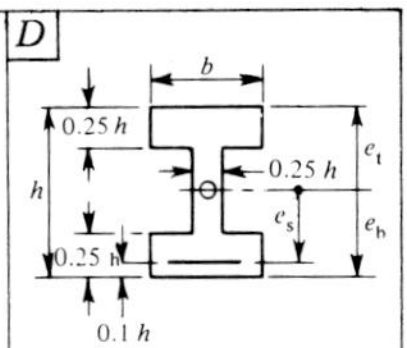
PROPERTIES Area A e_b e_t	bh $0.5\,h$ $0.5\,h$	$0.625\,bh$ $0.426\,h$ $0.574\,h$	$0.625\,bh$ $0.574\,h$ $0.426\,h$	$0.625\,bh$ $0.5\,h$ $0.5\,h$
I Z_b Z_t i^2	$\frac{bh^3}{12}$ $\frac{bh^2}{6}$ $\frac{bh^2}{6}$ $\frac{h^2}{12}$	$0.677\frac{bh^3}{12}$ $0.795\frac{bh^2}{6}$ $0.590\frac{bh^2}{6}$ $1.082\frac{h^2}{12}$	$0.677\frac{bh^3}{12}$ $0.590\frac{bh^2}{6}$ $0.795\frac{bh^2}{6}$ $1.082\frac{h^2}{12}$	$0.906\frac{bh^3}{12}$ $0.906\frac{bh^2}{6}$ $0.906\frac{bh^2}{6}$ $1.450\frac{h^2}{12}$
e_s k_b k_t	$0.4\,h$ 3.4 −1.4	$0.326\,h$ 2.54 −1.07	$0.474\,h$ 4.02 −1.24	$0.4\,h$ 2.66 −0.66
$\frac{k_t}{k_b}$	−0.411	−0.422	−0.308	−0.248
Stress Diagram ($f_g = \frac{P}{A}$ = stress at centroid)	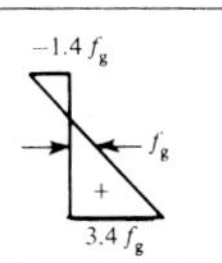	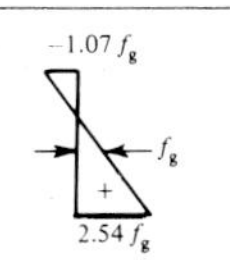	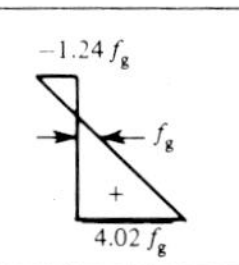	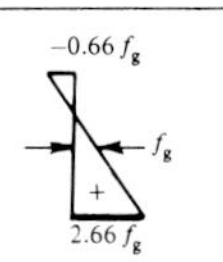

Table 8.3 Comparison of sections

ASSUMPTIONS: (1) No tensile stresses permitted (stress distribution assumed triangular)
(2) $f_{tr\ max} = \bar{f}_{cw} = 1.25 f_{br\ max} = 1.25 \bar{f}_{ct}$
(3) Effect of losses ignored
(4) No counter-action assumed

Section		A	B	C	D
Moment of resistance M_R	$Z_b f_{br\ max}$	$\frac{bh^2}{6} \bar{f}_{ct} = M_{R\ rect}$*	$0.795\ M_{R\ rect}$	$0.590\ M_{R\ rect}$*	$0.906\ M_{R\ rect}$*
	$Z_t f_{tr\ max}$	$1.25\ M_{R\ rect}$	$0.737\ M_{R\ rect}$*	$0.990\ M_{R\ rect}$	$1.133\ M_{R\ rect}$
Area required to make $M_{R\ min} = M_{R_1\ rect}$		$bh = A_{rect}$	$0.846 A_{rect}$	$1.06 A_{rect}$	$0.689 A_{rect}$
f_{bT}		$\bar{f}_{ct}$	†$1.25 \bar{f}_{ct} \times \frac{e_b}{e_t} = 0.927 \bar{f}_{ct}$	$\bar{f}_{ct}$	$\bar{f}_{ct}$
Centroidal stress $f_g = f_{bT} \times \frac{e_t}{h}$		$0.5 \bar{f}_{ct}$	$0.532 \bar{f}_{ct}$	$0.426 \bar{f}_{ct}$	$0.5 \bar{f}_{ct}$
Prestressing force $P = A f_g$		$0.5 A_{rect} \bar{f}_{ct} = P_{rect}$	$0.9 P_{rect}$	$0.9 P_{rect}$	$0.69 P_{rect}$

***Limiting value** *†Derived from limiting stress at top face*

greater than those required for working load. Consequently the sections intermediate between types C and D usually afford the best all-round efficiency.

In order to utilize the permissible stresses as fully as ultimate load requirements allow, a section which is unsymmetrical about the neutral axis, as shown in Figure 8.16, is advisable. The dimensions of the section should be of such magnitude that

$$\boxed{\frac{Z_t}{Z_{b\,min}} = \frac{e_b}{e_t} = \frac{R_0\bar{f}_{ct} + \bar{f}_{tw}}{\bar{f}_{cw} - R_0 f_{tt\,lim}} = \frac{f_{br}}{f_{tr}}} \quad (8.21)$$

The stress $f_{tt\,lim}$ may be tensile (that is, negative) or compressive. If this stress is tensile, and numerically greater than the permissible stress $-\bar{f}_{tt}$, tensioned wires or bars are required in the top and $-\bar{f}_{tt}$ must be substituted for $f_{tt\,lim}$ in equations 8.17 and 8.19.

If the section modulus for an I-section is related to that for a rectangular section by the expression $Z = \alpha \dfrac{bh^2}{6} = \alpha' bh^2$ then for a symmetrical I-beam the stress given by $\dfrac{M_w - R_0 M_t}{\alpha' bh^2}$ is either $(R_0 \bar{f}_{ct} + \bar{f}_{tw})$ or $(\bar{f}_{cw} - R_0 f_{tt\,lim})$. If tensile stresses are not permitted, and the value of f_{ct} is 1000 lbf/in^2 (70 kgf/cm^2; 7 N/mm^2) and R_0 is 0.85, the moment of resistance of a rectangular beam is 141 bh^2 lbf in. which is less than that of a reinforced concrete beam in which f_s = 18 000 lbf/in^2 (1260 kgf/cm^2; 125 N/mm^2) and $h = 1.16d$. Therefore, a greater moment of resistance cannot be obtained for the same stress in the concrete merely by prestressing a beam, unless the weight of the beam counteracts the effect of the prestressing force. The depth of a beam in which tensile stresses are not permitted is less than that of a reinforced concrete beam only if counter-action occurs; advantage is also obtained if the prestress can be applied in stages so that not only the weight of the beam but successive additions of dead load can be made to counteract the effect of the prestressing force.

The moment of resistance is considerably increased by increasing the permissible compressive stress in the concrete: for example, if the permissible stress at transfer is 2400 lbf/in^2 (170 kgf/cm^2; 17 N/mm^2) and R_0 = 0.80, the moment of resistance of a rectangular beam is 320 bh^2 lbf in. Advantage can also be taken of the great tensile resistance and resilience of prestressed concrete; if a tensile stress of 600 lbf/in^2 (42 kgf/cm^2; 4 N/mm^2) which will ensure absence of visible cracks even under conditions of fatigue in a member with well-bonded steel is permitted the moment of resistance becomes 420 bh^2 lbf in. and a tensile stress due to bending of 900 lbf/in^2 (63 kgf/cm^2; 6 N/mm^2), with a consequent moment of resistance of 470 bh^2 lbf in, is permissible with well-distributed pre-tensioned steel and high-strength concrete if the maximum load occurs rarely and permanent freedom from cracks is still required. With a nominal tensile stress of 1200 lbf/in^2 (84 kgf/cm^2; 8 N/mm^2) due to a maximum load which rarely occurs, harmless cracks may temporarily appear; the moment of resistance would then be 520 bh^2 lbf in. The economy of partial prestressing is clearly shown by these figures.

For the design of a beam, equation 8.18 can be written as follows. With regard to the bottom flange,

$$R_0\bar{f}_{ct} + \bar{f}_{tw} = \frac{M_w - R_0 M_t}{\alpha_b \dfrac{bh^2}{6}} = \frac{M_a + (1 - R_0)M_t}{\alpha_b \dfrac{bh^2}{6}} \quad (8.18a)$$

where $M_w = M_t + M_a$

Therefore

$$\boxed{\alpha_b Z_{rectangle} = \frac{M_a + (1 - R_0)M_t}{R_0 \bar{f}_{ct} + \bar{f}_{tw}}} \quad (8.18b)$$

in which $\alpha_b = \dfrac{Z_b}{Z_{rectangle}}$ is termed the 'shape factor', and M_a is the applied bending moment (excluding the bending moment M_t due to the weight of the beam). M_t depends on the value of b and h, and as $(1 - R_0)$ is small the effect of errors in the assumed value of M_t is also relatively small. Similarly

$$\boxed{\alpha_t Z_{rectangle} = \frac{M_a + (1 - R_0)M_t}{f_{cw} - R_0 f_{tt\,lim}}} \quad (8.18c)$$

this value is rarely critical, however, as the dimensions of the top flange of a beam are normally determined by conditions at ultimate load. If necessary, it may be conveniently obtained from the expression $\alpha_t = \alpha_b \dfrac{f_{br}}{f_{tr}}$

A fully elastic design for a beam is rarely needed; but if in some special cases it should be necessary the procedure is as follows.

(1) Determine allowable stresses.
(2) Calculate M_a. Determine R_0 as described in Chapter 6.
(3) Assume values of b and h; calculate M_t.
(4) Calculate $Z_{b\,min} = \alpha_b Z_{rectangle}$. Charts given in Part 2 describe whereby the most suitable section may be rapidly determined.
(5) Determine $Z_{t\,min} = \alpha_t Z_{rectangle}$.
(6) Determine the prestressing force from

$$P_t = \frac{A}{k_b}\left(\bar{f}_{ct} + \frac{M_t}{Z_{b\,min}}\right) \quad (8.22)$$

$$P_t = \frac{A}{k_b}\left(\bar{f}_{bt} + \frac{M_t}{Z_b}\right) \text{ when } Z_b > Z_{b\,min}$$

(Strictly speaking A in equation 8.22 should be A_0.)

For the special case when the stress f_{bn} at the bottom flange due to 'normal' loading is zero, then

$P_t = \dfrac{A}{k_b}\left(\dfrac{M_n + M_t}{Z_b}\right)$ in which M_n is the moment due to the 'normal' loading.

8.6 Application of general formulae to particular cases

For symmetrical sections, the following special cases can be obtained.

8.6.1 Rectangular sections

For a rectangular section, as noted previously,

$$\frac{M_a + (1 - R_0)M_t}{\frac{bh^2}{6}} = R_0 \bar{f}_{ct} + \bar{f}_{tw}$$

and hence

$$bh^2 = \frac{6(M_a + [1 - R_0]\, M_t)}{R_0 \bar{f}_{ct} + \bar{f}_{tw}} \quad \ldots \ldots \ldots \ldots \quad (8.23)$$

8.6.2 Sections symmetrical about two axes

The general expression

$$\frac{M_a + (1 - R_0)M_t}{Z_b} = R_0 \bar{f}_{ct} + \bar{f}_{tw} = f_{br} \quad \ldots \ldots \ldots \quad (8.24)$$

may be written in the form

$$\frac{M_a + (1 - R_0)M_t}{bh^2} = \alpha' f_{br} \quad \ldots \ldots \ldots \ldots \quad (8.25)$$

in which

$$\alpha' = \frac{Z_b}{bh^2}$$

Hence $bh^2 = \dfrac{M_a + (1 - R_0)M_t}{\alpha' f_{br}}$

Values of α' for symmetrical I-sections are given in Charts 8 and 9 in Part 2.

8.6.3 Sections unsymmetrical about the horizontal axis – I-sections and box sections

An unsymmetrical I-beam is shown in Figure 8.16. The general expression for the ratio is,

$$\frac{Z_t}{Z_b} = \frac{e_b}{e_t}$$

Also, as before, $\dfrac{M_a + (1 - R_0)M_t}{bh^2} = \alpha' f_{br}$ in which $\alpha' = \dfrac{Z_b}{bh^2}$

Values for α' for unsymmetrical sections with equal flange depths are given in Charts 12 and 13 in Part 2; these can be used to obtain suitable values of bh^2 for any particular case. Explanation for the use of these charts are given in Part 2.

8.6.4 Sections with tensioned steel at both faces

In general, the prestressing force required is obtained from the expression

$P_t = \frac{A}{k_b}\left(\bar{f}_{ct} + \frac{M_t}{Z_b}\right)$ in which $k_b = 1 + \frac{e_s e_b}{i^2}$. Hence it is necessary to obtain a value for e_s.

A method is often advocated in which the effect of the bending moment M_t due to the weight of the member is ignored. If it is also assumed that no tensile stresses are permitted, this leads to the expression

$$\boxed{P_t = \frac{M_a e_b e_t}{hi^2}} \qquad (8.26)$$

whence the calculated maximum eccentricity e_{sc} is given by

$$e_{sc} = e_k + \frac{M_t}{P_t}$$

where e_k is the kern distance. With large values of M_t this results in values of $e_{sc} \geqslant e_b$, which is clearly impossible. However, it is easy to obtain values of e_{sc} which are in agreement with the required limiting stresses and these should be used in preference. Three cases are considered here, corresponding to the number of limiting stresses. In all cases,

$$\frac{f_{bT}}{f_{tT}} = \frac{k_b}{k_t} = \frac{i^2 + e_{sc} e_b}{i^2 - e_{sc} e_t}$$

and hence

$$e_{sc} = \frac{i^2 (f_{bT} - f_{tT})}{e_b f_{tT} + e_t f_{bT}} \qquad (8.27)$$

(note that f_{tT} is always tensile (that is, negative) when k_t is negative. This is the case when the prestressing force is outside the kern.)

The values of f_{bT} and f_{tT} depend on the dimensions of the section and may be determined as follows.

8.6.5 Minimum section

$$Z_b = Z_{b\,min}, Z_t = Z_{t\,min}$$

In this case the maximum compressive and tensile stresses at transfer and working load must equal the maximum permissible stresses (Figure 8.17a). Hence at transfer,

$$f_{bT} = \bar{f}_{ct} + \frac{M_t}{Z_b} \text{ and } f_{tT} = -\bar{f}_{tt} - \frac{M_t}{Z_t}$$

This case rarely occurs in practice, since even if ultimate-load conditions can be satisfied with such a section some adjustment of calculated dimensions is usually necessary. Also since $Z_{b\,min}$ and $Z_{t\,min}$ are calculated from the applied bending moments and the permissible stresses it may be that the required value of e_{sc} exceeds $e_{s\,max}$; that is, the greatest possible eccentricity which still

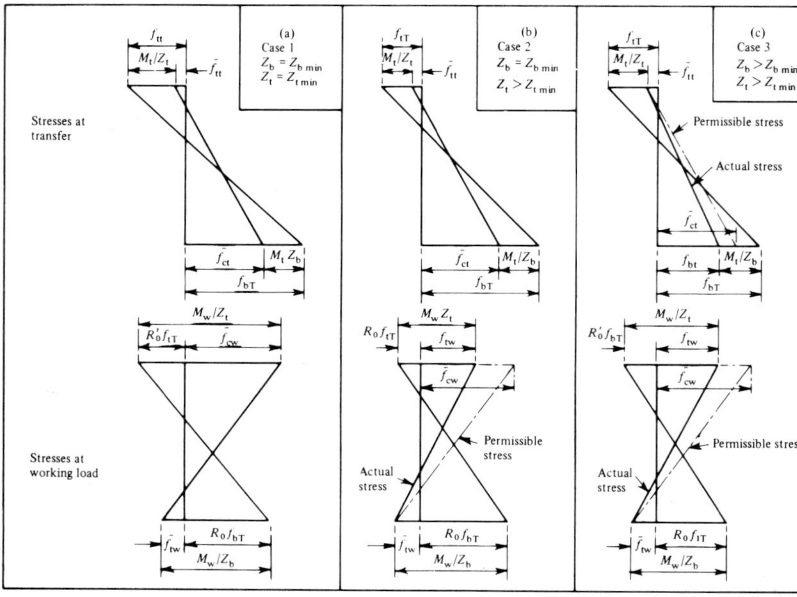

Figure 8.17 Limiting stresses at transfer and working load

provides sufficient cover. This is the preferred value, and is henceforward termed e_s. In general, a definite value of e_{sc} exists for each value of the ratio M_t/M_w and for large values of the ratio the calculated value will exceed the practical limit. The use of a section in which $Z_b = Z_{b\ min}$ and $Z_t = Z_{t\ min}$ is then impossible and in such a case it would be necessary to select a new section in which Z_t exceeds $Z_{t\ min}$

When no counter-action occurs $f_{bT} = \bar{f}_{ct}$ and $f_{tT} = -\bar{f}_{tt}$.

8.6.6 Section with enlarged compressive zone

$$Z_b = Z_{b\ min};\ Z_t > Z_{t\ min}$$

To ensure sufficient resistance at ultimate load, it is usually necessary for the area of the compressive flange to exceed that required for working load conditions (Figure 8.17b). In this case f_{bT} must equal $\bar{f}_{ct} + \dfrac{M_t}{Z_b}$, but f_{tT} need not equal $-\bar{f}_{tt} - \dfrac{M_t}{Z_t}$. However, the smallest prestressing force is obtained when $f_{tT} = -\bar{f}_{tt} - \dfrac{M_t}{Z_t}$ and the preferred value of e_s is therefore as in section 8.6.5, provided e_s does not exceed $e_{s\ max}$.

If the calculated value of e_{sc} exceeds the greatest possible value of e_s, the actual value of e_s is used to evaluate k_b and k_t and the value of f_{tT} is equal to $\dfrac{k_t}{k_b} f_{bT}$

When no counter-action occurs, $f_{bT} = \bar{f}_{ct}$ and $f_{tT} = -\bar{f}_{tt}$ $\left(\text{or } \dfrac{k_t}{k_b} f_{bT}\right)$

8.6.7 Oversized section

$Z_b > Z_{b\,min}$; $Z_t > Z_{t\,min}$

This is the usual case, in which the area of the compressive zone appreciably exceeds that required for working-load conditions and the tensile zone is slightly larger than the minimum. Consequently, only two permissible stresses can be attained (Figure 8.17c) and the smallest prestressing force is obtained when these are $-f_{tt}$ and $-f_{tw}$. Hence

$$f_{bT} = \frac{1}{R_0}\left[\left(\frac{M_a + M_t}{Z_b}\right) - \bar{f}_{tw}\right] \quad \ldots \ldots \quad (8.28)$$

and

$$f_{tT} = -\bar{f}_{tt} - \frac{M_t}{Z_t}$$

Again, if the calculated value of e_{sc} exceeds the actual value of e_s the actual value is used to calculate k_b and k_t and f_{tT} is then equal to $\frac{k_t}{k_b} f_{bT}$

If no counter-action occurs, f_{bT} is given by the expression $\frac{1}{R_0}\left(\frac{M_w}{Z_b} - \bar{f}_{tw}\right)$

and

$$f_{tT} = -\bar{f}_{tt} \left(\text{or } \frac{k_t}{k_b} f_{bT}\right)$$

A graphical method for determining stresses and e_s is also available[9].

8.6.8 Special cases

The general expression $e_{sc} = \dfrac{i^2 (f_{bT} - f_{tT})}{e_b f_{tT} + e_t f_{bT}}$ can be simplified in particular cases.

For a rectangular section $i^2 = \dfrac{h^2}{12}$ and $e_b = e_t = \dfrac{h}{2}$

Hence

$$e_{sc} = \frac{h}{6}\left(\frac{f_{bT} - f_{tT}}{f_{bT} + f_{tT}}\right)$$

Similarly for a symmetrical I-section or box-section, writing $i^2 = \dfrac{\gamma h^2}{12}$ and $e_b = e_t = \dfrac{h}{2}$,

$$e_{sc} = \frac{\gamma h}{6}\left(\frac{f_{bT} - f_{tT}}{f_{bT} + f_{tT}}\right) \quad \ldots \ldots \quad (8.29)$$

If it is intended that the maximum permissible tensile and compressive stresses should be reached at transfer, then this expression may be written in the form

$$e_{sc} = \frac{\gamma h}{6}\left(\frac{\bar{f}_{ct} + \dfrac{M_t}{Z_b} + \bar{f}_{tt} + \dfrac{M_t}{Z_t}}{\bar{f}_{ct} + \dfrac{M_t}{Z_b} - \bar{f}_{tt} - \dfrac{M_t}{Z_t}}\right) \qquad (8.30)$$

and since $Z_b = Z_t = Z$

$$e_{sc} = \frac{\gamma h}{6}\left(\frac{1 + \dfrac{\bar{f}_{tt}}{\bar{f}_{ct}} + \dfrac{2M_t}{Z\bar{f}_{ct}}}{1 - \dfrac{\bar{f}_{tt}}{\bar{f}_{ct}}}\right) = \frac{\gamma h}{6}\left(\frac{1 - \rho + 2K}{1 + \rho}\right)$$

in which

$$\rho = \frac{\bar{f}_{tt}}{\bar{f}_{ct}};\ K = \frac{M_t}{Z\bar{f}_{ct}}$$

Values of this expression corresponding to various values of ρ, K and γ are given in Table 8.4.

8.7 Position of prestressing steel

The values of e_{sc} obtained from the foregoing give the required position of the centroid of the prestressing steel. However, conditions at ultimate load are improved when the main prestressing steel has the greatest possible eccentricity. It is therefore advisable to divide the resultant prestressing force $\bar{P}_t$ into two components, one P_t at the bottom of the section and the other P_t' at the top, so that the lower tendons are close to the outer tensile face to control cracks. By replacing the distance e_{sc} obtained from the foregoing by the symbol $\bar{e}_s$, and defining e_s and e_s' as the actual distances of P_t and P_t' from the centroid, the following relations are derived (see Figure 8.18).

Table 8.4 Eccentricity of prestressing force for symmetrical sections

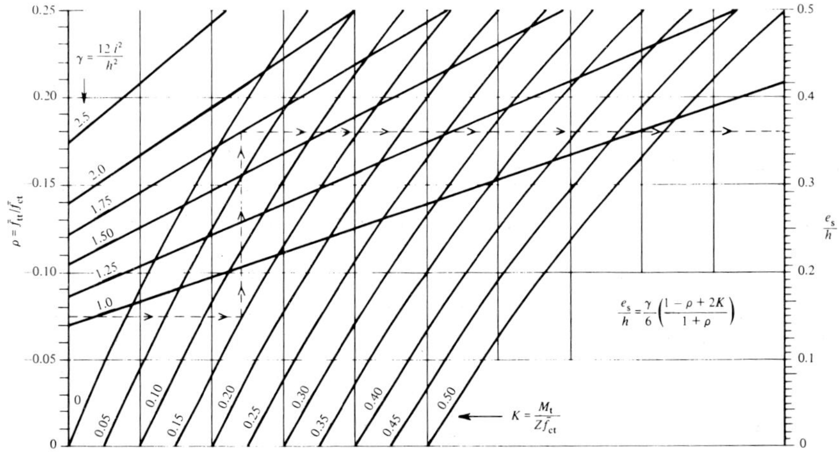

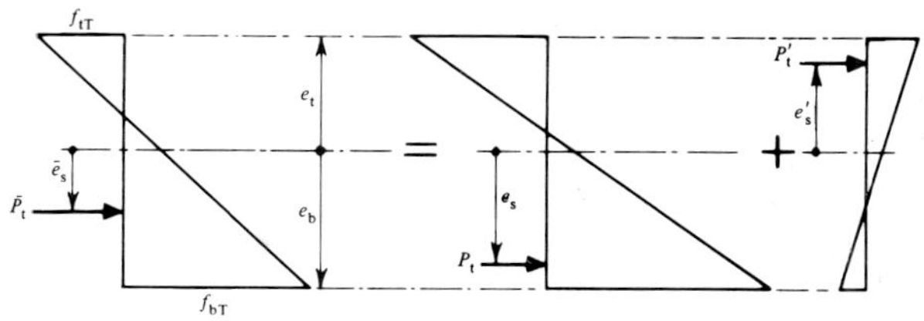

Figure 8.18 Position of prestressing tendons

$$P_t + P'_t = \bar{P}_t = \left(\frac{f_{bT} + f_{tT}}{2}\right) A \quad (8.31)$$

$$P_t e_s - P'_t e'_s = \bar{P}_t \bar{e}_s \quad (8.32)$$

Therefore

$$P_t = \left(\frac{\bar{e}_s + e'_s}{e_s + e'_s}\right) \bar{P}_t \quad (8.33)$$

and

$$P'_t = \bar{P}_t - P_t \quad (8.34)$$

When two such prestressing forces P_t and P'_t are applied the effect of different loss factors R_0 and R'_0 should be considered. These arise as a consequence of the difference between the stresses in the concrete at the centroids of P and P'. Hence a new resultant eccentricity e_{se} must be computed from the expression

$$\bar{e}_{se} = \frac{P'_e e'_s}{P_e + P'_e}$$

from which revised stress ratios $\bar{k}_{be}$ and $\bar{k}_{te}$ are obtained. Hence

$$\bar{f}_{be} = \frac{\bar{k}_{be} \bar{P}_e}{A} \quad (8.35)$$

and

$$\bar{f}_{te} = \frac{\bar{k}_{te} \bar{P}_e}{A} \quad (8.36)$$

8.8 Stress conditions at the ends of beams

At the ends of the member the bending moment M_t and M_a do not counteract and consequently the design formulae which have been derived are suitable only for the central portion. This may result in very high stresses due to prestress at the ends. It is therefore desirable, and indeed in certain cases essential to reduce the prestressing force and/or the eccentricity at the ends of the units to bring the resultant stresses within the permissible limits. This could be achieved either by keeping the tendons straight and debonding (also called blanketing) some of the cables to the required length or by draping (deflecting) the tendons and thus reducing the eccentricity towards the ends.

Sometimes tendons may be draped and at the same time some of them debonded.

From the manufacturing point of view, straight cables with debonding would be preferable to draping especially in long line production. In such a case, however, the spacing of the tendons will be the same throughout the beam and the minimum spacing will be governed by the requirement at anchorage. In the case of draping, the position of tendons towards the ends can be adjusted to suit the requirements of proper anchorage, and hence at the centre portion some of the tendons in different layers, if necessary, could be so positioned as to even touch one another.

In calculating the stresses at the ends of the beam the stress at the bottom face (tensile face) at transfer f_{bT} should not be greater than the tensile stress $\bar{f}_{tT}$ considered as permissible at that stage. At the service-load conditions the corresponding stress f_{bE} should not exceed the permissible $\bar{f}_{tw}$ (f_{bE} may be taken as $R_0 f_{bT}$). R_0 of course will be less than that obtained from the centre portion.

In draping cables it is preferable to continue the lowest layer of tendons straight and then deflect the cables of other layers. Depending on the design and manufacturing consideration the deflection of cables could be done from one central point or from two points, as shown in Figures 8.19 and 8.20. The profile could preferably be parabolic particularly with post-tensioned members.

In every case the designer should ensure that the vertical component of the prestressing forces are adequately contained during manufacture.

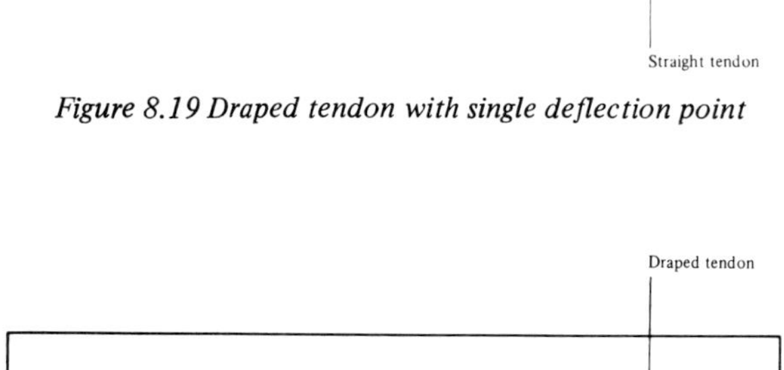

Figure 8.19 Draped tendon with single deflection point

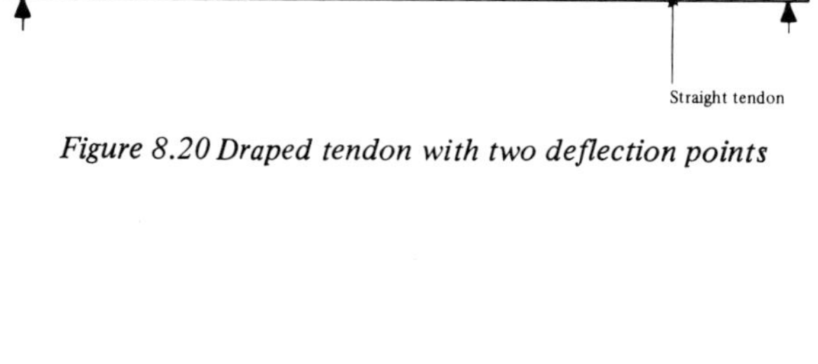

Figure 8.20 Draped tendon with two deflection points

8.9 Combined criteria for ultimate load and service load

As previously noted the area of the compressive flange and the total area of steel required are determined by ultimate load conditions, while the dimensions of the tensile flange and the magnitude and position of the prestressing force are obtained from the service load conditions appropriate to the particular class of the structure concerned.

It is sometimes suggested that all limit-states should be explicitly investigated in each design; this is laborious and often unnecessary, for example, there is clearly no need to consider crack widths in structures of Class 1 or Class 2. However, two methods of design are given in the following: the first is short, simple and reasonable, while the second allows for the case in which more detailed investigation might be required.

8.9.1 Simple design

(1) The compressive and tensile resistances required at ultimate load are obtained from:

$$M_R = F_{c\,max}\, z_{max} = F_{s\,max}\, z_{max} \quad \ldots \quad (8.37)$$

$$F_{s\,max} = K_u f_{su} A_{st} = F_{c\,max} = K_{cu}\, b d_n f_{cu} \quad \ldots \quad (8.38)$$

$K_{cu} f_{cu}$ is the limiting uniform compressive stress in accordance with section 8.4.
(K_u and d_n/d can be obtained from Table 8.1 if the design is based on CP 110 and CP 115.) This gives A_{st} and $b\,d_n$.

(2) Compute the size of the tensile flange (or check that given dimensions are satisfactory) from:

$$Z_{b\,min} = \frac{M_a + (1 - R_0)M_t}{R_0 \bar{f}_{ct} - \bar{f}_{tw}} \quad \ldots \quad (8.39)$$

If a suitable value is chosen for $\bar{f}_{ct}$ the collapse conditions at the tensile flange at transfer will not prove critical.

The order of steps 1 and 2 may be reversed if desired.

(3) Compute the effective prestressing force required (after losses) from:

$$P_e = \frac{A}{k_b}\left(\frac{M_w}{Z_b} - \bar{f}_{tw}\right) \quad \ldots \quad (8.40)$$

For Class 1:

$\bar{f}_{tw} = 0$;

$$\therefore \boxed{P_e = \frac{A M_w}{k_b Z_b}} \quad \ldots \quad (8.40a)$$

For Class 2: $f_{tw} > 0.75\, f_{tr}$ (or the specified limiting value), in which f_{tr} is the modulus of rupture of the section

$$P_e = \frac{A M_w}{k_b Z_b} - \frac{A \bar{f}_{tw}}{k_b} \quad \ldots \quad (8.40b)$$

For Class 1 + Class 3: As for Class 1, but M_w is then the bending moment at 'normal' loading. This method of design has been used in the Figures 8.2 and 8.3.

(4) Determine losses and compute the initial force P_i required.

(5) Investigate shearing, end blocks, and other aspects of design (see following chapters).

8.9.2 Combined criteria

In some cases it is possible to obtain a design in which the area of the compressive zone is suitable both for the ultimate load and service conditions, and a section may be selected to meet in full the requirements for transfer, service, and collapse conditions. This apparent 'efficiency' is often illusory, in that the economy of the section may not be appreciable enough both in terms of the quantities of materials used and their costs, and the goal itself is often impossible of attainment, as noted in section 8.1. However, a method of combining these criteria is as follows.

(1) Assuming counter-action occurs

For working load,

$$M_a + (1 - R_0)M_t = Z_{b\,min}(\bar{f}_{tw} + R_0 \bar{f}_{ct}) = Z_{b\,min} f_{br} \quad \ldots\ldots \quad (8.41)$$

For ultimate load,

$$M_{ult} = A_f z_{max} f_{cm} \quad \ldots\ldots \quad (8.42)$$

Where A_f = Area of compressive flange.

z_{max} = Lever arm at ultimate load.
f_{cm} = Average compressive stress at failure.

Hence,

$$\frac{M_{ult}}{M_a + (I - R_0)M_t} \frac{f_{br}}{f_{cm}} = \frac{A_f z_{max}}{Z_{b\,min}}$$

The right-hand expression relates solely to the dimensions of the section. The left-hand relates to the applied loading and permissible stresses. When these are known a suitable section is easily found.

When counter-action occurs, it is usually the case that the top flange is larger than the bottom flange. Values of $\dfrac{A_f z_{max}}{Z_{b\,min}}$ are therefore plotted in the lower parts of Charts 12 and 13 (in Part 2): the parts marked Z_t will then represent Z_b and the ratio b_t/b gives the bottom flange as a fraction of the top flange. The use of the charts in this manner is illustrated in Example 9.2.2.

(2) No counter-action assumed

When no counter-action occurs, the preceding analysis is slightly modified. In this case,

For working load

$$M_a + M_t = Z_{t\,min}(\bar{f}_{cw} + R_0 \bar{f}_{tt}) = Z_{t\,min} f_{tr}$$

$$= Z_{b\,min}(\bar{f}_{tw} + R_0 \bar{f}_{ct}) = Z_{b\,min} f_{br}$$

For ultimate load

$$M_{ult} = A_f z_{max} f_{cm}$$

Hence the distance of the centroid from the bottom flange is obtained from

$$\frac{e_b}{h} = \frac{Z_{t\ min}}{Z_{b\ min} + Z_{t\ min}} \qquad (8.43)$$

and

$$\frac{M_{ult}}{M_a + M_t} \frac{f_{tr}}{f_{cm}} = \frac{A_f z_{max}}{Z_{t\ min}} \qquad (8.44)$$

Values of $\frac{e_b}{h}$ and $\frac{A_f z_{max}}{Z_t}$ are plotted in the upper parts of Charts 12 and 13. These apply directly to sections in which the area of the top flange is equal to or less than that of the bottom flange, and in which the required value of $A_f z_{max}$ is approximately equal to the required value of $Z_{t\ min}$. A straight-edge is rotated about the required value of x until a section giving the appropriate value of $\frac{A_f z_{max}}{Z_{t\ min}}$ is found. When several such sections are available, their respective areas may be obtained from Charts 10 and 11 and that with the least area for a given depth is chosen. Alternatively, the section with the greatest value of x may be considered as the most efficient. The use of the charts (in Part 2) is illustrated in Example 9.2.1.

The area of steel required for the prestressing force will be sufficient for ultimate load when a large dead weight counteracts and tensile stresses are not permitted at working load. Otherwise, additional non-tensioned steel may be necessary.

8.10 Sections subjected to reversal of bending moments

Elastic conditions – If a section in a beam is subjected to bending moments of opposing signs and different magnitudes (Figure 8.21), and if it is desired to attain the greatest permissible stress in both flanges, both in tension and compression, it can be shown that the section must be symmetrical. In this case f_{bT} and f_{tT} are of minor importance, as it is clearly impossible to obtain the maximum stress at transfer.

Consider a section which may be subjected to a sagging bending moment M_{1a} or alternatively a hogging moment M_{2a}. If the permissible compressive and tensile stresses at working load are $\bar{f}_{cw}$ and $-\bar{f}_{tw}$, and the moment due to self weight at transfer is M_t, then for moment M_{1a}

$$R'_0 f_{tT} + \frac{M_t}{Z_t} + \frac{M_{1a}}{Z_t} = \bar{f}_{cw} \qquad (8.45)$$

$$R_0 f_{bT} - \frac{M_t}{Z_b} - \frac{M_{1a}}{Z_b} = -\bar{f}_{tw} \qquad (8.46)$$

Similarly for moment M_{2a}

$$R_0 f_{tT} + \frac{M_t}{Z_t} - \frac{M_{2a}}{Z_t} = -\bar{f}_{tw} \qquad (8.47)$$

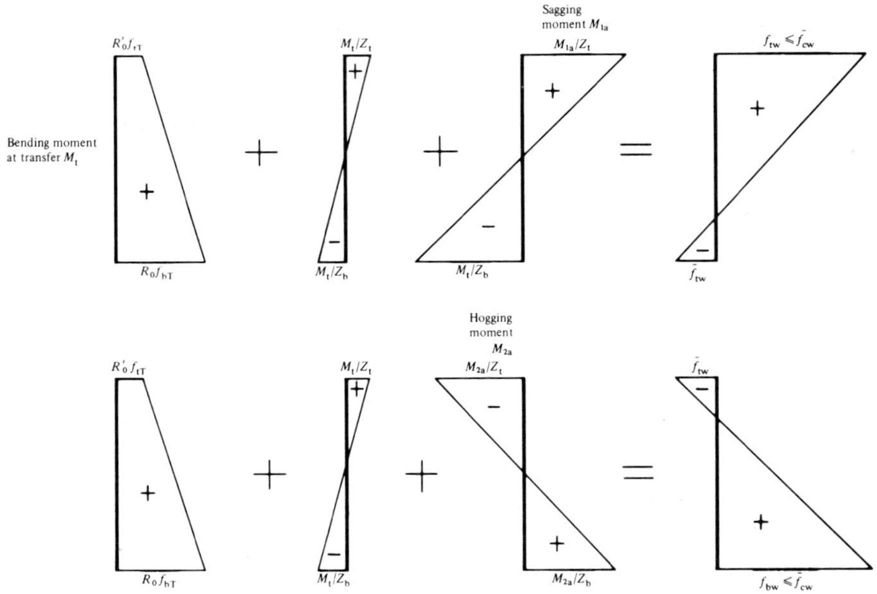

Figure 8.21 Effect of reversal of bending moment

$$R_0 f_{bT} - \frac{M_t}{Z_b} + \frac{M_{2a}}{Z_b} = \bar{f}_{cw} \qquad (8.48)$$

Assuming $R_0' = R_0$, expressions 8.14 and 8.17 may be combined to give equation 8.18, and 8.15 and 8.16 to give 8.19.

$$R_0\,(f_{bT} - f_{tT}) = \frac{(M_t + M_{1a})}{Z_t} + \frac{(M_{2a} - M_t)}{Z_b} \qquad (8.49)$$

$$R_0\,(f_{bT} - f_{tT}) = \frac{(M_t + M_{1a})}{Z_b} + \frac{(M_{2a} - M_t)}{Z_t}, \qquad (8.50)$$

Hence

$$\frac{M_{1a} - M_{2a} + 2M_t}{Z_b} = \frac{M_{1a} - M_{2a} + 2M_t}{Z_t}$$

or

$$Z_b = Z_t$$

In the special case when $M_t = 0, Z_b$ also equals Z_t. From equations 8.45 and 8.47

$$Z_b = Z_t = \frac{M_{1a} + M_{2a}}{\bar{f}_{cw} + \bar{f}_{tw}} \qquad (8.51)$$

and from equations 8.45 and 8.48

$$\frac{f_{tT}}{f_{bT}} = \frac{Z\bar{f}_{cw} - M_t - M_{1a}}{Z\bar{f}_{cw} + M_t - M_{2a}} \qquad (8.52)$$

The procedure for designing such a section may therefore be summarized as follows.

(1) Determine $Z_b = Z_t = \dfrac{M_{1a} + M_{2a}}{\bar{f}_{cw} + \bar{f}_{tw}}$

(2) Determine $f_{bT} = \dfrac{1}{R_0}\left(\bar{f}_{cw} + \dfrac{M_t - M_{2a}}{Z}\right)$

(3) Determine $f_{tT} = \dfrac{1}{R_0}\left(\bar{f}_{cw} - \dfrac{M_t + M_{1a}}{Z}\right)$

(4) Determine $\bar{P}_t = \dfrac{A}{2}(f_{bT} + f_{tT})$

(5) If $\bar{P}_t$ is the resultant prestressing force for a symmetrical section the components P_t and P'_t are easily calculated as follows.

$$k_b P_t + k_t P'_t = f_{bT} A$$

$$k_t P_t + k_b P'_t = f_{tT} A$$

Therefore,

$$P_t = \frac{(k_b f_{bT} - k_t f_{tT})}{(k_b{}^2 - k_t{}^2)} A \qquad (8.53)$$

and

$$P'_t = \frac{(k_b f_{tT} - k_t f_{bT})}{(k_b{}^2 - k_t{}^2)} \qquad (8.54)$$

ULTIMATE LOAD CONDITIONS

It will often be found that the dimensions determined for elastic conditions as described in the foregoing will be adequate also for ultimate conditions, and the following procedure will normally apply (Figure 8.22)

$$M_{1\,ult} = (f_{cm} - R_0 f_{tT}) A_f z_{max}$$

and

$$M_{2\,ult} = (f_{cm} - R_0 f_{bT}) A_f z_{max}$$

in which A_f is the area of each flange and Z_{max} the lever arm = $(h - d_f)$.

The stresses $f_{cm} - R_0 f_{tT}$ and $f_{cm} - R_0 f_{bT}$ are conservative values, since larger losses (and therefore smaller values of R_0) occur as failure is approached.

Since M_{1a} acts in the same direction as M_t, while M_{2a} is opposed by M_t, the value of $M_{1\,ult}$ is taken as $1.5 M_t + 2.5 M_{1a}$ as usual*, but $M_{2\,ult}$ should be assumed to be $2.5 M_{2a} - M_t$ (and not $2.5 M_{2a} - 1.5 M_t$). Therefore,

$$2.5 M_{1a} + 1.5 M_t = (f_{cm} - R_0 f_{tT}) A_f z_{max}$$

*CP 115 factors – for other Codes appropriate factors should be taken.

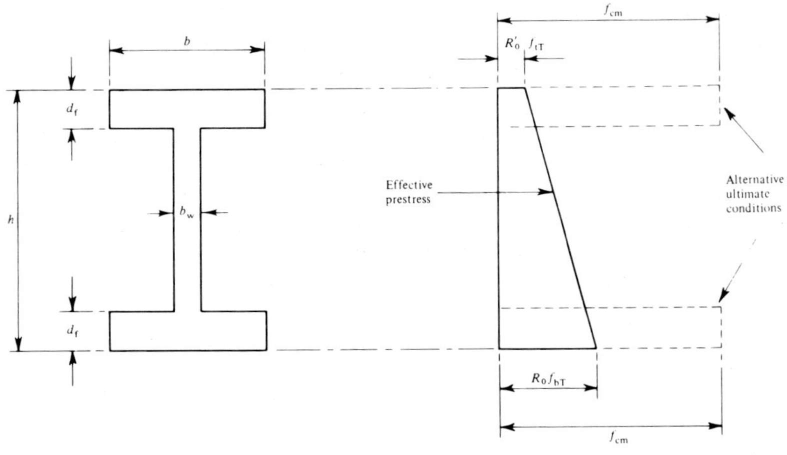

Figure 8.22 Reversal of moments: ultimate load conditions

from which

$$A_f z_{max} = \frac{2.5\,M_{1a} + 1.5\,M_t}{f_{cm} - R_0 f_{tT}}$$

Also

$$Z = \frac{M_1 + M_2}{\bar{f}_{cw} + \bar{f}_{tw}} = \frac{M_{1a} + M_{2a}}{f_{wr}}$$

in which f_{wr} is the range of stress at working load.

Hence,

$$\frac{A_f z_{max}}{Z} = \frac{2.5M_{1a} + 1.5M_t}{M_{1a} + M_{2a}} \frac{f_{wr}}{(f_{cm} - R_0 f_{tT})} = \frac{M_{1\,ult}}{M_{1a} + M_{2a}} \frac{f_{wr}}{(f_{cm} - R_0 f_{tT})} \qquad (8.55)$$

Similarly,

$$\frac{A_f z_{max}}{Z} = \frac{M_{2\,ult}}{M_{1a} + M_{2a}} \frac{f_{wr}}{(f_{cm} - R_0 f_{bT})} \qquad (8.56)$$

8.11 Design of sections unsymmetrical about vertical axis

In an unsymmetrical beam, torsional stresses and twisting can be avoided only if the load is applied through the shear centre; the position of the shear centre, for a number of typical sections, is given in Chart 6b in Part 2. However, this statement is true only for elastic sections; a plastic shear centre can be defined, but is hardly applicable to prestressed concrete, since in general the occurrence of cracking means that this material does not strictly comply with the assumption of the plastic theory and the extent of cracking varies with its load. Despite a considerable amount of research in recent years the knowledge of the behaviour of prestressed concrete members subjected to combined bending, shear and torsion has not yet reached a point at which simple design rules for ultimate-load conditions can be formulated. It therefore seems desirable to avoid the use of unsymmetrical sections where possible.

However, a different case arises with structural members which are only temporarily unsymmetrical; such as grandstand step units which are eventually supported by adjacent units (Figure 8.23). These can reasonably be designed as fully-elastic (Class 1) members during prestressing, transport, and erection, and hence the properties of an elastic section are appropriate for this case. It is also desirable to ensure that the difference in the resultant stresses near the ends and at mid-span, due to self weight is only small; thus the effects of shear and torsion can then be largely ignored. A more-or-less uniform prestress should be provided; no tensile stresses should be allowed, and the camber should be as small as possible. The design for ultimate-load conditions, on the other hand, should be based on the assumption that the structure is symmetrical for vertical loading. An example of the design of members of this type is given in Chapter 9 (Example 9.2.9).

The effects of shear are considered in Chapter 13.

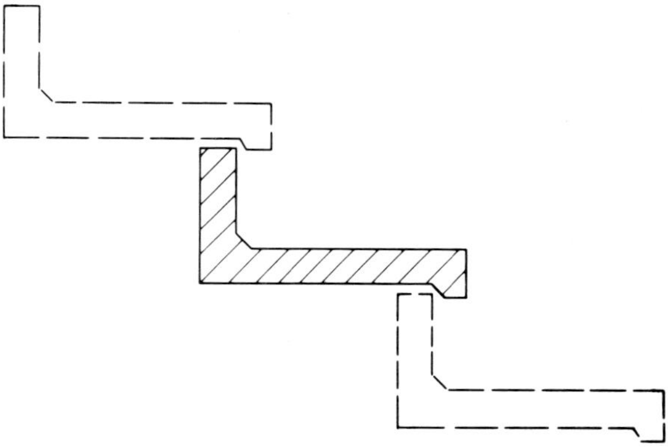

Figure 8.23 Typical step units for a grandstand structure

REFERENCES

1. FREYSSINET, E. Prestressed concrete: Principles and applications *Journal I.C.E.* Vol. 33, pp. 331–380, February 1950. No. 4, 1949/50.
2. FREYSSINET, E. The birth of prestressing, Translation No. 61.059 London 1956. pp. 1–38.
3. LIN, T. Y. Load-balancing method for design and analysis of prestressed concrete structures. *Journal A.C.I., proceedings* Vol. 60, No. 6, June 1963, pp. 719–742.
4. NILSON, A. H. Flexural design equations for prestressed concrete members. *Journal of Prestressed Concrete Institute,* Vol. 14, February 1969, pp. 54–62.
5. ABELES, P. W. Discussion to reference 4. *Journal of Prestressed Concrete Institute,* Vol. 14, February, 1969 pp. 78–81.
6. ABELES, P. W. Partial prestressing and its suitability for limit-state design. *Structural Engineer,* Vol. 49, No. 12, December 1971. pp. 529–541.
7. BOBROWSKI, J. and BARDHAN-ROY, B. K. A method of calculating the ultimate strength of reinforced and prestressed concrete beams in combined flexure and shear. *Structural Engineer,* Vol. 47, No. 5, May 1969. pp. 197–209.
8. KAAR, P. H., HANSON, N. W. and CAPELL, H. T., Stress-strain characteristics of high strength concrete. *Research and Development Bulletin,* Portland Cement Association.
9. BENNETT, E. W. A graphical method for the design of prestressed beams, *Concrete and Constructional Engineering,* Vol. 53, November 1958, pp. 399–403.

CHAPTER 9

EXAMPLES OF FLEXURAL DESIGN

9.1 Recommendations of CP 115, CP 116, CP 110 and CEB/FIP

As previously stated there is little difference in principle between CP 115[1], CP 116[2] and CP 110[3]. This book is based largely on the requirements of CP 115 (except for specific references to CP 110 and the CEB/FIP recommendations[4]), but the detailed calculations for design purposes can easily be adjusted to meet the requirements of any of these documents.

It has been stated, by the Chairman of the Drafting Committee for CP 110[5], that "the reason that very few changes were required to adapt the clauses for CP 110 was that CP 115 was effectively the first limit-state British Code". This is strictly correct: a similar basis was, however, already adopted in the *First report on prestressed concrete,* issued in 1951[6], but this document represented a set of recommendations rather than a Code of Practice.

The United Kingdom and many other countries have now adopted SI units for engineering and scientific work. In this book, however, the Imperial system has been retained, although corresponding metric and SI values have been added in virtually all cases. Further, since most of the values were originally obtained in Imperial units, the metric and SI values are given simply as conversions, and are not rounded-off to practical values.

The examples in this Chapter are, in fact, practical examples and are largely based on the requirements of the particular code of practice in use at the time the design was performed. If the design is done to comply with the recommendations of CP 115 or 116 it can easily be adjusted to accord with CP 110 if the following points are observed.

Strength. The strengths given in CP 115 are replaced by 'characteristic strengths', these are defined in CP 110 (2.3.2) as the values of the cube strength of the concrete, the yield or proof stress in the reinforcement, or the ultimate load of a prestressing tendon below which not more than 5 per cent of the test results fall.

Further, in CP 110, the design strength for a particular material and limit-state is obtained by dividing the characteristic strength f_k by an appropriate partial safety factor γ_m, so that

$$\text{design strength} = \frac{f_k}{\gamma_m} \text{(CP 110, 2.3.2)}$$

The factor γ_m is introduced to allow for the difference between the strength of a test specimen and that of the corresponding material in the structure or member itself (CP 110, 2.3.2). In general, the appropriate values of γ_m are 1.5 for concrete and 1.15 for steel; except that in the case of *excessive loads or localized damage* they may be taken as 1.3 for concrete and 1.0 for steel

(CP 110, 2.3.3.2). In the opinion of the authors, when the members are prefabricated under proper supervision and satisfactory quality control γ_m could be reduced to 1.2. Extensive study by Davis and Martin[7] on the quality control of concrete in the precast industry also suggests similar values.

Loading. CP 110 introduces the concept of characteristic loads; these are defined to be equal to the values given in CP 3, Chapter 8. The design load is obtained by multiplying the characteristic load by a partial safety factor γ_f.

The characteristic loads are denoted by G_k (dead load), Q_k (imposed design load), and W_k (design wind load). The design loads to be considered for the ultimate limit-state (CP 110, 2.3.3.1) are:

$1.4\,G_k + 1.6\,Q_k$ (dead + imposed)
$0.9\,G_k + 1.4\,W_k$ (dead + wind)
$1.2\,G_k + 1.2\,Q_k + 1.2\,W_k$ (dead + imposed + wind)

These compare with the requirements in CP 115 that 1.5 × dead load + 2.5 × live load (or 2.0 × total load, whichever is less) should be considered, without allowance for characteristic values and partial safety factors. The partial safety factor γ_f is usually taken as 1.0, but this is increased to 1.05 when the effects of misuse or accident are to be considered. On the other hand, only those loads likely to act simultaneously need to be considered. A value of 1.05 is also to be taken when considering the continued stability of a structure after it has sustained local damage. In this case, the loads should be those likely to occur before measures are taken to restrain the effect of the damage.

For the limit-state of serviceability, the loading combinations to be considered (CP 110, 2.3.4.1) are:

$G_k + Q_k$ (dead + imposed)
$G_k + W_k$ (dead + wind)
$G_k + 0.8Q_k + 0.8W_k$ (dead + imposed + wind)

In the CEB recommendations the single partial safety factor γ_f is replaced by three co-efficients, in which γ_{f1} allows for the probability that the characteristic load may be exceeded, γ_{f_2} for the chance that all loads will simultaneously reach the characteristic value, and γ_{f3} relates to inaccuracies in design assumptions and constructional tolerances.

From the foregoing, it is clear that the introduction of probability concepts into CP 110 and the CEB recommendations make it possible for skilled and experienced designers to achieve possibly (in some circumstances) more economical designs. However, as noted by Mill[5], CP 110 is "a further step in the direction of greater fineness in design. The grounds for such a step are questionable. The structural failures of recent years are more attributable to poor detail, or inadequate care in site construction, rather than to poor design". In the same discussion[5], it was generally accepted that CP 115 seems to be quite satisfactory for 90 per cent of design problems, and that a greater degree of refinement is warranted only for the remaining 10 per cent. Perhaps 10 per cent of this remainder (that is, 1 per cent of all designs) might require specialized knowledge and specific tests anyway.

9.2 Examples

9.2.1 Simply-supported beam with pre-tensioned steel

This first example relates to the preliminary design of a secondary roof beam; its results are used in Example 9.2.2 which is concerned with the primary beams for the same roof. This design is based on the roof structure at Yarmouth Station (British Rail, Eastern Region), which has given satisfactory service since 1953.

Data: The purlins are to be precast, with pre-tensioned steel. The design is to be to CP 115, and the basic design data are as follows:

Simply-supported purlins at 6 ft 3 in. (1905 mm) centres have a span of 40 ft (12.192 m) and are required to carry a live load of 15 lbf/ft^2 (73 kgf/m^2, 0.72 kN/m^2) in addition to a dead weight of 5 lbf/ft^2 (24 kgf/m^2 0.24 kN/m^2) from asbestos-cement roof sheeting. The prestressing steel is to comprise of pre-tensioned wires, of 0.2 in. (5 mm) dia., with a strength of 224 000 lbf/in^2 (15 749 kgf/cm^2, 1544 N/mm^2). The cube strength at transfer is to be 6000 lbf/in^2 (422 kgf/cm^2, 41.4 N/mm^2) and at working load 7500 lbf/in^2 (525 kgf/cm^2, 51.7 N/mm^2).

In the following, a design procedure suitable for preliminary design is given, and is explained at some length so that the successive steps may be clearly followed. More detailed examples are given later.

Permissible stresses:

Concrete $\bar{f}_{ct} = 0.4 \times 6000 = 2400$ lbf/in^2 (168.7 kgf/cm^2, 16.5 N/mm^2)
$\bar{f}_{tt} = 200$ lbf/in^2 (14.06 kgf/cm^2, 1.38 N/mm^2)
$\bar{f}_{cw} = 0.33 \times 7500 = 2500$ lbf/in^2 (175.8 kgf/cm^2, 17.2 N/mm^2)
$\bar{f}_{tw} = 500$ (for short-term loading) + 250 (assuming satisfactory test results) thus
$\bar{f}_{tw} = 750$ lbf/in^2 (52.7 kgf/cm^2, 5.17 N/mm^2)
f_{cu} (assuming equivalent rectangular stress block – see page 167).
$= 0.5 \times 7500 = 3750$ lbf/in^2 (263.7 kgf/cm^2, 25.85 N/mm^2)

Steel $f_{pi} = 0.7 \times 224\,000 = 157\,000$ lbf/in^2 (11 039 kgf/cm^2, 1082.5 N/mm^2)

Loading Roofing 5 lbf/ft^2
Live load 15 lbf/ft^2
$20 \times 6.25 = 125$ lbf/ft
Self-weight (assumed) 60 lbf/ft
185 lbf/ft (275 kgf/m, 2.7 kN/m)

Total dead load $= (5 \times 6.25) + 60 = 91.25$ lbf/ft (1.332 kN/m, 135.8 kgf/m)

Total live load $= 15 \times 6.25 = 93.75$ lbf/ft (1.368 kN/m, 139.5 kgf/m)

$$M_t = 60 \times 40^2 \times \frac{12}{8} = 144\,000 \text{ lbf in. } (1658.88 \text{ kgf m, } 16.267 \text{ kN m})$$

$$M_a = 125 \times 40^2 \times \frac{12}{8} = 300\,000 \text{ lbf in. } (3456 \text{ kgf m, } 33.89 \text{ kN m})$$

$$M_w = 185 \times 40^2 \times \frac{12}{8} = 444\,000 \text{ lbf in. } (5115 \text{ kgf m, } 50.16 \text{ kN m})$$

Assume f_{pt} = 140 000 lbf/in^2 (allowing 17 000 lbf/in^2 for initial losses) and f_{pe} = 110 000 lbf/in^2 (allowing 30 000 lbf/in^2 for later losses).

Hence

$$R_0 = \frac{110\,000}{140\,000} = 0.78$$

$$f_{br} = R_0\bar{f}_{ct} + \bar{f}_{tw} = (0.78 \times 2400) + 750 = 2620 \text{ lbf/in}^2$$

$$Z_{b\,min} = \frac{M_w}{f_{br}} = \frac{444\,000}{2620} = 169 \text{ in}^3 \text{ (2769 cm}^3\text{)}$$

$$f_{tr} = R_0\bar{f}_{tt} + \bar{f}_{cw} = (0.78 \times 200) + 2500 = 2656 \text{ lbf/in}^2$$

$$Z_{t\,min} = \frac{M_w}{f_{tr}} = \frac{444\,000}{2656} = 167 \text{ in}^3 \text{ (2737 cm}^3\text{)}$$

(No counter-action assumed)

Hence

$$Z_b \triangleq Z_t$$

$$M_{ult} = (1.5 \times \text{dead load} + 2.5 \times \text{live load}) \times \frac{l^2}{8}$$

$$M_{ult} = [(1.5 \times 91.25) + (2.5 \times 93.75)] \times 40^2 \times \frac{12}{8} = 890\,000 \text{ lbf in}$$

(10.253 kgf m, 100.54 kN m)

$$\frac{M_{ult}}{M_w}\frac{f_{tr}}{f_{cu}} = \frac{890\,000}{444\,000} \times \frac{2656}{3750} = 1.42$$

Now refer to Chart 13. Since $Z_b \triangleq Z_t$, a section with $b = b_t$ is obviously required; assuming also $b_w = 0.2b$, then a section in which $d_f = 0.25h$ gives the required value of

$$\frac{M_{ult}}{M_w}\frac{f_{tr}}{f_{cu}} \text{ and } Z_b = 0.155\,bh^2$$

Therefore

$$bh^2 = \frac{169}{0.155} = 1090 \text{ in}^3\text{, and assuming } b \triangleq 0.5h$$

$$h = \sqrt[3]{\frac{1090}{0.5}} = 13 \text{ in. (330.2 mm)}$$

Hence try section shown in Figure 9.1
Area = 0.65 bh (Chart 11) = 0.65 × 6 × 14 = 54.6 in^2 (352.26 cm^2)

$$e_b = e_t = 7 \text{ in.}$$

$$Z_b = Z_t = 0.155\,bh^2 \text{ (Chart 13)} = 0.155 \times 6 \times 14^2 = 182.2 \text{ in}^3 \text{ (2986 cm}^3\text{)}$$

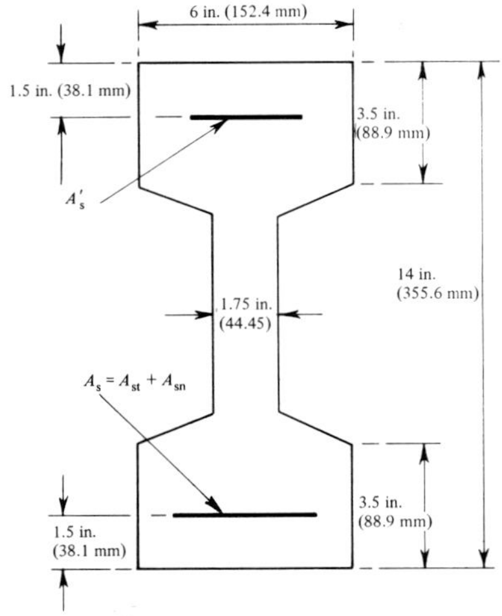

Figure 9.1 Cross-section of purlin (example 9.2.1)

$$I = Z_b \frac{h}{2} = 182.2 \times 7.0 = 1276 \text{ in}^4 \text{ (53.111 cm}^4\text{)}$$

$$i^2 = \frac{I}{A} = \frac{1276}{54.6} = 23.4 \text{ in}^2 \text{ (150.96 cm}^2\text{)}$$

$$e_s = e'_s = 7.0 - 1.5 = 5.5 \text{ in. (139.7 mm)}$$

Since both Z_b and Z_t exceed the minimum values the permissible stresses which give the lowest prestressing force are $\bar{f}_{tt}$ and $\bar{f}_{tw}$. Therefore, for no counter-action

$$f_{bT} = \frac{1}{R_0}\left[\frac{M_w}{Z_b} - \bar{f}_{tw}\right] = \frac{I}{0.78}\left[\frac{444\ 000}{182.2} - 750\right] = 2170 \text{ lbf/in}^2$$

(152 kgf/cm², 14.9 N/mm²)

$$f_{tT} = \bar{f}_{tt} = -200 \text{ lbf/in}^2 \text{ (14 kgf/cm}^2\text{, 1.37 N/mm}^2\text{)}$$

$$\bar{e}_s = \frac{i^2(f_{bT} - f_{tT})}{e_t f_{bT} + e_b f_{tT}} = \frac{23.4 \times (2170 + 200)}{7 \times 2170 - 7 \times 200} = \frac{55\ 500}{13\ 780} = 4.02 \text{ in.}$$

$$\bar{k}_b = 1 + \frac{\bar{e}_s e_b}{i^2} = 1 + \frac{4.02 \times 7}{23.4} = 2.205$$

$$\bar{k}_t = 1 - \frac{\bar{e}_s e_t}{i^2} = 1 - \frac{4.02 \times 7}{23.4} = -0.205$$

Hence

$$\bar{P}_t = \frac{A}{\bar{k}_b} f_{bT} = \frac{54.6}{2.205} \times 2170 = 53\ 700 \text{ lbf (24 358 kgf, 238.85 kN)}$$

Since possible counter-action of M_f is ignored

$$P_t = \left(\frac{\bar{e}_s + e'_s}{e_s + e'_s}\right) \bar{P}_t = \frac{4.02 + 5.5}{5.5 + 5.5} \times 53\ 700 = 46\ 600 \text{ lbf } (21\ 200 \text{ kgf, } 216 \text{ kN})$$

$$P'_t = \bar{P}_t - P_t = 53\ 700 - 46\ 600 = 7100 \text{ lbf } (3220 \text{ kgf, } 33 \text{ kN})$$

Therefore

$$A_{st} = \frac{46\ 600}{140\ 000} = 0.335 \text{ in}^2 \ (2.16 \text{ cm}^2)$$

Provide 11 No. 0.2 in. diameter wires (0.345 in^2)

$$A'_{st} = \frac{7100}{140\ 000} = 0.051 \text{ in}^2 \ (0.330 \text{ cm}^2)$$

Provide 2 No. 0.2 in. diameter wires (0.063 in^2)

ULTIMATE LOAD REQUIREMENTS

$$F_{s\,max} = \frac{M_{ult}}{z_{max}} = \frac{890\ 000}{14 - 1.5 - 2} = 85\ 000 \text{ lbf } (38\ 555 \text{ kgf, } 378 \text{ kN})$$

Therefore,

$$A_s = \frac{F_{s\,max}}{k_u f_{su}} = \frac{85\ 000}{1.0 \times 224\ 000} = 0.379 \text{ in}^2 \ (2.445 \text{ cm}^2)$$

$A_{st} = 0.345$ in^2. Hence additional non-tensioned steel required is given by

$$A_{sn} = 0.379 - 0.345 - 0.034 \text{ in}^2 \ (0.219 \text{ cm}^2)$$

Provide two 0.2 in. diameter wires (0.063 in^2)

If f_y is the yield or proof stress and $\frac{f_y}{f_{su}}$ is assumed to be 0.9

$$A_{s\,effective} = A_{st} + \left(A_{sn} \times \frac{f_y}{f_{su}}\right) = A_{st} + 0.9 \times A_{sn}$$

$$= 0.345 + 0.9 \times 0.063 = 0.401 \text{ in}^2 \ (2.587 \text{ cm}^2)$$

The properties of the section selected are known from Chart 13 to conform to ultimate-load requirements. The following demonstration (to show that this is in fact the case) is not normally required in practice, except in the final summary. It is included here simply to show the method employed for sections other than those tabulated.

$$\frac{A_{s\,effective}}{bd} \frac{f_{su}}{f_{cu}} = \frac{0.401 \times 224\ 000}{6 \times 12.5 \times 7500} = 0.16$$

Therefore, from Table 8.1, $k_u = 1.0$ and $d_{nu} = 0.4d = 0.4 \times 12.5 = 5$ in. (127 mm). For equivalent rectangular stress distribution $d_n = 0.8 \times 5 = 4$ in. (101.6 mm).

DESIGN SHEET NO 1: PRESTRESSED PURLINS

LOADING

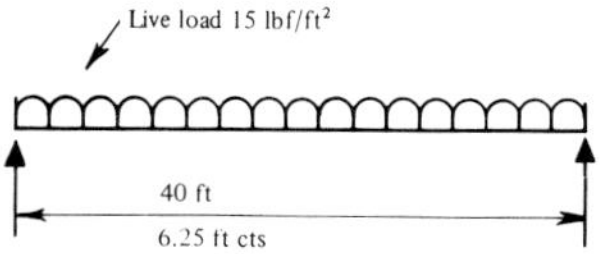

(a) Loading

Roofing: 5 lbf/ft²
Live: 15lbf/ft²

20 × 6.25 = 125 lbf/ft
Self-weight, say 60 lbf/ft

185 lbf/ft

Total dead load 91.25 lb/ft. Total live load 93.75 lbf/ft

BENDING MOMENTS $40^2 \times 1.5 = 2400$

$M_t = 60 \times 2400 = 144\ 000$ lbf in.
$M_a = 125 \times 2400 = 300\ 000$ lbf in.

$M_w = 444\ 000$ lbf in.

$M_{ult} = (1.5 \times 91.25 + 2.5 \times 93.75) \times 2400 = 890\ 000$ lbf in.

STRESSES (lb/in²)

$f_{ct} = 2400$ $f_{tt} = 200$ lbf/in²
$f_{cw} = 2500$ $f_{tw} = 500 + 250 = 750,\ f_{cu} = 0.5 \times 7500 = 3750$
$f_{pi} = 157\ 000$ $f_{pt} = 140\ 000$ (allowing initial loss of 17 000)
$f_{pe} = 110\ 000$ (allowing total loss of 47 000)
$f_{su} = 224\ 000,\ R_0 = 110/140 = 0.78$

SECTION

$$f_{br} = (0.78 \times 2400) + 750 = 2620;\ Z_{b\,min} = \frac{444\ 000}{2620} = 169 \text{ in}^3$$

$$f_{tr} = (0.78 \times 200) + 2500 = 2656;\ Z_{t\,min} = \frac{444\ 000}{2656} = 167 \text{ in}^3$$

$$\frac{M_{ult}}{M_w}\frac{f_{tr}}{f_{cu}} = 1.42. \text{ Hence try section sketched.}$$

Area = $0.65 \times 6 \times 14 = 54.6$ in² $Z_b = Z_t = 0.155 \times 6 \times 14^2 = 182.2$ in³

$$I = 182.2 \times 7 = 1276 \text{ in}^4,\ i^2 = \frac{1276}{54.6} = 23.4 \text{ in}^2,\ e_s' = e_s' = 5.5 \text{ in}$$

$$f_{bT} = \frac{1}{0.78}\left(\frac{444\ 000}{182.2} - 750\right) = 2170 \text{ lbf/in}^2\ f_{tT} = -200 \text{ lbf/in}^2$$

$$\bar{e}_s = \frac{23.4 \times 2370}{7 \times 1970} = 4.02 \text{ in. } \bar{k}_b = 1 + \frac{4.02 \times 7}{23.4} = 2.205.\ \bar{k}_t = -0.205$$

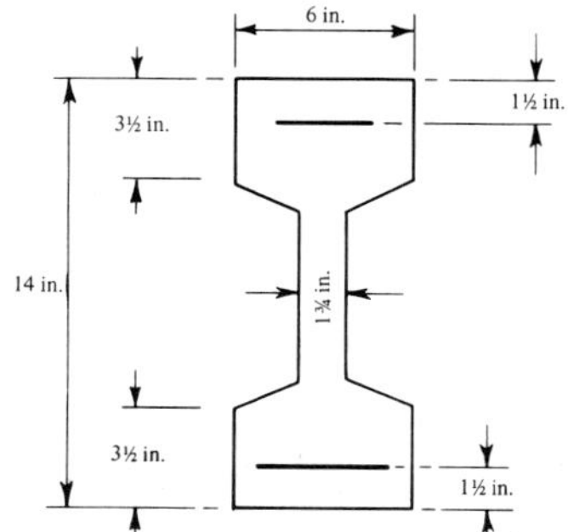

(b) Cross section

$$\bar{P}_t = \frac{54.6}{2.205} \times 2170 = 53\ 700 \text{ lbf}$$

$$P_t = \frac{4.02 + 6}{12} \times 53\ 700 = 45\ 600 \text{ lbf}$$

$$A_{st} = \frac{45\ 600}{140\ 000} = 0.325 \text{ in}^2$$

(11 No. 0.2 in. dia)

$P_t' = 8100$ lbf $A_{st}' = 0.058$ in² (2 No. 0.2 in. dia)

$$F_{su} = \frac{890\ 000}{12.5 - 2} = 85\ 000 \text{ lbf}$$

$$A_s = \frac{85\ 000}{224\ 000} = 0.379 \text{ in}^2 : A_{sn} = 0.379 - 0.345 = 0.034 \text{ in}^2 \text{ (2 No. 0.2 in. dia)}$$

Fig. 9.2 Typical calculation sheet

Actual flange thickness varies from 3.5 in. to 4.25 in.; therefore the design is suitable. Also

$$\begin{aligned} M_R &= F_{c\,max}\, z_{max} = f_{c\,max} \times \text{area of flange} \times z_{max} \\ &= 3750 \times 1.40 \times 0.155\, bh^2 \text{ (Chart 13)} \\ &= 3750 \times 0.216 \times 6 \times 14^2 = 952\,000 \text{ lbf in. (10 967 kgf m, 99.36 kN m)} \end{aligned}$$

M_{ult} = 890 000 lbf in.; hence the ultimate moment of resistance exceeds that required. A slight reduction would be possible, but since this is a preliminary design it is desirable that the section chosen should have some excess of strength. This allows for the possibility that more accurate calculations may lead to an increase in such assumed values as the dead weight and the losses; it is also possible that an increase in load may occur subsequent to the construction of the member.

The preliminary design procedure has been set out at some length in the foregoing. A more compact presentation, such as that shown in Figure 9.2 should be adopted in practice for record.

9.2.2 Roof beam with post-tensioned steel

Prepare a preliminary design for the midspan section of a roof beam of 80 ft span to support the purlins designed in Example 9.2.1. Post-tensioned grouted cables are to be used; the cube strength at transfer is to be 5500 lbf/in², and that at 28 days 7500 lbf/in² (525 kgf/cm², 51.48 N/mm²)

Permissible stresses:

Concrete $\bar{f}_{ct}$ = 0.4 × 5500 = 2200 lbf/in² (154 kgf/cm², 15.1 N/mm²)

$\bar{f}_{tt}$ = 190 lbf/in² (13.3 kgf/cm², 1.30 N/mm²)

$\bar{f}_{cw}$ = 0.33 × 7500 = 2500 lbf/in² (175 kgf/cm², 17.2 N/mm²)

$\bar{f}_{tw}$ = 325 (for short-term loading) + 250 (assuming satisfactory test results)

= 575 lbf/in² (40.3 kgf/cm², 3.95 N/mm²)

$\bar{f}_{cu}$ = 0.5 × 7500 = 3750 lbf/in² (263 kgf/cm², 25.74 N/mm²) (assuming equivalent rectangular stress block).

Steel f_{pi} = 0.7 × f_{su} = 0.7 × 224 000 = 157 000 lbf/in² (11 039 kgf/cm², 1082.5 N/mm²)

Loading

For beams at 40 ft centres, dead load (excluding self-weight)

$$= 91.25 \times \frac{40}{6.25} = 584 \text{ lbf/ft (869 kgf/m, 8.52 kN/m)}$$

$$\text{Live load} = 93.75 \times \frac{40}{6.25} = 600 \text{ lbf/ft (892.9 kgf/m, 8.76 kN/m)}$$

Self-weight, say 500 lbf/ft (744 kgf/m, 7.30 kN/m)

1684 lbf/ft (2506 kgf/m, 24.57 (kN/m)

Total dead weight = 500 + 584 = 1084 lbf/ft (1613.2 kgf/m, 15.82 kN/m)

$$M_a = (584 + 600) \times 80^2 \times \frac{12}{8} = 11\,400\,000 \text{ lbf in. (131 328 kgf m, 1287.8 kN m)}$$

$$M_t = 500 \times 80^2 \times \frac{12}{8} = 4\,800\,000 \text{ lbf in. } (55\,296 \text{ kgf m, } 542.2 \text{ kN m})$$

Assuming losses of 12 000 lbf/in^2 (due to elastic shortening and friction) before transfer and 37 000 lbf/in^2 at working load (for minimum prestressing force) then,

$$R_0 = \frac{157\,000 - 37\,000}{157\,000 - 12\,000} = \frac{120\,000}{145\,000} = 0.83$$

Hence

$$f_{br} = R_0 \bar{f}_{ct} + \bar{f}_{tw} = (0.83 \times 2200) + 575 = 2400 \text{ lbf/in}^2 \text{ (168 kgf/cm}^2\text{, } 16.47 \text{ N/mm}^2\text{)}$$

Therefore,

$$Z_{b\,min} = \frac{[M_a + (1 - R_0)M_t]}{f_{br}}$$

$$= \frac{11\,400\,000 + (0.17 \times 4\,800\,000)}{2400} = \frac{12\,216\,000}{2400} = 5080 \text{ in}^3 \text{ (83 246 cm}^3\text{)}$$

$$f_{tr} = R_0 \bar{f}_{tt} + \bar{f}_{cw} = (0.83 \times 190) + 2500 = 2658 \text{ lbf/in}^2 \text{ (186 kgf/cm}^2\text{, } 18.25 \text{ N/mm}^2\text{)}$$

Therefore,

$$Z_{t\,min} = \frac{[M_a + (1 - R_0)M_t]}{f_{tr}} = \frac{12\,216\,000}{2658} = 4600 \text{ in}^3 \text{ (75 381 cm}^3\text{)}$$

$$M_{ult} = (1.5 \times \text{dead load} + 2.5 \times \text{live load}) \times 80^2 \times \frac{12}{8}$$

$$= (1.5 \times 1084 + 2.5 \times 600) \times 9600 = 30\,000\,000 \text{ lbf in. (345 600 kgf m, 3389 kN m)}$$

Hence

$$A_f z_{max} = \frac{M_{ult}}{\bar{f}_{cu}} = \frac{30\,000\,000}{3750} = 8000 \text{ in}^3 \text{ (131 097 cm}^3\text{)}$$

Since $A_f z_{max}$ greatly exceeds $Z_{t\,min}$, it is clear that the top flange will have to be considerably larger than the bottom flange and it will thus not be possible to reach the maximum permissible stresses at working load at the top of the section. Charts 12 and 13 are prepared primarily for sections in which the top flange is smaller than the bottom; nevertheless they can easily be adapted for the opposite case if the section is imagined to be inverted.

Z_b required = 5080 in^3, $A_f z_{max}$ required = 8000 in^3

Therefore

$$\frac{A_f z_{max}}{Z_b} = \frac{8000}{5080} = 1.58$$

From Chart 12 a section with $b_w/b = 0.67$ (that is, the width of the bottom flange is 0.67 times that of the top flange) and a flange depth of $0.2h$ is suitable and Z_b (that is, Z_t on Chart 12) $= 0.101\ bh^2$. Hence $0.101\ bh^2 = 5080$, and assuming $b = 0.6h$ then

$$h = \sqrt[3]{\frac{5080}{0.6 \times 0.101}} = \sqrt[3]{83\ 600} = 43.7 \text{ in. } (111 \text{ cm})$$

Make $h = 42$ in. Then $b = \dfrac{5080}{0.101 \times 42^2} = 28.5$ in. and $b_w = 0.67 \times 28.5 = 19$ in.

(40.3 cm) and a suitable trial section is shown in Figure 9.3.

Area (Chart 11) $= 0.45\ bh = 0.45 \times 30 \times 42 = 570 \text{ in}^2\ (3677 \text{ cm}^2)$

The weight of the section is therefore $\dfrac{150}{144} \times 570 = 594$ lbf/ft (884 kgf/m, 8.67 kN/m). This is greater than the value assumed, but after multiplying by the factor $(1 - R_0)$ its effect is small.

$$Z_b\ (Z_t, \text{ Chart 12}) = 0.103\ bh^2 = 0.103 \times 30 \times 42^2 = 5440 \text{ in}^3$$
$$(89\ 150 \text{ cm}^3)$$

$$Z_t (Z_b, \text{ Chart 12}) = 0.13\ bh^2 = 0.13 \times 30 \times 42^2 = 6870 \text{ in}^3\ (112\ 580 \text{ cm}^3)$$

Distance of centroid (Chart 12) $= 0.41\,h$ (from top flange)

$$e_t = 0.41 \times 42 = 17.22 \text{ in. } (43.7 \text{ cm})$$
$$e_b = 42 - 17.22 = 24.78 \text{ in. } (62.9 \text{ cm})$$
$$I = Z_t e_t = 6870 \times 17.22 = 118\ 200 \text{ in}^4\ (4\ 919\ 850 \text{ cm}^4)$$
$$i^2 = \frac{I}{A} = \frac{118\ 200}{570} = 223 \text{ in}^2\ (1438.7 \text{ cm}^2)$$
$$e_s = 24.78 - 3 = 21.78 \text{ in. } (553.2 \text{ mm})$$
$$e_s' = 17.22 - 3 = 14.22 \text{ in. } (361.2 \text{ mm})$$

Since $Z_b > Z_{b\ min}$ and $Z_t > Z_{t\ min}$, the lowest prestressing force is obtained when the permissible tensile stresses at transfer and working load are reached.

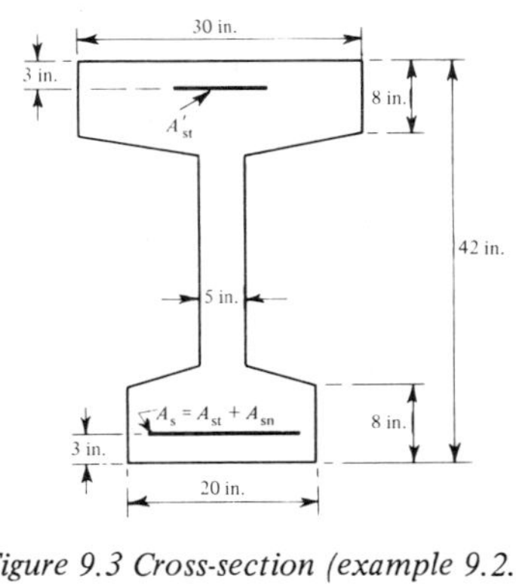

Figure 9.3 Cross-section (example 9.2.2)

However, in view of the relatively large value of M_t it is unlikely that the permissible tensile stress at transfer would be reached, and it would be advisable in a quick preliminary design to assume that the value of $\bar{e}_s$ is the greatest possible. Nevertheless, the computation is given in the following to illustrate the method.

$$f_{bT} = \frac{1}{R_0}\left[\left(\frac{M_a + M_t}{Z_b}\right) - \bar{f}_{tw}\right]$$

$$= \frac{1}{0.83}\left[\frac{11\,400\,000 + 4\,800\,000}{5440} - 575\right] = 2880 \text{ lbf/in}^2 \text{ (202 kgf/cm}^2\text{, 19.9 N/mm}^2\text{)}$$

$$f_{tT} = -\bar{f}_{tt} - \frac{M_t}{Z_t} = -190 - \frac{4\,800\,000}{6870} = -888 \text{ lbf/in}^2 \text{ (62.2 kgf/cm}^2\text{, 6.1 N/mm}^2\text{)}$$

Therefore, the value required is

$$\bar{e}_s = \frac{i^2\,(f_{bT} - f_{tT})}{e_t f_{bT} + e_b f_{tT}} = \frac{223\,[2880 - (-888)]}{17.22 \times 2880 - 24.78 \times 888} = 30.5 \text{ in. (77.3 cm)}$$

The value, as already expected, is inadmissible; hence $\bar{e}_s = e_s = 21.78$ in.

Therefore,

$$k_b = 1 + \frac{e_s e_b}{i^2} = 1 + \frac{21.78 \times 24.78}{223} = 3.32$$

and

$$k_t = 1 - \frac{e_s e_t}{i^2} = 1 - \frac{21.78 \times 18.22}{223} = -0.68$$

and

$$f_{tT} = \frac{k_t}{k_b} f_{bT} = -\frac{0.68}{3.32} \times 2880 = -590 \text{ lbf/in}^2$$

Hence

$$\bar{P}_t = P_t = \frac{A}{k_b} f_{bT} = \frac{570}{3.32} \times 2880 = 494\,000 \text{ lbf (224 000 kgf, 2195 kN)}$$

$$A_{st} = \frac{494\,000}{145\,000} = 3.406 \text{ in}^2 \text{ (21.9 cm}^2\text{)}$$

Provide 54 No. 0.276 in. (7 mm) diameter wires (3.23 in^2)

LIMIT-STATE OF COLLAPSE – M_{ult} = 30 000 000 lbf in. (345 600 kgf m, 3389 kN m) Assuming the centroid of the compressive zone to be 5 in. (127 mm) below the top face

$$F_{s\,max} = \frac{M_{ult}}{z_{max}} = \frac{30\,000\,000}{42 - 3 - 5} = 881\,000 \text{ lbf. (399 620 kgf, 3918 kN)}$$

$A_s = \frac{F_{s\,max}}{k_u f_{su}}$ and assuming $k_u = 1$, $A_s = \frac{881\,000}{1 \times 224\,000} = 3.94\ in^2\ (25.42\ cm^2)$

$A_{st} = 3.23\ in^2$ Hence additional non-tensioned steel A_{sn} required = $3.94 - 3.23 = 0.71\ in^2\ (4.58\ cm^2)$

Provide 12 No. 0.276 in. (7 mm) diameter wires ($0.718\ in^2$)

Assuming

$$\frac{f_y}{f_{su}} = 0.9, \text{ then } A_{s\,effective} = 3.23 + 0.9 \times 0.718 = 3.23 + 0.65 = 3.88\ in^2\ (25\ cm^2)$$

CHECK ON STRENGTH OF CONCRETE (unnecessary in practice)

$$\frac{A_{s\,effective}}{bd}\frac{f_{su}}{f_{cu}} = \frac{3.88}{30 \times 39} \times \frac{224\,000}{7500} = 0.099$$

Therefore from Table 8.1,

$$k_u = 1.0 \text{ (as assumed) and } d_{nu} = 0.25\,d = 0.25 \times 39 = 9.75 \text{ in.}$$

For equivalent rectangular stress distribution

$$d_n = 0.8 \times 9.75 = 7.8 \text{ in.}$$

The actual thickness of the flange varies from 8 in. to 10 in.; hence the section is adequate. As a further check taking $d_n = 8$ in. and $z = 35$ in.

$$M_R = \bar{f}_{cu} A_f z_{max} = 3750 \times 8 \times 30 \times 35$$
$$= 31\,500\,000 \text{ lbf in. } (365184 \text{ kgf m}, 3581 \text{ kN m})$$

$M_{ult} = 30\,000\,000$ lbf in. therefore, the ultimate resistance of the compressive flange of the section exceeds that required.

The preceding calculations should be set out in practice as shown previously in Figure 9.2.

9.2.3 Warehouse floor units

A warehouse floor, formed of precast prestressed units with a span of 50 ft (15.24 m) is required to support a live load of 3 tonf/yd² (3646 kgf/m², 35.75 kN/m²). The weight of each unit is not to exceed 15 tons (15 240 kgf, 149 kN) this being the maximum load for the cranes available. A complete calculation for the midspan section is required.

With a structure of this type it is convenient to consider a strip 1 ft (0.308 m) wide in the preliminary calculations, even though the final width of each unit may be several feet. It is also apparent, in view of the unusually severe loading, that the values of the stress ranges, f_{br} and f_{tr}, should be as great as possible and hence that the permissible stresses at transfer should approach, or equal, those at working load. Since the length is only 50 ft (15.24 m), pre-tensioning is quite practicable; this offers the advantages of factory conditions, with their better facilities for control and supervision, and also a better distribution of the prestressing force near the tensile face than is possible with post-tensioned cables. On the other hand, since straight wires are used, no allowance can be made for counter-action due to the weight of the member at transfer; it would be possible to overcome this disadvantage by deflecting the wires or by using additional post-tensioned cables, and this is considered in Example 9.2.4.

Assuming that the crushing strength of cubes at 28 days is to be 7500 lbf/in^2, the permissible working stress would be 2500 lbf/in^2. The permissible stress at transfer should preferably not exceed 40 per cent of the cube strength at transfer; assuming this stress is also required to be 2500 lbf/in^2, then the cube strength at transfer should not be less than 6250 lbf/in^2. The permissible tensile stress at transfer, for concrete with this cube strength, is slightly more than 200 lbf/in^2 and this value is therefore adopted. At working load, since the steel is pre-tensioned and assuming that the full working load will be applied for long periods, the permissible tensile stress is 325 lbf/in^2. According to the Code this may be increased by 250 lbf/in^2 to 575 lbf/in^2 provided that the maximum prestress is at least 1500 lbf/in^2. The prestressing steel is well distributed, and tests show that this enhanced stress is not more than 75 per cent of that at which the first crack appears. This should be achieved without difficulty since a much greater cracking stress, say of the order of 1000 lbf/in^2 would be expected.

The permissible initial tensile stress before transfer, f_{pi} in the steel is 70 per cent of the strength. Assuming the strength to be 100 tonf/in^2, f_{pi} = 70 tonf/in^2 = 157 000 lbf/in^2.

DESIGN OF MEMBERS SUBJECTED TO BENDING

Table 9.1 Preliminary estimation of losses

Loss due to	Minimum (at transfer)	Maximum (at working load)	Remarks
Elastic shortening	f_{bT} = 2500 lbf/in^2 (175 kgf/cm^2, 17.16 N/mm^2)	f_{bT} = 2500 lbf/in^2 (175 kgf/cm^2, 17.16 N/mm^2)	
Shrinkage	–	$0.3 \times 10^{-3} \times E_s$ = 8400 lbf/in^2 (588 kgf/cm^2, 57.66 N/mm^2)	$E_s = 28 \times 10^6$ lbf/in^2 (1.97×10^6 kgf/cm^2, 193 kN/mm^2) Assumes all shrinkage occurs after transfer
Creep	–	f_s equivalent = 0.95×2200 = 2100 lbf/in^2 (147 kgf/cm^2, 14.4 N/mm^2)	
Relaxation	–	10 000 lbf/in^2 (701 kgf/cm^2, 68.6 N/mm^2)	Assumes preliminary overstress of 10%

LOSSES. – The worst cases for design arise when the minimum losses occur at transfer and maximum losses at working load. The appropriate values in this case are obtained as shown in Table 9.1. Since the steel is pre-tensioned no losses due to slip or to friction need be considered. Hence at transfer, from Chart 4, f_{pt} = 145 000 lbf/in^2 and at working load f_{pe} = 109 000 lbf/in^2; R_0 = 0.74. Alternatively, using equation 6.11 Chapter 6, the losses are given by

$$\Delta f_{p\ total} = a + bx + cy + dz$$

in which a = 10 000 + 8400 = 18 400 lbf/in^2; b = 5.5; x = 2200 lbf/in^2; c = 0.33 × 28 = 9.33. For the losses due to creep an appropriate average stress must be taken into account (see Table 6.4). This is here termed $f_{s\ equiv}$ and is given by $y = f_{s\ equiv}$ = 0.95 × 2200 = 2100 lbf/in^2. d = 0. Therefore,

$$\Delta f_{p\ total} = 18\,400 + 12\,100 + 19\,600 = 50\,100 \text{ lbf/in}^2$$

Hence

$$R_0 = \frac{f_{pi} - \Delta f_{p\ total}}{f_{pi} - \Delta f_{pe}} = \frac{157\,000 - 50\,100}{157\,000 - 12\,100} = \frac{106\,900}{144\,900} = 0.737$$

From these data a preliminary investigation can be made to select a suitable depth. Assuming that 2 in. of screed will be added to the top of the floor and 1 in. of sprayed asbestos to the underside (giving fire resistances of 4 hours and 2 hours respectively – see Chapter 3), the following results are obtained.

Weight of added concrete = $\frac{2}{12}$ × 150	=	25 lbf/ft^2
Weight of concrete in joints	=	20 lbf/ft^2
Weight of asbestos and finishes	=	5 lbf/ft^2
Live load = $\frac{1}{3}$ ton per square foot	=	750 lbf/ft^2
Total		800 lbf/ft^2 (3906 kgf/m^2, 38.3 kN/m^2)

$$M_a = \frac{800 \times 50 \times 50 \times 12}{8} = 3\,038\,000 \text{ lbf in. width}$$

(34 998 kgf m, 343 kN m) per 0.308 m width

Assume dead weight is 20 per cent of added load = 0.20 × 800 = 160 lbf/ft^2. (This may differ appreciably from the actual weight, but since this is a preliminary design an incorrect assumption can be corrected later. When straight pre-tensioned steel is used, counter-action cannot be relied on and the entire self-weight must be considered). Hence assumed M_t = 0.20 × 3 038 000 = 608 000 lbf. in per foot. Therefore, ignoring counter-action,

$$Z_{b\ min} = \frac{M_a + M_t}{R_0 \bar{f}_{ct} - \bar{f}_{tw}}$$

$$= \frac{3\,038\,000 + 608\,000}{(0.74 \times 2500) + 575} = \frac{3\,646\,000}{2425} = 1505 \text{ in}^3 \text{ per foot width}$$

(24 663 cm^3 per 0.308 m width)

$$Z_{t\ min} = \frac{M_a + M_t}{R_0' \bar{f}_{tt} + \bar{f}_{cw}}$$ and assuming $R_0' = R_0$,

$$Z_{t\,min} = \frac{3\,646\,000}{(0.74 \times 200) + 2500} = \frac{3\,646\,000}{2650} = 1375 \text{ in}^3 \text{ per foot width}$$

(22 532 cm^3 per 0.308 m width)

Hence

$$\frac{Z_t}{Z_b + Z_t} = \frac{\bar{y}}{h} = \frac{1375}{1375 + 1505} = 0.478$$

The requirements given in CP 115 Code for the ultimate resistance of the compressive flange apply only to rectangular sections. In the present case an I- or box-section will be required, and the equivalent rectangular stress block is therefore used in the following.

$$\bar{f}_{cu} = 0.5 \times 7500 = 3750 \text{ lbf/in}^2$$

$$M_{ult} = (1.5 \times \text{dead load} + 2.5 \times \text{live load})\, l^2/8$$

$$= [1.5 \times (160 + 50) + 2.5 \times 750] \times 50^2 \times \frac{12}{8}$$

$$= 8\,210\,000 \text{ lbf in. per foot width (94 579 kgf m, 927.4 kN m per 0.308 m width)}$$

The alternative value of 2 M_w which is permitted by the Code would give a smaller result. However, the higher result is retained in the present example because of its instructive value.

$$M_w = 3\,646\,000 \text{ lbf in. per foot width}$$

Therefore,

$$\frac{M_u \bar{f}_{cw}}{M_w \bar{f}_{cu}} = \frac{8\,210\,000}{3\,646\,000} \times \frac{2500}{3750} = 1.50$$

Assuming that an I-section or box-section will be required, and that $b_w = 0.2\,b$ (because of the heavy loading and consequent large shearing forces) and also that $b_t \triangleq 0.85\,b$ (to provide a continuous surface for both floor and ceiling and to allow the formation of good joints between adjacent members), reference is made to Chart 13. It is clear that no section is available which meets all of these requirements; that is to say, no section which exactly fulfils the requirements for working load (with the consequent small top flange) can meet those for ultimate load (with the required larger top flange). An unsymmetrical section with a top flange deeper than the bottom flange will therefore be required.

Although Charts 12 and 13 are plotted for sections with flanges of equal depths it would be possible to adapt them to this case, by first imagining the section to be inverted and then by adjusting the dimensions of the bottom flange to suit the actual breadth required while retaining the same area. However, a simpler and better method is to select a section to suit the working load requirements and then adapt it to meet the ultimate-load conditions. From Chart 13, a section with $d_f = 0.2\,h$ (which gives a large value of d') is suitable for working-load conditions. For this case,

$\alpha' = \dfrac{Z}{bh^2} = 0.148$. Hence $bh^2 = 12h^2 = \dfrac{Z}{0.148} = 10\,150\ \text{in}^3$ per foot
$h^2 = 846\ \text{in}^2$ and $h = 29$ in. (736.6 mm).

The area of such a section is 0.579 *bh* (Chart 10) = 0.579 × 29.0 × 12 = 202 in^2 per foot width, and hence the weight per foot width is

$$\frac{202 \times 150 \times 50}{144 \times 2240} = 4.72 \text{ tons}$$

Since the limiting weight is 15 tons, the maximum suitable width for the section is about 3 ft, and this section is shown in Figure 9.4a

The moment of resistance of the concrete at ultimate load for this section is

$$M_R = 7 \times 32 \times 3750 \times 23 = 19\,300\,000 \text{ lbf in. } (222\,336 \text{ kgf m, } 2180 \text{ kN m})$$

The required moment of resistance is 8 210 000 × 3 = 24 630 000 lbf in. (according to CP 115). Hence sufficient additional concrete to provide a resistance of 5 330 000 lbf in. is required. Clearly a small increase in the thickness of the top flange of say 1 in. (25.4 mm) together with an increase in the effective depth will suffice. (This might also be achieved by means of added concrete cast in place – see Chapter 11). As a check, with the revised section shown in Figure 9.14b

$$M_u = 8 \times 32 \times 3750 \times 26 = 25\,010\,000 \text{ lbf in.}$$

which is sufficient. It will usually be the case that if the dimensions of the section are estimated carefully in this way little change will be necessary during the remainder of the design. However, in view of the large pre-tensioning force which will probably be required in the present case it is advisable to ensure that sufficient flange area is available to accommodate the steel. Taking A as 202 in^2 per foot width

$$i^2 = \frac{Z_b \bar{y}}{A} = \frac{1505 \times 0.478 \times 29}{202} = 103 \text{ in}^2$$

Hence at transfer (ignoring counter-action)

$$\bar{e}_s \simeq \frac{i^2\,(\bar{f}_{ct} + \bar{f}_{tt})}{-e_b \bar{f}_{tt} + e_t \bar{f}_{ct}} = \frac{103\,(2500 + 200)}{(29 \times -0.478 \times 200) + (0.522 \times 2500 \times 29)}$$

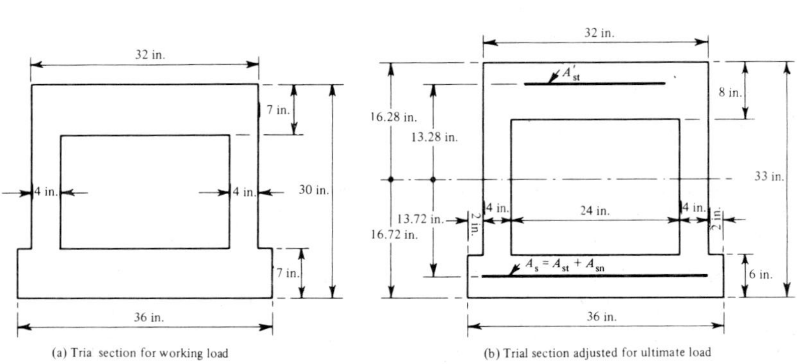

Figure 9.4 Trial sections (example 9.2.3)

$$= \frac{103 \times 2700}{29\,(1305 - 96)} = 7.92 \text{ in. } (201.17 \text{ mm})$$

Therefore,

$$\bar{k}_b = 1 + \frac{\bar{e}_s e_b}{i^2} = 1 + \frac{7.92 \times 0.478 \times 29}{103} = 2.07$$

and

$$\bar{P}_t = \frac{202}{2.07} \times 2500 = 244\,000 \text{ lbf } (110\,676 \text{ kgf, } 1085 \text{ kN})$$

Hence

$$P_t = \frac{\bar{e}_s + e_s'}{e_s + e_s'}\bar{P}_t = \frac{7.92 + 12.15}{13.85 + 12.15} \times 244\,000$$

$$= \frac{244\,000 \times 20.07}{26.0} = 188\,500 \text{ lbf } (85\,502 \text{ kgf, } 838 \text{ kN})$$

On the assumption that 0.276 in. (7 mm) diameter wires will be used the number required is

$$\frac{188\,500}{0.06 \times 145\,000} = 22 \text{ per ft}$$

These could be accommodated in three rows in a 6 in. flange; hence the flange area is adequate. However, this number is approaching the limit and in this case the use of strands could be considered as an alternative. If ½ in. (12.7 mm) diameter strands, with an area of 0.144 in^2 and a safe tensioning stress of 70 tonf/in^2 (156 800 lbf/in^2, 10 980 kgf/cm^2 = 1.08 kN/mm^2) are used, the number required would be $\dfrac{\pi \times 0.276^2 \times 22}{4 \times 0.144} = 9.1$ per foot.

The losses of the prestress would of course differ from those calculated for wires, and in view of the slightly greater relaxation which usually occurs with strands, a small increase in the number provided should be made in the final design unless stress-relieved strands are used. The requirements for ultimate load should also be checked.

The provisional section selected is now checked, the properties of the section being first calculated.

Area A (from Chart 10) = 0.53 bh = 0.53 × 36 × 33 = 629 in^2 (4058 cm^2)

Alternatively,

$$A = (8 \times 32) + (2 \times 4 \times 19) + (6 \times 36)$$
$$= 256 + 152 + 216 = 624 \text{ in}^2$$

Weight $= \dfrac{624}{144} \times 150 \times 50 = 32\,500$ lbf (14 742 kgf, 14.5 kN

Allowing 1000 lbf for diaphragms, blocks, etc., total weight = 33 500 lbf (15 195 kgf, 14.9 kN)

$$e_b = \frac{(8 \times 32 \times 29) + (2 \times 4 \times 19 \times 15.5) + (6 \times 36 \times 3)}{624}$$

$$= \frac{7420 + 2360 + 648}{624} = 16.72 \text{ in. } (424.69 \text{ mm})$$

$$e_t = 33 - 16.72 = 16.28 \text{ in. } (413.5 \text{ mm})$$

Second moment of area about centroid axis

$$I = \frac{36 \times 16.72^3}{3} - \frac{28 \times 10.72^3}{3} + \frac{32 \times 16.28^3}{3} - \frac{24 \times 8.28^3}{3} = 86\ 100 \text{ in}^4$$
$$(3\ 583\ 749 \text{ cm}^4)$$

$$Z_b = \frac{86\ 100}{16.72} = 5150 \text{ in}^3 \quad Z_t = \frac{86\ 100}{16.28} = 5290 \text{ in}^3 \ (86\ 688 \text{ cm}^3)$$

$$i^2 = \frac{I}{A} = \frac{86\ 100}{624} = 138 \text{ in}^2 \ (890.3 \text{ cm}^2)$$

Bending moments

$$M_t = 33\ 500 \times 50 \times \frac{12}{8} = 2\ 510\ 000 \text{ lbf in. } (28\ 915 \text{ kgf m, } 283.5 \text{ kN m})$$

Moment due to dead load $= 2\ 510\ 000 + 60 \times 3 \times 50^2 \times \frac{12}{8} = 3\ 118\ 000$ lbf in. (35 919 kgf m, 352.2 kN m)

Moment due to live load $= 750 \times 3 \times 50^2 \times \frac{12}{8} = 8\ 438\ 000$ lbf in. (97 206 kgf m, 953.2 kN m)

M_a = 8 438 000 + 608 000 = 9 046 000 lbf in. (104 210 kgf m, 1021.9 kN m)

M_w = 2 510 000 + 9 046 000 = 11 556 000 lbf in. (133 125 kgf m, 1305 kN m)

M_{ult} = (1.5 × dead load + 2.5 live load) moment = 4 670 000 + 21 095 000 = 25 765 000 lbf in. (296 812 kgf m., 2910.5 kN m)

Stresses – The top flange is larger than that required for working-load conditions; hence the greatest resultant eccentricity e_s of the prestressing force, and the minimum force required, are obtained when the permissible tensile stresses are reached at transfer and working load; therefore,

$$f_{bT} = \frac{1}{R_0}\left[\frac{M_w}{Z_b} + \bar{f}_{tw}\right] = \frac{1}{0.74}\left[\frac{11\ 556\ 000}{5150} - 575\right] = 2260 \text{ lbf/in}^2$$
$$(158 \text{ kgf/cm}^2, 15.5 \text{ N/mm}^2)$$

$$f_{tT} = -\bar{f}_{tt} = -200 \text{ lbf/in}^2 \ (14 \text{ kgf/cm}^2, 1.37 \text{ N/mm}^2)$$

$$\bar{e}_s = \frac{i^2 (f_{bT} - f_{tT})}{-e_b f_{tT} + e_t f_{bT}}$$

$$= \frac{138 \times (2260 - [-200])}{(16.72 \times -200) + (16.28 \times 2260)} = \frac{340\ 000}{33\ 450} = 10.15 \text{ in. } (257.81 \text{ mm})$$

$$\bar{k}_b = 1 + \frac{\bar{e}_s e_b}{i^2} = 1 + \frac{10.15 \times 16.72}{138} = 2.232$$

$$\bar{k}_t = 1 - \frac{\bar{e}_s e_t}{i^2} = 1 - \frac{10.15 \times 16.28}{138} = -0.2$$

Therefore,

$$\bar{P}_t = \frac{A}{\bar{k}_b} f_{bt} = \frac{624}{2.232} \times 2260 = 631\ 000 \text{ lbf } (286\ 215 \text{ kgf, } 2807 \text{ kN})$$

Assuming that the centroids of P_t and P_t' are 3 in. from the bottom and 3 in. from the top flanges respectively

$$P_t = \frac{\bar{e}_s + e_s'}{e_s + e_s'} \bar{P}_t = \frac{10.15 + 13.28}{13.72 + 13.28} \times \bar{P}_t = \frac{23.43}{27} \times 631\ 000 = 548\ 000 \text{ lbf}$$

(248 567 kgf, 2437 kN)

and

$$P_t' = 631\ 000 - 548\ 000 = 83\ 000 \text{ lbf } (37\ 648 \text{ kgf, } 369.2 \text{ kN})$$

Since M_t is zero at the ends of the beams, and the prestressing steel is straight, the stresses at transfer which govern the design are $f_{bt} = f_{bT} = 2260$ lbf/in^2 and $f_{tt} = f_{tT} = \frac{\bar{k}_t}{\bar{k}_b} f_{tT} = \frac{-0.2}{2.232} \times 2260 = -195$ lbf/in^2. (This figure gives a check on the accuracy of the arithmetic, since all the preceding results are obtained by inserting $f_{tT} = -200$ lb/in^2 in the appropriate formulae. Consequently, the same value should be obtained at this point.)

At the centre

$$f_{bt} = 2260 - \frac{M_t}{Z_b} = 2260 - \frac{2\ 510\ 000}{5150} = 1763 \text{ lbf/in}^2 \ (123 \text{ kgf/cm}^2,$$

12.10 N/mm^2)

$$f_{tt} = -195 + \frac{M_t}{Z_t} = -195 + \frac{2\ 510\ 000}{5290} = 279 \text{ lbf/in}^2 \ (19.5 \text{ kgf/cm}^2,$$

1.92 N/mm^2)

Check on losses – Assuming the stress at transfer in the steel f_{pi} is 150 000 lbf/in^2, the losses are now checked in detail as shown in Table 9.2. In this table the losses due to creep and elastic compression are based on the assumption that the permissible stresses are reached at transfer. This is a slightly worse condition than that actually occurring, and so gives slightly conservative results.

Hence

$$f_{pt} = 150\ 000 - 12\ 100 = 137\ 900 \text{ lbf/in}^2 \ (97 \text{ kgf/mm}^2, 950.8 \text{ N/mm}^2)$$

$f'_{pt} =$ 150 000 – 750 = 149 250 lbf/in² (105 kgf/mm², 1029 N/mm²)
$f_{pe} =$ 150 000 – 46 600 = 103 400 lbf/in² (727 kgf/mm², 712.9 N/mm²)
$f'_{pe} =$ 150 000 – 24 550 = 125 450 lbf/in² (875 kgf/mm², 864.8 N/mm²)

$$R_0 = \frac{103\,400}{137\,900} = 0.75 \qquad R'_0 = \frac{125\,450}{149\,250} = 0.84$$

$$R_e = \frac{103\,400}{150\,000} = 0.69 \qquad R'_e = \frac{125\,450}{150\,000} = 0.835$$

Hence the effective prestressing forces at working load are

$P_e =$ 548 000 × 0.75 = 411 000 lbf (186 426 kgf, 1828 kN)
$P'_e =$ 83 000 × 0.84 = 69 700 lbf (31 600 kgf, 311 kN)

Hence

$\bar{P}_e =$ 411 000 + 69 700 = 480 700 lbf (218 000 kgf, 2140 kN)

and

$$\bar{e}_{se} = \frac{(411\,000 \times 13.72) - (69\,700 \times 13.28)}{480\,700} = 9.81 \text{ in } (249.17 \text{ mm})$$

Stresses at working load – The stresses due to the effective prestressing force are

$$f_{bw} = \frac{k_{be}P_e}{A} - \frac{M_w}{Z_b}$$

and

$$f_{tw} = \frac{k_{te}P_e}{A} + \frac{M_w}{Z_t}$$

in which

$$k_{be} = 1 + \frac{\bar{e}_{se}e_b}{i^2} = 1 + \frac{9.81 \times 16.72}{138} = 2.189$$

and

$$k_{te} = 1 - \frac{\bar{e}_{se}e_t}{i^2} = 1 - \frac{9.83 \times 16.28}{138} = -0.157$$

Hence

$$f_{bw} = \frac{2.189 \times 480\,700}{624} - \frac{11\,556\,000}{5150} = 1686 - 2245$$

$$= -559 \text{ lbf/in}^2 \ (39.3 \text{ kgf/cm}^2,\ 3.86 \text{ N/mm}^2)$$

$(-\bar{f}_{tw} = -575 \text{ lbf/in}^2)$

and

$$f_{tw} = \frac{-0.157 \times 480\,700}{624} + \frac{11\,556\,000}{5290} = -121 + 2184$$

Table 9.2 Revised calculation of losses

	Bottom steel		Top steel		
Loss due to:	Minimum (at transfer)	Maximum (at working load)	Minimum (at transfer)	Maximum (at working load)	Remarks
Elastic shortening	f_{sT} = 2246 lbf/in² * Loss = 5.4 × 2246 = 12 100 lbf/in² (847 kgf/cm², 83 N/mm²)	12 100 lbf/in² (847 kgf/cm², 83 N/mm²)	f'_{sT} = 138 lbf/in² * Loss = 5.4 × 138 = 750 lbf/in² (52.7 kgf/cm², 5.15 N/mm²)	750 lbf/in² (52.7 kgf/cm², 5.15 N/mm²)	α = 5.4 when f_{cu} = 6250 lb/in² 42.9 N/mm²)
Shrinkage	–	8400 lbf/in² (588 kgf/cm², 57.7 N/mm²)	–	8400 lbf/in² (588 kgf/cm², 57.7 N/mm²)	No change
Creep	–	$f_{s\,equiv}$ = 0.95 × 1832 = 1742 lbf/in² Loss = 0.33 × 28 × 1742 = 16 100 lbf/in² (1127 kgf/cm², 110.5 N/mm²)	–	$f'_{s\,equiv}$ = 0.95⁺ × 509 + 0.05 × 2000 = 585 Loss = 0.33 × 28 × 585 = 5400 lbf/in² (380 kgf/cm², 37 N/mm²)	Assumes M_t acts alone for 12 months after which f_{tw} = 2000 lbf/in² (140 kgf/cm², 13.73, N/mm²)
Relaxation	–	10 000 lbf/in² (701 kgf/cm², 68.6 N/mm²)	–	10 000 lbf/in² (701 kgf/cm², 68.6 N/mm²)	No change
TOTALS:	12 100 (847 kgf/cm², 83 N/mm²)	46 600 (3276 kgf/cm², 327 N/mm³)	750 (52.7 kgf/cm², 5.15 N/mm²)	24 550 lbf/in² (1721 kgf/cm², 168.50 N/mm²)	

*

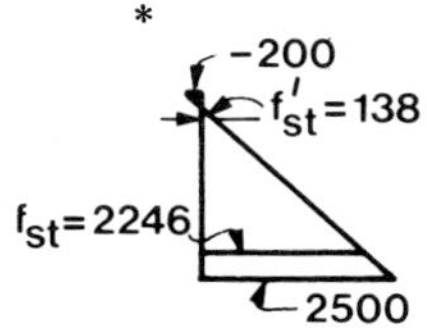

Stress diagram for elastic loss at transfer. (assuming permissible stresses are reached)

\+

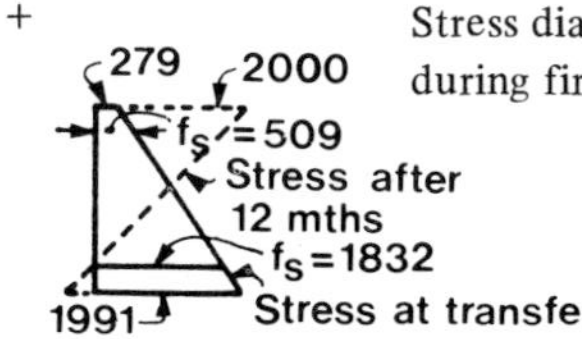

Stress diagram for loss due to creep during first 12 months and afterwards.

$= 2063\ \text{lbf/in}^2\ (\bar{f}_{cw} = 2500\ \text{lbf/in}^2)$

$(145\ \text{kgf/cm}^2,\ 14.22\ \text{N/mm}^2)$

Prestressing steel – At transfer

$P_t = 548\,000\ \text{lbf};\quad f_{pt} = 137\,900\ \text{lbf/in}^2$
$P'_t = 83\,000\ \text{lbf};\quad f'_{pt} = 149\,250\ \text{lbf/in}^2$

Hence

$$A_{st} = \frac{548\,000}{137\,900} = 3.98\ \text{in}^2\ (25.68\ \text{cm}^2)\ A'_{st} = \frac{83\,000}{149\,250} = 0.556\ \text{in}^2\ (3.58\ \text{cm}^2)$$

Assuming 0.276 in. diameter wires, the number of tensioned wires at the bottom

$$= \frac{3.98}{0.06} = 67\ (4.02\ \text{in}^2)\ \text{and at the top} = \frac{0.556}{0.06} = 10\ (0.60\ \text{in}^2)$$

LIMIT-STATE COLLAPSE: $M_u = 25\,765\,000$ lb in. Assuming $k_u = 1.0$, area of steel required is $= \dfrac{25\,765\,000}{1 \times 224\,000 \times 26} = 4.42\ \text{in}^2\ (28.52\ \text{cm}^2)$

$A_{st} = 4.02\ \text{in}^2$. Therefore $A_{sn} = 4.42 - 4.02 = 0.40\ \text{in}^2$. Provide 7 non-stressed wires (0.42 in^2). For determining the percentage/strength ratio, equivalent steel area $= 4.02 + 0.9 \times 0.42 = 4.02 + 0.38 = 4.40\ \text{in}^2$ (in which 0.9 is assumed to be the approximate ratio of the proof stress to the strength).

Check on concrete – For equivalent rectangular stress block, the mean stress in the concrete at ultimate load $= \dfrac{25\,765\,000}{8 \times 32 \times 26} = 3870\ \text{lbf/in}^2$ compared with the permissible value of 3750 lbf/in^2. The reason for this increase is the original underestimate of self-weight and dead load. Clearly it would be possible to increase either the top flange or the depth of the section; however, since the excess is small, it is worth while to make a more accurate assessment of the ultimate moment of resistance before doing so. The percentage/strength ratio is

$$\frac{A_{st}}{bd}\frac{f_{su}}{f_{cu28}} = \frac{4.40}{32 \times 30} \times \frac{224\,000}{7500} = 0.137$$

Hence $d_{nu} = 0.34d$. If the section is now considered as consisting of two solid rectangular portions (1,2) connected by two flanges (3) (Figure 9.5), the following results are obtained.

Rectangular portions 1, 2. –

Area of steel

$$A_{st1} = A_{st2} = \frac{4.40 \times 6}{36} = 0.733\ \text{in}^2\ (4.73\ \text{cm}^2)$$

Hence

$$\frac{A_{st1}}{b_1 d} \times \frac{f_{su}}{f_{cu}} = \frac{0.733}{4 \times 30} \times \frac{224\,000}{7500} = 0.1835$$

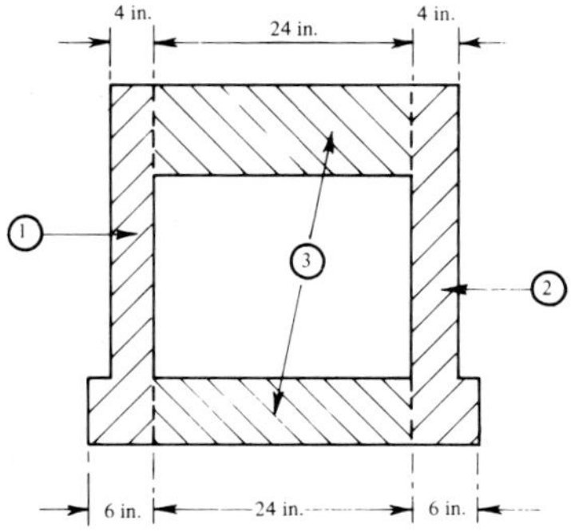

Figure 9.5 Revised cross-section (example 9.2.3)

and from Table 8.1, $d_{nu\,1} = 0.445 \times 30 = 13.35$ in. Therefore ultimate moment of resistance of rectangular portions = $2 \times (0.4 \times 7500) \times 4 \times 13.35 \times (30 - 0.4 \times 13.35) = 8\,500\,000$ lbf in. (97 920 kgf m, 960 kN m)

Flanges 3. –

Area of steel

$$A_{st3} = \frac{4.40 \times 24}{36} = 2.933 \text{ in}^2 \ (18.9 \text{ cm}^2)$$

$$\frac{A_{st3}}{b_3 d} \times \frac{f_{su}}{f_{cu}} = \frac{2.933}{24 \times 30} \times \frac{224\,000}{7500} = 0.122$$

Hence from Table 8.1, $d_{nu} = 0.31 \times 30 = 9.3$ in. This exceeds the actual depth of the flange, and the equivalent rectangular stress distribution is used instead.

Equivalent

$$d_{nu3} = 0.8 \times 9.3 = 7.44 \text{ in.}$$

$M_{R\,\text{flange}} = (0.5 \times 7500) \times 24 \times 7.44 \times (30 - 0.5 \times 7.44) = 17\,600\,000$ lbf in. Hence $M_{R\,\text{total}} = 17\,600\,000 + 8\,500\,000 = 26\,100\,000$ lbf in. (300 672 kgf m, 2948 kN m) which is greater than the required value of 25 765 000 lbf in.

Alternatively, using the formula given in the US Building Code

$$M_R = f'_c\,(0.25 \times 8 \times 30^2 + 0.85 \times 24 \times 8 \times 26)$$

$$= f'_c\,(1800 + 4250) = 6050 f'_c$$

and assuming

$$f'_c = 0.6 f_{cu\,28} = 4500 \text{ lbf in}^2 \text{ (which is a conservative value),}$$

$$M_R = 6050 \times 4500 = 27\,250\,000 \text{ lbf in. (313 920 kgf m, 3078 kN m)}$$

Because of the amount of work involved in the preparation of an accurate design, it is always worth while to extract for future reference the main checked data and properties, in a form such as that shown in Figure 9.6. This should preferably be prepared on a separate sheet, which can be filed apart from the main calculations.

DESIGN SHEETS FOR WAREHOUSE FLOOR UNITS:

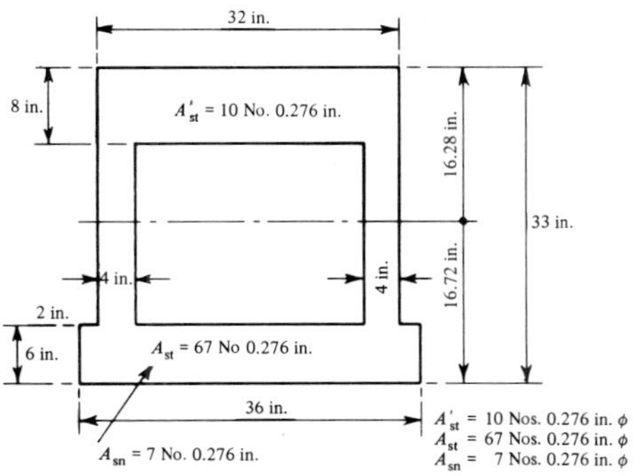

Span 50′–0″
Live load: 750 lbf/ft²
Weight: 220 lbf/ft²
Addl. dead wt: 50 lbf/ft²
TOTAL: 1020 lbf/ft²

Area = 624 in²
I = 86 100 in⁴ i^2 = 138 in²
Z_b = 5150 in³ Z_t = 5290 in³
A_{st} = 67/0.276 in. (4.02 in²)
A'_{st} = 10/0.276 in. (0.60 in²)
A_{sn} = 7/0.276 in. (0.42 in²)

STRESSES: (lb/in²)

$f_{bT} = 2260$; $f_{tT}\ -200$
$*f_{be} = 1690 - 497 = 1193$; $f_{te} = -132 + 474 = 342$
$f_{bd} = 1075$; $f_{td} = 457$
$f_{bw} = 555$; $f_{tw} = 2048$
$f_{pi} = 150\,000$
$f_{pt} = 137\,900$; $f'_{pt} = 149\,250$
$f_{pe} = 103\,400$; $f_{bn} = 118\,700$

$$*f_{be} = R_0 f_{bT} - \frac{M_t}{Z_b}$$

$$f_{te} = R'_0\, f_{tT} - \frac{M_t}{Z_t} \qquad Z_b M_t = 2\,510\,000 \text{ lbf in.}$$

Figure 9.6 Summary of design

9.2.4 Warehouse floor units with straight and deflected pre-tensioned steel

In large members with pre-tensioned steel, such as those designed in Example 9.2.3. some of the wires can be deflected to obtain counter-action of M_t. Assuming the dead-weight of the member to be 25 per cent of the added load and with the stresses and loadings given in Example 9.2.3, then

$R_0 = 0.74$; $1 - R_0 = 0.26$; $M_a = 3\,038\,000$ lbf in. per foot
$M_t = 0.25 \times 3\,038\,000 = 759\,500$ lbf in. per foot

$$Z_{b\,min} = \frac{3\,038\,000 + (0.26 \times 759\,500)}{2425} = 1322 \text{ in}^3 \text{ per foot } (21\,664 \text{ cm}^3 \text{ per } 0.308 \text{ m width})$$

$$Z_{t\,min} = \frac{3\,038\,000 + (0.26 \times 759\,500)}{2650} = 1221 \text{ in}^3 \text{ per foot } (20\,009 \text{ cm}^3 \text{ per } 0.308 \text{ m width})$$

The values obtained in Example 9.2.3. (without counter-action) are 1505 and 1375 in. ft respectively.

$$M_{ult} = [1.5 \times (200 + 50) + 2.5 \times 750] \times 50^2 \times \frac{12}{8} = 8\,430\,000 \text{ lbf in per foot}$$

(97 114 kgf m, 952 kN m per 0.308 m width)

$$A_f z_{max} = \frac{M_{ult}}{f_{cu}} = \frac{8\,430\,000}{3750} = 2250 \text{ in}^3 \text{ per foot} = 1.85\, Z_{t\,min}$$

A section is selected for conditions at working load and adapted to suit those at ultimate load. From Chart 13, a section with $b_t = 0.85\,b$, $b_w = 0.2\,b$ and $d_f = 0.2\,h$ is suitable, and hence $Z_{t\,min} = 1322 \text{ in}^3$ per foot $= 0.148\,bh^2$. Thus $bh^2 = 12h^2 = 1322/0.148$ and $h = 27.3$ in. (693.42 mm), giving the trial section shown in Figure 9.7

This section is smaller than that shown in Figure 9.6; however, the final dimensions of the section are still decided by the resistance required at ultimate load and the area required to accommodate the prestressing steel, and hence the section shown in Figure 9.4a will still apply.

Prestressing force – As before, the minimum prestressing force is obtained when the permissible tensile stresses are reached at transfer and working load. Since counter-action occurs in the present design, the value of e_s needed to obtain this condition would exceed the depth of the section. Hence the permissible tensile stress at transfer will not be reached, and

$$f_{bT} = \frac{1}{R_0}\left[\left(\frac{M_a + M_t}{Z_b}\right) + \bar{f}_{tw}\right]$$

$$= \frac{1}{0.74}\left[\left(\frac{9\,046\,000 + 2\,510\,000}{5150}\right) - 575\right] = 2320 \text{ lbf/in}^2 \text{ (162 kgf/cm}^2\text{,}$$

16 N/mm^2)

$$\bar{e}_s = 16.72 - 3 = 13.72 \text{ in. (348.49 mm)}$$

$$k_b = 1 + \frac{\bar{e}_s e_b}{i^2} = 1 + \frac{13.72 \times 16.72}{138} = 2.662$$

$$k_t = 1 - \frac{\bar{e}_s e_t}{i^2} = 1 - \frac{13.72 \times 16.28}{138} = -0.62$$

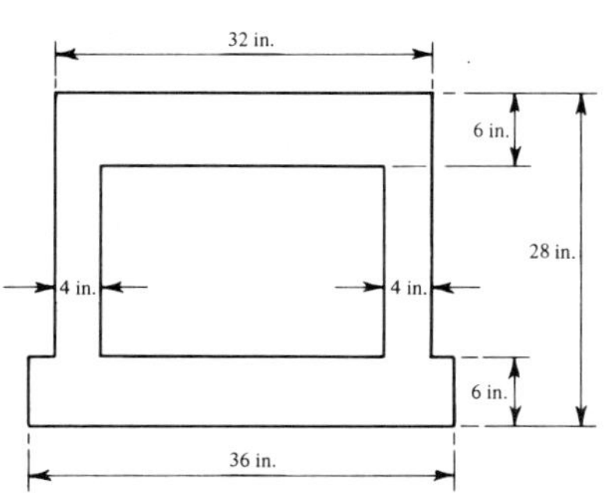

Figure 9.7 Trial section (example 9.2.4)

$$f_{tT} = \frac{k_t}{k_b} f_{bT} = \frac{-0.62}{2.662} \times 2320 = -540 \text{ lbf/in}^2 \text{ (38 kgf/cm}^2\text{, 3.7 N/mm}^2\text{)}$$

$$\bar{P}_t = \frac{A}{k_b} f_{bT} = \frac{624}{2.662} \times 2320 = 543\ 800 \text{ lbf (246 600 kgf, 2418 kN)}$$

This is the total prestressing force required at transfer, compared with 631 000 lbf (286 200 kgf, 2800 kN) required in Example 9.2.3.

Stresses – The figures in brackets are those from Example 9.2.3 and are repeated here for comparison (all in Imperial units only).

$$f_{bT} = 2320 \text{ lbf/in}^2 \text{ (2260)}$$

$$f_{tT} = -540 \text{ lbf/in}^2 \text{ (–195)}$$

$$f_{bt} = f_{bT} - \frac{M_t}{Z_b} = 2320 - \frac{2\ 510\ 000}{5150} = 1832 \text{ lbf/in}^2 \text{ (1763)}$$

$$f_{tt} = f_{tT} + \frac{M_t}{Z_t} = -540 + \frac{2\ 510\ 000}{5290} = -65 \text{ lbf/in}^2 \text{ (279)}$$

$$f_{be} = R_0 f_{bt} = 0.74 \times 1832 = 1355 \text{ lbf/in}^2 \text{ (83.4 kgf/cm}^2\text{, 8.18 N/mm}^2\text{)}$$

$$f_{te} = R_0' f_{tt} = 0.74 \text{ (approximately)} \times (-65) = -48 \text{ lbf/in}^2 \text{ (0.7 kgf/cm}^2\text{, 0.69 N/mm}^2\text{)}$$

$$f_{bw} = f_{be} - \frac{M_a}{Z_b} = 1355 - \frac{9\ 046\ 000}{5150} = 1355 - 1757 = -402 \text{ lbf/in}^2 \text{ (–555)}$$

$$f_{tw} = f_{te} + \frac{M_a}{Z_t} = -48 + \frac{9\ 046\ 000}{5290} = -48 + 1710 = 1662 \text{ lbf/in}^2 \text{ (2048)}$$

The method of calculating the number of deflected wires and ensuring that the permissible stresses are not exceeded is given in Chapter 14.

It will be noted that the prestressing force is appreciably smaller in the second case; this is because of the increased value of e_s made possible by counter-action. In consequence the stresses in the top flange are also reduced.

In a practical design it would be necessary to investigate more accurately the losses of prestress. In particular, if the wires are tensioned in the deflected position the consequent frictional losses should be considered. Alternatively, if the wires are deflected after tensioning, there are no frictional losses but a smaller tensioning stress should be allowed since additional stresses are imposed when the wires are deflected.

When the dimensions of the top flange and the depth of a section are determined by conditions at ultimate load, the presence of counter-action will in general lead to a small reduction in the dimensions of the bottom flange and an appreciable reduction in the prestressing force. When the dimensions are determined by conditions at working load, the presence of counter-action results in a reduction in the depth of the section and an appreciable increase in the prestressing force. An example illustrating this case is given in Chapter 11; in this the warehouse floor considered in the foregoing is designed for conditions at working load, the necessary additional ultimate resistance being provided by concrete cast in place. The statement made sometimes that in prestressed concrete the dead load carries itself for nothing, is incorrect; it requires either a greater depth (when the prestressing force is reduced) or a greater prestressing force (when the depth is not increased).

9.2.5 Beam of varying cross-section

If the cross-section varies along the beam, the point at which the bending moment is the maximum is not necessarily that at which the maximum stress occurs. The location of the point of maximum stress is a matter of trial and error, except in the particular case when the variation of the section is known in advance; this is considered in the following.

Design a beam with the profile shown in Figure 9.8 to support a live load of 175 lbf/ft (260 kgf/m, 2.55 kN/m) over a span of 56 ft (17.07 m) for conditions at working load. Allow $\bar{f}_{ct}$ = 2400 lbf/in^2, (166 kgf/cm^2, 16.5 N/mm^2) $\bar{f}_{tw}$ = 575 lbf/in^2 (40.4 kgf/cm^2, 3.96 N/mm^2) and R_0 = 0.8. The weight of the beam counteracts at transfer.

To locate the point of maximum stress, the section moduli and bending moments are calculated at points spaced 7 ft apart along the beam. (It is assumed that the variation of M_t is parabolic; this is sufficiently accurate for present purposes.) These values are given in Table 9.3, and plotted in Figure 9.9. It is seen that the greatest value of $\frac{M}{Z}\frac{Z_c}{M_c}$ occurs at a point 18 ft from the support, and this is therefore the critical section on which the remainder of the design is based.

Properties of critical section 18 ft from support.

$$h = h_c\left(0.6 + \left[\frac{18}{28} \times 0.4\right]\right) = 0.857\ h_c$$

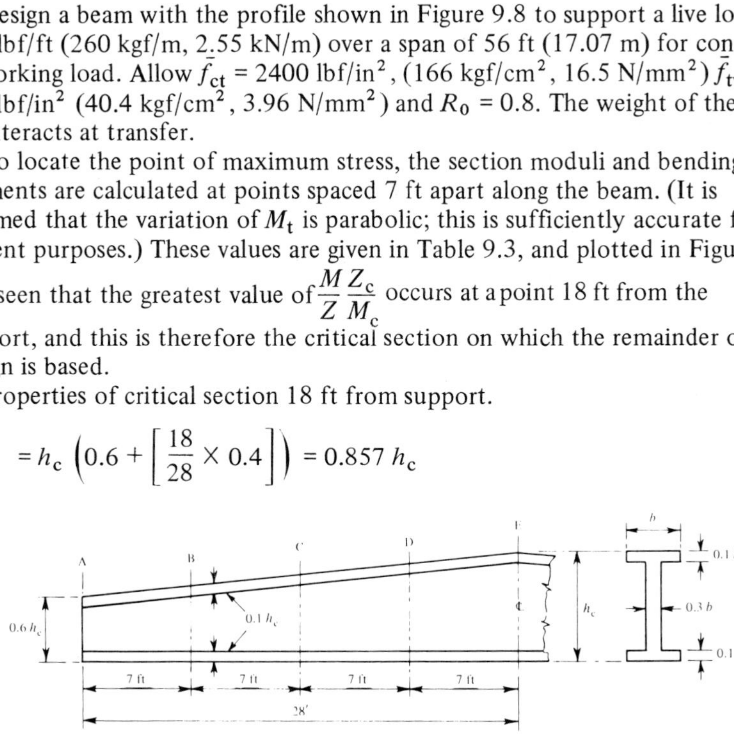

Figure 9.8 Beam of varying cross-section (example 9.2.5)

Table 9.3 Position of maximum stress

Point	A	B	C	D	E
$\frac{Z}{Z_c}$	0.445	0.569	0.705	0.849	1.000
$\frac{M}{M_c}$	0	0.438	0.75	0.938	1.000
$\frac{MZ_c}{ZM_c}$	0	0.769	1.065	1.105	1.000

$d_f = 0.1\,h_c = 0.1165\,h$

$$I = \frac{bh^3}{12}\,(1 - [0.7 \times 0.757^3]) = 0.057\,bh^3$$

$Z = 2 \times 0.057\,bh^2 = 0.114\,bh^2$

$$M_a = 175 \times 56^2 \times 1.5 \times \left(1 - \frac{10^2}{28^2}\right) = 718\,000\ \text{lbf in. (8271 kgf m, 81.1 kN m)}$$

Assume M_t = 350 000 lbf in. (4032 kg m, 39.5 kN m)

$f_{br} = R_0\bar{f}_{ct} + \bar{f}_{tw} = (0.8 \times 2400) + 575 = 2495$ lbf/in² (175 kgf/cm², 17.2 N/mm²)

$$Z_{b\,min} = \frac{M_a + (1 - R_0)M_t}{f_{br}} = \frac{718\,000 + 70\,000}{2495} = 317\ \text{in}^3\ (5195\ \text{cm}^3)$$

Assuming $b = \frac{h_c}{2} = \frac{h}{2 \times 0.857}$ then $\frac{0.114h^3}{2 \times 0.857} = 317$ and $h = 16.8$ in. (426.72 mm); $h_c = 19.6$ in. (497.85 mm)

Make $b = 10$ in. (254 mm); $h_c = 20$ in. (508 mm); $h = 0.857 \times 20 = 17.14$ in. (435.4 mm)

$I = 0.057 \times 10 \times 17.14^3 = 2875$ in⁴ (119 666 cm⁴)

$A = 10 \times 17.14\,(1 - [0.7 \times 0.767]) = 79.5$ in² (512.9 cm²)

$$i^2 = \frac{2875}{79.5} = 36.2\ \text{in}^2\ (233.5\ \text{cm}^2)$$

$$Z = \frac{2875 \times 2}{17.14} = 337.5\ \text{in}^3\ (5531\ \text{cm}^3)$$

$$e_b = e_t = \frac{17.14}{2} = 8.57\ \text{in. (217.7 mm)};\ e_s = 8.57 - 1 = 7.57\ \text{in. (192.3 mm)}$$

$$k_b = 1 + \frac{8.57 \times 7.57}{36.2} = 2.79;\ k_t = -0.79$$

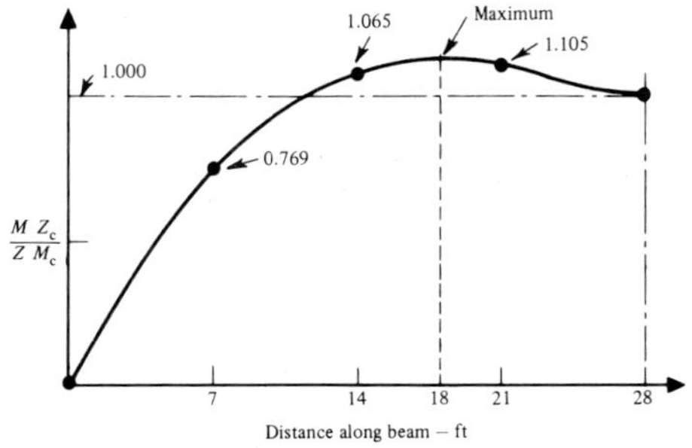

Figure 9.9 Position of maximum stresses (example 9.2.5)

Prestressing force – At the critical section, the exact value of M_t is 339 000 lbf in. and M_a = 718 000 lbf in.

Hence

$$f_{bT} = \frac{1}{R_0}\left[\left(\frac{M_a + M_t}{Z_b}\right) - \bar{f}_{tw}\right]$$

$$\therefore f_{bT} = \frac{1}{0.8}\left[\left(\frac{721\ 000 + 339\ 000}{337.5}\right) - 575\right] = 3200 \text{ lbf/in}^2 \text{ (224 kgf/cm}^2\text{, 22 N/mm}^2\text{)}$$

Hence

$$P_t = \frac{A}{k_b} f_{bT} = \frac{79.5}{2.79} \times 3200 = 91\ 000 \text{ lbf (41 270 kgf, 404 kN)}$$

These results are used in Chapter 13 in connection with the bending-up of the prestressing steel.

9.2.6 Example of lightweight concrete double tee beams

It is often necessary to design a beam for a definite shape. This applies particularly to T beams (either single-T or double-T beams) for which standardized moulds are available. As previously indicated T beams are very advantageous with regard to prestressing force and counter-action. It is possible to use existing shapes for different conditions, spans and functional requirements.

Usually T beams are used as composite members, i.e. precast section combined with added concrete. Such examples are shown in Chapter 11. In the present example a solution is shown in which the precast member itself is strong enough to carry the load and this is made of lightweight aggregate concrete in order to reduce the self-weight.

Figure 9.10 shows the typical cross-section of the double-T beam which is 27.6 in. (70 cm) deep with rib spacing at 47.52 in. (120 cm) centres. The total width of the present units is 94 in. (239 cm).

The unit spans 80 ft (24.4 m) and constitutes the roof of a factory building in Southern Ireland.

The density of the lightweight concrete used is 120 lb/ft^3 (1922 kg/m^3)

Concrete strength at 28 days = 7500 lbf/in^2 (620 kgf/cm^2, 62 N/mm^2) and at release 5000 lbf/in^2 (350 kgf/cm^2, 35 N/mm^2). $E_{c28} = 3.5 \times 10^6$ lbf/in^2 (2.45×10^5 kgf/cm^2, 2.4×10^4 N/mm^2) and $E_{cT} = 2.9 \times 10^6$ lbf/in^2 (2.00×10^5 kgf/cm^2, 2.0×10^4 N/mm^2). There are 8 No. 0.27 in. (7 mm) diameter wires in the top flange with a strength of 224 000 lbf/in^2 (155 kgf/mm^2, 1540 N/mm^2).

Each rib contains 8 No. ½ in. (12.7 mm) diameter Dyform strands with an ultimate stress of 270 000 lbf/in^2 (190 kgf/mm^2, 1870 N/mm^2).

Average $E_s = 28 \times 10^6$ lbf/in^2 (19.5×10^5 kgf/cm^2, 19.3×10^4 N/mm^2).

Modular ratio $\alpha_c = 8$ (final)

$\alpha_c \doteqdot 10$ (at transfer)

Cross-sectional area of ½ in. Dyform strand = 0.174 in^2 (112 mm^2)

Cross-sectional area of 0.276 in. diameter wire = 0.0598 in^2 (38 mm^2)

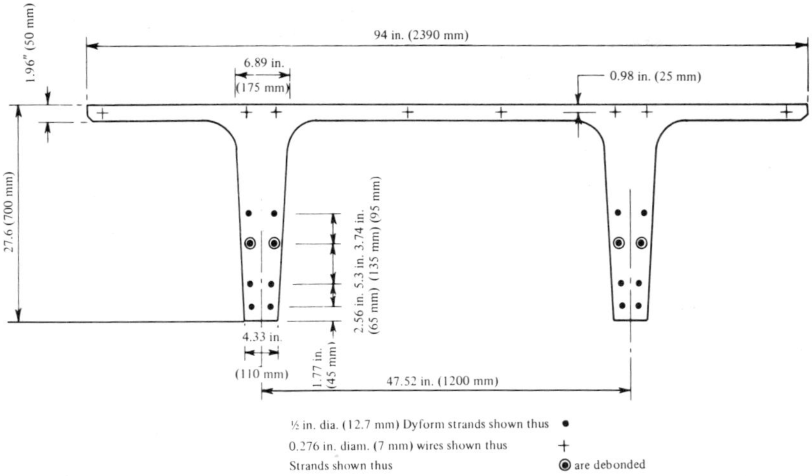

Figure 9.10 Cross-section of double-T beam (example 9.2.6)

Loading:

Own weight	= 405 lbf/ft
Screed @ 10 lbf/ft^2	= 79 lbf/ft
3-ply felt @ 5 lbf/ft^2	= 40 lbf/ft
Service pipes etc. @ 5 lbf/ft^2	= 40 lbf/ft
Dead load	= 564 lbf/ft (839 kgf/m, 8.23 kN/m)
Live load @ 15.0 lbf/ft^2	= 120 lbf/ft (178.5 kgf/m, 1.75 kN/m)

DESIGN FOR LIMIT-STATE OF COLLAPSE

Case 1: In accordance with CP 115

$$\frac{A_{st} \times f_{su}}{f_{cu} \cdot b.d} \qquad \text{Average } b = 2 \times 5\tfrac{1}{2} \text{ in} = 11 \text{ in (279.40 mm)}$$

$$= \frac{16 \times 0.174 \times 270\,000}{(1.2 \times 7500) \times 11 \times 20.31} = 0.37 \qquad \therefore k_u = 0.93 \text{ say (Table 8.1)}$$

$\therefore$ Ultimate force in steel $F_{su} = 0.93 \times 16 \times (0.174 \times 270\,000)$ (taking 16 Dyform strands in the ribs)
$= 699\,360$ lbf (317 225 kgf, 311 kN)

$$\text{C.G. of } F_{su} = \frac{2 \times (45 + 110 + 245 + 340)}{8} = 185 \text{ mm, i.e. 7.29 in. from}$$

soffit $\therefore d = 27.6 - 7.29 = 20.31$ in. (516 mm)

$\bar{f}_{cu} = 0.4 \times 7500 \times 1.20 = 3600 \text{ lbf/in}^2$

where 1.20 is the assumed age factor corresponding to six months.

$$\text{Area of concrete to balance } F_{su} = \frac{699\,360}{3600} = 194 \text{ in}^2 \text{ (125 334 mm}^2\text{)}$$

$\therefore$ Depth of neutral axis $= \dfrac{194}{94} = 2.06 >$ depth of flange

Assuming compression in flange only (ignoring web)

$F_{cu} = 94 \times 1.97 \times 3600 = 665\ 000$ lbf (301 639 kgf, 2958 kN)

$\therefore$ Lever arm $z = 27.6 - \left(7.29 + \dfrac{1.97}{2}\right) = 19.33$ in. (491 mm)

$\therefore$ Permissible $M_R = 665\ 000 \times 19.33 = 12\ 880\ 000$ lbf in. (148 120 kgf m, 1452.5 kN m)

Required $M_u = (564 \times 80^2 \times 1.5) \times 1.5 + (120 \times 80^2 \times 1.5) \times 2.5$
$= 11\ 000\ 000$ lbf in. (126 500 kgf m, 1240 kN m) $<$ Permissible M_R

Hence OK

Case 2: In accordance with CP 110

$F_{su} = \dfrac{699\ 360}{1.15} = 608\ 139$ lbf (275 848 kgf,)

As this is nearly equal to F_{cu} in Case 1, the lever arm = 19.33 in.

Permissible $M_R = 608\ 139 \times 19.33$
$= 11\ 755\ 000$ lbf in. (135 400 kgf m, 1328 kNm)

Required $M_{ult} = 1.4 \times (564 \times 80^2 \times 1.5) + 1.6 \times (120 \times 80^2 \times 1.5)$
$= 9\ 423\ 000$ lbf in. (108 500 kgf m, 1065 kN m)
$\therefore$ OK

Note: In accordance with CEB/FIP Manual(8) on Lightweight Aggregate Concrete, $f_{cu} = 0.5 \times 7500 \times$ age factor, which obviously leads to higher ultimate moment of resistance, due to less compression area required, and greater lever arm, resulting in even higher factor of safety.

CHECK FOR SERVICEABILITY CONDITIONS

Properties of concrete section

$A_0 = 485$ in^2 (3134 cm^2) (area of concrete section)
$e_t = 8.6$ in (21.8 cm)
$e_b = 19$ in (48.2 cm)
$I_0 = 34\ 400$ in^4 (1 427 800 cm^4)
$i^2 = 71$ in^2 (458 cm^2)
C.G. of steel in the rib = 7.29 in. from the bottom surface
$\therefore\ e_s = 19 - 7.29 = 11.71$ in.
C.G. of steel in flange from top surface = 0.98 in. (25 mm)
$\therefore\ e_s' = 8.6 - 0.98 = 7.62$ in.

$$k_b = 1 + \frac{e_s e_b}{i^2} = 4.14$$

$$k_t = 1 - \frac{e_s e_t}{i^2} = 0.42$$

$$k'_b = 1 - \frac{e'_s e_b}{i^2} = -1.03$$

$$k'_t = 1 + \frac{e'_s e_t}{i^2} = 1.93$$

Properties of entire section (taking into account bonded steel)

A_e = 494 in^2 (3280 cm^2)
e_{be} = 18.6 in. (47.2 cm)
e_{te} = 9 in. (22.8 cm)
I_e = 37 500 in^4 (1 559 800 cm^4)
Z_t = 4160 in^3 (68 300 cm^3)
Z_b = 2015 in^3 (33 000 cm^3)

Initial prestressing forces

P_i = 16 × 32 900 = 526 400 lbf (238 700 kgf, 2341 kN)
(70 per cent of ultimate)

P'_i = 8 × 4500 = 36 000 lbf (16 300 kgf, 160 kN)
(33 per cent of ultimate)

Therefore nominal initial stresses

$$f_{bI} = \frac{k_b P_i + k'_b P'_i}{A_0} = +4410 \text{ lbf/in}^2 \qquad \text{(compression)}$$

$$f_{tI} = \frac{k_t P_i + k'_t P'_i}{A_0} = -315 \text{ lbf/in}^2 \qquad \text{(tension)}$$

Elastic shortening at transfer:

Nominal initial stress at C.G. of steel (total)

$$f_{sI} = \frac{4410 + 315}{27.6} \times 19.10 - 315 = 2955 \text{ lbf/in}^2$$

Assume instantaneous loss due to elastic shortening = 10 per cent then

f_{sT} = 0.9 × 2955 = 2659 lbf/in^2

Δf_{pe} = 10 × 2659 = 26590

This is greater than 10 per cent of steel stress 188 000 lbf/in^2 i.e. 18 800 lbf/in^2.

Try 15 per cent

f_{sT} = 0.85 × 2955 = 2510 lbf/in^2
Δf_{pe} = 10 × 2510 = 25 100 lbf/in^2
15% of 188 000 = 28 200 lbf/in^2

Try as 3rd approximation loss = 13.5 per cent

f_{sT} = 0.865 × 2955 = 2550 lbf/in^2
Δf_{pe} = 10 × 2550 = 25 500 lbf/in^2
13.5% × 188 000 = 25 400 ∴ OK

(This could be directly calculated without trial and error by using the formula on page 99.)

Therefore the stresses at transfer are

$$f_{bT} = 0.865 \times 4410 = 3810 \text{ lbf/in}^2$$

$$f_{tT} = 0.865 \times 315 = -270 \text{ lbf/in}^2$$

Part of the relaxation loss likely to occur at this stage is ignored.

Self-weight bending moment $= M_t = 405 \times 80^2 \times 1.5$ lbf in. = 3 880 000 lbf in.

$$f_{bs} = \frac{3\,880\,000}{2015} = -1930 \text{ lbf/in}^2$$

$$f_{ts} = \frac{3\,880\,000}{4160} = 935 \text{ lbf/in}^2$$

∴ Resultant stresses at transfer (at midspan)

$$f_{bt} = 3810 - 1930 = 1880 \text{ lbf/in}^2 < \frac{5000}{2}$$

$$f_{tt} = 935 - 270 = 665 \text{ lbf/in}^2$$

Stress at C.G. of steel

$$f_{st} = \frac{(1800 - 665)}{27.6} \times 19.10 + 665 = 1505 \text{ lbf/in}^2 \qquad \text{(compression)}$$

Since there is no counter-action at the ends, the compressive stress level is higher than the permissible. Therefore debonding of four strands per rib is necessary at the ends.

Stresses at full service load

Maximum losses

	(a)	Elastic shortening	= 25 500 lbf/in²
*	(b)	Relaxation @ 2½ per cent of initial stress	= 4600 lbf/in²
**	(c)	Creep = $0.9 \times 0.5 \times 10^{-6} \times 28 \times 10^6 \times 1505$	= 18 960 lbf/in²
***	(d)	Shrinkage = $400 \times 10^{-6} \times 28 \times 10^6$	= 11 200 lbf/in²
			60 260 lbf/in²

$$\therefore \% \text{ loss} = \frac{60\,260 \times 100}{188\,000} = 32\%$$

* Relaxation in accordance with manufacturers' specification for low relaxation Dyform strands.

** Creep for lightweight aggregate concrete taken as 0.5×10^{-6} based on material properties. Coefficient 0.9 could be reduced if more particulars of loading are known.

*** Shrinkage – average taken as 400×10^{-6}, is also based on material properties.

Effective prestress

$$f_{bE} = 0.68 \times 4410 = 2998 \text{ lbf/in}^2 \text{ (210 kgf/cm}^2\text{, 20.7 N/mm}^2\text{)}$$
$$f_{tE} = -0.68 \times 315 = -214 \text{ lbf/in}^2 \text{ (15 kgf/cm}^2\text{, 1.47 N/mm}^2\text{)}$$

M_w = B.M. due to maximum service load
= $(564 + 120) \times 80^2 \times 1.5 = 6\,566\,000$ lbf in. (75 640 kgf m, 742kN m)

∴ Maximum resultant stresses at midspan

$$f_{bw} = 2998 - \frac{6\,566\,000}{2015}$$

$= -260$ lbf/in² (18 kgf/cm², 1.8 N/mm²) (tension)

$$f_{tw} = -214 + \frac{6\,566\,000}{4160} = 1364 \text{ lbf/in}^2 \text{ (95 kgf/cm}^2\text{, 9.4 N/mm}^2\text{)}$$

(compression)

As the modulus of rupture of this grade of lightweight concrete(9) is in the order of 550 lbf/in² the maximum tensile stress is less than ¾ of the modulus of rupture and hence OK as a Class 2 structure. If the limiting tensile stresses for lightweight aggregate concrete are taken as 80% of the values given in CP 110 for normal weight concrete, then permissible tensile stress will be 371 lbf/in² (26.10 kgf/cm², 2.56 N/mm²) for this grade of concrete. According to CEB/FIP Manual allowable flexural tensile stress is even higher.

9.2.7 Example of cantilever double tee beams

An example is given here showing the use of double -T beam for a cantilever construction of 39.4 ft (12 m). The total length of the double -T unit is 80.35 ft (24.50 m). The span between the supports is 32.8 ft (10 m) and the cantilever on the rear side is 8.15 ft (2.48 m) (see Figure 9.11). Figure 9.12 shows the cross-section of the unit.

In this case, in addition to the required tendons at the top, tendons at the bottom are provided, which are needed during transport and at a definite loading which occurs only betwen the supports, without extending onto the cantilever. At cantilever bending moment, the compressive zone is in the web of the T beams and thus *the beam is over-reinforced* at the support which cannot be avoided in such a case if the most economical solution has to be established.

Simplified method of calculation based on assumptions that the concrete stress is uniform and the effective depth does not exceed ½*d*, has been made for ultimate moment of resistance.

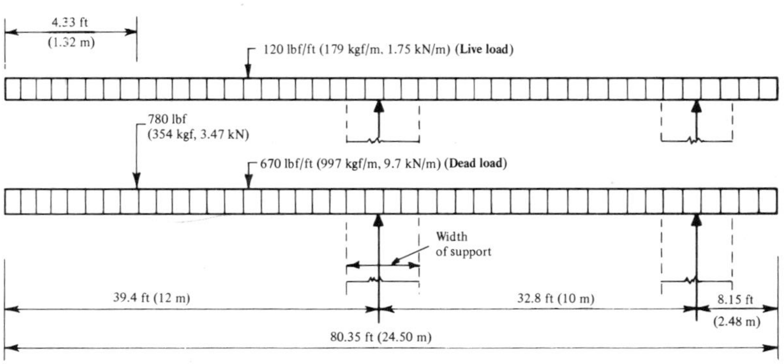

Figure 9.11 Load diagram (example 9.2.7)

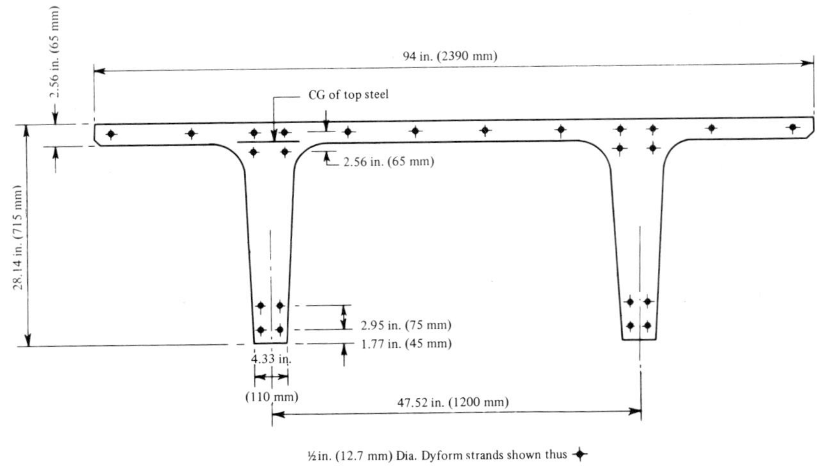

Figure 9.12 Cross-section of double-T beam (example 9.2.7)

At the end of the example, however, calculation has been presented to obtain the moment of resistance by the strain compatability method for comparison of results.

Characteristic strength at 28 days 9000 lbf/in² (630 kgf/cm², 62 N/mm²) and at release 7500 lbf/in² (520 kgf/cm², 52 N/mm²) $E_c = 5.75 \times 10^6$ lbf/in² (4×10^5 kgf/cm², 3.96×10^4 N/mm²)

There are 16 No. 0.5 in. (12.7 mm) diameter Dyform strands in the top flange each stressed to 70 per cent of its characteristic strength of 270 000 lbf/in² (190 kgf/mm², 1861 N/mm²), i.e. up to 189 000 lbf/in² (133 kgf/mm²; 1300 kN/mm²).

The relaxation loss of such steel, in accordance with the manufacturer's specification is 2½ per cent of the initial stress.

Eight Nos. of the same kind of strand have been provided in the ribs but with an initial prestress of 34 500 lbf/in² (2425 kgf/cm²; 237 N/mm²) only.

E value of the steel is 28×10^6 lbf/in² (19.6×10^5 kgf/cm², 19.3×10^4 N/mm²)

Consequently the modular ratio $\alpha_e = 4.87$ say 4.9

Sectional area of each strand = 0.174 in² (112 mm²)

LOADING

Referring to Figure 9.11 the loading over 7.68 ft wide unit (full service load) is as follows:

(a)	Self-weight of the unit	=	565 lbf/ft
(b)	Screeding between units (say)	=	25 lbf/ft
(c)	Felt and waterproofing	=	80 lbf/ft
	Total dead load	=	670 lbf/ft (997 kgf/m, 9.7 kN/m)

(d) Concentrated dead load due to upstand = 780 lbf acting at 4.33 ft from free end.

(e) Live load = 120 lbf/ft (179 kgf/m, 1.75 kN/m)

Critical section for flexure is 38 ft from the free end (at face of support).

DESIGN FOR THE LIMIT-STATE OF COLLAPSE (TO CP 110)
Using simplified method of calculation

Average $b = 2 \times 5 = 10$ in (254 mm)

$$\frac{A_{st} \cdot f_{su}}{f_{cu} \cdot b.d} = \frac{16 \times 0.174 \times 270\,000}{(9000 \times 1.15) \times 10 \times 26.21} = 0.28$$

$\therefore k_u = 1$ (see Table 8.1)

$$\therefore\ F_{su} = 16 \times \frac{(0.174 \times 270\,000)}{1.15} = 653\,635 \text{ lbf } (296\,482 \text{ kgf}, 2907 \text{ kN})$$

C.G. of $F_{su} = 1.93$ in. from top (49 mm)
$h = 28.14$ in. (715 mm)
$\therefore\ d = 26.21$ in. (666 mm)

$f_{cu} = 0.4 \times f_{cu} \times$ age factor (corresponding to six months)

Assume age factor = 1.15

$\therefore\ f_{cu} = 0.4 \times 9000 \times 1.15 = 4140$ lbf/in^2 (291 kgf/cm^2, 28.55 N/mm^2)

Width of the rib at ½ d (from the bottom) = 4.34 + 1.32 = 5.66 in.

$$\therefore\ \text{Average width of the rib } \tfrac{1}{4}\, d = \frac{4.34 + 5.66}{2} = 5 \text{ in.}$$

$\therefore\ F_{cu} = 2 \times (5 \times 13.11) \times 4140 = 542\,754 \text{ lbf} < F_{su}$

Ignoring the tendons in compressive zone

$$\therefore\ z = 26.36 - \frac{13.11}{2} = 19.70 \text{ in.}$$

$\therefore$ Permissible $M_R = 542\,754 \times 19.70 = 10\,692\,250$ lbf in
(123 175 kgf m, 1207 kN m)

Maximum cantilever moment due to dead load

$$= \frac{670 \times 38.00^2 \times 12}{2} + 780 \times 33.67 \times 12 = 6\,116\,000 \text{ lbf in.}$$

Live-load moment at the same section = $120 \times 38.00^2 \times 6 = 1\,039\,000$ lbf in.

M_u required (according to CP 110) = $1.4 \times 6\,111\,000 + 1.6 \times 1.039\,000$
= 10 524 800 lbf in. (121 200 kgf m, 1188 kNm) $M_R > M_u$ $\therefore$ o.k.

Calculation by strain-compatability method is given after the serviceability conditions:

CHECK FOR SERVICEABILITY CONDITIONS

Properties of the concrete section

A_0 = 540.5 in^2 (3494 cm^2)
e_t = 8.3 in. (21.1 cm)
e_b = 19.84 in. (50.4 cm)
I_0 = 38 283 (1 593 475 cm^4)
Z_t = 4600 in^3 (75 100 cm^3)
Z_b = 1930 in^3 (31 600 cm^3)

Properties of the entire section (taking into account bonded steel)

A_e = 557 in^2 (3598 cm^2)
I_e = 40 220 in^4 (1 674 165 cm^4)
Z_{te} = 4825 in^3 (79 000 cm^3)
Z_{be} = 2030 in^3 (33 200 cm^3)

In this example since the compression and tensile zones are reversed, standard notation with regard to k value has not been used in order to avoid confusion. Direct compression and bending component of the eccentric prestress have been separately calculated.

Initial prestressing force

16 × 32 900 + 8 × 6000 = 573 100 lbf (260 000 kgf, 2549 kN)

$$\text{Average steel stress} = \frac{573\ 400}{24 \times 0.174} = 137\ 100 \text{ lbf/in}^2 \text{ (9590 kgf/cm}^2\text{, 946 N/mm}^2\text{)}$$

Distance of the centre of gravity of prestressing force from the top face = 3.86 in. (9.8 cm)

∴ Eccentricity = 8.3 – 3.86 = 4.44 in (11.27 cm)

Nominal stresses due to initial prestressing force:

$$f_{tI} = \frac{573\ 400}{540.5} + \frac{573\ 400 \times 4.44}{4600} = 1614 \text{ lbf/in}^2 \text{ compression (113 kgf/cm}^2\text{, 11.0 N/mm}^2\text{)}$$

$$f_{bI} = \frac{573\ 400}{540.5} - \frac{573\ 400 \times 4.44}{1930} = -260 \text{ lbf/in}^2 \text{ tension (18 kgf/cm}^2\text{, 1.79 N/mm}^2\text{)}$$

$$\therefore f_{sI} = \frac{(1.614 + 260)}{28.14} \times 24.28 - 260 = 1360 \text{ lbf/in}^2 \text{ (96 kgf/cm}^2\text{, 9.4 N/mm}^2\text{)}$$

Elastic shortening at transfer

Assume 5 per cent loss at first approximation, then

f_{st} = 0.95 × 1360 = 1292 lbf/in^2

∴Δ f_{pe} = 4.90 × 1292 = 6292 lbf/in^2

5% of 137 100 = 6855 lbf/in^2

Since the difference is small 5 per cent loss has been assumed being on the safe side. However, more accurate calculations for the loss due to elastic shortening can be done by using the formula in page 99, considering top steel and bottom steel separately.

As the nominal initial stresses are relatively low, there is no need to check the stresses at transfer. It will suffice to check the resultant stresses under service load.

Calculation of total losses:

(a)	Elastic shortening = 5% of 137 100	= 6850 lbf/in^2
(b)	Relaxation = 2½% of 70% stress	= 4750 lbf/in^2
(c)	Creep = $0.9 \times 0.33 \times 10^{-6} \times 28 \times 10^6 \times 1292$	= 10740 lbf/in^2
(d)	Shrinkage = $300 \times 10^{-6} \times 28 \times 10^6$	= 8400 lbf/in
		30 740 lbf/in^2 (2161 kgf/cm^2, 212 N/mm^2)

$$\therefore \Delta f_{\text{p total}} = \frac{30\,740}{137\,100} = 0.224$$

Therefore effective prestress

$f_{\text{tE}} = 0.776 \times 1614 = 1254$ lbf/in^2 (87 kgf/cm^2, 8.65 N/mm^2)

$f_{\text{bE}} = -0.775 \times 260 = -202$ lbf/in^2 (14 kgf/cm^2, 1.40 N/mm^2)

Total service load bending moment = 6 116 000 + 1 039 000
= 7 155 000 lbf in.

Therefore resultant stresses

$$f_{\text{tw}} = 1254 - \frac{7.155 \times 10^6}{4825} = -228 \text{ lbf/in}^2 \text{ (16 kgf/cm}^2\text{, 1.6 N/mm}^2\text{)}$$

(tension)

$$f_{\text{bw}} = -202 + \frac{7.155 \times 10^6}{2030} = 3302 \text{ lbf/in}^2 \text{ (232 kgf/cm}^2\text{, 22.5 N/mm}^2\text{)}$$

(compression)

Note: The maximum compressive stress under working load is within the permissible limit if an age factor of only 12 per cent is taken, i.e., $< 0.33 \times 9000 \times 1.12$. According to CP 116, however, permissible compressive stress is $\frac{f_{\text{cu}}}{2.73}$. Therefore this section is o.k. without any age factor.

ULTIMATE STRENGTH CALCULATION BY STRAIN COMPATIBILITY METHOD (Figure 9.13)

In this example the stress-strain relationship for concrete (Figure 9.14) and steel (Figure 9.15) have been taken from CP 110. Calculation has been done in SI units as the Figures 9.14 and 9.15 are based on these units.

ϵ_{se} = average strain in steel due to effective prestress
ϵ_{su} = average strain in steel at failure of member
f_{pE} = effective prestress in steel

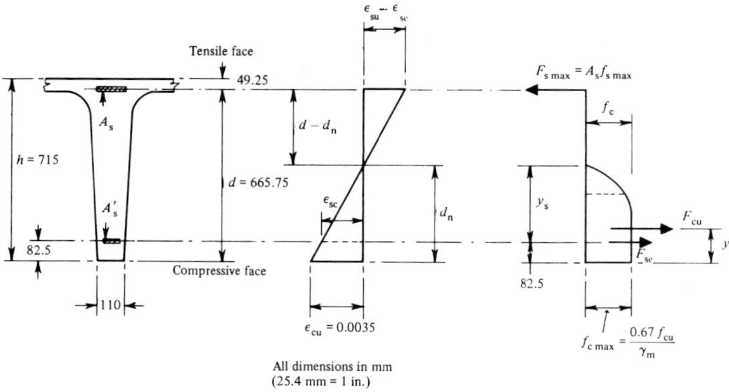

Figure 9.13 Strain compatibility at ultimate condition

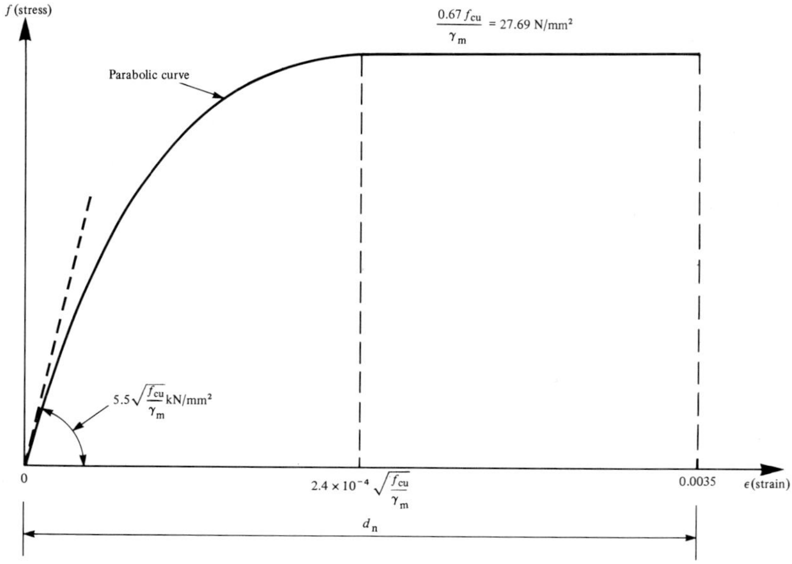

Figure 9.14 Compressive stress–strain diagram for concrete (example 9.2.7)

f'_{pE} = tensile stress in prestressing steel in the compression zone
f_{su} = average stress in tension steel at failure of member
f_{sc} = compressive stress in steel

$$f_{c\,max} = \frac{0.67 \times f_{cu}}{\gamma_m} = 27.69 \text{ N/mm}^2 \text{ taking } \gamma_m = 1.5$$

A'_s = 8 × 112 = 896 mm²
A_s = 16 × 112 = 1792 mm²
Loss of prestress = 23%

ϵ_{se} = strain in tension steel due to effective prestress

$$= 0.77 \times \frac{146.3 \times 10^3}{112} \times \frac{1}{E} = \frac{1006}{E} = 0.00503$$ (Initial prestressing force in each top strand = 146.3 kN)

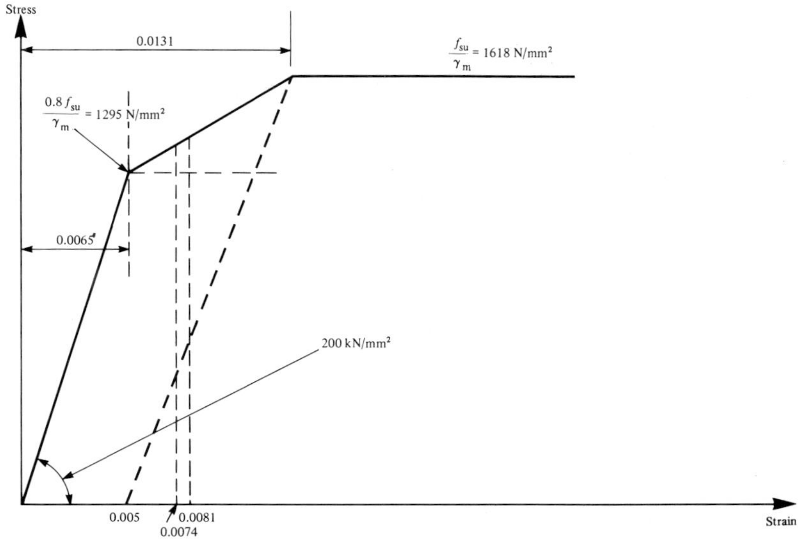

Figure 9.15 Stress–strain diagram for strand (example 9.2.7)

Stress in the bottom strands (compression side) due to effective prestresses

$$f'_{pE} = 0.77 \times \frac{26.7 \times 10^3}{112} = 183.5625 \text{ N/mm}^2$$ (Initial prestressing force in each bottom strand = 26.7 kN)

$$\frac{\epsilon_{su} - \epsilon_{se}}{665.75 - d_n} = \frac{0.0035}{d_n}$$

or

$$\epsilon_{su} = \epsilon_{se} + 0.0035 \frac{(665.75 - d_n)}{d_n}$$

or

$$\epsilon_{su} = 0.00503 + 0.0035 \frac{(665.75 - d_n)}{d_n}$$

Concrete stress (Figure 9.14)
The maximum compressive stress, 27.69 N/mm² commences at a strain = $2.4 \times 10^{-4} \frac{f_{cu}}{\gamma_m}$, i.e., at a strain = 0.0015

Let d_n = Distance of N.A. from the extreme compression fibre

The Parabolic curve of the stress-strain diagram then extends up to $\frac{0.0015}{0.0035} d_n$ i.e. $0.44\,d_n$

∴ The rectangular part of the diagram extends for $0.56\,d_n$

$$\therefore\ F_{cu} = [27.69\ b.d_n - \tfrac{1}{3} \times b.\ (0.44\ d_n).\ 27.69]$$

$$= [27.69\ b.d_n - 4.061\ b.d_n] = 23.629\ b.d_n$$

$$\text{C.G. of } F_{cu} = \frac{[27.69\ b.d_n.d_n/2 - 4.061.b.d_n\ (0.25 \times 0.44\ d_n)]}{23.629\ b.d_n}$$

$$= 0.5675\ d_n \text{ from N.A.}$$

$$\therefore\ \text{y (see Figure 9.13)} = (1 - 0.5675)\ d_n = 0.4325\ d_n$$

For trapizoidal section it is sufficiently accurate to take b as the average of the width at N.A. and that at the extreme compression fibre.

First trial: Assume d_n = 400 mm

$$\therefore\ \epsilon_{su} = 0.00503 + 0.0035 \times \frac{(665.75 - 400)}{400} = 0.0074$$

$$\therefore\ f_{s\,max}\ \text{(Figure 9.15)} = 1295 + \frac{(1618 - 1295) \times (0.0074 - 0.0065)}{(0.0131 - 0.0065)}$$

$$= 1339\ \text{N/mm}^2$$

$$\therefore\ F_{su} = 1792 \times 1339 = 2\ 399\ 500\ \text{N}$$

$$F_{cu} = 23.629 \times \frac{(110 + 150)}{2} \times 2 \times 400 = 2\ 457\ 400\ \text{N}$$

$$F_{sc} = 896 \times \left[0.0035 \times \frac{(400 - 82.5)}{400} \times 200\ 000 - 183.5625\right] = 333\ 000\ \text{N}$$

$$\therefore\ F_c = F_{cu} + F_{sc} = 2\ 790\ 400\ \text{N} > F_{su}$$

$\therefore$ try different value for d_n

Second trial: Assume d_n = 350 mm

$$\therefore\ \epsilon_{su} = 0.00503 + 0.0035 \times \frac{(665.75 - 350)}{350} = 0.0082$$

$$\therefore\ f_{s\,max}\ \text{(Figure 9.15)} = 1295 + \frac{(1618 - 1295) \times (0.0082 - 0.0065)}{(0.0131 - 0.0065)}$$

$$= 1378.20\ \text{N/mm}^2$$

$$\therefore\ F_{su} = 1792 \times 1378.20 = 2\ 469\ 729\ \text{N}$$

$$F_{cu} = 23.629 \times \frac{(110 + 145)}{2} \times 2 \times 350 = 2\ 108\ 900\ \text{N}$$

$$F_{sc} = 896 \times \left[0.035 \times \frac{(350 - 82.5)}{350} \times 200\ 00 - 183.5625\right] = 314\ 900\ \text{N}$$

$$\therefore\ F_c = F_{cu} + F_{sc} = 2\ 423\ 800\ \text{N} < F_{su}$$

Try d_n = 355 mm

$$\epsilon_{su} = 0.00503 + 0.0035 \times \frac{(665.75 - 355)}{355} = 0.0081$$

$$f_{s\,max} = 1295 + \frac{(1618 - 1295) \times (0.0081 - 0.0065)}{(0.0131 - 0.0065)} = 1373.30 \text{ N/mm}^2$$

$$\therefore F_{su} = 1792 \times 1373.30 = 2\,461\,000 \text{ N}$$

$$F_{cu} = 23.629 \times \frac{(110 + 145.50)}{2} \times 2 \times 355 = 2\,143\,210 \text{ N}$$

$$F_{sc} = 896 \times \left[0.0035 \times \frac{(355 - 82.5)}{355}\,200\,000 - 183.5625\right] = 316\,970$$

$$\therefore F_c = F_{cu} + F_{sc} = 2\,460\,200 \approx F_{su}$$

$$\therefore M_R = F_{sc}\,(d - 82.5) + F_{cu}\,(d - y)$$

$$y = 0.4325 \times 355 = 153.5375 \approx 153.54 \text{ mm}$$

$$\begin{aligned} \therefore M_R &= 316\,970\,(665.75 - 82.5) + 2\,143\,210 \times (665.75 - 153.54) \\ &= 184\,872\,752 + 1\,097\,773\,594 \\ &= 1\,282\,646\,346 \text{ N mm} \\ &= 1282.646 \text{ kNm } (11\,353\,134 \text{ lbf in}, 130\,802 \text{ kgf m}) \end{aligned}$$

$M_R > M_u$ $\therefore$ o.k.

M_R calculated by strain compatability method compares well with M_R obtained by simplified method.

9.2.8 Example of long cantilever in lightweight concrete

Cantilever beams of inverted T-shape of a length of about 78ft (23.80 m) were used at the roof of the Grandstand of Doncaster Racecourse, England, built 1968/69. Lightweight concrete was used, as it was essential to reduce the weight of the beam to the extent that the capacity of the crane available for lifting was not exceeded.

This practical example has been included to demonstrate the possibilities of the use of prestressed lightweight concrete in a special structure(8).

Figure 9.16 and 9.17 show the construction. Each unit was designed as a cantilever beam the curved flange forming the compression zone. As the width – radius of curvature is small the shell effect was negligible and therefore ignored.

Although the structure was designed as Class 2, in actual fact only a small tensile stress under maximum service load was allowed and a greater than usual overall factor of safety against collapse conditions was ensured for the following reasons.

(a) No waterproofing would be applied to the central upstand rib, the top surface of which would be in tension and exposed to the weather.

(b) Possibilities of vibration due to wind-gust effect or other sources of excitation.

(c) To reduce the stress range in consideration of fatigue life as a result of fluctuating wind loading.

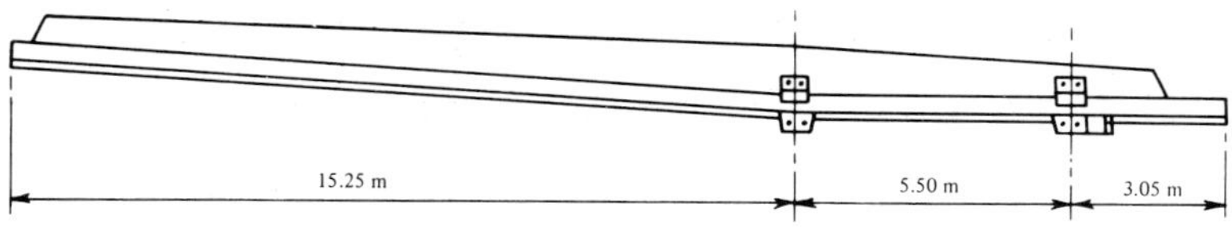

Figure 9.16 Elevation of roof unit (example 9.2.8)

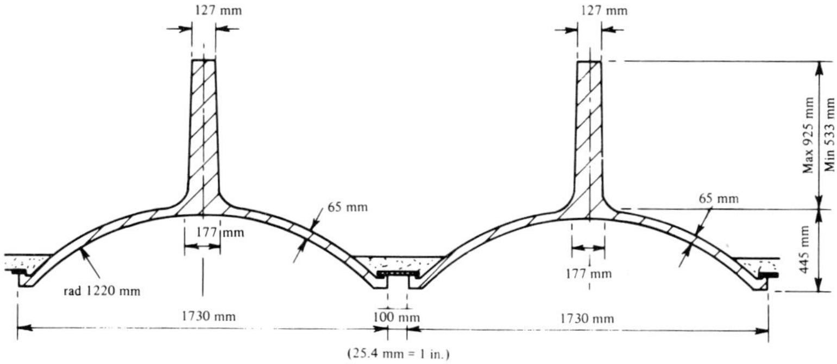

Figure 9.17 Cross-section of roof construction (example 9.2.8)

Concrete quality

(a)	Characteristic strength at release	35 N/mm^2
(b)	Characteristic strength at 28 days	52.5 N/mm^2
(c)	Density	1900 kg/m^3
	E at release	20×10^3 N/mm^2
	E at 28 days	24.5×10^3 N/mm^2

Steel

12.7 mm diameter Dyform strands with a minimum breaking load of 210 kN per strand. Nominal cross-sectional area of each strand = 112 mm^2.

E for Dyform strand	= 19.3×10^4 N/mm^2
Relaxation	= maximum 2½% of 70% of ultimate stress (manufacturers' specification)
α_e	= 10 (at transfer)
α_e	= 8 (final)

Loading (Figure 9.18)

(1) Own weight

(a) Uniformly distributed load due to curved flange = 2.60 kN/m

(b) Uniformly distributed load (except for 0.61 m and 1.22 m at front end and rear end respectively due to the minimum height of the rib) = 1.530 kN/m

(c) Triangular load (see diagram) due to variation in the depth of the central rib = 0 to 1.10 kN/m.

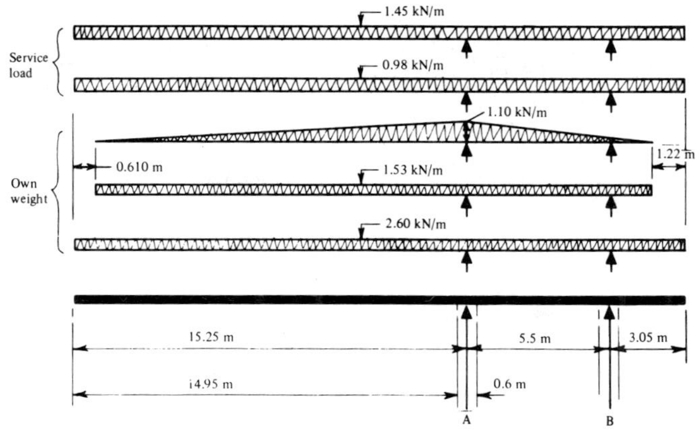

Figure 9.18 Load diagram (example 9.2.8)

(2) Permanent service load
Due to screeding and waterproofing etc. uniformly distributed load = 0.98 kN/m

(3) Snow load
Uniformly distributed load = 1.45 kN/m

Bending moment – at point A

(a) Own weight: $2.60 \times \dfrac{15.25^2}{2} + \dfrac{1.53 \times 14.64^2}{2} + 1.1 \times \dfrac{14.64^2}{6}$

$= 302 + 164 + 39.2 = 505.2$ kNm $\quad = 505.2$ kN m

(b) Permanent service load $= \dfrac{0.98 \times 15.25^2}{2}$ $\quad = 114$ kN m

(c) Temporary service load (snow load) $= \dfrac{1.45 \times 15.25^2}{2}$ $= 169$ kN m

Total = 788.2 kN m

Bending moment – at edge of support A (cantilever end)

(a) $2.60 \times \dfrac{14.95^2}{2} + \dfrac{1.53 \times 14.34^2}{2} + 1.1^* \times \dfrac{14.34^2}{6}$

$= 290 + 157 + 37.7 = 484.7$ kN m

(b) $0.98 \times \dfrac{14.95}{2}$ $\quad = 109.5$ kN m

*approximate

(c) $1.45 \times \frac{14.95^2}{2}$ = 162.0 kN m

Total = 756.2 kN m

Bending moment – at support B

(a) $\frac{2.60 \times 3.05^2}{2}$ = 12.10 kN m

(b) $1.53 \times \frac{1.83^2}{2}$ = 2.56 kN m

(c) $0.98 \times \frac{3.05^2}{2}$ = 4.55 kN m

Total due to dead load 19.21 kN m

(d) $145 \times \frac{3.05^2}{2}$ = 6.75 kN m

Total including live load = 25.96 kN m

Maximum bending moment – at the edge of support A (span side)

$$25.96 + \frac{(788.2 - 25.96)}{5.5} \times 5.20 - \left(20.1 \times 0.3 - \frac{6.56 \times 0.3^2}{2} - \frac{1.10 \times 0.3^2}{2}\right)^*$$

$$= 25.96 + 720 - 6.04 + 0.35 = 745.96 - 5.69 = 739.27 \text{ kN m}$$

Maximum shear

$$\frac{(788.2 - 25.96)}{5.5} + [20.1 - (6.56 + 1.10^*) \times 0.3] = 138.5 + 20.1 - 2.3$$

$$= 156.30 \text{ kN}$$

LIMIT-STATE OF COLLAPSE CONDITION (near support A).
For simplification it has been assumed that the entire curved flange would be in uniform compression at the ultimate condition.

$$\frac{A_{st} \times f_{su}}{b.d.f_{cu}} = \frac{12 \times 210\,000}{127 \times 990 \times 50} = 0.40$$

$\therefore\ k_u = 0.9$

Maximum height of N.A. = 445 mm; O.K. $<d/2$
Ultimate tensile force = 0.9 × 12 × 210 = 2268 kN
C.G. of tensile force = 990 mm from bottom.
C.G. of compression area = 220 mm above bottom line.
∴ lever arm = 990 – 220 = 770 mm

$$\therefore\ M_R = \frac{2268 \times 770}{1000} = 1746.36 \text{ kN m}$$

*approximate

$$\therefore \text{ factor of safety} = \frac{1746.36}{756.2} = \underline{2.30} \text{ o.k.}$$

It is also necessary to investigate the combined effect of bending and shear which has been shown in Chapter 13

CHECK FOR SERVICE CONDITION

Properties of the precast section

(A) At cantilever end
- A_0 = Area = 2220 cm^2
- e_t = 55.4 cm
- e_b = 40.8 cm
- I_0 = 1 585 000 cm^4
- Z_t = 28 600 cm^3
- Z_b = 38 800 cm^3

(B) At support section (Figure 9.19)
- A_0 = Area = 2820 cm^2
- e_t = 80.60 cm
- e_b = 56.40 cm
- I_0 = 4 310 000 cm^4
- Z_t = 54 500 cm^3
- Z_b = 76 400 cm^3

$$\text{Modular ratio} = \frac{E_{\text{steel}}}{E_{\text{concrete}}} = 8 = \alpha_{e28}$$

$$\text{and } 10 = \alpha_{eT}$$

12 No.12.7 mm diameter Dyform strands each stressed to 70 per cent of ultimate, i.e. 147 kN.

12.7 mm diameter Dyform strands shown thus ● and 7 mm diameter wire shown thus + in Figure 9.19. Each 7 mm diameter wire (ultimate stress 1.54 kN/mm^2) stressed to 41.42 kN (70 per cent of ultimate).

Total prestressing force (initial) at support section

$$12 \times 147 + 7 \times 41.42 = 2055 \text{ kN}$$

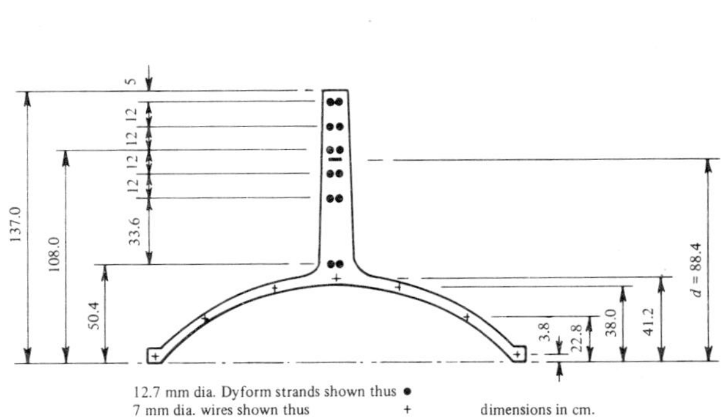

Figure 9.19 Cross-section of roof unit at support B (example 9.2.8)

Total steel area $= 12 \times 112 + 7 \times 38.5 = 1611.5\ \text{mm}^2$

$$\therefore\ f_s = \text{average steel stress} = \frac{2055 \times 10^3}{1611.5} = 1270\ \text{N/mm}^2$$

C.G. of prestress = 884 mm from bottom

$\therefore$ eccentricity = 884 – 564 = 320 mm

In this example again, as in the previous one, the stresses due to prestressing force have been computed by the combination of direct and bending effect; since the tensile zone is at the top.

Initial stresses (support section)

$$f_{tI} = \frac{2055 \times 10^3}{2820 \times 10^2} + \frac{2055 \times 10^3 \times 320}{54\,500 \times 10^3} = 7.37 + 12.03$$

$$= 19.40\ \text{N/mm}^2\ \text{(compression)}$$

$$f_{bI} = 7.37 - \frac{2055 \times 10^3 \times 302}{76\,400 \times 10^3} = (-)\ 1.25\ \text{N/mm}^2\ \text{(tension)}$$

$$\text{Stress at C.G. of steel} = f_{sI} = \frac{(19.40 + 1.25)}{137} \times 88.4 - 1.25$$

$$= 13.32 - 1.25 = 12.07\ \text{N/mm}^2\ \text{(compression)}$$

Let loss due to elastic shortening = 8.5 per cent

then $f_{sT} = 0.915 \times 12.07 = 11\ \text{N/mm}^2$

$\Delta f_{pe} = 10 \times 11 = 110\ \text{N/mm}^2$

8.5% of 1270 = $107.8\ \text{N/mm}^2$

Therefore, to be on the safe side assume loss due to elastic shortening = 9 per cent

Losses

(i)	Elastic shortening $= 10 \times (0.91 \times 12.07)$	=	$109.8\ \text{N/mm}^2$
+(ii)	Relaxation @ 2½% of 70% of ultimate	=	$31.8\ \text{N/mm}^2$
*(iii)	Shrinkage $= 600 \times 10^{-6} \times 19.3 \times 10^4$	=	$116.0\ \text{N/mm}^2$
*(iv)	Creep $= 0.85 \times 11.0 \times 8 \times 10^{-5} \times 19.3 \times 10^4$	=	$144.0\ \text{N/mm}^2$
			$401.6\ \text{N/mm}^2$

$$\therefore\ \text{Loss} = \frac{401.6 \times 100}{1270} = 31.6\%\ \text{say}\ 32\%$$

Final stresses after losses (due to prestress only)

$f_{tE} = 0.68 \times 19.40 = 13.19\ \text{N/mm}^2$ (compression)

+Manufacturers specification

*Data re: shrinkage strain and specific creep per unit stress were obtained from the manufacturer. They appear to be on the high side by comparison with the corresponding data in CEB manual.

$f_{bE} = 0.68 \times (-1.25) = -0.85$ N/mm² (tension)

B.M. due to service load = 756.2 kNm

$$\therefore\ f_{tw} = \frac{756.2 \times 10^3 \times 10^3}{54\,500 \times 10^3} = 13.90 \text{ N/mm}^2 \text{ (tension)}$$

$f_{bw} = 9.90$ N/mm² (compression)

$\therefore$ Resultant stresses $f_{tw} = -0.71$ N/mm² (tension) (103 lbf/in², 7.2 kgf/cm²)

O.K.

$f_{bw} = 9.05$ N/mm² (compression) (1310 lbf/in², 92 kgf/cm²)

As the section varies along the span and some strands are debonded near the ends, the ultimate as well as serviceability conditions should be investigated at every 1 m distance.

9.2.9 Example of an unsymmetrical section

This example deals with precast members of unsymmetrical section which are used as seatings in stadiums or grandstands as already generally described in Chapter 7. They are unsymmetrical only during transport at which state they are preferably in compression or with very small tensile stress. As soon as they are placed in position they are supported along their entire length by the adjacent beams and do not act any more as unsymmetrical beams. It is important to avoid extensive camber. This example relates to the grandstand in Calgary, Canada, built in 1973/74 and this type of construction is called bleacher in Canada. The members are about 40 ft (12 m) long and about 3 ft (0.90 m) wide, with a depth of approximately 2 ft (0.60 m).

For this example special acknowledgement is due to the designer, Dr. S. Chakrabarty.

Figure 9.20 shows the cross-section at the centre, and Figure 9.21 at the support of the beam having a span of 39.5 ft (12.0 m). It is seen that the ½ in. (12.7 mm) strands of the beam are grouped together at the centre, the centroid being 3.25 in. (82.5 mm) above the bottom surface, but are spaced at 3 in. (75 mm) centres at the ends, the centroid being 8 in. (203.2 mm) above the bottom surface.

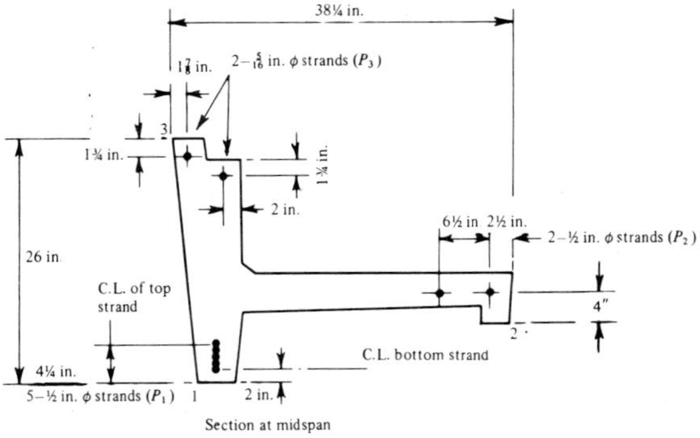

Figure 9.20 Cross-section at midspan (example 9.2.9)

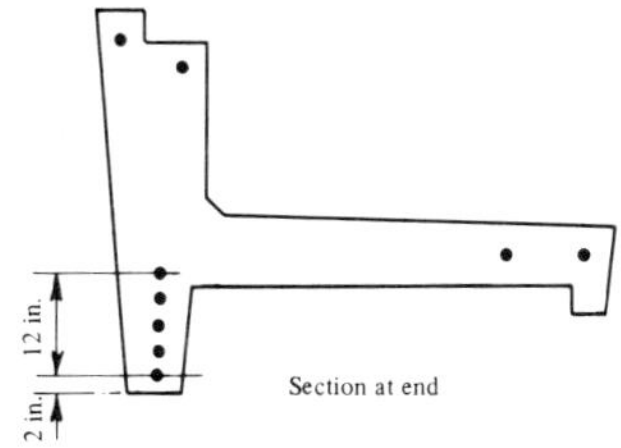

Figure 9.21 Cross-section at ends (example 9.2.1)

The slope is

$$\frac{(8-3.25)}{12} \times \frac{39.5}{2} = 0.02$$

This is very small. Hence the cosine may be taken as 1. The strength of a 0.5 in. diameter strand is 270 000 lbf/in^2 (1862 N/mm^2, 189 kgf/mm^2). Initial prestressing force at 0.7 of the strength is 28 900 lbf (128 kN, 13 109 kgf).

The section values from the individual areas (Figure 9.22), as shown in previous examples, are obtained with the origin at 0_0.

The following results are obtained:

$A = 280.2$ in^2 (180 800 mm^2)

Centroid, $\bar{x} = 12.9$ in. (327.7 mm) and $\bar{y} = 12.2$ in. (310.0 mm) from 0_0

$I_x = 8607$ in^4 (3583 × 10^6 mm^4)
$I_y = 40\,685$ in^4 (16 934 × 10^6 mm^4)

In order to obtain the principal axes the value of I_{xy} has to be computed. Care has to be taken with regard to the signs. The same areas are used again.

$$I_{xy} = \sum_{i=1}^{i=n} A_i x_i Y_i$$

The resulting $I_{xy} = -5149$ m^4 (2143 × 10^6 mm^4)

The slope of the principal axes (Figure 9.23) is given by

$$\tan 2\theta = \frac{2I_{xy}}{I_y - I_x} = -0.321$$

or

$$2\theta = -17.8°$$

and

$$\theta = -8.9°$$

The principal axes X_p and Y_p are indicated in Figure 9.22, and the principal I_p values amount to:

$I_{xp} = I_x \cos^2\theta + I_y \sin^2\theta - I_{xy} \sin 2\theta = 7800$ in^4 (3246 × 10^6 mm^4)

$I_{yp} = I_x \sin^2\theta + I_y \cos^2\theta + I_{xy} \sin 2\theta = 41\,492$ in^4 (17 270 × 10^6 mm^4)

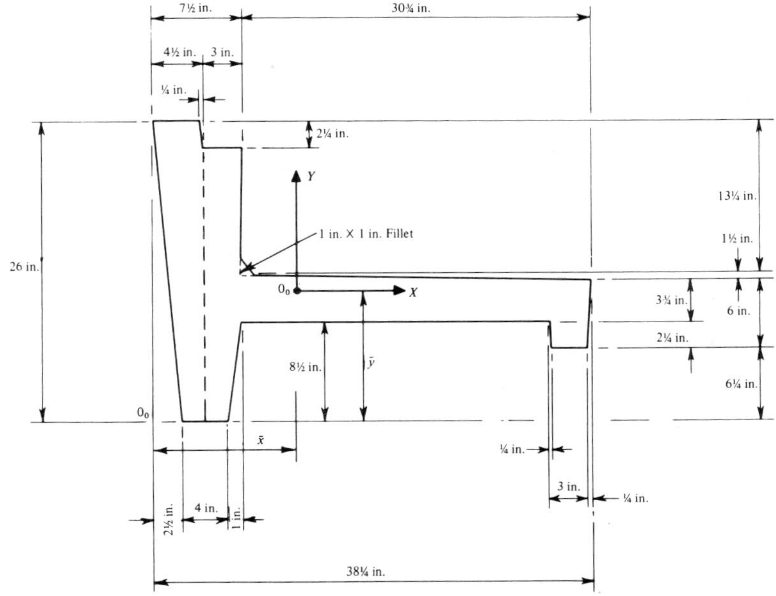

Figure 9.22 Cross-section showing concrete dimensions and orthogonal co-ordinates (example 9.2.9)

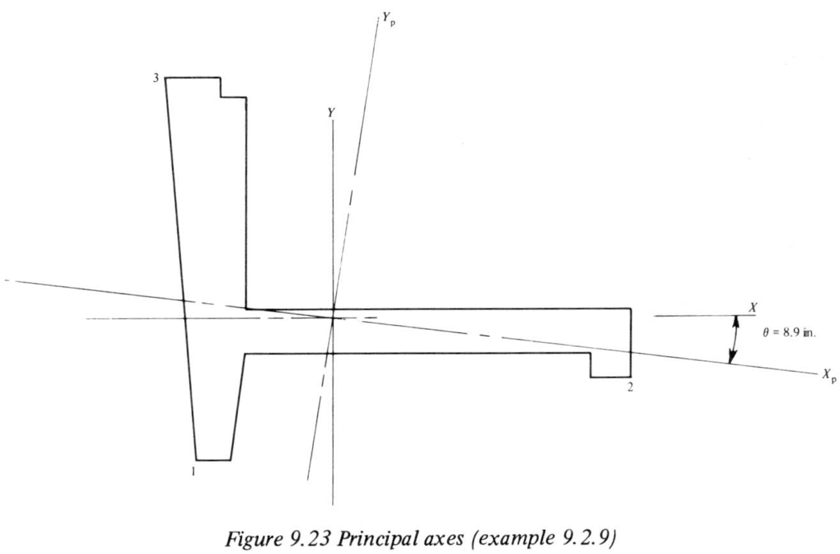

Figure 9.23 Principal axes (example 9.2.9)

The x and y co-ordinates of the three corner points (Figure 9.23) are shown in Table 9.4

Table 9.4

	1	2	3
x	−10.4 in. (264.2 mm)	+25.1 in. (637.5 mm)	−12.9 in. (327.7 mm)
y	−12.2 in. (310.0 mm)	−5.95 in. (151.1 mm)	+13.8 in. (350.5 mm)

Their appropriate co-ordinates with respect to the principal axes can be obtained from

$$x_p = x \cos\theta + y \sin\theta \quad y_p = -x \sin\theta + y \cos\theta$$

This results in

Table 9.5

	1	2	3
x_p	−8.39 in. (213 mm)	+25.72 in. (653 mm)	−14.88 in. (378 mm)
y_p	−13.66 in. (347 mm)	−2.06 in. (52 mm)	+11.64 in. (295 mm)

There are three prestressing forces, P_1, P_2 and P_3. First the two prestressing forces P_1 and P_2 are considered which produce compression at the points 1 and 2. It is assumed that $R_0 = 0.75$

Hence

$$P_{1E} = 0.75 \times 5 \times 28\,910 = 108\,400 \text{ lbf } (482.16 \text{ kN}, 49\,170 \text{ kgf})$$

$$P_{2E} = 0.75 \times 2 \times 28\,910 = 43\,400 \text{ lbf } (193.0 \text{ kN}, 19\,700 \text{ kgf})$$

where, the initial prestressing force per strand is 28 900 lbf (128.5 kN).

With $e_{s1} = 8.95$ in. (227 mm) and $e_{s2} = 2.075$ in. (52 mm) the bending moment causing compression of the lower face is obtained as

$$M_x = 108\,400 \times 8.95 + 43\,400 \times 2.075 = 1\,060\,300 \text{ lbf in.}$$

The horizontal distance of P_1 from the centroid of the combined forces $(P_1 + P_2)$ is

$$\frac{43\,400\,(8.4 + 19.6)}{43\,400 + 108\,410} = 8 \text{ in. } (203.2 \text{ mm})$$

Hence the bending moment about the y axis is

$$M_y = (43\,400 + 108\,410) \times 0.4 = 60\;700 \text{ lbf in.}$$

This bending moment causes compression at the left side. Figure 9.24 shows the actual directions of the moments M_x and M_y.

The principal moments are:

$$\begin{aligned} M_{xp} &= M_x \cos\theta + M_y \sin\theta \\ &= 1\,047\,550 + 9390 = 1\,056\,940 \text{ lbf in.} \end{aligned}$$

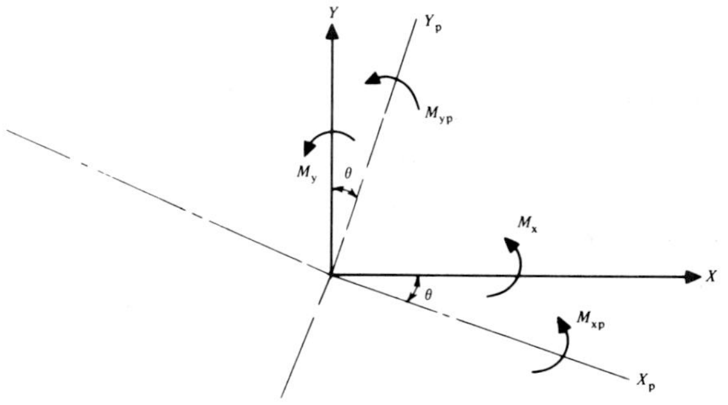

Figure 9.24 Principal moments due to P_1 and P_2 (example 9.2.9)

$$M_{yp} = M_x \sin\theta - M_y \cos\theta$$
$$= 164\,020 - 59\,980 = 104\,040 \text{ lbf in.}$$

Stresses at points 1, 2 and 3 are computed below.
Direct stress:

$$\frac{P}{A} = \frac{151\,800}{280.2} = +542 \text{ lbf/in}^2 \; (3.74 \text{ N/mm}^2)$$

Bending stresses:

At 1

$$f_{1xp} = \frac{1\,056\,940}{7800} \times 13.6 = +1843 \text{ lbf/in}^2$$

$$f_{1yp} = -\frac{104\,040}{41\,492} \times 8.4 = -21 \text{ lbf/in}^2$$

i.e. $f_1 = +1822 \text{ lbf/in}^2$

At 2

$$f_{2xb} = +\frac{1\,056\,940}{7800} \times 2.075 = +281 \text{ lbf/in}^2$$

$$f_{2yp} = +\frac{104\,040}{41\,492} \times 25.72 = 64 \text{ lbf/in}^2$$

i.e. $f_2 = +346 \text{ lbf/in}^2$

At 3

$$f_{3xp} = -\frac{1\,056\,940}{7800} \times 11.64 = -1577 \text{ lbf/in}^2$$

$$f_{3yp} = -\frac{104\,040}{41\,492} \times 14.9 = -37 \text{ lbf/in}^2$$

i.e. $f_3 = -1614 \text{ lbf/in}^2$

Therefore, the resultant stresses due to prestress are:

$$f_{p1} = 542 + 1822 = 2364 \text{ lbf/in}^2$$

$$f_{p2} = 542 + 346 = +888 \text{ lbf/in}^2$$

$$f_{p3} = 542 - 1615 = -1073 \text{ lbf/in}^2$$

It is seen that both point 1 and point 2 are in compression. But the tensile stress at point 3 is very large. Consequently 2 upper tendons $\frac{5}{8}$ in. (16 mm) diameter strands P_3 (Figure 9.20 and 9.21) are stressed.

Effect of P_3:

Initial prestress is

$$P_{3t} = 2 \times 9840 = 19\,680 \text{ lbf}$$
$$P_{3E} = 0.75 \times 19\,680 = 14\,760 \text{ lbf}$$

Bending moments M_x and M_y are:

$$M_x = 14\,760 \times (26 - 12.2 - 3.25) = 155\,720 \text{ lbf in.}$$
$$M_y = 14\,760 \times (12.9 - 3.75) = 135\,050 \text{ lbf in.}$$

The actual directions of moments M_x and M_y are shown in Figure 9.25. The principal moments M_{xp} and M_{yp} are:

$$M_{xp} = M_x \cos\theta - M_y \sin\theta$$
$$= 132\,950 \text{ lbf in.}$$
$$M_{yp} = M_x \sin\theta + M_y \cos\theta$$
$$= 157\,520 \text{ lbf in.}$$

Computed stresses at points 1, 2 and 3 are as follows:

$$f_1 = -147 \text{ lbf/in}^2$$
$$f_2 = -80 \text{ lbf/in}^2$$
$$f_3 = +307 \text{ lbf/in}^2$$

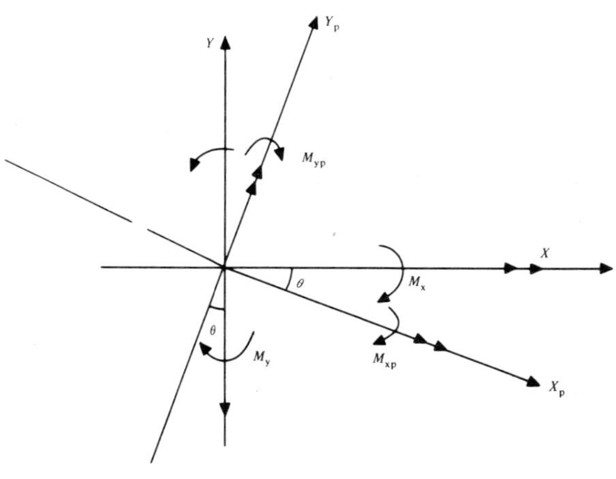

Figure 9.25 Principal moments due to P_3 (example 9.2.9)

Therefore, the resultant stresses at points 1, 2 and 3 due to total prestress are:

$$f_{p1} = 2364 - 147 = 2217 \text{ lbf/in}^2$$
$$f_{p2} = +888 - 80 = 808 \text{ lbf/in}^2$$
$$f_{p3} = -1073 + 307 = -766 \text{ lbf/in}^2$$

$$\text{Self-weight at } 153 \text{ lbf/ft}^3 = 280.2 \times \frac{153}{144} = 298 \text{ lbf/ft}$$

Maximum moment at midspan due to self-weight

$$M_d = \frac{wl^2}{8} = \frac{298 \times 39.5^2}{8}$$

$$= 58\ 120 \text{ lbf ft}$$

The moment M_d is about x-axis.

The principal moments are:

$$M_{xp} = M_d \cos\theta$$
$$= 58\ 120 \text{ lbf ft}$$

$$M_{yp} = M_d \sin\theta$$
$$= 58\ 120 \times 0.1547$$
$$= 8990 \text{ lbf ft}$$

As before, the stresses were computed at points 1, 2 and 3. The results are:

$$f_1 = -1180 \text{ lbf/in}^2$$
$$f_2 = -250 \text{ lbf/in}^2$$
$$f_3 = +1067 \text{ lbf/in}^2$$

Resultant stresses at points 1, 2 and 3 due to prestress and self-weight are:

$$f_1 = 2217 - 1180 = +1037 \text{ lbf/in}^2 \ (72 \text{ kgf/cm}^2, 7.17 \text{ N/mm}^2)$$
$$f_2 = 808 - 250 = +558 \text{ lbf/in}^2 \ (+3.85 \text{ N/mm}^2, +39 \text{ kgf/cm}^2)$$
$$f_3 = -766 + 1067 = +301 \text{ lbf/in}^2 \ (+2.09 \text{ N/mm}^2, 21 \text{ kgf/cm}^2)$$

9.3 General notes on the design of simply supported long-span precast beams

In this section general considerations are given on two examples about the economy and suitability of the design of long-span precast members. The first example relates to a beam of 85 ft (25.5 m) span of the Victoria Station in Sheffield designed by The Civil Engineering Department of British Railways Eastern Region in 1954. Figure 9.26 is an elevation of the roof beams and Figure 9.27

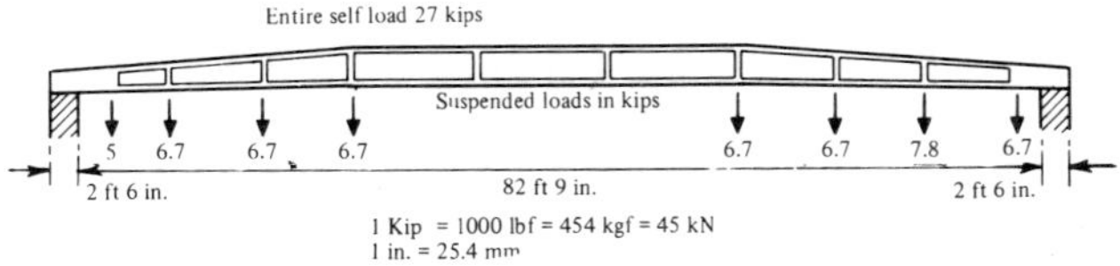

Figure 9.26 Elevation of roof beam at Sheffield Victoria station

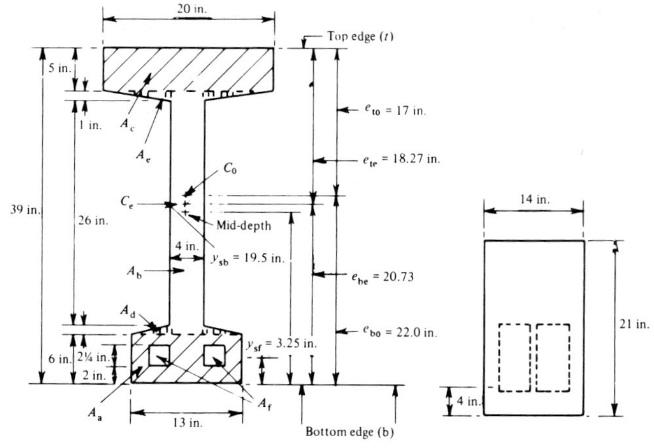

Figure 9.27 Cross-section at midspan and end elevation of beam in Figure 9.26

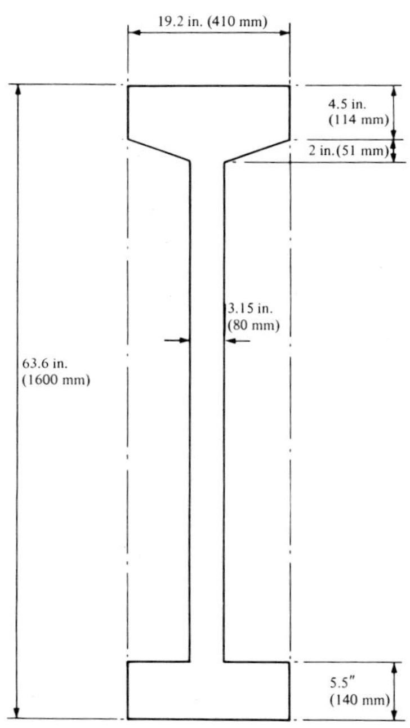

Figure 9.28 Slender roof beam

show the cross-sections at the centre and the end elevation. These beams carry the glass canopies which are suspended from the beams with appreciable loads as indicated in Figure 9.27. The beams themselves are sturdy enough to act without transverse connection. In the first author's book[10], an analysis of this construction is published based on CP 115 although the original design was carried out before detailed particulars of this Code about the losses were known

Table 9.6 Particulars of girder shown in Figure 9.28 (built in 1966)

Thickness of web at centre	8 cm		3.15 in
Entire length	23.88 m		76.33 ft
Self weight	15 000 kgf	147.1 kN	33 000 lbf
Max live load per unit length	2190 kgf/m	21.47 kN/m	1471 lbf/ft
Cube strength of concrete	500 kgf/cm^2	49 N/mm^2	7250 lbf/in^2

and the design was mainly carried out on the basis of the First Report[(6)]. It should be noted that with the particulars given in CP 115, the design for collapse load as well as that for service load agreed very well with the original assumption, according to which the maximum tensile stress did not exceed 650 lbf/in^2 (45.5 kgf/cm^2, 4.5 N/mm^2).

In this case the tendons consist of two straight Magnel cables of 16 wires 0.276 in. (7 mm) dia., some non-tensioned wires were added. The web is 4 in. (10 cm) wide which allowed satisfactory vibration of the concrete, the precast members being cast at the site.

Another example is shown from a publication[(11)] relating to a prestressed lightweight concrete beam of slightly shorter span in which a much deeper I-shaped beam was manufactured in a factory. In this case the thickness of the web was only 3.15 in. (80 mm). However, in the paper in reference[(11)] slightly shorter beams of webs having half the widths of the above example are described. The cross-section of this example is shown in Figure 9.28 and a comparison of span, self-weight and additional service load is given in the Table 9.6. Obviously with a design of precast members with very thin slabs, having stiffener ribs, sufficient vertical shear reinforcement is necessary which was not the case with the roof beams at Sheffield.

REFERENCES

1. BRITISH STANDARDS INSTITUTION. CP 115: 1959 *The structural use of prestressed concrete in buildings.* London.
2. BRITISH STANDARDS INSTITUTION. CP 116: 1965 *The structural use of precast concrete.* London.
3. BRITISH STANDARDS INSTITUTION. CP 110: 1972 *Unified code for structural concrete.* London.
4. CEB/FIP. *CEB/FIP Model code for concrete structures.* Comité Euro International du Beton, Bulletin d'information, N/124/125E. 1978.
5. INSTITUTION OF STRUCTURAL ENGINEERS AND CONCRETE SOCIETY, *Unified code symposium.* London, September 1973.
6. INSTITUTION OF STRUCTURAL ENGINEERS. *First report on prestressed concrete,* 1951.
7. DAVIS, S. G. and MARTIN, S. J. *The quality of concrete and its variation in structure.* Cement and Concrete Association, Technical Paper 42.487 November 1973, pp. 1–24.

8. CEB/FIP. *CEB/FIP Manual of lightweight aggregate concrete – design and technology.* The Construction Press, 1977.
9. BOBROWSKI J. and BARDHAN-ROY B. K. Structural assessment of lightweight aggregate concrete. *Concrete,* July 1971. Vol. 5. No 7, pp 229–34.
10. ABELES, P. W. *Introduction to prestressed concrete,* Vol. 2.
11. REIFFENSTUHL, H. *Development of prestressed lightweight concrete beams.* Beton and Stahlbeton November 1968.

CHAPTER 10

END BLOCKS

10.1 Post-tensioned steel

10.1.1 General

When post-tensioned cables are used the prestressing force is applied to the concrete at the points of anchorage of the cables, and it is obvious that the stresses in the immediate vicinity of the points of application of the prestressing force differ from those at sections remote from the anchorage. The accurate determination of these stresses is difficult and usually unnecessary; when their nature is understood, approximate methods may be used to calculate the reinforcement required.

The distribution of forces resembles that which is usually assumed to occur in a deep column footing (Figure 10.1), and the distribution of stresses is qualitatively similar; the analogy is by no means exact, however. Figure 10.2 shows the distribution of stress and the directions of the principal tensile and compressive stresses which actually occur, and which should be allowed for in design. It is of interest to note that this diagram was published by Coker in 1921 in connection with the effects of concentrated loads[1]. It is obvious that the

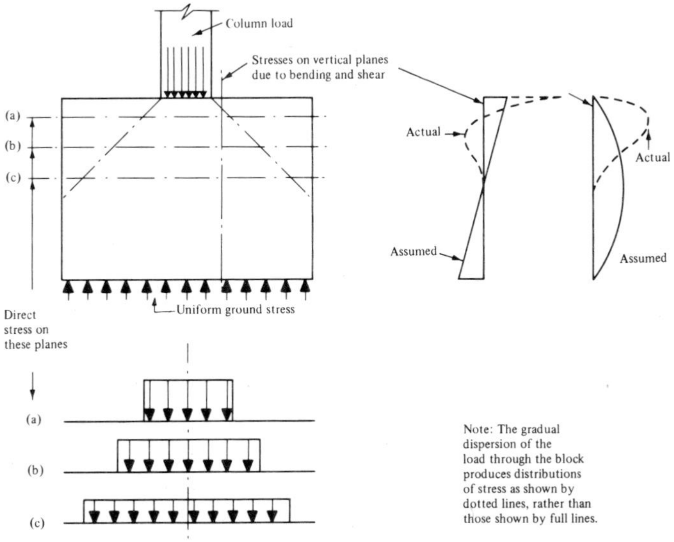

Figure 10.1 Stresses in column footing

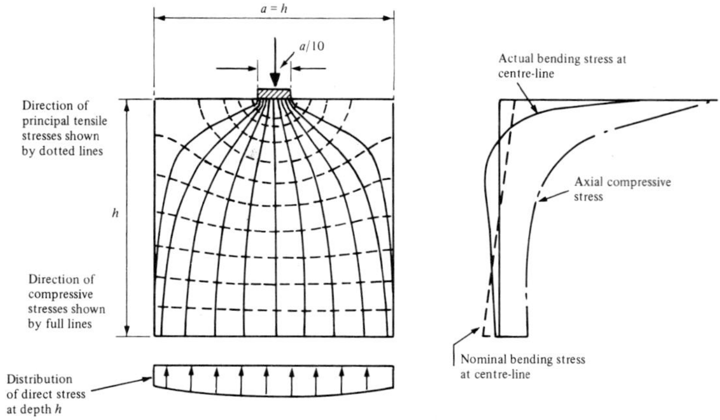

Figure 10.2 Stresses in cube due to concentrated load

more closely the distribution of the prestressing force approaches the resultant distribution of stress, the smaller are the local tensile stresses in the anchorage zone. From the stress diagram (Figure 10.2) it will be noted that there are large transverse compressive stresses close to the point of loading and transverse tensile stresses at a greater depth. The compressive stresses tend to cause bursting of the concrete and helical reinforcement and/or welded mats are required to resist this tendency; the tensile stresses may cause cracks along the axis of loading, and straight transverse reinforcement (usually in the form of stirrups) should be provided to control this cracking. It should be noted also that cracking cannot always be completely eliminated since the stress diagram shows that the presence of tensile stresses (and hence in some cases of micro-cracks) is unavoidable. The transverse reinforcement should therefore be sufficient to ensure that no visible cracks develop.

The case shown in Figure 10.2 corresponds only approximately to the end-block of a prestressed beam. A closer analogy (which may be considered as equivalent to a deep strip footing or continuous wall) is shown in Figure 10.3; in this case the distribution of stress along lines X – X and Y – Y are as shown in Figures 10.3b and 10.3a. The particular values given in this diagram are due to Bay[2] and correspond to the particular case in which the width of the loaded area is one-tenth of the distance between the centre-lines of the loads. It is clear that midway between the loads the nature of the stresses is the reverse of that previously described; this is still the case when the distribution of stress is trapezoidal or triangular (as in most prestressed beams) instead of uniform (as in Figure 10.3).

10.1.2 Calculation of stresses

Apart from recommendations in the national codes of practice, four methods of calculating the stresses (and hence the reinforcement required) in the end-blocks of beams with post-tensioned steel are in general use. The first is due to Magnel[3], the second to Guyon[4], the third to Zielinski and Rowe[5,6] and the fourth to Mörsch[7]. The methods are described in the following. The recommendations quoted are those of the originators of the methods; the recommendations of the authors are given later. All of these methods are

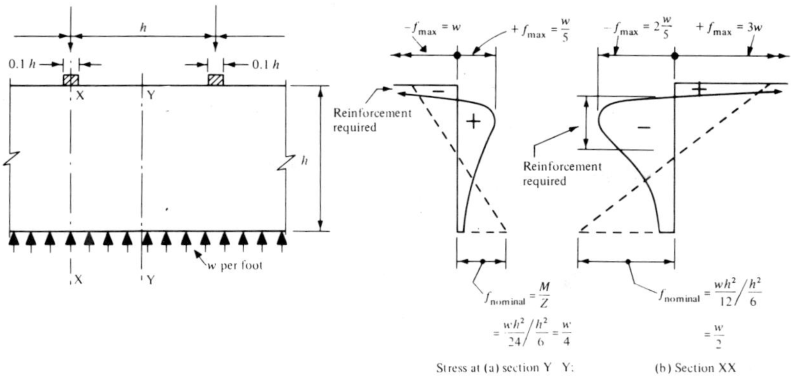

Figure 10.3 Stresses in deep strip footing

approximate and can be applied equally to lightweight aggregate concrete, provided the value of allowable principal tensile stress or the permissible tensile stress of concrete as the case may be, is adjusted.

Magnel's method. It is assumed that on any plane AB (Figure 10.4) there are a bending moment M and a shearing force V. The bending moment causes stresses f_M, the distribution of which is arbitrarily assumed to be a cubic parabola, given by the expression $K\dfrac{M}{ba^2}$ and the shearing force causes stresses f_V given by $K_1\dfrac{V}{ab}$ in which b is the nett width of the member at the level of the plane AB after deducting cable holes or other openings, a is the length of the anchorage zone, and K and K_1 are the coefficients given in Figure 10.5.

The direct stresses f_D (Figure 10.6) are evaluated from the assumption that the loads from the anchors disperse at an angle of 45 deg. These stresses are calculated for the gross effective area, ignoring the cable holes; the reduced value so obtained has the effect of increasing the calculated value of the principal tensile stress and the result is therefore conservative.

When f_M, f_V and f_D have been calculated in this way the principal tensile stress f_P is obtained from the formula

$$f_P = \frac{f_M + f_D}{2} - \sqrt{f_V{}^2 + \frac{(f_D - f_M)^2}{4}}$$

f_P is negative for tensile stresses, and a negative result corresponds to a tensile principal stress. A graphical method is also available and is given in Chart I.

When calculating f_M, f_V and f_D, the length of the anchorage zone is usually assumed to be equal to its depth. The procedure suggested by Magnel is as follows (Figure 10.4).

(1) Draw diagrams of loading, bending moment and shearing force.
(2) Select possible critical planes AB by inspection. These would normally occur at the top or bottom of a cable duct, or at the level which includes the greatest number of cables.
(3) Calculate f_M, f_V, f_D, and the principal tensile stresses, for value of $\dfrac{x}{a}$ (Figure 10.4) between -0.2 and $+0.3$.

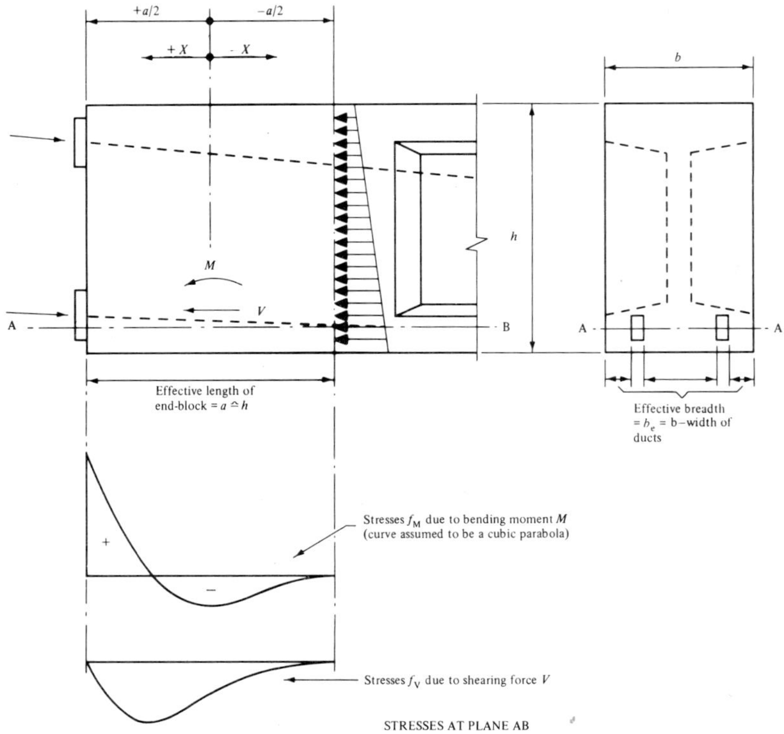

Figure 10.4 Magnel's method

(4) Provide reinforcement to resist the greatest principal tensile stress if this exceeds the permissible value, assuming this stress to be operative over the whole of the middle-half of the anchorage (Figure 10.7).

This procedure is often laborious and as its sole object is the determination of the greatest principal tensile stress the following shorter method may be used.

(1) Draw diagrams of loading.

(2) Calculate the nominal shearing stress $f_{V\,nom} = \dfrac{\text{load}}{\text{nett area}}$ at the top and bottom of each duct or series of ducts, and select the plane at which the greatest value occurs.

EITHER

(3A.) If the value of $f_{V\,nom}$ is large, say about 100 lbf/in^2 (7 kgf/cm^2, 0.7 N/mm^2), calculate the bending moment on this plane; determine f_M, f_V, f_D, and the principal tensile stress at sections distant 0.1a and 0.2a from the centre line of the anchorage towards the anchor end. If the value obtained at 0.4a exceeds that at 0.2a, calculate the principal tensile stress at the centre of the anchorage to ensure that the maximum is determined.

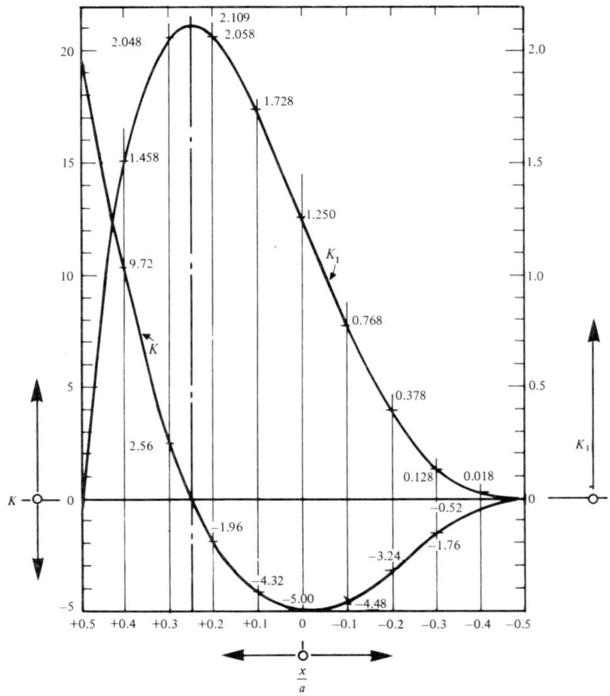

Figure 10.5 Curves for Magnel's method

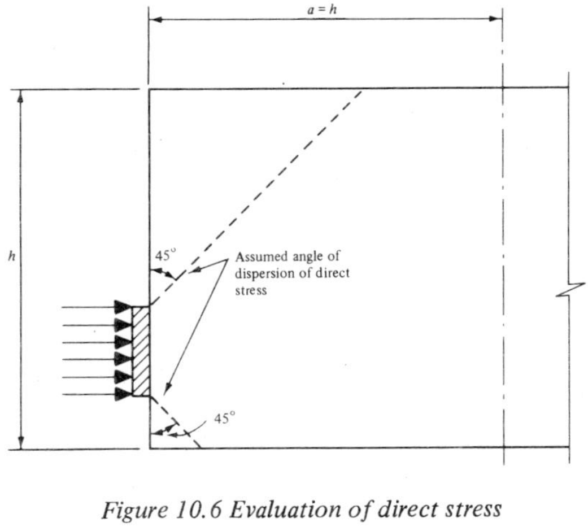

Figure 10.6 Evaluation of direct stress

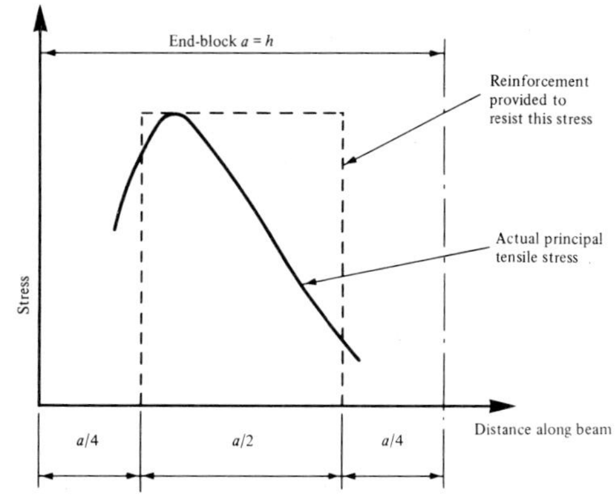

Figure 10.7 Principal tensile stress according to Magnel

OR

(3B.) If the value of $f_{V\,nom}$ is small, the same procedure is applied, but only to the plane at which the greatest bending moment occurs.

(4) Provide reinforcement, if required (as before), within the middle-half of the anchorage.

The angle of inclination of the principal tensile stress is usually small and it is normally sufficient to provide vertical reinforcement only. Alternatively, the horizontal and vertical components of the principal stress can be calculated, the angle of inclination θ being obtained from the formula

$$\theta = \tfrac{1}{2}\tan^{-1}\left(\frac{2f_V}{f_M - f_D}\right)$$

or from Chart 1.

Experience has shown the method to be conservative, provided that part of the reinforcement is placed close to the end-face when the anchorages are widely spaced, to resist tensile stresses occurring between the anchorages, as indicated in Figure 10.3. Good compaction is the most important factor; indeed in the case of a beam of 18 ft span which was made for the Stationery Building in Edinburgh in 1951, in which the principal tensile stress was more than 300 lbf/in^2 (21 kgf/cm^2, 2.1 N/mm^2), the substantial secondary reinforcement determined in accordance with this method was intentionally omitted from one beam and the cable force was increased; nevertheless, no visible cracking occurred(8). The entire quantity of reinforcement involved is not usually excessive, although it may be rather concentrated, but if economy is of particular importance Magnel recommends that the area of reinforcement be determined by adopting a permissible stress of 25 000 lbf/in^2 (17.5 kgf/mm^2, 172 N/mm^2).

Guyon's method. In the method devised by Guyon, which is based on a two-dimensional mathematical analysis of the stresses in the end-block, with particular (though not exclusive) references to the Freyssinet system, the anchorage is considered to consist of a number of separate prisms. Two general cases are distinguished.

(1) LINEAR DISTRIBUTION OF PRESTRESSING FORCE

If the prestressing force is applied by anchors which are linearly distributed along the end of the member in a manner corresponding to the distribution of stress at the inner end of the end-block (Figure 10.8) each anchorage is considered to consist of a prism which is in equilibrium under the action of the prestressing force at one end and the linear stress in the concrete at the other. If the horizontal component of each prestressing force does not act in the same line as the resultant force representing the section of the stress diagram (which is usually the case) each force is then assumed to act on a prism the height ($2d$) of which is taken as twice the distance from the boundary of the appropriate section of the stress diagram or from the free edge (Figure 10.9). The mean compressive stress f_D is calculated for each prism; the prestressing force is applied over a depth $2d'$, and the ratio d'/d is termed the 'concentration ratio'. The stresses along the horizontal line of action of each force are determined from the curves given in Figure 10.10.

(2) NON-LINEAR DISTRIBUTION OF PRESTRESSING FORCE (SUCCESSIVE RESULTANTS)

If the anchors are arranged in groups, it is assumed that the stresses f_M reach their greatest values first on the lines of action of each separate force, then on the line of action of the resultant of each group, and finally on the line of the

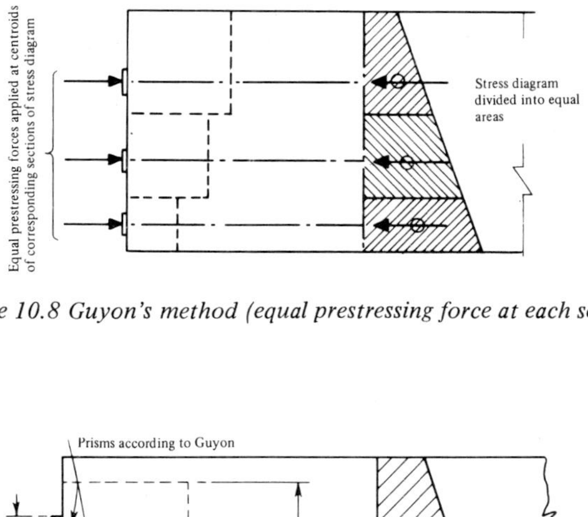

Figure 10.8 Guyon's method (equal prestressing force at each section)

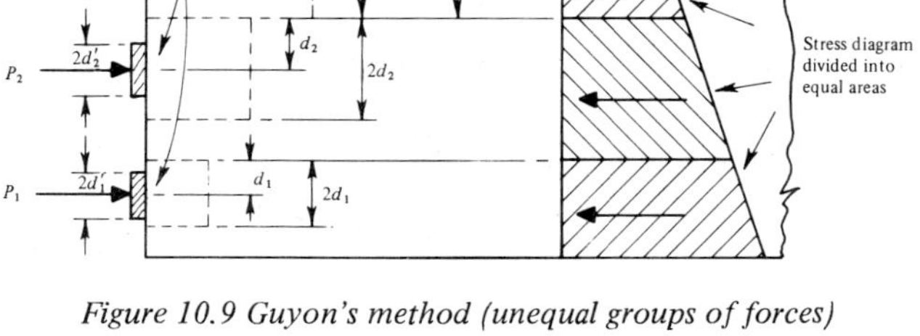

Figure 10.9 Guyon's method (unequal groups of forces)

total resultant (Figure 10.11). Guyon terms this the 'law of successive resultants', and the reinforcement is provided in corresponding stages.

The total resultant is considered first and it is assumed at all stages that the force under consideration acts axially on a symmetrical prism with a height of twice the distance of the force from the nearest face of the prism previously considered, or from the nearest edge of the anchorage.

The stresses f_M, f_V and f_D at the axis of each prism are then determined from Table 10.1.

Zielinski and Rowe's method. The third method, due to Zielinski and Rowe, also adopts the concept of prisms, though their definitions differ slightly from those given by Guyon. A square anchor of side $2d'$ is assumed to act on a square prism of side $2d$. The distance $2d$ is assumed to be either the distance between the centre-lines of anchors or twice the distance from the centre-line of the anchor to the edge of the concrete, whichever is less.

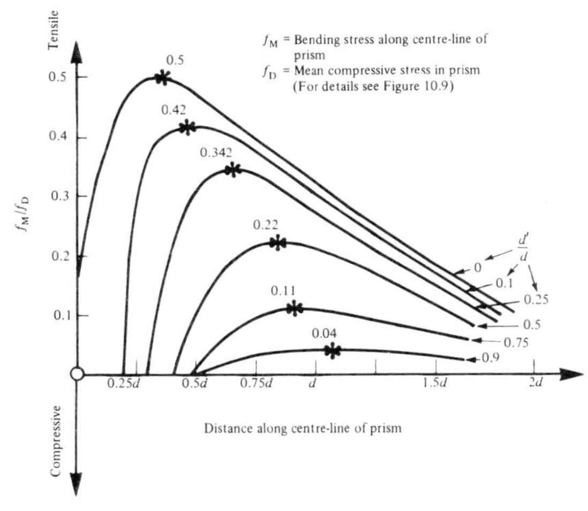

Figure 10.10 Curves for Guyon's method

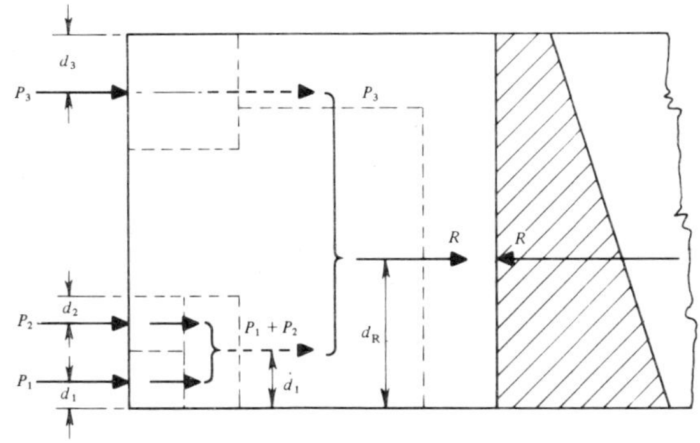

Figure 10.11 Guyon's method (successive resultants)

Table 10.1 Guyon's Method

VALUES OF $\frac{f_M}{f_D}$; in which f_M = AXIAL bending stress due to LONGITUDINAL force P
f_D = Mean compressive stress $\frac{P}{BD}$

Eccentricity of LONGITUDINAL force P	0	$h/12$	$h/6$	$h/4$	$h/3$	$h/2$	$3h/4$
$\frac{h}{2}$	−2.187	−0.913	−0.428	−0.014	+0.307	+0.399	+0.192
$\frac{3h}{8}$	+1.610	−0.800	−0.130	+0.120	+0.280	+0.325	+0.166
	−1.222	−0.601 Tension	+0.125	+0.192	+0.250	+0.242	+0.122
	−0.950	−0.380 Compress.	+0.210	+0.220	+0.174	+0.130 ↑ Compression	+0.062
$\frac{h}{4}$	−0.758	−0.250	+0.238	+0.154	+0.062	−0.024	−0.016
	−0.642	+0.550	+0.190	+0.04	−0.090	−0.170 Tension ↓	−0.074
$\frac{h}{8}$	−0.566	+1.004	+0.074	−0.144	−0.266	−0.262	−0.128
	−0.540	+1.120	−0.320	−0.340	−0.36	−0.300	−0.158
0	−0.523	−0.448	−0.500	−0.462	−0.423	−0.314 ℄ BEAM	−0.161

Distance along beam →

Eccentricity of TRANSVERSE force V							
0	±∞	0	0	0	0	0	0
	−10.44	−1.02	−0.15	−0.05	+0.05	+0.08	+0.04
$\frac{h}{8}$	−5.66	−2.96	−0.87	−0.19	+0.05	+0.14	+0.07
	−4.32	−3.05	−1.36	−0.35	+0.20	+0.16	+0.08
$\frac{h}{4}$	−4.00	−3.10	−1.52	−0.44	+0.22	+0.20	+0.10
	−4.32	−3.05	−1.36	−0.35	+0.20	+0.16	+0.08
$\frac{3h}{8}$	−5.66	−2.96	−0.87	−0.19 Tension	Compression +0.05	+0.14	+0.07
	−10.44	−1.02	−0.15	−0.05	+0.05	+0.08	+0.04
$\frac{h}{2}$	±∞	0	0	0	0	0	0

VALUES of $\frac{f_M}{f_V}$; in which f_M = Axial bending stress due to TRANSVERSE load V

f_V = Mean shearing stress $\frac{V}{bh}$

Similarly, a rectangular anchor with dimensions of $2b'$ and $2d'$ acts on a rectangular prism of $2b \times 2d$. Circular anchors are assumed equivalent to square anchors of side $2d'$.

For each prism, the direct stress f_D is obtained from

$$f_D = \frac{P_i}{A_p}$$

in which P_i is the maximum (initial) prestressing force and A_p is the nett cross-sectional area of the prism after subtracting the area of the duct. For anchors comprising steel plates bearing directly on the concrete, with cones or trumpets forming the ends of the ducts, the area to be subtracted is that at the point of maximum tension, taken as $d/2$ from the end-face of the concrete. With embedded anchors, which distribute prestressing force along their lengths, the area of the duct immediately behind the anchor is subtracted.

The maximum tensile splitting stress, f_T, and the total splitting force T_{sp} in the direction normal to the axis of the prism, are then obtained from

$$f_T = B f_D$$

and

$$T_{sp} = C P_i$$

in which the parameters B and C, derived from tests, are as given in Figures 10.12 a and 10.12 b. The distribution of tensile stress along the axis of the prism is very nearly triangular (as found by Guyon), with the apex of the triangle at a distance of $d/2$ from the anchor face, and diminishing to zero at $2d$ from the anchor face.

The concrete is assumed to resist part of the total tension T_{sp}, and the amount T_{spr} to be resisted by the reinforcement is given by

$$T_{spr} = T_{sp}\left[1 - \left(\frac{\bar{f}_t}{f_T}\right)^2\right]$$

in which $\bar{f}_t$ is the permissible tensile stress in the concrete.

Zielinski and Rowe found that the strains which occur in end-blocks prior to cracking correspond to apparent tensile strength in excess of the splitting strength of the concrete. The ratio of apparent strength to splitting strength is

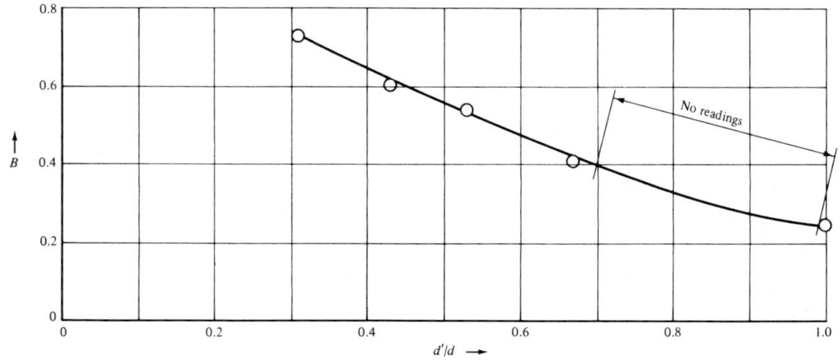

Figure 10.12(a) *Parameter 'B' (Zielinski and Rowe)*

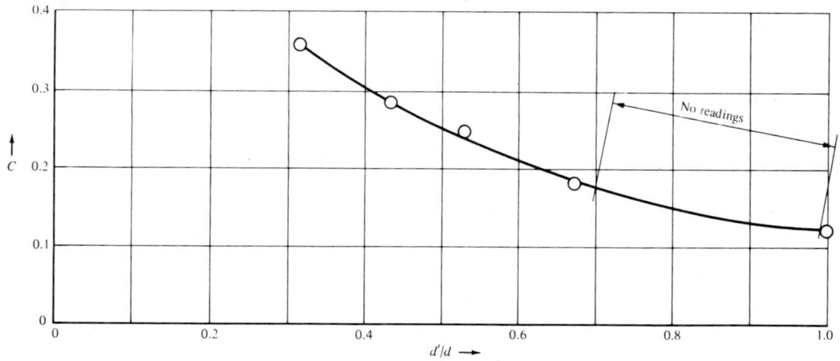

Figure 10.12(b) *Parameter 'C' (Zielinski and Rowe)*

denoted by K and is also given in Figure 10.13. The permissible tensile stress $\bar{f}_t$ is therefore assumed to have a value of $K f_{tens}$ in which f_{tens} is the tensile splitting strength. Because of the complex nature of stresses in end-blocks Zielinski and Rowe state that 'it would be logical to increase the apparent permissible stresses in tension' in concrete by the factor K the value of which for different d'/d ratio have been given in Figure 10.13. A_s is then obtained from the relationship

$$A_s = \frac{T_{spr}}{\bar{f}_s}$$

where $\bar{f}_s$ is the permissible tensile stress in steel.

In CP 110 a simplified variation of the method is recommended, which ignores the tensile strength of the concrete but allows the steel to be used at a stress of f_k/γ_m. The dimensions of the prism are assessed in the manner described above, and the splitting force to be resisted is obtained from the following values:

d'/d (or b'/b)	0.3	0.4	0.5	0.6	0.7
T_{spr}/P_i	0.23	0.20	0.17	0.14	0.11

According to CEB/FIP Model Code the transverse reinforcement required can be obtained from the relationship

$$A_s = 0.3 \,.\, P_i \,.\, \left(1 - \frac{d'}{d}\right) \Big/ \left(f_y/\gamma_m\right).$$

The reinforcement can consist of stirrups, ties or loops. The total tensile force $(A_s f_y/\gamma_m)$ can be assumed constant over a distance between $0.2d$ and $2d$.

Mörsch's method. A fourth method which involves much less calculation than any of the foregoing methods and which was originally devised in 1920 in connection with the design of stone hinges for masonry arches, is illustrated in Figure 10.14. The splitting force T_{sp} is assumed to act midway along the end-block, and is given by

$$T_{sp} = \frac{P_t}{4}\left(1 - \frac{d'}{d}\right)$$

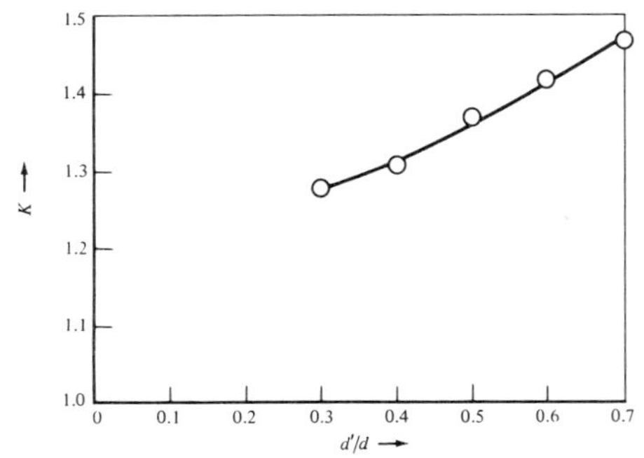

Figure 10.13 Parameter K (Zielinski and Rowe)

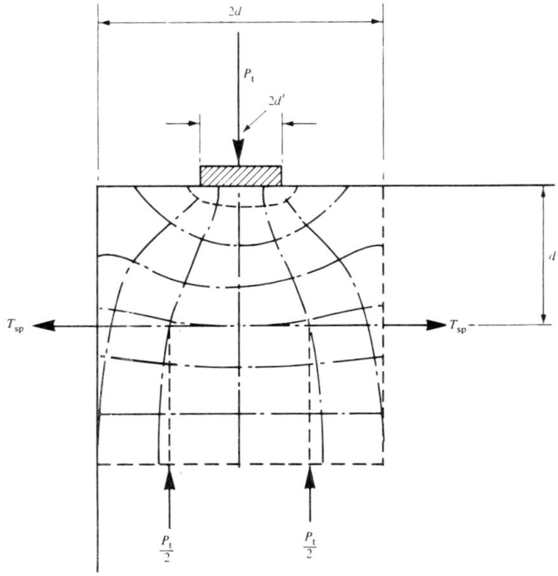

Figure 10.14 Mörsch's method

10.1.3 Examples

In order to compare the calculations involved and the results arising from each of these methods, an example (originally due to Magnel) is given in the following. In Figure 10.15 is shown the end of a slab to which a total prestressing force of 372 000 lbf is applied by means of two cables per foot.

(1) *Magnel's method.* The shearing force and bending moment diagrams are shown in full, but only the values at the sections required need be calculated in practice. The appropriate values of f_M, f_D, and f_V for sections I, II, and III are given in Table 10.2. It should be noted that in this example, the assumption that the axial force is dispersed at an angle of 45 deg. through the slab leads to a constant value of f_D of $\frac{372\,000}{12 \times 38} = 815$ lbf/in^2 (87 kgf/cm^2,

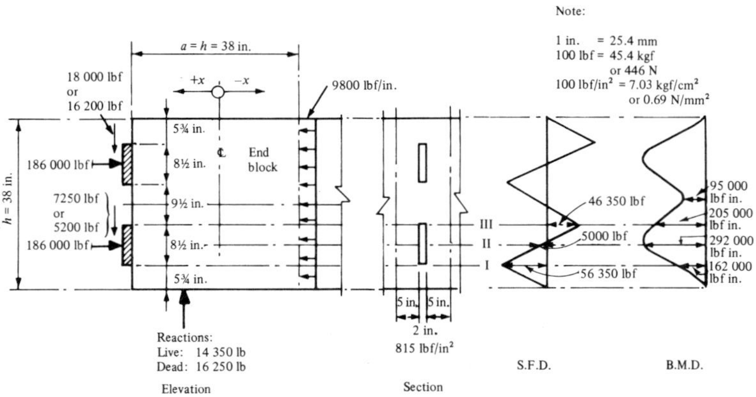

Figure 10.15 Example 10.1.3 (Magnel's method)

5.7 N/mm²) at all critical values of x/a. For the bending stress f_M, $ba^2 = 10 \times 38^2 = 14\,400$ in³ (2.36×10^5 cm³) for the shearing stress f_V, $ba = 10 \times 38 = 380$ in² (2450 cm²).

In addition to the calculated value of f_V the additional shearing stresses due to the dead and live loads and to the vertical component of the prestressing force should be included. In the present example, the shearing force due to dead load and prestress (before losses) is $-16\,250 + 18\,000 + 7250 = 9000$ lbf and that due to the total load and prestress is $-16\,250 - 14\,350 + 16\,200 + 5200 = -9200$ lbf. The greater of these forces gives rise to a maximum shearing stress of $\dfrac{9200}{12 \times 38 \times 0.67} = 30$ lbf/in² (2.1 kgf/cm², 0.2 N/mm²) and all the calculated values of f_V are increased by this amount.

The greatest value of the principal tensile stress is 139 lbf/in² (9.6 kgf/cm², 0.95 N/mm²), and hence no reinforcement is required if the specified works cube strength exceeds 5250 lbf/in² (367 kgf/cm², 36 N/mm²). In cases where reinforcement is required, however, the method of calculation is as follows. It is assumed that the maximum principal tensile stress occurs throughout half the length of the end-block (Figure 10.7). In the present case the tensile force to be related would therefore be $139 \times 10 \times 38 = 52\,800$ lbf. If the permissible stress in the steel is 25 000 lbf/in², as recommended by Magnel, stirrups with a total cross-sectional area of $\dfrac{52\,800}{25\,000} = 2.11$ in² (1360 mm²) would be required (Figure 10.22a). It will be seen later that even using this high permissible stress the calculated area exceeds those obtained by other methods. The senior author successfully employed a nominal stress of 40 000 lb/in² (or more accurately half the calculated area) in a number of beams designed in 1952 without any trouble afterwards

(2) *Guyon's method.* The end-block is divided into prisms as shown in Figure 10.16.

Effective height of prism $= 2 \times 9 = 18$ in.

Effective breadth of prism $= 12$ in.

Table 10.2 Calculations for Magnel's method

Section	$\frac{x}{a}$	f_D	$f_M = \frac{KM}{a^2b}$	$f_V = \frac{K_1 V}{a\ b}$	f'_V	$f_V + f'_V$	$\frac{f_D + f_M}{2}$	$\left[\frac{f_D - f_M}{2}\right]^2$	$(f_V + f'_V)^2$	$\sqrt{\bar{f}_V^2 + \left(\frac{f_D - f_M}{2}\right)^2}$	f_p
I	+0.3	↑	+29	303	↑	333	422	156 200	110 800	517	−95
V = 56 350 lbf	+0.2		−22	305		335	397	175 100	112 100	536	−139
M = 162 000 lbf in.	+0.1	815	−49	255	30	285	383	186 500	81 200	517	−134
$\frac{V}{ab}$ = 148 lbf/in²	0		−56	185		215	380	190 000	46 200	487	−107
	−0.1		−50	114		144	382	187 000	20 750	456	−74
$\frac{M}{a^2b}$ = 11.3 lbf/in²	−0.2	↓	−36	56	↓	86	390	181 200	7 400	433	−43
II	+0.3	↑	+52	28	↑	58	434	149 900	3400	387	+47
V = 5000 lbf	+0.2		−40	28		58	388	183 200	3400	431	−43
M = 292 000 lbf in.	+0.1	815	−88	23	30	53	364	204 300	2800	456	−92
$\frac{V}{ab}$ = 13.3 lbf/in²	0		−101	17		47	358	209 800	2200	460	−102
	−0.1		−91	11		41	363	205 200	1700	455	−92
$\frac{M}{a^2b}$ = 20.2/lbf/in²	−0.2	↓	−66	5	↓	35	375	194 500	1200	442	−67
III	+0.3	↑	+37	250	↑	280	426	152 100	78 400	480	−54
V = 46 350 lbf	+0.2		−28	251		281	394	178 100	79 000	507	−113
M = 205 000 lbf in.	+0.1	815	−62	211	30	241	377	192 700	58 100	500	−123
$\frac{V}{ab}$ = 122 lbf/in²	0		−71	153		183	373	196 200	33 500	479	−106
	−0.1		−64	94		124	376	193 600	15 400	457	−81
$\frac{M}{a^2b}$ = 14.2 lbf/in²	−0.2	↓	−46	47	↓	77	385	185 800	5900	437	−52

1000 lbf/in² = 70 kgf/cm² = 7 N/mm² 100 lbf = 45 kgf = 446 N, f'_V = shear stress due to vertical load and vertical component of prestressing force
$\bar{f}_V = f_V + f'_V$

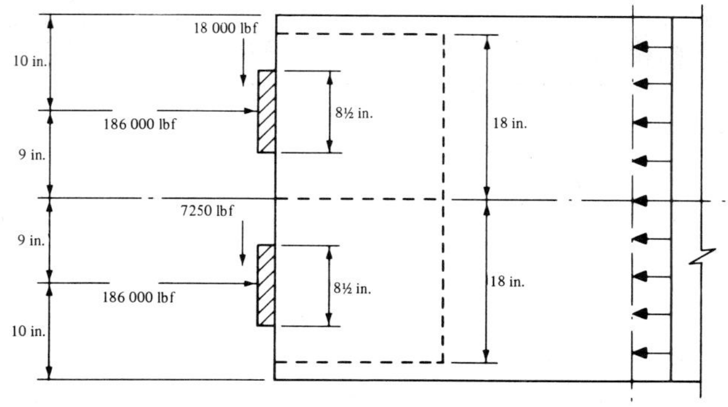

Figure 10.16 Example 10.1.3 (Guyon's method)

$$\text{Mean compression} = \frac{186\ 000}{18 \times 12} = 860 \text{ lbf/in}^2 \ (60.2 \text{ kgf/cm}^2,\ 5.9 \text{ N/mm}^2)$$

$$\text{Concentration ratio} = \frac{8.5}{18} = 0.473$$

Therefore, from Figure 10.10 (by interpolation)

$f_{M\ max} = 0.23 f_D = 0.23 \times 860 = 198$ lbf/in^2 and distance of $f_{M\ max}$

from end of beam $= \dfrac{5d}{6} = \dfrac{5 \times 9}{6} = 7.5$ in. (19 cm)

The distribution of stress is sketched in Figure 10.17; this may be considered as equivalent to a triangle with a peak value of 230 lbf/in^2. If it is assumed that reinforcement must be provided to resist all tensile stresses greater than the permissible value (for example, 150 lbf/in^2), then the force to be resisted (that is, the shaded area in Figure 9.17) is given by

$$bT \left[1 - \left(\frac{\bar{f}_t}{f_T}\right)^2\right]$$

in which T is the area of the stress triangle (that is, tensile force per unit width). In the present case therefore the force is

$$12 \times \frac{230 \times 14.6}{2} \times \left[1 - \left(\frac{150}{230}\right)^2\right] = 20\ 150 \times 0.575$$

$$= 11\ 600 \text{ lbf } (5200 \text{ kgf},\ 51.6 \text{ kN})$$

Hence if the permissible stress in the steel is 20 000 lbf/in^2 (14 kgf/mm^2, 138 N/mm^2), the area required is $\dfrac{11\ 600}{20\ 000} = 0.58$ in^2 (374 mm^2). Guyon also recommends that additional reinforcement should be placed as near as possible to the anchorage to resist a force of $0.04P = 0.04 \times 186\ 000 = 7500$ lbf; the area required is therefore $\dfrac{7500}{20\ 000} = 0.375$ in^2 (245 mm^2).

Resultant stresses

(a) Stresses due to normal forces. Particular eccentricities of the prestressing forces are given in Table 10.1. Because of this, and also to allow for the effect of the concentration ratio, Guyon recommends that the actual forces should be replaced by equivalent forces acting at the points shown in Figure 10.18. The stresses due to these forces are given in Table 10.3. They occur on the centre-line of the resultant prism; that is, at the centre-line of the member.

(b) Stresses due to transverse forces. In cases where these stresses are sufficiently large to be worth considering, Guyon proposes the following approximate method.

If a transverse force is applied at an anchor distant y from the axis of the beam, the maximum stress occurs at the axis and the distribution of stress is

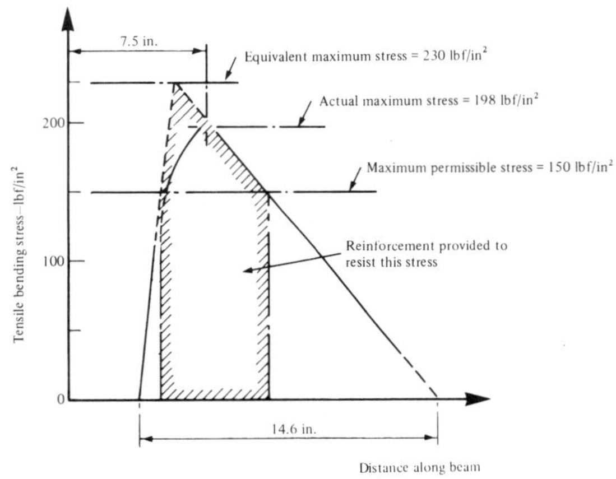

1 in. = 25.4 mm

100 lbf/in² = 7 kgf/cm² or 0.7 N/mm²

Figure 10.17 Distribution of stresses (Guyon's method)

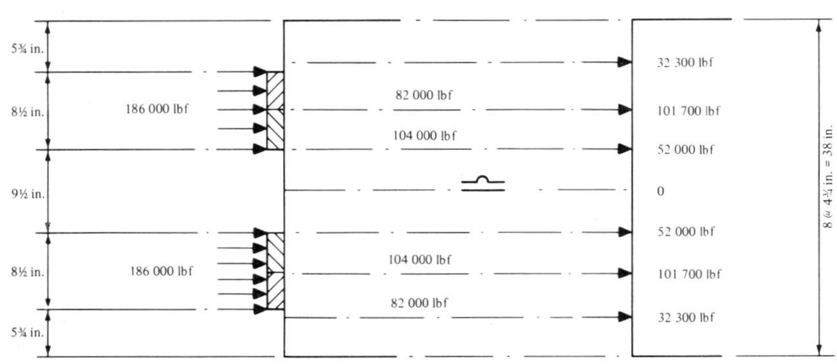

Figure 10.18 Equivalent forces according to Guyon

Table 10.3 Calculations for Guyon's method

Prestressing force and eccentricity	Distance along end-block						
	0	$a/12$	$a/6$	$a/4$	$a/3$	$a/2$	$3a/4$
32 000 lbf at $\frac{3h}{8}$	−39 500	−19 400	+4030	+6200	+8070	+7820	+3940
101 700 lbf at $\frac{h}{4}$	−77 000	−2540	+24 200	+15 650	+6300	−2400	−1630
52 000 lbf at $\frac{h}{8}$	−29 500	+52 200	+3850	−7500	−13 840	−13 620	−6650
Total for half-block	−146 000	+30 260	+32 080	+14 350	+530	−8440	−4340
Total for block	−292 000	+60 520	+64 160	+28 700	+1060	−16 880	−8680
$f_M = \frac{\text{Total}}{12 \times 38}$	−640	+133	+141	+63	+3	−37	−19
$f_{M\ \text{transverse}}$	+100	+67	+33	0	0	0	0
$f_{M\ \text{total}}$	−540	+200	+174	+63	+3	−37	−19

approximately triangular over a length l along the axis (Figure 10.19).

The greatest value occurs at the end of the beam, and the area of the triangle is obtained from the requirement that the forces acting on each half of the beam must balance. Therefore the force given by this stress diagram must be equal to the difference between the applied force and the shearing force representing the distributed shearing stress at the inner end of the end-block.

In the present case $\frac{y}{h} = \frac{9}{38} = 0.237$. Hence the length over which the stress is distributed (Figure 10.19b) = $0.235h = 0.235 \times 38 = 8.9$ in. (226 mm).

Transverse force = 18 000 + 7250 = 25 250 lbf (Figure 10.20)

Therefore the balancing force required = $18\,000 - \frac{25\,250}{2}$

= 5375 lbf (2439 kgf, 23.3 kN)

$$\frac{f_{M\ \max}}{2} \times 8.9 \times 12 = 5375 \text{ and } f_{M\ \max} = +100\ \text{lbf/in}^2\ (7\ \text{kgf/cm}^2, 0.69\ \text{N/mm}^2)$$

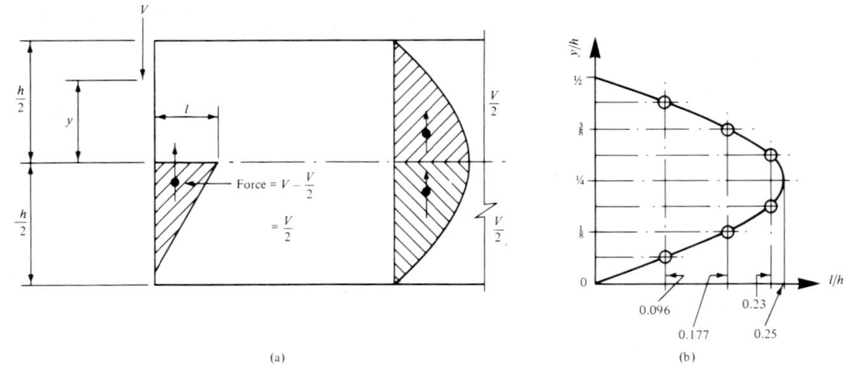

Figure 10.19 Distribution of transverse forces (Guyon)

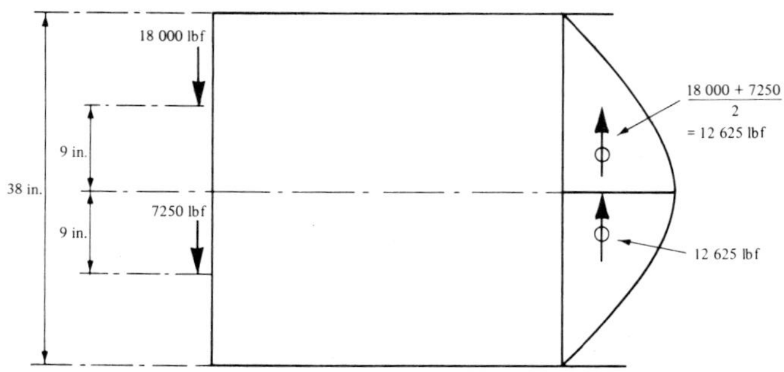

1 in. = 25.4 mm
100 lbf = 45.4 kgf or 446 N

Figure 10.20 Numerical illustration of Figure 10.19

(c) Total stresses. The total stresses due to the normal and transverse forces are shown in Table 10.3. Assuming the point of zero stress to be 1 in. from the surface, the total tensile force at the end of the beam is $0.5 \times 540 \times 12 \times 1 = 3240$ lbf (1460 kgf, 14.42 kN) and the area of reinforcement is therefore

$$\frac{3240}{20\,000} = 0.16 \text{ in}^2 \ (1 \text{ mm}^2).$$

The total reinforcement required is shown in Figure 10.22b

(3) *Zielinski and Rowe's method.* For this method, the sequence of calculation is simple. With the arrangement of anchors shown in Figure 10.16, the computations are as follows:

$$\begin{aligned} P_i &= 186\,000 \text{ lbf } (84\,000 \text{ kgf, } 828 \text{ kN}) \\ \text{Prism area} &= 18 \times 12 = 216 \text{ in}^2 \ (135\,000 \text{ mm}^2) \\ \text{Duct area} &= 5 \times 2 = 10 \text{ in}^2 \ (6250 \text{ mm}^2) \\ A_p &= 206 \text{ in}^2 \ (128\,750 \text{ mm}^2) \end{aligned}$$

$f_D = P_i/A_p = 186\ 000/206 = 903$ lbf/in² (63.3 kgf/cm², 6.25 N/mm²)

$\dfrac{d'}{d} = \dfrac{8.5}{18} = 0.472$

$B = 0.54$, $C = 0.26$ — See Figures 10.12a and 10.12b

$K = 1.37$ — See Figure 10.13

$f_T = Bf_D = 0.54 \times 903 = 488$ lbf/in² (34.2 kgf/cm², 3.37 N/mm²)

$T_{sp} = CP_i = 0.26 \times 186\ 000 = 48\ 300$ lbf (22 000 kgf, 215 kN)

Assuming $f_{ten} = 400$ lbf/in² (28 kgf/cm², 2.76 N/mm²)

$\bar{f}_t = 1.37 \times 400 = 548$ lbf/in² (38.36 kgf/cm²; 3.78 N/mm²)

$\therefore\ \bar{f}_t > f_T$

$\therefore$ No reinforcement necessary

Using the further simplification given in CP 110, the corresponding calculations are:

$d'/d = 0.472$	$b'/b = 0.83$
$T_{spr}/P_i = 0.18$	0.07 (by extrapolation)
$\therefore\ T_{spr} = 0.18 \times 186\ 000 = 33\ 500$ lbf (15 200 kgf, 149 kN)	$0.07 \times 186\ 000 = 13\ 020$ lbf (5920 kgf, 58 kN)

Assuming a design strength of $f_K/1.15$ equals 30 000 lbf/in² (2100 kgf/cm², 207 N/mm²) then

$A_s = \dfrac{33\ 500}{30\ 000} = 1.12$ in² (700 mm²)	$\dfrac{13\ 020}{30\ 000} = 0.434$ in² (271 mm²)

This reinforcement is distributed throughout a length from 0.2 d' (or b') to $2d'$ (or b') from the end-face of the beam (Figure 10.22c)

(4) *Mörsch's method.* From Figure 10.21, the splitting force is given by

$$T_{sp} = \frac{186\ 000}{4}\left(1 - \frac{8.5}{18}\right) = 24\ 500 \text{ lbf (11 120 kgf, 109 kN)}$$

The reinforcement required is therefore $\dfrac{24\ 500}{20\ 000} = 1.23$ in² and the corresponding arrangement of reinforcement is shown in Figure 10.22d.

The differences in the amounts and the distributions of reinforcement obtained by different methods are seen from Figure 10.22. All of the methods are based on more or less simplified theories; Magnel assumed the curve of bending stress to be a cubic parabola, Guyon's theory is based on a two-dimensional elastic analysis of a three-dimensional system of stress, and Mörsch's method replaces a complex system of principal stress by a simple linear system. Zielinski and Rowe's method is based on experimental work.

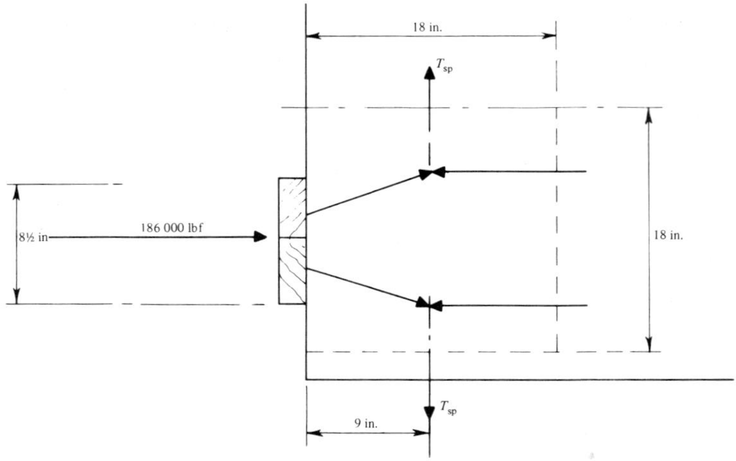

Figure 10.21 Example 10.1.3 (Mörsch)

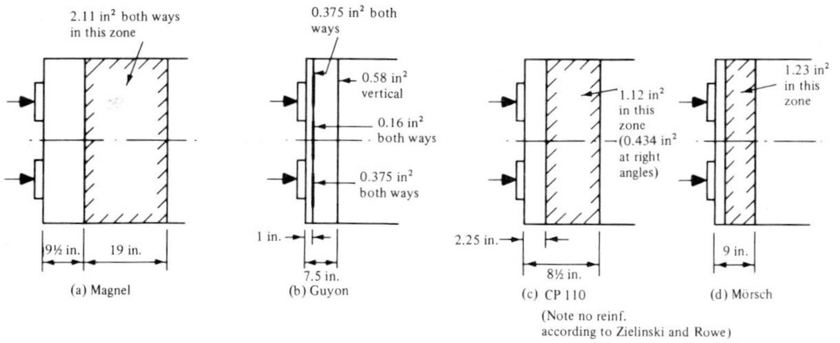

Figure 10.22 Comparison of four methods (example 10.1.3)

The recommendations given in the following are based on the experience and observation of the authors. The most important aspects of the design of end blocks have been found to be as follows.

(1) The reinforcement must be placed in the correct position; that is, partly between the anchorages and as close as possible to the end-face, and partly on the axis of the anchorage and farther away from the end-face, in the form of transverse stirrups. This is the most suitable arrangement of reinforcement to resist stresses of the type shown in Figure 10.3.

(2) The reinforcement should not be so congested that consolidation of the concrete is difficult to achieve. The attainment of full compaction is probably the most important requirement for end-blocks. It has been shown that if this is guaranteed no trouble occurs in individual cases even if the secondary reinforcement is entirely omitted; on the other hand, the presence of secondary steel of itself will not prevent the collapse of a poorly-compacted end-block.

(3) Subject to (2), the quantity of reinforcement provided is not the governing criterion. Because of this, Mörsch's method appears to be the simplest solution.

(4) The provision of helical reinforcement and/or welded steel mats around the anchorages, close to the end-face, is advisable to prevent bursting. This may be omitted if large anchor-plates are provided, as the bursting stresses are then small.

10.1.4 Bearing stresses under anchorage plates

In accordance with ACI's tentative recommendations[9] the average bearing stresses on the concrete created by the anchorage plates shall not exceed the values allowed by the following equations:

At service load:

$$f_{cp} = 0.6 f_{cyl} \sqrt{A_b'/A_b}$$

but not greater than f_{cyl}.

At maximum jacking load:

$$f_{cp} = 0.8 f_{cyl} \sqrt{(A_b'/A_b)} - 0.2$$

but not greater than 1.25 f_{cyli},

where

f_{cp} = Permissible compressive concrete stress in bearing
f_{cyl} = Specified compressive strength of the concrete
$f_{cyl\,i}$ = Specified compressive strength of the concrete at the time of initial prestress
A_b' = Maximum area of the portion of the concrete anchorage surface that is geometrically similar to and concentric with the area of the anchorage
A_b = Bearing area of the anchorage

As used in the preceding equations f_{cp} is the average bearing stress, P/A_b in the concrete computed by dividing the force P of the prestressing steel by the net projected area, A_b, between the concrete and the bearing plate or other structural element of the anchorage which has the function of transferring the force to the concrete.

In the CEB/FIP Model Code the maximum value of f_{cp} or $\frac{P}{A_b}$ is limited to $\frac{f_{cyl}}{\gamma_m} \times \sqrt{\frac{A_b'}{A_b}}$ but $\not> 3.3 \frac{f_{cyl}}{\gamma_m}$ for normal weight concrete. For the lightweight aggregate concrete the square root in the above equation is replaced by the cube root.

Post-tensioning anchorages which do not utilize bearing plates may be used and need not comply with the above provisions for bearing stresses if the anchorage is one of recognized proven performance, and test data satisfactory to the engineer is provided.

10.2 Pre-tensioned steel

10.2.1 Transmission length

When pre-tensioned steel is used it is essential to obtain satisfactory bond, and to avoid any slipping of the steel. The length in which the prestress is transmitted to the concrete, termed the 'transmission length', depends partly on the strength of the concrete (which in turn depends on the proportions and properties of the constituents and the degree of compaction) and partly on the surface condition and the diameter of the steel.

It was at one time assumed that the bond by means of which the prestress is transferred to the concrete was solely due to the tendency for the diameter of the steel to increase slightly at the ends when it is released. It is now known to be due to a combination of adhesion, friction, and wedging effect; a very small slip occurs at transfer, which allows the friction to develop. When pre-tensioning was introduced it was thought that only piano wire, with its small diameter, was suitable for the purpose and piano wires in long-lines were used exclusively for some years. (The long-line system, which is sometimes incorrectly credited to Hoyer, was in fact suggested by Freyssinet; the use of piano wire was proposed by Hoyer.)

Smooth wires of 0.2 in. and 0.276 in. diameter were mainly used in this country for pre-tensioning; 0.276 in. wires were introduced by British Railways (Eastern Region) in 1951–1952.

The E.M.P.A. of Switzerland proposed that indented wires should be used for pre-tensioning; these may be advantageous when the working load is static but otherwise their bond resistance does not exceed that of smooth wires, as the concrete embedded in the grooves may be sheared off. Crimped wire may also be used; this has a bond resistance which is appreciably greater than that of plain wire of equal diameter but the relaxation is also greater. Strands with diameters up to 0.5 in. were introduced in the USA. These possess a greater bond resistance than smooth wires of equal area, due to the interlocking between the helices of the separate wires forming the strand and the concrete. They are now commonly used in this country and diameters up to 0.75 in. have been successfully employed for pre-tensioning in connection with high-strength concrete.

In every case it is essential to know the profile and length of the transmission curve for the particular type of steel used. Random tests[(10)] made in pre-tensioning works showed great variations in these values, particularly between wires at the top and bottom of the members. The efficiency of the bond is greatly influenced by the degree of compaction, and unless particular care is taken fully to compact the concrete at the top of a member there may be no bond at all for the first 6 in. of the wires and the transmission of stress may thereafter be only linear; with good bond the transmission curve may be parabolic or exponential (Figure 10.23).

Values for the transmission length for wires can vary between 50 and 160X diameter. A suitable transmission curve is shown in Figure 10.23, in which 50 per cent of the prestressing force is assumed to be transferred in the first quarter of the length and 80 per cent in the first half. The transmission length for stranded cables is relatively shorter, but the test results which are at present available show considerable variations and as the degree of compaction is not known it is difficult to draw firm conclusions from them. In a test made in 1953 the transmission length of a stranded cable of 0.5 in. diameter was found to be only 12 in. (that is, $24d$), while a cable of $\frac{7}{16}$ in. diameter had a greater transmission length. On the other hand, transmission lengths from 6 in. to 20 in. have been observed with strands of $\frac{5}{16}$ in. diameter. The following extract in connection with transmission length is quoted from CP 110.

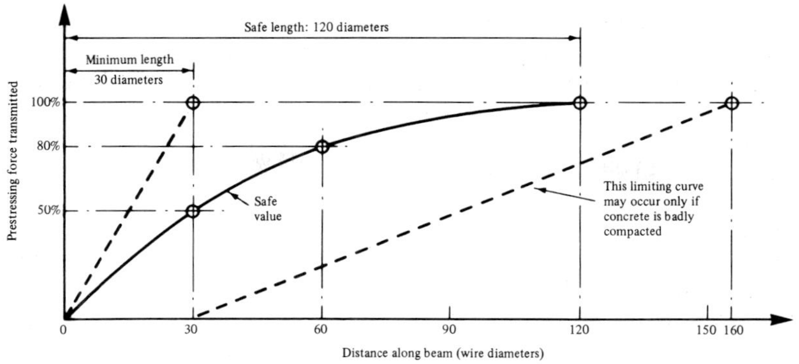

Figure 10.23 Transmission curves

"The transmission length is defined as being the length of member required to transmit the initial prestressing force in a tendon to the concrete.

The transmission length depends on a number of variables, the most important being:

the degree of compaction of the concrete,
the size and type of tendon,
the strength of the concrete,
the deformation, e.g. spiral effect of strand or crimp of the tendon.

The transmission length can vary a great deal for different factory or site conditions; for example, it has been shown that the transmission length for wire can vary between 50 and 160 diameters. As far as possible, therefore, the engineer should base the transmission length on experimental evidence for known site or factory conditions.

The following general recommendations, based on research, should be considered in relation to the known site or factory conditions.

(1) For factory produced units using plain or indented wire with a small offset crimp (e.g. 0.3 mm offset, 40 mm pitch), a transmission length of 100 diameters may be assumed when the ends of the units are fully compacted and the cube strength of the concrete at transfer is not less than 35 N/mm^2.

(2) For units using wire of a considerable crimp (e.g. 1.0 mm offset, 40 mm pitch), a bond length of 65 diameters may be assumed when the ends of the units are fully compacted and the cube strength of the concrete at transfer is not less than 35 N/mm^2.

(3) The development of stress from the end of the unit to the point of maximum stress is such that it may be assumed that 80 per cent of the maximum stress is developed in a length of 70 diameters for the conditions mentioned in (1) and in a length of 54 diameters for the conditions mentioned in (2).

(4) When the cube strength of the concrete at transfer is less than 35 N/mm (5075 lbf/in^2), the transmission lengths are likely to be greater.

(5) The transmission length for tendons near the top of a beam may well be greater than that for identical tendons placed lower in the beam, since the concrete near the top is less likely to be as well compacted.

(6) The sudden release of tendons leads to a great increase in the transmission lengths in the units near the releasing end of the bed.

(7) From the available experimental data, the transmission length for small diameter strand is not proportional to the diameter of the tendon, nor is the scatter of results so great as it is for wire. Table 10.4 gives values for the transmission length for strand; in the absence of more exact data, these values may be used in design.

Table 10.4 Transmission lengths for small diameter strand to BS 5896

Diameter of strand	Transmission length (range of results given in brackets)
mm	mm
9.3	200 (±25)
12.5	330 (±25)
18.0	500 (±50)

(8) If the tendons are prevented from bonding to the concrete near the ends of the members by the use of sleeves or tape, the values given in Table 10.4 for the transmission length may be used, it being assumed that the transmission zone starts at the point where the de-bonding process has been stopped."

Hot-rolled wires of oval or rectangular section, with diagonal ribs, are used in Germany (see Chart 2). For these, transmission lengths of 15½ in. (for wires with cross-sectional areas of 0.062 in^2) and 18½ in. (for wires with areas of 0.093 in^2) have been approved in Germany, the required minimum strength of the concrete at transfer being 4250 lbf/in^2. The corresponding prestressing forces are 7750 lbf/in^2 and 11 000 lbf/in^2. These transmission lengths are less than those in the foregoing.

The effect of time on the transmission length has been studied by Evans(11). The results obtained indicate that the transmission length increases with time; as shrinkage occurs the points of commencement and completion of the curve move away from the end of the member. The final transmission length given by Evans for wires of 0.08 in., 0.20 in., and 0.276 in. diameter are $84d$, $120d$, and $187d$ respectively; the last-mentioned value appears to relate to a concrete of limited strength.

Tests at the Building Research Station with lightweight concrete using 'foamed slag', Lytag and expanded clay aggregates show that the transmission length in high strength lightweight concrete – 5000 lbf/in^2 (35 N/mm^2) or above is not substantially different from that of gravel concrete of similar strength(12,13,14). For well compacted concrete of cube strength 5000 lbf/in^2 (35 N/mm^2) at transfer a transmission length of 100 diameter for plain or indented wire and 50 diameter for strands has been suggested(14).

10.2.2 The transmission zone

As in the case of post-tensioning, it is of major importance when pre-tensioned steel is used to ensure that there is no splitting or bursting of the concrete. Some danger of splitting arises whenever a large force is transferred from the steel to the concrete by bond, and it is advisable (and often necessary) when

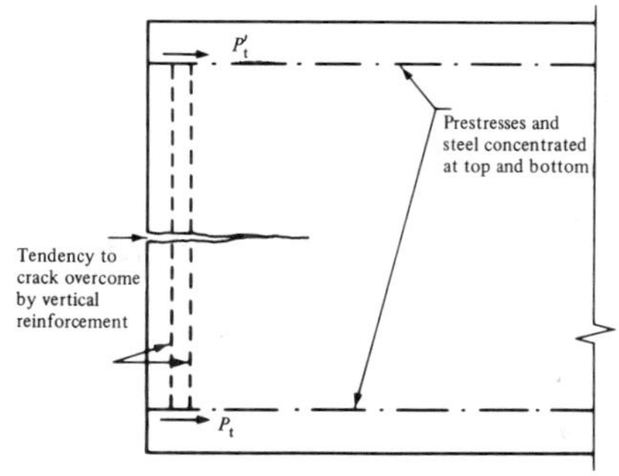

Figure 10.24 Splitting tendency at ends of pretensioned units

concentrated forces occur to avoid this risk by providing stirrups around the tensioned steel. With separate forces at the top and bottom, horizontal splitting between the forces may occur in deep beams even in rectangular sections (Figure 10.24). The forces P_t and P_t' are gradually developed in the concrete at the top and bottom; they can be transmitted to the central part of the section (which is initially without stress) only by the action of horizontal shearing stresses, and there is therefore a tendency for horizontal cracks to form. The tendency itself can be eliminated only by distributing the pre-tensioned steel over the whole of the cross-section; this, however, is not recommended by the authors as it would reduce the resistance to the formation of cracks and also the ultimate load due to bending, and vertical stirrups should be provided in preference.

No satisfactory theory is at present available for the calculation of these stresses; however, a simple method based on test results which is suitable for the common case of an I-beam with end-stiffeners is described in the following. For a beam such as that shown in Figure 10.25, the end-stiffener and part of the adjacent web can be considered as a vertical beam with a horizontal 'depth' equal to the vertical depth h. The stiffening ribs themselves should have a width a less than h, as otherwise the stress conditions in the rectangular portion would be worse than those in the I-section. The value of a is arbitrary; it may reasonably be taken as equal to $\frac{b - b_w}{2}$, but much narrower stiffeners have been used with success or they may be omitted altogether, provided that sufficient reinforcement (in the form of vertical stirrups) is placed to resist the total tension T due to bending.

If f_{st} is the stress in the concrete at the level of the lower prestressing force P_t and f_{st}' is that at the level of the upper force P_t', then the bending moment is given by

$$M = b_w z_1^2 \left[\frac{f_{st}'}{8} + \frac{\sqrt{3}}{27}(f_{st} - f_{st}')\right] = 0.64\, b_w z_1^2 (f_{st} + f_{st}')$$

in which b_w is the breadth of the web and z_i is the distance between P_t and P_t'.

To illustrate this method, the end of a roof beam with a span of 40 ft is shown in Figure 10.26. These beams were designed by British Railways (Eastern Region) and erected at Bury St. Edmunds in 1952, at a time when

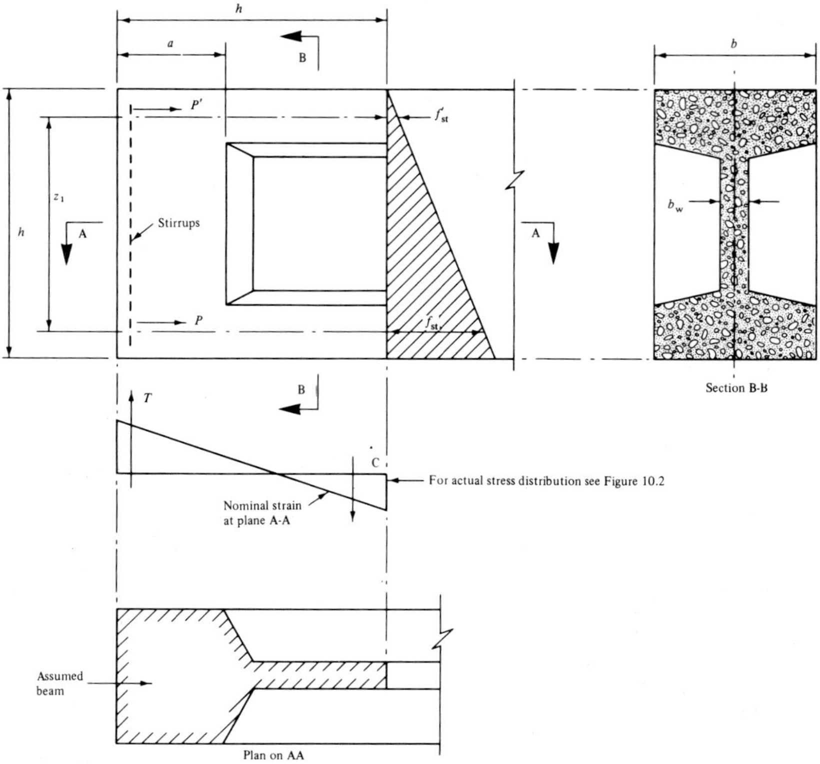

Figure 10.25 Reinforcement in transmission zone

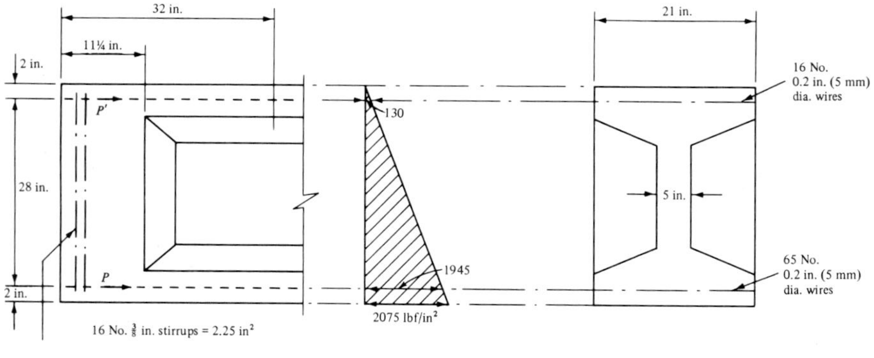

Figure 10.26 Reinforcement at end face of pretensioned roof beams

the sufficiency of the method had not been proved; additional steel was therefore provided as a safety measure.

The bending moment is
$0.064 \times 5 \times 28^2 \times (130 + 1945) = 521\ 000$ lbf in. (6000 kgf m, 58.85 kN m)

The total tension $T = \frac{M}{z} = \frac{521\ 000}{\text{say } 20} = 26\ 000$ lbf (11 800 kgf, 115.8 kN)

The area of reinforcement $= \frac{26\ 000}{18\ 000} = 1.45\ \text{in}^2\ (935\ \text{mm}^2)$
and this is concentrated at the end of the beam.

REFERENCES

1. COKER E. G. and FILON, L. N. G. Photo-elasticity. *British Association for the Advancement of Science.* 1921.
2. BAY, H. *Der Wandârtige Träger* und Bogenscheibe. Stuttgart, Wittwer, 1960.
3. MAGNEL, G. *Prestressed concrete.* 2nd edition London, Concrete Publications 1950. pp. 300.
4. GUYON, Y. Prestressed concrete. London, *Contractors' Record,* 1957. pp. 543.
5. ZIELINSKI, J. and ROWE, R. E. *An investigation of the stress distribution in anchorage zones of post-tensioned concrete members.* London, Cement and Concrete Association Research Report No. 9, September 1960. pp. 1–32. Publication 41.009.
6. ZIELINSKI, J. and ROWE, R. E. *The stress distribution associated with groups of anchorages in post-tensioned concrete members* London, Cement and Concrete Association Research Report No. 13, June 1961. pp. 1–39. 41.013.
7. MÖRSCH, E. *Uber die Berechnung der Gelenkquader.* Beton-und-Eisen, 1924.
8. WALLEY, F. W. *Prestressed concrete design and construction.* London. H.M.S.O. 1953. pp. 279.
9. ACI-ASCE COMMITTEE Tentative recommendations for prestressed concrete flat plates *Journal of the American Concrete Institute Proceedings* Vol. 71 No. 2 February 1974 pp. 61–71.
10. BASE, G. D. *An investigation of transmission length in pre-tensioned concrete,* London, Cement and Concrete Association, 1957. pp. 1–32. Publication 42.260.
11. EVANS, R. H. and BENNETT, E. W. *Prestressed concrete theory and design,* London, Chapman and Hall, 1958. pp. 294.
12. SWAMY, R. N. *Prestressed Lightweight Concrete Developments in Prestressed Concrete.* 1. Applied Science Publishers Ltd., Barking, Essex. 1978.
13. BATE, S. C. C. *"Prestressed Concrete – Civil Engineering Reference Book".* Volume 3, Second Edition, Butterworth, London. 1961.
14. SHORT, A. and KINNIBURG, W. *Lightweight Concrete.* Third edition. Applied Science Publishers, 1978, pp. 252.

CHAPTER 11

COMPOSITE MEMBERS

11.1 General principle

In modern design the cost-effectiveness does not entirely depend on the quantity of materials but to a great extent on time related operation and the construction plant used. Consequently it is often of great advantage to use factory made components in combination with in-situ structural topping in actual construction. Such factory made components could either be in normal weight or lightweight concrete.

11.2 Historical development

In a composite prestressed member the concrete is placed in at least two separate stages, some or all of the prestressing force being applied between the first and second stages. In the first members of this type, which were made and tested in Switzerland in 1943[1], the compressive zones of prestressed beams were obtained by casting an additional concrete slab in place (Figure 11.1a). Other applications based on the same principle, including one in which prestressed concrete rods are used to replace mild steel reinforcement, are shown in Figures 11.1b to 11.1e. All of these alternatives (except the one employing concrete rods) have been tested, and full co-operation between the prestressed and added concrete has been proved to occur.

Some more recent developments are shown in Figure 11.2. Figure 11.2a depicts a type of bridge deck developed and used by British Railways (Eastern Region)[2]. This design is based on a system proposed by Hajnal-Kónyi[3], in which the prestressed beams are designed to carry their own dead weight, together with that of the additional concrete, without props, and the composite slab is designed to support all additional dead loads and live loads. A further development of this principle is the 'wafer' slab bridge deck[4], also designed and used by British Railways (Eastern Region), shown in Figure 11.2b, in which further prestress is applied to the composite section after the added concrete has hardened. A composite beam devised by Samuely[5], in which castellated prestressed planks are used at the tensile face is shown in Figure 11.2c

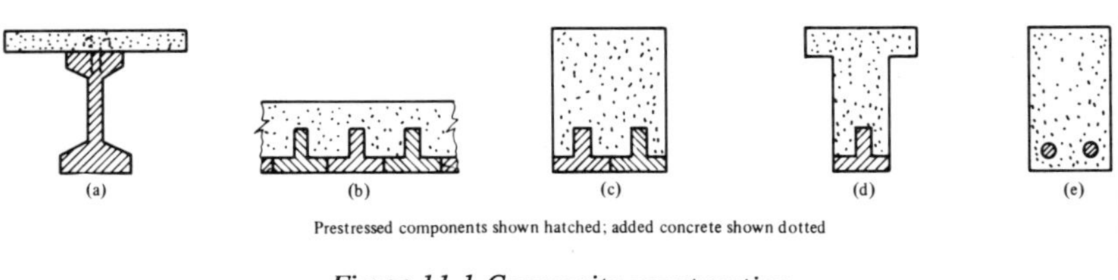

Figure 11.1 Composite construction

and a type of composite floor in which prestressed planks are used to support the weight of hollow tiles and the additional concrete during construction is shown in Figure 11.2d[6]. During the last ten years a combination of precast-T and double-T units with in-situ added concrete has been extensively used (Figure 11.2e).

11.3 Analysis and design of composite sections

The general formulae for the properties of composite sections given below are based on the following assumptions.

(1) Linear distribution of strain across the section; this entails the corollary that the connection between the two components of the section is strong

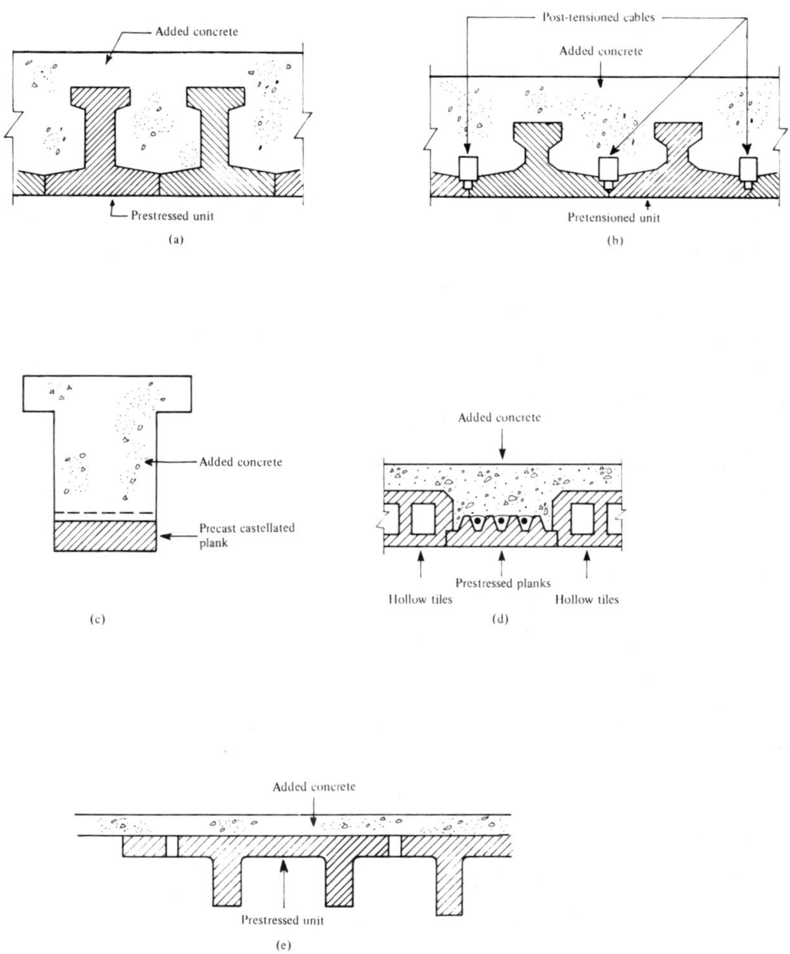

Figure 11.2 Developments in composite construction

enough to ensure that the longitudinal shearing forces cause no relative movement.

(2) Stress is directly proportional to strain, but the values of Young's modulus for the prestressed component and the added concrete may differ.

The most convenient method of analysis is to replace the composite section by an equivalent section consisting of a material with a Young's modulus equal to that of one of the materials of the composite section. In the following, this equivalent section is assumed to have the same Young's modulus as the prestressed component (Figure 11.3b), and the strain at any point in the section is required to be equal to that at the same point in the composite member. The stresses in the prestressed portion of the composite member may be calculated directly using the properties of the equivalent section, but those in the added concrete are decreased in the ratio of the Young's moduli of the two components.

A typical composite section is shown in Figure 11.3a. In the notation, suffixes 1 and 2 relate to the lower and upper faces of the prestressed component and suffixes 3 and 4 to the lower and upper faces of the additional concrete. The suffixes 'p', 'a', and 'c' relate to the properties of the prestressed component, the additional concrete, and the equivalent composite section respectively. The actual section shown in Figure 11.3a may be replaced by the equivalent section shown in Figure 11.3b. Assuming that Young's modulus for the added concrete E_{ca} is equal to α_a times that for the prestressed component E_{cp}, then $A_{pc} = A_p + \alpha_a A_a$, and hence the properties of the combined section are given by

$$e_{1c} = \frac{A_p e_{1p} + \alpha_a A_a e_{1a}}{A_{pc}} = \frac{A_p e_{1p} + \alpha_a A_a e_{1a}}{A_p + \alpha_a A_a}$$

and

$$I_{pc} = I_p + A_p (e_{1c} - e_{1p})^2 + \alpha_a [I_a + A_a (e_{1a} - e_{1c})^2]$$

To determine the stresses in the prestressed component, the appropriate section moduli are

$$Z_{1c} = \frac{I_{pc}}{e_{1c}} \text{ and } Z_{2c} = \frac{I_{pc}}{e_{2c}}$$

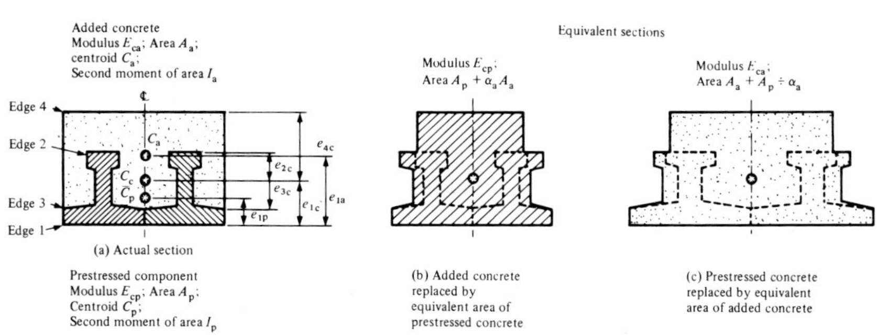

Figure 11.3 Equivalent sections

For stresses in the added concrete, I_{pc} is replaced by $I_{ac} = \dfrac{I_{pc}}{\alpha_a}$

Hence

$$Z_{3c} = \frac{I_{pc}}{\alpha_a e_{3c}} = \frac{I_{ac}}{e_{3c}} \text{ and } Z_{4c} = \frac{I_{ac}}{e_{4c}}$$

In the particular case when $e_{2c} = e_{3c}$ (Figure 11.4)

then

$$\frac{Z_{2c}}{Z_{3c}} = \frac{I_{pc}}{e_{2c}} \times \frac{\alpha_a e_{3c}}{I_{pc}} = \alpha_a$$

The resultant stresses (Figure 11.5) in a composite member are composed of the following.

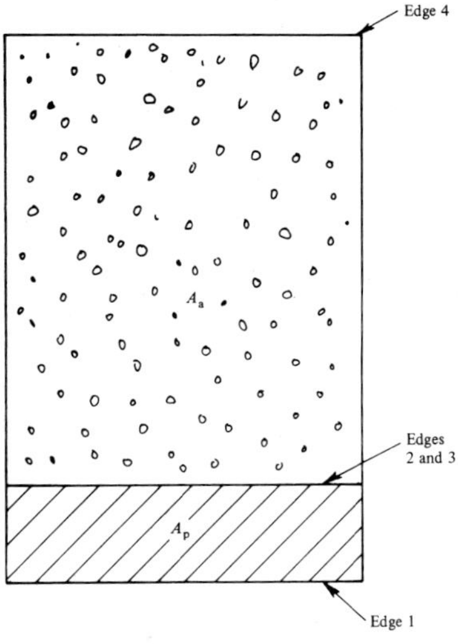

Figure 11.4 Composite section (added concrete above precast concrete)

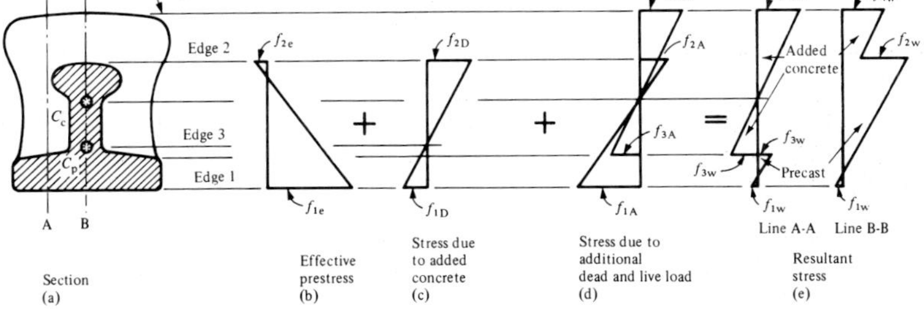

Figure 11.5 Stress distributions in a composite section

(1) The effective prestress in the prestressed component.
(2) The further stresses occurring in the prestressed component alone, before the added concrete has hardened and is able fully to co-operate with the prestressed component.
(3) The stresses in the composite section due to additional dead load and live load. Stresses of this type occurring at the same level in the prestressed component and the added concrete will be equal if $\alpha_a = 1$. If there is little difference in the properties of the two concretes, α_a may be assumed to be unity; otherwise the stress in the added concrete will be less than that in the prestressed concrete, as shown in Figure 11.5d.

11.4 Design of composite members

If M_t is the bending moment due to the weight of the prestressed component at transfer, M_d is the total bending moment due to dead weight supported by the prestressed member alone and M_a is the bending moment due to the additional dead and live loads supported by the composite construction, then the following design equations are obtained. In the following k_1 and k_2 correspond to k_b and k_t respectively of prestressed member.

At transfer

$$\frac{k_1 P_t}{A_p} - \frac{M_t}{Z_{1p}} = f_{1t} \qquad (11.1)$$

$$\frac{k_2 P_t}{A_p} + \frac{M_t}{Z_{2p}} = f_{2t} \qquad (11.2)$$

At working load

$$\frac{R_0 k_1 P_t}{A_p} - \frac{M_d}{Z_{1p}} - \frac{M_a}{Z_{1c}} = f_{1w} \qquad (11.3)$$

$$\frac{R_0 k_2 P_t}{A_p} + \frac{M_d}{Z_{2p}} + \frac{M_a}{Z_{2c}} = f_{2w} \qquad (11.4)$$

$$\frac{M_a}{Z_{3c}} = f_{3w} \qquad (11.5)$$

$$\frac{M_a}{Z_{4c}} = f_{4w} \qquad (11.6)$$

If $f_{1t} = \bar{f}_{ct}$ and $f_{1w} = -\bar{f}_{tw}$, then multiplying equation 11.1 by R_0 and subtracting it from equation 11.3 gives

$$\frac{M_d - R_0 M_t}{Z_{1p}} + \frac{M_a}{Z_{1c}} = R_0 \bar{f}_{ct} + \bar{f}_{tw}$$

and hence, if Z_{1p} is known, the Z_{1c} required is given by

$$Z_{1c} = \frac{M_a}{R_0 \bar{f}_{ct} + \bar{f}_{tw} - \dfrac{1}{Z_{1p}}(M_d - R_0 M_t)} \qquad (11.7)$$

Similarly, multiplying equation 11.2 by R_0 and subtracting it from equation 11.4,

$$\frac{M_d - R_0 M_t}{Z_{2p}} + \frac{M_a}{Z_{2c}} = f_{2w} - R_0 f_{2t}$$

and hence required

$$Z_{2c} = \frac{M_a}{R_0 f_{2t} + f_{2w} - \frac{1}{Z_{2p}}(M_d - R_0 M_t)} \qquad (11.8)$$

In equation 11.8, f_{2t} may be $\bar{f}_{tt}$ and f_{2w} may equal or even exceed $\bar{f}_{cw}$ (see Chapter 7). If the prestressed component is propped in position until after the added concrete has hardened, the terms $\frac{M_d}{Z_{1p}}$ and $\frac{M_d}{Z_{2p}}$ in equations 11.3 and 11.4 become

$$\frac{M_d}{Z_{1c}} \text{ and } \frac{M_d}{Z_{2c}}$$

When the sections have been chosen the stresses in the prestressed component must be investigated. The prestress must be within the limits

$$\bar{f}_{ct} + \frac{M_t}{Z_{1p}} \geqslant f_{1t} \geqslant \frac{M_d}{R_0 Z_{1p}} + \frac{M_a}{R_0 Z_{1c}} - \frac{\bar{f}_{tw}}{R_0} \qquad (11.9)$$

and

$$-\bar{f}_{tt} - \frac{M_t}{Z_{2p}} \leqslant f_{2t} \leqslant -\frac{M_d}{R_0 Z_{2p}} - \frac{M_a}{R_0 Z_{2c}} + \frac{\bar{f}_{cw}}{R_0} \qquad (11.10)$$

The considerations governing the conditions at ultimate load are the same as those for ordinary prestressed concrete members, except that particular care must be taken to ensure that adequate resistance to the ultimate shearing forces is provided. It must be borne in mind that the principal tensile stresses in the added concrete are much greater in most cases than those in the prestressed component, because of the high tensile stress in the additional concrete. Hence it is often necessary to provide shear reinforcement in the added concrete.

If flexural tensile stresses are induced in the added concrete by imposed service loading the magnitude of these stresses in the in-situ concrete at interface in accordance with CP 110 should be limited to the values given in the Table 11.1. In CP 116, however, the corresponding values are much less – even less than half of the allowable limit in Table 11.1.

Both CP 116 and CP 110 permit the limits to be increased by up to 50 per cent provided the permissible tensile stress in the prestressed concrete units is reduced by the same numerical amount. These values are applicable at contact surface only.

Similarly the allowable compressive stress in the added concrete at interface may be increased by up to 50 per cent provided it can be ensured that the failure of the composite member will be due to the excessive elongation of the steel.

Table 11.1 Flexural tensile stresses in added concrete

Grade of in-situ concrete	25	30	40	50	N/mm²
	253.7	304.5	406	507.5	kgf/cm²
	3625	4350	5800	7250	lbf/in²
Maximum allowable tensile stress	3.2	3.6	4.4	5.0	N/mm²
	32.5	36.5	44.7	50.8	kgf/cm²
	464	522	638	725	lbf/in²

11.5 Losses of prestress

It is difficult to make an exact assessment of the losses of prestress in composite members. In addition to shrinkage, creep and relaxation in the prestressed component, the effects of differential shrinkage and creep between the two concretes must be considered. On the one hand, nearly all of the losses in the prestressed concrete may have taken place before the added concrete is placed. On the other hand, if little or no loss has taken place in the prestressed component before the additional concrete is placed, the shortening of the prestressed component will be restrained by the added concrete to some extent, depending on the shapes of the sections and the quantity of added concrete. For example, if no losses occur before the added concrete has hardened, a change in strain of $\Delta\epsilon$, measured at the level of the prestressing steel, between transfer and the time when the maximum losses have taken place, would result in a loss of stress of Δf_p in the tensioned steel. Consequently, the prestressing force would be reduced by an amount ΔP and the effective prestress f_{1E} would be reduced by

$\Delta f_{1E} = \dfrac{-k_{1c}\,\Delta P}{A_c}$ in which $k_{1c} = 1 + \dfrac{e_{sc}\,e_{1c}}{i^2}$; and $i^2 = \dfrac{I_{pc}}{A_{pc}}$.

The effects of differential shrinkage and creep may be determined as follows. In Figure 11.6b, ϵ_{pf} represents the unrestrained 'free' shrinkage and creep which would occur at the centroid of the prestressed components alone, and ϵ_{af} the unrestrained shrinkage and creep at the centroid of the added concrete alone, under the same conditions of stress and environment. The difference between the two values is $\Delta\epsilon$; this is therefore the unrestrained differential movement which would occur if there were no connection between the added concrete and the prestressed component. No such relative movement occurs in fact, because of the bond between the two components, and stresses occur in each component as a consequence of this restraint.

In the analysis (due to Evans and Parker)[7] which follows, a perfect bond between the sections and a linear distribution of strain are assumed. Since no external forces occur, the tensile force in the additional concrete must be equal to and co-linear with the compressive force in the prestressed concrete. From the similar triangles shown in Figure 11.6c and 11.6d.

$$\frac{F_{sp}\,e_p^{\,2}}{e_p\,I_p\,E_{cp}} = \frac{F_{sa}\,e_a^{\,2}}{e_a\,I_a\,E_{ca}} \qquad (11.11)$$

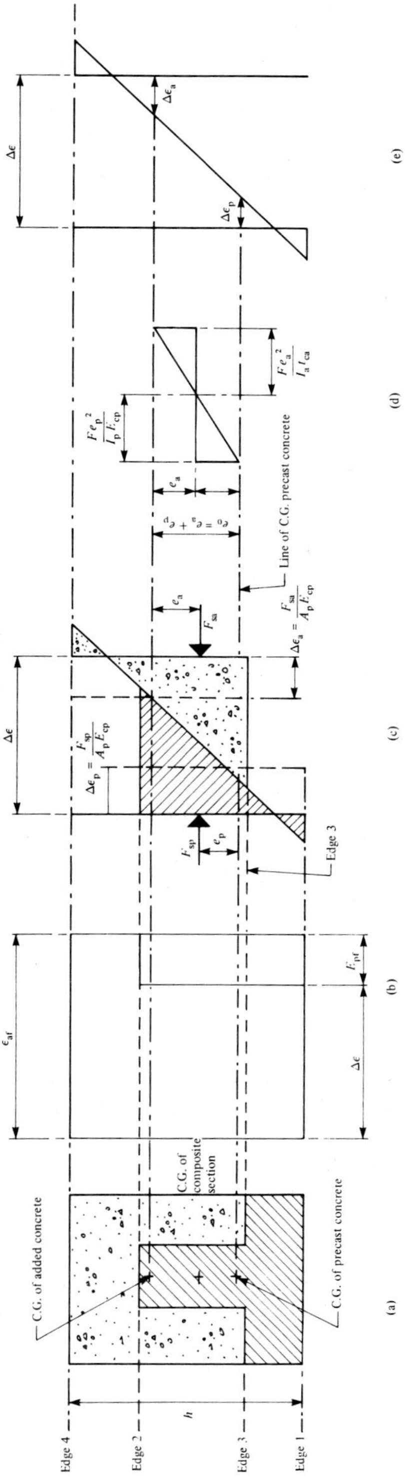

Figure 11.6 Differential shrinkage and creep

But $F_{sp} = F_{sa}, = F_s$, and writing $E_{ca} = \alpha_a E_{cp}$,

$$\frac{e_a}{e_p} = \frac{\alpha_a I_a}{I_p}; e_a = \frac{\alpha_a I_a e_0}{I_p + \alpha_a I_a}; e_p = \frac{I_p e_0}{I_p + \alpha_a I_a}$$

Also, from the strain diagram shown in Figure 11.6c

$$\Delta\epsilon = \Delta\epsilon_p + \Delta\epsilon_a + F_s\left(\frac{e_p^2}{I_p E_{cp}} + \frac{e_a^2}{I_a E_{ca}}\right) \qquad (11.12)$$

Substituting $\Delta\epsilon_p = \dfrac{F_s}{A_p E_{cp}}$, $\Delta\epsilon_a = \dfrac{F_s}{A_a E_{ca}}$ and the relationships obtained in the foregoing, and simplifying;

$$\Delta\epsilon_p = \frac{\Delta\epsilon}{\left(\dfrac{A_{pc}}{\alpha_a A_a} + \dfrac{A_p e_0^2}{I_p + \alpha_a I_a}\right)} \qquad (11.13)$$

and

$$\Delta\epsilon_a = \frac{\Delta\epsilon}{\left(\dfrac{A_{pc}}{A_p} + \dfrac{\alpha_a A_a e_0^2}{I_p + \alpha_a I_a}\right)} \qquad (11.14)$$

in which A_{pc} is the area of the equivalent section. Hence, when a suitable value is obtained for $\Delta\epsilon$, the values of $\Delta\epsilon_p$ and $\Delta\epsilon_a$ can be calculated and the strains and stresses at all other planes are then determinate (Figure 11.6d).

The equations 11.13 and 11.14 agree well with other methods which have been based on similar basic assumptions but derived differently.

Another method, due to Mattock[8] gives the following simplified equation to obtain the bending moment induced in a composite section as a result of differential shrinkage.

$$M_\Delta = \Delta\epsilon E_{ca} A_a \left(\frac{d_{fa}}{2} + e_{c2}\right) \times \left(\frac{1 - e^{-\phi}}{\phi}\right) \qquad (11.15)$$

where

M_Δ	= Moment induced due to differential shrinkage
A_a	= Area of added concrete
d_{fa}	= Depth of added concrete
e_{c2}	= Distance of C.G. of composite section from the top face of precast section
$\left(\dfrac{d_{fa}}{2} + e_{c2}\right)$	= Distance from C.G. of added concrete to C.G. of composite section
ϕ	= Creep factor usually taken as 2.

It should be borne in mind that the moment due to differential shrinkage effects the elastic stress conditions only and does not influence the ultimate behaviour.

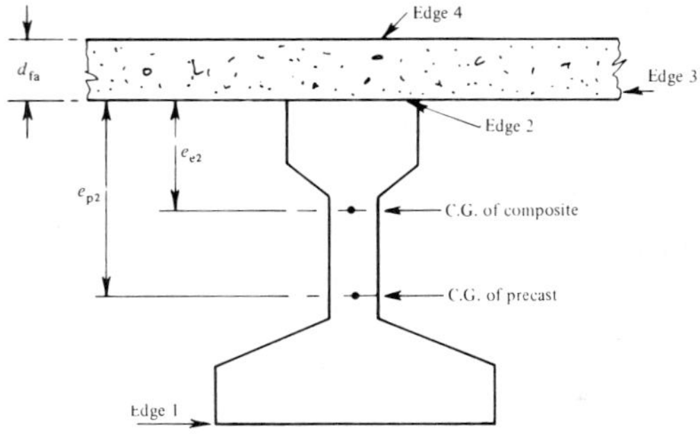

Figure 11.7 Parameter for calculations of effects of differential shrinkage and creep (Mattock's method)

$$\text{Stresses induced due to } M_\Delta \text{ at edge 1} = \frac{M_\Delta}{Z_{1c}}$$

$$\text{Stresses induced due to } M_\Delta \text{ at edge 2} = \frac{M_\Delta}{Z_{2c}}$$

$$\text{Stresses induced due to } M_\Delta \text{ at edge 3} = \frac{M_\Delta}{Z_{3c}}$$

$$\text{Stresses induced due to } M_\Delta \text{ at edge 4} = \frac{M_\Delta}{Z_{4c}}$$

In the Code it is stated 'Differential shrinkage of the added concrete and the prestressed concrete may in certain circumstances lead to increased tensile stress in the member. The effect, however, is not generally of great importance.' This is also true for the effect of differential creep.

The designer usually has no exact knowledge of the conditions under which the separate components will be cured, and it is therefore difficult for him to make a realistic assessment of the probable differential deformations.

In general, the effect of differential deformation can be ignored if it causes no tensile stress at the bottom face. The examples given later show that when the differential shrinkage exceeds the differential creep this condition is usually present, except when the added concrete forms the compressive flange of a T-section or where a composite member is formed from approximately equal components.

It seems to be advisable to compute the losses of the prestressed concrete component in the usual manner, disregarding the effects both of the composite action and of differential shrinkage and creep; in most cases this may be expected to give slightly conservative results. An exception to this can occur if the added concrete is placed immediately after transfer, so that the differential shrinkage is negligible; the differential creep of the prestressed component would then induce stresses of opposite signs to those induced by differential shrinkage. See also the concluding paragraph of section 11.6.

11.6 Horizontal shear at interface

For a composite member to act successfully, a good connection between the two components is essential. Roughening of the common surface has generally proved sufficient; the provision of castellations as a 'shear key' is also useful. Protruding stirrups are useful but not essential; if they are badly detailed or awkwardly placed they can be an embarrassment. Where stirrups are not available, it is essential to ensure satisfactory bond either by roughening or by castellation or both.

According to CP 110 calculations for horizontal shear between the two components of a composite section are only required for the serviceability limit-state. The methods given below ensure that composite action does not break down for the serviceability limit-states and that the shear strength is adequate for the ultimate limit-state.

Shear connector. A design approach for particular types of shear connection between the component parts of a composite member is given here. In other composite members where evidence of the suitability of the shear connection is not available it should be established by tests to failure.

The horizontal shear stress due to the design service loads ($G_k + Q_k$) at the contact surface of the precast and in-situ components at any point along the length of the member may be calculated as:

$$v_h = \frac{V_d S_c}{I b_e}$$

where

v_h is the horizontal shear stress per unit area of the contact surface,
V_d is the total vertical shear at the point considered due to the design service load,
S_c is the first moment of the concrete to one side of the contact surface, about the neutral axis of the transformed composite section,
I is the second moment of area of the composite section,
b_e is the width of the contact surface.

The shear stresses along the contact surface under maximum shear conditions calculated as above should not exceed the values given in Table 11.2. for beam and slab construction and for the three types of surface considered. The types of surface are defined as follows.

(1) Where links are not provided and the contact surface has been prepared in the following manner: when the concrete in the precast member has set but not hardened, the surface which will subsequently receive the in-situ concrete should be sprayed with a fine spray of water or brushed with a stiff brush, just sufficient to remove the outer mortar skin and expose the larger aggregate without disturbing it.

(2) Where the contact surface is not as described in (1) and where links are provided having a minimum cross-sectional area of 0.15 per cent of the contact area: the spacing of links should not be excessive; for composite-T beams with an in-situ flange, the spacing should not exceed 4 times the minimum thickness of the in-situ concrete nor 600 mm.

(3) Where the contact surface has been prepared as described in (1) (or where this treatment proved impracticable and the surface skin and laitence has been removed by sand blasting or the use of a needle gun and not by hacking) and when links are provided in excess of the minimum, the allowable shear stresses given in Table 11.2 for this type of surface may be increased by 72.5 lbf/in^2 (5 kgf/cm^2 0.5 N/mm^2) for each additional area of links equal to 1 per cent of the contact area.

Table 11.2 Horizontal shear stress for composite beam and slab sections

Grade of in-situ concrete		25	30	40	50	60	N/mm^2
		253.7	304.5	406	507.5	609	kgf/cm^2
		3625	4350	5800	7250	8700	lbf/in^2
Maximum shear stress	Surface type 1	0.38	0.45	0.54	0.59	0.64	N/mm^2
		3.85	4.57	5.48	5.99	6.49	kgf/cm^2
		55	65.25	78.3	85.6	92.8	lbf/in^2
	Surface type 2	0.36	0.38	0.42	0.46	0.50	N/mm^2
		3.65	3.85	4.26	4.67	5.07	kgf/cm^2
		52.2	55	61	66.7	72.5	lbf/in^2
	Surface type 3	1.22	1.25	1.32	1.38	1.45	N/mm^2
		12.38	12.69	13.4	14	14.7	kgf/cm^2
		177	181.2	191.4	200	210.3	lbf/in^2

Shear reinforcement provided in the composite section to resist vertical shear (see CP 110 5.4.2) should extend into the in-situ slab and can be deemed to resist horizontal shear at the precast/in-situ interface.

In composite slabs when links are not provided, the horizontal shear stresses should be limited as follows:

(1) For surface type 1 the stress should not exceed 101.5 lbf/in^2 (0.7 N/mm^2) nor 1.2 times the appropriate value given in Table 11.2.

(2) Where the top surface of the precast unit has not been so treated, the stress should be limited to 0.8 times the appropriate value given in Table 11.2 for surface type 2.

According to CEB/FIP Model Code a composite member may be designed as a monolithic component if the horizontal shear stress ν_h per unit of contact length is such that

$$\nu_h \leqslant \frac{A_s}{s} . X . \frac{f_y}{\gamma_m} . (1 \times \cot \alpha) . \sin \alpha + Y . \nu_c . b.$$

A_s = area of reinforcement passing across the interface and completely anchored on each side of the interface (in mm^2)

s = spacing (mm)
α = inclination of reinforcement
b = width of the interface (mm)
ν_c = permissible shear stress (as given in the Code) for the weaker concrete (N/mm^2)

X and Y are coefficients, values of which are given in Table 11.3.

Table 11.3 Values of coefficients X and Y

$A_s/s.b$	X	Y
$\leqslant 0.002$	0	1.25
$\geqslant 0.005$	0.9	2.5

It should be noted that tabulated values of X and Y are applicable only when adequate roughness is present at interface. NO values are given for smooth surface.

The above relationship is equally applicable if the in situ concrete is in lightweight aggregate concrete.

In the ACI Building Code the approach is to calculate horizontal shear stress at interface by the formula

$\nu_h = \dfrac{V_u}{\phi_1 b_e.d.}$ where b_e = width of contact surface and d relates the full composite section. ϕ_1 factor specified for shear is to be used (see Table 5.5).

If ν_h so obtained does not exceed 80 lbf/in^2 (5.62 kfg/cm^2, 0.55 N/mm^2) composite action is ensured even without any reinforcement passing through the interface provided the contact surface is adequately rough (i.e. full amplitude of approximately 5 mm or ¼ in) and clean.

If in addition to clean and adequately rough contact surfaces, the minimum tie reinforcement as required by ACI standard, passes through the interface being anchored on either side, the permissible shear stress is 350 lbf/in^2 (24.61 kgf/cm^2, 2.41 N/mm^2). These limiting values relate to the ultimate load.

From a number of 'girder tests' and push-off tests Hanson[9] concluded that the critical slip at which the section fails to act as a composite member is 0.005 in. (0.12 mm). The corresponding horizontal shear stress at the interface obtained by the conventional formula $\nu_h = \dfrac{V_d S_c}{I b_e}$ is 500 lbf/in^2 (39 kgf/cm^2, 3.44 N/mm^2) for a rough bonded surface and 300 lbf/in^2 (21 kgf/cm^2, 2.06 N/mm^2) for a smooth bonded surface. The above limiting stresses were obtained for f_{cyl} 3000 lbf/in^2 (210 kgf/cm^2, 20.6 N/mm^2) and 5000 lbf/in^2 (390 kgf/cm^2, 34.4 N/mm) respectively of the added concrete and precast concrete. In addition to these values approximately 175 lbf/in^2 (12.25 kgf/cm^2, 1.2 N/mm^2) shear capacity may be added for each per cent stirrup reinforcement crossing the joint.

Tests by Evans and Chung[10] on composite-T beams in which the added slab was in 'Lytag' lightweight concrete and the precast section of normal weight concrete showed a horizontal shear strength of 400 lbf/in^2 (28 kgf/cm^2, 2.76 N/mm^2) at the interface for a rough bonded surface.

In the opinion of the authors, the interface should always be roughened. This was found to be more effective than even castellated connection with smooth surface. Comparative fatigue tests on three types of composite members similar to Figure 11.4 were carried out by British Railways[11]. The prestressed bottom planks had a short rib 2½ in. (6.3 cm) at two sides forming a channel. With the first type there were only roughened surfaces; with the second type surfaces were smooth with alternating castellations 0.25 in. deep. In the third type surfaces were also smooth but connecting stirrups were provided. All three types proved satisfactory at failure even after one million pulsating loading. However, as far as cracking is concerned the rough surface proved to be a much better interlocking medium than either castellations or projecting stirrups with smooth concrete surfaces.

The tests also showed that there was a need for sufficient shear reinforcement to resist the high principal tensile stresses in the added non-stressed concrete which occured as a combination of shear and bending due to point loads. These stirrups are required solely in the added concrete and are unnecessary as shear connections between the precast and added concrete, provided that satisfactory co-operation is obtained by roughening, castellations or separate short stirrups.

Further tests were carried out at Duke University[12] on beams similar to Figure 11.4 but without any side rib in the precast plank. The conclusions of these tests are given below:

(1) A roughened surface was sufficiently strong to avoid horizontal shear failure which occurred only as a secondary effect after yielding resulted in primary failure; but when the added concrete was cast immediately on transfer of the prestress, horizontal separation occurred first at a load of 86 per cent of that at which flexural failure was expected.

(2) With composite beams in which the two precast components were glued together no separation took place after flexural failure.

(3) When the added concrete was placed onto the prestressed concrete after considerable shrinkage and creep had already taken place, compressive stresses occurred in the prestressed component which increased the effective prestress, but considerable tensile stresses appeared in the added concrete which may have reached or exceeded the tensile strength.

(4) When creep of the prestressed component was the major factor, compressive stresses occurred in the lower part of the added concrete and tensile stresses in the upper part which acted favourably as far as the non-prestressed component was concerned.

(5) When both shrinkage and creep occurred and the age of the prestressed component varied, the problem became more complex.

(6) Microcracks occurred as soon as the concrete stresses reached the tensile strength. Thus with differential shrinkage, microcracks developed in the added concrete at an early age, even under dead load.

(7) When the nominal tensile stresses in the non-prestressed concrete, after microcracking, became excessive, microcracks developed at the outer tensile face of the prestressed component at relatively small tensile stresses or even at nominal compressive stresses.

(8) When the nominal stresses in the added concrete further increased, these microcracks sometimes became visible, although the nominal concrete stress was still below the tensile strength. Visible cracks occurred in both the prestressed and the non-prestressed components without uniting, there being still nominal compression in the upper part of the prestressed component. The cracks in the prestressed and non-prestressed components combined only at further increase of loads.

(9) Stress re-distribution must have taken place in the cases at which cracking occurred at an earlier stage as described in (6) to (8) above. However, due to the limitation of tests, it has not yet been possible to obtain a theory concerning this stress redistribution.

(10) With a good connection obtained either by roughening or by glueing, the actual failure moments agreed well with the calculated values computed for a stress in the steel of 263×10^3 and 280×10^3 lbf/in^2 (i.e. 2 per cent proof stress and ultimate strength) for the 'R' and 'S' beams respectively.

(11) Formulae have been derived based on the measured strain in the composite beam, covering all shrinkage and creep which occurred in the composite member.

(12) It seems advisable to consider two cases, i.e. either differential shrinkage or differential creep, which cover the extreme practical limits.

11.7 Examples

11.7.1 Example: Warehouse floor units

In Example 9.2.3 of Chapter 9, the dimensions of the prestressed member required for conditions at working load are increased in order to obtain the necessary ultimate resistance. No allowance was made for the resistance of the added concrete, and this is investigated in the following.

The trial section obtained for the prestressed component in Example 9.2.3 is used as the basis for the trial section shown in Figure 11.8. The only appreciable difference is that part of the bottom flange of the prestressed component which is adjacent to the added concrete is increased in thickness, to ensure that the tensile stress in the added concrete is less than the permissible value given in Chapter 5.

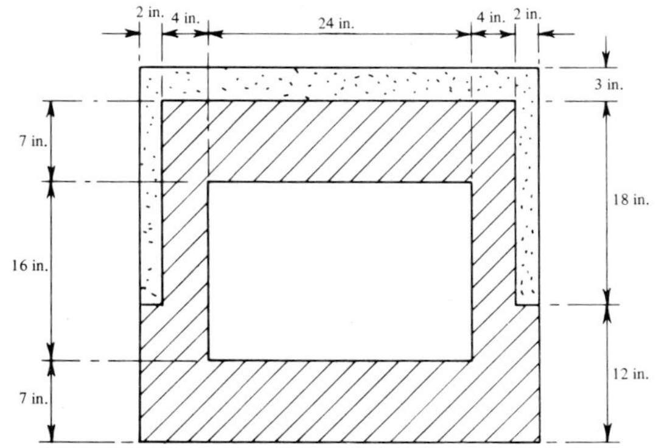

Figure 11.8 Composite section in example 11.7.1

The ultimate resistance of the prestressed concrete alone is found in Example 9.2.3 to be 19 300 000 lbf in., (222 336 kgf m, 2180.2 kN m) and the value required is about 25 000 000 lbf in. (288 000 kgf m, 2824 kN m). If it is assumed that the added concrete will have a cube strength of 4000 lbf/in^2 (281.24 kgf/cm^2, 27.5 N/mm^2) it can be assumed that the mean compressive stress in this concrete at ultimate load will be 2000 lbf/in^2 and the thickness required is therefore equal to

$$\frac{25\,000\,000 - 19\,300\,000}{2000 \times 36 \times \text{say } 28} = 2.8 \text{ in. } (71.12 \text{ mm})$$

A trial thickness of 3 in. is therefore shown in Figure 11.8.

SECTION PROPERTIES

Prestressed component:

$$A_p = (7 \times 36) + (2 \times 4 \times 16) + (7 \times 32) + (2 \times 5 \times 2) = 624 \text{ in}^2 \ (4026 \text{ cm}^2)$$

$$e_{1p} = \frac{(36 \times 7 \times 3.5) + (2 \times 4 \times 16 \times 15) + (32 \times 7 \times 26.5) + (2 \times 5 \times 2 \times 9.5)}{624}$$

$$= 14.3 \text{ in. } (363.22 \text{ mm})$$

$$I_p = \frac{36 \times 14.3^3}{3} - \frac{24 \times 7.3^3}{3} + \frac{32 \times 15.7^3}{3} - \frac{24 \times 8.7^3}{3} + \frac{4 \times 7^3}{12}$$

$$+ 4 \times 7 \times (14.3 - 9.5)^2 = 67\,900 \text{ in}^4 \ (2\,826\,209 \text{ cm}^4)$$

$$Z_{1p} = \frac{67\,900}{14.3} = 4750 \text{ in}^3 \ (77\,839 \text{ cm}^3)$$

$$Z_{2p} = \frac{67\,900}{15.7} = 4330 \text{ in}^3 \ (70\,956 \text{ cm}^3)$$

$$i_p^2 = \frac{I_p}{A_p} = \frac{67\,900}{624} = 109 \text{ in}^2 \ (703 \text{ cm}^2)$$

Added concrete:

$$A_a = (2 \times 2 \times 18) + (3 \times 36) = 180 \text{ in}^2 \ (1161 \text{ cm}^2)$$

$$e_{4a} = \frac{(36 \times 3 \times 1.5) + (18 \times 4 \times 12)}{180} = 5.69 \text{ in. } (144.53 \text{ mm})$$

$$I_a = \frac{36 \times 5.69^3}{3} - \frac{32 \times 2.69^3}{3} + \frac{4 \times 15.31^3}{3} = 6790 \text{ in}^4 \ (282\,621 \text{ cm}^4)$$

Composite section:

From Example 9.2.3 the cube strength of the prestressed concrete is 7500 lbf/in^2, hence $E_{cp} = 5.75 \times 10^6$ lbf/in^2 (4.04×10^5 kgf/cm^2, 39.6 kN/mm^2). Assuming the cube strength of the added concrete is 4000 lbf/in^2, the corresponding value for E_{ca} is 4×10^6 lbf/in^2 (2.81×10^5 kgf/cm^2, 27.58 kN/mm^2).

$$\alpha_a = \frac{E_{ca}}{E_{cp}} = \frac{4}{5.75} = 0.695$$

Therefore

$$e_{1c} = \frac{(624 \times 14.3) + (0.695 \times 180 \times 27.31)}{624 + (0.695 \times 180)} = 16.5 \text{ in. } (419.1 \text{ mm})$$

$$I_{pc} = 67\,900 + 624\,(16.5 - 14.3)^2 + 0.695\,[6790 + 180\,(27.31 - 16.5)^2]$$
$$= 90\,275 \text{ in}^4 \;(843\,908 \text{ cm}^4)$$

$$Z_{1c} = \frac{90\,275}{16.5} = 5470 \text{ in}^3 \;(89\,637 \text{ cm}^3)$$

$$Z_{2c} = \frac{90\,275}{13.5} = 6690 \text{ in}^3, (109\,630 \text{ cm}^3)$$

$$Z_{3c} = \frac{90\,275}{0.695 \times 4.5} = 28\,850 \text{ in}^3, (472\,768 \text{ cm}^3)$$

$$Z_{4c} = \frac{90\,275}{0.695 \times 16.5} = 7870 \text{ in}^3 \;(128\,966 \text{ cm}^3)$$

Bending moments:

$$M_t = 624 \times \frac{150}{144} \times 50^2 \times 1.5 = 2\,440\,000 \text{ lbf in. } (28\,109 \text{ kgf m}, 275.6 \text{ kN m})$$

$$M_d = 180 \times \frac{150}{144} \times 50^2 \times 1.5 + M_t = 3\,143\,000 \text{ lbf in.} (36\,207 \text{ kgf m}, 335 \text{ kN m})$$

$$M_a = 2255 \times 50^2 \times 1.5 = 8\,460\,000 \text{ lbf in. } (97\,459 \text{ kgf m}, 955.7 \text{ kN m})$$

$$M_{ult} = 1.5 \times 3\,141\,000 + 2.5 \times 8\,460\,000 = 25\,860\,000 \text{ lbf in.}$$
$$(297\,907 \text{ kgf m}, 2921.3 \text{ kN m})$$

No test results are available for the ultimate strength of composite members composed of concretes of different strengths which act together in the compressive zone; the following method, which is based on equality of strains, appears to give reasonable results although it has as yet no experimental support. Assuming mean compressive stresses of 2000 lbf/in^2 in the added concrete and 3750 lbf/in^2 in the prestressed concrete at ultimate load (that is, stresses of $0.5 f_{cu}$ in each case as previously discussed in Chapter 8), the resistance available (Figure 11.9) is

$$36 \times 2000 \times 3 \times (33 - 1.5 - 3.5) + 4 \times 2000 \times 7 \times 23 +$$
$$32 \times 3750 \times 7 \times (30 - 3.5 - 3.5) = 26\,460\,000 \text{ lbf in. } (304\,819 \text{ kgf m}, 2989 \text{ kN m})$$

which is greater than that required.

An alternative, and equally empirical, method is to assume that the mean compressive stress at ultimate load is 0.8 times that which is permissible for the prestressed component alone. With this method, the calculated resistance is

$$0.8 \times 3750 \times 36 \times 10 \times (33 - 3.5 - 5) = 26\,500\,000 \text{ lbf in. } (305\,280 \text{ kgf m}, 2993.6 \text{ kN m})$$

which again is greater than required.

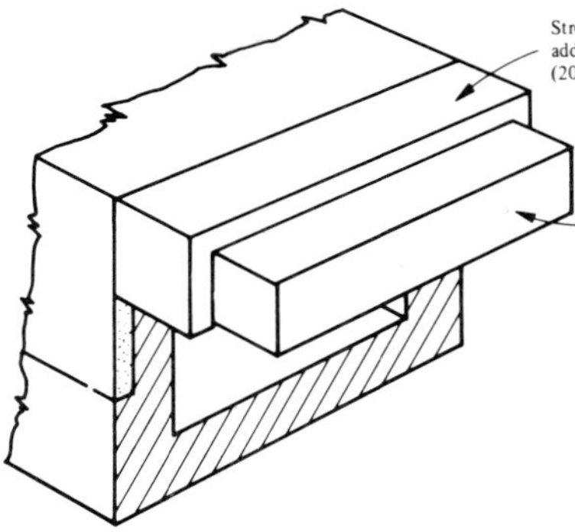

Figure 11.9 Assumed stress-block of a composite section at ultimate condition (example 11.7.1)

PRESTRESSING FORCE

As in ordinary prestressed construction, the lowest prestressing force is obtained when the permissible tensile stresses are reached at transfer and working load. If $\bar{f}_{tt}$ = 200 lbf/in^2 (14.06 kgf/cm^2, 1.38 N/mm^2) and $\bar{f}_{tw}$ = 575 lbf/in^2, (40.42 kgf/cm^2, 3.96 N/mm^2).

$$f_{1T} = \frac{1}{R_0}\left[\frac{M_d}{Z_{1p}} + \frac{M_a}{Z_{1e}} + \bar{f}_{tw}\right]$$

$$= \frac{1}{0.74}\left[\frac{3\,143\,000}{4750} + \frac{8\,460\,000}{5470} - 575\right] = 2210 \text{ lbf/in}^2 \ (155.4 \text{ kgf/cm}^2, 15.24 \text{ N/mm}^2)$$

$$f_{2T} = \bar{f}_{tt} = -200 \text{ lbf/in}^2. \ (14.06 \text{ kgf/cm}^2, 1.38 \text{ N/mm}^2)$$

$$\bar{e}_{sp} = \frac{i_p^{\,2}(f_{1T} - f_{2T})}{e_{2p} f_{1T} + e_{1p} f_{2T}} = \frac{109\,(2210 + 200)}{15.7 \times 2210 - 14.3 \times 200} = 8.25 \text{ in. } (209.55 \text{ mm})$$

$$k_1 = 1 + \frac{\bar{e}_{sp}\, e_{1p}}{i_p^{\,2}} = 1 + \frac{8.25 \times 14.3}{109} = 2.082$$

$$k_2 = 1 - \frac{\bar{e}_{sp}\, e_{2p}}{i_p^{\,2}} = 1 - \frac{8.25 \times 15.7}{109} = -0.190$$

$$\bar{P}_t = \frac{A_p}{k_1} f_{1T} = \frac{624}{2.082} \times 2210 = 632\,000 \text{ lbf } (286\,669 \text{ kgf, } 2811 \text{ kN})$$

$$P_t = \frac{\bar{e}_{sp} + e'_{sp}}{e_{sp} + e'_{sp}}\, \bar{P}_t = \frac{8.25 + 10.8}{12.2 + 10.8} \times 632\,000 = 523\,000 \text{ lbf } (237\,228 \text{ kgf, } 2326 \text{ kN})$$

$$P'_t = \bar{P}_t - P_t = 632\,000 - 523\,000 = 109\,000 \text{ lbf } (49\,441 \text{ kgf, } 485 \text{ kN})$$

At this stage, the assessment of the losses should be rechecked, as in Example 9.2.3. However, this presents no new problems and is therefore omitted here; hence assuming $R_0 = R_0' = 0.74$ the stresses are as follows (Figure 11.10).

At end of member:

$$f_{1T} = f_{1t} = 2210 \text{ lbf/in}^2 \ (155.4 \text{ kgf/cm}^2, 15.24 \text{ N/mm}^2)$$

$$f_{2T} = f_{2t} = \frac{k_2}{k_1} f_{1T} = \frac{-0.19}{2.082} \times 2210 = -202 \text{ lbf/in}^2 \ (14.2 \text{ kgf/cm}^2, 1.39 \text{ N/mm}^2)$$

At centre of member:

$$f_{1t} = 2210 - \frac{2\,440\,000}{4750} = 1697 \text{ lbf/in}^2 \ (119.3 \text{ kgf/cm}^2, 11.7 \text{ N/mm}^2)$$

$$f_{2t} = -202 + \frac{2\,440\,000}{4330} = 360 \text{ lbf/in}^2 \ (25.3 \text{ kgf/cm}^2, 2.48 \text{ N/mm}^2)$$

$$f_{1e} = (0.74 \times 2210) - 513 = 1125 \text{ lbf/in}^2 \ (79.1 \text{ kgf/cm}^2, 7.76 \text{ N/mm}^2)$$

$$f_{2e} = (0.74 \times -202) + 562 = 412 \text{ lbf/in}^2 \ (28.97 \text{ kgf/cm}^2, 2.84 \text{ N/mm}^2)$$

$$f_{1d} = 1125 - \frac{703\,000}{4750} = 977 \text{ lbf/in}^2 \ (68.7 \text{ kgf/cm}^2, 6.74 \text{ N/mm}^2)$$

$$f_{2d} = 412 + \frac{703\,000}{4330} = 575 \text{ lbf/in}^2 \ (40.4 \text{ kgf/cm}^2, 3.96 \text{ N/mm}^2)$$

$$f_{1w} = 977 - \frac{8\,460\,000}{5470} = -571 \text{ lbf/in}^2 \ (40.1 \text{ kgf/cm}^2, 3.94 \text{ N/mm}^2)$$

$$f_{2w} = 575 + \frac{8\,460\,000}{6690} = 1838 \text{ lbf/in}^2 \ (129.2 \text{ kgf/cm}^2, 12.67 \text{ N/mm}^2)$$

The stresses in the added concrete are,

$$f_{3w} = \frac{-8\,460\,000}{28\,850} \times 0.695 = -204 \text{ lbf/in}^2 \ (14.34 \text{ kgf/cm}^2, 1.41 \text{ N/mm}^2)$$

$$f_{4w} = \frac{8\,460\,000}{7870} \times 0.695 = 746 \text{ lbf/in}^2 \ (52.5 \text{ kgf/cm}^2, 5.14 \text{ N/mm}^2)$$

For added concrete with a cube strength of 4000 lbf/in^2, the permissible tensile stress given in the Code is 233 lbf/in^2. Hence this value of f_{3w} is suitable and the design is satisfactory for working-load conditions.

ULTIMATE LOAD

$M_{ult} = 25\,860\,000$ lbf in. (297 907 kgf m, 2921 kN m)

Assuming $k_u = 1.0$, the area of steel required is $\dfrac{25\,860\,000}{1 \times 224\,000 \times 24} = 4.81$ in^2 (31.03 cm^2). The tensioned steel required (assuming the stress at transfer to be

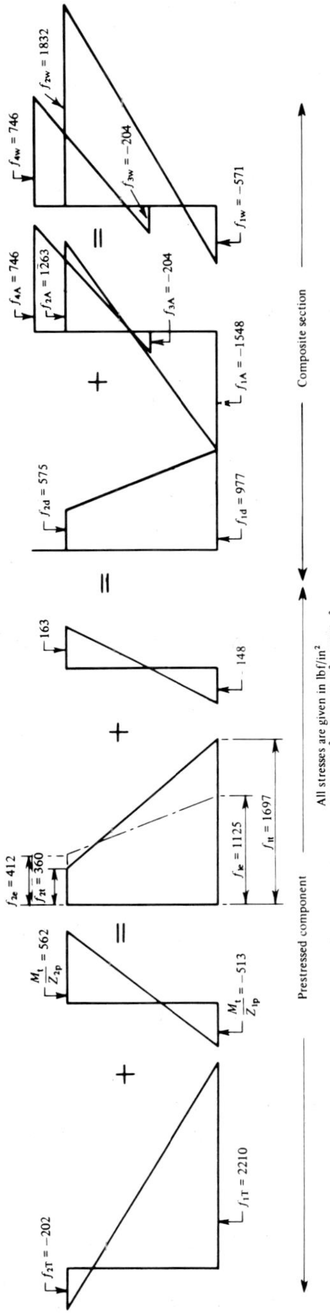

Figure 11.10 Stress diagram for composite section (example 11.7.1)

as in Example 9.2.3) is $\frac{523\ 000}{137\ 900}$ = 3.80 in^2 (24.52 cm^2) Hence the area of additional non-tensioned steel required is 1.01 in^2 (6.52 cm^2). The sufficiency of the concrete has already been proved.

Comparing the results obtained with those of Example 9.2.3 the depth of the composite section is 2 in. (50.8 mm) less than before; the area of the tensioned steel is the same, but the area of the non-tensioned steel is slightly increased.

11.7.2 Example: Propped composite floor

Design a composite floor slab to support a superimposed load of 200 lbf/ft^2 (976 kgf/m^2, 9.58 kN/m^2) over an effective span of 14 ft 6 in. (4.420 m). The structure will be propped during the placing of the added concrete at suitable intervals along the span such that the bending moment due to the weight of added concrete during construction can be ignored.

STRESSES

Assuming that the cube strengths at transfer and 28 days for the prestressed component are 4000 lbf/in^2 (to allow early release) and 6000 lbf/in^2 respectively, and that for the added concrete is 4000 lbf/in^2 when the load is applied, the permissible stresses are as follows.

Prestressed component:

Compressive stress at transfer: 0.4 × 4000 = 1600 lbf/in^2
(112.5 kgf/cm^2, 11.03 N/mm^2)

Compressive stress at 28 days: 0.33 × 6000 = 2000 lbf/in^2
(140.6 kgf/cm^2, 13.79 N/mm^2)

Tensile stress at transfer: For a cube strength of 4000 lbf/in^2
the permissible tensile stress is 167 lbf/in^2 (11.74 kgf/cm^2, 1.14 N/mm^2)

Tensile stress at working load: It is recommended in the Code that the tensile stress should not exceed three-quarters of that at which cracks become visible. In beams with well-distributed steel close to the tensile face this could be expected to occur when the stress is about 1000 lbf/in^2, however, in a shallow plank the tensioned steel is placed at about the centroid, and the tensile stress at which cracks become visible may be only two-thirds of this value (that is, 670 lbf/in^2). Hence the permissible stress is 0.75 × 670 = 500 lbf/in^2, since this is less than the value 300 + 250 = 550 lbf/in^2 which is otherwise obtained.

Added concrete:

Compressive stress at working load: 0.33 × 4000 = 1333 lbf/in^2

Tensile stress at working load: For a cube strength of 4000 lbf/in^2 the permissible tensile stress is 233 lbf/in^2.

Composite section:

At the interface with the added concrete the compressive stress in the prestressed component may be increased by up to 50 per cent for under-reinforced beams. The tensile stress in the added concrete may also be increased by up to 50 per cent, if the tensile stress in the prestressed component is reduced by the same numerical amount.

Table 11.4 Calculation of losses

Cause of loss	Minimum (at transfer)	Maximum (at working load)	Notes
Elastic compression	$\frac{28}{4} \times 1600$ $= 11\,200$ lbf/in² (787 kgf/cm², 77 N/mm²)	11 200 lbf/in² (787 kgf/cm², 77 N/mm²)	$f_{cut} = 4000$ $E_c = 4 \times 10^6$ $E_s = 28 \times 10^6$
Relaxation	0	10 000 lbf/in² (703.1 kgf/cm², 68.95 N/mm²)	–
Creep	0	$0.33 \times \frac{6}{4} \times 28 \times 1600$ $= 22\,400$ lbf/in² (1574.9 kgf/cm², 154.4 N/mm²)	Full creep loss allowed because of low strength at transfer
Shrinkage	0	$300 \times 28 = 8400$ lbf/in² (590.6 kgf/cm², 57.9 N/mm²)	–
Δf_p total	11 200 lbf/in² (787 kgf/cm², 77 N/mm²)	52 000 lbf/in² (3636 kgf/cm², 358.5 N/mm²)	–

LOSSES

The losses are obtained as shown in Table 11.4. Since the steel is pre-tensioned, no losses due to slip or friction need be considered, and since the prestress is uniform no counter-action occurs; the full stress is therefore allowed in calculating losses due to creep. Hence if the permissible initial stress in the steel is 157 000 lbf/in² (11 039 kg/cm², 1082 N/mm²), $f_{pt} = 145\,800$ lbf/in², $f_{pe} = 105\,000$ lbf/in² and,

$$R_0 = \frac{f_{pe}}{f_{pt}} = \frac{105\,000}{145\,800} = 0.72$$

BENDING MOMENTS

Assuming the weight of the prestressed component and added concrete to be 35 lbf/ft² each.

Weight of prestressed component = 35 lbf/ft² (170.9 kgf/m², 1.675 kN/m²)
Weight of added concrete = 35 lbf/ft² (170.9 kgf/m², 1.675 kN/m²)
Superimposed load = 200 lbf/ft² (976 kgf/m², 9.575 kN/m²)

$$M_t = 35 \times 14.5^2 \times 1.5 = 11\,000 \text{ lbf in. (127 kgf m, 1247 N m)}$$

$$M_d = 2 \times 11\,000 = 22\,000 \text{ lbf in. (254 kgf m, 249.4 N m)}$$

It is assumed that before the added concrete is placed the prestressed component will be propped in such a way that M_t is reduced to zero, and the weight of the added concrete causes only negligible stresses in the prestressed component. For example, if props are placed at three intermediate points the effective span becomes $\frac{14.5}{4}$ ft and the bending moment due to the total dead weight is

$$70 \times \left(\frac{14.5}{4}\right)^2 \times \frac{12}{12} = 920 \text{ lbf in.}$$

The corresponding stress is

$$\frac{920 \times 6}{12 \times 2.5^2} = 74 \text{ lbf/in}^2 \text{ which can be neglected.}$$

It is therefore necessary to consider only the bending moment at full load. This is given by

$$M_w = 270 \times 14.5^2 \times 1.5 = 85\,100 \text{ lbf in. (980 kgf m, 9.61 kN m)}$$

$$M_a = 85\,100 - 22\,000 = 63\,100 \text{ lbf in. (727 kgf m, 7.13 kN m)}$$

PROPERTIES OF SECTION

The effective prestress, after losses, is 0.72 × 1600 = 1155 lbf/in² (81.2 kgf/cm², 7.96 N/mm²).

Hence for the tensile zone of the composite section

$$Z_{1c} = \frac{M_w}{f_{1r}} = \frac{85\,100}{1155 + 500} = 51.5 \text{ in}^3 \text{ (844 cm}^3\text{)}$$

and for the compressive zone

$$Z_{4c} = \frac{M_a}{f_{ca}} = \frac{63\,100}{1333} = 47.3 \text{ in}^3 \text{ (775 cm}^3\text{)}$$

Assuming the section to be rectangular, and ignoring for the moment the difference between the properties of the prestressed and added concretes,

$$Z_1 = \frac{bh^2}{6} \text{ and } h = \sqrt{\frac{6Z_1}{b}} = \sqrt{\frac{6 \times 51.5}{12}} = 5.1 \text{ in. (134.5 mm)}$$

Consequently, a suitable trial section is that shown in Figure 11.11. Its properties are as follows.

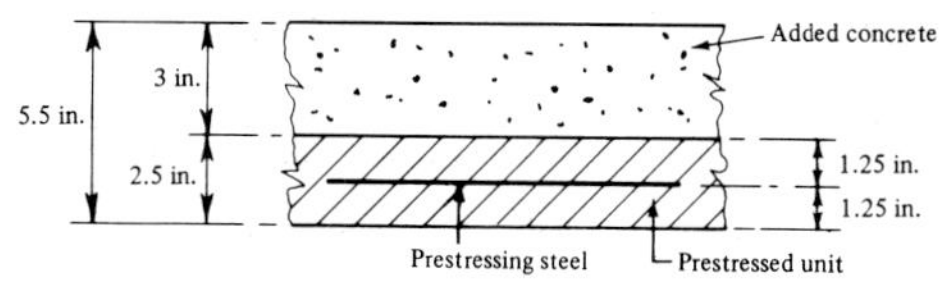

Figure 11.11 Composite section in example 11.7.2

Prestressed component:

$$A_p = 2.5 \times 12 = 30 \text{ in}^2 \ (193.55 \text{ cm}^2)$$

$$e_{1p} = 1.25 \text{ in. } (31.75 \text{ mm})$$

$$I_p = \frac{12 \times 2.5^3}{12} = 15.63 \text{ in}^4 \ (650.6 \text{ cm}^4)$$

$$Z_{1p} = Z_{2p} = \frac{15.63}{1.25} = 12.5 \text{ in}^3 \ (204.8 \text{ cm}^3)$$

$$i_p^{\,2} = \frac{15.63}{30} = 0.521 \text{ in}^2 \ (3.36 \text{ cm}^2)$$

Added concrete:

$$A_a = 3 \times 12 = 36 \text{ in}^2 \ (232.3 \text{ cm}^2)$$

$$e_{4a} = 1.5 \text{ in. } (38.1 \text{ mm})$$

$$I_a = \frac{12 \times 3^3}{12} = 27 \text{ in}^4 \ (1123.8 \text{ cm}^4)$$

Composite section:
At working load, the concrete strength of the prestressed component is 6000 lbf/in^2 and that of the added concrete is 4000 lbf/in^2 Therefore $E_{cp} = 5 \times 10^6$ lbf/in^2, $E_{ca} = 4 \times 10^6$ lbf/in^2 and $\alpha_a = 4/5 = 0.8$.

$$e_{1c} = \frac{(30 \times 1.25) + (0.8 \times 36 \times 4)}{30 + (0.8 \times 36)} = 2.6 \text{ in. } (66.04 \text{ mm})$$

$$I_{pc} = 15.63 + [30 \times (2.6 - 1.25)^2] + 0.8\ [27 + 36 \times (4.0 - 2.6)^2]$$
$$= 148.3 \text{ in}^4 \ (6172.7 \text{ cm}^4)$$

$$Z_{1c} = \frac{148.3}{2.6} = 57.1 \text{ in}^3 \ (935.7 \text{ cm}^3)$$

$$Z_{2c} = Z_{3c} = \frac{148.3}{0.1} = 1483 \text{ in}^3 \ (24\,302 \text{ cm}^3)$$

$$Z_{4c} = \frac{148.3}{2.9} = 51.1 \text{ in}^3 \ (837.4 \text{ cm}^3)$$

Bending moments:

$$M_t = 2.5 \times 12 \times \frac{150}{144} \times 14.5^2 \times 1.5 = 9850 \text{ lbf in.}$$
$$(113.5 \text{ kgf m, } 1113 \text{ N m})$$

$$M_d = \frac{5.5}{2.5} \times 9850 = 21\,700 \text{ lbf in. } (250 \text{ kgf m, } 2451 \text{ N m})$$

$$M_a = 200 \times 14.5^2 \times 1.5 = 63\,100 \text{ lbf in. } (727 \text{ kgf m, } 7128 \text{ N m})$$

$$M_w = 63\,100 + 21\,700 = 84\,800 \text{ lbf in. } (977 \text{ kgf m, } 9579 \text{ N m})$$

$$M_u = 2 \times 84\,800 = 169\,600 \text{ lbf in. } (1954 \text{ kgf m, } 19\,159 \text{ N m})$$

PRESTRESSING FORCE

Since the prestressed slab is only 2½ in. deep, it is clear that the prestressing steel must be placed along its centre line, and hence

$$P_t = A_p f_{1T} = A_p \times \frac{I}{R_0}\left(\frac{M_w}{Z_{1c}} + \bar{f}_{tw}\right)$$

$$= 30 \times \frac{I}{0.72}\left[\frac{84\,800}{57.1} - 500\right] = 40\,950 \text{ lbf } (18\,575 \text{ kgf, } 182 \text{ kN})$$

STRESSES

$$f_{1T} = f_{2T} = \left[\frac{84\,800}{57.1} - 500\right] = 1365 \text{ lbf/in}^2 \ (95.97 \text{ kgf/cm}^2, 9.41 \text{ N/mm}^2)$$

$$f_{1t} = 1365 - \frac{9850}{15.63} = 735 \text{ lbf/in}^2 \ (51.7 \text{ kgf/cm}^2, 5.07 \text{ N/mm}^2)$$

$$f_{2t} = 1365 + \frac{9850}{15.63} = 1995 \text{ lbf/in}^2 \ (140.3 \text{ kgf/cm}^2, 13.75 \text{ N/mm}^2)$$

Since the maximum permissible stress at transfer is 1600 lbf/in^2, this value for f_{2t} is not acceptable. However, the calculated value of M_t occurs only if the member is lifted at each end, and this can be avoided by specifying the position of lifting points so that the stress due to the weight of the member will not exceed 1600 – 1365 = 235 lbf/in^2. The least stress is obtained when the lifting points are about $0.2l$ ($0.207l$, to be exact) from each end; in this case the bending moments are as shown in Figure 11.12 and the corresponding maximum stress is 0.2 × 630 = 126 lbf/in^2 (8.86 kg/cm^2, 0.87 N/mm^2).

Therefore

$$f_{1t} = 1365 - 126 = 1239 \text{ lbf/in}^2 \ (87.1 \text{ kgf/cm}^2, 8.54 \text{ N/mm}^2)$$

$$f_{2t} = 1365 + 126 = 1491 \text{ lbf/in}^2 \ (104.8 \text{ kgf/cm}^2, 10.28 \text{ N/mm}^2)$$

$$f_{1e} = (0.72 \times 1365) - 126 = 856 \text{ lbf/in}^2 \ (60.2 \text{ kgf/cm}^2, 5.9 \text{ N/mm}^2)$$

$$f_{2e} = (0.72 \times 1365) + 126 = 1108 \text{ lbf/in}^2 \ (77.9 \text{ kgf/cm}^2, 7.64 \text{ N/mm}^2)$$

As the member is propped until the added concrete has hardened, it is necessary to consider only the stresses due to dead and live load acting on the composite section.

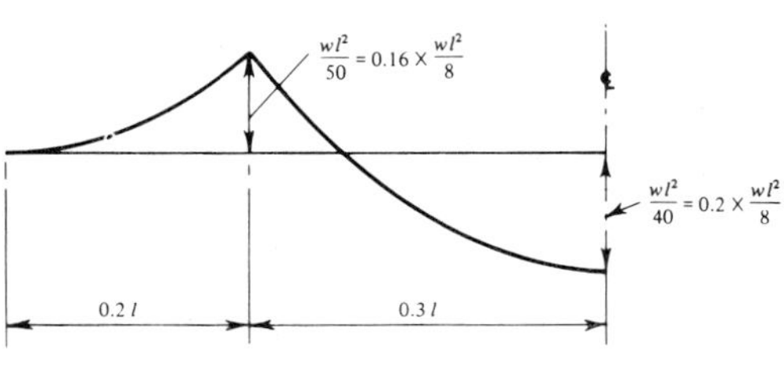

Figure 11.12 Bending moment diagram at lifting with lifting points 0.2 l from ends

Hence

$$f_{1d} = R_0 f_{1T} - \frac{M_d}{Z_{1c}} = 982 - \frac{21\,700}{57.1} = 602\,\text{lbf/in}^2\ (42.3\,\text{kgf/cm}^2, 4.15\,\text{N/mm}^2).$$

$$f_{2d} = R_0 f_{2T} - \frac{M_d}{Z_{2c}} = 982 - \frac{21\,700}{1483} = 967\ \text{lbf/in}^2\ (68\ \text{kgf/cm}^2, 6.67\ \text{N/mm}^2)$$

$$f_{3d} = -\frac{\alpha_a M_d}{Z_{3c}} = -\frac{0.8 \times 21\,700}{1483} = -12\ \text{lbf/in}^2\ (0.84\ \text{kgf/cm}^2, 0.08\ \text{N/mm}^2)$$

$$f_{4d} = +\frac{\alpha_a M_d}{Z_{4c}} = \frac{0.8 \times 21\,700}{51.1} = 340\ \text{lbf/in}^2\ (23.9\ \text{kgf/cm}^2, 2.34\ \text{N/mm}^2)$$

$$f_{1w} = f_{1d} - \frac{M_a}{Z_{1c}} = 602 - \frac{63\,100}{57.1} = -503\ \text{lbf/in}^2\ (35.4\ \text{kgf/cm}^2, 3.47\ \text{N/mm}^2)$$

$$f_{2w} = f_{2d} - \frac{M_a}{Z_{2c}} = 967 - \frac{63\,100}{1483} = 924\ \text{lbf/in}^2\ (65.0\ \text{kgf/cm}^2, 6.37\ \text{N/mm}^2)$$

$$f_{3w} = f_{3d} - \frac{\alpha_a M_a}{Z_{3c}} = -12 - \frac{0.8 \times 63\,000}{1483} = -46\,\text{lbf/in}^2\ (3.23\ \text{kgf/cm}^2, 0.31\ \text{N/mm}^2)$$

$$f_{4w} = f_{4d} + \frac{\alpha_a M_a}{Z_{4c}} = 340 + \frac{0.8 \times 63\,000}{51.1} = 1327\,\text{lbf/in}^2\ (93.3\ \text{kgf/cm}^2, 9.15\ \text{N/mm}^2)$$

The stress diagrams are shown in Figure 11.13

ULTIMATE LOAD

Resistance of steel

The prestressing force required is 40 950 lbf, and if the permissible stress at transfer in the steel is 150 000 lbf/in² the area required is $\frac{40\,950}{150\,000} = 0.273$ in² (1.76 cm²). Assuming k_u = 1.0 and the lever arm is 3 in. (76.2 mm), the area of steel required for ultimate load is

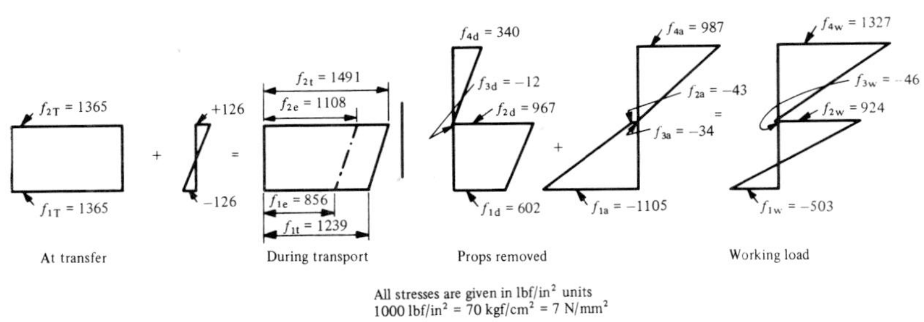

Figure 11.13 Stress diagram of the composite section (example 11.7.2)

$$\frac{169\ 600}{224\ 000 \times 3 \times 1.0} = 0.253 \text{ in}^2 \ (1.63 \text{ cm}^2)$$

which is less than that provided.

Resistance of concrete
The percentage/strength ratio is

$$\frac{224\ 000}{4000} \times \frac{0.273}{12 \times 4.25} = 0.3$$

Hence from Table 8.4

$$d_n = 0.75 \times 4.25 = 3.19 \text{ in. } (81.03 \text{ mm})$$

and the mean compressive stress in the concrete is

$$\frac{169\ 600}{(4.25 - 0.4 \times 3.19) \times 12 \times 3.19} = 1495 \text{ lbf/in}^2 \ (105 \text{ kgf/cm}^2, 10.3 \text{ N/mm}^2)$$

compared with a permissible value of $0.4 \times 4000 = 1600$ lbf/in^2 (112.5 kgf/cm^2, 11.0 N/mm^2). Hence the design is suitable.

EFFECT OF INACCURATE POSITIONING OF PRESTRESSING WIRES
In deep beams the effect of small inaccuracies in the positioning of the prestressing steel is usually negligible, but in shallow members such as that considered in this example this is no longer the case. A reasonable tolerance would be $\pm \frac{1}{16}$ in. and the effect of this is considered in the following.

$$e_{sp} = \pm \tfrac{1}{16} \text{ in. } (1.59 \text{ mm})$$

$$k_{1p} = 1 + \frac{e_{sp}\, e_{1p}}{i^2}$$

$$= 1 \pm \frac{1.25}{16 \times 0.521} = 1.15 \text{ or } 0.85$$

Similarly $k_{2p} = 0.85$ or 1.15

If the prestressing force is increased to ensure that the required prestress at the bottom face is obtained, the amount required is $40\ 950 \times 1.15 = 47\ 100$ lbf (21 364 kgf, 209.5 kN) for which

$$f_{1T} = \frac{0.85 \times 47\ 100}{30} = 1365 \text{ lbf/in}^2 \ (95.97 \text{ kgf/cm}^2, 9.41 \text{ N/mm}^2)$$

as before, and

$$f_{2T} = \frac{47\ 100 \times 1.15}{30} = 1850 \text{ lbf/in}^2 \ (130 \text{ kgf/cm}^2, 12.75 \text{ N/mm}^2)$$

At the opposite tolerance the stresses would be $f_{1T} = 1850$ lbf/in^2 and $f_{2T} = 1365$ lbf/in^2.

Stress diagrams for these two conditions are shown in Figure 11.14. The change of stress is clearly appreciable, although in this case probably not serious enough to warrant a change in design. If uniform deflection is desired,

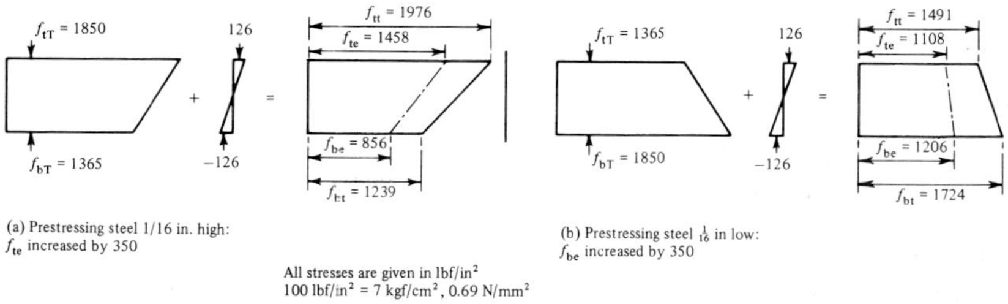

Figure 11.14 Variation of stresses due to tolerance ($\pm\frac{1}{16}$ in.) in steel position (example 11.7.2)

however, variations of this order can lead to comparatively large differences in deformation and this is considered in Chapter 12.

11.7.3 Example: Composite floor without props

Design a composite floor slab for the same conditions as Example 11.7.2, except that propping during construction is no longer possible.

Assuming the permissible stresses and the losses to be as in Example 11.7.2, then for the composite section Z_{1c} = 51.5 in³ (844 cm³) and Z_{4c} = 47.3 in³ (775 cm³) as before. However, the prestressed component must now support its own weight and that of the added concrete over a span of 14 ft 6 in., and a stiffer member will therefore be required. A suitable trial section for these conditions is shown in Figure 11.15.

PROPERTIES OF SECTION

Prestressed component:

$$A_p = (2 \times 12) + (2.5 \times 4) = 34 \text{ in}^2 \ (219.4 \text{ cm}^2)$$

$$e_{1p} = \frac{(2 \times 12 \times 1) + (2.5 \times 4 \times 3.25)}{34} = 1.67 \text{ in. } (42.4 \text{ mm})$$

$$I_{1p} = \frac{12 \times 1.67^3}{3} + \frac{8 \times 0.33^3}{3} + \frac{4 \times 2.83^3}{3} = 49 \text{ in}^4 \ (2039 \text{ cm}^4)$$

$$Z_{1p} = \frac{49}{1.67} = 29.5 \text{ in}^3 \ (483.4 \text{ cm}^3);\ Z_{2p} = \frac{49}{2.83} = 17.3 \text{ in}^3 \ (283.5 \text{ cm}^3)$$

Added concrete:

$$A_a = (2 \times 4 \times 3.5) + (4 \times 1) = 32 \text{ in}^2 \ (206.5 \text{ cm}^2)$$

$$e_{4a} = \frac{(2 \times 4 \times 3.5 \times 1.75) + (4 \times 1 \times 0.5)}{32} = 1.6 \text{ in. } (40.6 \text{ mm})$$

$$I_a = \frac{8 \times 1.9^3}{3} + \frac{12 \times 1.6^3}{3} - \frac{4 \times 0.6^3}{3} = 34.3 \text{ in}^4 \ (1428 \text{ cm}^4)$$

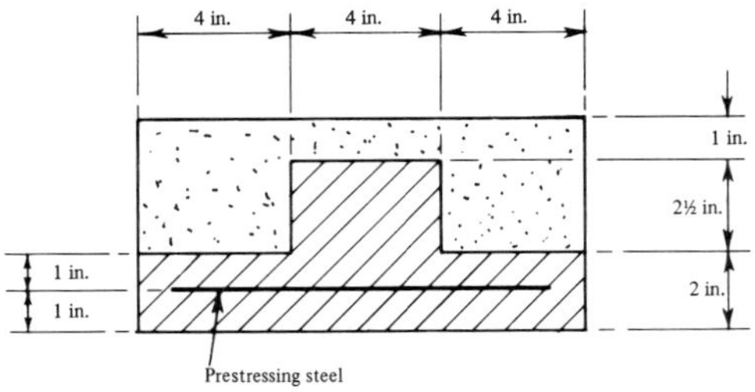

Figure 11.15 Composite section for example 11.7.3

Composite section:

$$\alpha_a = 0.8$$

$$e_{1c} = \frac{(34 \times 1.67) + [0.8 \times 32 \times (5.5 - 1.6)]}{34 + (0.8 \times 32)} = 2.63 \text{ in. } (66.8 \text{ mm})$$

$$I_{pc} = 49 + 34 \times (2.63 - 1.67)^2 + 0.8 \times [34.3 + 32\,(3.9 - 2.63)^2] = 149.6 \text{ in}^4 \ (6227 \text{ cm}^4)$$

$$Z_{1c} = \frac{149.6}{2.63} = 56.8 \text{ in}^3 \ (930.8 \text{ cm}^3); Z_{2c} = \frac{149.6}{1.87} = 79.9 \text{ in}^3 \ (1\,309.3 \text{ cm}^3)$$

$$Z_{3c} = \frac{149.6}{0.63} = 237 \text{ in}^3 \ (3884 \text{ cm}^3); \quad Z_{4c} = \frac{149.6}{2.87} = 52.1 \text{ in}^3 \ (853.8 \text{ cm}^3)$$

BENDING MOMENTS

$$M_t = 34 \times \frac{150}{144} \times 14.5^2 \times 1.5 = 11\,150 \text{ lbf in. } (128.4 \text{ kgf m, } 1260 \text{ N m})$$

$$M_d = \left(\frac{34 + 32}{34}\right) \times 11\,150 = 21\,700 \text{ lbf in. } (250 \text{ kgf m, } 2\,451 \text{ N m})$$

$$M_a = 63\,100 \text{ lbf in. } (727 \text{ kgf m, } 7\,128 \text{ N m})$$

$$M_w = 84\,800 \text{ lbf in. } (977 \text{ kgf m, } 9\,579 \text{ N m})$$

$$M_u = 169\,600 \text{ lbf in. } (1954 \text{ kgf m, } 19\,159 \text{ N m})$$

PRESTRESSING FORCE

Again, because of the shallow flange of the prestressed section, the prestressing steel must be placed on the centre line of the flange, 1 in. from the tensile face.

Hence

$$e_{sp} = 1.67 - 1 = 0.67 \text{ in. } (17 \text{ mm})$$

$$i_p^2 = \frac{49}{34} = 1.44 \text{ in}^2 \ (9.29 \text{ cm}^2)$$

$$k_{1p} = 1 + \frac{0.67 \times 1.67}{1.44} = 1.77$$

$$k_{2p} = 1 - \frac{0.67 \times 2.83}{1.44} = -0.31$$

$$\frac{M_d}{Z_{1p}} = \frac{21\,700}{29.5} = 737 \text{ lbf/in}^2 \text{ (51.8 kgf/cm}^2\text{, 5.08 N/mm}^2\text{)}$$

$$\frac{M_d}{Z_{2p}} = \frac{21\,700}{17.3} = 1255 \text{ lbf/in}^2 \text{ (88.2 kgf/cm}^2\text{, 8.65 N/mm}^2\text{)}$$

$$\frac{M_a}{Z_{1c}} = \frac{63\,100}{56.8} = 1112 \text{ lbf/in}^2 \text{ (78.2 kgf/cm}^2\text{, 7.67 N/mm}^2\text{)}$$

$$\frac{M_a}{Z_{2c}} = \frac{63\,100}{79.9} = 790 \text{ lbf/in}^2 \text{ (55.5 kgf/cm}^2\text{, 5.45 N/mm}^2\text{)}$$

Assuming in this case that counter-action is ensured by bending up some of the wires then $R_0 f_{1T} - 737 - 1112 = -500$ and $f_{1T} = \frac{1112 + 737 - 500}{0.72} =$ 1870 lbf/in^2 (131.5 kg/cm^2, 12.89 N/mm^2). The prestressing force P_t is therefore

$$\frac{A_p}{k_{1p}} f_{1T} = \frac{34}{1.77} \times 1870 = 35\,900 \text{ lbf (16 284 kgf, 160 kN)}$$

STRESSES

$$f_{1T} = 1870 \text{ lbf/in}^2 \text{ (131.5 kgf/cm}^2\text{, 12.89 N/mm}^2\text{)}$$

$$f_{2T} = \frac{k_2}{k_1} f_{1T} = \frac{-0.31}{1.77} \times 1870 = -330 \text{ lbf/in}^2 \text{ (23.2 kgf/cm}^2\text{, 2.28 N/mm}^2\text{)}$$

$$f_{1t} = 1870 - 379 = 1491 \text{ lbf/in}^2 \text{ (104.8 kgf/cm}^2\text{, 10.28 N/mm}^2\text{)}$$

$$f_{2t} = -330 + 646 = 316 \text{ lbf/in}^2 \text{ (22.2 kgf/cm}^2\text{, 2.18 N/mm}^2\text{)}$$

These values are acceptable if counter-action is ensured.

$$f_{1d} = 0.72 \times 1870 - 737 = 608 \text{ lbf/in}^2 \text{ (42.7 kgf/cm}^2\text{, 4.19 N/mm}^2\text{)}$$

$$f_{2d} = 0.72 \times -330 + 1255 = 1017 \text{ lbf/in}^2 \text{ (71.5 kgf/cm}^2\text{, 7.01 N/mm}^2\text{)}$$

$$f_{1w} = 608 - 1112 = -504 \text{ lbf/in}^2 \text{ (35.4 kgf/cm}^2\text{, 3.47 N/mm}^2\text{)}$$

$$f_{2w} = 1017 + 790 = 1807 \text{ lbf/in}^2 \text{ (127 kgf/cm}^2\text{, 12.46 N/mm}^2\text{)}$$

$$f_{3w} = -0.8 \times \frac{63\,100}{237} = -213 \text{ lbf/in}^2 \text{ (14.98 kgf/cm}^2\text{, 1.47 N/mm}^2\text{)}$$

$$f_{4w} = 0.8 \times \frac{63\,100}{52.1} = 969 \text{ lbf/in}^2 \text{ (68.18 kgf/cm}^2\text{, 6.68 N/mm}^2\text{)}$$

The stress diagrams are shown in Figure 11.16.

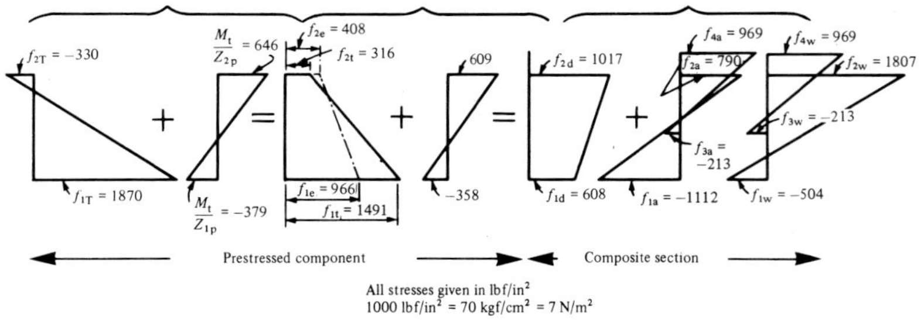

Figure 11.16 Stress diagrams: example 11.7.3

ULTIMATE LOAD

Resistance of steel

$$A_{st} = \frac{35\ 900}{150\ 000} = 0.239 \text{ in}^2 \text{ (1.54 cm}^2\text{)}$$

Assuming k_u = 1.0 and the lever arm is 3 in., then A_s = 0.253 in^2, as before.

Hence

$$A_{sn} = 0.14 \text{ in}^2 \text{ (0.9 cm}^2\text{)}$$

Resistance of concrete

The percentage/strength ratio is

$$\frac{224\ 000}{4000} \times \frac{0.253}{12 \times 4.5} = 0.263$$

From Table 8.4, d_n = 0.65 × 4.5 = 2.93 in. (74.4 mm) and the mean compressive stress in the concrete is $\frac{169\ 600}{(4.5 - 0.4 \times 2.93) \times 12 \times 2.93}$ = 1450 lbf/in^2 (101.94 kg/cm^2, 10 N/mm^2). The permissible value for the added concrete is 1600 lbf/in^2; hence the design is suitable.

EFFECT OF INACCURATE POSITIONING OF PRESTRESSING WIRES

As before, a tolerance of ± $\frac{1}{16}$ in. is assumed and therefore

$$e_{sp} = 0.667 \pm 0.063 = 0.73 \text{ in. (18.5 mm) or } 0.604 \text{ in. (15.3 mm)}$$

$$k_{1p} = 1 + \frac{\left.\begin{matrix}0.73\\0.604\end{matrix}\right\} \times 1.67}{1.44} = 1.844 \text{ or } 1.699$$

$$k_{2p} = 1 - \frac{\left.\begin{matrix}0.73\\0.604\end{matrix}\right\} \times 2.83}{1.44} = -0.44 \text{ or } -0.19$$

The increased prestressing force is 35 900 × $\frac{1.771}{1.699}$ = 37 500 lbf (17 010 kgf,

166.8 kN) for which

$$f_{1T} = \frac{1.699}{34} \times 37\,500 = 1870 \text{ lbf/in}^2 \ (131.5 \text{ kgf/cm}^2, 12.89 \text{ N/mm}^2)$$

and

$$f_{2T} = 1870 \times \frac{-0.19}{1.699} = -209 \text{ lbf/in}^2 \ (14.7 \text{ kgf/cm}^2, 1.44 \text{ N/mm}^2)$$

At the opposite tolerance

$$f_{1T} = \frac{1.844}{34} \times 37\,500 = 2030 \text{ lbf/in}^2 \ (142.7 \text{ kgf/cm}^2, 14.0 \text{ N/mm}^2)$$

and

$$f_{2T} = 2030 \times \frac{-0.44}{1.844} = -485 \text{ lbf/in}^2 \ (34.1 \text{ kgf/cm}^2, 3.34 \text{ N/mm}^2)$$

Diagrams of stress are given in Figure 11.17.

The effect on deflection is considered in Chapter 12.

11.7.4 Example: Differential shrinkage

Investigate the effect of a differential shrinkage of the added concrete of 50×10^{-6} on the sections obtained in Examples 11.7.2 and 11.7.3.

For the section of Example 11.7.2

$$\Delta\epsilon_p = \frac{\Delta\epsilon}{\left(\dfrac{A_{pc}}{\alpha_a A_a} + \dfrac{A_p e_0{}^2}{I_p + \alpha_a I_a}\right)}$$

$$= \frac{50 \times 10^{-6}}{\left(\dfrac{30 + 28.8}{28.8} + \dfrac{30 \times 2.75^2}{15.63 + 0.8 \times 27}\right)} = 6.15 \times 10^{-6}$$

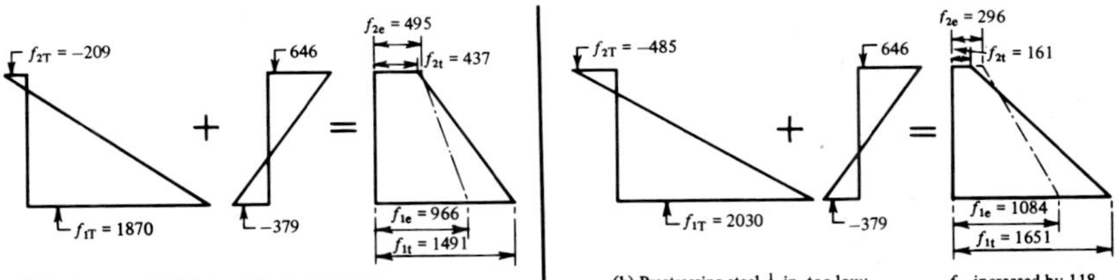

(a) Prestressing steel $\frac{1}{16}$ in. too high: f_{2e} increased by 87, Tolerance allowed in design.

(b) Prestressing steel $\frac{1}{16}$ in. too low: f_{1e} increased by 118, f_{2c} decreased by 112

All stresses are given in lbf/in^2

Figure 11.17 Variation of stresses due to $\pm\frac{1}{16}$ in. tolerance (example 11.7.3)

$$\Delta\epsilon_a = \frac{\Delta\epsilon}{\left(\frac{A_{pc}}{A_p} + \frac{\alpha_a A_a e_0{}^2}{I_p + \alpha_a I_a}\right)} = \frac{50 \times 10^{-6}}{\left(\frac{30 + 28.8}{30} + \frac{28.8 \times 2.75^2}{15.63 + 0.8 \times 27}\right)} = 6.40 \times 10^{-6}$$

Hence, from Figure 11.18

$$\Delta\epsilon_1 = 10^{-6} \times (-37.5 \times \frac{4}{2.75} + 43.6) = 10.9 \times 10^{-6}$$

$$\Delta\epsilon_2 = 10^{-6} \times [2 \times (10.9 + 6.15) - 10.9] = 23.2 \times 10^{-6}$$

$$\Delta\epsilon_3 = 10^{-6} \times (50 - 23.2) = 26.8 \times 10^{-6}$$

$$\Delta\epsilon_4 = [2 \times (26.8 - 6.4) - 26.8] \times 10^{-6} = 14 \times 10^{-6}$$

$$f_{1s} = -10.9 \times 10^{-6} \times 5 \times 10^6 = -55 \text{ lbf/in}^2 \ (3.87 \text{ kgf/cm}^2, 0.38 \text{ N/mm}^2)$$

$$f_{2s} = 23.2 \times 5 = 116 \text{ lbf/in}^2 \ (8.16 \text{ kgf/cm}^2, 0.8 \text{ N/mm}^2)$$

$$f_{3s} = -26.8 \times 4 = -107 \text{ lbf/in}^2 \ (7.52 \text{ kgf/cm}^2, 0.74 \text{ N/mm}^2)$$

$$f_{4s} = 14 \times 4 = 56 \text{ lbf/in}^2 \ (3.94 \text{ kgf/cm}^2, 0.39 \text{ N/mm}^2)$$

The stress diagram is shown in Figure 11.18b

For the section of Example 11.7.3, inserting the appropriate values in the same formulae

$$\Delta\epsilon_1 = 7.3 \times 10^{-6}$$

$$\Delta\epsilon_2 = 42 \times 10^{-6}$$

$$\Delta\epsilon_3 = 36 \times 10^{-6}$$

$$\Delta\epsilon_4 = 3 \times 10^{-6}$$

These values are shown in Figure 11.19a. The stresses are

$$f_{1s} = -7.3 \times 5 = -36.5 \text{ lbf/in}^2 \ (2.57 \text{ kgf/cm}^2, 0.25 \text{ N/mm}^2)$$

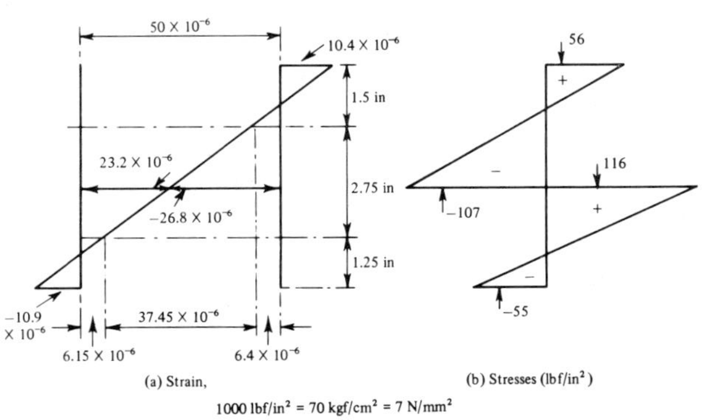

Figure 11.18 Strain and stress diagram for this section in example 11.7.2 due to differential shrinkage

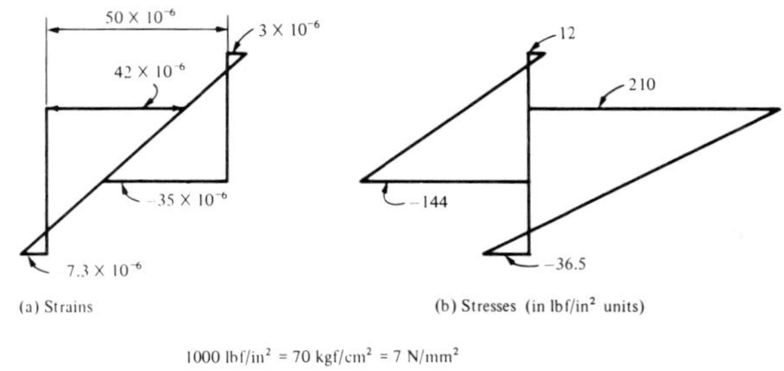

Figure 11.19 Strain and stress diagram for this section in example 11.7.3 due to differential shrinkage

f_{2s} = 42 X 5 = 210 lbf/in² (14.77 kgf/cm², 1.45 N/mm²)

f_{3s} = −36 X 4 = −144 lbf/in² (10.12 kgf/cm², 0.99 N/mm²)

f_{4s} = 3 X 4 = 12 lbf/in² (0.84 kgf/cm², 0.08 N/mm²)

The stress diagram is shown in Figure 11.19b

The preceding calculations give typical values for the magnitude and distribution of the stresses due to shrinkage. In extreme cases, different distribution of stresses can occur. One such extreme is shown in Figure 11.20 for the case in which the added concrete is concentrated at the top flange of a deep section, and the opposite case in which the prestressed concrete is concentrated at the bottom of a deep section. It is also possible for the shrinkage and creep of the prestressed component to exceed that of the added concrete, if the latter is placed very soon after transfer. In this case the nature of the strains and stresses is reversed.

11.7.5 Example: Composite double tee construction

In Examples 9.2.6 and 9.2.7 double-T units with pre-tensioned tendons have been used in the construction consisting of precast members only. It is often economical to use such units in conjunction with in-situ structural topping to carry the entire load as a composite member.

Stresses are calculated in stages consistent with the sequence of construction such that the precast part would carry only its own weight and the weight of the added concrete while the additional loads (superloads) will be supported by the composite section. The resultant stress at any point due to loading will be the algebraic sum of the stresses at that point in different stages as shown in the previous examples of this chapter.

In this particular example an additional alternative solution is presented for serviceability condition at which the precast member is propped at midspan when the in-situ concrete is added, the props being removed after the concrete has matured and the composite section becomes operative. Long-span floor: A 55 ft (16.764 m) effective span classroom constructed of prestressed 28 in. (715 mm) deep 8 ft (2438 mm) wide double-T units (ribs at 4 ft (1219 mm) centres) with a minimum of 2½ in. (63.5 mm) thick in-situ structural topping to act compositely under service load.

Figure 11.21 shows the cross-section of the composite construction.

For the serviceability conditions two sets of calculations are presented: one

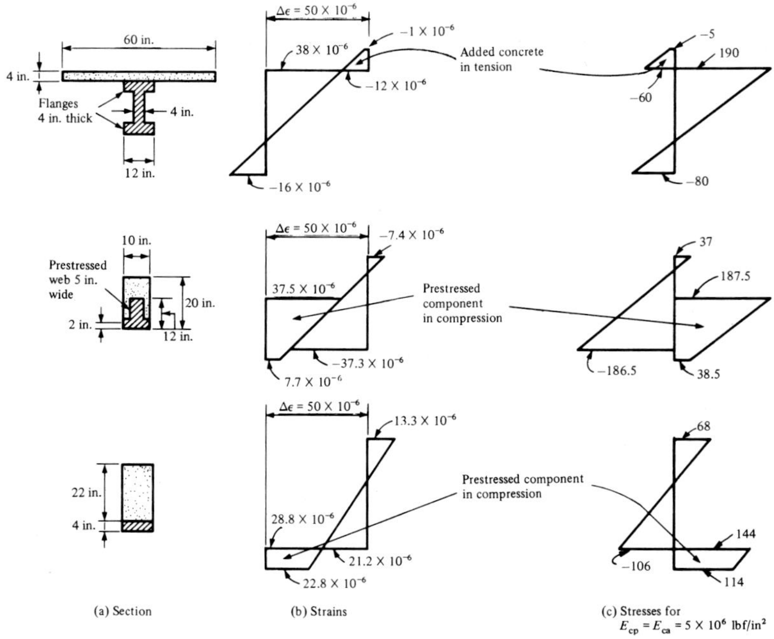

Figure 11.20 Different distribution of stress and strain depending on position of added concrete in sections

using temporary prop at midspan when in-situ concrete is cast, another without use of any prop. Although the ultimate condition is not effected, it is obvious from the following calculation that the case with the prop shows some advantage as far as elastic stress levels are concerned.

The service load for the design is as follows:

(a) Self weight of precast unit = 525 lbf/ft (781.28 kgf/m, 7.66 kN/m) run of TT

(b) Structural topping average 3¼ in. thick* = 320lbf/ft(476.21 kgf/m,4.67 kN/m) run of TT

(c) Finishes at 24 lbf/ft^2 = 24 X 8 = 192 lbf/ft (285.73 kgf/m, 2.80 kN/m) run of TT

(d) Partitions at 40 lbf/ft^2 = 40 X 8 = 320 lbf/ft (476.21 kgf/m, 4.67 kN/m) run of TT

(e) Live at 60 lbf/ft^2 = 60 X 8 = 480 lbf/ft (714.32 kgf/m, 7.01 kN/m) run of TT

Concrete quality:

(a) For precast concrete
6000 lbf/in^2 at release (421.86 kgf/cm^2, 41.37 N/mm^2)

*(4 in. at support and 2½ in. at midspan to allow for camber)

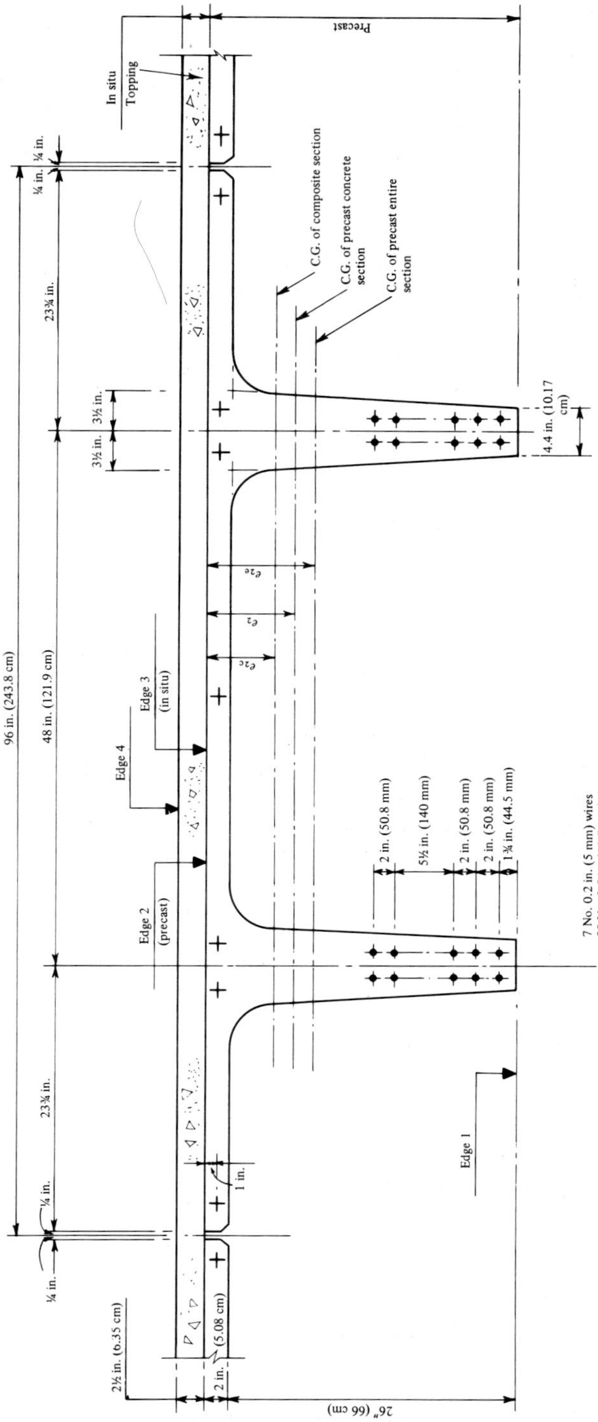

Figure 11.21 Composite double-T construction (example 11.7.5)

9000 lbf/in^2 at 28 days (632.79 kgf/cm^2, 62.05 N/mm^2)

(b) For in-situ topping

4500 lbf/in^2 at 28 days (316.40 kgf/cm^2, 31.03 N/mm^2

Steel:
20 No. ½ in. (12.7 mm) diameter stabilized strands (10 per rib) having a guaranteed minimum ultimate strength of 37 000 lbf (16 783 kgf, 164.57 kN) each and 7 No. 0.2 in. (5 mm) diameter wires in the flange with a guaranteed minimum ultimate stress of 224 000 lbf/in^2 (15 750 kgf/cm^2, 1544.4 N/mm^2)

Area of each 0.5 in. diameter strand = 0.144 in^2 (92.9 mm^2)

Area of each 0.2 in. diameter wire = 0.0314 in^2 (20.3 mm^2)

$E_s = 28 \times 10^6$ lbf/in^2 (1.97×10^6 kgf/cm^2, 193 kN/mm^2)

Relaxation of stabilized strand in accordance with the manufacturer's specification 2–3 per cent of stress when stressed to 70 per cent of the ultimate.

Ultimate condition (to CP 115)

M_R = Moment of resistance of the unit in flexure:

$$\frac{A_{st} \cdot f_{su}}{b.d.f_{cu}} = 0.3 \qquad \therefore K_u = 1$$

F_{su} = Ultimate tensile force (taking all strands in the rib) = $20 \times 37\,000$
= 740 000 lbf (335 659 kgf, 3291 kN)

C.G. of tensile force = 7.15 in. (181.6 mm) from the bottom.

Uniform concrete compression at failure = $0.4 \times 4500 = 1800$ lbf/in^2

Therefore, area of concrete required to balance tensile force

$$= \frac{740\,000}{1800} = 411 \text{ in}^2$$

$$\therefore d_n = \frac{411}{96} = 4.28 \text{ in. (i.e. within the flange)}$$

$$\therefore z = 30.5 - \left(7.15 + \frac{4.28}{2}\right) = 21.21 \text{ in.}$$

∴ Permissible $M_R = 740\,000 \times 21.21 = 15\,695\,000$ lbf in. (180 800 kgf m, 1773 kN m)

Load for ultimate condition

$1.5 \times (525 + 320 + 192 + 320) + 2.5 \times 480$
$= 2035.5 + 1200 = 3235.5$ lbf/ft (4814.9 kgf/m, 472.2 kN/m)

$$M_u \text{ required} = \frac{3235.5 \times 55^2}{8} \times 12 = 14\,681\,000 \text{ lbf in. (169 125 kgf m,}$$

1658 kN m) < Permissible M_R ; O.K.

According to CP 110 the ultimate tensile force is $F_{su} = \dfrac{\text{Characteristic strength}}{1.15}$

where characteristic strength in prestressing strands and wires is to be taken as the minimum specified strength. The uniform compressive stress of concrete = $0.4 f_{cu}$

∴ the ultimate moment of resistance M_R = 13 828 000 lbf in. (159 300 kgf m, 1562 kN m)

Factor of safety required is 1.4 × dead load and 1.6 × live load

$$\therefore\ M_u \text{ required} = [1.4 \times (525 + 320 + 192 + 320) + 1.6 \times 480] \times \frac{55^2}{8} \times 12$$

$$= 12\ 100\ 000 \text{ lbf in. } (139\ 400 \text{ kgf m, } 1367 \text{ kN m}) < \text{permissible } M_R$$

CHECK FOR SERVICEABILITY CONDITION

Properties of the precast section

E_{cp} = 5.75×10^6 lbf/in^2 (0.4×10^6 kgf/cm^2, 39.6 kN/mm^2)

A_p (Cross-sectional area) = 487.9 in^2 (3147.73 cm^2)

e_2 = 8.9 in. (22.61 cm)

e_1 = 19.1 in. (48.51 cm)

I_p = 36 167.8 in^4 (1 505 416 cm^4)

Z_2 (section modulus top) = 4061.69 in^3 (66 559 cm^3)

Z_1 (section modulus bottom) = 1894.06 in^3 (31 038 cm^3)

Properties of entire precast section (taking into account bonded steel)

$$\alpha_e = \frac{E_s}{E_c} = 4.87 \simeq 4.9$$

$$A_{ep} = 487.9 + 20 \times (4.9 - 1) \times 0.144 + 7 \times (4.9 - 1) \times 0.0314$$
$$= 487.9 + 11.15 + 0.85 = 499.9 \text{ in}^2\ (3225.15 \text{ cm}^2)$$

$$e_{1e} = \frac{(4 \times 3.9 \times 0.144) \times (1.75 + 3.75 + 5.75 + 11.25 + 13.25) +}{499.9}$$

$$\frac{(0.85 \times 27 + 487.9 \times 19.1)}{499.9}$$

$$= 18.8 \text{ in. } (47.752 \text{ cm})$$

e_{2e} = 9.2 in. (23.368 cm)

I_{ep} = 38 000 in^4 (1 581 600 cm^4)

Z_{2e} = 4125 in^3 (67 600 cm^3)

Z_{1e} = 2020 in^3 (33 100 cm^3)

Initial prestressing force

Each wire stressed to 70 per cent of ultimate, i.e. to 4900 lbf.

Each strand stressed to 70 per cent of ultimate i.e. 25 900 lbf.

P_i = 20 × 25 900 + 7 × 4900 = 552 300 lbf (250 000 kgf 2457 kN)

C.G. of P_i = 8.37 in. from edge 1 (21.26 cm)

f_{sp} = 178 000 lbf/in² (12 515 kgf/cm², 1227 N/mm²)

e_s = 10.73 in. (27.25 cm)

Initial stresses solely due to prestress

$$f_{1I} = \frac{552\,300}{488.0} + \frac{552\,300 \times 10.73}{1894} = 4662 \text{ lbf/in}^2 \ (300 \text{ kgf/cm}^2, 29 \text{ N/mm}^2)$$

$$f_{2I} = \frac{552\,300}{488} - \frac{552\,300 \times 10.73}{4062} = -328 \text{ lbf/in}^2 \ (23 \text{ kgf/cm}^2, 2.26 \text{ N/mm}^2)$$

$$f_{SI} = \frac{(4262 + 328)}{28} \times 19.63 - 328 = 2892 \text{ lbf/in}^2 \ (203 \text{ kgf/cm}^2, 20 \text{ N/mm}^2)$$

Assume loss due to elastic shortening = 7.5 per cent

$\therefore f_{sT}$ = 0.925 × 2892 = 2675 lbf/in²

Δf_{pe} = 4.9 × 2675 = 13 100 lbf/in²

7.5 % f_{sI} = 0.075 × 178 000 = 13 350 lbf/in²

As the difference is negligible take loss due to eastic shortening = 7.5 per cent

Consequently

f_{1T} = 0.925 × 4262 = 3942 lbf/in² (compression)

f_{2T} = −0.925 × 328 = −303 lbf/in² (tension)

Since $f_{1T} > \frac{6000}{2}$, and there is no counter-action at the ends, debonding of four strands done to the required length.

Stress due to self-weight at x ft from support

$$M_x = \left[525 \times 27.5\,x - \frac{525}{2}x^2\right] \times 12 \text{ lbf in.}$$

$$= (173\,000\,x - 3150\,x^2) \text{ lbf in.}$$

$$f_{1d} = \frac{173\,000\,x}{2020} - \frac{3150\,x^2}{2020} = (85.5\,x - 1.56\,x^2) \text{ lbf/in}^2 \qquad \text{(tension)}$$

$$f_{2d} = (42\,x - 0.764\,x^2) \text{ lbf/in}^2 \qquad \text{(compression)}$$

∴ stress at C.G. of steel

$$f_{sd} = (-)\left[\frac{(85.5x - 1.56x^2 + 42x - 0.764x^2)}{28} \times 19.63 - 42x + 0.764x^2\right]$$

$$= (-)(47.4x - 0.866x^2) \qquad \text{(tension)}$$

Since propping at midspan will shift the critical section for stresses under service load, the losses and final stresses have been determined at any section x. When the critical section x is determined the value is substituted. In the unpropped or continuously propped case where the midspan is critical these (i.e. losses and stresses) could be determined directly as in the previous examples.

CALCULATION OF FINAL LOSSES

(1) Elastic shortening = 13 100 lbf/in^2

(2) Relaxation at 2½ per cent of 178 000 (Manufacturers' specification) = 4450 lbf/in^2

(3) Shrinkage = $300 \times 10^{-6} \times 28 \times 10^6 = 8400$ lbf/in^2

(4) Creep = $0.33 \times 10^{-6} \times 28 \times 10^6 \times [2675 - 47.4x + 0.866x^2]$
$= 24\,760 - 437x + 8.0x^2$

∴ $\Delta f_{\text{p total}} = (50\,710 - 437x + 8x^2)$ lbf/in^2

∴ % loss at a section x ft from support

$$= \frac{(50\,710 - 437x + 8x^2)}{178\,000} \times 100 = (28.46 - 0.246x + 0.0045x^2)$$

∴ Final stress at a section x ft from support (due to prestress only)

$$f_{1p} = \left[\frac{100 - (28.46 - 0.246x + 0.0045x^2)}{100}\right] \times 4262 \text{ lbf/in}^2$$

$$= [0.715 + 0.00246x + 0.000045x^2] \times 4262 \text{ lbf/in}^2 \quad \text{(compression)}$$

$$f_{2p} = [0.715 + 0.00246x + 0.000045x^2] \times 328 \text{ lbf/in}^2 \qquad \text{(tension)}$$

Properties of the composite section

$$\alpha_p = \frac{E_p}{E_a} = \frac{5.75}{4} = 1.41$$

$$A_c = \text{Area of composite section} = \frac{96 \times 2.5}{1.41} + 499.9 = 170 + 499.9$$

$$= 669.9 \text{ in}^2 \ (4321.92 \text{ cm}^2)$$

$$e_{1c} = \frac{14\,378.76}{699.9} = 20.6 \text{ in. } (52.32 \text{ cm})$$

$$I_c \text{ composite} = 170 \times 8.65^2 + 499.9 \times 1.8^2 + \frac{1}{12} \times \frac{96}{1.41} \times 2.5^2 + 37\,999$$

$$= 52\,403 \text{ in}^4 \ (2\,181\,175.3 \text{ cm}^4)$$

$$\therefore\ Z_{1c} = \frac{52\,403}{20.6} = 2560 \text{ in}^3 \ (41\,950.98 \text{ cm}^3)$$

$$Z_{2c} = (\text{interface} - \text{precast}) = \frac{52\,403}{7.4} = 7060 \text{ in}^3 \ (115\,693 \text{ cm}^3)$$

$$Z_{3c} \ (\text{interface} - \text{in-situ}) = Z_{2c} \times \alpha_p = 9800 \text{ in}^3 \ (160\,594 \text{ cm}^3)$$

$$Z_{4c} \ (\text{in-situ}) = \frac{52403}{9.9} \times 1.41 = 7450 \text{ in}^3 \ (122\,084 \text{ cm}^3)$$

Stages of construction and flexural stress

CASE 1: UNPROPPED CONDITION

Stage 1. Precast unit placed on bearings and in-situ topping cast (no composite action)

Loading = 525 + 320 = 845 lbf/ft (1257.5 kgf/m, 12.33 kN/m)

$$M_{max} = \frac{845 \times 55^2}{8} \times 12 = 3\,840\,000 \text{ lbf in. } (44\,200 \text{ kgf m, } 433.8 \text{ kN m})$$

$$\therefore\ f_{1d} = -\frac{3\,840\,000}{2020} = -1900 \text{ lbf/in}^2 \ (133.59 \text{ kgf/cm}^2, 13.1 \text{ N/mm}^2)$$

(tension)

$$f_{2d} = \frac{3\,840\,000}{4125} = 930 \text{ lbf/in}^2 \ (65.39 \text{ kgf/cm}^2, 6.41 \text{ N/mm}^2)$$

(compression)

Stage 2. Finishes, partitions and live load added: composite action between precast and in-situ concrete

Added load = 192 + 320 + 480 = 992 lbf/ft (1476.3 kgf/m, 14.48 kN/m)

$$M_{max} = \frac{992 \times 55^2}{8} \times 12 = 4\,500\,000 \text{ lbf in. } (51\,800 \text{ kgf m, } 508.34 \text{ kN m})$$

$$f_{1a} = \frac{4\,500\,000}{2560} = -1758 \text{ lbf/in}^2 \ (123.6 \text{ kgf/cm}^2, 12.12 \text{ N/mm}^2)$$

(tension)

$$f_{2a} = \frac{4\,500\,000}{7060} = 635 \text{ lbf/in}^2 \ (44.65 \text{ kgf/cm}^2, 4.38 \text{ N/mm}^2)$$

(compression)

$$f_{3a} = \frac{635}{1.41} = 450 \text{ lbf/in}^2 \ (31.64 \text{ kgf/cm}^2, 3.10 \text{ N/mm}^2)$$

(compression)

$$f_{4a} = \frac{4\,500\,000}{7450} = 605 \text{ lbf/in}^2 \ (42.54 \text{ kgf/cm}^2, 4.17 \text{ N/mm}^2)$$

(compression)

At the point of maximum stresses (here midspan section) stresses due to prestress only (after all losses)

f_{1p} = $(0.715 + 0.00246x - 0.000045x^2) \times 4262 = 3190$ lbf/in^2 (224 kgf/cm^2, 22 N/mm^2)

f_{2p} = $-0.7485 \times 328 = -246$ lbf/in^2 (17.3 kgf/cm^2, 1.7 N/mm^2)

(tension)

Therefore, resultant stresses under service load

f_{1w} = $(3190 - 1900 - 1758) = (-)\,468$ lbf/in^2 (32.9 kgf/cm^2, 3.23 N/mm^2)

(tension)

f_{2w} = $(930 + 635 - 246) = 1319$ lbf/in^2 (92.74 kgf/cm^2, 9.09 N/mm^2)

(compression)

f_{3w} = 450 lbf/in^2 (31.64 kg/cm^2, 3.1 N/mm^2) (compression)

f_{4w} = 605 lbf/in^2 (42.54 kgf/cm^2, 4.17 N/mm^2) (compression)

Stress diagram is shown in Figure 11.22.

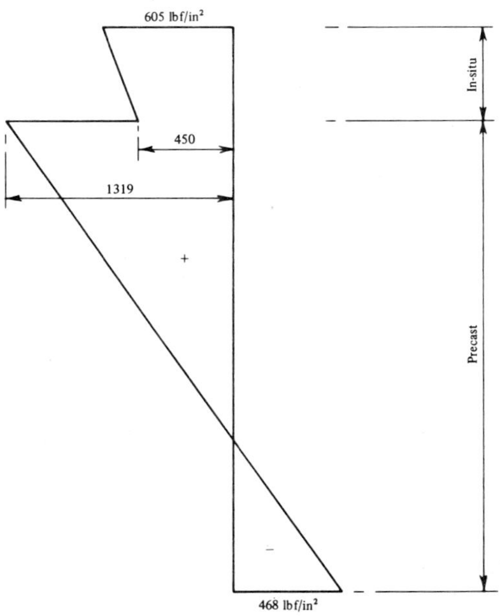

Figure 11.22 Stress diagram under service load (example 11.7.5 – upropped case)

CASE 2: PROPPED CONDITION

Let the critical section be x ft from the support. Since the tensile stress is the governing factor, only the bottom stress (i.e. stress at edge 1) is considered.

Stage 1. Precast unit placed on bearing (Figure 11.23)

Self-weight = 525 lbf/ft (781.3 kgf/m, 7.66 kN/m)

$$M_{x1} = \left(525 \times 27.5x - \frac{525x^2}{2}\right) \times 12 \text{ lbf in.}$$

$$f_{1d} = \frac{M_{x1}}{2020} = (85.5x - 1.56x^2) \text{ lbf/in}^2 \qquad \text{(tension)}$$

Stage 2. Prop added at midspan in such a way that it will prevent deflection at that point due to further loading. No stresses. (Figure 11.24).

Stage 3. In-situ topping cast: No composite action (Figure 11.25)

Weight of added concrete = 320 lbf/ft (476.2 kgf/m, 4.67 kN/m)

Reaction at supports:

$$R_L = \tfrac{3}{16} \times 320 \times 55.0 = 3300 \text{ lbf}$$

$$R_R = 3300 \text{ lbf}$$

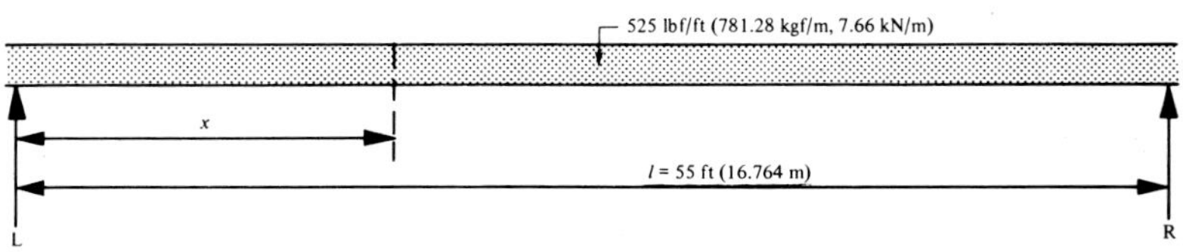

Figure 11.23 Loading due to self-weight (example 11.7.5, case 2)

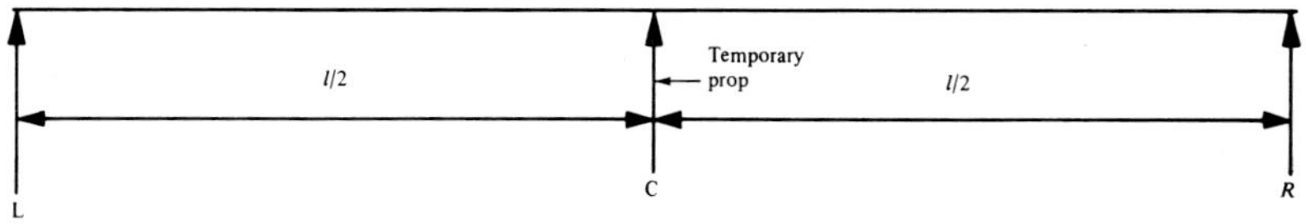

Figure 11.24 Temporary prop placed at midspan (example 11.7.5 – case 2)

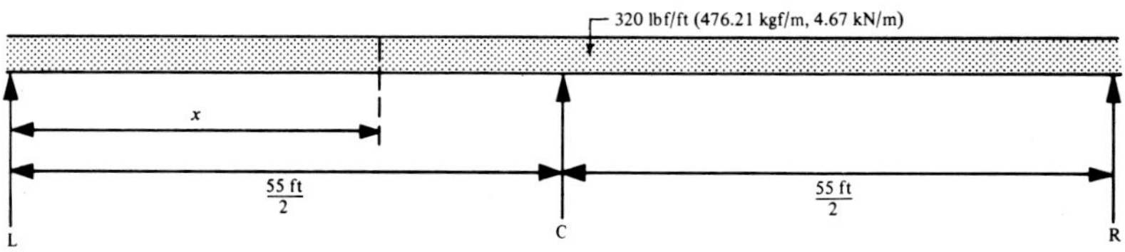

Figure 11.25 Load diagram in stage 3 (example 11.7.5 – case 2)

Reaction at temporary support C = $R_C = \frac{5}{8} \times 320 \times 55.0 = 11\ 000$ lbf

$$M_{x3} = \left(3300x - \frac{320x^2}{2}\right) \times 12 \text{ lbf in.}$$

$$f_{1d}' = \frac{3300 \times 12x}{2020} - \frac{160 \times 12x^2}{2020} = (19.6x - 0.95x^2) \text{ lbf/in}^2 \qquad \text{(tension)}$$

Stage 4. Prop removed when in-situ concrete is matured: composite action between precast and in-situ concrete. (See Figure 11.26)

$$M_{x4} = (5500\,x) \times 12 = 66000x \text{ lbf in.}$$

$$\therefore f_{1d}'' = \frac{66000x}{2560} = 25.8x \text{ lbf/in}^2 \qquad \text{(Tension)}$$

Stage 5. Superload operative (See Figure 11.27)

Added load as in Case 1 = 992 lbf/ft (1476 kgf/m, 14.48 kN/m)

$$M_{x5} = \left(992 \times 27.5x - \frac{992x^2}{2}\right) \times 12$$

$$f_{1a} = \frac{992 \times 27.5 \times 12}{2560}x - \frac{496 \times 12x^2}{2560}$$

$$= (128x - 2.33x^2) \text{ lbf/in}^2 \qquad \text{(tension)}$$

Since tensile stress is the critical factor only the bottom fibre stresses are considered.

Resultant bottom fibre stress (service load and prestress)

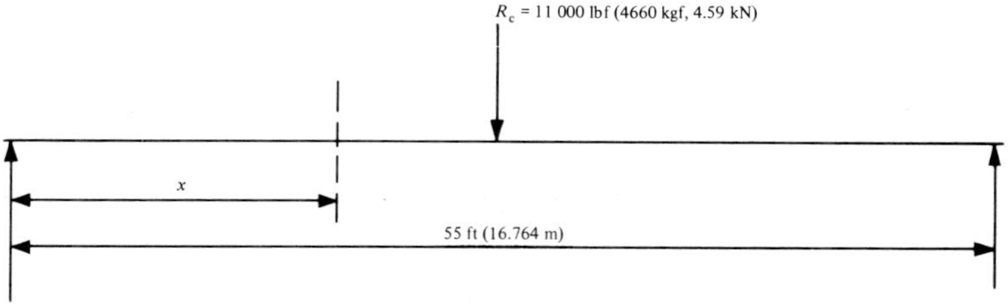

Figure 11.26 Load diagram in stage 4 (example 11.7.5 – case 2)

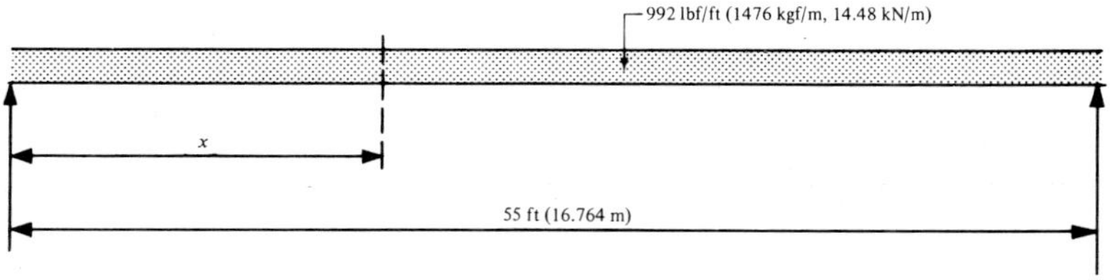

Figure 11.27 Superload stage 5 (example 11.7.5, case 2)

$$f_{1w} = f_{1p} - (f_{1d} + f_{1d}' + f_{1d}'' + f_{1a}) = (0.715 + 0.00246x - 0.000045x^2) \times 4262 - (85.5x - 1.56x^2 + 19.6x - 0.95x^2 + 25.8x + 128x - 2.33x^2)$$

or

$$f_{1w} = -(248.4x - 4.647x^2 - 3050) \text{ lbf/in}^2 \qquad \text{(tensile)}$$

For maximum f_{1w}

$$\frac{d f_{1w}}{dx} = 248.4 - 4.647 \times 2x = 0$$

or

$$x = \frac{248.4}{4.647 \times 2} = 26.8 \text{ ft } (8.17 \text{ m})$$

$$\therefore \text{ maximum } f_{1w} = -(248.4 \times 26.8 - 4.647 \times 26.8^2 - 3050)$$

$$= -(6650 - 3325 - 3050) = -275 \text{ lbf/in}^2 \ (19.34 \text{ kgf/cm}^2,\ 1.90 \text{ N/mm}^2) \qquad \text{(tensile)}$$

Therefore comparing maximum tensile stress obtained in case 1 (unpropped condition) it is evident that addition of prop during casting in-situ concrete improves the serviceability state of design.

REFERENCES

1. RITTER, M. and LARDY, P. Vorgespannter Beton: thorie und Berechung, Schweizerische Versuche und Ausfuhrunger (Prestressed concrete: theory and calculations, Swiss tests and achievements). Zurich Leemann and Co., 1946. pp. 118.
2. SADLER, R. E. Development in overhead electrification of railways as it affects the civil engineer *Proceedings of the Institute of Civil Engineers.* Vol. 12. February 1959. pp. 125–156.
3. HAJNAL-KÓNYI, K. Discussion to the paper 'Fully and partly stressed concrete', by P. W. Abeles, *Journal of the American Concrete Institute.* Vol. 41. 1945. pp. 216–225.
4. ABELES, P. W. *A new bridge deck (the wafer slab) used by British Railways, Eastern Region.* Third Congress F.I.P. Berlin 1958. Session III, paper No. 12 *
5. SAMUELY, F. J. Some recent experience in composite precast and in-situ concrete construction with particular reference to prestressing. *Proceedings of the Institute of Civil Engineers.* Vol. 1. 1952. Pt. 1. No. 30. pp. 222–259.
6. BIRKENMAIER, M. Vorgespannte Ziegalkonstruktionen (Prestressed Clay Construction) *Schw. Bauzeitung* Vol. 68. 1950. pp. 141–144 and 166–68.
7. EVANS, R. H. and PARKER, A. S. Behaviour of prestressed concrete composite beams. *Journal of the American Concrete Institute.* 1955. Vol. 51. pp. 861–881.
8. MATTOCK, A. H. Precast prestressed concrete bridges, creep and shrinkage studies. *Journal of the PCA Research and Development Laboratories.* Vol. 3, No. 2 May 1961. pp. 32–66, Development Department Bulletin D46.
9. HANSON, N. W. Precast prestressed concrete bridges, horizontal shear connections. PCA 1960.
10. EVANS, R. H. and CHUNG, Horizontal shear failure of prestressed composite-T beams with cast-in-situ lightweight concrete deck. *Concrete.* April 1969. pp. 124–126.
11. ABELES, P. W. Restraint and stress redistribution in composite prestressed concrete beams. *IABSE, Congress* Stockholm, 1960, Section V12.
12. ABELES, P. W., BROWN, E. I. and HU, C. H. Tests of composite concrete beams with prestressed planks. *Materiaux et Construction.* Vol. 5. No. 5, 1972. pp. 31–40.

CHAPTER 12

DEFLECTION

12.1 General notes

The deflection of a prestressed beam is the sum of the deformation due to the prestress and that due to the loading. For a given loading and prestress, its magnitude changes with time as a consequence primarily of the creep of the concrete and also, to a lesser degree, the increase in the elastic modulus of the concrete and the gradual reduction of the prestressing force by the losses described in Chapter 6, and these effects should be taken into account in calculating the deflection.

Usually the prestressing steel is placed symmetrically about the centre-line of the beam, and if the prestress is applied at both ends of the cable the prestressing force is also symmetrical. In this case the simplest method of calculation is the moment–area method. If the cables are unsymmetrical, or if the prestress is applied from one end only (leading to frictional losses and a consequent variation in the prestressing force) Macaulay's method can be employed.

In Figure 12.1 is shown the case of a straight cable, with a prestressing force P_t applied at an eccentricity e_s to a simply-supported beam. The bending moment $M_p = -P_t e_s$ causes hogging and the deflection can be obtained by considering the ratio M_p/EI as a load and calculating the central bending moment produced by this fictitious load (Figure 12.1). This and other standard cases are given in Table 12.1.

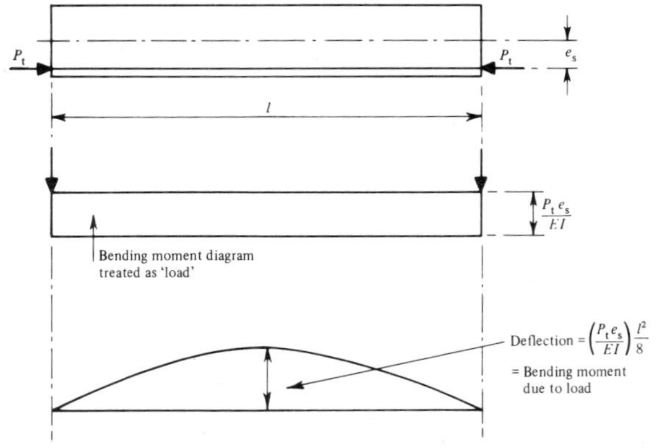

Figure 12.1 Moment-area method

Table 12.1 Deflection due to various profiles of cables

Symmetrical cases

Diagram of prestressing moment	Deflection	M_p for zero camber
	$\frac{M_p l^2}{8EI}$	$M_p = \frac{5M_t}{6}$
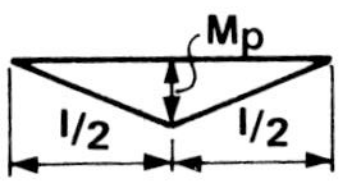	$\frac{M_p l^2}{12\,EI}$	$M_p = \frac{5M_t}{4}$
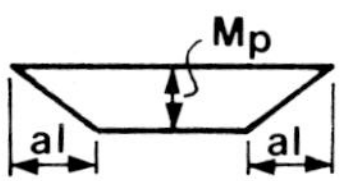	$\frac{M_p l^2\ (3 - 4a^2)}{24\ EI}$	$M_p = \frac{5M_t}{2\ (3 - 4a^2)}$
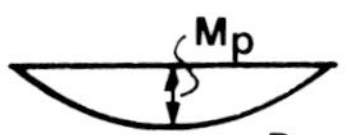	$\frac{5M_p l^2}{48\ EI}$	$M_p = M_t$
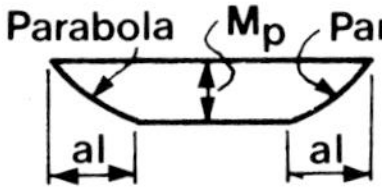	$\frac{M_p l^2\ (3 - 2a^2)}{24\ EI}$	$M_p = \frac{5M_t}{2\ (3 - 2a^2)}$

Unsymmetrical cases

Diagram of prestressing moment	Deflection	M_p for zero central camber
	Central: $\frac{M_p l^2}{16\,EI}$ Maximum: $\frac{M_p l^2}{9\sqrt{3}EI}, \left(\frac{l}{\sqrt{3}} \text{ from } A\right)$	$M_p = \frac{5M_t}{3}$
	Central: $\frac{7M_p l^2}{192\ EI}$ Maximum: $\frac{0.0384\,M_p l^2}{EI}$ (0.63 l, from A)	$M_p = \frac{20M_t}{7}$
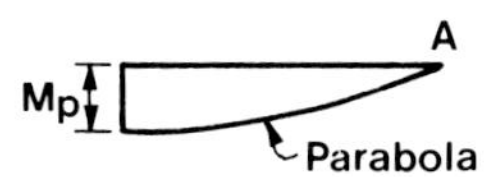	Central: $\frac{17M_p l^2}{192\ EI}$ Maximum: $\frac{9M_p l^2}{100\,EI}$, 0.58 l from A	$M_p = \frac{20M_t}{17}$
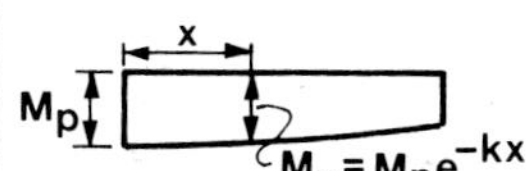	Maximum (varies) ≏ Central: $\frac{M_p}{2k^2 EI}\ (e^{-kl/2} - 1)^2$	$M_p = \frac{5\ k^2 M_t}{24\ (e^{-kl/2} - 1)^2}$

**Zero camber throughout the length of beam is obtained only if the prestressing moment is equal to the bending moment (here assumed parabolic) at all points. Otherwise, formulae gives zero central camber.*

12.2 Unsymmetrical load

Unsymmetrical load may be caused by unsymmetrical cables or by frictional losses in symmetrical cables tensioned at one end. As an example, the case of frictional loss is given in the following; this and other standard cases are included in Table 12.1.

The prestressing force P_{tx} at any point distant x from the end of the beam (Figure 12.2) is given by

$$P_{tx} = P_{t0}\, e^{-kx}$$

in which P_{t0} is the force at the jacking end. The bending moment is therefore

$$M_{px} = -P_{t0}\, e_s\, e^{-kx} = E_c I_c \frac{d^2 y}{dx^2}$$

in which y is the deflection, and by Macaulay's method

$$E_c I_c \frac{dy}{dx} = -P_{t0}\, e_s \left(-\frac{1}{k} e^{-kx} + A \right)$$

and

$$E_c I_c y = -P_{t0}\, e_s \left(\frac{1}{k^2} e^{-kx} + Ax + B \right)$$

When $x = 0$, $y = 0$; hence

$$-P_{t0}\, e_s \left(\frac{1}{k^2} + B \right) = 0 \text{ and } B = -\frac{1}{k^2}$$

When $x = 1$, $y = 0$; hence

$$-P_{t0}\, e_s \left(\frac{1}{k^2} e^{-kl} + Al - \frac{1}{k^2} \right) = 0$$

and

$$A = \frac{1}{k^2 l} \left(1 - e^{-kl} \right)$$

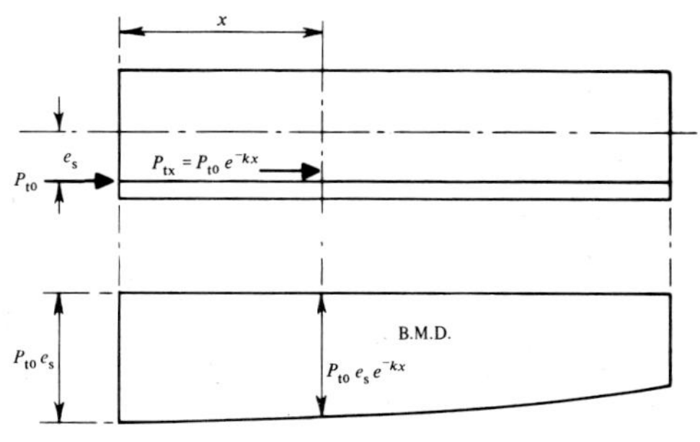

Figure 12.2 Bending moment due to unsymmetrical prestressing force

The upward deflection y at any point x is therefore

$$y = -\frac{P_{to}\, e_s}{E_c I_c}\left(\frac{1}{k^2}\, e^{-kx} + \left[\frac{x}{k^2 l}\,(1 - e^{-kl})\right] - \frac{1}{k^2}\right)$$

$$= \frac{P_{to}\, e_s}{k^2 E_c I_c}\left[(1 - e^{-kx}) - \frac{x}{l}\,(1 - e^{-kl})\right] \quad . \; . \; . \; . \; . \; . \; . \; . \; . \; . \; . \quad (12.1)$$

Since the lack of symmetry is small, the maximum deflection is assumed to occur at midspan; hence

$$y_{max} \triangleq \frac{P_{to}\, e_s}{k^2 E_c I_c}\,[(1 - e^{-kl/2}) - \tfrac{1}{2}(1 - e^{-kl})]$$

$$= \frac{P_{to}\, e_s}{2k^2 E_c I_c}\,(1 - e^{-kl/2})^2 \quad . \; . \; . \; . \; . \; . \; . \; . \; . \; . \; . \; . \; . \; . \quad (12.2)$$

As a numerical example consider the case when $l = 50$ ft and $k = 1 \times 10^{-3}$ per ft length

The maximum deflection is then

$$y_{max} = \frac{P_{to}\, e_s}{2 \times 10^{-6}\, E_c I_c}\,(1 - e^{-0.025})^2$$

$$= \frac{P_{to}\, e_s}{2 \times 10^{-6}\, E_c I_c}\,(1 - 0.9753)^2 = \frac{305\, P_{to}\, e_s}{E_c I_c} \text{ upward}$$

The effect of friction on the deflection is usually negligible. In the present case, ignoring friction, the calculated upward deflection would then be

$$\frac{P_{to}\, e_s\, l^2}{8 E_c I_c} = \frac{312.5\, P_{to}\, e_s}{E_c I_c}.$$

The deflection due to the prestressing force is counteracted by that due to the weight of the beam. When this is uniformly distributed the corresponding downward deflection is $-\dfrac{5}{384}\dfrac{wl^4}{E_c I_c}$.

12.3 Variation of deflection with time

The elastic modulus E_c depends greatly on the strength of the concrete. The values given in CP 115 are 4×10^6, 5×10^6, 5×10^6, and 6×10^6 lbf/in^2 (2.8×10^5; 3.5×10^5 and 4.2×10^5 kgf/cm^2 or 2.7×10^4; 3.4×10^4 and 4.1×10^4 N/mm^2) for concrete with cube strengths of 4000, 6000, and 8000 lbf/in^2 (280, 420 and 560 kgf/cm^2 or 27, 41 and 55 N/mm^2) at transfer respectively. Sometimes the strength is less than that specified, and the elastic modulus is then reduced also. With certain aggregates, such as sandstone, chalk, and lightweight aggregates, the elastic modulus is appreciably less than the values given in the Code.

The deflection will increase in the course of time, by an amount which depends largely on the ratio of strength at transfer to that at 28 days; the greater the ratio the smaller is the effect of creep. This normally increases the deflection by about 25 to 50 per cent and in unfavourable circumstances by 100 per cent or even more, when the prestress is applied at a very early age. On the other hand, the counter-action of the live load occurs usually at a later stage when the concrete has matured and the elastic modulus has increased. In this case, the effect of creep is likely to be small.

Tests described by Walley* have shown that the increase of deflection in a cracked beam, even with large and sustained loads, is not excessive; a load of 85 per cent of the ultimate load, sustained for 282 days, produced an increase of deflection of less than 100 per cent. Similar results were obtained during tests by the German Federal Railways on cracked masts, which were loaded so as to produce a nominal tensile bending stress of 1500 lbf/in^2 (105 kgf/cm^2; 10 N/mm^2). In this case, however, the duration of load was somewhat limited. Hence it is clear that there is little increase of deflection when the load is applied after the concrete has aged. On the other hand, early loading, particularly if the prestress is applied too soon, may cause excessive deflection.

An approximate indication of the probable increase in deflection with time can be obtained by means of a method given in CP 115. In this method it is assumed that the immediate deflection at transfer is elastic, and is proportional to the elastic modulus at transfer E_{ct}; this in turn is obtained from Figure 12.3. By the time the working load is applied however, the increase in the elastic modulus and the effect of creep will have caused the deflection to change. The ratio of the total deflection to the original deflection can be seen from Figure 12.3; it is given by the formula

$$\frac{\left(\dfrac{1}{E_{c\infty}} + \epsilon_{c\infty}\right)}{\dfrac{1}{E_{ct}}} = \frac{E_{ct}}{E_{c\infty}} (1 + \epsilon_{c\infty} E_{c\infty}), \qquad (12.3)$$

in which $E_{c\infty}$ is the elastic modulus when the working load is applied and $\epsilon_{c\infty}$ is the creep coefficient described in Chapter 6. For example, if $E_{ct} = 4.5 \times 10^6$ lbf/in^2 (corresponding to a cube strength of 5000 lbf/in^2), $E_{c\infty} = 6 \times 10^6$ lbf/in^2 (corresponding to 8000 lbf/in^2), and $\epsilon_{c\infty} = 0.33 \times 10^{-6} \times \frac{6000}{5000}$ per lbf/in^2, the ratio is

$$\frac{4.5}{6}\left(1 + 6 \times 10^6 \times 0.33 \times 10^{-6} \times \frac{6}{5}\right) = 2.53$$

that is, the initial deflection may be expected to increase by about 153 per cent before the load is applied. Obviously the method is approximate, and should be regarded as giving an indication only of the increase to be expected. A more realistic approach is to calculate the range of deflections which can be expected, by means of two sets of extreme values for E_{ct}, $E_{c\infty}$, and $\epsilon_{c\infty}$.

*F. W. Walley, *Prestressed concrete design and construction.* H.M.S.O., 1953.

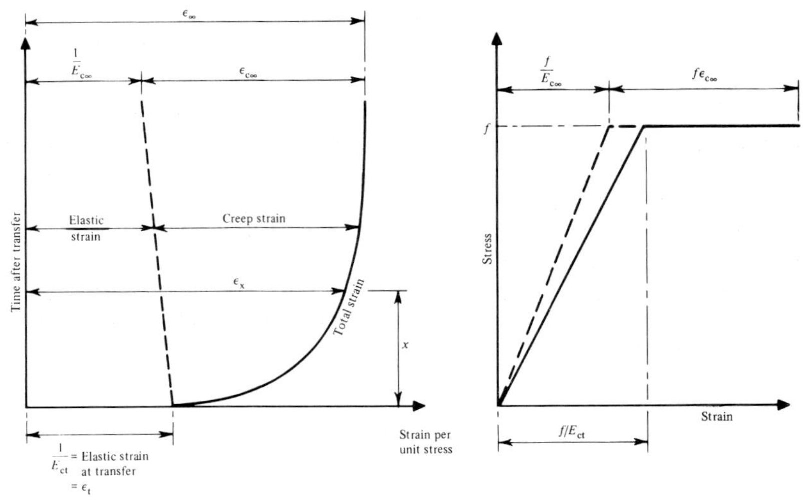

Figure 12.3 Variation of strain with time

12.4 Design for no camber

From the point of view of appearance, it is usually desirable that members such as precast floor units should have equal cambers. This can be achieved only if all units have identical properties, and it is impossible in practice. The major causes of unequal deflections include variations in the elastic modulus, the strength at transfer (and consequently the subsequent deformation due to creep), and the curing history prior to transfer; the effect of small displacements of the prestressing steel can also be appreciable, and this is considered later.

The effects of variations in the properties of the concrete can be minimized by reducing the prestress without decreasing the strength of the concrete. A particular solution is often adopted in which the deflection caused by the prestress is equal and opposite to that due to the weight of the member, thereby eliminating the effect of variations in the elastic modulus.

Formulae for this case, with respect to different cable profiles, are given in Table 12.1. These relate to conditions at transfer; with the passage of time the stress f_{pt} at transfer decreases to $f_{pe} = R_0 f_{pt}$, and the initial elastic modulus is in effect reduced by creep. However, because of the equality between the bending moment M_t and that due to the prestress the latter effect is small, and the reduction of prestress can be allowed for approximately as follows. If it is desired that the combined deflection due to the weight of the member and the effective prestressing moment should be zero, an approximate reduction factor between 1 and $\frac{1 + R_0}{2}$ can be applied; for instance, in the first case of Table 12.1, at one limit

$$\frac{(1 + R_0)}{2} \frac{(P_t e_s - P'_t e'_s)}{8} - \frac{5}{48} \frac{wl^2}{8} = 0 \quad . \quad . \quad . \quad . \quad . \quad . \quad . \quad . \quad . \quad . \quad . \quad . \quad (12.4)$$

and hence $P_t e_s - P_t' e_s' = \frac{5}{3} \times \frac{M_t}{1 + R_0}$. At the other limit, the tabulated value is unchanged.

If it is desired that the camber should be zero when the full dead load is present the calculations become more complicated and the result is less definite, since the effect of the stress at transfer and the subsequent creep must be allowed for. The elastic deflection of a beam may increase by less than 50 per cent or more than 100 per cent, depending on the strength at transfer and the time of application of the additional dead load. If the ratio of the eventual deformation due to dead load to the elastic deformation due to dead load is denoted by λ, and if it is desired that the eventual deflection due to the full dead load is to be zero, then in the first case of Table 12.1, at one limit,

$$\frac{(1 + R_0)}{2} \frac{(P_t e_s - P_t' e_s')}{8} - \frac{5}{48} M_t - \frac{5}{48\lambda} M_{ad} = 0 \qquad (12.5)$$

and hence

$$(P_t e_s - P_t' e_s') = \frac{5}{3(1 + R_0)} \left(M_t + \frac{M_{ad}}{\lambda} \right) \qquad (12.6)$$

At the other limit, the tabulated value is unchanged.

The accuracy of this result depends of course on the accuracy with which λ is predicted; this in turn depends on the accuracy with which the strength at transfer is known.

12.5 Effect of variation in position of prestressing steel

Some tolerance must always be allowed in the positioning of prestressing steel, and in large members this usually has little effect. In shallow members such as prestressed planks however the effect of a tolerance of, for example, $\pm \frac{1}{16}$ in. (1.6 mm) can be appreciable and should be taken into account in the calculations. The effect on the stresses is considered in Examples 11.6.2 and 11.6.3 of Chapter 11; the results obtained are used in the following to calculate the deflection.

12.5.1 Example: Composite floor propped during construction (Figure 12.4)

From Example 11.6.2, Chapter 11, the prestressing force is 47 100 lbf (21 360 kgf; 210 kN) and the eccentricity $\pm \frac{1}{16}$ in. The bending moment due to the prestress is therefore

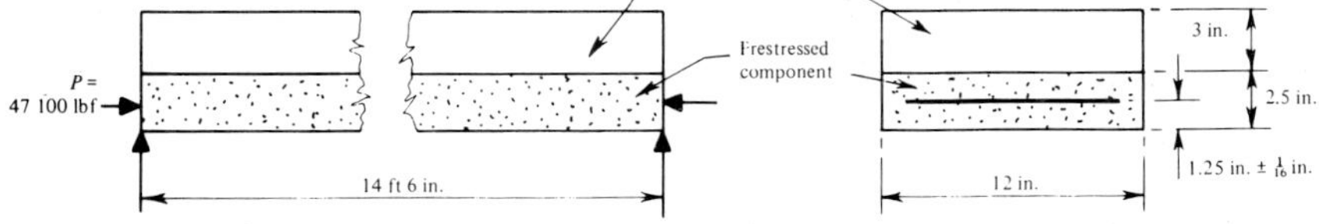

Figure 12.4 Composite floor in example 12.5.1

$\pm 47\,100 \times \frac{1}{16} = 2950$ lbf in. (34 kgf m; 333 Nm)

$I = \dfrac{12 \times 2.5^3}{12} = 15.6 \text{ in}^4$ (649 cm^4) for a cube strength at transfer of 4000 lbf/in^2 (280 kgf/cm^2; 27.5 N/mm^2), $E_{ct} = 4 \times 10^6$ lbf/in^2; $l = 174$ in. The elastic deflection due to the prestress at transfer is therefore $\pm \dfrac{2950 \times 174^2}{8 \times 4 \times 10^6 \times 15.6}$ $= \pm 0.18$ in. (4.5 mm) and this is caused solely by errors of $\pm \frac{1}{16}$ in. in the placing of the prestressing steel.

This can be compared with the elastic deflection due to the weight of the plank (assuming it to be supported initially at each end). The weight of a 2½ in. plank is 32 lbf/ft, and the deflection is therefore

$$\frac{5 \times 32 \times 174^4}{384 \times 4 \times 10^6 \times 15.6 \times 12} = 0.51 \text{ in. (13 mm)}$$

The effect of a tolerance of $\pm \frac{1}{16}$ in the position of the steel is therefore to produce nett deflections ranging from 0.33 in. to 0.69 in. (8 mm to 18 mm).

12.6 Effect of creep

If the time that will elapse between transfer and propping is known, the effect of creep can also be estimated. Assuming for example that this period will be 28 days, and that the creep strain is then one-half of the total creep, then if $E_{ct} = 4 \times 10^6$ lbf/in^2 (2.8×10^5 kgf/cm^2; 2.75×10^4 N/mm^2) (corresponding to a strength at transfer of 4000 lbf/in^2), $E_{c28} = 4.75 \times 10^6$ lbf/in^2 (corresponding to 5500 lbf/in^2 at 28 days), and $\epsilon_{c28} = 0.5 \times 0.33 \times 10^{-6} \times \dfrac{6000}{4000}$ (see Chapter 6), the increase of deflection is

$$\frac{4}{4.75}\left[1 + \left(0.5 \times 0.33 \times 4.75 \times \frac{6}{4}\right)\right] = 1.84$$

The range of probable deflection at 28 days is therefore increased, the values being $1.84 \times 0.33 = 0.61$ in. (15 mm) and $1.84 \times 0.69 = 1.27$ in. (32 mm).

As the prestressed component is considered in this example to be propped during the placing of the added concrete, no further deflection due to this cause can take place. If an attempt is made to eliminate unequal deflections of adjacent planks by adjusting the height of the props, this should be done by raising the units with large deflections rather than by allowing further deflections of those with small cambers; any additional tensile stress will then occur at the top face of the prestressed unit and will not adversely affect the strength.

12.6.1 Example: Composite member without props (Figure 12.5).

From Example 11.6.3, Chapter 11, the prestressing force required is 37 500 lbf (17 000 kgf; 167 kN) and the eccentricity is 0.67 ± 0.06 in.; that is, 0.73 in. or 0.61 in. (18 mm or 15 mm). If $E_{ct} = 5.5 \times 10^6$ lbf/in^2 (3.85×10^5 kgf/cm^2; 3.8×10^4 N/mm^2) and $I = 49$ in^4 (2040 cm^4) the maximum upward deflection at transfer is therefore

$$\frac{37\,500 \times 0.73 \times 174^2}{8 \times 5.5 \times 10^6 \times 49} = 0.39 \text{ in. (10 mm)}$$

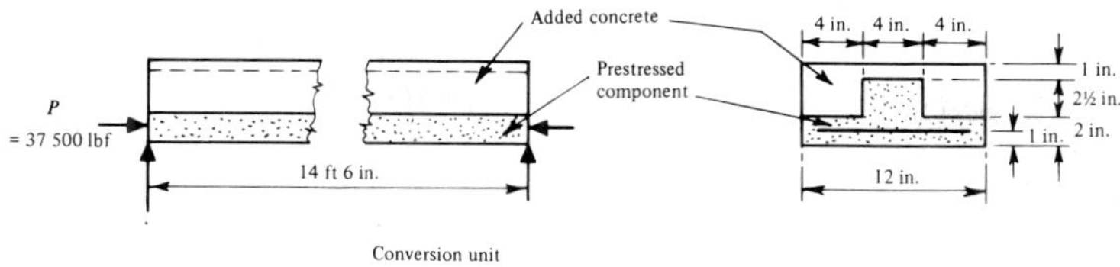

Figure 12.5 Composite section in example 12.6.1

and the minimum is $\frac{0.61}{0.73} \times 0.39 = 0.32$ in. The weight per foot of the prestressed unit is 36.5 lbf; hence the deflection due to M_t is $\frac{5 \times 36.5 \times 14.5 \times 174^3}{384 \times 5.5 \times 10^6 \times 49} = 0.13$ in. There is therefore a nett upward deflection at transfer of between 0.26 in. (7 mm) and 0.19 in. (5 mm).

EFFECT OF CREEP

Assuming again that the additional concrete is placed 28 days after transfer, and that the cube strength is then 8000 lbf/in^2 (560 kgf/cm^2; 55 N/mm^2) (for which $E_{c28} = 6 \times 10^6$ lbf/in^2), then $\epsilon_{c28} = 0.5 \times 0.33 \times 10^{-6}$, and the increase in deflection is $\frac{5.5}{6}$ $[1 + (0.5 \times 0.33 \times 6)] = 1.83$; the deflection immediately before the added concrete is placed will therefore be between 0.48 in. (12 mm) and 0.35 in. (9 mm) upward.

It is apparent that the deflection can vary considerably due to the different influences.

CHAPTER 13

OTHER DESIGN CONSIDERATIONS

13.1 Bending-up of cables

13.1.1 General principle

When the design of a beam requires that counter-action should occur immediately the prestress is applied, the stresses f_{bT}, and f_{tT} at transfer are nominal only and care must be taken that the actual stresses never exceed computed $f_{bt} = f_{bT} - \dfrac{M_{t\,max}}{Z_b}$ and $f_{tt} = f_{tT} + \dfrac{M_{t\,max}}{Z_t}$. In this case f_{bt} or f_{tt} or both, are usually made equal to the maximum permissible value at the section where M_t is also a maximum. It is therefore necessary to ensure that the permissible stresses are not exceeded at sections where M_t is small or zero; this is usually accomplished by bending-up some of the prestressing cables, thereby reducing the eccentricity and hence the prestressing moment. The calculations are slightly more complicated for beams of varying cross-section, and the simpler case of beams with uniform cross-section is considered first.

13.1.2 Beams of uniform cross-section

Figure 13.1 shows the stresses at the top and bottom of a simply-supported beam. To ensure that the permissible values are not exceeded, at the bottom face

$$f_{bt} = \bar{f}_{ct} = \frac{k_b P_t}{A} - \frac{M_t}{Z_b} = \left(1 + \frac{e_s e_b}{i^2}\right)\frac{P_t}{A} - \frac{M_t}{Z_b}$$

and hence

$$e_s = \frac{i^2}{e_b}\left[\left(\bar{f}_{ct} + \frac{M_t}{Z_b}\right)\frac{A}{P_t} - 1\right]$$

For the top face,

$$f_{tt} = -\bar{f}_{tt} = \frac{k_t P_t}{A} + \frac{M_t}{Z_t} = \left(1 - \frac{e_s e_t}{i^2}\right)\frac{P_t}{A} + \frac{M_t}{Z_t}$$

and hence

$$e_s = \frac{i^2}{e_t}\left[\left(\bar{f}_{tt} + \frac{M_t}{Z_t}\right)\frac{A}{P_t} + 1\right]$$

The smaller of these two values gives the greatest permissible value of e_s at any section. If $M_t = 0$ (for example, at the supports of simply-supported beams), then

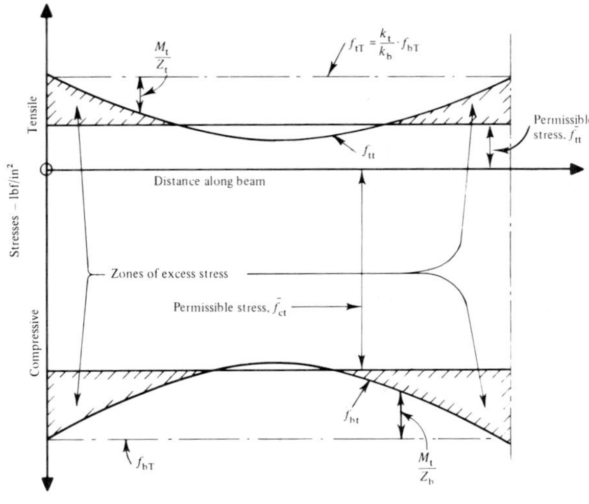

Figure 13.1 Stresses at transfer: beam of uniform section

$$e_s = \frac{i^2}{e_b}\left(\frac{\bar{f}_{ct} A}{P_t} - 1\right) \text{ or } \frac{i^2}{e_t}\left(\frac{\bar{f}_{tt} A}{P_t} + 1\right)$$

In theory an upper limit for e_s can also be obtained. In practice it is never necessary to deflect cables so that the centroid of the prestressing force is higher than that of the beam, and this can therefore be taken as the upper limit. If the effective eccentricity of the prestressing force is between these two limits the permissible stresses at transfer are not exceeded. The method of calculation for a uniform beam is shown in the following example.

13.1.3 Example: Investigate the bending-up of the prestressing steel for the beam designed in Example 9.2.2.

At the centre,

f_{bT} = 2880 lbf/in² (202 kgf/cm²; 19.9 N/mm²)

f_{bt} = 2880 – 882 = 1998 lbf/in² (140 kgf/cm²; 14 N/mm²)

f_{tT} = –888 lbf/in² (62 kgf/cm²; 6 N/mm²)

f_{tt} = –888 + 698 = –190 lbf/in² (13 kgf/cm²; 1.3 N/mm²)

$\bar{f}_{ct}$ = 2200 (154 kgf/cm²; 15.2 N/mm²) and

$\bar{f}_{tt}$ = 190 lbf/in² (13.3 kgf/cm²; 1.3 N/mm²)

At the support,

$$\frac{M_t}{Z_b} = \frac{M_t}{Z_t} = 0$$

Therefore, $f_{bT} = f_{bt} \ngtr 2200$ and $f_{tT} = f_{tt} \ngtr -190$ lbf/in² (13 kgf/cm²; 1.3 N/mm²).

For the first condition

$$e_s \leqslant \frac{i^2}{e_b}\left[\frac{\bar{f}_{ct}A}{P_t} - 1\right] = \frac{223}{24.78} \times \left[\frac{2200 \times 570}{496\ 000} - 1\right] = 13.75 \text{ in. (35 cm)}$$

For the second condition $e_s = 15.8 > 13.75$

Therefore the eccentricity, which is 21.78 in. at the centre should not be more than 13.75 in. at the support. Hence, some of the steel will remain straight and some will be deflected. The prestressing steel comprises 54 wires of 0.276 in. (7 mm) diameter; assuming that the deflected wires will be anchored 3 in. (76 mm) from the top face of the beam, and that the number of bent-up wires is x then

$$[(54 - x) \times 21.78] - (x \times 14.22) = 54 \times 13.75$$

Hence $x = 12$, that is, 12 wires should be deflected and the profile of the deflected wires should be approximately parabolic.

The following alternative method of calculation, which gives similar results, is also possible. From the foregoing analysis, it is apparent that if the maximum prestressing moment is reduced by M_t between the point at which M_t is a maximum and the support, the permissible stresses will not be exceeded. The prestressing moment can therefore be considered as made up of two components. The first, due to the straight wires, is constant; the second, due to the deflected wires, is equal and opposite to M_t at any point. Consider, for example, the beam designed in Example 9.2.4. The total prestressing force at the centre is 543 800 lbf (246 800 kgf; 2420 kN) and $M_{t\ max}$ is 2 510 000 lbf in. (28 900 kgf m; 283 kN m). The greatest distance through which the wires can be deflected is 27 in. and hence, taking moments about point 0 (Figure 13.2), $P \times 27 = M_t$, or

$P = \dfrac{2\ 510\ 000}{27} = 93\ 000$ lbf. This is 17 per cent of the total prestressing force.

The total width of the webs of the section is 8 in., or 22 per cent of the width of the section; therefore the wires immediately under the webs may be deflected to give the required counter-action, the profile of the deflected wires being the same as the shape of the diagram for M_t.

With pretensioning it is obviously more suitable to bend up (or deflect) the tendon in straight lines rather than in any curved profile. It is normally achieved by pressing down the tendon at the centre (see Figure 13.3a) or at two points (Figure 13.3b). If the angle of deviation θ (Figure 13.4) of the cable is small

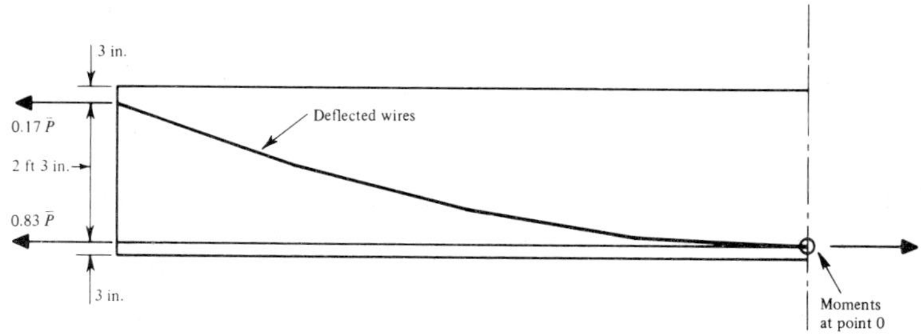

Figure 13.2 Profile of deflected wire

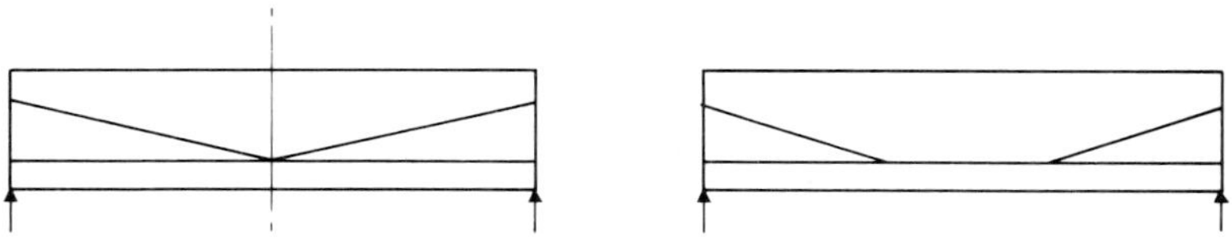

(a) Single point deflection (b) Two point deflection

Figure 13.3 Deflection of pre-tensioned tendons

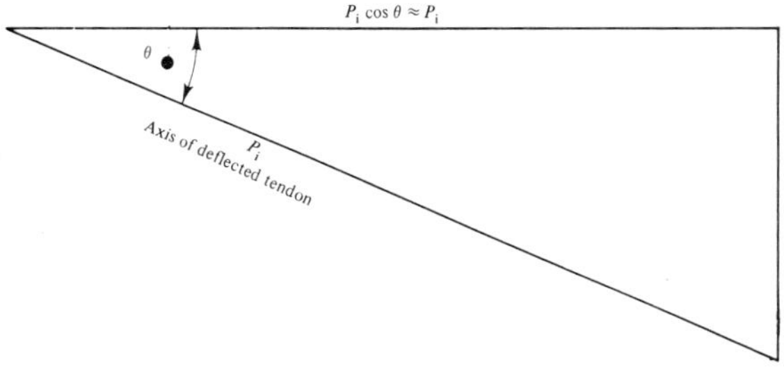

Figure 13.4 Angle of deviation of deflected tendon

then for all intents and purposes the prestressing forces along the axis of cable and the horizontal direction may be considered equal. It is necessary to investigate the maximum tensile stresses in concrete under service load which may govern the distance between the two points where the deflection starts.

For simplicity's sake many designers prefer to provide straight tendons at the bottom with a few added at the top to control the stresses. (This no doubt simplifies the manufacturing process considerably and helps during handling and transport). If necessary some of the straight tendons at the bottom may be debonded.

13.1.4 Beams of varying cross-section

The method is the same for beams of varying cross-section, except that the point of maximum stress can no longer be selected by inspection (see Example 9.2.5). In Figure 13.1 the curves for the maximum values of f_{bT} and f_{tT} are proportional to those for M_t, since Z_b and Z_t are constant; if Z_b and Z_t are not constant (Figure 13.5) it is sufficient in practice to calculate their values, together with those of M_t at several intermediate points in order to be certain that the worst condition has been allowed for. In Figure 13.5 the different shapes of the diagrams for M_t and $1/Z$ and hence of M_t/Z, are shown for a beam of the type considered in Example 9.2.5. It is apparent that the stress at the central section is less than the maximum.

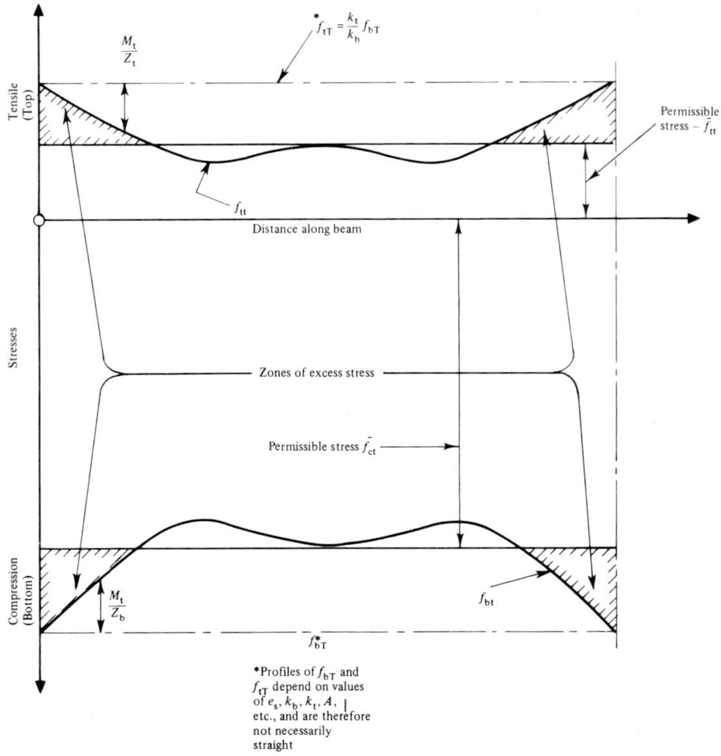

Figure 13.5 Stresses at transfer: beam of varying section

A similar procedure is necessary for the primary design for such a beam, and the additional work required to determine the deflected profiles of the cables is shown in the following example.

13.1.5 Example: Investigate the bending-up of the prestressing steel for the beam designed in Example 9.2.5.

The stresses at points 7 ft apart along the beam, assuming straight cables, are given in Figure 13.6 and plotted in Figure 13.7. If $\bar{f}_{ct}$ = 2400 lbf/in^2 and $\bar{f}_{tt}$ = 575 lbf/in^2, it is apparent that these stresses would be exceeded near the ends of the beam. The calculations for determining the number of wires to be bent-up are as given in the preceding section and are not repeated here.

It should be noted that the length of beam over which the prestressing steel is to be deflected is smaller than in the case of a uniform beam. It is sometimes necessary to terminate cables at intervals throughout this distance, rather than attempt to deflect them through sharp angles.

13.2 Shearing and stirrups

The principal tensile stress caused by shear is much more favourable in a prestressed beam under working-load conditions than in a reinforced concrete

Section	A	B	C	D	E
e_s (in.)*	5	6	7	8	9
A (in²)	64	70	76	82	88
$Z_b = Z_t$ (in³)	190	243	301	362	427
$\frac{K_b}{A} = \left(\frac{1}{A} + \frac{e_s}{Z}\right)$ (in⁻²)	0.0420	0.0390	0.0365	0.0343	0.0324
$\frac{k_t}{A} = \left(\frac{1}{A} - \frac{e_s}{Z}\right)$ (in⁻²)	−0.0108	−0.0104	−0.0102	−0.0099	−0.0097
$f_{bT} = \left(\frac{k_b P_t}{A}\right)$ lbf/in²	3822	3550	3320	3120	2950
$f_{tT} = \left(\frac{k_t P_t}{A}\right)$ lbf/in²	−982	−946	−928	−900	−888
M_t (lbf in.)	0	165 400	288 000	364 000	389 800
$\frac{M_t}{Z}$ (lbf/in²)	0	680	957	1005	913
$f_{bt} = \left(f_{bT} - \frac{M_t}{Z}\right)$ lbf/in²*	3822	2870	2363	2115	2037
$f_{tt} = \left(f_{tT} + \frac{M_t}{Z}\right)$ lbf/in²*	−982	−266	+29	+105	+25

**Values for straight tendons (100 lbf/in² = 7 kgf/cm² = 0.7 N/mm²)*

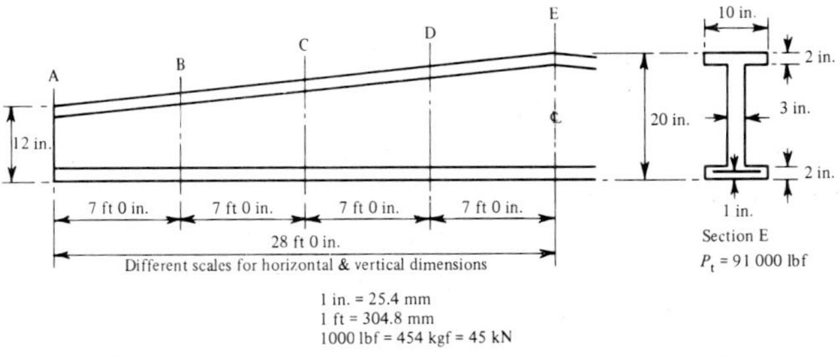

Figure 13.6 Data for bending-up of tendons in a beam of varying cross-section (example 13.1.5)

beam, because of the reduction made by the prestress itself. Consequently, it is usually sufficient to base the primary design of a member on its resistance to bending and then to check that the principal tensile stresses do not exceed the permissible values. On the other hand, care should be taken to ensure that there is a sufficient factor of safety against failure due to shear also. The method of calculating these stresses is given in Chapter 7, and the permissible values are given in Chapter 5. The procedure is illustrated in the following.

13.2.1 Example: Calculate the reinforcement required to resist shearing, for the beam designed in Example 9.2.3.
The relevant dimensions and properties of the member are shown in the upper part of Figure 13.8. The principal stresses at planes a, b, and c at the end of the beam are obtained as shown in the table in the lower part of Figure 13.8. From Chapter 5, reinforcement is required if the principal tensile stress exceeds 175 lbf/in^2 (12.25 kgf/cm^2; 1.2 N/mm^2) at working load or 400 lbf/in^2 (28 kgf/cm^2; 2.75 N/mm^2) at ultimate load.

WORKING LOAD – The maximum principal tensile stress = 225 lbf/in^2 = 1.3 × 175 lbf/in^2. Therefore, from Chapter 5, the shearing force to be resisted by the reinforcement is 0.6 × 76 500 = 45 900 lbf. The area of reinforcement is calculated in the same way as for a reinforced concrete beam. Assuming that ½ in. (12.7 mm) diameter stirrups with four legs are required, the permissible stress being 20 000 lbf/in^2, then the pitch is, in accordance with equation 7.31:

$$s = \frac{A_s f_s z}{V} = \frac{(4 \times 0.196) \times 20\,000 \times 28}{45\,900} = 9½ \text{ in. (24 cm)}$$

COLLAPSE LOAD – The maximum principal tensile stress is 637 lbf/in^2 (44.6 kgf/cm^2; 4.4 N/mm^2) hence reinforcement is required to resist a shearing force of 170 500 lbf at a stress of 80 per cent of the yield stress or 0.2 per cent proof stress. Assuming the yield point of the mild steel reinforcement to be 35 000 lb/in^2 its resistance at ultimate load is

$$\frac{0.8 \times 35\,000}{20\,000} \times 45\,900 = 64\,200 \text{ lbf}$$

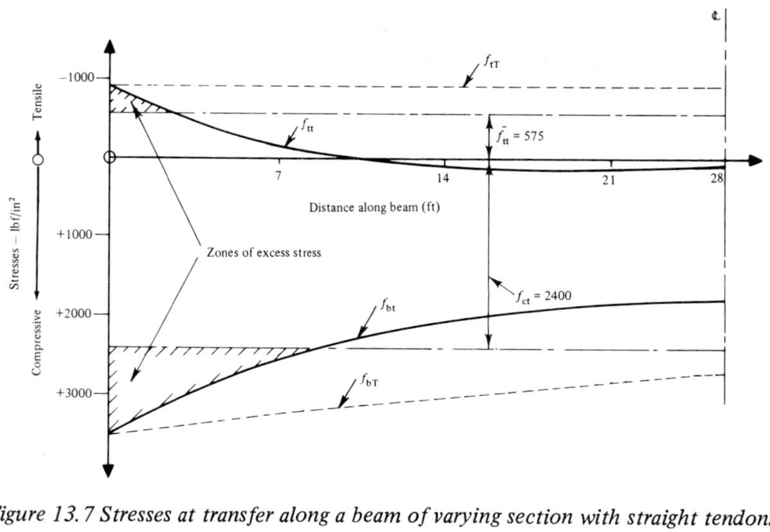

Figure 13.7 Stresses at transfer along a beam of varying section with straight tendons

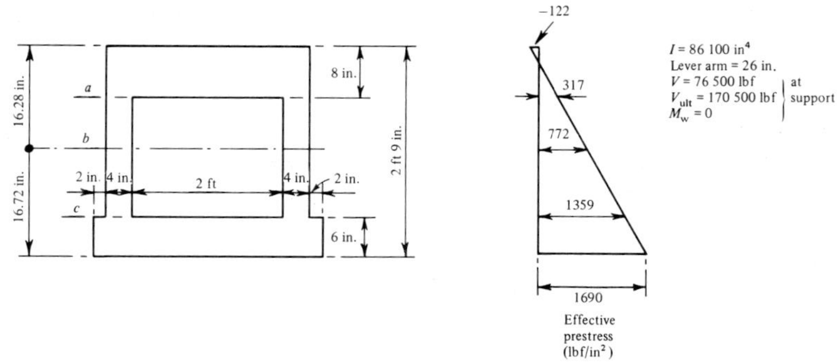

Plane	$\frac{A\,y}{b\,I}$	f_{sh}	$f_{sh\ ult}$	f_{pe} = Effective prestress	f_{pt}	$f_{pt\ ult}$
a	4.56×10^{-3}	349	778	317	−225	−637
b	4.96×10^{-3}	380	848	772	−155	−578
c	4.29×10^{-3}	329	733	1359	−76	−321
	(in^{-2})	lbf/in²				
		100 lbf/in² = 7 kgf/cm² or 0.7 N/mm²				

Figure 13.8 Design for shear (example 13.2.1)

Additional reinforcement is therefore needed to resist

170 500 − 64 200 = 106 300 lbf

Assuming the spacing is 9½ in. (as for the mild steel) and the permissible stress to be 150 000 lbf/in², the area required is

$$A_{sv} = \frac{s\,V}{z\,f_{sv}} = \frac{9.5 \times 106\,300}{28 \times 150\,000} = 0.24 \text{ in}^2 \ (155 \text{ mm}^2)$$

that is, 4 No. 0.276 in. wires (7 mm dia.).

The total reinforcement required to resist the effects of shear at the end of the beam (in addition to that required for the end block) is therefore 4 No. 0.5 in. mild steel bars and 4 No. 0.276 in. high-tensile wires, at 9½ in. spacing.

In the collapse state *v* the shearing stress in the cracked section rather than $f_{sh\ ult}$ (and consequent $f_{pt\ ult}$) is relevant and so reinforcement should be determined accordingly. Also in view of the difficulty of bending hard-drawn wire the use of such wires as stirrups is rare.

If the principal tensile stresses are such that stirrups are not required to resist them (see Chapter 7), it is nevertheless usually desirable to provide nominal stirrups throughout the length of the beam. If it is known at the time of the design that the placing, compacting, and curing of the concrete will be carefully done, under good supervision, there is some practical gain in omitting stirrups since this allows better compaction of the concrete. When it is not certain that these conditions will be fulfilled, nominal stirrups should be provided.

The vertical reinforcement in the end block is, of course, required whether the permissible principal tensile stresses due to shear are exceeded or not. Stirrups may also be required at points where the slope of a cable changes, to resist the vertical component of the force in the cable, and at the points of application of concentrated loads. If the principal stress exceeds the permissible tensile stress, sufficient stirrups should be provided for conditions at ultimate load. Bent-up tendons serve to reduce the shearing force under working load, but are of little value at ultimate load. Some designers favour bending-up of a large proportion of the tendons; this, however, may unfavourably affect the bending stresses when the losses are higher than assumed.

13.2.2 Example: Investigate shearing stress and reinforcement required in the beam designed in Example 9.2.8.

Shear at edge of support **A** (cantilever end) (Service load)

(a) Due to own weight

$$2.60 \times 14.95 + 1.53 \times 14.34 + 1.1 \times \frac{14.34}{2}$$

$$= 38.9 + 21.9 + 7.87 = \quad 68.67 \text{ kN}$$

(b) Due to permanent service load

$$0.98 \times 14.95 = \quad 14.65 \text{ kN}$$

(c) Due to snow load

$$1.45 \times 14.95 = \quad 21.70 \text{ kN}$$

$$\text{Total} = 105.02 \text{ kN}$$

Bending moment at support B (Service load)

(a) $\dfrac{2.60 \times 3.05^2}{2} = 12.10 \text{ kN m}$

(b) $1.53 \times \dfrac{1.83^2}{2} = 2.56 \text{ kN m}$

(c) $0.98 \times \dfrac{3.05^2}{2} = 4.55 \text{ kN m}$

Total 19.21 kN m

(d) $1.45 \times \dfrac{3.05^2}{2} = 6.75 \text{ kN m}$

Total 25.96 kN m

Maximum bending moment at the edge of support **A** (span-side) (Service load)

$$25.96 + \frac{(788.2 - 25.96)}{5.5} \times 5.20 - \left(20.1 \times 0.3 - \frac{6.56 \times 0.3^2}{2} - \frac{1.10 \times 0.3^2}{2}\right)$$

$$= 25.96 + 720 - 6.04 + 0.35$$

$= 735.96 - 5.69$

$= 739.27$ kN m

$$\text{Maximum shear} = \frac{(788.2 - 25.96)}{5.5} + [20.1 - (6.56 + 1.19) \times 0.3]$$
$$= 156.30 \text{ kN}$$

(See also Chapter 9 page 240)

Shear check (to CP 110 – formula 7.29)
Ultimate load = $1.4\, G_K + 1.6\, Q_K$
(Referring to Figure 9.18 example 9.2.8, page 239)
Maximum M_A due to ultimate load = $(505.20 + 114) \times 1.4 + 169 \times 1.6$
$= 1137.280$ kN m

Minimum $M_B = 25.96$ kN m

Ultimate shear at A (span A – B)

$$V = \frac{(1137.28 - 25.96)}{5.50} + (2.60 + 1.53 + 0.98) \times 1.4 \times \frac{5.5}{2} + (\tfrac{1}{2} \times 1.10 \times 5.5)$$

$$\times 1.4 \times \frac{2}{3} + 1.45 \times 1.6 \times \frac{5.5}{2} = 230.9342 \text{ kN}$$

A_{st} = area of tensile steel = $12 \times 112 = 1344$ mm^2
b = minimum width = 127 mm
d = 990 mm

$$\therefore \frac{100 A_{st}}{b.d} = \frac{100 \times 1344}{127 \times 990} = 1.069 \approx 1$$

$\therefore$ v_c (for lightweight concrete) = 0.60 N/mm^2 (Table 25 of CP 110).

Losses of prestress = 32% of initial prestress
Initial prestress = 70% of f_{sk} or f_{su}
$\therefore$ f_{pe} (effective prestress) = $[(1 - 0.32) \times 0.70] f_{su} = 0.476 f_{su}$

$$\therefore \left(1 - 0.55 \frac{f_{pe}}{f_{su}}\right) . v_c . b.d = (1 - 0.55 \times 0.476) \times 0.60 \times 127 \times 610 = 34\,313 \text{ N}$$

(only rib considered – curved portion of flange ignored in taking d here).

At y = 426 mm

$$f_{yE} = \frac{(13.19 + 0.85)}{1370} \times 990 - 0.85 = 9.2957 \text{ N/mm}^2$$

$$\therefore M_o = 0.8 f_{yE} . \frac{I}{y} = 0.8 \times 9.2957 \times \frac{4\,310\,000 \times 10^4}{426}$$

$$= 752.3844 \times 10^6 \text{ N mm} = 752.3844 \text{ kN m}$$

$$\therefore V_{cr} = 34\,313 + \frac{M_o \times V}{M} = 34\,313 + \frac{752.3844}{1137.28} \times 230\,934.20$$

$= 187\,100$ N $< V$ $\therefore$ Reinforcement necessary.

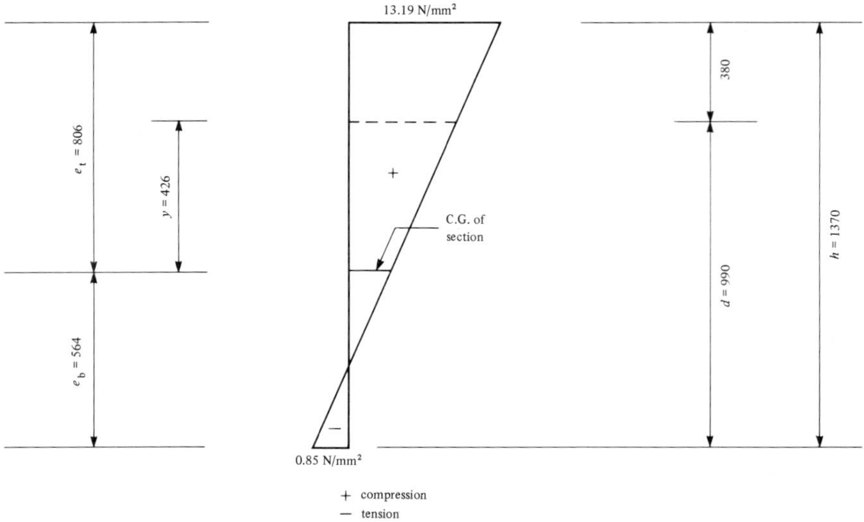

Figure 13.9 Stress diagram at support section due to prestress only after all losses

From equation 7.30:

$$\frac{A_{sv}}{s} = \frac{(V - V_{cr})}{0.87 f_y d^*} = \frac{(230\,934 - 187\,100)}{0.87 \times 250 \times 950^*} = 0.2121$$

f_y for mild steel = 250 N/mm^2

(*In calculating shear reinforcement in accordance with this formula for rectangular [also nearly rectangular] members, CP 110 permits value of 'd' to be taken as the distance from the extreme compression fibre [in this case the soffit of the rib] to the corner longitudinal steel in the tensile zone round which the stirrups pass.)

Assuming s = 150 mm
A_{sv} required = 0.2121 × 150 = 31.85 mm^2
Two legs of 6 mm stirrups provide 56.54 mm^2.

However, if adequate factor of safety is obtained by checking in accordance with formula (7.37a) as shown in the following, only nominal transverse reinforcement would suffice.

Check by formula (7.37a)

$$M = 0.875\, d\, l_s \left(0.342 \times b_1 + 0.3 \times \frac{M_R}{d^2} \sqrt{\frac{z}{l_s}}\right) \times K_1$$

where $K_1 = 4\sqrt{\dfrac{16.66}{\phi f_{su}}}$

$$l_s = \text{Shear span} = \frac{M}{\text{shear}} = \frac{739.27}{156\ 30} = 4.74 \text{ m} = 4740 \text{ mm}$$

d = 990 mm

$\frac{l_s}{d} = 4.788 \therefore$ the formula is applicable

b_1 = say 200 mm

z = 770 mm

Area of tensile steel = $12 \times 112 = 1344 \text{ mm}^2$

Area of concrete up to effective depth

= Total area − $(1370 - 990) \times 125$

= $282\ 000^* - 47\ 500 = 234\ 500 \text{ mm}^2$

$$\therefore \quad \phi = \frac{1344}{234\ 500} = 5.74 \times 10^{-3}; f_{su} = 1.87 \times 10^3 \text{ N/mm}^2$$

$$\therefore \quad K_1 = \sqrt[4]{\frac{16.6}{5.74 \times 10^{-3} \times 1.87 \times 10^3}} = \sqrt[4]{1.55} = 1.12$$

$$\therefore \quad M = 0.875 \times 990 \times 4740 \times \left(0.342 \times 200 + 0.3 \times \frac{1940 \times 10^6}{990^2} \times \sqrt{\frac{770}{4740}}\right) \times 1.12$$

$$= 4.1 \times 10^6 \times 308.4 \times 1.12$$

$$= 1420 \times 10^6 \text{ N mm}$$

$$= 1420 \text{ kN m}$$

$$\therefore \quad \text{Factor of safety} = \frac{1420}{739.27} = 1.92 \text{ O.K.}$$

On the cantilever side

$$l_s = \frac{756.2}{105.02} = 7.2 \text{ m } 7200 \text{ mm}$$

Therefore M is much greater. Adequate safety factor is ensured.

13.3 Stability problems

13.3.1 General notes

Buckling in columns has so far not been of great importance with prestressed concrete because the prestressing force itself does not cause buckling provided that the tendons are bonded or at least gripped at definite distances. However, with further development of prestressed concrete, structures of lesser thickness are being produced and similar conditions as with steel structures may occur. In the present section three cases are investigated.

*See Chapter 9 Page 241.

Stability problems may occur,

(a) as instability with individual members,
(b) as overall instability (with entire structures, in particular with a structure which is composed of individual, not well connected members) and
(c) as a combination of (a) and (b).

Reinforced and prestressed concrete were previously less subject to instability because they were generally not so slender as e.g. steel structures, and stability examinations were made only for thin plates and shells, but gradually more slender constructions are being designed. It seems therefore important to deal with these problems briefly. The lateral stability is important during handling, transport and erection of prestressed concrete members as usually they are weakest at that time in the lateral direction.

13.3.2 The basic theory

The column buckling theory can be taken as the basis for all stability investigations. The well-known classical Euler formula (13.1) is based on the elastic conditions of bending and compression, as illustrated in Figure 13.10 in which the Euler's buckling load N_E is plotted against the slenderness ratio l/i

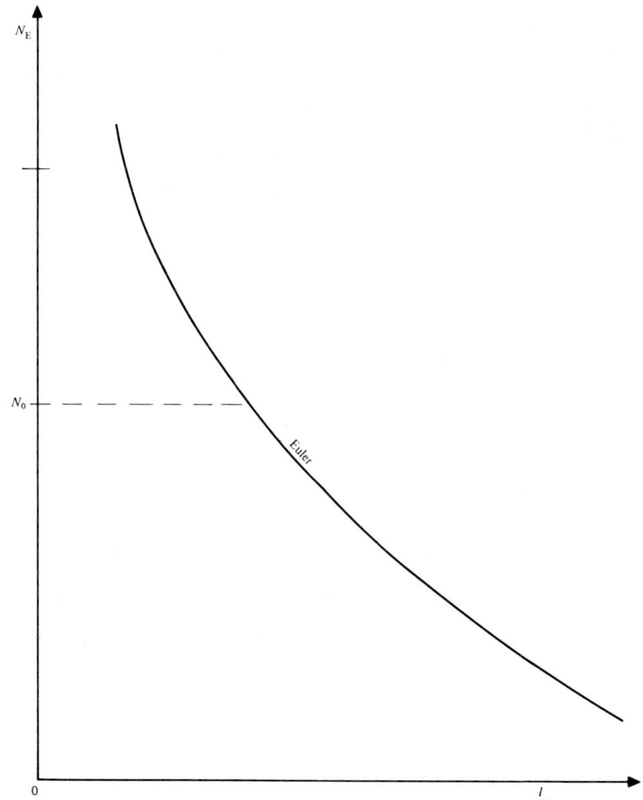

Figure 13.10 Ideal (Euler) buckling diagram

$$N_E = \frac{\pi^2 EI}{l^2} = \frac{k_0}{\left(\frac{l}{i}\right)^2} \quad (13.1)$$

This formula applies only to slender columns. For a stocky column, the failure load is N_0 as indicated in Figure 13.10. If the parameter $\bar{N}_0 = \frac{N}{N_0}$ is introduced, the simpler graph 13.11 with dimensionless co-ordinates can be set-up with the abscissa.

$$\bar{\lambda} = \frac{N_0}{N_E} = k\frac{l}{i}$$

In this case the failure load ratio $\bar{N}_0$ up to $\bar{\lambda} = 1$, and the elastic Euler load N_E for higher values of $\bar{\lambda}$ is the upper boundary of the failure load. As a possible practical lower limit, the Rankine formula as modified by Merchant(1) may be considered.

$$\frac{N}{N_0} + \frac{N}{N_E} = 1 \quad (13.2)$$

These considerations are based on a study by Barta(2) of University College, London which was presented at the Column Colloquium for Steel Structures; but it is of basic importance, since it presents a simple solution, which can be applied also to other materials, and only a few research data are necessary, as shown in the following.

Merchant(1) suggested that the special form of the Rankine formula $\bar{N}_0 + \bar{N}_E = 1$ should be taken as the practical lower limit, as shown in Figure 13.12

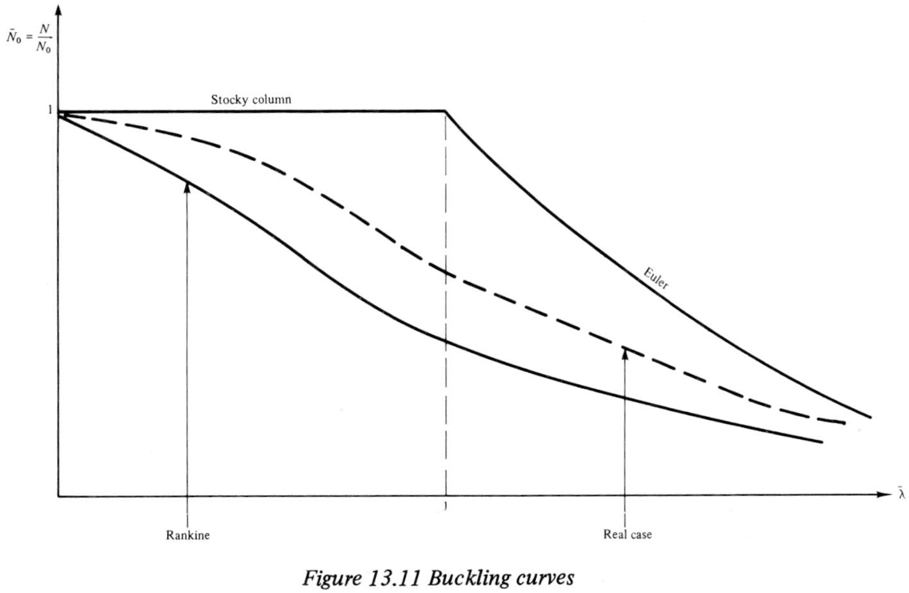

Figure 13.11 Buckling curves

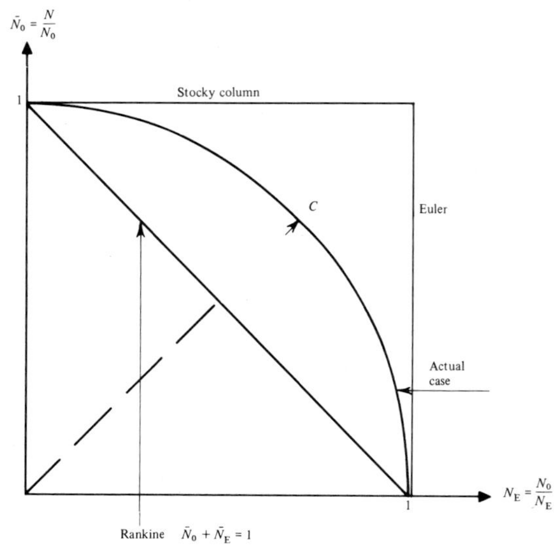

Figure 13.12 Buckling interaction curves

according to Barta. This presents a very simple solution for explanation of the problem showing the upper limit: $(\bar{N}_0 - 1)(\bar{N}_E - 1) = 0$, the lower limit $\bar{N}_0 + \bar{N}_E - 1 = 0$ and between the equation of the actual column formula passing through the point 'c' which depends on the degree of imperfection sensitivity. It is therefore only required to carry out relatively few tests to obtain the actual buckling curve. For the symmetrical steel section the equation for the relation covering the actual case is $(\bar{N}_E + \bar{N}_0) + (1 - c)\,\bar{N}_0\bar{N}_E = 0$. This equation, plotted also in Figure 13.11, can be expressed by

$$\bar{\lambda} = \frac{1 - \bar{N}_0}{\bar{N}_0\,[1 - (1 - c)\,\bar{N}_0]}$$

For reinforced concrete, research is in progress at University College, London.

13.3.3 Lateral stability

In the first edition the following was stated about lateral stability:

> 'In the British Code C.P. 115 it is specified that when the ratio of the span to the minimum breadth exceeds 30, the stresses may have to be reduced, unless stiffeners are provided or adequate lateral supports are provided. An example of a design incorporating stiffeners is that of the roof beams of Victoria Station at Sheffield, in which the ratio of span/breadth of compressive flange is 52. In this structure, a canopy is suspended from beams which have no lateral support. On the other hand, no stiffeners were provided for the purlins at Yarmouth Station, which have a span of 40 ft. The roofing consists of corrugated asbestos sheeting, which affords no lateral support, and the ratio of span to breadth in this case is as high as 80; yet this roof was built in 1953 and has proved to be entirely satisfactory. Thus the provision of stiffeners or other means of improving the lateral strength does not seem to be essential even for span/breadth ratios of up to 80, provided that high-strength

prestressed concrete is used. The design of these purlins forms the basis of Example 9.2.1.'

In the Code of Practice CP 110 no definite limitation regarding the span to width ratio is given, but the designer is advised to investigate the problem.

For homogeneous elastic materials like steel the problem of lateral buckling has been extensively studied both theoretically and experimentally and accurate analytical solutions are possible for evaluation of critical loads causing elastic instability.

Such solutions may be adopted for application to prestressed concrete beams with reasonable accuracy as long as it can be ensured that the member is uncracked and elastic-rigid. Even then perhaps it is advisable that a higher than usual factor of safety (not less than three) should be kept since some of the basic material properties (like modulus of elasticity, shear modulus, etc.) used in the formulae cannot be accurately determined for concrete.

For a prismatic beam of narrow rectangular cross-section, the critical buckling load can be expressed, according to Timoshenko(3), by the following general equation:

$$W_{cr} = \frac{K\sqrt{B_1 C}}{l^2} \qquad \text{(13.3)}$$

where

W_{cr} = critical load (total) causing buckling

K = a numerical constant the value of which depends on the nature of support (i.e., simply supported or cantilever or fixed-end) and the nature and position of the load (i.e., whether a point load or uniformly distributed load).

B_1 = flexural rigidity about Y.Y axis $\left(\text{i.e. } \frac{h\,b^3}{12} E_c \text{ for a rectangular section}\right)$

C = torsional rigidity of the section

$= h\,b^3 \left[\frac{1}{3} - 0.21\frac{b}{h}\cdot\left(1 - \frac{b^4}{12\,h^4}\right)\right]. G$ for a rectangular section.

G = modulus of rigidity of concrete = $\dfrac{E_c}{2(1+\nu)}$

ν being the poisson's ratio

l = span

In reference (3) the following values of 'K' have been worked out for different conditions:

Case 1: Cantilever beams (rectangular)

(a) $K = 4.013$ if W_{cr} is a point load at the end of the cantilever

(b) $K = 12.85$ if W_{cr} is the total load uniformly distributed for the full length.

Case 2: Simply supported beam (rectangular)

(a) $K = 16.93$ for the point load at midspan

$= 19.04$ for the point load at 0.31 from support

$= 24.10$ for the point load at 0.25 from support

(b) K = 28.3 if W_{cr} is the total load uniformly distributed for the full length.

Case 3: Fixed-end beam (rectangular)
$K = 26.6$ for the point load at midspan

The above values are true as long as the stress level remains within the limit of proportionality. Otherwise E value (and consequently G) should be reduced to take account of varying stress. In the case of concrete even within the elastic range E should be taken as modified by the creep co-efficient.

In the equation for critical buckling load it has been assumed that the load is acting at the centroid of the section. If the position of the load happens to be above the centroid, the magnitude of the critical load will be reduced. Conversely, if the position is below the centroid the magnitude of the critical load will be increased.

The equation (13.3) is also applicable to I section when appropriate B_1 and C are introduced. The numerical values of the constant 'K' however, are quite different. For symmetrical I section values of K for various ratios of

$$\frac{l^2}{a^2} \qquad \left[\text{where } a^2 = \frac{B_{f1} \cdot h^2}{2C} = \left(\frac{\text{flexural rigidity of one flange} \times h^2}{2C}\right)\right]$$

have been reproduced from reference (3) for simply-supported and fixed-end beams in Table 13.1 and for cantilever beams in Table 13.2.

$$B_{f1} = \frac{(d_{ft} \times b_t^3) \cdot E_c}{12}$$

Table 13.1 Values of K for simply-supported and fixed-end beams

$\frac{l^2}{a^2}$	Values of K			
	Simply-supported beam		Fixed-end beam	
	Point load at mid span	*u.d.l.*	Point load at mid span	*u.d.l.*
0.40	86.40	143.00	268.00	488.00
4.00	31.90	53.00	88.80	161.00
8.00	25.60	42.60	65.50	119.00
16.00	21.80	36.30	50.20	91.30
24.00	20.30	33.80	43.60	–
32.00	19.60	32.60	40.20	73.00
48.00	18.80	31.50	–	–
64.00	18.30	30.50	34.10	–
80.00	18.10	30.10	–	–
128.00	–	29.40	30.70	55.80
160.00	17.50	–	–	–
200.00	–	29.00	29.40	53.50
240.00	17.40	–	–	–
280.00	–	28.80	–	–
320.00	17.20	–	28.40	–
360.00	17.20	28.60	–	–
400.00	17.20	28.60	–	51.20

Table 13.2 Values of K for cantilever beams with a point load at the end

$\frac{l^2}{a^2}$	Value of K
0.10	44.30
1.00	15.70
2.00	12.20
4.00	9.76
8.00	8.03
16.00	6.73
24.00	6.19
32.00	5.87
40.00	5.64
above 40.00	$\dfrac{4.01}{\left(1-\frac{a^2}{l}\right)}$

The torsional rigidity C of a symmetrical I section can be assumed as

$$C = 2.G.\left[b_t . d_{ft}^3\left(\tfrac{1}{3} - 0.21\frac{d_{ft}}{b_t}\right) \times \left(1 - \frac{d_{ft}^4}{12.b_t^4}\right) + \tfrac{1}{3}(h - 2.b_t).b_w^3\right]$$

Because of symmetry $b_t = b_b$ and $d_{ft} = d_{fb}$ (see figure on page xx).

Another formula of approximation (13.4) is shown, as derived by Csonka[(4)(5)], giving the critical elastic uniformly distributed load g_k according to Figure 13.13:

$$g_k = \frac{EI_y e}{l^4}\,\frac{960}{3 - 30c^2 + 40c^3 - 5c^4} \qquad (13.4)$$

This results for lifting from the two ends in

$$g_k = 120\frac{EI_y e}{l^4} \qquad (13.5)$$

and for lifting from the centre in

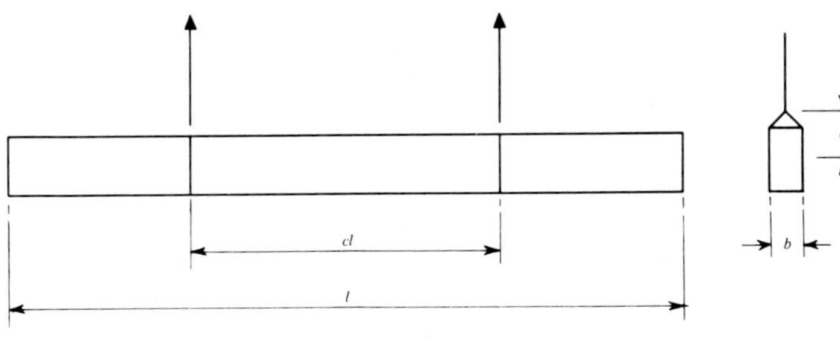

Figure 13.13 Lateral stability during lifting of beams

$$g_k = 320 \frac{E I_y e}{l^4} \qquad (13.6)$$

This shows that obviously lifting from the ends is preferable and an increase in the eccentricity *e*, also increases the safety. This formula ignores the other terms of the series. When '*e*' is sufficiently large, other terms may not be discarded. Therefore increase of '*e*' increases the safety only to a limited extent.

References may also be made to a paper by Anderson(6) on the subject and on the correspondence by Swann to this paper(7).

Finally a paper by Mehlborn(8) should be mentioned in which a method of approximation is described to estimate the lateral stability of slender members with an example of the trough shaped 'Kalle' bridge in Wiesbaden made of precast prestressed lightweight concrete which is discussed in Chapter 15.

13.3.4 Dynamic instability as a consequence of resonance

Reference has already been made in Chapter 3 that the imposed frequency of fatigue or vibration should not be equal to the natural frequency. If this cannot be avoided artificial damping is necessary.

13.3.5 Instability of building constructions due to insufficient connection between precast members

In the British Code CP 116 it has been expressly stated that construction composed of precast members should be connected in such a manner that they are equivalent to a monolithic construction. Generally speaking, the construction produced from individual precast members well connected in all directions to frame systems should be as good as a well designed construction where the entire structure is built in-situ. This is shown in Chapter 14. However, satisfactory connection between the assembled members to form a certain degree of rigidity is essential to obtain a stable structure. An example of a construction which was unsatisfactory and unstable was the case of Ronan Point(9) where a partial collapse occurred owing to an internal explosion. The individual members were not properly connected and the whole construction failed without any resistance to lateral forces, and part of it collapsed like a pack of cards.

Consequent to this event the consideration of 'progressive collapse' was introduced in the Building Regulations.

13.4 Special thin-walled members

The classical theory of design of thin-walled members is based on a number of assumptions: (1) plane sections remain plane, which excludes warping (see also Chapter 7); (2) the cross-section does not deform which excludes distortion, and (3) uniform stress distribution which excludes shear lag.

Generally the effect of non-uniform shear-stress distribution is considered to be negligible. This is, however, incorrect with thin-walled structures, e.g. box shaped sections as indicated in Figure 13.14. According to Oden(10) experiments have shown that the true normal stress distribution lags behind that given by the elementary theory. This deviation from the elementary theory is known as *shear lag*; the phenomenon is also referred to as stress diffusion. According to Saint Venant's principle the stress is uniform on sections sufficiently far removed from the point of load. A certain distance is required for the loads, including the reactions, to 'diffuse' into a uniform stress distribution. Consequently the intensity of stress is substantially higher at the longitudinal edges than in the centre as indicated in Figure 13.14.

Such problems occurred first with aircraft constructions and appeared later in civil engineering with steel structures.

They are also of importance with prestressed concrete where thin-walled constructions are being introduced.

In fact this problem occurs also with a construction composed of thin slab and webs which may require stiffening. The problem of the co-operation of the slab in T-beams is a simple example for this.

13.5 Fulfilment of design assumptions

When preparing a design, it is necessary to ensure that the conditions assumed to occur at transfer and during transport can in fact be obtained. If pre-tensioned steel is used, cracks may occur if counter-action is assumed by the designer but is not obtained in practice. Counter-action should not always be assumed to occur automatically in members with pre-tensioned steel. Even if counter-action is not relied on, it is still possible that cracks occur if the prestressing force is fully applied at the top of the beam before any is applied at the bottom, and at the same time the bottom sticks to the mould. This can be overcome either by the uniform application of the upper and lower prestressing forces, to the concrete, or by partial application of each alternately, at transfer.

With post-tensioning, the beam may stick to the shutters; it is often advisable to apply part of the prestress and then jack the beam clear of the bottom shutter before applying the remainder. Alternatively, portions of the bottom shutter may be removed before the stress is applied. The presence of scaffolding or supports for the shuttering can interfere with the deformation as well as horizontal movement caused by the prestress; for example, in beams with cantilevered ends the prestressing of the cantilevers may produce sagging of the central span, and this may be prevented if the scaffolding is present at this stage.

The suitability of available methods of lifting and transport of precast members should always be investigated. Any restrictions imposed by the methods or facilities available should be allowed for in the design. When precast members are used the capacity of the crane available should be considered at the design

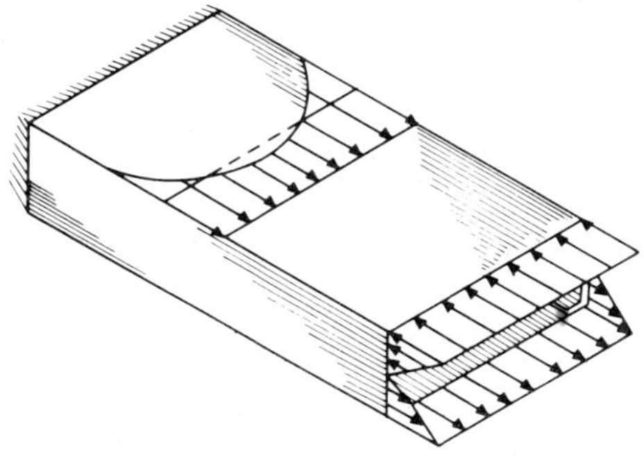

Figure 13.14 Stress distribution in a thin-walled box section (shear lag)

stage. This may affect the choice of material, i.e.whether lightweight aggregate or normal weight aggregate should be used. This may also influence the choice of composite construction.

13.6 Specification and drawings

In preparing the specification and drawings care should be taken to ensure that all relevant information is given. The necessity for this may appear so obvious as to make the statement unnecessary; nevertheless, it is by no means unusual for this rule to be broken. Apart from such obvious items as concrete strengths at transfer and working load, the information given should include the allowances for initial overstress and slip of the prestressing steel, the expected strains of cable for given jack loads when friction is present, the conditions required at transfer (particularly when immediate counter-action is required), the method of lifting and transport, together with the positions of points of lifting and support, and the sequence of loading. The last item may be particularly important in the case of beams with cantilevered ends; if the loading of a cantilever before the remainder of the beam is not allowed for in the design this should be made clear to those concerned with the construction.

REFERENCES

1. MERCHANT, W. The failure load of rigidly jointed frameworks, as influenced by stability. *Structural Engineering,* Vol. 32 No. 7 1954, pp. 185–190.
2. BARTA, T. A. Some simple thoughts of column buckling. *International Column Colloquium*, Paris 1972. IABSE Report of the Working Committee Vol. 23, 1975 pp. 277–291.
3. TIMOSHENKO, S. *Theory of Elastic Stability.* McGraw Hill Book Co Inc, New York, 1936. pp. 245–268.
4. CSONKA, P. Die Stabilitaet des an einem Punkte aufgehaengten geraden Balkens (Stability of a rectangular beam suspended at one point), *Acta Technica* Vol. VIII Nos. 3–4 1952. pp. 389–397.
5. CSONKA, P. Stabilitaet des an 2 Punktenaufgehaengten Recteckbalken (Stability of a beam rectangular suspended in 2 points). *Baupl. u. Bautechnik* 1954 p. 297.
6. ANDERSON, A. P. Lateral stability of long prestressed concrete beams. *Journal of prestressed concrete Institute* Vol. 16,No. 3 May-June 1971, pp. 7–9.
7. SWANN, R. A. Lateral stability of long prestressed concrete beams. *Journal of Prestressed Concrete Institute* Vol. 16,No. 6 November-December 1971. pp. 85–86.
8. MEHLBORN, G. Naeherungsverfahren zur Abschaetzung der Kippstabilitaet vorgespannter Traeger (Approximate calculation of transverse stability of prestressed concrete beams). *Beton u. Stahlbeton,* Vol. 69, No. 1 January 1974. pp. 7–12.
9. GRIFFITHS, H. (Chairman) Report of the enquiry into the collapse of flats at Ronan Point, Canning Town, London. H. M. Stationary Office, 1968, London and New York pp. 84.
10. ODEN, J. T. *Mechanics of elastic structures.* McGraw Hill, 1964.

CHAPTER 14

STATICALLY-INDETERMINATE STRUCTURES

14.1 General notes

Reinforced concrete structures are usually statically indeterminate for several reasons. The bending moments are almost always numerically smaller than those in equivalent statically-determinate structures, and monolithic behaviour is often advantageous because of the re-distribution of stress which would occur if the structure were to be overloaded. Also, methods of analysis based solely on the free bending moments and giving rapid and reasonably accurate results are now available. It should be noted that it is usual to ignore the deformations due to axial loads as these are important only in very stiff frames (that is, frames in which the ratios of the second moments of area of the members to their length are large) and the effects of shrinkage and creep are usually ignored. The possible settlement of supports is considered only rarely. Most structures which are cast in place become statically indeterminate; when statically-indeterminate structures are required (as is often the case with bridges, for example), particular care is often necessary to ensure that this condition is achieved.

Smaller bending moments are also obtained in statically-indeterminate prestressed concrete structures but the bending due to eccentric prestressing force usually causes secondary reactions and bending moments which increase or reduce the primary effects of the eccentric forces. Moreover, it is not possible to ignore the effects of elastic shortening, shrinkage and creep.

While statically-indeterminate prestressed structures possess some advantages, there are also disadvantages which do not occur with reinforced concrete structures. The major disadvantage is that portions of a continuous member are usually subjected to reversal of bending moment; in the case shown in Figure 14.1 for example, more than half of the member may be subjected to positive as well as negative bending moments, depending on the distribution of the load. This causes no difficulty in reinforced concrete or steel, but in prestressed concrete such a condition may make the use of a highly eccentric prestressing force impossible, thereby eliminating one of the major advantages and also affecting the ultimate resistance. Practical difficulties may also be created by the need to use undulating cables (Figure 14.2) which may involve excessive frictional losses; also it is usually simpler to make a prestressed structure statically determinate. For these reasons, the benefits of statically-indeterminate structures in prestressed concrete are often less than in other forms of construction; however, there are many occasions when their use is to be preferred.

14.2 Secondary bending moments

A member to which an eccentric longitudinal force is applied is in consequence

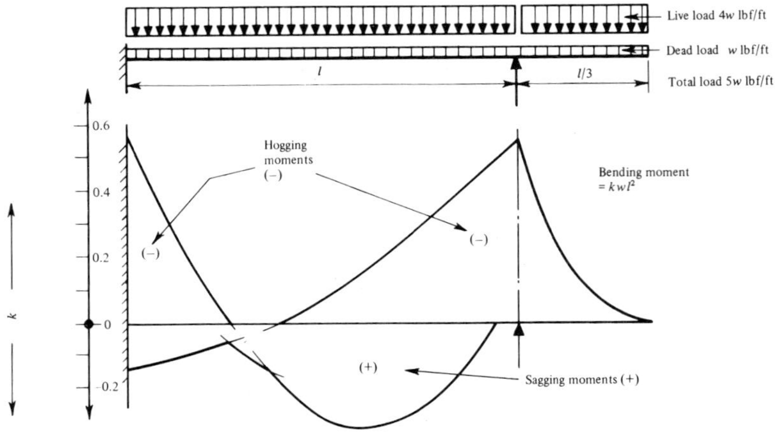

Figure 14.1 Envelope of bending moments: propped cantilever

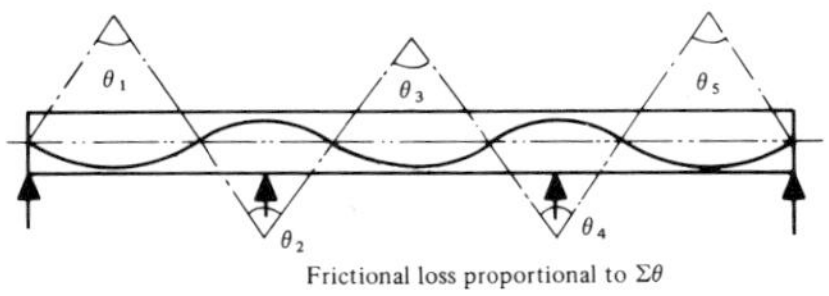

Figure 14.2 Friction of tendons in a continuous member

subjected to a bending moment, and deflects. If there are more than two supports, or if the member is not simply supported, the deformations which would otherwise occur at the supports are modified or prevented, and additional reactions are induced; these cause secondary bending moments which may be of the same order as the primary prestressing moments but of the opposite sign. In the following it is assumed that all sections are symmetrical about a vertical line through the centroid and are subjected to prestressing forces which are constant in a direction parallel to the axis of the member; that is, no frictional losses are allowed for.

If in the case shown in Figure 14.3 there was no support at B, the prestressing force near the lower edge of the member would cause a downward deflection at B. The reaction R created by the pressure on the support is therefore equal to the force necessary to impose an equal and opposite deflection at B, and the primary prestressing moment Pe_a is modified by the secondary moment due to R. The net prestressing moment is shown in Figure 14.3. The secondary moment is sometimes termed a 'parasitic' moment, but as its magnitude and distribution depend on the nature of the supports as well as the primary prestressing moment this term is not entirely accurate and is avoided here. It can conveniently be combined with the moment due to the load; it is more instructive, however, to combine it with the primary prestressing moment (Figure 14.4) as both moments

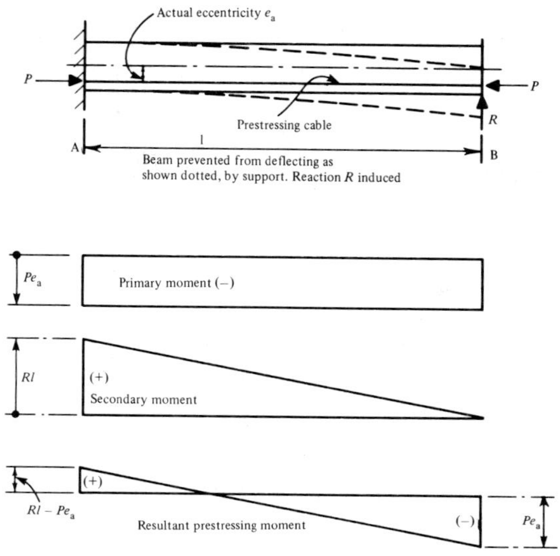

Figure 14.3 Effect of secondary moment of a parallel cable

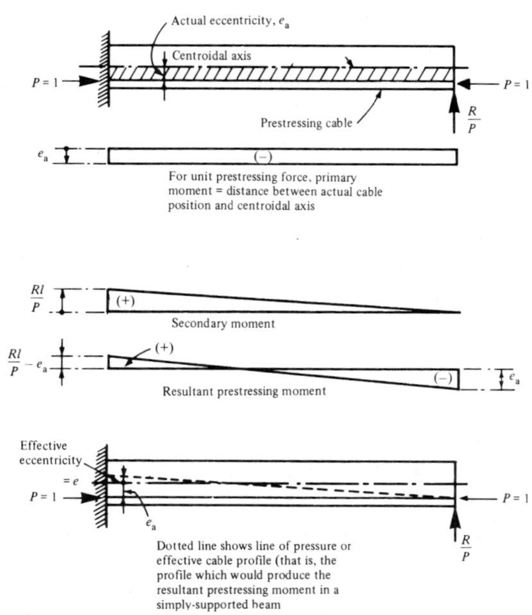

Figure 14.4 Effective eccentricity of tendon due to secondary moment

are proportional to the prestressing force. The algebraic sum of the primary and secondary moments then represents the effective profile of the prestressing cable, from which the effective eccentricity e may be obtained (Figure 14.4). Secondary moments corresponding to various cable profiles are given in Chart 14.

It is unwise merely to ensure that the primary and secondary prestressing moments are equal and opposite to the bending moments caused by the dead and live loads, and to make no allowance for the axial stress due to the prestressing force. The design should be such that suitable resultant stresses are ensured, and not that the bending moments are exactly counteracted.

14.3 Analysis of statically-indeterminate structures

14.3.1 Three and four moment theorems

Many methods are available for the analysis of statically-indeterminate structures; the Moment-Distribution method due to Hardy Cross is probably the most widely used. When considering prestressed structures, however, a method is required in which the effects of side-sway and axial compression can easily be allowed for, and therefore the Theorem of Three Moments, and the general Theorem of Four Moments, of which the Three-Moment Theorem is a special case is recommended. The effects of axial external loads are ignored, but those due to prestressing forces are taken into account.

The Theorem of Three Moments is known to most engineers in the following form, in which the symbols are as defined in Figure 14.5.

$$M_K \frac{l_{KL}}{I_{KL}} + 2M_L \left(\frac{l_{KL}}{I_{KL}} + \frac{l_{LM}}{I_{LM}} \right) + M_M \frac{l_{LM}}{I_{LM}} = -\frac{6A_{KL}\bar{x}_{KL}}{I_{KL}\,l_{KL}} - \frac{6A_{LM}\bar{x}_{ML}}{I_{LM}\,l_{LM}}$$

The expressions $\dfrac{A_{KL}\bar{x}_{KL}}{l_{KL}}$ and $\dfrac{A_{LM}\bar{x}_{ML}}{l_{LM}}$

in this equation denote the reactions at support L of the free bending moment diagrams for spans KL and LM treated as imaginary loads. These reactions are henceforward denoted by R_{LK} and R_{LM}, so that

$$R_{LK} = \frac{A_{KL}\bar{x}_{KL}}{l_{KL}} \text{ and } R_{LM} = \frac{A_{LM}\bar{x}_{ML}}{l_{LM}}$$

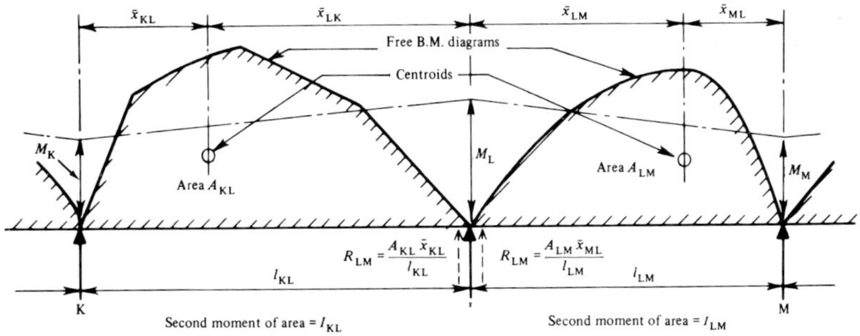

Figure 14.5 Theorem of three moments

The double-suffix notation should be carefully noted; in general, the first letter of the suffix denotes the support to which the property relates, and the two letters together denote the span. Thus R_{LK} is the reaction at support L of the free bending moment diagram for the span LK, treated as an imaginary load, and in order to obtain this value, the lever arm $\bar{x}_{KL}$ is required. The same notation may conveniently be used for A_{KL} and l_{KL}; but as these two properties relate exclusively to the span KL, $A_{KL} = A_{LK}$ and $l_{KL} = l_{LK}$ and consequently the first letter of the suffix has no special significance.

The equation may therefore be written as follows:

$$M_K \frac{l_{KL}}{I_{KL}} + 2M_L \left(\frac{l_{KL}}{I_{KL}} + \frac{l_{LM}}{I_{LM}} \right) + M_M \frac{l_{LM}}{I_{LM}} = -\frac{6R_{LK}}{I_{KL}} - \frac{6R_{LM}}{I_{LM}}$$

In this form the theorem applies to adjacent spans of a continuous beam in which no settlement of the supports occurs; each span is assumed to be prismatic, and if the second moment of area is constant throughout the expression simplifies to:

$$M_K l_{KL} + 2M_L \ (l_{KL} + l_{LM}) + M_M l_{LM} = -6R_{LK} - 6R_{LM}$$

The inclusion of terms to allow for the settlement of supports (see, for example *The analysis of engineering structures,* by A. J. S. Pippard and J. F. Baker[1]) gives the equation in the following form:

$$M_K \frac{l_{KL}}{I_{KL}} + 2M_L \left(\frac{l_{KL}}{I_{KL}} + \frac{l_{LM}}{I_{LM}} \right) + M_M \frac{l_{LM}}{I_{LM}} - \frac{6E\delta_L}{l_{KL}} + \frac{6E\delta_M}{l_{LM}} = -\frac{6R_{LK}}{I_{KL}} - \frac{6R_{LM}}{I_{LM}}$$

The additional symbols are defined in Figure 14.6. If more than two members meet at joint L, it is no longer the case that $M_{LK} = M_{LM}$ and it can be shown[2] that the appropriate equation is then

$$M_{KL} \frac{l_{KL}}{I_{KL}} + 2M_{LK} \frac{l_{KL}}{I_{KL}} + 2M_{LM} \frac{l_{LM}}{I_{LM}} + M_{ML} \frac{l_{LM}}{I_{LM}} - \frac{6E\delta_L}{l_{KL}} +$$
$$\frac{6E\delta_M}{l_{LM}} = -\frac{6R_{LK}}{l_{KL}} - \frac{6R_{LM}}{l_{LM}}$$

This is the Theorem of Four Moments, which differs from the Theorem of Three Moments only in that M_{LK} and M_{LM} are assumed to have different numerical values.

Writing $K = \frac{I}{l}$, $\theta = \frac{\delta}{l}$ and $N = -\frac{6R}{l}$

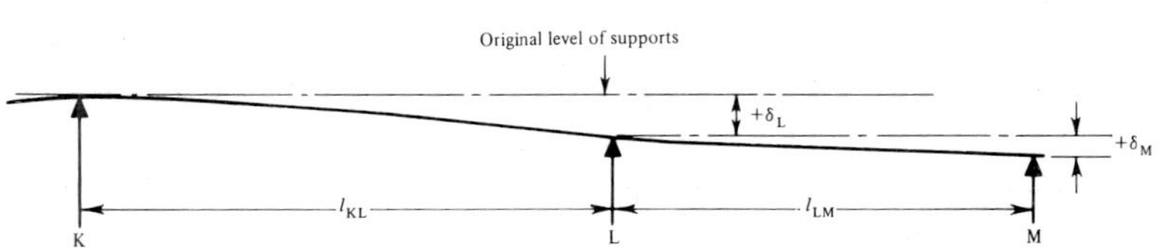

Figure 14.6 Settlement of support

$$\frac{(M_{KL} + 2M_{LK})}{K_{KL}} + \frac{(2M_{LM} + M_{ML})}{K_{LM}} - 6E\,(\theta_{KL} - \theta_{LM}) = \frac{N_{LK}}{K_{KL}} + \frac{N_{LM}}{K_{LM}}$$

Finally, if $k_L = \dfrac{K_{KL}}{K_{LM}}$, the equation may be rewritten in the form

$$M_{KL} + 2M_{LK} + k_L\,(2M_{LM} + M_{ML}) - 6EK_{KL}\,(\theta_{KL} - \theta_{LM}) = N_{LK} + k_L N_{LM}$$

and if $M_{LK} = M_{LM} = M_L$, this becomes

$$M_K + 2M_L\,(1 + k_L) + M_M k_L - 6EK_{KL}\,(\theta_{KL} - \theta_{LM}) = N_{LK} + k_L N_{LM}$$

In this equation θ_{KL} is the angle of distortion of the member KL, and is positive when the member distorts in a clockwise direction. Bending moments which cause tension on the underside of a beam or cantilever, or the inside of a frame are positive (see also p. 271). Values of N for various loadings are given in Chart 15; in this chart

$$N_{AB} = -\frac{6R_{AB}}{l_{AB}} = -\frac{6A_{AB}\bar{x}_{BA}}{(l_{AB})^2}$$

and

$$N_{BA} = -\frac{6R_{BA}}{l_{AB}} = -\frac{6A_{AB}\bar{x}_{AB}}{(l_{AB})^2}$$

It is necessary to obtain sufficient additional information to eliminate the angles of distortion from the equations, as the number of unknowns is otherwise excessive. Consider the following simple example (Figure 14.7).

$$A_{BC} = 40 \times 15 \times 0.5 = 300 \text{ tonf ft}^2 \text{ (28 317 kgf m}^2\text{; 278 kN m}^2\text{)}$$

$$\bar{x}_{BC} = \frac{20}{3};\ \bar{x}_{CB} = \frac{25}{3};\ N_{CB} = -\frac{6 \times 300 \times 20}{3 \times 15^2} = -\frac{800}{15}$$

$$N_{BC} = -\frac{6 \times 300 \times 25}{3 \times 15^2} = -\frac{1000}{15}$$

Members AB, BC:

$$\frac{10}{0.8}(M_{AB} + 2M_{BA}) + \frac{15}{1}(2M_{BC} + M_{CB}) - 6EI_{BC}\,(\theta_{AB} - \theta_{BC}) = -0 - 1000$$

Members BC, CD:

$$\frac{15}{1}(M_{BC} + 2M_{CB}) + \frac{10}{0.8}(2M_{CD} + M_{DC}) - 6EI_{BC}\,(\theta_{BC} - \theta_{CD}) = -800 - 0$$

As A and D are pinned, $M_{AB} = M_{DC} = 0$, and as only two members meet at joints B and C, $M_{BA} = M_{BC}$ and $M_{CB} = M_{CD}$. Also, since the nett horizontal force acting on the frame is zero, the horizontal reactions at A and D must be equal and opposite, and hence $M_{BA} = M_{CD}$. Therefore,

$$70M_B - 6EI_{BC}\,(\theta_{AB} - \theta_{BC}) = -1000$$

$$70M_B - 6EI_{BC}(\theta_{BC} - \theta_{CD}) = -800$$

In order to obtain a solution the angles must be eliminated from the equations. From the deflection diagram (Figure 14.7b) it is apparent that $\theta_{AB} = \theta_{CD}$, and $\theta_{BC} = 0$. Therefore,

$$70M_B - 6EI_{BC}(\theta_{AB} - 0) = -1000$$

$$70M_B - 6EI_{BC}(0 - \theta_{AB}) = -800$$

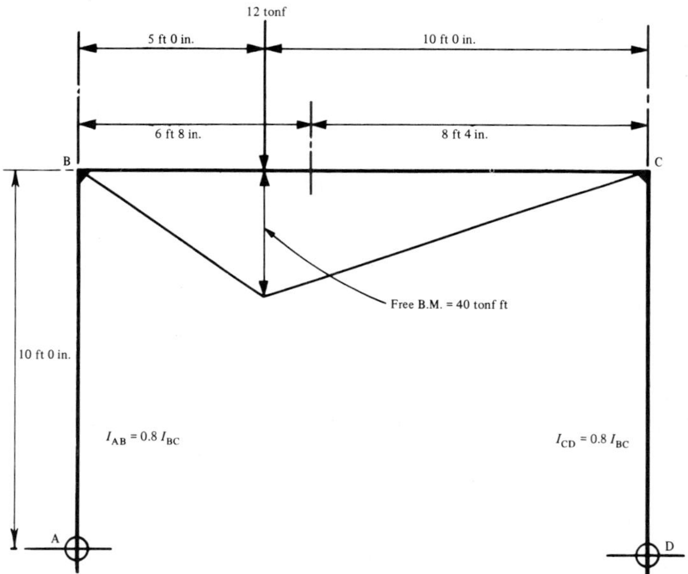

(*a*) Bending moment

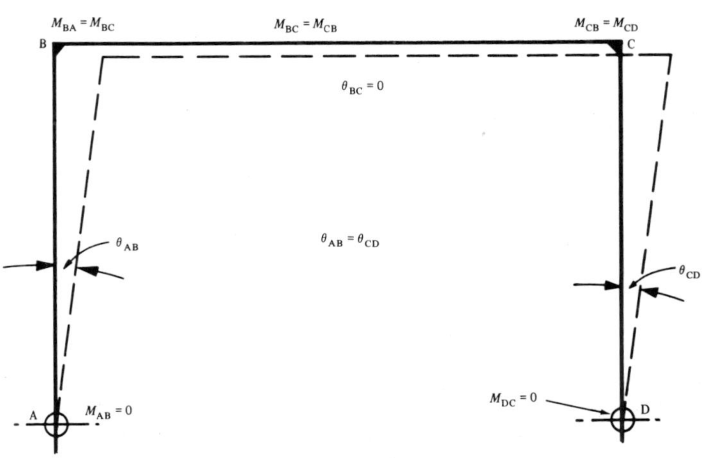

(*b*) Angles of distortion

Figure 14.7 Bending moment and deformation of a two-hinged portal frame

Hence $140\,M_B = -1800$ and $M_B = M_C = -\dfrac{1800}{140} = 12.86$ tonf ft. (3983 kgf m; 39 kN m).

The method can also be used to obtain a more general solution, as in the following (Figure 14.8).

From Chart 15

$$N_{CB} = -\,W\alpha\,(1-\alpha^2)\,l_{BC}$$

$$N_{BC} = -\,W\alpha\,(1-\alpha)\,(2-\alpha)\,l_{BC}$$

Members AB, BC:

$$\frac{(M_{AB} + 2M_{BA})}{K_{AB}} + \frac{(2M_{BC} + M_{CB})}{K_{BC}} - 6E\,(\theta_{AB} - \theta_{BC}) = \frac{N_{BC}}{K_{BC}}$$

Members BC, CD:

$$\frac{(M_{BC} + 2M_{CB})}{K_{BC}} + \frac{(2M_{CD} + M_{DC})}{K_{CD}} - 6E\,(\theta_{BC} - \theta_{CD}) = \frac{N_{CB}}{K_{BC}}$$

As before, $M_{AB} = M_{DC} = 0$; $K_{AB} = K_{CD}$; $M_{BA} = M_{BC} = M_{CB} = M_{CD}$

$\theta_{AB} = \theta_{CD}$

$\theta_{BC} = 0$

Let $K_{AB} = K_v, K_{BC} = K_h, \theta_{AB} = \theta$

Then

$$M_B\left(\frac{2}{K_v} + \frac{3}{K_h}\right) - 6E\theta = \frac{N_{BC}}{K_h}$$

$$M_B\left(\frac{3}{K_h} + \frac{2}{K_v}\right) + 6E\theta = \frac{N_{CB}}{K_h}$$

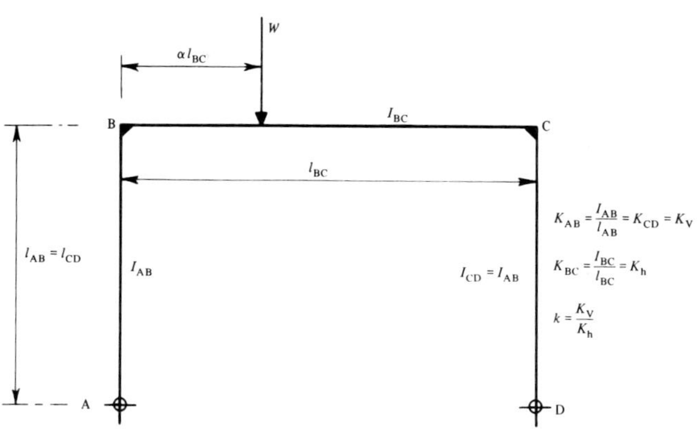

Figure 14.8 Two-hinged portal frame: general equation

Hence

$$2M_B\left(\frac{2}{K_v}+\frac{3}{K_h}\right)=\frac{N_{BC}+N_{CB}}{K_h}$$

If $k=\dfrac{K_v}{K_h}$, $\qquad 2M_B\,(2+3k)=k\,(N_{BC}+N_{CB})$

from which

$$M_B=M_C=\left(\frac{k}{2+3k}\right)\left(\frac{N_{BC}+N_{CB}}{2}\right)=-\left(\frac{k}{2+3k}\right)\frac{3}{2}Wl_{BC}\,\alpha(1-\alpha)$$

It should be noted that the two initial equations in these examples apparently contain nine unknowns; four are eliminated by considerations of compatibility of moments imposed by the nature of the structure, two by considerations of the compatibility of deformations, and one by the known conditions at the supports, leaving two to be eliminated between the two equations. In any structure, sufficient conditions can always be obtained to reduce the number of apparent unknowns to the number of equations; the work is made much simpler, and a more reliable result obtained more quickly, when the degree of indeterminacy is first determined and unnecessary equations are eliminated. Methods of achieving these objects are described in the following.

14.3.2 Degree of indeterminacy

A structure in equilibrium is statically indeterminate if the forces and moments at any point are dependent on the physical properties of the cross-sections of the members of the structure. Two cases, here termed 'external indeterminacy' and 'internal indeterminacy', shall be considered.

External indeterminacy. A simple beam or unclosed structure (that is, not forming a completely closed frame) is statically determinate if it is pinned at one support and has a roller as the other support (Figure 14.9a), or if it is fixed in position and direction at one support only (Figure 14.9b). In either case there are three reactions only, which can be calculated from the three general equations of equilibrium (that is, Σ Horizontal forces = Σ Vertical forces = ΣMoments = 0).

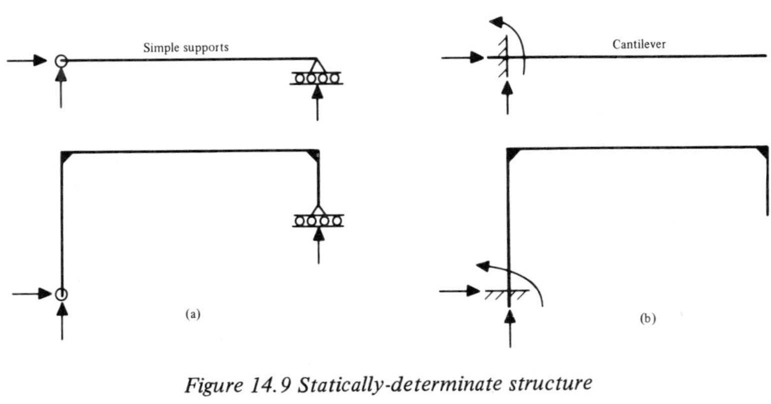

Figure 14.9 Statically-determinate structure

If more supports are present, or if they are of such a nature as to impose additional restraints (Figure 14.10), the additional reactions are unknown quantities which may be determined by imagining the supports or restraints to be equivalent to forces and moments of unknown magnitude acting on a simply-supported or cantilevered structure. These forces and moments must therefore be of such magnitudes that the structure is maintained in its original position or that any expected settlements or rotations of the supports are allowed for. If hinges or other discontinuities are to be provided, additional equations of equilibrium are obtained which sometimes enable the values of more than three reactions to be calculated. The two-hinged arch shown in Figure 14.11a has four reactions. The replacement of the horizontal reaction at one support by an unknown force *H* acting on a statically-determinate structure allows the remaining reactions to be obtained in terms of *H*, but the value of *H* must be determined from the condition that the displacement of the support is zero. However, if there is an additional hinge (at the crown, for example) the structure becomes statically determinate, the value of *H* being calculated from the condition that the moment at this hinge must be zero (Figure 14.11b).

It should be noted in this connection that the position of the hinges is of importance. The reactions for the structure shown in Figure 14.12a may be calculated from the equations of equilibrium and the additional condition that the moment at the hinge must be zero; but this is not the case if the hinge is placed as shown in Figure 14.12b, as the condition that the moment at the hinge in the upper closed frame is zero does not lead directly to a reduction in the number of external unknown quantities. Such a structure is both internally and externally indeterminate, and is subject to the considerations described in the following.

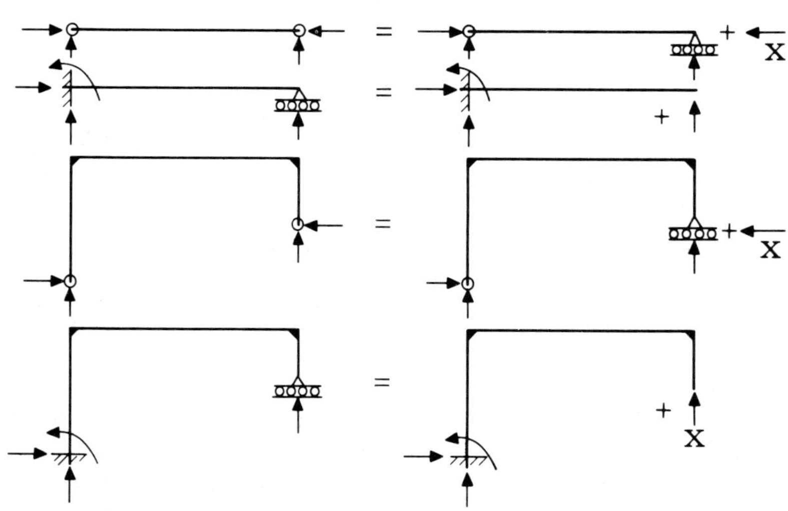

Figure 14.10 External indeterminacy

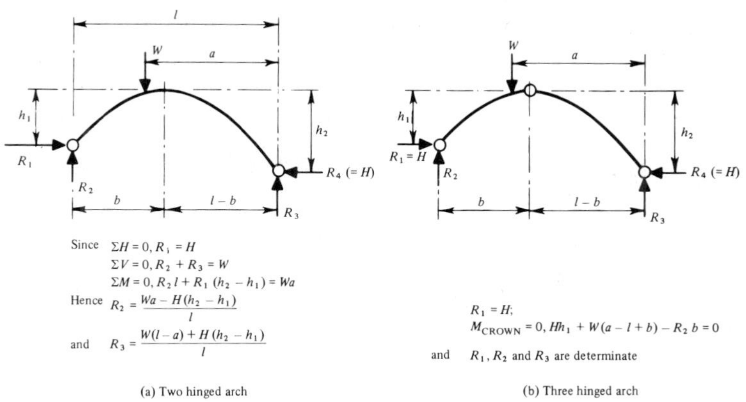

Figure 14.11 Two and three-hinged arches

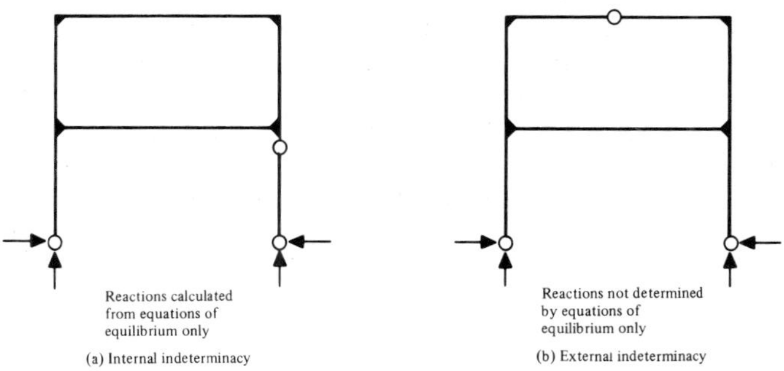

Figure 14.12 Two storey frame with hinges

Internal indeterminacy. The closed frame shown in Figure 14.13 is pinned to a support at one corner, and rests on a roller at another; there are therefore three reactions only, and these may be calculated from the general equations of equilibrium. The structure is still statically indeterminate, however, since a cut could be made in any one member without causing the collapse of the structure. If such a cut is imagined, then in order that the distortion of the structure should remain unaltered and the cut should remain closed, equal and opposite axial forces, shearing forces, and bending moments, of appropriate magnitudes must be considered to act at each side of the cut.

A structure may be externally indeterminate only (Figures 14.10 and 14.11a), internally indeterminate only (Figures 14.12a and 14.13), or both (Figure 14.12b). In the case shown in Figure 14.14, the degree of external indeterminacy is 2 and of internal indeterminacy 3; the total degree of indeterminacy is therefore 5.

Three unknown quantities (the direct force, shearing force, and bending moment) are released at every cut in a rigid member, and one unknown quantity

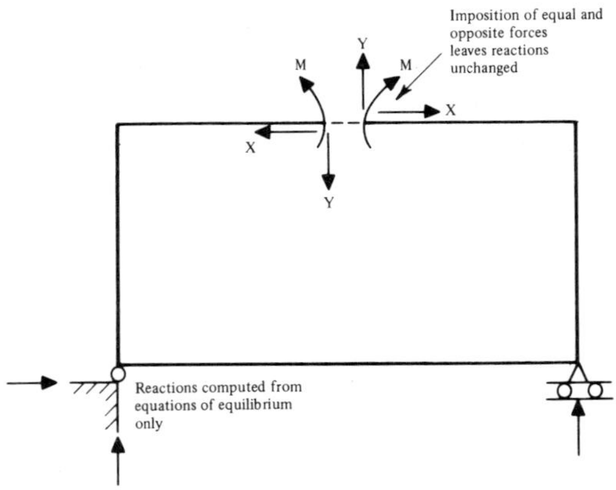

Figure 14.13 Closed rigid frame

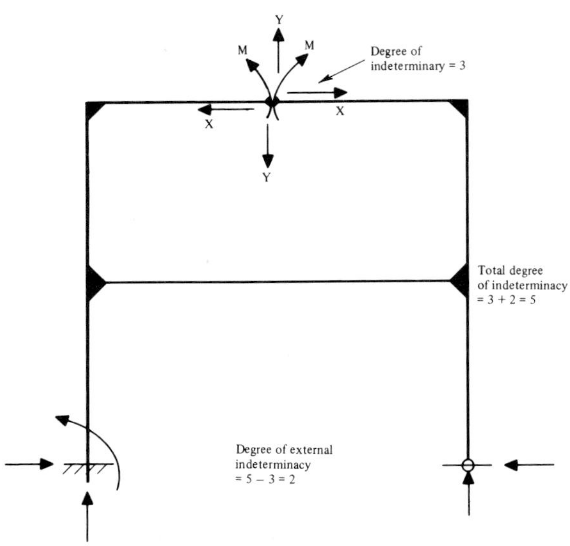

Figure 14.14 External and internal indeterminacy

(the direct force) in a tie; one, two, or three unknown quantities are eliminated when a support is removed, depending on whether it is a roller (normal force only), pin (direct and shearing forces), or fixed support (direct force, shearing force, and bending moment). One unknown is released if a pin is replaced by a roller, and two if a fixed end is replaced by a roller. Similarly, if a pin is inserted in a rigid member, one unknown is removed. In the preceding examples, the structure is indeterminate in the first degree, since the removal of one unknown would render it statically determinate. For example, if a hinge were inserted in the horizontal member BC (Figure 14.7) a three-hinged frame would be obtained, or if one pinned support were replaced by a roller the frame would act as a simply-supported beam. Other examples are shown in Figure 14.15.

14.3.3 Symmetry and negative symmetry

If a symmetrical structure is subjected to unsymmetrical loading, the analysis is often simplified by considering the loading to consist of two components, one being symmetrical and the other having negative symmetry, and combining the two results. In the previous example, for instance, the loading could have been divided as shown in Figure 14.16. In Case (a), no displacement of the supports occurs, and the Theorem of Three Moments may therefore be applied directly.

From Chart 15

$$N_{CB} = N_{BC} = -3W\alpha l(1-\alpha) = -3 \times \frac{12}{2} \times 5 \times \frac{2}{3} = -60$$

Therefore

$$0 + 2M_B\left(\frac{10}{0.8} + \frac{15}{1}\right) + M_C\left(\frac{15}{1}\right) = -\frac{60 \times 15}{1}$$

and by symmetry

$$M_B = M_C = -\frac{900}{70}$$

In Case (b), the bending moment at the mid-point of BC must be zero, and therefore a hinge may be inserted (Figure 14.16b) without altering the system, which thereby becomes statically determinate, and $M_B = M_C = 0$.

Hence

$$M_{B\,\text{total}} = M_{C\,\text{total}} = -\frac{900}{70} \text{ as before}$$

It is helpful to note that the sum of the degrees of indeterminacy of the two subdivided cases obtained in this manner must be the total degree of indeterminacy of the original. In the preceding example, the degree of indeterminacy of the original structure is one; after division, the degree of indeterminacy of Case (a) is also one, and Case (b) is determinate.

As a further example, the frame previously considered is now assumed to have fixed supports (Figure 14.17). By removing one support completely, a statically-determinate system would be obtained; the degree of indeterminacy of the structure is therefore three. The solution for a structure with fixed ends is obtained by including the fixed-end moment for an imaginary span A_a (Figure 14.17) of zero length in the equation. The division of the load is shown in Figure 14.18.

Case (a) In order to obtain symmetry, the shearing force at the mid-point of BB′ must be zero. Hence, the degree of indeterminacy of case (a) is two. $\theta_v = \theta_h = 0$, by symmetry. From Chart 15

$$N'_{BB'} = N'_{B'B} = -\tfrac{3}{2} W\alpha(1-\alpha) l_h$$

The equation for the imaginary span A_a and span AB is

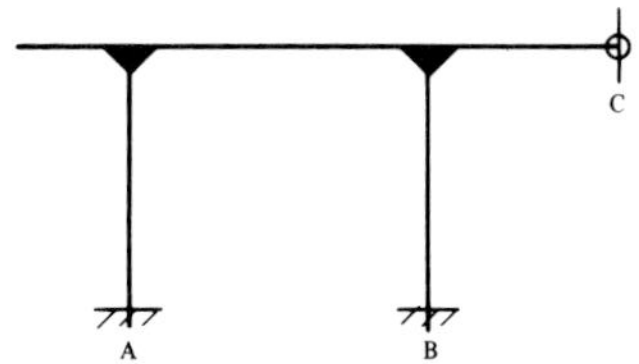

Remove support B: 3 Releases
Remove support C: 2 Releases

Total: 5 Releases

Result: cantilever (statically determinate) removal of 5 unknown forces and moments renders structure statically determinate; hence degree of indeterminacy of original structure = 5.

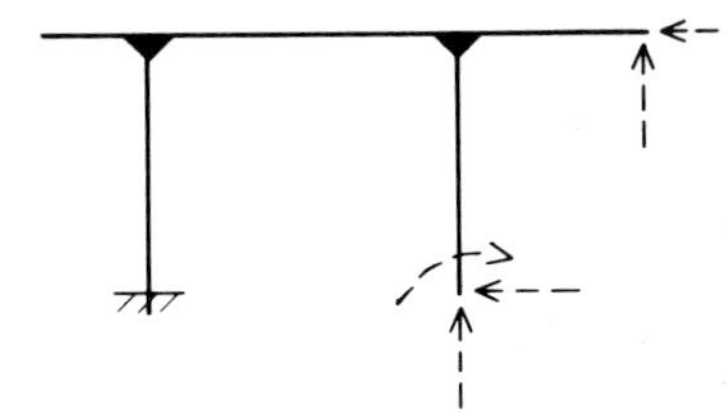

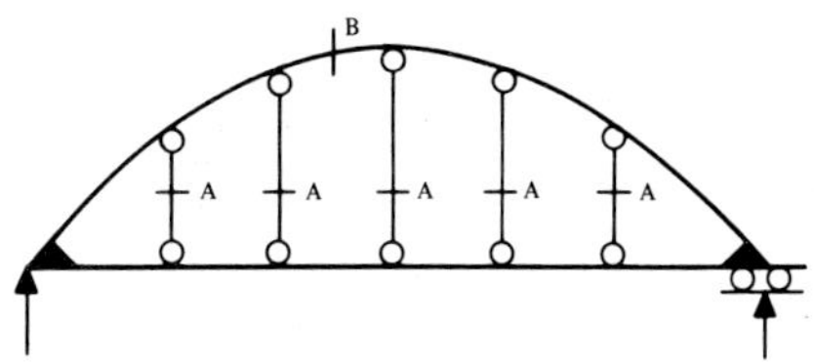

Cut ties A: 5 × 1 = 5 Releases
Cut arch at B: 3 Releases

Total: 8 Releases

Result: Simply-supported beam with cantilevered ends. Therefore original structure 8 times indeterminate.

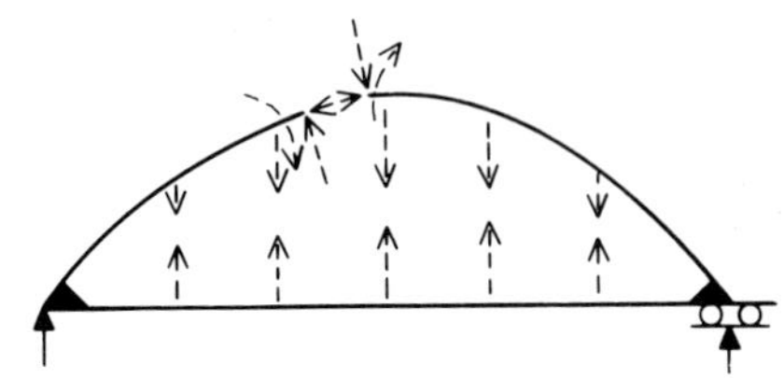

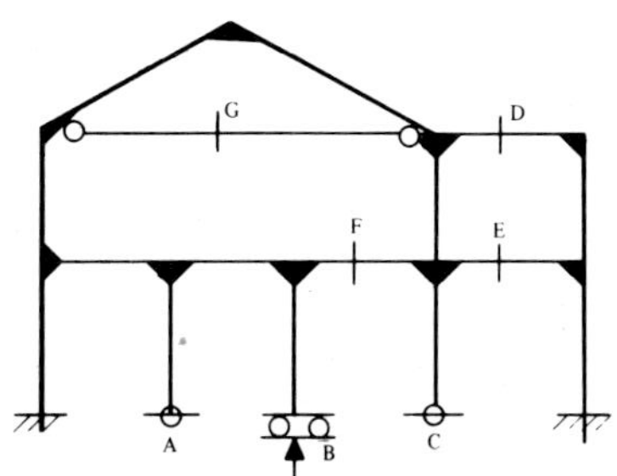

Remove supports A, B, C: 2 + 1 + 2 = 5 Releases
Cut beams D, E, F: 3 × 3 = 9 Releases
Cut tie G: 1 Release

Total: 15 Releases

Result: two cantilevers therefore original structure 15 times indeterminate.

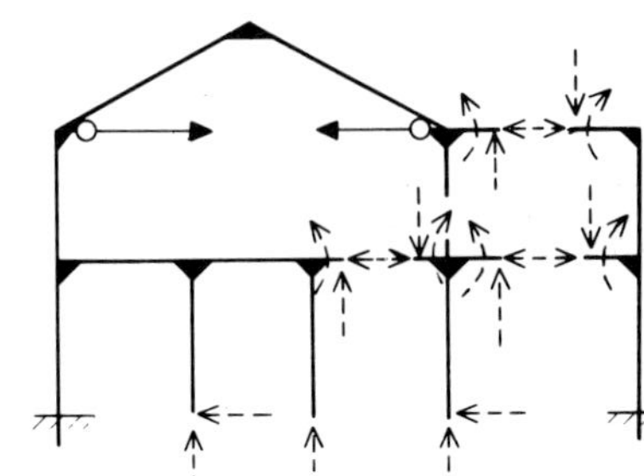

Figure 14.15 Degree of indeterminacy

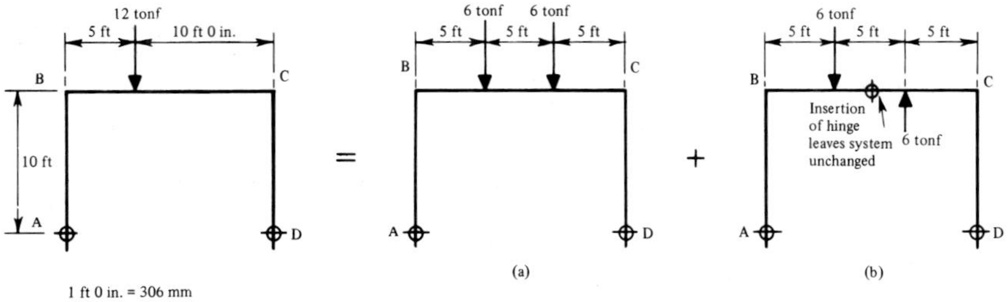

Figure 14.16 Symmetry and negative symmetry

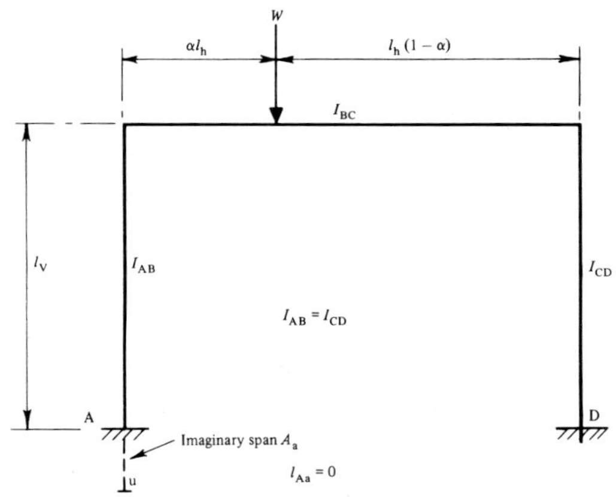

Figure 14.17 Fixed end portal

$$0 + 2M'_{AB} + M'_{BA} + 6EK_{AB}\,\theta'_{AB} = N_{AB}$$

In the present case,

$$\theta'_{AB} = N'_{AB} = 0$$

Hence

$$2M'_{AB} + M'_{AB} = 0, \text{ and } M'_{AB} = \frac{-M'_{BA}}{2}, \text{ or } M'_{A} = \frac{-M'_{B}}{2}$$

For members AB, BB′, the moment equation is

$$M'_A + 2M'_B\,(1 + k_B) + M'_{B'}\,k_B = 0 + k_B\,N'_{BB'}$$

and since $M'_B = M'_{B'}$, by symmetry,

$$M'_A + M'_B\,(2 + 3k_B) = k_B\,N'_{BB'}$$

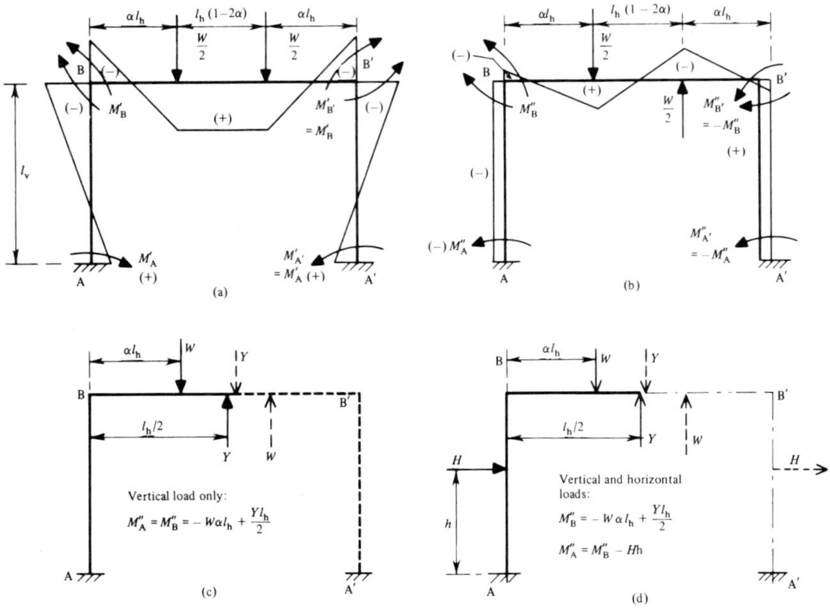

Figure 14.18 Solution of fixed end portal for loading shown in Figure 14.17

Therefore

$$M'_A = \frac{-k_B N'_{BB'}}{3(1+2k_B)} = \frac{+k_B W\alpha(1-\alpha)l_h}{2(1+2k_B)}$$

and

$$M'_B = \frac{+2k_B N'_{BB'}}{3(1+2k_B)} = \frac{-k_B W\alpha(1-\alpha)l_h}{(1+2k_B)}$$

Case (b) Since Case (a) is two times indeterminate, Case (b) must be indeterminate in the first degree. The only unknown at the middle of the horizontal member is therefore a shearing force, since the bending moment and axial force at this point must be zero when the loading has negative symmetry. It follows that, when vertical forces only are present (Figure 14.18c) then,

$$M''_A = M''_B = \frac{Yl_h}{2} - \frac{W}{2}\alpha l_h$$

If horizontal forces are also present (Figure 14 18d), then $M''_A = M''_B - Hh$, and the two moments are unequal. In the present case, as there is no horizontal force,

$M''_A = M''_B$. Also, $\theta_h = 0$

and from Chart 15

$$N''_{BB'} = -\frac{W\alpha}{2}(1-2\alpha)(1-\alpha)\,l_h$$

For the vertical member:

$$2M''_A + M''_B + 6EK_{AB}\,\theta''_{AB} = 0 \qquad (14.1)$$

For members AB, BB':

$$M''_A + 2M''_B(1+k_B) + M''_{B'}k_B - 6EK_{AB}(\theta''_{AB} - 0) = k_B N''_{BB'}$$

and since $M''_{B'} = -M''_B$

$$M''_A + M''_B(2+k_B) - 6EK_{AB}\,\theta''_{AB} = k_B N''_{BB'} \qquad (14.2)$$

Adding (14.1) and (14.2)

$$3M''_A + (3+k_B)M''_B = k_B N''_{BB'}$$

and therefore

$$M''_A = M''_B = \frac{k_B N''_{BB'}}{6+k_B} = \frac{-k_B}{2(6+k_B)}\,W\alpha(1-2\alpha)(1-\alpha)\,l_h$$

The total moments are therefore

$$M_A = M'_A + M''_A = +\,W\alpha(1-\alpha)\,l_h\left[\frac{k_B}{2(1+2k_B)} - \frac{k_B(1-2\alpha)}{2(6+k_B)}\right]$$

$$M_B = M'_B + M''_B = -\,W\alpha(1-\alpha)\,l_h\left[\frac{k_B}{1+2k_B} + \frac{k_B(1-2\alpha)}{2(6+k_B)}\right]$$

$$M_C = M'_B - M''_B = -\,W\alpha(1-\alpha)\,l_h\left[\frac{k_B}{1+2k_B} - \frac{k_B(1-2\alpha)}{2(6+k_B)}\right]$$

$$M_D = M'_A - M''_A = +\,W\alpha(1-\alpha)\,l_h\left[\frac{k_B}{2(1+2k_B)} + \frac{k_B(1-2\alpha)}{2(6+k_B)}\right]$$

For the frame shown in Figure 14.19,

$W = 12$ tonf (12 200 kgf; 119 kN), $\alpha = \frac{1}{3}$, $l_h = 15$ ft (4.57m),

$$k_B = \frac{I_v}{l_v}\frac{l_h}{I_h} = \frac{0.8}{10} \times \frac{15}{1} = 1.20$$

Therefore

$W\alpha(1-\alpha)l_h = 12 \times \frac{1}{3} \times \frac{2}{3} \times 15 = 40$ tonf ft ; (12 387 kgf m, 121.5 kNm)

$$\frac{k_B}{2(1+2k_B)} = \frac{1.2}{2(1+2.4)} = \frac{1.2}{6.8} = 0.1765$$

$$\frac{k_B}{I+2k_B} = 2 \times 0.1765 = 0.353$$

$$\frac{k_B(1-2\alpha)}{2(6+k_B)} = \frac{1.2 \times \frac{1}{3}}{2(6+1.2)} = \frac{0.4}{14.4} = 0.0278$$

Hence

$$M_A = +40\,(0.1765 - 0.0278) = +5.95 \text{ tonf ft (1842 kgf m; 18 kN m)}$$

$$M_B = -40\,(0.353 + 0.0278) = -15.23 \text{ tonf ft (4717 kgf m; 46.25 kN m)}$$

$$M_C = -40\,(0.353 - 0.0278) = -13.01 \text{ tonf ft (4029 kgf m; 39.5 kN m)}$$

$$M_D = +40\,(0.1765 + 0.0278) = +8.17 \text{ tonf ft (2530 kgf m; 24.8 kN m)}$$

14.3.4 Closed-frame structures

In any closed frame (Figure 14.20) the total horizontal and vertical components of the deformation, both before and after loading, must be zero. Hence

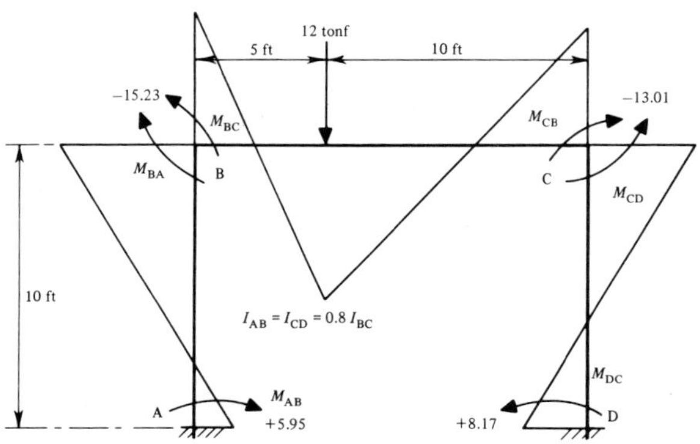

Figure 14.19 Numerical solution of a fixed-end portal

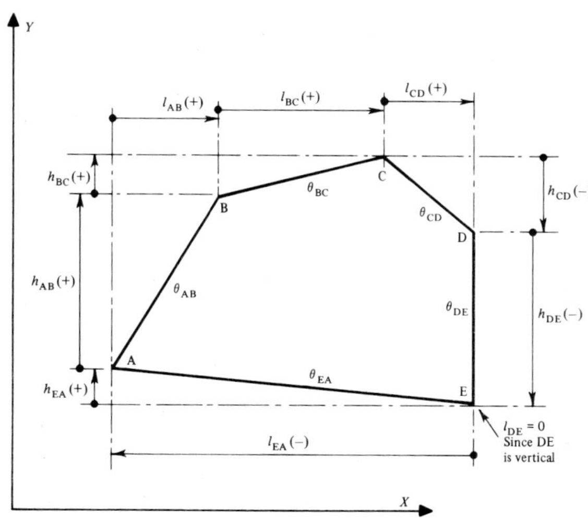

Figure 14.20 Closed frames (distortion angles)

$\Sigma\theta h = 0$ and $\Sigma\theta l = 0$

in which h is the vertical component of the length of each member and l the horizontal component. The structure is traversed in a clockwise direction and values of h and l are assumed to be positive when the direction of travel is parallel to the X and Y axes: in Figure 14.20, for example, the equations are

$$+\theta_{AB}h_{AB} + \theta_{BC}h_{BC} - \theta_{CD}h_{CD} - \theta_{DE}h_{DE} + \theta_{EA}h_{EA} = 0$$

$$+\theta_{AB}l_{AB} + \theta_{BC}l_{BC} + \theta_{CD}l_{CD} + 0 - \theta_{EA}l_{EA} = 0$$

It should be noted that in this statement the deformations due to the longitudinal forces are neglected. The application of these equations to the particular case of a rectangular frame (Figure 14.21a) leads to the useful result that all columns in any storey have equal angles of distortion and so also do the beams in any bay (Figure 14.21b). If all the columns are supported on unyielding foundations, the angles of distortion of all the beams are zero (Figure 14.21c). If the structure shown in Figure 14.21b and its loading are both symmetrical, then $\theta_2 = 0$, $\theta_1 = -\theta_3$, and all values of θ relating to the verticals (θ_4, θ_5, and θ_6) become zero. If the loading has negative symmetry $\theta_1 = \theta_3$.

If the frame is not fully closed, an imaginary rigid member may be assumed to connect the supports and the equations may still be used, provided that no settlement of the supports occurs. Otherwise the movement of the support must be considered, the equations then becoming

$\Sigma\theta h = \Delta x$ and $\Sigma\theta l = \Delta y$

In the example shown in Figure 14.22, $\Delta y = 0$ and $\Delta x = 0.1$ ft (3 cm).

Therefore

$$\Sigma\theta h\text{: } 10\,\theta_{AB} - 10\,\theta_{CD} = 0.1$$

$$\Sigma\theta l\text{: } 15\,\theta_{BC} - 15\,\theta_{DA} = 0$$

The displacement in any direction of one point relative to another must equal the sum of the displacements in that direction of the separate members connecting the points. The displacement Δy of D relative to A is zero, and as the deformations of members AB and CD in this direction are zero (ignoring

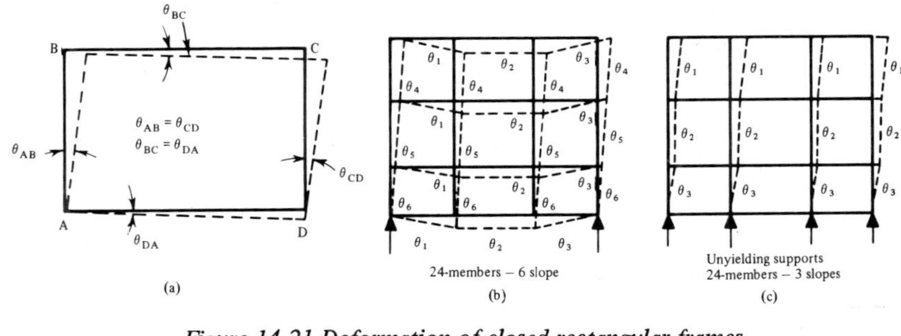

Figure 14.21 Deformation of closed rectangular frames

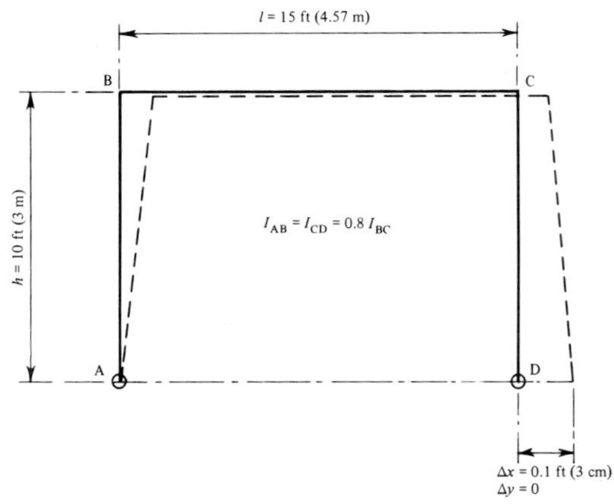

Figure 14.22 Solution of two-hinged portal introducing imaginary closed frame

deformations due to axial forces) the distortion of members BC in the same direction must also be zero. Similarly $10\theta_{AB} - 10\theta_{CD} = \Delta x = 0.1$ and, as the structure is symmetrical, $\theta_{AB} = -\theta_{CD}$ and $M_B = M_c$. Hence $10\theta_{AB} - 10\theta_{CD} = 20\theta_{AB} = 0.1$. Therefore, $\theta_{AB} = 0.005$.

The moment equation is therefore

$$0 + 2M_B\left(\frac{10}{0.8} + \frac{15}{1}\right) + M_B\frac{15}{1} - 6\,EI_{BC} \times 0.005 = 0$$

Hence

$$70M_B - 0.03\,EI_{BC} = 0$$

and

$$M_B = \frac{0.03EI_{BC}}{70}$$

14.3.5 Sign convention

In the formation of the equations in the preceding examples, the support moments are initially assumed to be positive (that is, causing tension on the underside of a beam or cantilever or the inside of a frame), and when this assumption is incorrect a negative value is obtained upon solution of the equations. This procedure is therefore equivalent to making an initial arbitrary assumption that the deformation is such that the underside of a beam or the inside of a frame is in tension.

This is quite satisfactory for continuous beams or single-bay frames. In the case of multiple-bay frames, however, the deformation of an interior column which causes tension on the inside face of one bay must cause compression on the inside face of the adjacent bay (Figure 14.23), for which the support moments must be negative. In the example shown in Figure 14.24, the assumed deformations lead to the equations given. (Note that, if advantage is taken of

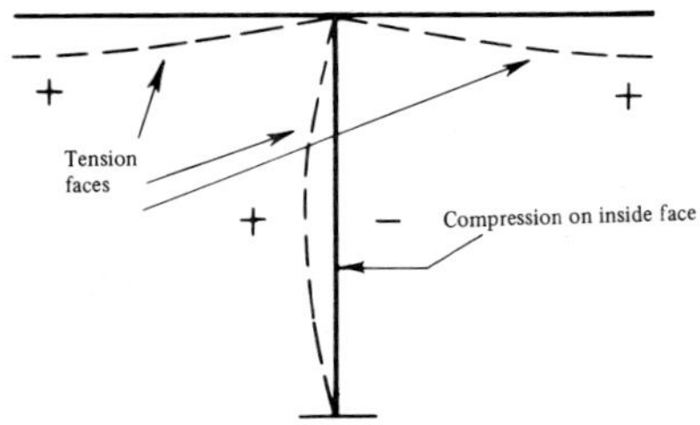

Figure 14.23 Sign convention for moments

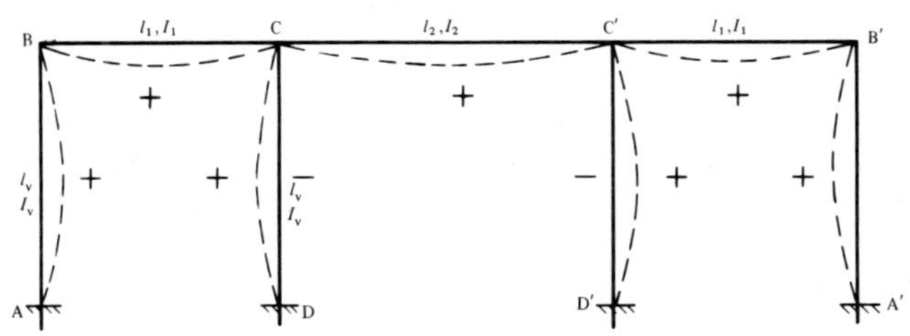

Loading vertical
assumed deflection
shown dotted

$$k_1 = \frac{l_1}{l_v} \cdot \frac{I_v}{I_1}$$

$$k_2 = \frac{l_2}{l_v} \cdot \frac{I_v}{I_2}$$

$$k_3 = \frac{l_1}{l_2} \cdot \frac{I_2}{I_1} = \frac{k_2}{k_1}$$

SYMMETRICAL LOADING – CASE (a)

AB: $2M'_A + M'_B = 0$ 1
ABC: $M'_A + 2(1 + k_1)M'_B + k_1 M'_{CB} = k_1 N'_{BC}$ 2
BCD: $k_1 M'_B + 2k_1 M_{CB} + 2M'_{CD} + M'_D = k_1 N'_{CB}$ 3
DC: $2M'_D + M'_{CD} = 0$ 4
BCC′: $k_3 M'_B + 2k_3 M'_{CB} + 3M'_{cc'} = k_3 N'_{CB} + N'_{cc'}$ 5
Check: DCC′: $-M'_D - 2M'_{CD} + 3k_2 M'_{CC'} = k_2 N'_{CC'}$ 6
Joint C: $M'_{CB} = M'_{CC'} + M'_{CD}$ 7

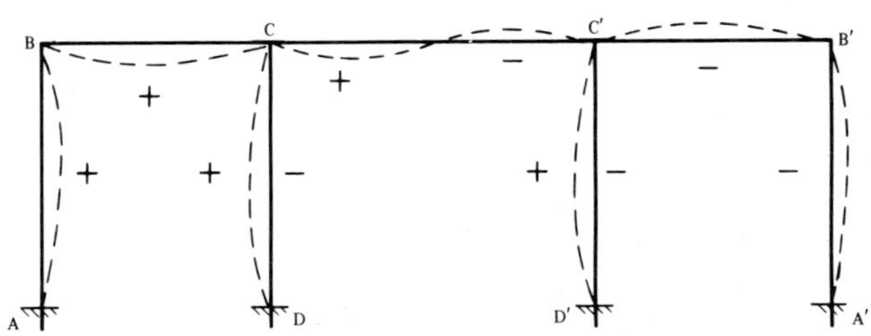

LOADING WITH NEGATIVE SYMMETRY – CASE (b)

AB: $2M''_A + M''_B - 6EI_v\theta = 0$ 1
ABC: $M''_A + 2(1 + k_1)M''_B + k_1 M''_{CB} + 6EI_v\theta = k_1 N''_{BC}$ 2
BCD: $k_1 M''_B + 2k_1 M''_{CB} + 2M''_{CD} + M''_D - 6EI_v\theta = k_1 N''_{CB}$ 3
DC: $2M''_D + M''_{CD} - 6EI_v\theta = 0$ 4
BCC′: $k_3 M''_B + 2k_3 M''_{CB} + M''_{CC'} = k_3 N''_{CB}$ 5
DCC′: $-M''_D - 2M''_{CD} + k_2 M''_{CC'} + 6EI_v\theta = 0$ 6
Joint C: $M''_{CB} = M'_{CC} + M''_{CD}$ 7
Equilibrium: $M''_A + M''_B - M''_{CD} - M''_D = 0$ 8

Figure 14.24 Example of forming equations from assumed deformation

symmetry or negative symmetry to reduce the number of equations, the assumed deformations should be consistent with the loading case). The joint equations [No. 7, Case (a) and No. 7, Case (b)] are obtained by considering the members to be cut near to the joint, and subjected to positive moments as shown in Figure 14.25. It should also be noted that equations 6 of Case (a) and 6 of Case (b) can be obtained from the two preceding equations in each case; they are therefore not independent, but may be used for checking the accuracy of the solution.

The assumed directions of deformation are thus chosen quite arbitrarily, provided that they are consistent with the nature of the loading, but the signs of the terms in the equations must be consistent with the directions chosen. If they are assumed to be as shown in Figure 14.26, for example, the equations will be as shown in that figure. However, it is desirable to use the same assumptions and the same sign convention wherever possible, in order to reduce the chances of mistakes which might occur.

14.4 Applications to prestressed concrete

14.4.1 General

When the Theorem of Three Moments is applied to prestressed concrete at working load, the area between the prestressing cable and the longitudinal axis is similar in shape to the diagram of the primary bending moment. A visual indication of the magnitudes of the loading terms in the equations may be obtained, since these are directly proportional to the reactions of the imaginary load represented by the primary-moment diagrams. For example, the cable shown in Figure 14.27a produces a primary bending moment $M_p = Pe_a$; this is an upward or hogging moment and is therefore negative. The reaction at either support, of this diagram treated as a load, would therefore be $-\frac{1}{2}M_p\,l$, and the loading term N, which is $-6/l$ times the reaction, is $+\,3M_p$ for either support. Similarly the cable in Figure 14.27b would cause a downward, or sagging, moment, and the loading term for either support is $N = -3M_p$. Loading terms for a number of primary-moment diagrams are given in Chart 14.

14.4.2 Concordant cables

In Figure 14.27c a straight inclined cable is shown, and the reaction, and consequently the loading term, for support L is required to be zero. The following condition therefore applies.

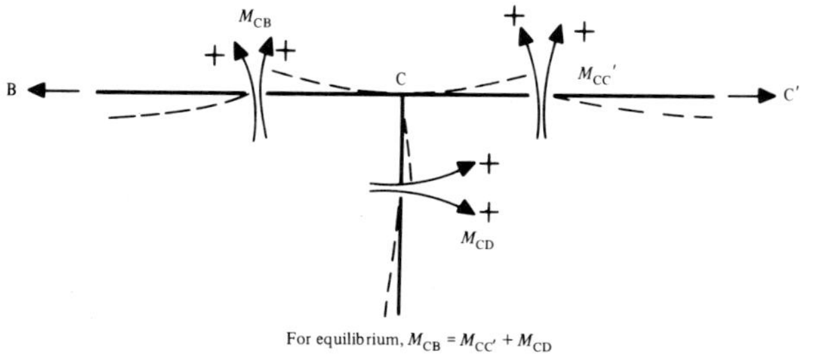

Figure 14.25 Equilibrium at a joint

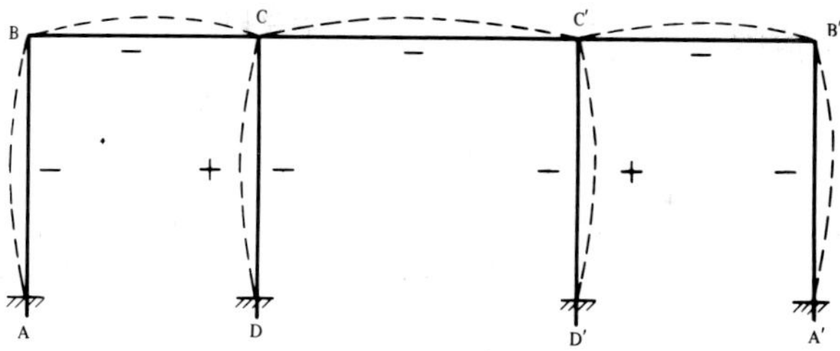

LOADING SYMMETRICAL – CASE (a)

AB:	$2M'_A - M'_B = 0$	1
ABC:	$M'_A - 2(1 + k_1)M'_B - k_1 M'_{CB} = k_1 N'_{BC}$	2
BCD:	$-k_1 M'_B - 2k_1 M'_{CB} + 2M'_{CD} + M'_D = k_1 N'_{CB}$	3
CD:	$-M'_{CD} - 2M'_D = 0$	4
BCC′:	$-k_3 M'_B - 2k_3 M'_{CB} - 3M'_{CC'} = k_3 N'_{CB} + N'_{CC'}$	5
Check DCC′:	$-M'_D - 2M'_{CD} - 3k_2 M'_{CC'} = k_2 N'_{CB} + N'_{CC'}$	6
Joint C:	$M'_{CB} = M'_{CC'} + M'_{CD}$	7

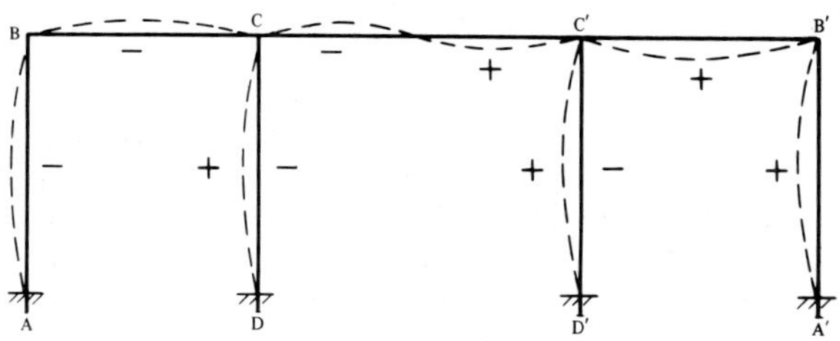

LOADING WITH NEGATIVE SYMMETRY – CASE (b)

AB:	$-2M''_A - M''_B - 6EI_v\theta = 0$	1
ABC:	$-M''_A - 2(1 + k_1)M''_B - k_1 M''_{CB} + 6EI_v\theta = k_1 N''_{BC}$	2
BCD:	$-k_1 M''_B - 2k_1 M''_{CB} + 2M''_{CD} + M''_D - 6EI_v\theta = k_1 N''_{CB}$	3
CD:	$-M''_{CD} - 2M''_D + 6EI_v\theta = 0$	4
BCC′:	$-k_3 M''_B - 2k_3 M''_{CB} - 3M''_{CC'} = k_3 N''_{CB}$	5
Check DCC′:	$-M''_D - 2M''_{CD} - 3k_2 M''_{CC'} + 6EI_v\theta = 0$	6
Joint C:	$M''_{CB} = M''_{CC'} + M''_{CD}$	7
Equilibrium:	$-M''_A - M''_B - M''_{CD} - M''_D = 0$	8

Figure 14.26 Variation to Figure 14.24

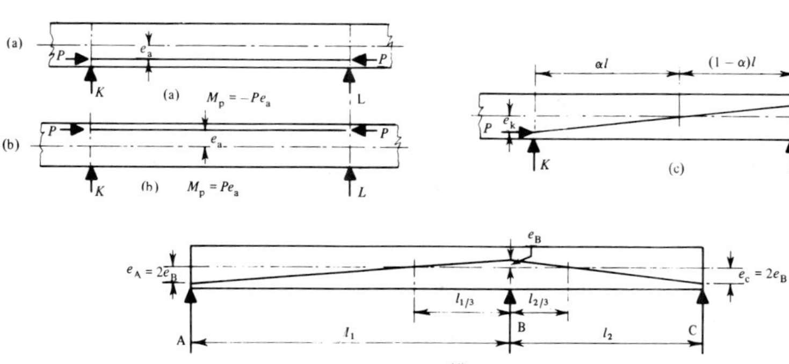

Figure 14.27 Effect of profile of cable

Reaction at L is proportional to

$$\frac{e_K \alpha l}{2} \frac{\alpha l}{3} - \frac{(1-\alpha)\, l e_L}{2} \left(l - \frac{(1-\alpha)}{3} l \right)$$

If the reaction is zero, then

$$\frac{e_K \alpha^2}{6} = e_L \left(\frac{1-\alpha}{2}\right)\left(\frac{2+\alpha}{3}\right)$$

Also

$$\frac{e_K}{e_L} = \frac{\alpha}{1-\alpha} \text{ and hence } \frac{\alpha^3}{1-\alpha} = (1-\alpha)(2+\alpha)$$

from which $\alpha = \frac{2}{3}$, and $e_K = 2e_L$.

If this cable profile were used for the continuous beam shown in Figure 14.27 d no secondary bending moment would be caused; such a profile, which produces no tendency to deflect at an intermediate support, is termed 'concordant'. More generally, when a cable is so placed that no change in the reactions at intermediate or end supports or distortions at fixed supports occurs at tensioning, then secondary bending moments do not develop. The profile of such a cable was termed 'concordant' by Guyon(3) as the line of pressure which it produces coincides with the cable itself.

The examples which follow illustrate a method for the determination of secondary bending moments and do not necessarily show the best position of the cables in the structures.

14.4.3 Beam with one end fixed

The beam shown in Figure 14.28 is fixed at one end and simply supported at the other. A straight prestressing cable is placed near the lower face. The effective cable profile is to be determined.

The prestressing force is P, and its eccentricity is e_a. The primary moment everywhere is equal to $-Pe_a$. Applying the Theorem of Three Moments: at the fixed end

$$2M_A\, l = +\, 3M_p\, l = 3Pe_a l$$

Hence

$$M_A = +\, 3/2\, Pe_a$$

The bending moment diagrams and the effective cable profile are given in Figure 14.28. It is clear that some counter-action is obtained at the support A despite the fact that the cable is apparently placed parallel.

If the eccentricity of the tendon varies uniformly from e_a at B to ke_a at A (Figure 14.29) then

$$2M_A\, l = +\, (2Pke_a + Pe_a) l$$

and

$$M_A = Pe_a\, (k + \tfrac{1}{2})$$

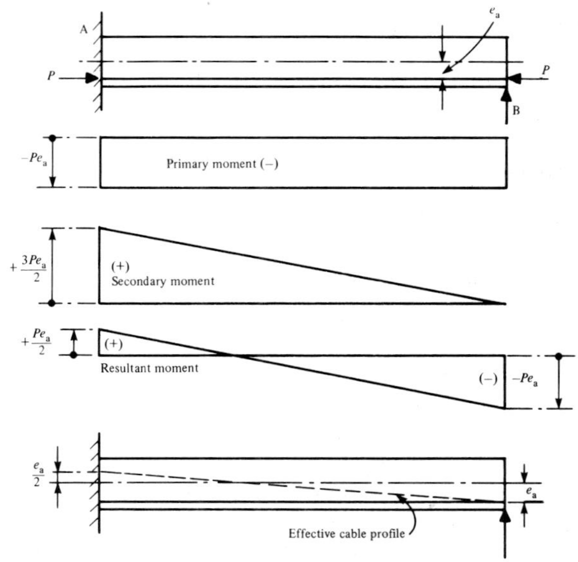

Figure 14.28 Secondary moment of parallel tendon (apparent constant eccentricity)

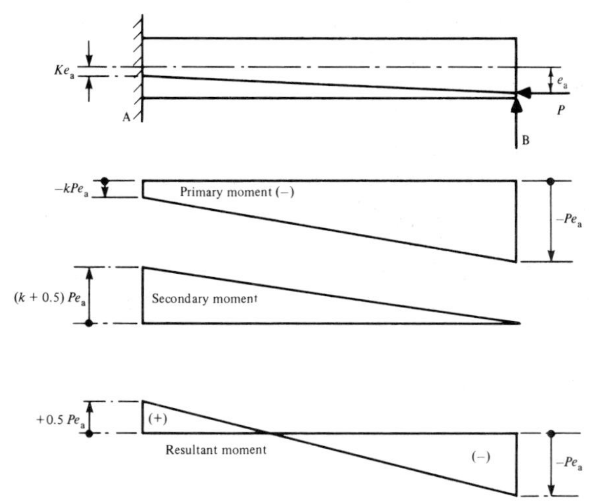

Figure 14.29 Resultant prestressing moment with cable of uniformly varying eccentricity (one end fixed the other simply supported)

The resultant prestressing moment at A is therefore $Pe_a\,(k + \frac{1}{2} - k) = + Pe_a/2$ and is independent of the value of k. In other words, whatever the eccentricity of the cable at A, the position of the effective cable profile is constant. It follows that, when $k = \frac{1}{2}$ (that is, when the actual and effective cable profiles coincide), $M_A = 0$ and the cable is concordant.

14.4.4 Beam with both ends fixed
A beam is fixed at both ends, and has a straight prestressing cable near the lower edge. The effective profile of the cable is required.

The primary moment is $-Pe_a$, and applying the Theorem of Three Moments,

$$2M_A l + M_B l = +3\,Pe_a l$$

Hence $M_A = M_B$ (by symmetry) $= +Pe_a$. The secondary bending moment therefore counteracts the primary prestressing moment, and the beam behaves as if the cable were coincident with the centroidal axis, the total prestressing moment being everywhere zero.

14.4.5 Continuous beam
The beam shown in Figure 14.30 is continuous over two spans, and the prestressing cable is straight in each span and is anchored on the centroidal axis. The effective profile of the cable is to be determined.

The primary moment at any point is $+Pe_x$, as shown in Figure 14.30, and applying the Theorem of Three Moments,

$$2M_B\,(l_1 + l_2) = -2Pe_B\,(l_1 + l_2)$$

Hence $M_B = -Pe_B$, and again the secondary moment counteracts the primary moment, the beam behaving as though the cable were everywhere coincident with the centroidal axis. It can be shown that this result is valid for any number of spans, with any linear variation of the position of the cable, provided that the anchorages are located at the centroidal axis.

14.4.6 The law of linear transformation
The examples given in the foregoing are illustrations of the Law of Linear Transformation, which may be stated as follows. A cable profile which changes direction only at supports, and has no eccentricity at simple end-supports, produces no resultant prestressing moment. If the position of a cable is altered in accordance with these conditions, no change occurs in the prestress.

This law was first recognized by Guyon[3], who published in 1945 a theory of design based on the conditions at working load only, and making use of this

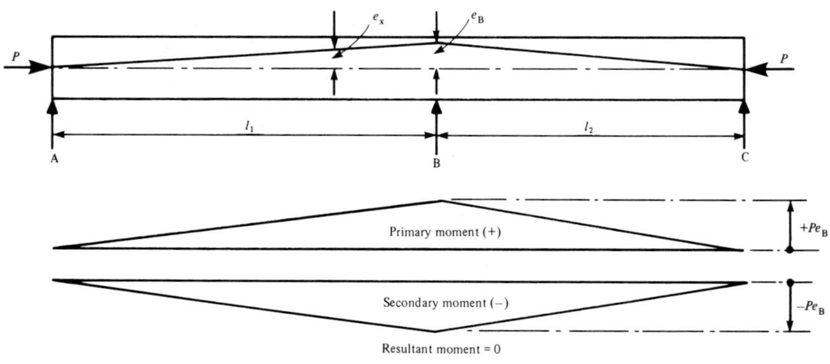

Figure 14.30 Law of linear transformation

Law of Linear Transformation, allowance being made for the frictional losses in curved cables and the probability of reversal of bending moment at any section. At that time the importance of conditions at ultimate load was not always appreciated, but in 1952 Guyon expressed the view that a structure designed in accordance with these principles automatically possesses suitable properties at ultimate load, because of the development of plastic hinges. While this might be the case in some instances, it is by no means certain that such hinges always develop. In structures subjected to rolling loads, for example, failure due to shearing may occur before plastic hinges can develop, as extensive cracking may occur near the supports at an early stage of overloading, leading to the development of diagonal cracks due to the higher shearing stress occurring in the reduced section. On the other hand, ultimate-load design based on ensuring the development of plastic hinges at pre-determined points, with a check on conditions at working load, has been advocated by Baker[4].

The author's opinion is that it is essential to consider conditions at ultimate load when selecting the profile of the cable; a cable placed solely in accordance with the law of linear transformation is not entirely satisfactory, as it is not related to ultimate load and does not entirely satisfy conditions at working load since the apparent counter-action indicated by the eccentricity of the cable is not in fact obtained, the resultant effect on the member being simply axial compression. The conditions at the supports, due to the load causing failure, may be improved by the linear transformation of a centroidal cable but those at mid-span are made worse, and transformed curved cables are sometimes provided instead (Figure 14.31). Since the design is based on the conditions at working load, the eccentricity of the curved cable after transformation is usually small, and the resistance to cracking is consequently greatly reduced in the event of overloading; sufficient tensioned steel should therefore be provided near the top and bottom of the member. As the curved cable acts as if it were anchored at the centroid, it is preferable, in the author's opinion, to provide two separate cables near the top and bottom instead of one near the centre, together with curved cables of greater eccentricity.

Reduction of the friction on a cable is important and may be achieved by haunching (Figure 14.32a) or by widening the upper flange (Figure 14.32b)

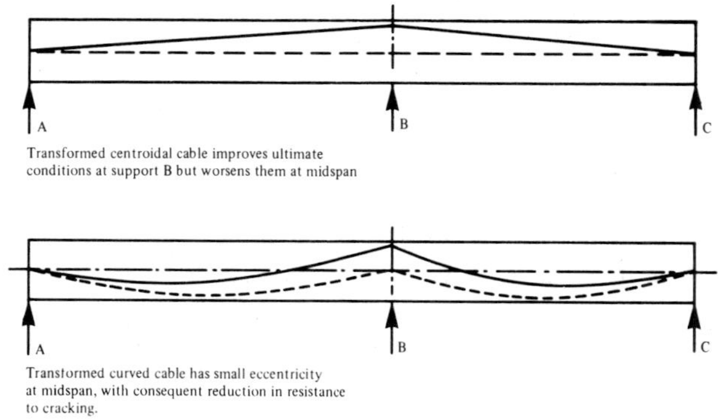

Figure 14.31 Effects of transformation of tendons

enabling a flatter profile to be used. Cap cables (Figure 14.32c) may also be provided for ultimate-load conditions when the large friction on the cable, and consequent reduction in the effective prestress are of minor importance. The profile of the beam can also be suitably adjusted (Figure 13.32d).

14.4.7 Selection of cable profile

The choice of profile depends on the following considerations in order of importance.

(1) The eccentricity should be as large as possible where the largest bending moment occurs at incipient failure.

(2) The total prestressing moment at any section should be sufficient to counteract to a certain extent the average bending moment at that section due to conditions at working load. Since a prestressing cable causes axial stresses in addition to bending stresses, the resultant stress, not the bending moment, is the criterion of design.

(3) The profiles of curved cables should be such as to produce the smallest possible frictional losses.

The following procedure is recommended.

Consider the bending moment envelope due to external loads. Note where a highly eccentric cable is required and sketch a suitable curve. This should be as flat as possible, to reduce friction.

Examine the area included between the cable profile and the centroidal axis, and improve the shape and magnitude of the secondary moment diagram if necessary by adjusting the profile of the cable.

In a continuous beam, the secondary moments should then be evaluated and they should preferably be added to the bending moments produced by the external loads in order to simplify the design procedure. If the profile has been well chosen, the secondary moment should not be excessive. In most cases, the stresses due to the combined bending moments are calculated and added to those due to the eccentric prestressing force causing the primary moment, to obtain the conditions at working load. Finally, the design is checked for ultimate conditions.

In a frame, the shortening of the separate members due to prestressing causes, in addition to the primary and secondary moments, tertiary moments which must be taken into account. If the separate members are not able to shorten freely when the prestress is applied, allowance for the restrained elastic

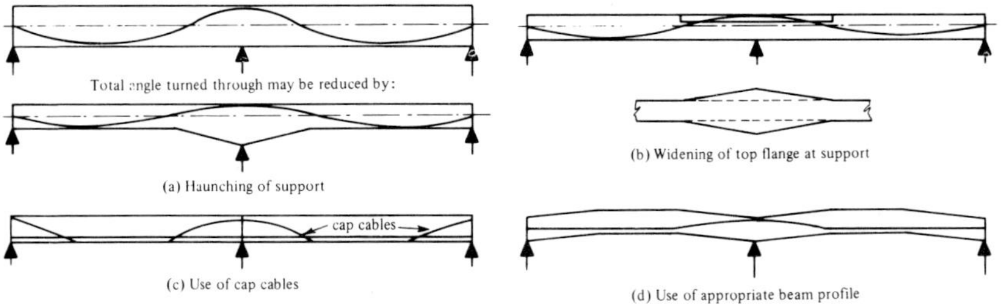

Figure 14.32 Reduction of friction

shortening must be made. Otherwise only the shortening due to creep need be considered. Under unfavourable conditions, the effect may give rise to moments as large as the bending moment due to the working load. In the case of continuous beams, longitudinal shortening must be allowed to take place when the prestress is applied or no prestress will exist in the beam, the prestressing force being transferred directly to the supports.

It is sometimes preferable to use precast freely-supported beams with straight prestressing wires near the bottom face to resist the dead load, and additional steel at the top over the supports to resist the live load. The steel over the supports may be tensioned if desired. Such an arrangement has the advantage that cracking in the area in tension is prevented; this is important when rolling loads are present, in order to avoid the early development of inclined cracks due to shear which might lead to early failure due to shearing forces.

14.5 Examples of the calculation of secondary bending moments

14.5.1 Beams with fixed ends

As the loading term N is directly proportional to the area of the bending moment diagram (when this is symmetrical), no secondary bending moment will occur when the area of a symmetrical primary moment diagram is zero. In the case of a prestressed beam with fixed ends, this condition can easily be arranged and the resulting cable profile is satisfactory. However, as the moments at the supports due to the load exceed that at midspan, it is better to arrange the cable profile so that a secondary moment is present which will increase the counter-action at the supports. This is obtained when the area of the primary moment diagram at midspan exceeds that near the supports (Figure 14.33). The total effect will then comprise (a) the counter-action due to the actual eccentricity of the cable, and (b) a secondary moment which increases the counter-action at the support and reduces it at midspan. Such a profile is shown in Figure 14.33. From Chart 14 for loading No. 6 ($b = c = 0.2; a = 0.6$),

$$N_{AB} = N_{BA} = +2a(a + 2b)Pe_a = +1.2\,Pe_a$$

and from loading No 10 ($a = 0.2$)

$$N_{AB} = N_{BA} = -a\left(4 - \frac{3a}{2} + \frac{3a}{2}\right)Pe_a = -0.8Pe_a$$

The combined values are therefore

$$N_{AB} = N_{BA} = +(1.2 - 0.8)\,Pe_a = +0.4Pe_a$$

By the Theorem of Three Moments,

$$2M_A\,l + M_B\,l = +0.4Pe_a\,l$$

By symmetry

$$M_A = M_B$$

and hence $M_A = M_B = +0.133\,Pe_a$.

The combined bending moment diagrams are shown in Figure 14.33.

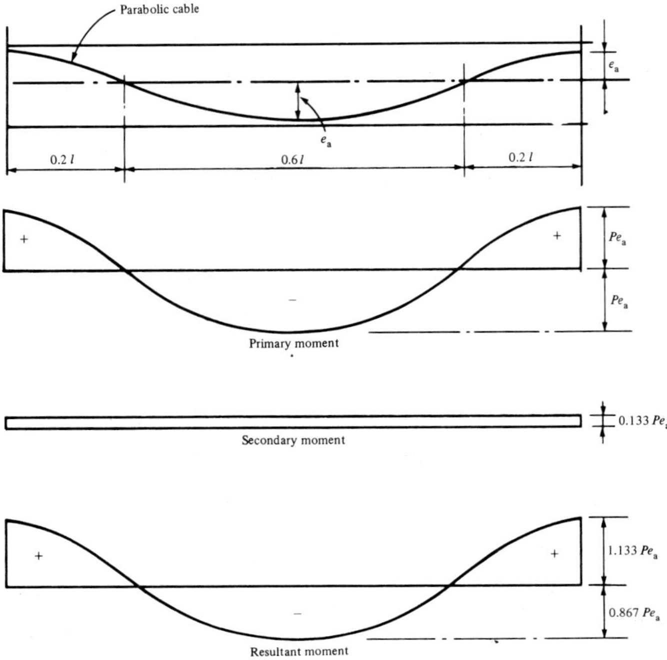

Figure 14.33 Prestressing moments on beams with fixed ends

14.5.2 Continuous beams

The bending moment diagram for a continuous beam with two equal spans is shown in Figure 14.34a and the average bending moment at working load is shown by the chain-dotted line. Alternative cable profiles with equal upper and lower eccentricities e_a which appear suitable are shown in Figure 14.34b.

From Chart 14, for profile No. 1, loadings 16 (for N_{AB}) and 8 (for N_{BA}),

$$N_{BA} = N_{BC} = -Pe_a \left[4 \times 0.25 \times \left(1 - \frac{0.75}{8}\right) - 2 \times (0.75)^2\right]$$

$$= +0.219\, Pe_a$$

and

$$M_B = +0.109\, Pe_a$$

For the alternative profile No. 2 shown in chain dots

$$N_{BA} = N_{BC} = -Pe_a \left[4 \times 0.3 \times \left(1 - \frac{0.9}{8}\right) - 2 \times (0.7)^2\right]$$

$$= -0.085\, Pe_a$$

Hence

$$2M_B = -0.085\, Pe_a \text{ and } M_B = -0.0425\, Pe_a$$

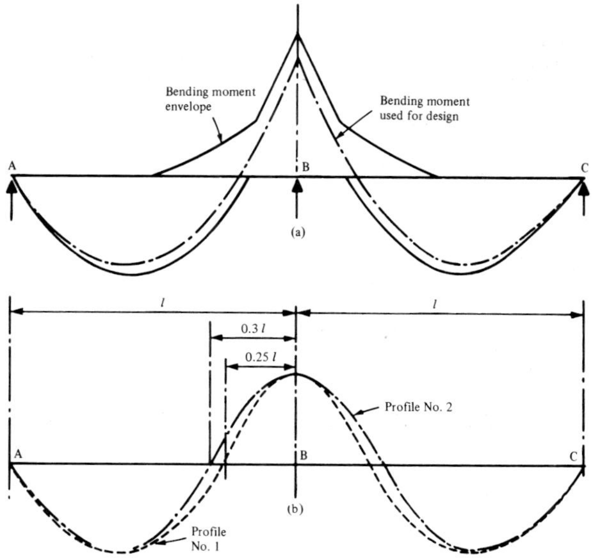

Figure 14.34 Alternative profiles of bending moment envelope and tendons for continuous beams of two equal spans

The effects of the secondary bending moments, when added to those due to the load, are shown in Figure 14.35a; and when added to the primary moments, in Figure 14.35b.

14.5.3 Single-bay frames: General solutions

In Figure 14.36 are shown two symmetrical single-bay frames, one having columns with fixed ends and the other with hinges at the bottoms of the columns. In the following, general solutions are obtained for the frames assuming the loading to be symmetrical; these are used to investigate the effects of uniform vertical load, lateral earth pressure, and secondary and tertiary moments.

$$k = \frac{l_v}{l_h}\frac{I_h}{I_v} \text{ (Figure 14.36)}$$

Frame with fixed ends

The general equations are

$$2k M_A + k M_B + \frac{6EI_h\theta}{l_h} = k N_{AB} \quad \ldots \quad (14.3)$$

$$k M_A + (2k + 3) M_B - \frac{6EI_h\theta}{l_h} = k N_{BA} + N_{BB'} \quad \ldots \quad (14.4)$$

Subtracting equation 14.3 from 2 × equation 14.4 gives

$$3(k + 2) M_B - \frac{18 EI_h\theta}{l_h} = -k N_{AB} + 2k N_{BA} + 2N_{BB'}$$

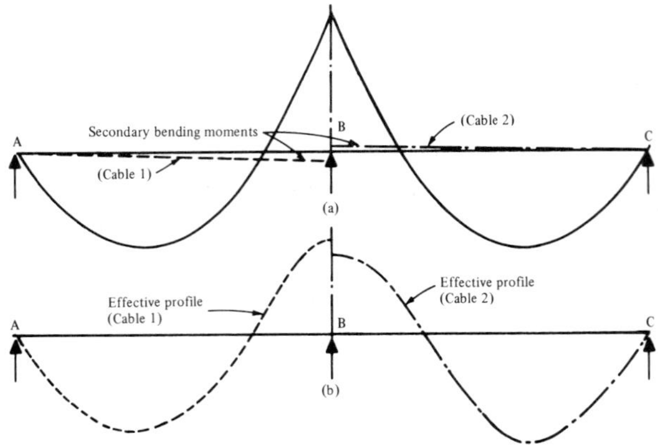

Figure 14.35 Effect of alternative tendon profiles

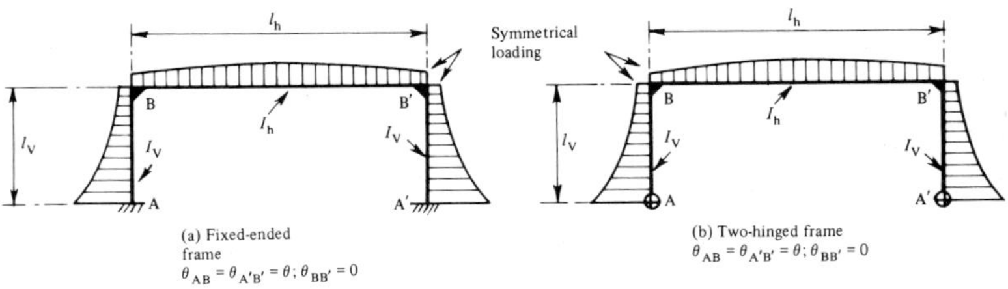

Figure 14.36 Single-bay frames

from which

$$M_B = \frac{6EI_h\theta}{(k+2)l_h} + \frac{1}{3(k+2)}(-kN_{AB} + 2kN_{BA} + 2N_{BB'})$$

Similarly, subtracting $k \times$ equation 14.4 from $(2k + 3) \times$ equation 14.3 gives

$$3k(k+2)M_A = -18(k+1)\frac{EI_h\theta}{l_h} + k(2k+3)N_{AB} - k^2 N_{BA} - kN_{BB'},$$

from which

$$M_A = -\frac{6(k+1)}{k(k+2)}\frac{EI_h\theta}{l_h} + \frac{2k+3}{3(k+2)}N_{AB} - \frac{(kN_{AB} + N_{BB'})}{3(k+2)}$$

Frame with two hinges

$$(2k+3)M_B - \frac{6EI_h\theta}{l_h} = kN_{BA} + N_{BB'}$$

Hence

$$M_B = \frac{6EI_h\theta}{(2k+3)l} + \frac{kN_{BA} + N_{BB'}}{(2k+3)}$$

These general solutions can be simplified by putting $\theta = 0$, which is true for all cases of symmetrical loading; that is, in which the vertical displacement of one end of the beam relative to the other is zero. This condition is satisfied in all of the following cases.

14.5.4 Single bay frame: Uniform vertical load

Refer to Figure 14.37a

$$N_{AB} = N_{BA} = 0; N_{BB'} = -\frac{wl_h^2}{4}$$

Frame with fixed ends [Case II (a)]

$$M_B = \frac{-2}{k+2}\frac{wl_h^2}{12}; M_A = +\frac{1}{k+2}\frac{wl_h^2}{12}$$

Frame with two hinges [Case II (b)]

$$M_B = \frac{-3}{2k+3}\frac{wl_h^2}{12}$$

14.5.5 Single-bay frame: Lateral earth pressure

Refer to Figure 14.37b.

$$N_{BB'} = 0; N_{AB} = -\frac{w_h l_v^2}{4} - \frac{2w_e l_v^2}{15}; N_{BA} = -\frac{w_h l_v^2}{4} - \frac{7w_e l_v^2}{60}$$

Frame with fixed ends [Case III(a)]

$$M_B = \frac{-k}{3(k+2)}\left(\frac{w_h l_v^2}{4} + \frac{w_e l_v^2}{10}\right)$$

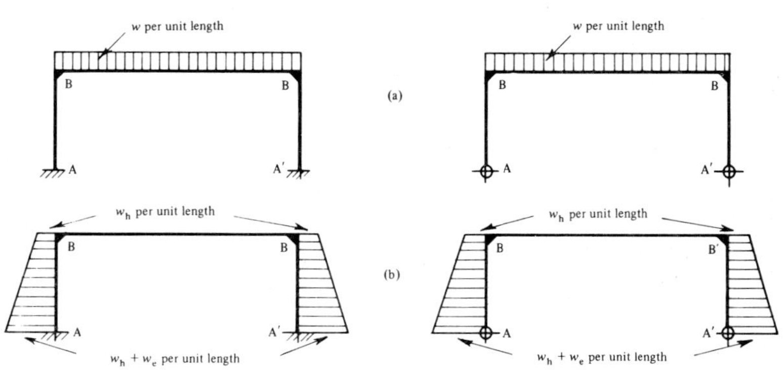

Figure 14.37 Symmetrical, vertical and horizontal loading for single bay fixed and two-hinged frame

$$M_A = +\frac{1}{3(k+2)}\left(\frac{(k+3)\,w_h\,l_v^2}{4} + \frac{(3k+1)\,w_e\,l_v^2}{20}\right)$$

Frame with two hinges [Case III (b)]

$$M_B = -\frac{k}{2k+3}\left(\frac{w_h\,l_v^2}{4} + \frac{7w_e\,l_v^2}{60}\right)$$

When the span of the horizontal member of the frame is large and its depth is small, for instance when $l_v/l_h = 1/10$ and $l_v \geqslant I_h$ the factor k approaches zero and the bending moments approach the following values.

Case II (a) $M_B \to -\frac{wl_h^2}{12}$; $M_A \to +\frac{wl_h^2}{24}$

Case II (b) $M_B \to -\frac{wl_h^2}{12}$

Case III (a) $M_B \to 0; M_A \to +\frac{w_h\,l_v^2}{8} + \frac{w_e\,l_v^2}{40}$

Case III (b) $M_B \to 0$

14.5.6 Single-bay frame: Secondary bending moments

Frame with fixed ends

To determine the secondary bending moments, the profiles of the cables must be assumed (Figure 14.38). The cable in the beam is assumed to be concordant; those in the columns are positioned to resist the pressure of the earth and the expected tertiary moment caused by the shortening of the beam.

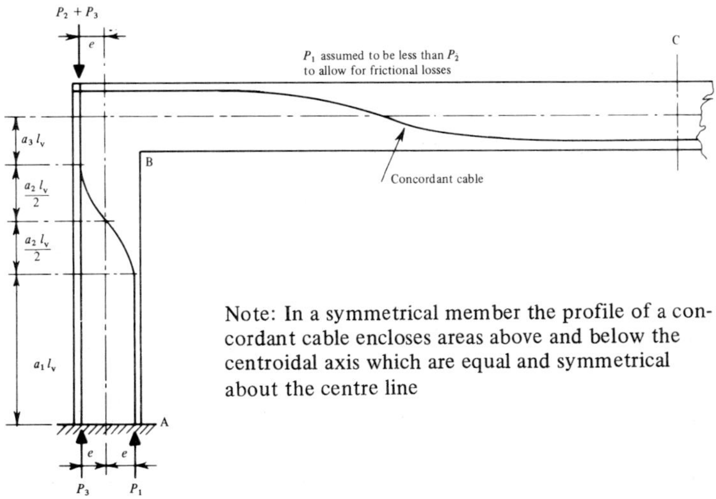

Figure 14.38 Cable profile in a fixed single-bay frame

This would cause tensile stresses at the outer faces of the vertical members. A cable is therefore positioned to counteract these stresses. An approximate method of allowing for the effects of friction is also shown in Figure 14.38. The force P_2 at the top of the curved cable in the column is assumed to exceed the force P_1 at the bottom by an amount equal to the anticipated frictional loss.

The resulting primary-moment diagrams and the corresponding loading terms are shown in Figure 14.39. Since θ and $N_{BB'}$ are both zero, it is apparent that as k approaches zero, M_B tends to zero and M_A to $N_{AB}/2$.

Frame with two hinges

$$\theta = 0; N_{BB'} = 0$$

Hence as $k \to 0, M_B \to 0$

14.5.7 Single-bay frame: Tertiary bending moments

Assuming the change in length of the horizontal member is Δl, the horizontal deflection at the top of each column is $\Delta l/2$; hence $\theta = \Delta l/2l_v$ (Figure 14.40). The strain in the horizontal member is $\Delta l/l = \epsilon_{BB'}$.

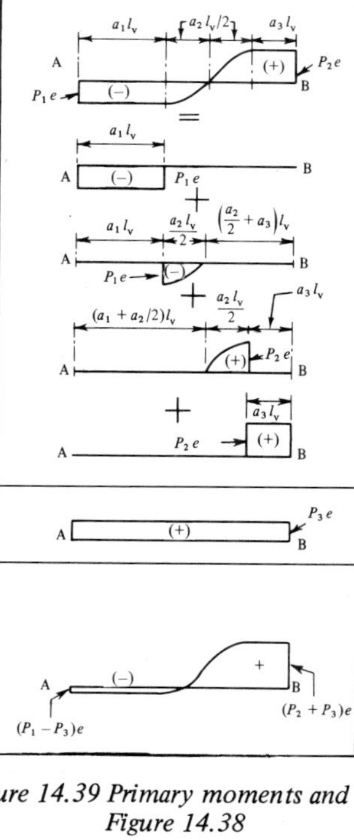

Case No	Primary moments	Loading terms N_{AB}	Loading terms N_{BA}
5		$+3a_1(2-a_1)P_1e$	$+3a_1^2P_1e$
12		$+a_2\left(2a_3+\frac{13a_2}{8}\right)P_1e$	$+a_2\left(2a_1+\frac{3a_2}{8}\right)P_1e$
12		$-a_2\left(2a_3+\frac{3a_2}{8}\right)P_2e$	$-a_2\left(2a_1+\frac{13a_2}{8}\right)P_2e$
5		$-3a_3^2P_2e$	$-3a_3(2-a_3)P_2e$
9		$-3P_3e$	$-3P_3e$
Total		$P_1e\left(6a_1-3a_1^2+2a_3a_2+\frac{13a_2^2}{8}\right)$ $-P_2e\left(\frac{3a_2^2}{8}+2a_2a_3+3a_3^2\right)$ $-3P_3e$	$P_1e\left(3a_1^2+2a_2a_1+\frac{3a_2^2}{8}\right)$ $-P_2e\left(\frac{13a_2^2}{8}+2a_1a_2+6a_3-3a_3^2\right)$ $-3P_3e$

Figure 14.39 Primary moments and loading terms due to cable profile in Figure 14.38

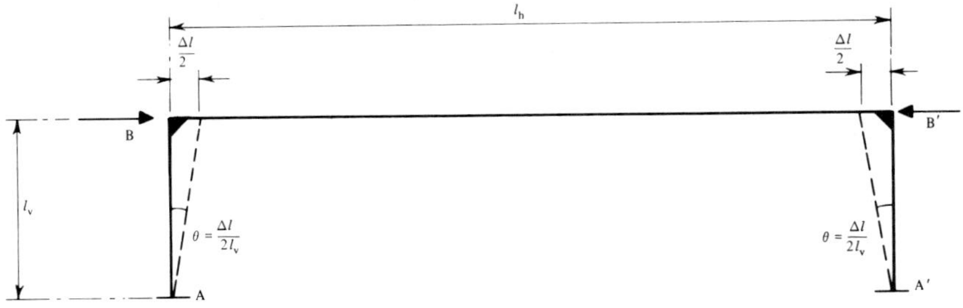

Figure 14.40 Tertiary moment due to axial deformation

Frame with fixed ends

$$N_{AB} = N_{BA} = N_{BB'} = 0$$

$$M_B = \frac{6}{k+2}\frac{EI_h\theta}{l_h} = \frac{6}{k+2}\frac{EI_h}{l_h}\frac{\Delta l}{2l_v} = \frac{3}{k+2}\frac{EI_h}{l_v}\epsilon_{BB'}$$

Similarly

$$M_A = -\frac{3(k+1)}{k(k+2)}\frac{EI_h}{l_h}\epsilon_{BB'}$$

As $k \to 0$, $M_B \to \frac{1.5EI_h}{l_v}\epsilon_{BB'}$ and $M_A \to \infty$; the reasons for this result and its effect on the structure are considered in 14.5.8.

Frame with two hinges

$$N_{AB} = N_{BA} = N_{BB'} = 0; M_A = 0$$

$$M_B = \frac{6}{2k+3}\frac{EI_h\theta}{l_h} = \frac{6}{2k+3}\frac{EI_h}{l_h}\frac{\Delta l}{2l_v} = \frac{3}{2k+3}\frac{EI_h}{l_v}\epsilon_{BB'}$$

As $k \to 0$, $M_B \to \frac{EI_h}{l_v}\epsilon_{BB'}$

14.5.8 Effect of horizontal reactions on prestressing force

General notes

The tertiary moments due to the longitudinal shortening caused by the prestressing force give rise to horizontal reactions at the bottoms of the columns. A portion of each frame is shown in Figure 14.41, from which it is clear that, to produce a prestressing force P in the longitudinal member, a force of $P + \Delta P$, in which ΔP is the horizontal reaction, must be applied to the frame. The magnitude of the horizontal reaction is given by the following expressions.

Frame with fixed ends

$$\Delta P = \frac{M_B - M_A}{l_h} = \frac{EI_h}{l_v^2}\epsilon_{BB'}\left(\frac{3}{k+2} + \frac{k+3}{k(k+2)}\right) = \frac{3(2k+1)}{k(k+2)}\frac{EI_h}{l_v^2}\epsilon_{BB'}$$

As $k \to 0$, $\Delta P \to \infty$; so that as the stiffness of the vertical members, relative to the horizontal members, approaches infinity, the frame tends towards the system shown in Figure 14.42.

Frame with two hinges

$$\Delta P = \frac{M_B}{l_v} = \frac{3}{2k+3}\frac{EI_h}{l_v^2}\epsilon_{BB'}$$

As $k \to 0$, $\Delta P \to \frac{EI_h}{l_v^2}\epsilon_{BB'}$

Example

The tertiary moments and reactions may, in severe cases, exceed the primary and secondary moments and reactions. Consider the case shown in Figure 14.43 for which $k = 0.1$, $l_h/l_v = 1/10$, $I_v = I_h$. If the frame has fixed ends, the tertiary moments and reactions are

$$M_A = -\frac{3.3}{0.21} \times \frac{10EI_h}{l_h} \times \epsilon_{BB'} = -157.1\frac{EI_h}{l_h}\epsilon_{BB'}$$

$$M_B = \frac{3}{2.1} \times \frac{10EI_h}{l_h} \times \epsilon_{BB'} = +\,14.3\frac{EI_h}{l_h}\epsilon_{BB'} \quad \left(\simeq \frac{M_A}{k}\right)$$

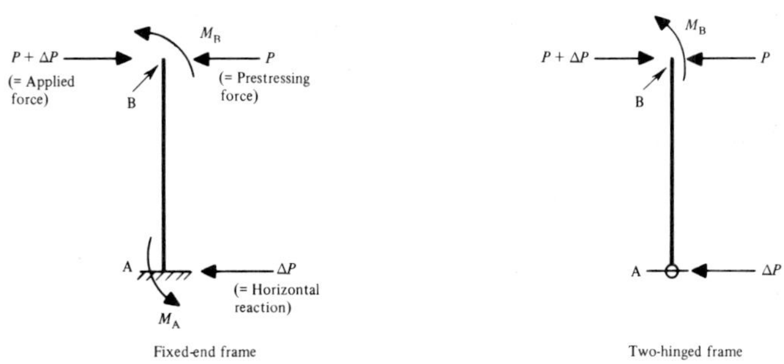

Figure 14.41 Effect of horizontal reactions on prestressing force

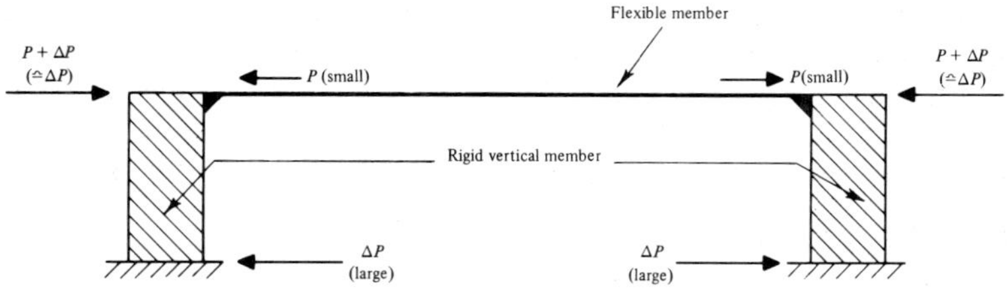

Figure 14.42 The effect of secondary and tertiary bending moments

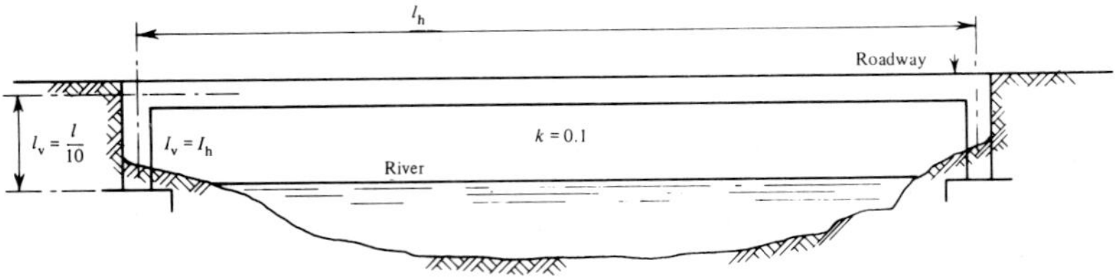

Figure 14.43 Example to Figure 14.42

$$\Delta P = \frac{3.6}{0.21} \times \frac{100EI_h}{l_h^2} \times \epsilon_{BB'} = 1714\frac{EI_h}{l_h^2}\epsilon_{BB'}$$

The values for the frame with two hinges are

$M_A = 0$

$$M_B = \frac{3}{3.2} \times \frac{10EI_h}{l_h} \times \epsilon_{BB'} = +9.4\frac{EI_h}{l_h}\epsilon_{BB'}$$

$$\Delta P = \frac{3}{3.2} \times \frac{100EI_h}{l_h^2} \times \epsilon_{BB'} = 9.4\frac{EI_h}{l_h^2}\epsilon_{BB'}$$

If the bending moment at working load at the centre of the beam is M_C, the greatest stress in the concrete at working load is 3000 lbf/in^2 (213 kgf/cm^2; 20.5 N/mm^2) the mean stress is 1200 lbf/in^2 (84 kgf/cm^2; 8.27 N/mm^2), the depth of the beam $h = l_h/30$, and E is 5×10^6 lbf/in^2 (3.5×10^5 kgf/cm^2; 34.5 kN/mm^2), then

$$I_h = \frac{M_C}{3000} \times \frac{h}{2} = \frac{M_C h}{6000} \text{ and } P = 1200A = 1200\frac{I_h}{i^2} \text{ or } \frac{I_h}{i^2} = \frac{P}{1200}$$

in which $i = \sqrt{h^2/12}$ is the radius of gyration. For a rectangle, $i^2 = h^2/12$; for an I-section the value is greater and is here assumed to be $h^2/7.5$.

Therefore

$$\frac{I_h}{l_h^2} = \frac{I_h}{900h^2} = \frac{I_h}{6750i^2} = \frac{P}{8.1 \times 10^6}$$

and

$$\frac{I_h}{l_h} = \frac{M_C h}{6000\, l_h} = \frac{M_C}{180\,000}$$

Substituting these values in the expressions given in the foregoing,

For the frame with fixed ends

$$M_A = -\frac{157.1 \times 5 \times 10^6 \times M_C \times \epsilon_{BB'}}{180\,000} = -4363\,\epsilon_{BB'} M_C$$

$$M_B = +\frac{14.3 \times 5 \times 10^6 \times M_C \times \epsilon_{BB'}}{180\ 000} = +389\ \epsilon_{BB'} M_C$$

$$\Delta P = \frac{1714 \times 5 \times 10^6 \times P \times \epsilon_{BB'}}{8.1 \times 10^6} = 1055\ \epsilon_{BB'} P$$

For the frame with two hinges

$$M_A = 0$$

$$M_B = \frac{9.4 \times 5 \times 10^6 \times M_C \times \epsilon_{BB'}}{180\ 000} = +261\ \epsilon_{BB'} M_C$$

$$\Delta P = \frac{94 \times 5 \times 10^6 \times P \times \epsilon_{BB'}}{8.1 \times 10^6} = 58\ \epsilon_{BB'} P$$

The strain $\epsilon_{BB'}$ will be the sum of the elastic shortening and the movements due to shrinkage and creep and can be assessed as follows.

LEAST PROBABLE STRAIN

Elastic shortening	$\frac{1200}{5 \times 10^6}$	$= 2.4 \times 10^{-4}$
Shrinkage	0.02%	$= 2.0 \times 10^{-4}$
Creep = 0.5 × elastic shortening		$= 1.2 \times 10^{-4}$
	$\epsilon_{BB'}$ minimum	$= 5.6 \times 10^{-4}$

GREATEST PROBABLE STRAIN

Elastic shortening: $\frac{1200}{5 \times 10^6}$	$= 2.4 \times 10^{-4}$
Shrinkage (moist curing before stressing): 0.03%	$= 3.0 \times 10^{-4}$
Creep = 1.5 × elastic shortening	$= 3.6 \times 10^{-4}$
$\epsilon_{BB'}$ maximum	$= 9.0 \times 10^{-4}$

For the frame with fixed ends the maximum and minimum tertiary moments and reactions are given by

$M_A = -3.66 M_C$ (min) to $-5.89 M_C$ (max)

$M_B = +0.33 M_C$ (min) to $+0.54 M_C$ (max)

$\Delta P = 0.59P$ (min) to $0.95P$ (max)

and for the frame with two hinges by

$M_A = 0$

$M_B = +0.22 M_C$ (min) to $+0.35 M_C$ (max)

$\Delta P = 0.033P$ (min) to $0.052P$ (max)

In order to produce a prestressing force of P in the horizontal member of the frame with fixed ends, it would therefore be necessary to apply a force of $1.59P$ to $1.95P$, depending on the conditions, and tertiary moments would be between $0.33\,M_C$ and $0.54\,M_C$ at the support B and throughout the member BB$'$ and between $-3.66M_C$ and $-5.89\,M_C$ at support A. The magnitudes and ranges of uncertainty of these values are both so great that alternative methods of construction (for example, prestressing the longitudinal member before casting joints B and B$'$) should be adopted. In the case of the frame with two hinges, the tertiary moment M_B is two-thirds of that for the fixed frame and a force of between $1.033P$ and $1.052P$ would be needed to obtain a prestressing force P. These effects must be taken into account in designing the structure, but the magnitudes and ranges of uncertainty are not such as to necessitate a change in the method of construction.

It is clear that the magnitudes of the tertiary effects are such that it would be dangerous to ignore them, particularly as they involve both an increase in the bending moment and a reduction in the available prestressing force. While the frames considered in the foregoing are an extreme case, the effects should always be included in the analysis of statically-indeterminate prestressed framed structures, and proper allowance made at all stages of design and construction.

14.6 Continuity with precast elements

Up till now statically-indeterminate structures formed by post-tensioning cast in place concrete have been discussed. The same principles also apply when precast segments are post-tensioned together to form the statically-indeterminate system. No load comes on the structure until it is completed.

In practice it is often advantageous and economic to design a structure in such a way that for dead load the construction is statically determinate and the continuity is provided only for super-load. Standard units with pre-tensioned steel can then be suitably utilized to form statically-indeterminate member. Normally the continuity is achieved either by adding suitably reinforced in-situ concrete and/or by introducing rigid connections (such as welding) at supports.

The flexural design for the limit-state of collapse condition is virtually the same for all three constructions above. The critical negative and positive bending moments due to external loading are determined either by plastic-hinge or yield-line method; or sometimes even by using conventional elastic method with or without allowing suitable redistribution of moment due to plastic flow. The moment of resistance of the member at any section can be calculated using any standard method given in Chapter 7. As the high bending moment is invariably accompanied by high shear the sections (specially near the supports) need to be checked for the combined effect of flexure and shear as discussed in Chapter 7.

The design for serviceability conditions is, however, different for this type of construction. The design is done in two stages: first as a simply supported member supporting the self-weight and construction load, and then in the completed form as a statically-indeterminate structure supporting the super-load. As pre-tensioned units are used the problem of parasitic moment due to cable profile does not arise. Additional secondary moments, however, do develop due to

(a) creep rotation restraint of pre-tensioned elements, and
(b) differential shrinkage in case of in-situ topping.

The effect of these two over the support are of opposite nature; the creep rotation restraint introducing positive moment (bottom tension) and differential shrinkage, when present, causing negative moment (top tension) at the support. Figure 14.44 which is reproduced from reference (5) illustrates the deformation and restraint moment due to creep in a two-span beam.

The simplified formula for calculation of restraint moments in a two-span continuous girder due to creep is

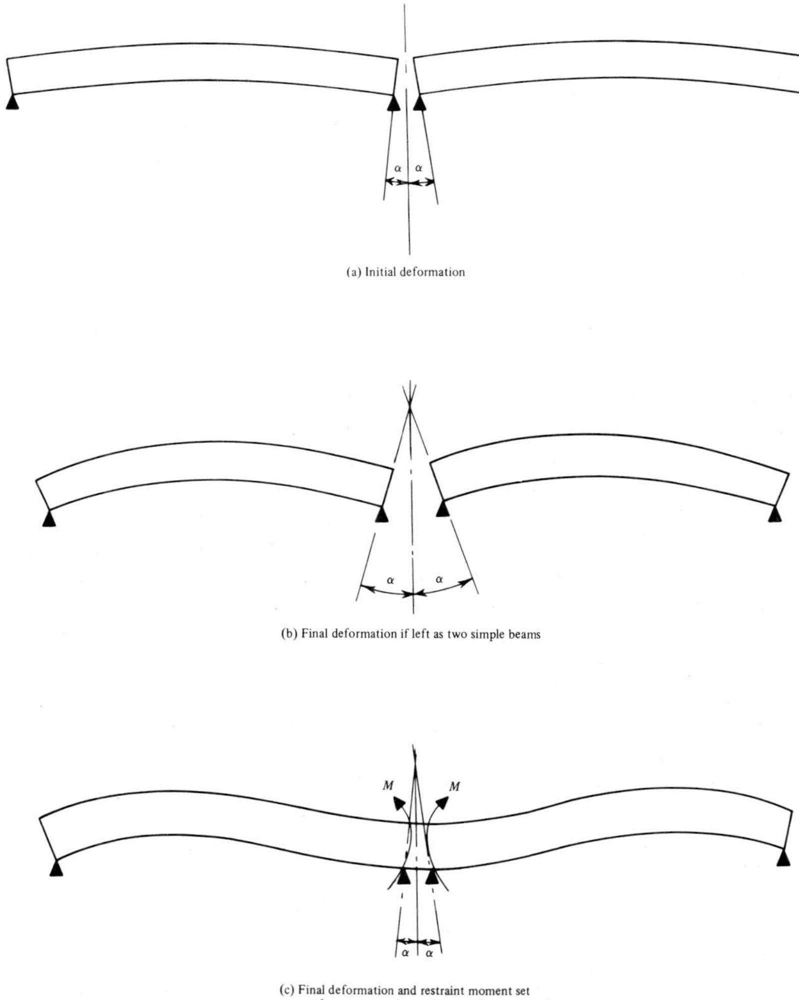

Figure 14.44 Effect of creep rotation restraint

$$(3/2\,M_p - M_d)\,(1 - e^{-\phi})$$

where M_p is the product of effective prestressing force and its eccentricity with respect to the centroid of the full composite section; M_d is the dead-load moment at the support considering continuous girder; e is the base of Naperian logarithm and ϕ is the creep factor usually 2.

The formula for the restraint moment in continuous girder due to differential shrinkage is

$$3/2\,M_\Delta\left(\frac{1 - e^{-\phi}}{\phi}\right)$$

where M_Δ is the differential shrinkage moment as given in equation 11.15.

14.6.1 Example

It is often necessary to use standard pre-tensioned units to form statically indeterminate structures. In the following example a design of a two-hinged portal of 90 ft span (27.44 m) consisting of standard double-T vertical and horizontal members has been presented. The vertical double-T units are placed in position plumbed levelled and adequately propped. Horizontal double-T units are then placed which span in a freely supported condition for their own weight. The next operation is to form rigid connections at supports by site-welding specially cast well-anchored plates in the horizontal and vertical double-T units. For any subsequent loading (such as finishes, live, wind, thermal, creep and shrinkage) the construction acts as a two-hinged portal.

Figure 14.45a and 14.45b show respectively the cross-section of the horizontal and vertical double-T units. The vertical members are generally uniformly prestressed to avoid the problem of differential camber between adjacent units.

Figure 14.46 shows the portal construction.

Horizontal double-T units

These are 32 in. (81.3 cm) deep units, 95½ in. (2.426 m) wide with ribs at 4 ft (1.22 m) centres. The units are prestressed with 24 No. ½ in. (12.7 mm) Dyform strands, each initially stressed to 70 per cent of its ultimate, i.e. to 32 900 lbf (14 920 kgf; 146 kN) and 7 No. 0.2 in. (5 mm) diameter wires each initially stressed to 4900 lbf (2220 kgf; 21.8 kN) i.e. 70 per cent of ultimate.

Initial prestressing force = 823 900 lbf (373 700 kgf; 3664 kN)

Average initial stress in steel = 187 800 lbf/in^2 (13 160 kgf/cm^2, 1297 N/mm^2)

Concrete strength is 7500 lbf/in^2 (525 kgf/cm^2; 52 N/mm^2) at transfer and 9000 lbf/in^2 (630 kgf/cm^2; 62 N/mm^2) at 28 days. $E_c = 5.75 \times 10^6$ lbf/in^2 (4×10^5 kgf/cm^2; 39.6 kN/mm^2). The difference between E_{ct} and E_{c28} has been ignored being very small.

Sectional properties

A_0 = 534.6 in^2 (3449 cm^2)

e_t = 10.1 in. (25.65 cm)

e_b = 21.9 in. (55.65 cm)

I_0 = 50 720 in^4 (2 111 000 cm^4)

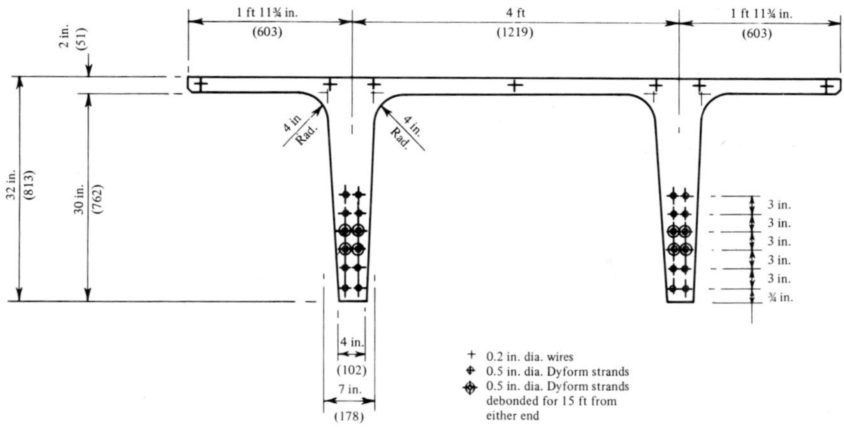

Figure 14.45(a) Cross-section of horizontal double-T unit (example 14.8.1)

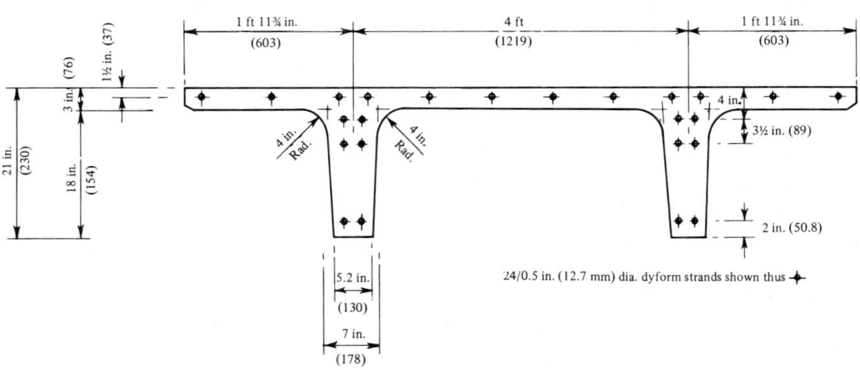

Figure 14.45(b) Cross-section of vertical double-T unit (example 14.8.1)

Z_t = 5030 in^3 (82 427 cm^3)

Z_b = 2318 in^3 (37 985 cm^3)

C.G. of prestressing force = 10.2 in. (25.63 cm) from the bottom face
Therefore eccentricity = 11.7 in. (29.72 cm)

Nominal stresses due to initial prestressing force (at midspan)

f_{bI} = +5700 lbf/in^2 (399 kgf/cm^2; 39.3 N/mm^2)

f_{tI} = −373 lbf/in^2 (26 kgf/cm^2; 2.57 N/mm^2)

$$\therefore\ f_{sI} = \frac{(5700 + 373)}{32} \times 21.8 - 373 = 3769 \text{ lbf/in}^2\ (264 \text{ kgf/cm}^2;\ 26 \text{ N/mm}^2)$$

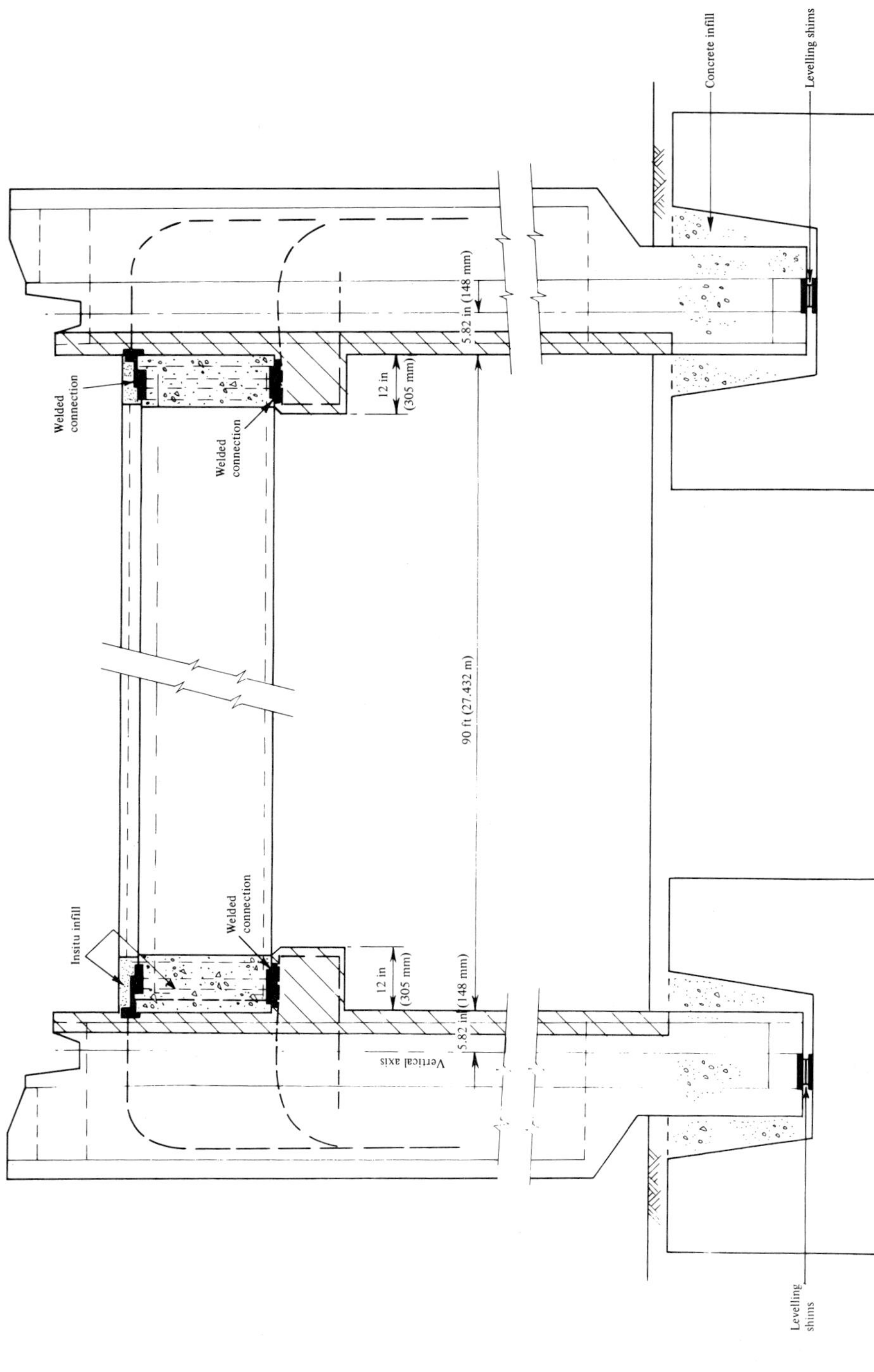

Figure 14.46 Portal frame

Assume 9 per cent loss due to elastic shortening, then

$f_{st} = 0.91 \times 3769 = 3430 \text{ lbf/in}^2$

$\therefore \; \Delta f_{pe} = 4.90 \times 3430 = 16\,810 \text{ lbf/in}^2$

9% of 187 800 = 16 890 lbf/in²

As the difference is very small 9 per cent loss due to elastic shortening has been taken.

Therefore the stresses at transfer (due to prestress only):

$f_{bT} = 0.91 \times 5700 = 5187 \text{ lbf/in}^2$ (363 kgf/cm²; 35.8 N/mm²)

$f_{tT} = -0.91 \times 373 = -339 \text{ lbf/in}^2$ (23.7 kgf/cm²; 2.34 N/mm²)

Self-weight bending moment (at midspan)

$M_d = 575 \times 90^2 \times 1.5 = 6\,986\,300 \text{ lbf in.}$

$$f_{bG} = -\frac{6\,986\,300}{2318} = -3013 \text{ lbf/in}^2 \text{ (211 kgf/cm}^2\text{; 20.8 N/mm}^2\text{)}$$

$$f_{tG} = \frac{6\,986\,300}{5030} = 1389 \text{ lbf/in}^2 \text{ (97 kgf/cm}^2\text{; 9.6 N/mm}^2\text{)}$$

Therefore resultant stresses at transfer (at midspan)

$$f_{bt} = 5187 - 3013 = 2174 \text{ lbf/in}^2 < \frac{7500}{2} \text{ Hence O.K.}$$

$f_{tt} = 1389 - 339 = 1050 \text{ lbf/in}^2$

$$f_{st} = \frac{(2174 - 1050)}{32} \times 21.8 + 1050 = 1816 \text{ lbf/in}^2$$

At debonded section: 8 strands debonded for 15 ft length
Prestressing force (initial) = 16 × 32 900 + 7 × 4900 = 526 400 lbf

Average initial stress in steel = 175 000 lbf/in²
C.G. of prestressing force = 10.58 in. from bottom
Therefore eccentricity = 11.32 in.

Nominal initial stresses are

$f_{bI} = 3556 \text{ lbf/in}^2$

$f_{tI} = -200 \text{ lbf/in}^2$

$f_{sI} = 2310 \text{ lbf/in}^2$

Assume 6 per cent loss due to elastic shortening then
$f_{st} = 0.94 \times 2310 = 2170 \text{ lbf/in}^2$

$\therefore\ \Delta f_{pe} = 4.90 \times 2065 = 10\,630$ lbf/in²

6% of 175 000 = 10 500 lbf/in²

Therefore the stresses at transfer (due to prestress only)

$f_{bT} = 0.94 \times 3556 = 3340$ lbf/in²

$f_{tT} = -0.94 \times 200 = -188$ lbf/in²

$f_{sT} = 2173$ lbf/in²

f_{bT} in fact will be less than 3340 as there will be small counter-action at the end of the transmission length and some relaxation loss also will occur.

Maximum losses (at midspan section)

(a) Elastic shortening	= 16 890 lbf/in²
(b) Relaxation at 2½% of initial stress	= 4 700 lbf/in²
(c) Creep $0.33 \times 10^{-6} \times 28 \times 10^{6} \times 1816$	= 16 780 lbf/in²
(d) Shrinkage $300 \times 10^{-6} \times 28 \times 10^{6}$	= 8 400 lbf/in²
	46 770 lbf/in²

Therefore percentage loss $= \dfrac{46\,770 \times 100}{187\,800} = 24.9\%$, say 25%

The same final losses of prestress have been assumed in the central 60 ft of the beam (where all strands are bonded).

Therefore effective prestress (at midspan)

$f_{bE} = 0.75 \times 5700 = 4275$ lbf/in² (299 kgf/cm²; 29.5 N/mm²)

$f_{tE} = -0.75 \times 373 = -280$ lbf/in² (19.6 kgf/cm²; 1.93 N/mm²)

Maximum losses at debonded section (15 ft from either end),

(a) Elastic shortening	= 10 500 lbf/in²
(b) Relaxation at 2½% of 175 000	= 4 400 lbf/in²
(c) Creep = $0.9 \times 0.33 \times 10^{-6} \times 28 \times 10^{6} \times 2173$	= 18 000 lbf/in²
(d) Shrinkage	= 8 400 lbf/in²
	41 300 lbf/in²

Therefore percentage loss $= \dfrac{41\,300 \times 100}{175\,000} = 23.6\%$, say 24%

Therefore effective prestress (at debonded section),

$f_{bE} = 0.76 \times 3556 = 2700$ lbf/in² (177 kgf/cm²; 17.4 N/mm²)

$f_{tE} = -0.76 \times 200 = -152$ lbf/in^2 (5 kgf/cm^2; 0.5 N/mm^2)

Vertical double-T units

These 21 in. (53.34 cm) double-tees are made from the standard 20 in. (50.8 cm) deep mould adding 1 in. (2.54 cm) extra thickness at the top. The units are uniformly prestressed with 24 No. ½ in. (12.7 mm) diameter Dyform strands each stressed to 32 900 lbf (14 920 kgf; 146.00 kN)

Initial prestressing force = 789 600 lbf (358 200 kgf; 3512 kN)

Initial stress in steel = 189 000 lbf/in^2 (13 230 kgf/cm^2; 1303 N/mm^2)

Concrete strength similar to that in horizontal double-T units.

Sectional properties

A_0 = 519.83 in^2 (3353.74 cm^2)

e_t = 5.82 in. (14.78 cm)

e_b = 15.18 in. (38.56 cm)

I_0 = 18 703 in^4 (778 500 cm^4)

Z_t = 3213 in^3 (52 650 cm^3)

Z_b = 1232 in^3 (20 190 cm^3)

Here e_b and Z_b relate to outer face whereas e_t and Z_t relate to inside face of vertical double-T units

$$\text{C.G. of prestress} = \frac{(12 \times 1.5 + 4 \times 16 + 4 \times 7.5 + 4 \times 19)}{24}$$

= 5.83 in. from inside face

∴ eccentricity = 0.01 in. – ignored

Therefore nominal initial stress

$$f_{sI} = f_{tI} = f_{bI} = \frac{789\,600}{519.73} = 1519 \text{ lbf/in}^2 \text{ (106 kgf/cm}^2\text{; 10.5 N/mm}^2\text{)}$$

Assume loss due to elastic shortening 4 per cent then,

f_{st} = 0.960 × 1519 = 1460 lbf/in^2

f_{pe} = 4.9 × 1460 = 7150 lbf/in^2

4% of 189 000 = 7560 lbf/in^2

Therefore assumption is correct for all practical purposes.

Checking stresses at transfer is not necessary as the stress level is low.

Maximum losses

(a) Elastic shortening $= 7\,500$ lbf/in²

(b) Relaxation 2½% of initial $= 4\,700$ lbf/in²

(c) Creep $0.33 \times 10^{-6} \times 28 \times 10^{6} \times 1460 = 13\,490$ lbf/in²

(d) Shrinkage $300 \times 10^{-6} \times 28 \times 10^{6} = 8\,400$ lbf/in²

$34\,090$ lbf/in²

$$\text{Therefore percentage loss} = \frac{34\,090 \times 100}{189\,000} = 18\%$$

Therefore effective uniform prestress $= 0.82 \times 1519 = 1246$ lbf/in² (87 kgf/cm²; 8.6 N/mm²)

DESIGN OF PORTAL FOR SERVICEABILITY CONDITION

Since the main purpose of this example is to show the secondary effect due to creep and shrinkage etc. only the serviceability condition is considered.

Stage 1. Horizontal double-T unit is placed in position: Freely supported condition

Figure 14.47 shows the loading

$$M_G \text{ at midspan} = \frac{575 \times 89^2}{8} \times 12 = 6\,832\,000 \text{ lbf in. (78 704 kgf m; 771 kN m)}$$

$$\text{Reaction on bracket} = \frac{575 \times 89}{2} = 25\,590 \text{ lbf (11 600 kgf; 114 kN)}$$

Stage 2. Finishes applied: Portal action

Loading: (Figure 14.48)

(a) Pointing at joints of double-T units $=$ 6 lbf/ft²

(b) Insulation $=$ 2 lbf/ft²

(c) Asphalt $=$ 12 lbf/ft²

$20 \text{ lbf/ft}^2 \times 8$

$= 160$ lbf/ft (238 kgf/m; 2.33 kN/m)

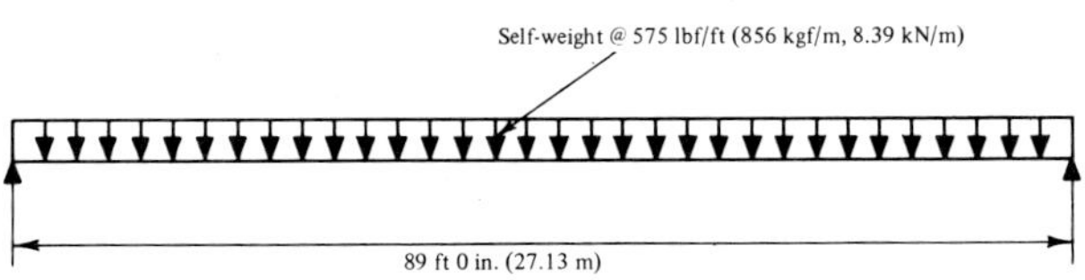

Figure 14.47 Loading due to own weight of double T (example 14.8.1 – stage 1)

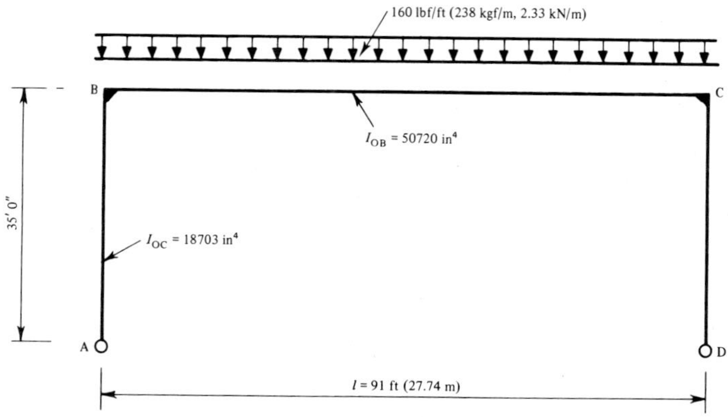

Figure 14.48 Permanent superload on the portal (example 14.8.1 – stage 2)

Let $R = \dfrac{I_{OB}}{I_{OC}} \times \dfrac{H}{l} = \dfrac{50\,720}{18\,703} \times \dfrac{35}{91} = 1.04$

$X = 2R + 3 = 2 \times 1.04 + 3 = 5.08$

Therefore, M at B and C = $(-)\ \dfrac{160 \times 91^2 \times 12}{4 \times 5.08} = -782\,400$ lbf in. (9.013 kgf m; 88 kN m) (Tension outer face)

Assume 10 per cent redistribution

M at B and C = 0.9 × 782 400 = –704 160 lbf in. (8110 kgf m; 79.5 kN m)

Therefore moment at midspan

$= \dfrac{160 \times 91^2 \times 12}{8} - 704\,160 = 1\,283\,300$ lbf in. (14 783 kgf m; 145 kN m)

Reaction at vertical supports

$= \dfrac{160 \times 91}{2} = 7280$ lbf (3300 kgf; 32.38 kN)

Stage 3. Live load operative

Loading = 15 lbf/ft^2

Therefore, load per ft run

= 120 lbf/ft (179 kgf/m; 1.75 kN/m)

Moment at B and C after 10 per cent adjustment

$= -\dfrac{704\,160}{160} \times 120 = -528\,100$ lbf in. (6 084 kgf m; 59.6 kN m)

Moment at midspan

$$= \frac{120 \times 91^2 \times 12}{8} - 528\ 100 = 962\ 480 \text{ lbf in. } (11\ 100 \text{ kgf m, } 109 \text{ kN m})$$

$$\text{Reaction on vertical support} = \frac{120 \times 91}{2} = 5460 \text{ lbf } (2477 \text{ kgf; } 24.3 \text{ kN})$$

Stage 4. Wind load

Let 113 lbf/ft be the total wind loading of which 66 lbf/ft represents pressure on the windward face and 47 lbf/ft represents suction on the leeward side [the wind loading has been derived in accordance with reference (6) for Class C for a basic wind gust speed of 44 m/sec (98.5 mph)]. Figure 14.49 shows the loading.

Moment at B due to w_1

$$= \frac{66 \times 35^2}{4} \left[-\frac{1.04}{2 \times 5.08} + 1 \right]$$

= 18 140 lbf ft (tension inside face or bottom)

Moment at B due to w_2

$$-\frac{47 \times 35^2}{4} \times \left[-\frac{1.04}{2 \times 5.08} - 1 \right]$$

$$= \frac{47 \times 1225}{4} \times 1.102$$

= 15 860 lbf ft (tension inside or bottom face)

Moment at C due to w_1

$$= \frac{66 \times 35^2}{4} \times \left(-\frac{1.04}{2 \times 5.08} - 1 \right)$$

= −22 270 lbf ft. (tension at outer or top face)

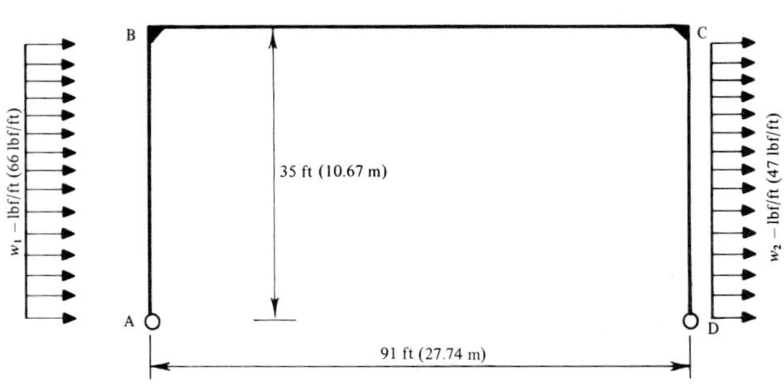

Figure 14.49 Wind loading on the portal (example 14.8.1 stage 4)

Moment at C due to w_2

$$= -\frac{47 \times 35^2}{4}\left(-\frac{1.04}{2 \times 5.08} + 1\right)$$

$= -12\ 900$ lbf ft. (tension at outer or top face)

Reaction at A or D $= \pm \dfrac{113 \times 35^2}{2 \times 90} = \pm 760$ lbf (344 kgf; 3.4 kN)

$\therefore$ Maximum (+)ve moment at joint = (18 140 + 15 860)

= 34 000 lbf ft.

= 408 000 lbf in. (4700 kgf m; 46.1 kN m)

Maximum (–)ve moment at joint = – (22 270 + 12 900) = –35 170 lbf ft

= –422 000 lbf in. (4860 kgf m; 48 kN m)

SECONDARY EFFECTS

(1) *Effect of creep rotation restraint*

Assume the units will be erected one month after casting. At that time two thirds of shrinkage and creep will most probably occur.

Loss of prestress after one month (fully prestressed area)

(a) Elastic shortening	=	16 890 lbf/in^2
(b) Relaxation, say 75%	=	3 520 lbf/in^2
(c) Creep = 16 780 $\times \frac{2}{3}$	=	11 200 lbf/in^2
(d) Shrinkage = 8400 $\times \frac{2}{3}$	=	5 600 lbf/in^2
		37 210 lbf/in^2

Therefore percentage loss $= \dfrac{37\ 210}{187\ 800} \times 100 = 19.8\%$, say 20%

In the debonded zone probable losses after one month

(a) Elastic shortening	=	10 500 lbf/in^2
(b) Relaxation 75%	=	3 300 lbf/in^2
(c) Creep = 18 000 $\times \frac{2}{3}$	=	12 000 lbf/in^2
(d) Shrinkage = 8400 $\times \frac{2}{3}$	=	5 600 lbf/in^2
		31 400 lbf/in^2

Percentage loss $= \dfrac{31\ 400 \times 100}{175\ 000} \doteqdot 18\%$

Therefore effective prestress (in fully prestressed area)

$= 0.80 \times 823\ 900 = 659\ 100$ lbf at erection

and

$= 0.75 \times 823\ 900 = 617\ 900$ lbf after final losses

Effective prestress in debonded area

$= 0.82 \times 526\ 400 = 431\ 600$ lbf at erection

and

$0.76 \times 526\ 400 = 400\ 000$ lbf after all losses

Modified E value of concrete

At erection (i.e. after one month)

$$\text{Elastic strain per unit stress} = \frac{1}{5.75 \times 10^6} = 0.173 \times 10^{-6}$$

$$\text{Specific creep strain} = \tfrac{2}{3} \times 0.33 \times 10^{-6} = \underline{0.220 \times 10^{-6}}$$

$$\text{Total strain per unit stress} = 0.393 \times 10^{-6}$$

$$\therefore\ E_{c1} = \frac{1}{0.393 \times 10^{-6}} = 2.54 \times 10^6 \text{ lbf/in}^2$$

Modified E value after final losses

$$E_{c2} = \frac{1}{(0.173 + 0.33) \times 10^{-6}} = 1.98 \times 10^6 \text{ lbf/in}^2$$

Referring to Figures 14.50 and 14.51

$$y_1 \text{ (after one month)} = \frac{431\ 600 \times 11.32}{2.54 \times 10^6\, I_0} = \frac{1.92}{I_0}$$

$$y_1 \text{ (after final losses)} = \frac{400\ 000 \times 11.32}{1.98 \times 10^6\, I_0} = \frac{2.28}{I_0}$$

$$y_2 \text{ (after one month)} = \frac{659\ 100 \times 11.7}{2.54 \times 10^6\, I_0} = \frac{3.03}{I_0}$$

$$y_2 \text{ (after final losses)} = \frac{617\ 900 \times 11.7}{1.98 \times 10^6\, I_0} = \frac{3.65}{I_0}$$

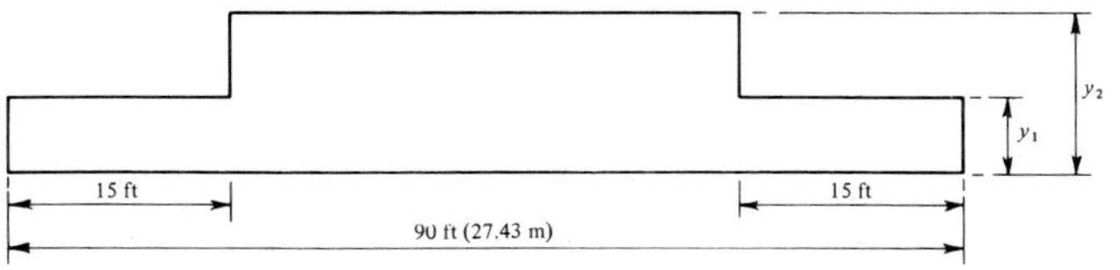

Figure 14.50 Prestressing moment diagrams (example 14.8.1)

$$y_3 \text{ (after one month)} = \left(\frac{575 \times 90^2 \times 12}{8}\right) \times \frac{1}{2.54 \times 10^6\, I_0} = \frac{2.75}{I_0}$$

$$y_3 \text{ (after final losses)} = \left(\frac{575 \times 90^2 \times 12}{8}\right) \times \frac{1}{1.98 \times 10^6 \times I_0} = \frac{3.52}{I_0}$$

Therefore slope at B or C at the time of erection (one month)

$$= (1.92 \times 15 \times 12 + 3.03 \times 30 \times 12 - 2.75 \times 45 \times 12 \times \tfrac{2}{3}) \times \frac{1}{I_0} = \frac{446.4}{I_0}$$

Total slope at B and C (if full creep can take place)

$$(2.28 \times 15 \times 12 + 3.65 \times 30 \times 12 - 3.52 \times 45 \times 12 \times \tfrac{2}{3}) \times \frac{1}{I_0} = \frac{457.4}{I_0}$$

Therefore rotation restrained by welding $= \dfrac{11}{I_0}$

Let M_c = moment required to restrain the rotation then

$$\frac{M_c \times 45 \times 12}{E I_0} = \frac{11}{I_0}$$

or

$$M_c = \frac{11 \times 5.75 \times 10^6}{45 \times 12} = 117\,100 \text{ lbf in. (1350 kgf m; 13.2 kN m) here 28 day}$$

E value will be appropriate

(2) *Effect of axial deformation due to creep and shrinkage*

(a) The equivalent axial deformation due to creep ϵ_{ec} occurring after the restraint is applied is given by the simplified equation

$$\epsilon_{ec} = (1 - e^{-\phi})\, \epsilon$$

where $\epsilon = \dfrac{Pl}{AE}$, P being the prestressing for after appropriate losses

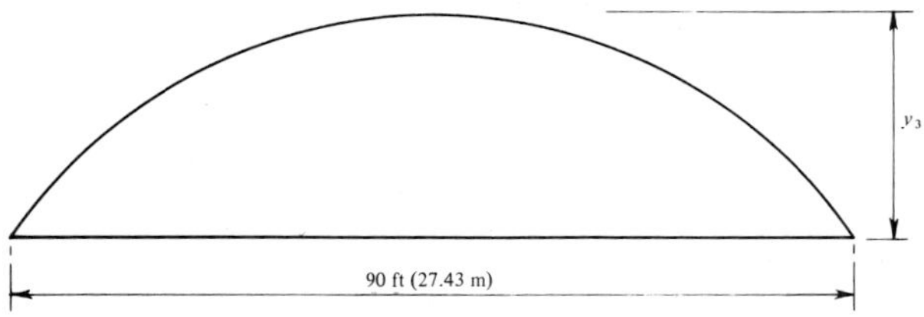

Figure 14.51 Self weight bending moment diagram (example 14.8.1)

$$\frac{Pl}{AE} = \frac{617\,900 \times 90 \times 12}{534.6 \times 5.75 \times 10^6} = 0.217 \quad E = \text{Young's modulus of concrete} -$$

generally 28-day value will suffice

$\therefore \ \epsilon_{ec} = 0.87 \times 0.217 = 0.189$ in.

(b) Equivalent axial deformation due to shrinkage is given by the equation

$\epsilon_{es} = \left(\frac{1 - e^{-\phi}}{\phi}\right) \epsilon_s l$; ϵ_s = shrinkage strain between the time of restraint and final

$= 0.43 \times 100 \times 10^{-6} \times 90 \times 12$

$= 0.0464$ in.

Therefore total axial deformation due to creep and shrinkage

$= (0.189 + 0.464) = 0.235$ in.

Therefore at either end $\epsilon = 0.1175$ in.

Fixed-end moment at B (span AB) = F.E.M at C (span CD)

$$= \frac{3\,EI\,\epsilon}{l^2} = \frac{3 \times 5.75 \times 10^6 \times 18\,700 \times 0.1175}{35^2 \times 144}$$

= 214 900 lbf in.

Either by Hardy Cross method or by Three-Moment Theory the actual moment at joint B and C works out to be +93 200 lbf in. (1075 kgf m; 1.05 kN m). This moment will produce tensile stress at the inside face of vertical member and bottom face of the horizontal member.

RESULTANT STRESSES AT MIDSPAN SECTION

Moment at stage 1 = 6 832 000 lbf in.

Moment at stage 2 = 1 283 300 lbf in.

Moment at stage 3 = 962 480 lbf in.

9 077 780 lbf in.

Moment at stage 4 (due to wind) – negligible

Secondary moments

(1) Creep rotation restraint	= 117 100 lbf in.
(2) Axial deformation due to creep and shrinkage	= 93 200 lbf in.
Total moment	9 288 080 lbf in.

$$\therefore f_{bw} = 4275 - \frac{9\,288\,080}{2318} = 4275 - 4020$$

= 255 lbf/in^2 (compression) (18 kgf/cm^2; 1.75 N/mm^2)

Bending moment of horizontal member at supports ignoring relief due to secondary effect

Stage 1 = 0 lbf in.

Stage 2 = –704 160 lbf in.

Stage 3 = –528 000 lbf in.

Stage 4 = –422 000 lbf in.

–1 654 260 lbf in.

This bending moment is at support where the section is virtually reinforced concrete. The prestress section is effective at the end of the transmission length, i.e. about 2 ft away from the point where these moments are calculated. In accordance with CP 110, however, 80 per cent of Stage 3 and Stage 4 (i.e. imposed and wind) moments need to be taken into account.

Without any reduction of moment within the transmission length resultant stresses are as follows:

$$f_{tw} = -\frac{1\,654\,260}{5030} - 152 = -481 \text{ lbf/in}^2 \text{ (tension) } (33.6 \text{ kgf/cm}^2;\ 3.32 \text{ N/mm}^2)$$

$$f_{bw} = 2700 + \frac{1\,654\,260}{2318} = 3413 \text{ lbf/in}^2 \text{ (compression)} (239 \text{ kgf/cm}^2;\ 23.5 \text{ N/mm}^2)$$

This is O.K. allowing temporary overtress due to wind.

Reinforcement necessary at support section $z \doteqdot 30$ in.

$$A_s = \frac{1\,654\,260}{(30\,000 \times 1.25) \times 30} = 1.47 \text{ in}^2 \ (948 \text{ mm}^2)$$

2–25 mm diameter bars (one per rib) will do. Area provided = 1.52 in^2 (491 mm^2)

STRESSES IN VERTICAL DOUBLE-T UNIT

Stage	*Direct load*	*Eccentricity*	*Bending moment*
1	25 590 lbf	× 12 =	307 000 lbf in.
2	7 280 lbf		704 160 lbf in.
3	5 460 lbf		528 100 lbf in.
4	760 lbf		422 000 lbf in.
Total =	39 090 lbf		1 961 260 lbf in
	(17 700 kgf; 173 kN)		(22 600 kgf m; 222 kN in.

Therefore resultant stress

$$f_{bw} \text{ (outer face)} = 1246 + \frac{39\,090}{519.83} - \frac{1\,961\,260}{1232}$$

$$= -271 \text{ lbf/in}^2 \text{ (tension) } (19 \text{ kgf/cm}^2; 1.89 \text{ N/mm}^2)$$

$$f_{tw} \text{ (inner face)} = 1246 + \frac{39\,090}{519.83} + \frac{1\,961\,260}{3213}$$

$$= 1932 \text{ lbf/in}^2 \text{ (compression) } (135 \text{ kgf/cm}^2; 13 \text{ N/mm}^2)$$

14.6.2 Example of a continuous bridge over two spans

In the following, an interesting example is shown relating to a continuous bridge over two spans, composed of two precast prestressed standard I-girders, spaced at 6 ft (1.83 m), each 75 ft (22.8 m) long, and with a third member of the same type 64 ft (19.55 m) long to form together with an additional concrete slab made in place and post-tensioning tendons, a continuous two-span construction. According to Figures 14.52 and 14.53 there are two temporary supports. The beams are 4 ft (1.22 m) deep and the cast in-situ slab is 7 in. (178 mm) thick. The connections between the individual parts are formed by cast-in-place concrete joints. The outer precast members have pre-tensioned tendons (12 strands 0.5 in. (12.7 mm) dia.), whereas the centre member contains only ordinary reinforcement. The post-tensioning tendons comprise 28 single strands. This is

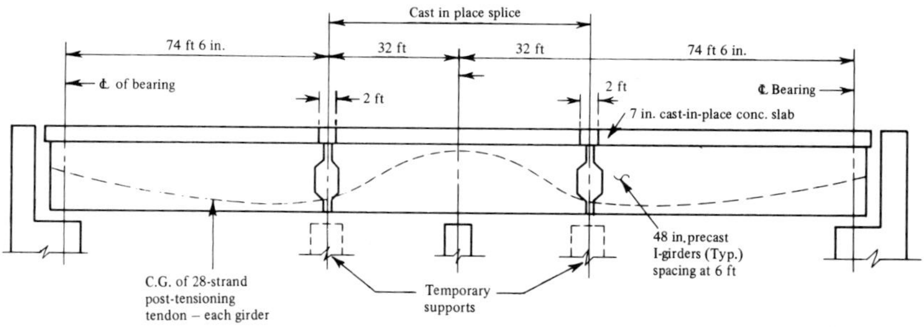

Figure 14.52 Prestressed segmental continuous girder (example 14.8.2)

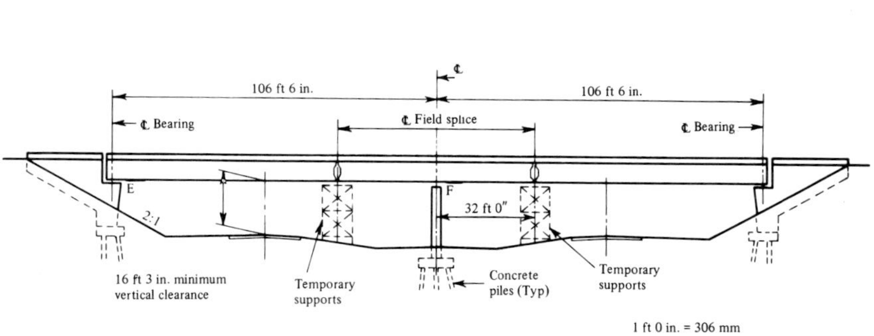

Figure 14.53 Two-span continuous girder formed by three precast segments post-tensioned together (example 14.8.2)

based on a report in the PCI Bridge Bulletin, First Quarter 1974 and relates to a bridge for Platt County for the Illinois Department of Transportation. As seen in the figures two spans of bridge each 106 ft 6 in. (32.50 m) long are formed. This design is based on a succesful test on a prototype construction of a similar construction[6].

REFERENCES

1. PIPPARD, A. J. S. and BAKER, J. F. *The analysis of engineering structures.* Edward Arnold Publ. Ltd, London 3rd Ed. 1957.
2. ABELES, P. W. *The four moment theorem.* Concrete and Constructional Eng, April, May and June 1942, London, based on a paper in *Beton und Eisen,* 1924.
3. GUYON, Y. A Study of continuous beams and of statically redundant systems in prestressed concrete. (Circ. J. No. 8 20th Sept. 1945 of Institut Technique du Bâtiment et des Travaux Publiques) CACC Translation No. 33, London 1951.
4. BAKER, A. L. L. A Plastic theory of design for ordinary reinforced and prestressed concrete including moment redistribution in continuous members. *Magazine of Concrete Research,* June 1949, Vol. 1 No. 2.
5. MATTOCK, A. H. Precast prestressed concrete bridge: creep and shrinkage studies. Bulletin D46. Journal of PCA Research and Development Laboratories, Vol. 3, No. 2, 1961, p. 33.
6. BRITISH STANDARDS INSTITUTION CP3: Part 3: 1972 Wind loads.
7. Giant precast prestressed segmental composite girder load tested successfully. Journal of the Prestressed Concrete Institute, September-October 1973, Vol. 18 No. 5.

CHAPTER 15

APPLICATIONS OF PRESTRESSING

Prestressing can be applied with benefit to many types of structure. In this chapter some brief notes are given on some of the applications which are commonly encountered; these are intended to be indicative rather than exhaustive, as complete descriptions of many of these types of structure would fill several books. In addition, there are a number of applications of a highly specialized nature which most designers are unlikely to encounter; these include cases like prestressing of dams, roads, and airport runways, and are not covered here.

15.1 Structural members in building

15.1.1 Floors

The design of beams and floors was discussed and illustrated in previous chapters, and is not repeated here. There is also a large range of proprietary floor and roof units – but they are subject to so much variation, that no attempt has been made to describe them in detail. Some of the types which are widely used are shown in Figures 15.1 and 15.2; they comprise hollow sections (15.1a and 15.1b), inverted channels 15.1c, and joist and tile construction 15.1d, as well as the double T, single T, and wide inverted channel units. (Figure 15.2.)

These units may be used to span between supporting beams of prestressed concrete, reinforced concrete, or steel. They can often be installed cheaply, but

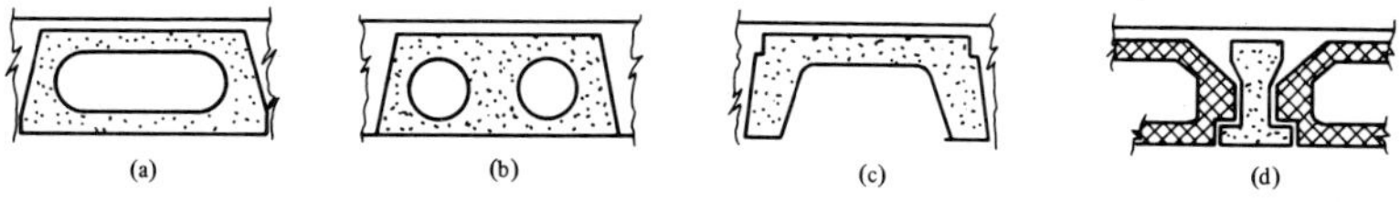

Figure 15.1 Typical floor units spanning normally up to 30 ft 0 in. (9.2 m)

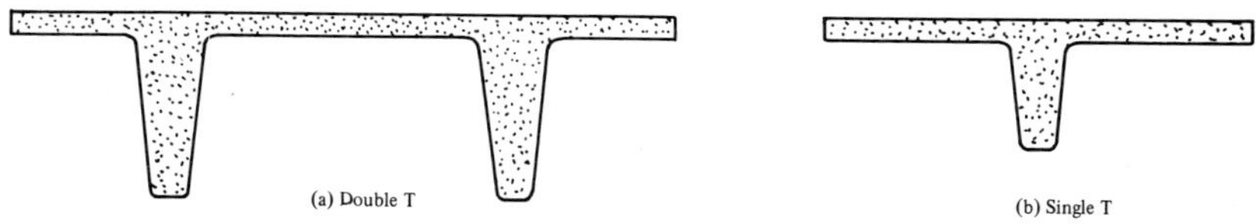

Figure 15.2 Typical double and single T units

suffer from the disadvantage common to all systems of precast floors, whether prestressed or reinforced, that if openings are subsequently required it may be an awkward and costly business to provide them. With some of these types it is also necessary to give special consideration to the accommodation of services. On the other hand with proper planning, the use of these units may result in considerable economy. Standard (pre-tensioned) single and double-T units have been successfully used for long-span floor construction. In the USA single-T unit of width 8 ft–12 ft (2.4 m–3.6 m) and depth 36 in. to 48 in. (900–1200 mm), and double-T units of width 8 ft–10 ft (2.4 m–3 m) having depth 20 in.–36 in. (500 to 900mm) have been standardized. These units could be manufactured in normal weight or lightweight aggregate concrete and can be used with or without structural topping. In Great Britain 8 ft (2.4 m) wide double-T units and 4 ft (1.2 m) wide single-T units of depth from 12 in. (300 mm) to 32 in. (800 mm) are commonly used.

15.1.2 Roofs

Sloped roof beams (Figure 15.3) may be advantageously made on a pre-tensioning bed, since with such a profile bending up of the prestressing steel may be dispensed with. The tensile stresses can be avoided or reduced by ensuring that the dead weight counteracts immediately the prestress is applied. For this purpose special precautions are required to prevent the concrete from adhering to the bottom of the mould, and it is advisable to insist that an appreciable part of the support be withdrawn before transfer; otherwise cracks will occur in the compressive zone as a consequence of the presence of high tensile stresses before the self-weight counteracts. It is advisable also to allow rather low resultant tensile stresses in the concrete at the top of simply-supported beams.

Shrinkage cracks may develop where permanent tensile stresses occur; the upper faces of simply supported roof beams of this type seem particularly prone to this trouble.

It is obvious that the stress conditions can be improved by the provision of draped prestressed steel (Figure 15.3b), whether the steel is pre-tensioned or post-tensioned. The conditions at the ends are thereby greatly improved and the depth can then be reduced to a minimum. The development of permanent tensile stresses at the top face can easily be prevented by the adoption of a suitable profile for the draped prestressing steel (see Chapter 13). The development of temporary tensile stresses at the top face can be avoided by the employment of the methods described in the foregoing. Double or single-T units are also successfully used in long-span flat roof construction, either as simply-supported member or cantilever or part of portal frames (examples given in

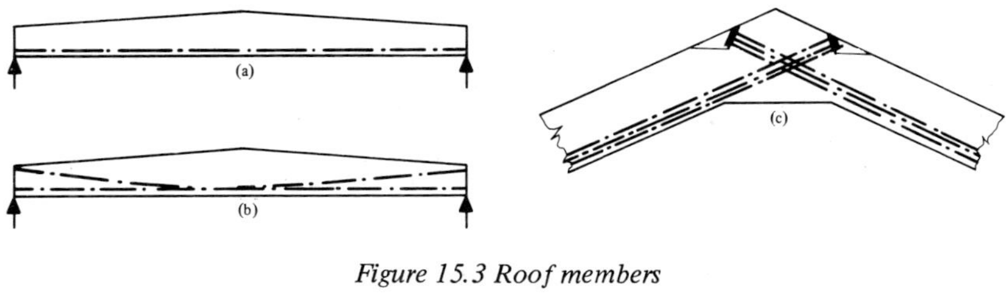

Figure 15.3 Roof members

Chapters 9 and 14). In the Rembrandt dress factory at Stockton-on-Tees (England) 8 ft (2.4 m) wide double-T units (42 in. deep) have been used to span 122 ft (37 m). Pre-tensioned double or single-T unit can also be used to form external load-bearing wall. In such a case it is desirable that the prestress should be more or less uniform to avoid differential camber between adjacent units. (See example 14.8.1.)

For steeper roofs it is sometimes preferable to use two inclined members, the ridge being constructed as indicated in Figure 15.3c.

15.1.3 Trusses

In the analysis of a triangulated framework it is usually assumed that the members are pin-jointed, and the assessment of secondary stresses due to the rigidity of the joints is often either ignored or allowed for by means of an arbitrary safety margin. Experience has shown this procedure to be generally safe (although sometimes wasteful) for steelwork, but often inadequate for reinforced concrete.

Prestressing may be used to improve the resistance and reduce the size of the tensile members of such a framework, and as cracking is largely or completely eliminated the sectional properties of the members are much less uncertain than in reinforced concrete. The secondary stresses can therefore be calculated with a greater degree of confidence, and at the same time can be kept to a minimum by the use of an arrangement of tendons such as that shown in Figure 15.4, which was suggested by Rüsch. In this arrangement, care is taken to ensure that the prestress is applied uniformly to all tension members, and the curtailment of the tendon in the bottom member is so arranged that no unsymmetrical force occurs. In detailing, care should be taken so that the internal and external forces meet at a point as shown in Figure 15.4.

15.1.4 Arches

The first application of prestressing to reinforced concrete arches is due to Freyssinet, who employed jacks at the crown of a bridge as early as 1908 to push the two halves apart temporarily, thereby allowing the falsework to be lowered.

Another form of construction where prestressing can be successfully applied is the bow-string arch (Figure 15.5). This is usually a highly-indeterminate structure, but this difficulty can be eliminated by stipulating the tensile forces to be provided in the slings and main tie, and ensuring (by prestressing) that these forces do in fact occur.

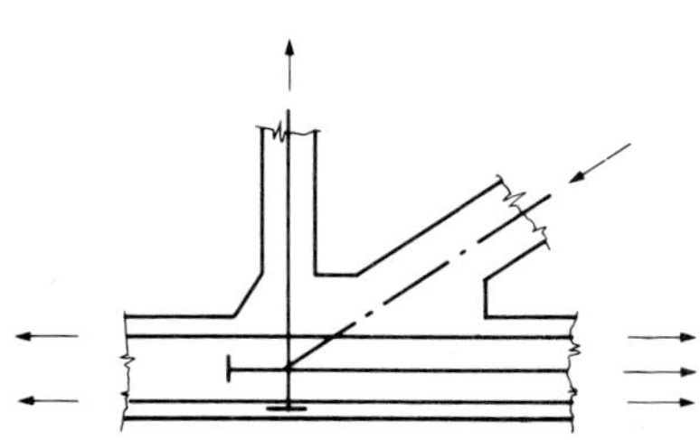

Figure 15.4 Joint in truss

An alternative use of prestressing to produce a structure intermediate between an arch and a frame, introduced by Finsterwalder, is shown in Figure 15.6.

15.2 Bridges

In the case of a bridge the dead weight usually constitutes a large proportion of the total design load, and as it is a feature of prestressed concrete that dead load can be supported with relatively little increase in depth, economy is often obtained by its use for this purpose. This is particularly significant for long spans.

15.2.1 Road bridges (up to 100 ft, 30 m)

The use of precast beams with pre-tensioned steel for bridges with spans up to about 100 ft and subjected to Ministry of Transport loading has become so widespread that standard sizes have been introduced. Figure 15.7 shows details of standard bridge beams adopted by British Railways (Eastern Region) in 1954, as a result of experience gained during the previous six years. In these, a range of three widths of bottom (18 in., 24 in.and 27 in., i.e. 450 mm, 600 mm and 680 mm) and top flanges (6½ in., 8½ in. and 12 in., i.e. 165, 215 and 300 mm respectively) are combined with suitable depths between 10 in. to 27 in. (250 mm to 680 mm) to produce a range of composite sections of considerable versatility. Web thickness may vary between 3 in. (75 mm) to 4¼ in. (108 mm). The advantages of this type of construction include elimination of propping and soffit shuttering, the economy obtained by the use of comparatively small prestressed members in conjunction with larger quantities of concrete cast in place, the use of smaller cranes to place the beams, and the elimination of the

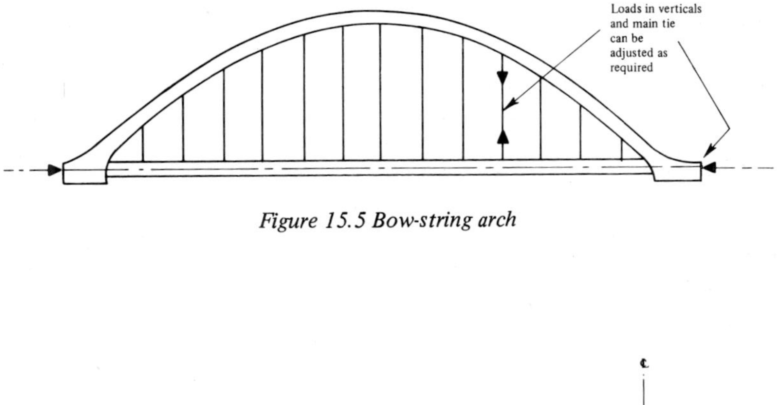

Figure 15.5 Bow-string arch

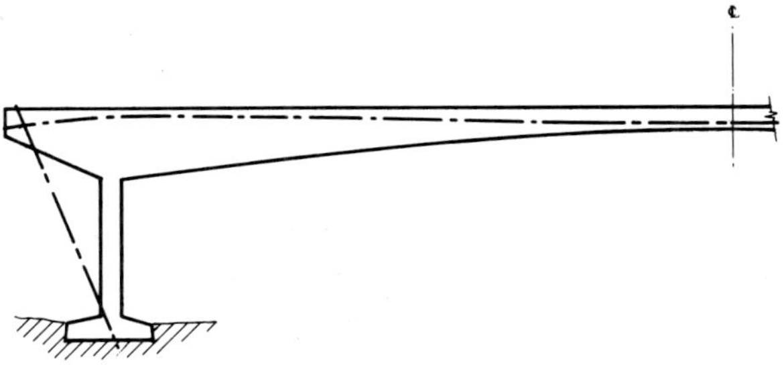

Figure 15.6 Tied frame

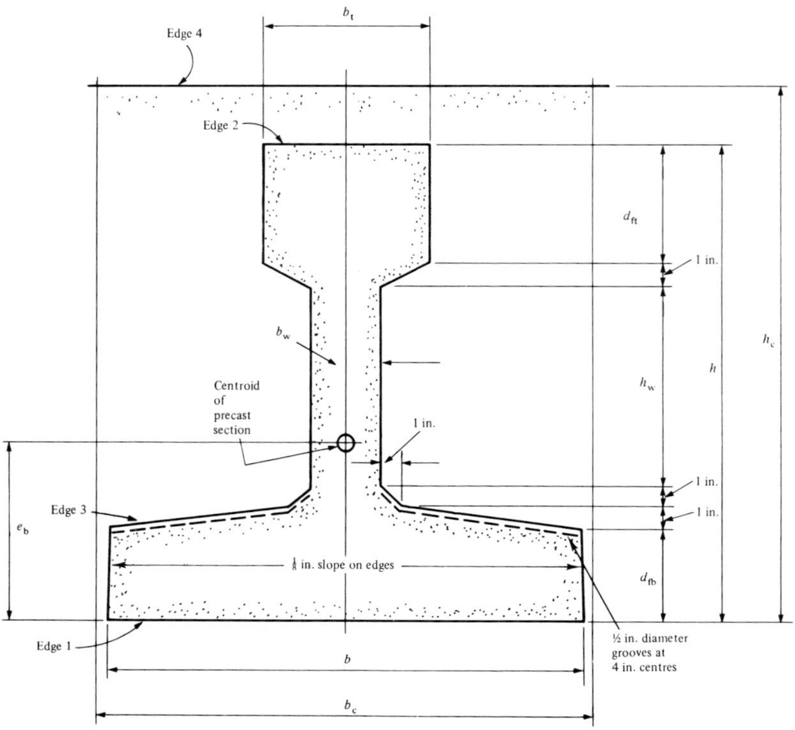

Figure 15.7 Standard bridge beams (British Railways, Eastern region)

need for transverse prestressing as a result of the good transverse distribution of load. A range of similar types was introduced by the Prestressed Concrete Development Group (now known as The Concrete Society) in 1961; these are shown in Figure 15.8. In this case, only the depths of the web and top flange are varied in order to simplify the moulds. The Concrete Society also introduced Standard box sections as shown in Figure 15.9 to span up to 85 ft (26 m).

In the USA the range of standard beams shown in Figure 15.10 was introduced by the American Association of State and Highway Officials and the Prestressed Concrete Institute in 1957. These are intended for use in conjunction with a deck slab whereas both series of British beams are for use with infilling concrete. Other sections have been standardized by individual manufacturers and authorities in America; examples are given in Figure 15.11.

In the case of multiple-span bridges, advantage can be taken of the reduced bending moments which occur in continuous beams, provided that the supports are rigid. Standard precast beams can be adapted for this purpose, either by providing cap cables (see Chapter 14) or by means of the system shown in Figure 15.12. In this method, precast prestressed beams are first placed in position and support (without continuity) the weight of the additional concrete in the deck. Non-tensioned reinforcement is placed over the supports of the

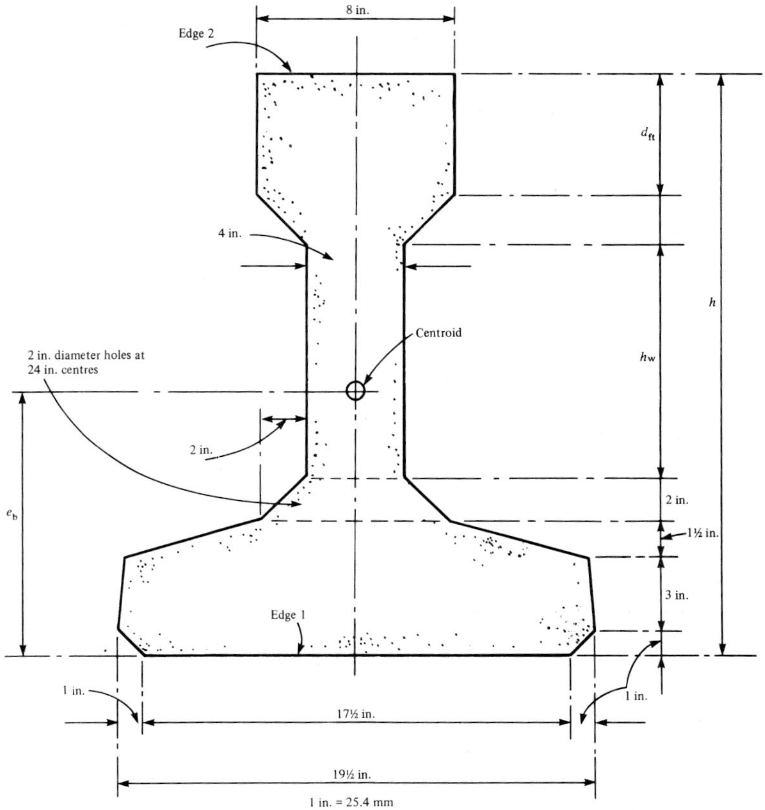

Precast section only

h	h_w	df_t	e_b	A	Z_b	Z_t
Inches				in²	in³	
15	3½	2	5.47	150.5	544	312
16½	3½	3½	6.23	162.5	665	403
21	9½	2	7.60	174.5	1002	565
22½	9½	3½	8.50	186.5	1158	703
24	9½	5	9.40	198.5	1310	843
25½	9½	6½	10.28	210.5	1458	983
27	9½	8	11.14	222.5	1605	1125

Figure 15.8 Standard bridge beams (The Concrete Society)

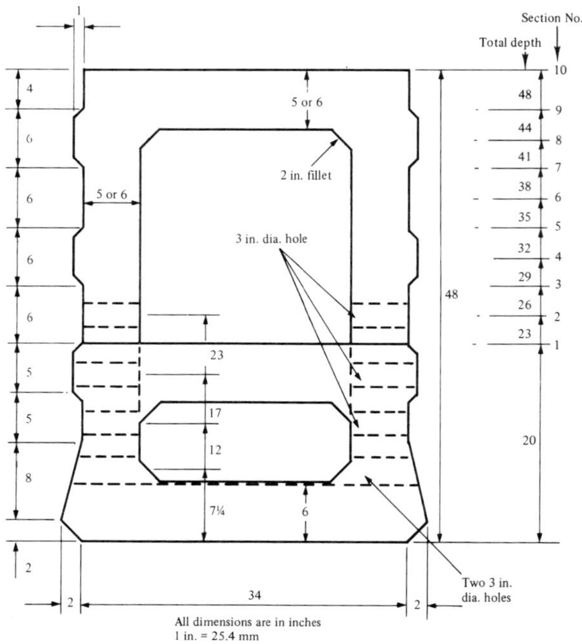

Figure 15.9 Standard box sections to span up to 85 ft 0 in. (The Concrete Society)

beams, so that once the concrete has hardened the composite structure becomes continuous. Tests by the Portland Cement Association USA have shown that this system is very satisfactory[1].

15.2.2 Precast road bridges over 100 ft (30 m) span

Standard beam range shown in Figure 15.10 has been extended in the USA by two further types as illustrated in Figure 15.13. These sections can span up to 140 ft (43 m). Figures 15.14 and 15.15 show respectively Concrete Society's standard I-beam and box-beam which can span up to 120 ft (36.5 m). It is possible to use these sections, and even shallower sections as illustrated before, as parts of longer beam to be spliced and post-tensioned together at construction site. Such judicious application of pre-tensioning and post-tensioning often result in economy. For example, Type 4 unit (Figure 15.10) can be used in a similar way with 8 in. (200 mm) composite deck 60 in. (1500 mm) wide for a span of 165 ft (49.3 m) capable of carrying AASHO loading[2].

15.2.3 Long-span bridges

Where long spans are necessary the problems involved, and their solutions, are usually peculiar to the structure. Three types of cross-section which have been successfully employed are sketched in Figure 15.16. The first of these (Figure 15.16a) is a cellular structure which exhibits a smooth soffit and allows the prestressing steel to be accommodated within the thickness of the concrete without difficulty, while at the same time ample provision for services can be made in the

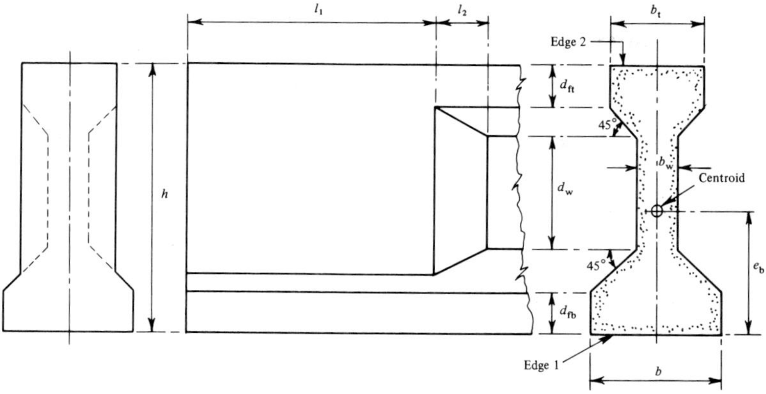

Precast section only

h	b	b_w	b_t	df_b	d_w	df_t	l_1	l_2	e_b	A	Z_b	Z_t	Type
Inches										In²	In³		
28	16	6	12	5	11	4	24	6	12.59	276	1 810	1 477	1
36	18	6	12	6	15	6	24	6	15.83	369	3 220	2 530	2
45	22	7	16	7	19	7	42	9	20.27	560	6 185	5 065	3
54	26	8	20	8	23	8	54	12	24.73	789	10 545	8 910	4

Figure 15.10 Standard bridge beam (U.S.A.)

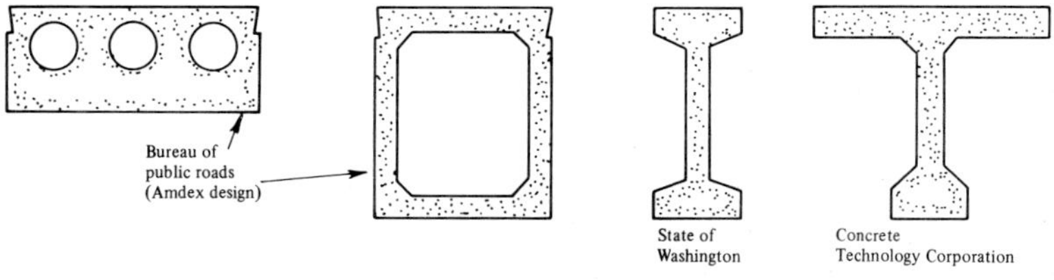

Figure 15.11 Some typical bridge beams (U.S.A.)

spaces between the ribs. Another type of structure is shown in Figure 15.16b. In this case the soffit is replaced by a series of flanges, accommodation for services being sometimes provided by separate bays, or sometimes by direct suspension from the deck slab. It may be difficult with this type to place the large cables required inside the thickness of the concrete, and external cables are then used and subsequently grouted. Cables of this type are preferably connected to the prestressed beams by means of transverse steel connectors, which resist any

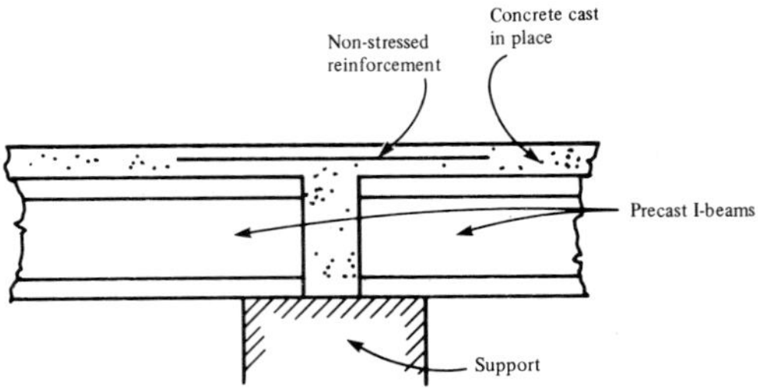

Figure 15.12 Intermediate support of continuous bridge

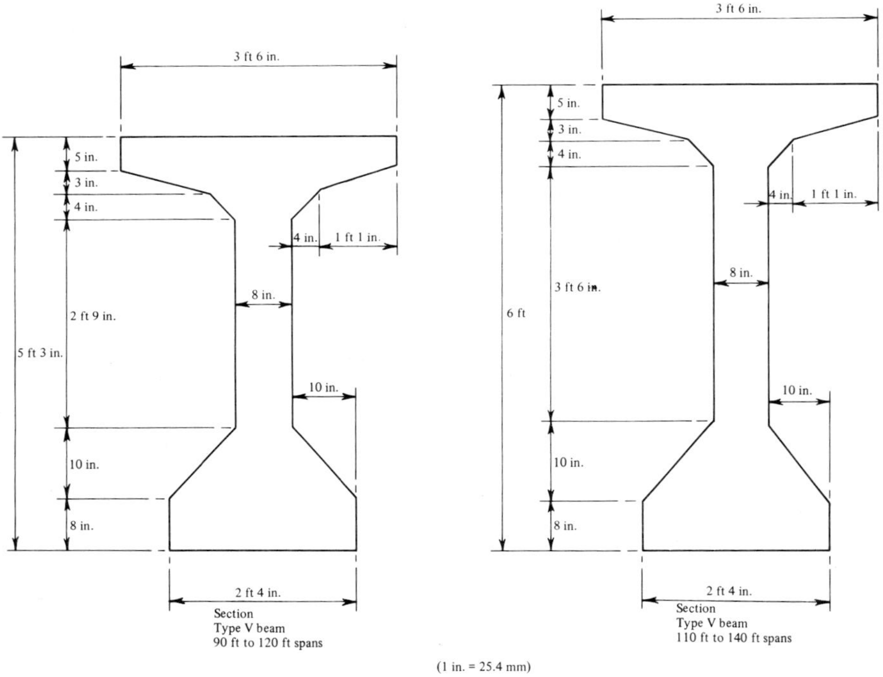

Figure 15.13 Standard bridge beam (U.S.A.) to span up to 140 ft (43 m)

differential shearing stresses between the grout and the beam and ensure uniformity of deformation. This method was introduced by Gifford for the Narrows Bridge at Perth, Australia(3). A third type (Figure 15.16c) combines the convenience of Type (a) with the efficiency of Type (b) but the transverse distribution of the load is likely to lead to higher stresses and deformations. Box-shaped girders are often constructed in segments, either using 'free cantilever method' of construction (Finsterwalder) or using temporary supports and precast elements (Lee).

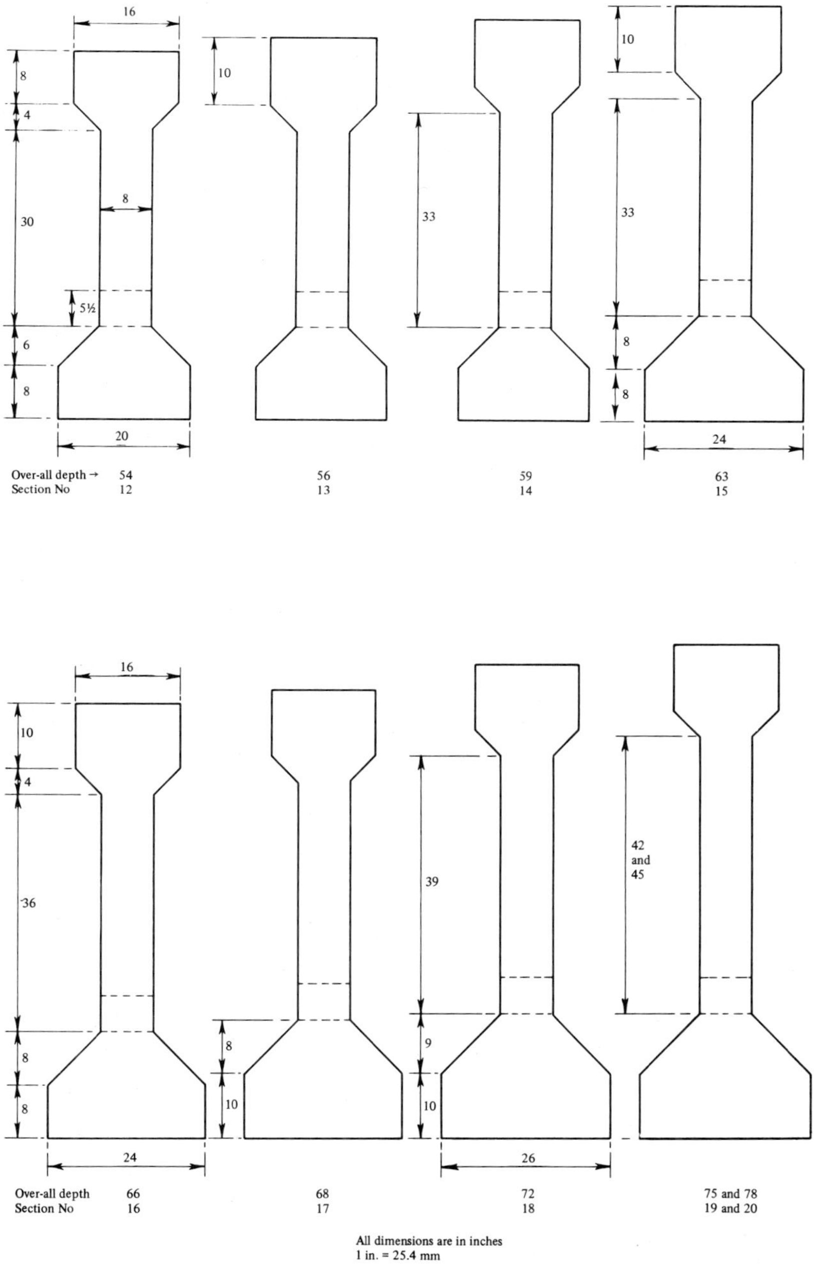

Figure 15.14 Standard I sections up to span up to 120 ft (The Concrete Society)

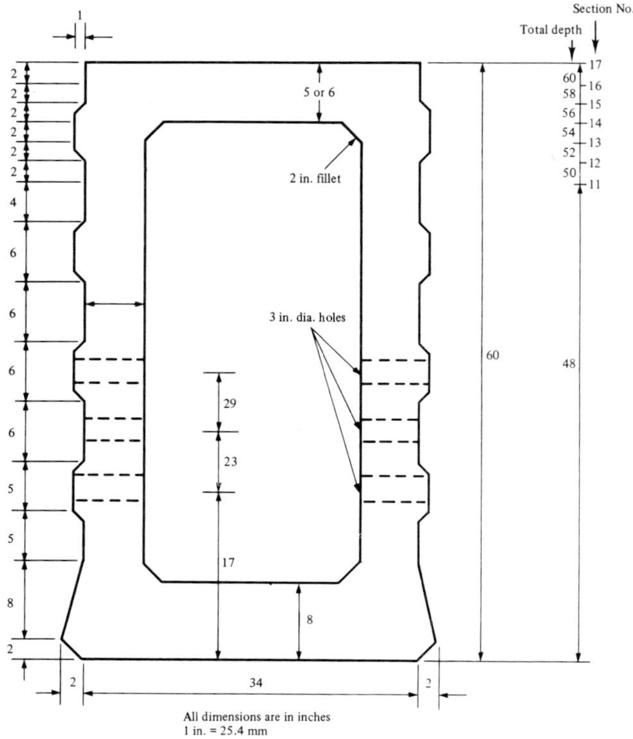

Figure 15.15 Standard box beams (The Concrete Society) to span up to 120 ft 0 in.

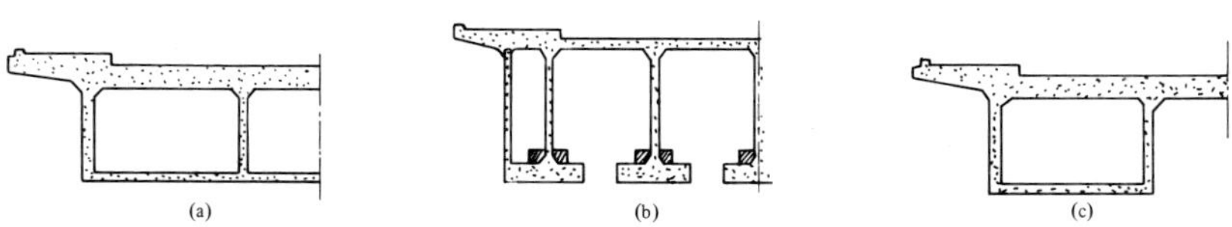

Figure 15.16 Cross-section of long-span bridges

15.2.4 Footbridges

Footbridges and service bridges are subjected to much smaller loads than road and railway bridges, and the use of a number of precast components is often possible. This may involve the disadvantage of erecting a structure to support the components prior to the formation of joints cast in place and the application of the prestress, but this is often offset by the advantages accruing from the use of standard factory-made units.

15.3 Columns and struts

As the main effect of prestressing is the creation of compressive stresses, it is obviously useless to prestress an axially-loaded column. If a column is loaded eccentrically, or if it is required to resist both bending and direct load, the use of prestressing will be advantageous.

The benefit to be gained becomes more pronounced as the ratio of bending stress to direct stress increases. For example, if the stress due to bending is one-tenth of the direct stress (Figure 15.17a), and if the design is based on Code of Practice CP 115, the permissible direct compressive stress may be used without exceeding the permissible bending stress. On the other hand, if the stress due to bending equals or exceeds the direct stress (Figure 15.17b), the use of prestressing may lead to an appreciable reduction in the size of the column.

A distinction must be drawn between columns in which the applied bending moment is approximately constant in direction and amount (for example, the external columns of a building frame) and those with bending moments of varying direction. In the first case the prestressing force is designed to provide a counter-acting moment; in the second a uniform prestress is usually more appropriate, as discussed in the section of masts.

The requirements given in Code of Practice CP 115 are intended to apply only to members in which the mean prestress exceeds 400 lbf/in^2 (28 kgf/cm^2; 2.7 N/mm^2). This proviso is introduced to prevent the use of concrete with a very small prestress in place of ordinary reinforced concrete, as a means of avoiding the requirements of the Code for reinforced concrete. In this Code (CP 114) the maximum permissible direct stress is limited to 1520 lbf/in^2 (108 kgf/cm^2; 10.6 N/mm^2) and without a restriction of this kind it would be possible, by applying a small prestress to a reinforced column made of high-strength concrete, to obtain a greater load-carrying capacity from the requirements of Code CP 115 than from CP 114. The exclusion of columns with a prestress of less than 400 lbf/in^2 (28 kgf/cm^2; 2.7 N/mm^2) means that the concrete in a prestressed column must have a strength in excess of 7680 lbf/in^2 (538 kgf/cm^2; 53 N/mm^2) before any advantage could be gained in this way over a reinforced concrete column with a strength of 6000 lbf/in^2 (420 kgf/cm^2; 42 N/mm^2). This restriction may be justifiable in principle, but seems conservative in amount; an effective prestress of 200 lbf/in^2 (14 kgf/cm^2; 1.4 N/mm^2) if losses are properly allowed for, might be a more reasonable value. According to CP 110, the minimum effective prestress should be 370 lbf/in^2 (26 kgf/cm^2; 2.5 N/mm^2).

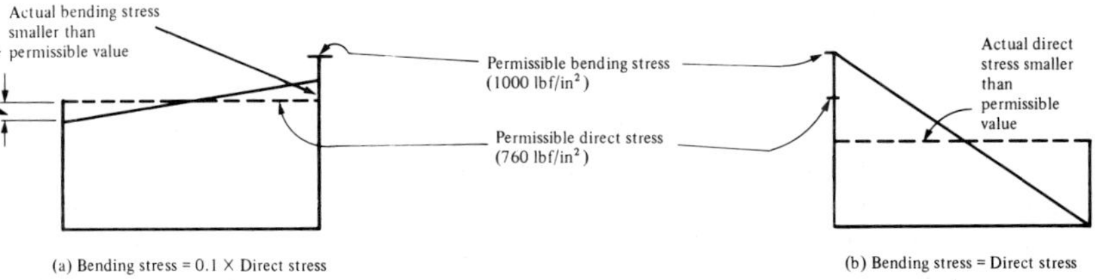

Figure 15.17 Columns with bending and direct stresses

The requirements given in Code of Practice CP 115 are as follows.

(1) When the stress due to bending is more than four times the stress due to the axial load, the greatest stress in the concrete should not exceed that permitted for pure bending; Otherwise the stress should not exceed that permitted for direct compression. Compared with the requirements of CP 114, in which no such restriction is imposed, this limitation seems to be unduly severe.

(2) For long columns, the load reduction factors given in CP 114 should be used. This requirement is adopted empirically, in the absence of test results; compared with reinforced columns, prestressed columns might be expected to offer a greater resistance to buckling but a smaller resistance to crushing. It is similarly possible that the lateral stability of slender prestressed beams may exceed that of reinforced concrete beams, but again insufficient test results are available to confirm this.

(3) Where longitudinal non-stressed reinforcement is provided, transverse stirrups or helical binders should also be adopted in accordance with the recommendations of CP 114. The cross-sectional area of the non-stressed steel is of particular importance; it is usually substantial in amount, and the redistribution of stress due to creep should be considered in calculating the losses of the prestress and the cracking strength of the column, otherwise the entire prestress may become ineffective. There is, however, no specific requirement given in CP 110 regarding the design of prestressed columns.

The provision of some prestress for any precast column is usually desirable since it increases the resistance to cracking during handling and transport and causes cracks to close after these loads are removed. If necessary, the prestress can be withdrawn after the column is erected. Similar considerations apply also to piles.

Prestressed walls can be designed on the same principle as the columns.

In the large wall-panel system of construction it is often advantageous to use vertical post-tensioning to ensure total and integrated action of the walls and floors.

15.4 Masts

It is often the case that structures of this type are required to sustain comparatively light loads, but may be subject to bending moments (often due to wind) of varying amounts and directions. The basic requirement is usually that a high degree of resilience is required; this, in conjunction with a relatively low prestress, ensures that the mast has the ability to absorb excess load without difficulty. On the other hand, highly-prestressed masts have been proved to be liable to brittle failure and their use has been abandoned in various countries. Masts designed for high resilience were introduced by the German National Railways in which the nominal tensile stresses are about 700 lbf/in^2 (49 kgf/cm^2; 4.8 N/mm^2) due to the permanent load and about 1500 lbf/in^2 (105 kgf/cm^2; 10.4 N/mm^2) due to the rarely-occurring full load, together with a factor of safety against failure of 1.75. Obviously, at the higher loads, visible cracks will occur.

The method of designing structures subject to reversal of bending moments is described in Chapter 8, and this can be applied directly to the design of masts. In addition to wind loading, the possibility of a different distribution of load during construction should also be allowed for. A further possibility in the case of certain masts supporting electric cables is that a breakage of the cable may impose a torsional moment on the mast; this should then be considered as a possible design case. Generally an arrangement is adopted in which anchor masts, designed to resist the full torque which may occur, are placed at intervals, the intermediate masts providing only vertical support to the cables. It is then necessary to design the intermediate masts to allow for the smaller torque caused by the drag of a broken cable; this may be done by providing sufficient torsional strength to resist the effect, or sufficient flexibility to nullify it. In the first case, masts of tubular or octagonal section offer the greatest resistance to torsion. If other sections, such as I-beams, are used, rotating cable supports can be provided.

15.5 Piles

The cross-sectional dimensions of precast reinforced concrete piles are often determined by the desirability of avoiding cracking during lifting and transport (so far as this is possible with reinforced concrete) since any cracks tend to remain open, leading to corrosion of the reinforcement after the pile is driven. The use of prestressing enables these dimensions to be reduced; at the same time the consequences of cracking are likely to be much less severe since flexural cracks tend to close under the action of the prestress. These advantages are present with all types of prestressed piles, the extent of the reduction in area being greater for long piles, but additional benefits can be obtained in particular cases. If the pile is to be subjected to a bending moment only (as in the case of sheet piles) or a bending moment and direct load, without cracking, or if a high degree of resilience is required (as in the case of fender piles) the use of prestressing usually leads to a further improvement in economy and performance.

In general, since cracks due to bending close when the load is removed, occasional cracks do not lead to corrosion of the prestressing steel. If impact occurs, then with good bond the cracks may be opened and closed many times with no consequent ill-effects. In sheet-pile retaining walls, if a slip should cause an increase of the working load with consequent cracking, a redistribution of the load takes place, leading to the closure of the cracks. There is therefore no need to design piles to be free from cracks at the maximum possible load.

15.5.1 Stresses

The stresses to be considered in the design of prestressed piles are the prestress itself; the stresses due to lifting and transport; those occurring during driving; and those due to the final design loads.

Prestress

In the case of long axially-loaded piles the prestress is required primarily to resist stresses due to lifting and transport, and to a lesser degree those occurring during driving. Any increase over the minimum required may reduce the capacity of the pile to resist direct load and is therefore undesirable as well as uneconomical. The use of temporary prestressing may be worth considering, in which all or part of the prestress is withdrawn after the pile is driven; this may be advantageous for point-bearing piles. With friction piles it often happens that the surface area, and hence the cross-sectional area, is determined by the coefficient of friction between the pile and the ground, and in this case the direct stress which the pile can safely support may be less than that which the

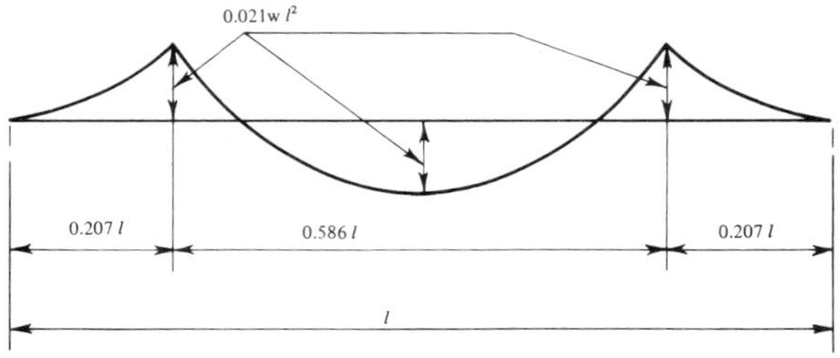

Figure 15.18 Minimum bending moments (Two-point lifting)

concrete could resist. If this area is greater than that required to resist handling stresses no prestress is needed, although its use is still desirable to prevent cracking due to shrinkage.

A uniform distribution of the prestress is usually suitable.

Stresses during lifting and transport

As shown in Chapter 11, the bending moment due to the weight of a member of uniform section is the least when the points of lifting or support are distant $0.207l$ from each end; the bending moment is then $0.021\,wl^2$ lbf ft in which w is the weight (per unit length) and l the length (Figure 15.18).

Assuming the cross-section of the pile to be a square of side D ft, the stress due to bending is then

$$f = \frac{M_t}{Z} = \frac{0.021\,wl^2 \times 6}{D^3} = 0.126\,\frac{wl^2}{D^3} \quad \text{lbf/ft}^2$$

When l and D are in metres and w is in N/m (or kgf/m) the unit of the bending moment $0.021\,wl^2$ will be in Nm (or kgf m). The expression for $f = 0.126\,\dfrac{wl^2}{D^3}$ will then be in N/m² (or kgf/m²) unit.

If the resultant tensile stress is to be zero, the prestress must equal this value and the prestressing force is therefore

$$P_t = 0.126\,\frac{wl^2}{D^3}\,D^2 = 0.126\,\frac{wl^2}{D}$$

Assuming

$w = 150D^2$, this becomes

$P_t = 18.9\,Dl^2$ lbf $= 18.9l^2$ lbf/ft width.

This formula gives the prestressing force required to avoid tensile stresses during lifting immediately after transfer. After losses take place some tensile stress would develop and it is advisable to check the resultant tensile stress during lifting prior to driving, allowing an appropriate impact factor (of, say, 33 to 50 per cent) for the entire weight of the pile.

Stresses during driving

Driving stresses often exceed those occurring at any other time during the life of the pile. They cannot be predicted with much accuracy; of the several formulae available for the purpose, that due to Hiley appears to give more consistent results than most and is recommended here. The formula is

$$R = \frac{\eta WH}{s + 0.5c} + P + W$$

in which R is the resistance to settlement (tonf); η a coefficient indicating the efficiency of the hammer blow; W is the weight of the moving parts of the hammer (tonf); s the set (in. per blow); c the elastic compression of the pile and cap (in.); P the weight of the pile, helmet, and stationary parts of the hammer (tonf) and H the drop of hammer (in.).

In a paper[4] published in 1938 it is shown that the stresses caused by the impact of the hammer are propagated as a shock-wave along the length of the pile; the wave is reflected back upon itself at the toe and then travels back to the head of the pile. The effects of a wave of this type can be calculated with the aid of a suitably-programmed electronic computer, and if the variables involved are known with sufficient accuracy (which is rarely the case) an accurate assessment of driving stresses can be made. Smith has shown[5] that Hiley's formula gives results generally in accordance with the wave equation, but that in the case of long and heavy piles the compressive stress produced at the toe of the pile may be almost twice that at the head, as a consequence of the reflection of the stress wave. The reflected wave can also cause tensile stresses as a consequence of the inertial 'rebound' of the pile, and Strobel and Heald[6] state that a prestress of 600 to 700 lbf/in^2 (42–49 kgf/cm^2; 4.1–4.8 N/mm^2) is usually sufficient to eliminate cracking due to this cause. Secondary reinforcement should also be provided at the head and toe, where the compressive stresses are most severe, as in reinforced concrete piles, and is also desirable throughout the length of the pile.

It is sometimes overlooked that during driving the pile is temporarily called upon to store large amounts of energy, and the ability of concrete to perform this duty increases with time. It is therefore undesirable to achieve a large early compressive strength, in the mistaken belief that in this case the pile can be driven at an early age; a concrete with smaller strength but greater resilience is to be preferred, and a reasonable curing period (often taken as 28 days) should be allowed, otherwise failure may occur during driving.

Stresses due to design loading

(1) Direct stress. The load which a point-bearing pile can support may be determined from a recognized formula (such as that due to Hiley), or from tests on a driven pile, or from both. In the case of a friction pile, the load is calculated from the frictional resistance per square foot of pile surface, again supplemented by tests; Smith has shown[5] that Hiley's formula gives a conservative estimate of the driving stresses in friction piles. In both cases, a factor of safety of between 2 and 3 is usually employed to obtain the safe working load.

When a considerable length of the pile is unsupported by the ground (as, for example, in the supports of many marine structures) the pile should be considered as a slender column; the permissible stress so obtained may be less than that derived from the foregoing, and is then used instead.

(2) Bending stresses. When prestressed piles are required to resist bending moments applied through the pile cap, it is desirable to limit the magnitude of the prestress, as described previously. If additional bending resistance is required, this can sometimes be obtained by increasing the prestress after driving is completed, or (when only the top of the pile is highly stressed) by providing additional non-stressed reinforcement. In either case, careful detailing of the top of the pile is necessary, to allow for the possible removal of excess length and the transfer of the bending moment (for example, by stripping the reinforcement).

15.5.2 General notes

Piles with pre-tensioned steel

All piles should be standardized, particularly when they are made on pre-tensioning beds, as standard sets of moulds can be used. As the prestress is then transferred by bond the problems of cutting-off and stripping mentioned in the previous paragraph are simplified. It is usually difficult to add a prestressed extension to such a pile, and this sometimes leads to waste by encouraging the use of over-long piles. Nevertheless, production on pre-tensioning beds offers great technical and economic advantages, which may be sufficient to offset this drawback.

Piles with post-tensioned steel

When post-tensioning is used it is often easier to extend piles than to shorten them. In consequence uncertainties regarding the lengths of the piles can be reduced, and this may result in appreciable economy. It is also possible to cast long piles in shorter sections, which can then be assembled and prestressed, or to employ piles of varying section. For example, a solid head and toe may be employed in conjunction with a hollow shaft, to reduce the weight of the pile. Typical details of such a pile are shown in Figure 15.19, and as the satisfactory design of piles of this type depends on many features, the following points should be noted.

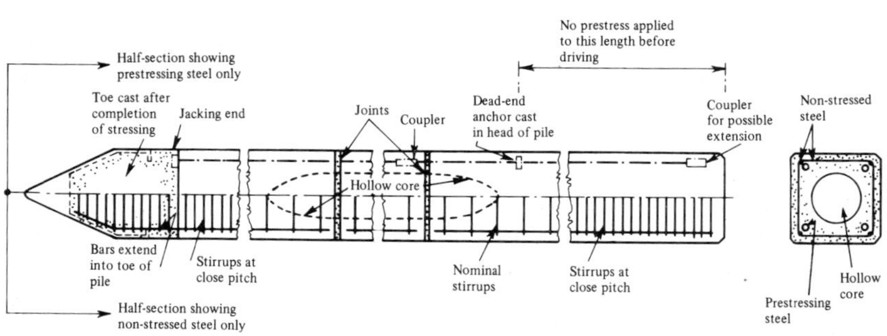

Figure 15.19 Hollow pile with post-tensioned steel

(1) If the wall thickness of the hollow portion of the pile is small, any eccentricity of the hollow core can lead to failure during driving as a consequence of the increased stresses which occur due to prestressing and to driving. For normal structural tolerances, the diameter of a circular core should not exceed one-half of the side of a square pile; this value might be increased by 25 per cent if careful positioning can be guaranteed.

(2) The change of section from the hollow to the solid portions should be accomplished gradually.

(3) The joints between the precast units should be wide enough to permit satisfactory compaction of the concrete in the joint.

(4) Little or no prestress should be applied to the portions of the pile immediately adjacent to the head and toe; non-stressed reinforcement in these zones is usually more satisfactory.

(5) Torsional failure of piles during driving sometimes occurs as a consequence of rotational restraint. This has been observed with both prestressed and reinforced concrete piles(7) and is caused primarily by the rigidity of the driving rig and the restraint imposed by the helmet. It seems probable, however, that the reduced torsional resistance of a hollow pile renders it more liable to this type of failure than a solid pile, and it may be advisable to allow for this by adopting a circular section for the top 12 in. (305 mm) of the pile, and by ensuring that the pile is free to twist in the helmet.

15.5.3 Piles in the earthquake zone

Prestressed concrete piles can be safely and economically used for foundations in weak soils even when subjected to strong earthquakes(8). In the paper by Bertero *et al* (8) it has been shown from dynamic analysis of layered soils in specific areas of the San Francisco Bay that the displacements and curvatures in piles subject to earthquake forces are not uniform and that in many cases S-curves develop giving rise to curvature of 1 in 1200 as shown in Figure 15.20b.

In order to allow rotation to take place it is necessary to provide hinge connections between the pile head and the structure (Figure 15.21a and 15.21b as well as at the interface of weak soil and firm soil/clay 15.21c. It is also essential to use sufficient steel with adequate anchorage at the hinges. In addition it is recommended to use moderate size piles and increase the spiral reinforcement through the body of the pile.

15.5.4 Sheet piles

Where sheet piles are used as retaining walls, the bending moments imposed on them are generally as shown in Figure 15.22. There is clearly some economy to be gained by using a non-uniform distribution of prestress, and if the piles are less than about 20 ft (6 m) in length the resulting camber has no adverse effect on the driving properties. In addition, since sheet piles are normally designed to resist a nominal distribution of load which readjusts itself favourably as soon as the piles deflect, a tensile stress of about 750 lbf/in^2 (52.5 kgf/cm^2; 5.2 N/mm^2) can safely be allowed in the concrete. Any cracks which may be caused by any unforeseen excess loading (due to a slip, for example) close again when this redistribution occurs. In Figure 15.23 are shown details of sheet piles designed in accordance with these considerations for a small cantilever retaining wall. In this case the nature of the ground enabled metal driving shoes to be dispensed with, but this is the exception rather than the rule. The unsymmetrical

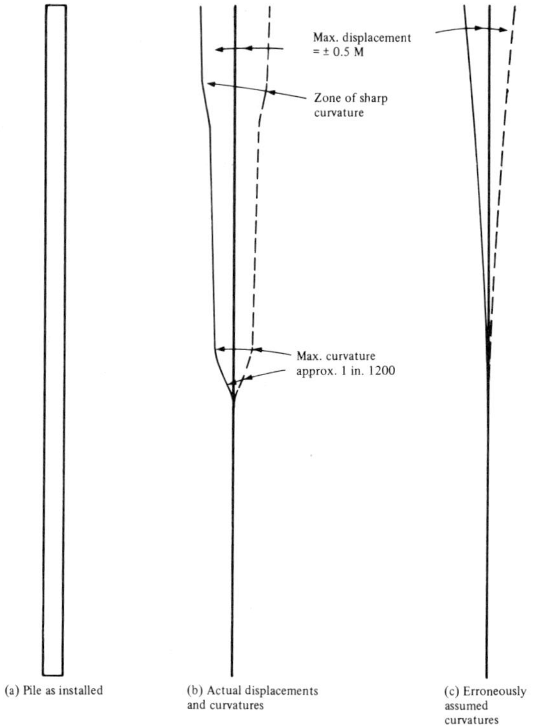

Figure 15.20 Displacement and curvatures of piling in very strong earthquake according to Ben Gerwick Jn.

chamfer at the toe of the pile is also of interest; while no generalization is possible, it has been the authors' experience that chamfers of this type give a driving performance superior to either the symmetrical chamfer shown in Figure 15.23a or the single chamfer shown in Figure 15.23b.

Stirrups should be provided at close centres at the head and toe of a sheet pile, to resist the stresses due to driving. When the ground is heavy, metal shoes should be provided and some transverse reinforcement should be present throughout the length of the pile.

An example of in-situ post-tensioned sheet pile (diaphragm wall) is shown in Figure 15.24a and 15.24b. This wall was built in 1974 as a retaining structure for the deep basement construction of the West German Embassy Building in London.

The wall was designed as a free cantilever and after excavation (for about 18 months or so) until the slabs of the substructure floors of the main building provided lateral support at different levels it did act as one. The maximum extent of the cantilever was 34 ft (10 m) and the total depth of diaphragm wall 57 ft (17.4 m) with a thickness of 2ft 11½ in. (900 mm). Because of the condition of the ground and the surrounding area ground anchorage could not be introduced.

ICOS-flex system was used in the construction of the diaphragm wall. The stability of the trench was maintained by Bentonite slurry – the amount of bentonite was controlled to suit the nature of the surrounding soil.

An interesting point about prestressing such construction is that the limitation of initial stresses in concrete (tensile or compressive) is of little significance, since the counter-action as a result of mobilisation of passive pressure of earth owing to deformation tendency of the structure will balance the stresses.

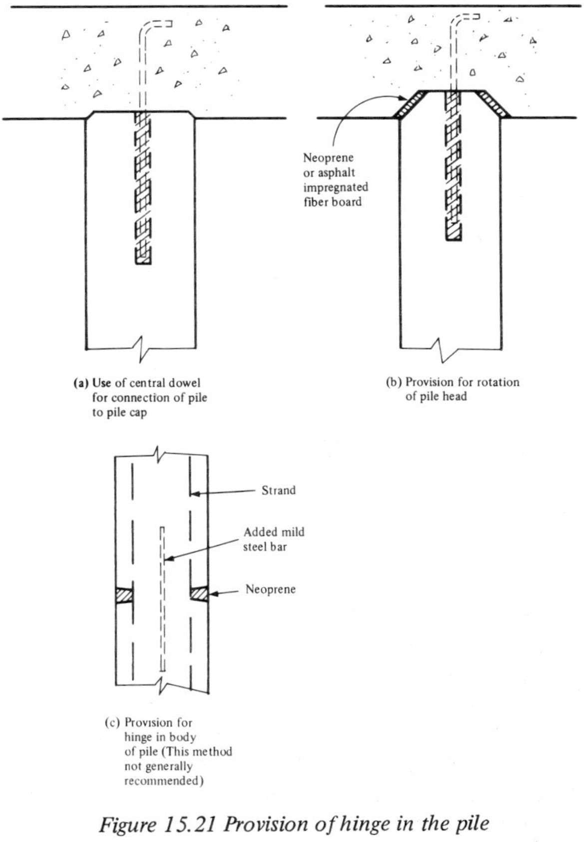

Figure 15.21 Provision of hinge in the pile

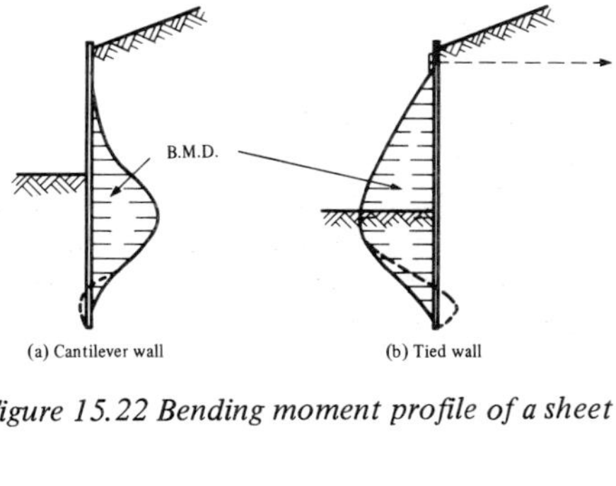

Figure 15.22 Bending moment profile of a sheet pile

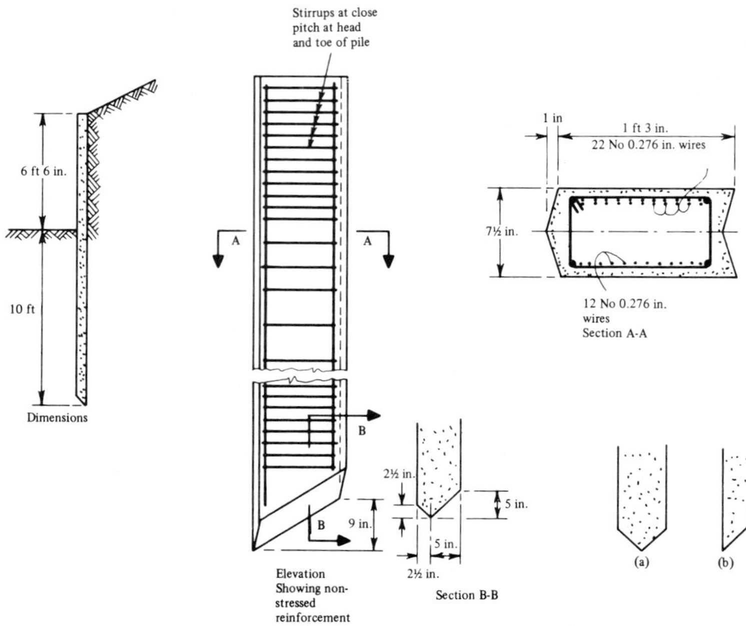

Figure 15.23 Details of precast sheet piles

15.5.5 Fender piles

In marine structures such as wharves and jetties, the kinetic energy of a berthing ship must be absorbed without damage to the structure or the ship, and some form of flexible construction is often provided for this purpose. Until recently, the most suitable materials possessing sufficient resilience were timber, rubber, and steel. Reinforced concrete piles are less suitable than these, and highly-prestressed concrete is even more brittle than reinforced concrete. However, if the prestress is limited to a low value to allow sufficient deformation, tests have proved that large impacts can be absorbed without fracture, and this type of prestressed concrete has been used with success for fender construction.

A sketch of a typical fendering system is shown in Figure 15.25. When the weight of the ship and its velocity of approach are known, the force exerted on the structure is computed from the travel of the upper flexible support, and the horizontal members are designed to spread this force over a sufficient number of piles. The piles themselves are designed to resist the bending moment and deflection so caused, the upper end being supported elastically and the lower end either simply supported or fixed, depending on the penetration and the ground conditions.

The resistance of such a pile to impact forces is improved if non-stressed longitudinal and transverse steel is provided in addition to the tensioned steel throughout the length of the pile.

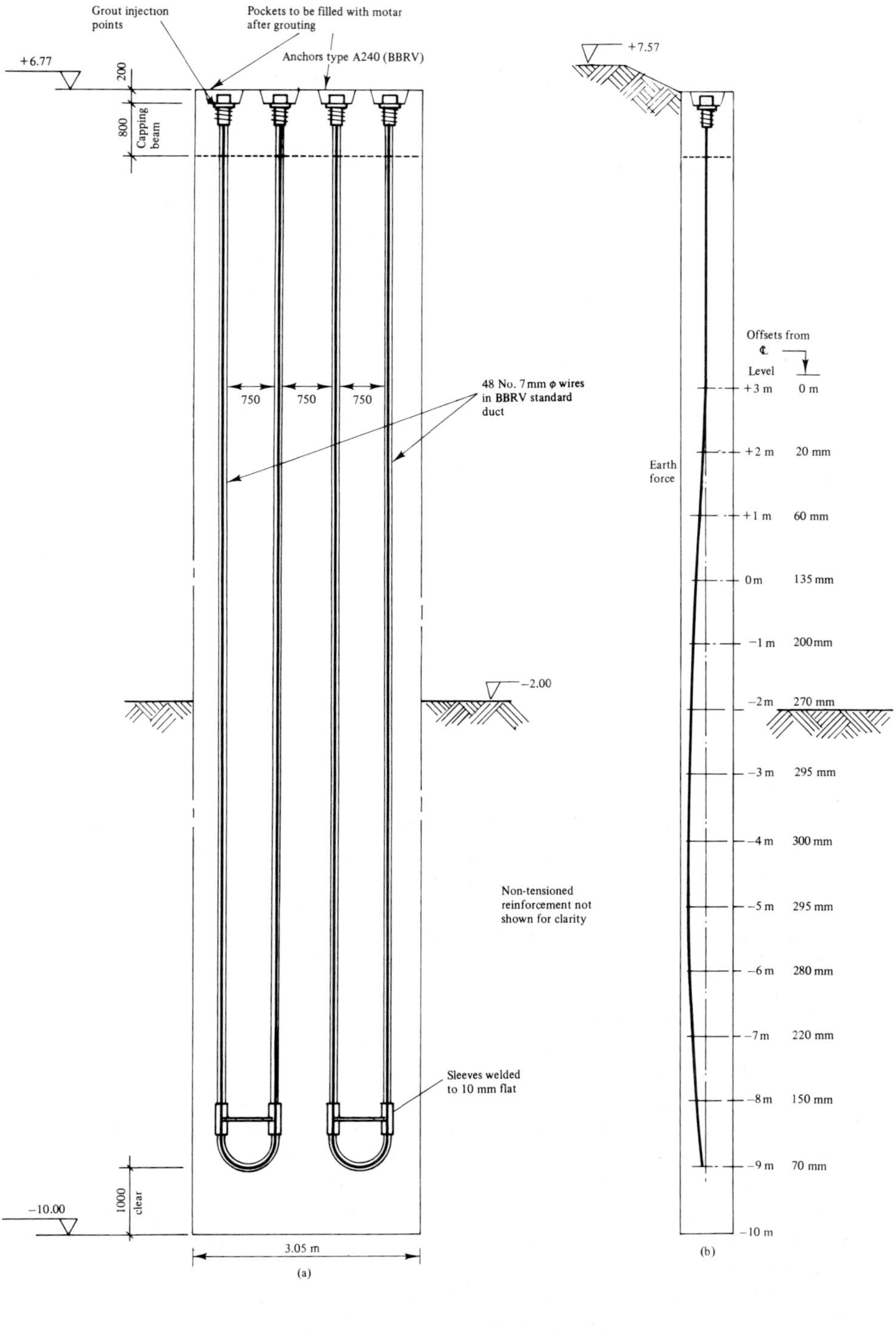

Figure 15.24 Cable profile in diaphragm wall

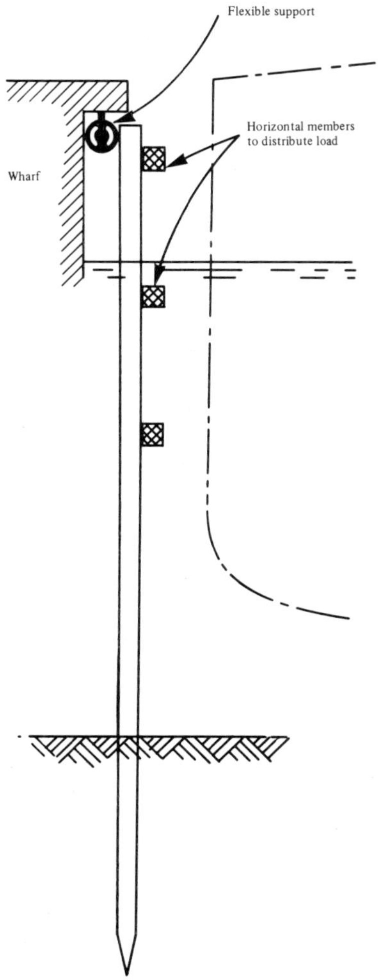

Figure 15.25 Fender piles

15.5.6 Anchor piles

It is sometimes advantageous to use prestressing cables to provide an anchorage in rock. This type of anchorage was used by Coyne for dams as early as 1929, and a type of anchor pile employing the same principle is sketched in Figure 15.26.

15.6 Shells

The theory of shells is too complex to be summarized in a short space, and in the following notes consideration is given primarily to the advantages and effects of prestressing. The methods of design described are approximate, and are suitable only for preliminary calculations. In all cases a more accurate calculation should be made to confirm the preliminary design.

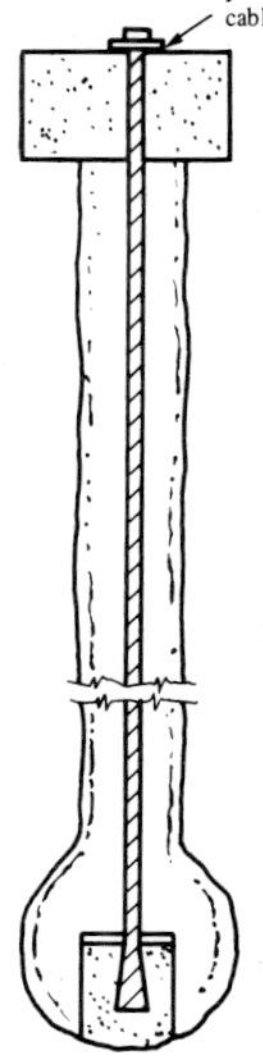

Figure 15.26 Anchor pile

15.6.1 Advantages of prestressing

The advantages claimed for prestressing in various papers and discussions of the International Association of Shell Structures and the Fédération Internationale de la Précontrainte may be summarized as follows.

Doubly-curved shells and domes

(1) A prestressed circular ring beam improves the conditions of stress at the edge of a dome by counteracting the tensile stresses due to bending.
(2) Similar effects are obtained in the case of doubly-curved domes of non-circular base by the provision of prestressed edge members.
(3) Rigid prestressed tie members may be used advantageously when doubly-curved shells (for example, shells of negative curvature, such as hyperbolic paraboloids) are to be connected together.
(4) Straight pre-tensioned wires can be employed along the generators of hyperbolic paraboloids. This method has been used for water containers and for precast roof members of inverted curvature.

Cylindrical shells

(1) The longitudinal span of shells can be appreciably increased by prestressing the edge beams.
(2) Cracking cannot be usually avoided in reinforced concrete shells as a consequence of tensile stresses due to shrinkage, creep and bending, but can be avoided or at least minimized by providing additional bent-up prestressing steel in the membrane of the shell adjacent to the edge beam.

(3) If all the prestressing steel is placed in the membrane of the shell, edge beams can be dispensed with altogether. Sufficient safety against buckling at ultimate load can still be obtained if transverse stiffening ribs are provided. Similar advantages against cracking are also obtained as in (2).
(4) Greater uniformity of alignment and deformation of adjacent shells can be achieved when prestressing is used.
(5) If the membrane is prestressed the trajectories of stress are modified to such an extent that the provision of a rectangular mesh of non-stressed reinforcement is quite satisfactory, whereas in a reinforced concrete shell the reinforcement is better adjusted to conform to the trajectories of the principal tensile stress, which are curved.

15.6.2 Domes

The calculation of the stresses in a thin spherical dome is not difficult if the dome is simply-supported (as shown in Figure 15.27b). If the reactions are not statically-determinate (for example, if the dome is rigidly connected to an edge beam or tank wall, as shown in Figure 15.27c), the redundant reactions and moments impose additional direct and bending stresses near the springing of the dome. The method of calculating these is given in many textbooks[9]; it is not repeated here as it is too complex to be treated satisfactorily in a few pages. It is emphasized that the following discussion is directly applicable only to simply-supported domes; the prestressing of statically-indeterminate domes is discussed briefly on page 437.

Simply-supported domes

The stresses in a simply-supported spherical dome, the thickness of which is so small that the resistance to bending is negligible, are given in Figure 15.28. At any point they consist of a direct compression acting along the meridian, denoted

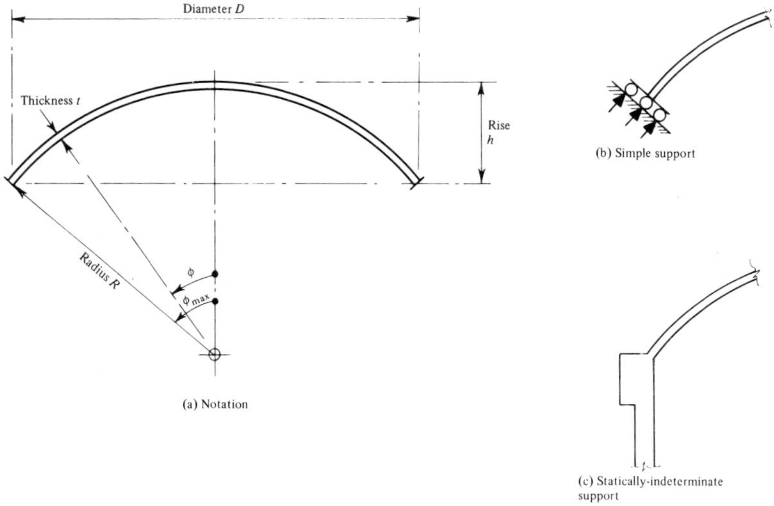

Figure 15.27 Domes

by N_ϕ, and a direct force (which may be tensile or compressive, depending on the value of ϕ) acting along the horizontal small circle through the point, denoted by N_θ (Figure 15.28).

These forces are given by the expressions

$$N_\phi = R\left(\frac{w_s}{1+\cos\phi} + \frac{w_l}{2}\right)$$

$$N_\phi = R\cos\phi\,(w_s + w_l\cos\phi) - N_\phi$$

in which w_s is the weight of the shell (weight per unit surface area) and w_l the uniform live load (per unit projected area). Compressive stresses are positive.

If the dome is hemispherical the thrust N_ϕ at the springing is vertical; in all other cases, a horizontal force of $N_\phi \cos\phi$ is exerted at the springing, and this is conveniently resisted by the prestressing, the force required being given by

$$P = \frac{N_\phi \cos\phi\, D}{2} \qquad (15.1)$$

If this prestressing force can be applied directly to the rim of the dome, the stresses in the dome will be those stated in the foregoing. This is not usually possible, however, as the amount of prestressing steel required is usually too great to be accommodated in so small a space. An edge-beam is therefore provided, making the dome statically-indeterminate.

Prestressing of statically-indeterminate domes

The simplest case occurs with the arrangement shown in Figure 15.29 in which the edge beam itself is unrestrained and the thrust from the dome passes through the centroid of the beam. If an imaginary cut is made along line A.A. (Figure 15.29) it can be shown that the horizontal thrust $N_\phi \cos\phi$ causes the edge of the dome to move inwards by an amount

$$y_d = \frac{D}{2Et}\left(N_\theta - \nu N_\phi\right)$$

in which ν is Poisson's ratio.

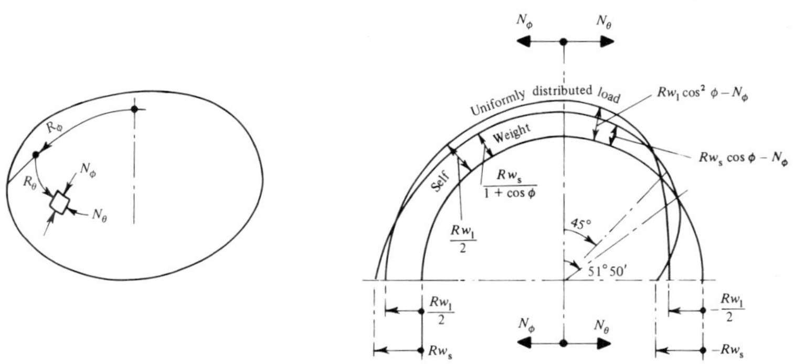

Figure 15.28 Stresses in statically determinate domes

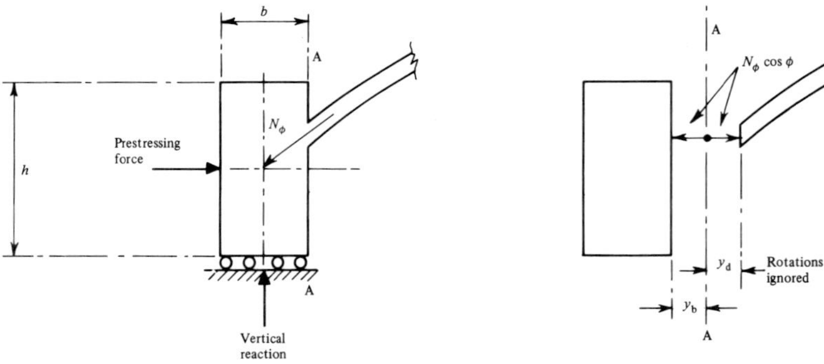

Figure 15.29 Effect of ring beam

On the other hand, the thrust causes the ring beam to move outwards by an amount

$$y_b = \frac{N_\phi \cos\phi D^2}{4Ebh}$$

The prestressing force must therefore be sufficient to move the ring beam a total distance of $y_b + y_d$ in order to close the gap. The movement due to a prestressing force P is

$$y_p = \frac{PD}{2Ebh} = y_b + y_d$$

and hence

$$P = \frac{bh}{t}(N_\theta - \nu N_\phi) + \frac{N_\phi \cos\phi D}{2} \quad . \; . \; . \; . \; . \; . \; . \; . \; . \; . \; . \; . \; . \quad (15.2)$$

By comparing equations 15.1 and 15.2, it is seen that the effective prestressing force required is greater in equation 15.2; the magnitude of the increase is usually between 5 and 10 per cent. The effect of varying the prestressing force is shown in Figure 15.30 for a particular case.

This is also true for the more general case shown in Figure 15.31, in which the line of thrust from the dome does not pass through the centroid of the beam and the beam is itself rigidly attached to a supporting wall. The effective prestressing force required can be obtained by increasing the value given in equation 15.1 by 10 per cent. In such a case, however, the stresses at the springing of the dome may differ greatly from those predicted by the membrane theory given in the foregoing, and should be the subject of a separate calculation. The results of such a computation, based on formulae given by Timoshenko(9), are compared with those obtained by the approximate method in Figure 15.32.

It should also be noted that the application of a prestressing force larger than required is detrimental to the structure as well as wasteful. The effect of such a force is shown (exaggerated) in Figure 15.33. It is clear that, while the value of N_ϕ is not greatly increased, much larger value of N_θ will occur, together with significant bending stresses, at the edge of the dome.

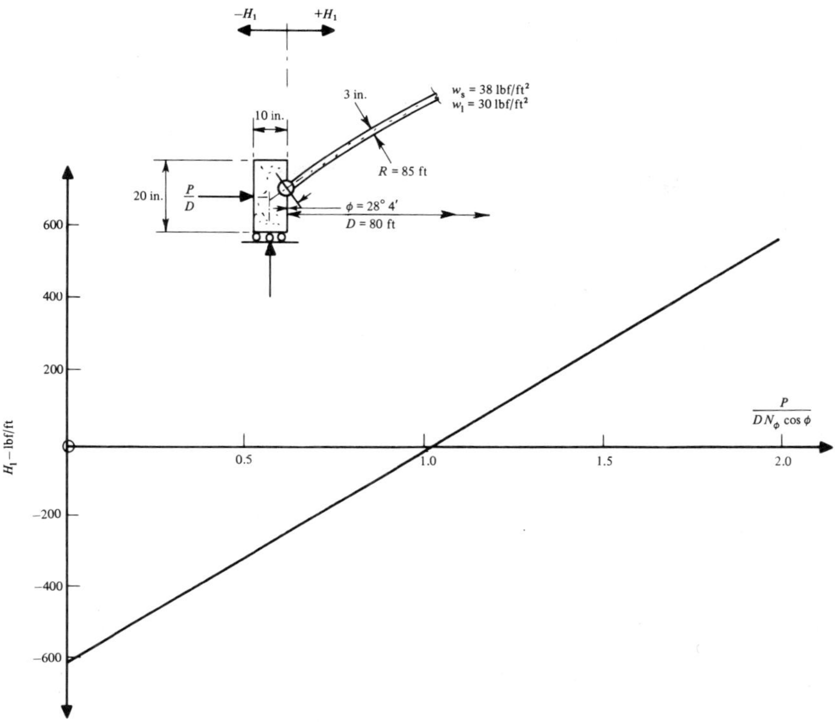

Figure 15.30 Effect of variation of prestressing force

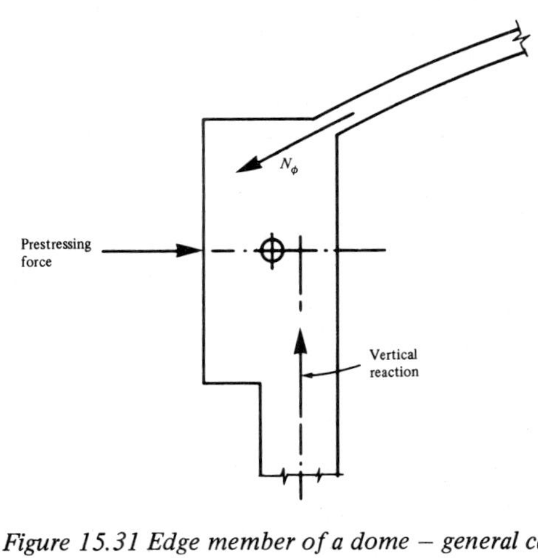

Figure 15.31 Edge member of a dome – general case

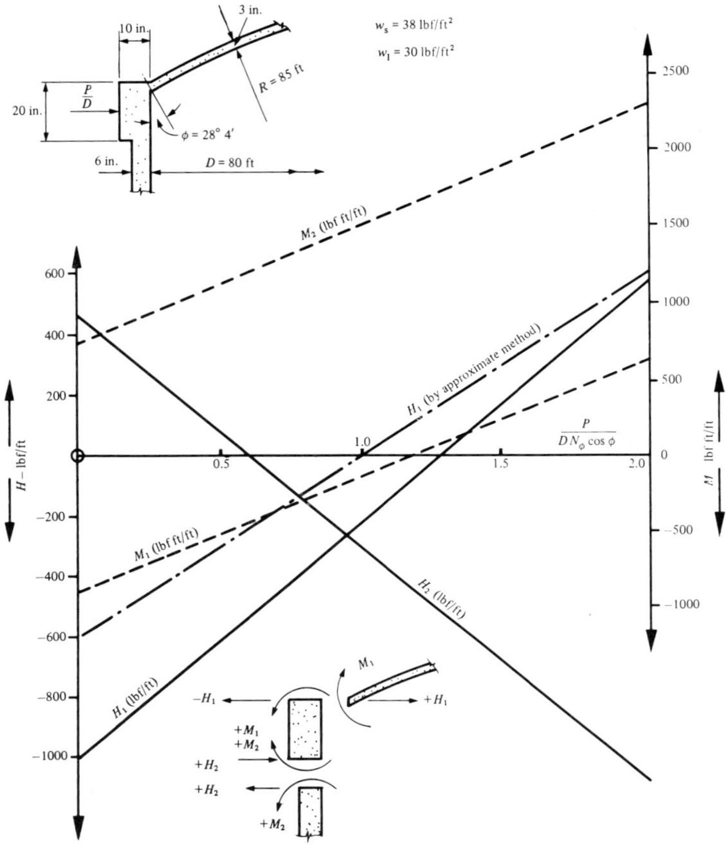

Figure 15.32 Comparison of Timoshenko's method and approximate method

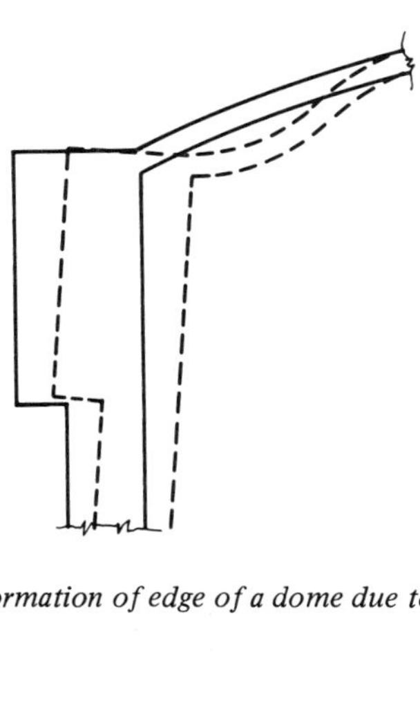

Figure 15.33 Deformation of edge of a dome due to over prestressing

15.6.3 Cylindrical shells

As in the case of domes, no difficulty occurs in the calculation of the stresses in a simply-supported cylindrical shell without edge-beams. When edge-beams are present however, the calculation of the redundant reactions and moments is too complicated to be given here, and an approximate method, applicable only to 'long' shells is given instead.

Shell without edge beams

Unlike a dome, the properties of a cylindrical shell are not the same in all directions, and in order to simplify the mathematics it is assumed that the distribution of loading along the shell is sinusoidal; more exactly, a uniform load of w is replaced by $w' = \frac{4}{\pi} w \cos \frac{\pi x}{l}$, in which x is measured from the centre of the span and the constant $\frac{4}{\pi}$ is obtained from the first term of the equivalent Fourier series.

With this assumption, the stresses in a simply-supported cylindrical shell of small thickness are as given in Figure 15.34. They consist of a tangential thrust N_ϕ, a longitudinal thrust N_x, and a longitudinal shearing force V_0 (all per unit length) at any point, and are given by the expressions

$$N_\phi = R \cos \phi \cos kx \, (w_s' + w_l' \cos \phi)$$

$$N_x = \frac{\cos kx}{Rk^2} \left(2w_s' \cos \phi + 3w_l' (\cos \phi^2 - \sin \phi^2) \right)$$

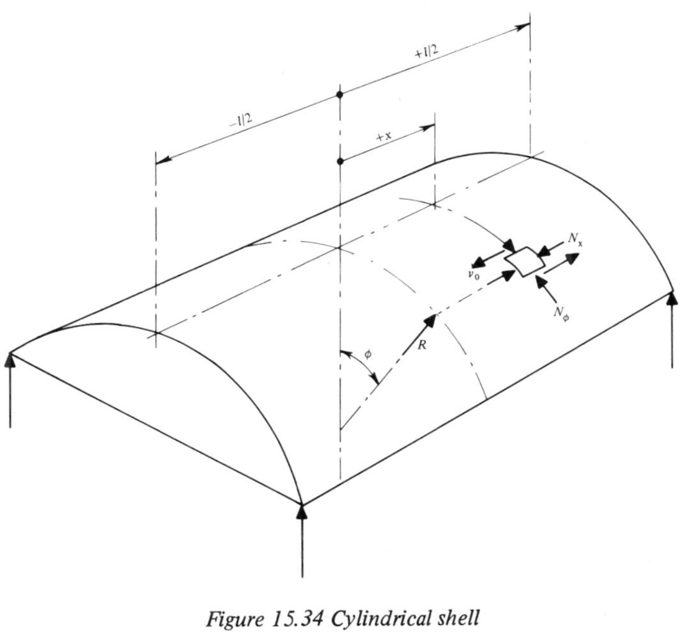

Figure 15.34 Cylindrical shell

$$V_0 = \frac{\sin \phi \sin kx}{k} (2w_s' + 3w_1' \cos \phi)$$

in which

$$w_s' = \frac{4w_s}{\pi}; w_1' = \frac{4w_1}{\pi}; \text{ and } k = \frac{4\pi}{l}$$

These can be maintained only if reactions equal and opposite to N_ϕ and V_0 are provided at the edge of the shell. It is not usually possible to provide an arrangement of cables which will provide the exact value of both N_ϕ and V_0. However, a prestressing force which satisfies the requirement for N_ϕ is greater than that needed for V_0, and this case is considered here (Figure 15.35). When the curvature of the prestressing cables around the shell is ignored, then from the force diagram it follows that

$$\frac{P\pi}{N_{\phi 0}\, l} = \sqrt{\frac{h_0^2 + 0.319^2 l^2}{h_0}}$$

and hence

$$P = \frac{l N_{\phi c}}{\pi} \sqrt{1 + 0.319^2 \left(\frac{l}{h_0}\right)^2} \quad \ldots \ldots \ldots \ldots \quad (15.3)$$

In this equation h_0 is the distance through which the centroid of the prestressing steel is deflected between midspan and support; the greater this value is, the smaller is the prestressing force required. The resistance to shear will exceed that required provided that $\frac{V_{0\,max}}{N_{\phi 0}} < 0.319 \frac{l}{h_0}$. However, this excess will affect the stresses in the shell, and the prestressing force given by Equation 15.3 should not be adopted without checking that this is acceptable.

Shells with edge beams

In considering the prestressing of domes it was noted that the redundant reaction and moment introduced when an edge-beam is provided modify the stresses in the dome. In the case of a cylindrical shell four redundants are introduced and the problem becomes too complex for solution in general terms.

The action of a cylindrical shell may be thought of as two-fold. Longitudinally the behaviour is similar to that of a beam, and transversely of an arch. The longitudinal effects therefore comprise stresses due to bending and shear, while the transverse stresses are due to direct thrust and transverse bending. In the case of long shells, in which the length is more than about twice the chord, the distribution of stress due to bending and shear is nearly the same as that given by simple-beam theory, and the transverse stresses (although not necessarily the transverse deflections) are not usually of great importance. In short shells, on the other hand, the distribution of the longitudinal stresses is similar to that in a deep beam (see Chapter 10) and the transverse stresses may be of major importance. The variation in the distribution of the longitudinal stress is shown in Figure 15.36, which is from A.S.C.E. Manual No. 31. Shells with edge-beams when prestressed, often have the tendons placed in the edge-beams only since they are in tension. However, it is advisable to provide tendons in the shell as well. In the case of continuous-shell this is essential.

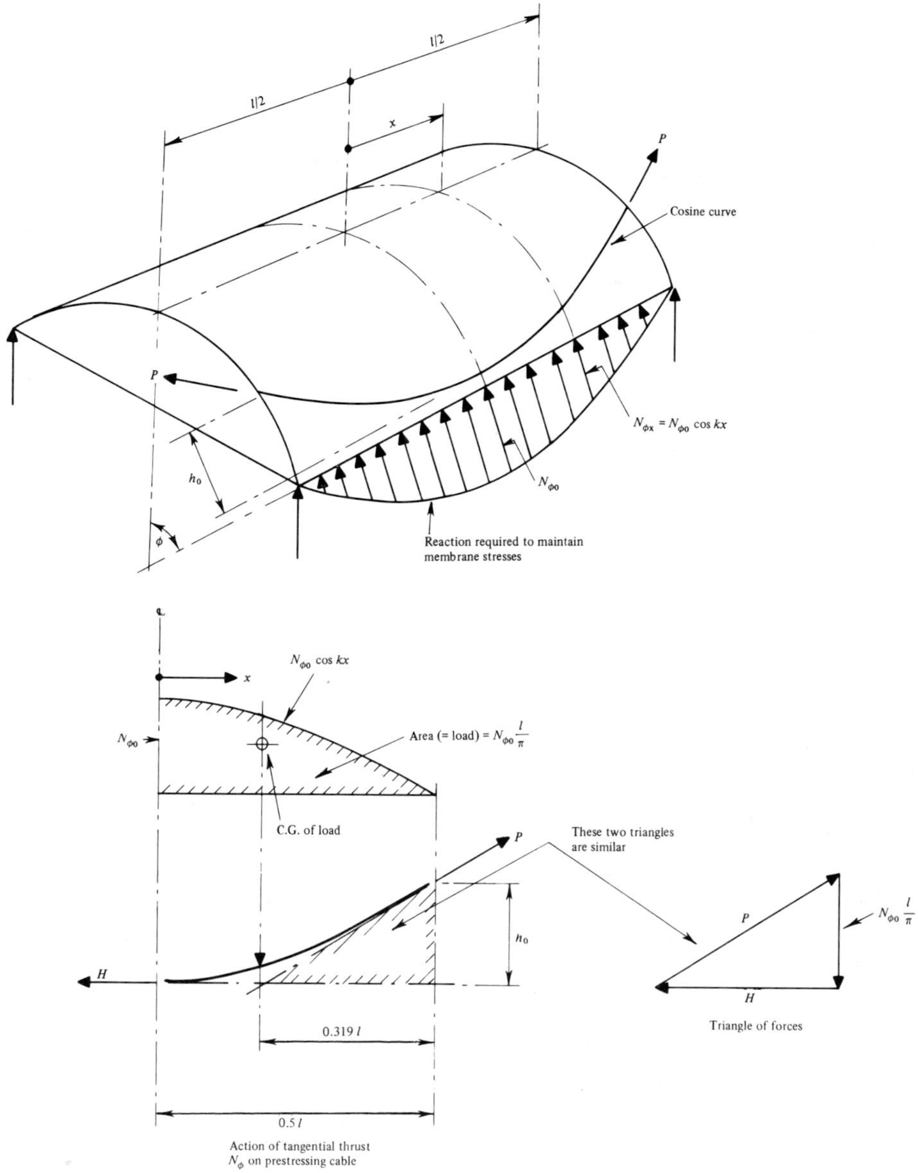

Figure 15.35 Prestressing of statically determinate cylindrical shell

A simple approximation, which is valid for long shells only, is given in the following. This is suitable for preliminary calculations but should not be used for a final design. In this method, due to Lundgren(10), the structure is regarded as a simple beam, the properties of which are given in Figure 15.37.

The bending moment M and shearing force V are calculated, and the stresses at any point are obtained from the usual expressions

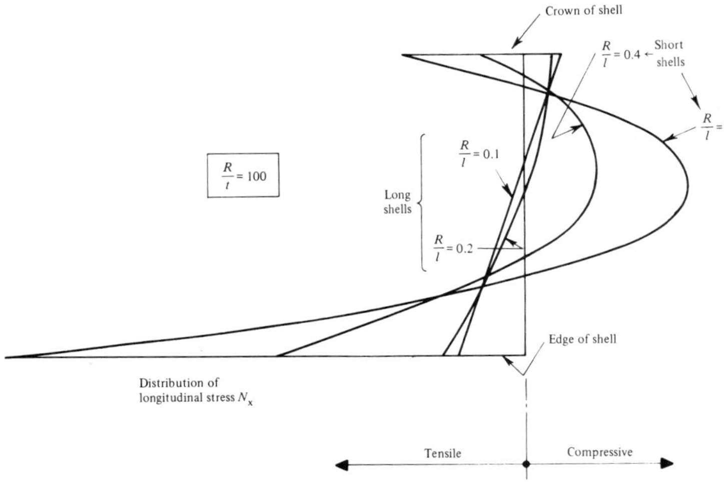

Figure 15.36 Longitudinal stresses in long and short shells

F_x = Longitudinal bending force per unit arc

= Longitudinal bending stress × thickness

$= \dfrac{M_x}{Z} t$

V_0 = Maximum shearing force per unit length

$= \dfrac{VS'}{I}$ per unit length

The effect of the prestressing force is computed in the same way.

The lateral thrust N_ϕ and bending moment M_ϕ can also be obtained approximately. These are given by considering a small length of the shell as an arch, subjected to the loading shown in Figure 15.38.

15.6.4 North-light shell roofs

If a number of North-light shell roofs (Figure 15.39) are arranged one beside the other they are capable of taking up the thrust between the adjacent rows of shells except at the edge members. Each shell presents in general a complex problem in view of its asymmetry. However, if they are long (i.e. the length being at least twice the width) they may be computed as approximation by the beam theory. Transverse bending will occur which has to be taken up by individual cross ribs unless this transverse bending is counteracted by prestressing. The upper edge of the shell is supported by a number of window posts which transfer the reactions to the adjacent shell.

Generally, the resultant of the shear stresses obtained from the membrane state of the shell does not coincide with the resultant of the loading. In consequence, additional torsional moments occur. The approximate calculation of such an inside shell as a long beam is based on conditions illustrated in Figure

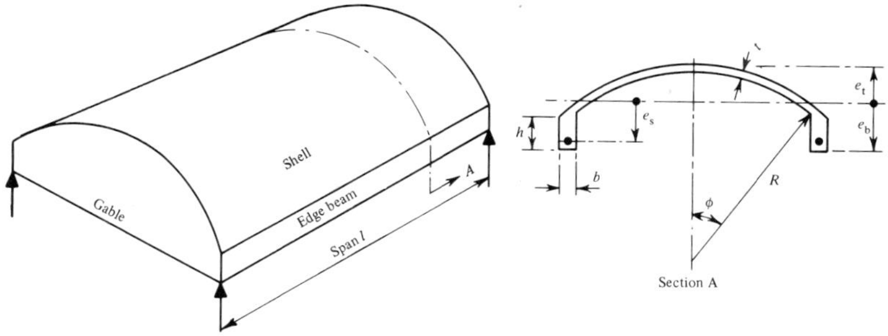

Properties of section

$$\text{Area } A = 2\,(R\,t\,\phi + bh)$$

$$e_t = R\left[\frac{\phi^2}{6}\left(1-\frac{\phi^2}{20}\right)+\frac{C_1\,C_2}{1+C_2}\right]$$

$$I = 2R^3t\left[\frac{\phi^5}{45}\left(1-\frac{\phi^2}{7}\right)+\frac{C_1{}^2\,C_2\,\phi}{1+C_2}+\frac{C_2\,\phi\,h^2}{12R^2}\right]$$

$$2S' = \frac{4}{3}te_t\sqrt{2e_tR}\left(1+\frac{e_t}{20R}\right) \text{ when } e_b > h$$

$$= e_b^2 \text{ when } e_b < h$$

$$C_1 = \frac{\phi^2}{3}\left(1-\frac{\phi^2}{10}\right)+\frac{h}{2R}$$

$$C_2 = \frac{b\,h}{R\,t\,\phi}$$

Figure 15.37 Lundgren's method: longitudinal stresses

15.40. It is necessary first to obtain the principal axes and the components of the bending moment and normal forces along these axes. The shear force multiplied by the distance from the shear centre give the torsional moment.

In order to obtain the basic conditions for the design the following steps have to be made. First, the shell cross-section has to be sub-divided in a number of preferably equal parts assuming an origin '0', say at the middle of the circular part of the shell (see Figure 15.40). The two axes $0Y$ and $0Z$ have to be assumed and the y and z co-ordinates for the centroids of the individual members to be tabulated. Then the values y^2, z^2, yz, $\sin\theta$, $\cos\theta$ (θ being the angle between y-axis and the connection between centroid of the section and origin 0) as well as the loads for self-weight of each segment for additional load due to insulation and roofing for segment and of the superimposed live load per segment have to be computed. From the total weight of shell per unit span the components of load per unit length in direction of Y and Z axes have to be evaluated and the centre of gravity of the shell has to be computed.

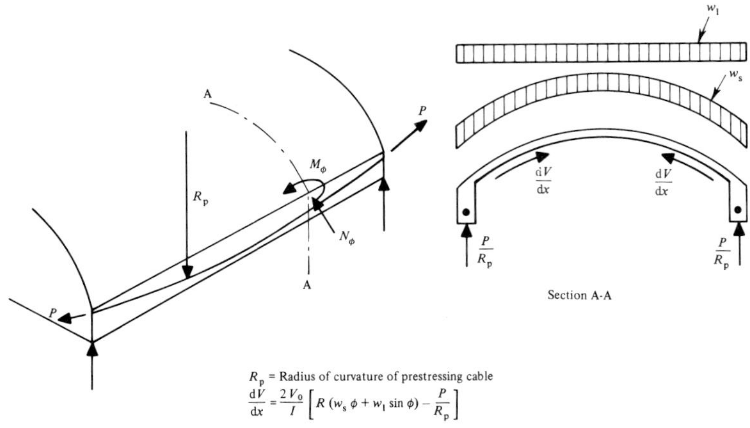

Figure 15.38 Lundgren's method: transverse stresses

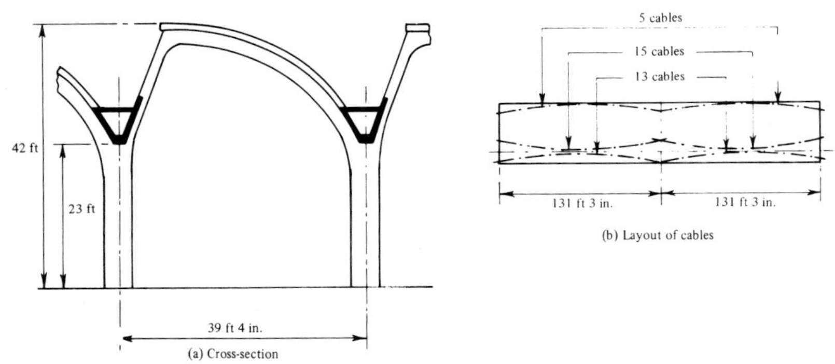

Figure 15.39 North-light shell roofs

The co-ordinates of the various segments have to be translated to the new origin 0′ at the centre of gravity of the main axes and the new values $I_{y'y'}$, $I_{z'z'}$ and $I_{z'y'}$ computed. Parabolic cables have to be arranged in such a way that the resulting force is vertical. Specific shear (assumed as force tangential to the element) and effective loads due to prestressing have to be in equilibrium with the reactions induced by glazing posts and acting along their length. Therefore the values of the reactions in the glazing posts can be determined and the transverse moment at any point of the shell.

The unit length of the shell is subjected to the following forces: the vertical load of the shell, the specific shear, reactions from glazing posts and effective load due to the prestressing forces in parabolic tendons. The latter P_e can be obtained from the relationship $P_e = \frac{8Ph_n}{l^2}$ where P is the prestressing force, l is the span of the shell, and h_n is the rise of the cable.

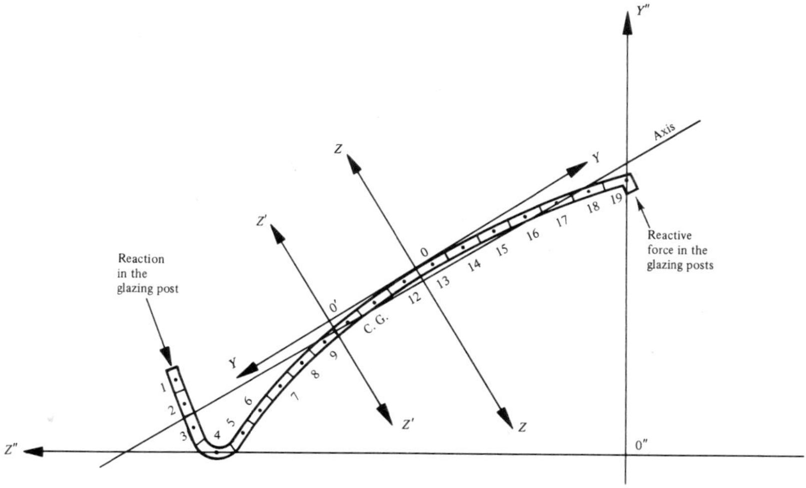

Figure 15.40 North-light shell roof: approximate method of calculation

This type of structure can be most suitably used for an assembly of precast members as introduced by Hossdorf(11). Such a solution was used as a roof construction of a shed, and model tests were carried out to compare its resistance with the actual behaviour under ultimate condition.

The system 'Hossdorf' is characterized by the fact that precast members have upper end ribs with holes through which curved tendons placed in a duct tube are threaded in a shape as indicated in Figure 15.41. The cross-section being shown in Figure 15.42. With this type of shell construction, an upper roofing is provided (e.g. asbestos-cement) which on the one hand protects the encased tendons (which are pressure grouted) and acts as insulation for the room beneath the roof structure. In a model test(12) (one-tenth of actual size) one shell structure consisted of 25 units corresponding to the main type of 117 ft (37.5 m) length and 35.7 ft (10.7 m) width (Figure 15.42). The tendons are above the relatively thin membrane and are curved so as to offset or minimize the transverse bending under working load. The great number of precast units involved (approximately 2500 units) made this an attractive proposition. However, the roof structure based on this model test was not built, although test results were very satisfactory.

A series of loadings applied within the elastic range of the shell based on the assumption that much higher loads would occur at the valley than at the ridge, owing to the accumulation of snow. In order to obtain with one model two different conditions before failure, arrangements with and without anti-twist provisions were made, the latter relating to the edge member of the adjacent row of shells. The effect of the glazing posts to take up the thrust was duly taken into account at the loading. The final loading was carried out without anti-twist arrangements and the failure load corresponded to 8.5 times the live load, when a kink occurred in the ridge beam without the tensioned steel reaching the yield point. Cracks developed at about 4½ times the live load and

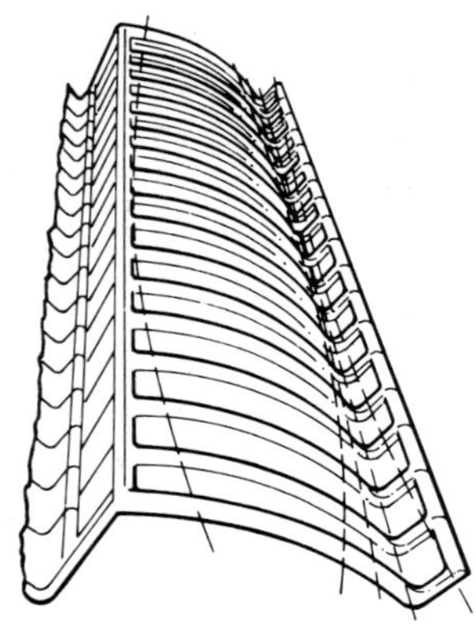

Figure 15.41 Hossdorf system

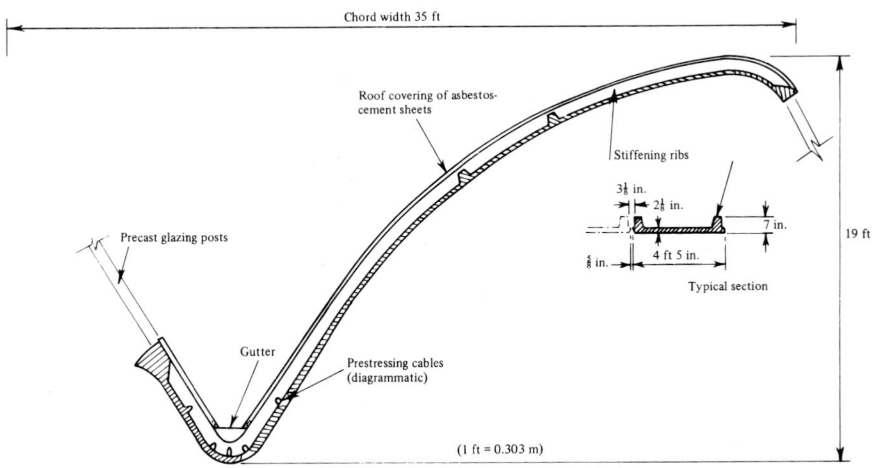

Figure 15.42 Cross-section of Hossdorf north-light shell

failure occurred at about 8.5 times live load; it was noted that the cracks extended to a line parallel to the principal axis. Although the maximum load was almost nine times the live load or 2.2 times the overall load, almost complete recovery of deflection occurred and all cracks associated with bending in the membrane closed[12]. This seems to be, therefore, a very suitable solution if a large area with North-light has to be built.

15.6.5 Hyperbolic paraboloids

Hyperbolic paraboloid shells offer exciting and interesting structures with possibilities of a wide range of form. They are surfaces of translation and make effecient use of materials by relying on shape or form for strength rather than for the mass(13). They could be supported by one, two or four columns, or sometimes by a number of columns along the periphery which could be circular, elliptical, square or rectangular in plan. The form can be achieved by moving vertical parabola along the axis of another parabola at right-angles to it and having an opposite curvature (see Figure 15.43). This can also be generated by moving along the y axis a straight line that remains parallel to zx plane at all times but pivots while sliding along straight line ABC (Figure 15.44).

The membrane stresses are relatively small and the thickness of connection is generally determined from the consideration of cover to reinforcement. The edge members are subjected to tensile or compressive thrusts which equal the sum of shear forces acting along its length. Prestressing is mainly applied to the tensile-edge member.

Examples of hyperbolic paraboloid on elliptical or circular base having a number of supports along the perimeter have been shown in reference (14). The analysis was done on the principle of suspension cables where the forces have been balanced within the ring beam.

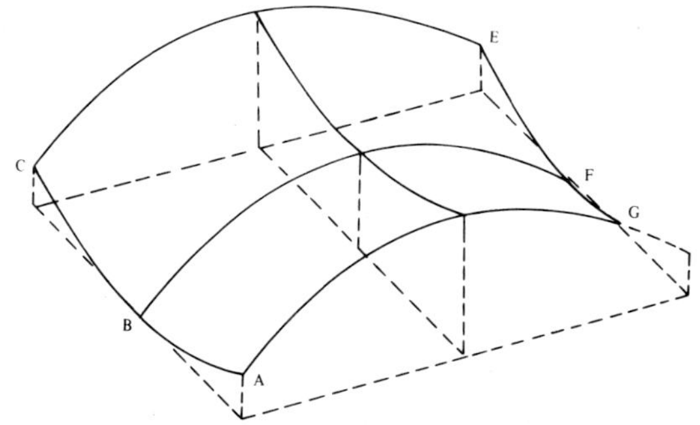

Figure 15.43 Hyperbolic paraboloid shells

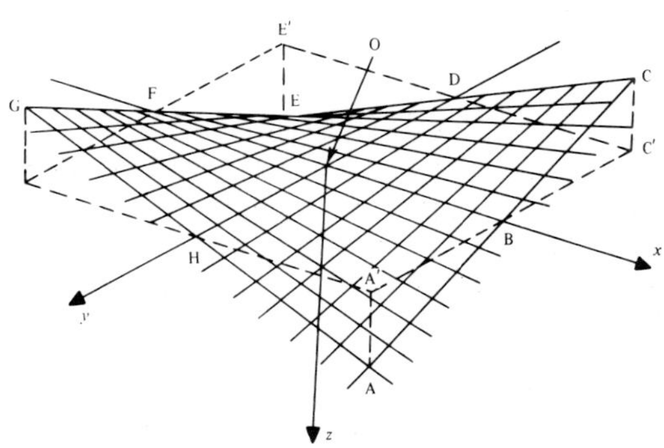

Figure 15.44 Surfaces of translation generated by moving straight lines

The sagging suspension cables run along the major axis, while the hogging cables (at right angles to major axis), stiffening the suspension cables, are prestressed in stages and the magnitude of prestress is controlled in such a way as to reduce the resultant bending moment in the ring beams.

Riblath (expanded metal) or similar permanent shuttering is placed on intersecting grids of cables to support insulation screed and waterproofing material. The cables are protected inside by spraying with fire-resistant material.

Hyperboloid of rotation

This type of construction is particularly well adapted for prestressing since the tendons are laid along the straight generatrix. This method is particularly suitable for pre-tensioning as employed by the Silverkuhl system developed in Germany and used for roofing members (see Figure 15.45). This type of precast roof member is produced in a standard width of 2.5 m (8 ft 2 in.) and in lengths up to 20 m (66 ft 2 in.). In order to obtain sufficient fire-resistance, preferably a lower layer of Perlite of a thickness of 2–7 cm (0.8 in. to 2.8 in.) is placed into the mould before casting and fire-resistance tests have proved that a thickness of 4.5 cm (1.8 in.) suffices to ensure a fire rating of at least two hours.

15.7 Folded slabs

The behaviour of folded plates is very similar in principle to that of shell structures. The main difference between the two is the more pronounced transverse bending of the folded plate, which acts as a continuous frame in the transverse direction, whereas the action of a cylindrical shell is more like that of an arch. Shearing stresses proportional to the bending moments occur along the edge connections of the separate plates. These can usually be resisted by a triangular thickening of the concrete along the joints.

The considerations which apply with regard to prestressing are similar to those for cylindrical shells. When a simple two-fold plate is used as a roof structure, with or without edge-beams, the advantages of prestressing are the same as in the case of a cylindrical shell. Bending-up of cables within the inclined plates is of great advantage (Figure 15.46), but if the structure is cast in place this may require the plate to be thicker than if no prestress were used; thus

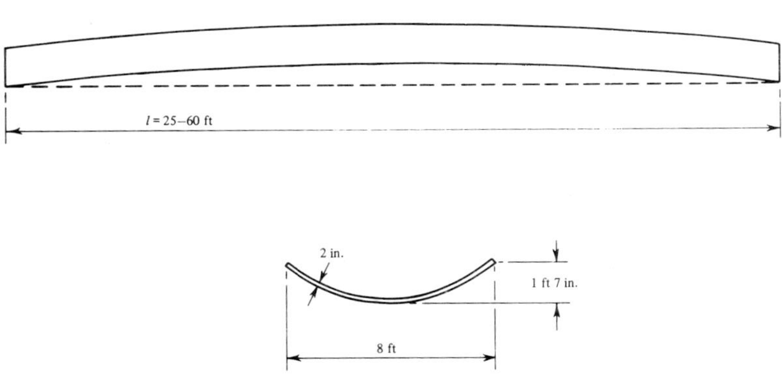

Figure 15.45 Silverkuhl system

the advantage is greater when the size of the plate is sufficient to warrant such a thickening. The position is different when precast members of limited size are used, since in this case pre-tensioned wires can be provided without increasing the thickness over that required for ordinary reinforced concrete.

15.8 Pipes and tanks

15.8.1 Pipes

If a pipe is required to transmit a fluid under pressure, this causes a circumferential tensile stress in the pipe. In the simplest case, when the thickness of the pipe is small compared with the diameter and the variations of pressure between the top and bottom of the pipe due to the density of the fluid can be neglected, the stress is given by

$$f = \frac{pD}{2t}$$

in which p is the pressure, D the diameter and t the thickness of the pipe. Assuming that no tensile stresses are to be present under working load, then the effective prestress f_{pe} must not be less than $\frac{pD}{2t}$ and the prestress at transfer is therefore,

$$f_{pt} = \frac{pD}{2R_0 t}$$

Hence the prestressing force per unit length P is

$$2f_{pt}t = \frac{pD}{R_0}$$

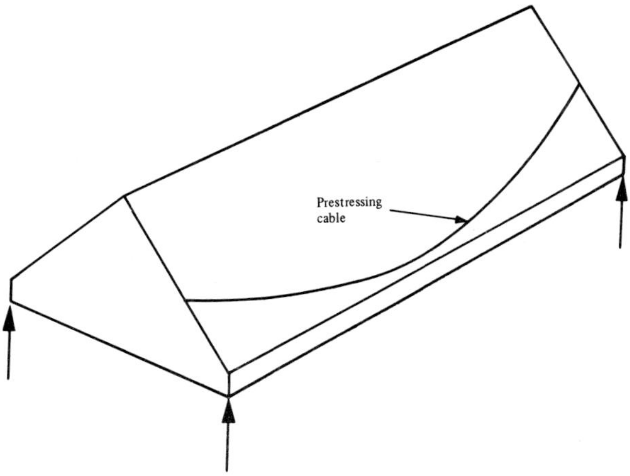

Figure 15.46 Prestressing of folded plate

In assuming the losses, particularly for pipes containing water, it is sometimes suggested that as the resultant stress at working load (and hence the loss due to creep) is small, and the presence of the water will reduce losses due to shrinkage, a reduced value of R_0 should be used. However, since the pipes may be stored for some time before use and may not be subjected to pressure for some time after installation, it is preferable to allow for the full losses which might occur.

If the pipe is required to act as a beam, supporting its own weight and that of the fluid, longitudinal stresses due to bending will occur. The structure can then be designed as a beam of annular cross-section and the prestressing force calculated in the usual way. Longitudinal stresses are also caused at points of discontinuity of prestress; an approximate method of analysis, due to Davies[(15)], is available and again prestressing can be employed to improve the distribution of stress.

In the British Standard BS 4625 prestressed concrete pressure pipes have been classified into the following two categories:

(1) *Prestressed concrete cylinder pipe:*
This is a welded sheet-steel cylinder with steel sockets and spigot rings welded to its ends, lined with concrete suitably compacted and circumferentially prestressed to withstand internal pressure and external design load, they are subsequently coated with cement mortar to protect the steel cylinder and prestressing wire (Figure 15.47).

(2) *Prestressed concrete non-cylinder pipe:*
This is a suitably compacted concrete core longitudinally prestressed with pre-tensioned high-tensile steel wire embedded in the concrete, circumferentially prestressed to withstand internal pressure and external design loads and subsequently coated with cement mortar to protect the circumferential prestressing wire (Figure 15.48).

The requirement for the second category pipes is that the longitudinal pretension should be provided throughout its length, the concrete strength at transfer being not less than 3000 lbf/in^2 (210 kgf/cm^2; 21 N/mm^2). Circumferential prestressing shall not be applied before the concrete has attained a strength of 4600 lbf/in^2 (322 kgf/cm^2; 32 N/mm^2).

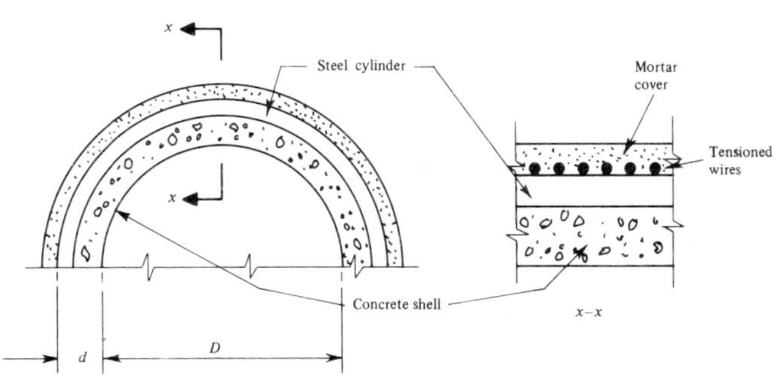

Figure 15.47 Prestressed concrete pipe cylinder type (BS 4625)

Initial compression induced in the concrete shall not exceed 55 per cent of the concrete strength at the time of stressing.

For prestressed concrete pressure pipes it is specified that there should be no tensile stress due to backfill, working pressure and the prestress. However, limited tensile stresses are allowed when additional transient surface load is applied.

For cylinder pipes, areas to take up the internal pressure would be modified and the concrete area including the area of steel cylinder multiplied by the modular ratio. This should be taken into account in determining the prestressing force in accordance with the above formula.

For prestressed concrete non-pressure pipes (between 400–3000 mm diameter) the recommendation contained in the draft British Standard document 74/10221 regarding longitudinal and/or circumferential prestress is similar.

15.8.2 Tanks

If the junction between the wall and base of a tank is such as to permit free relative movement, the stresses in the wall of the tank are calculated in exactly the same way as for a pipe and no longitudinal stresses occur. If the junction is such that the base of the wall is either hinged or fully fixed, however, the circumferential stresses are modified and longitudinal stresses are caused. Figure 15.49 shows the base details of three types. The variation in the distribution of ring forces and vertical moments in the three types are shown in Figure 15.50[16]. The method of computing these stresses is available elsewhere[16,17] and is not repeated here. If the resultant stresses at working

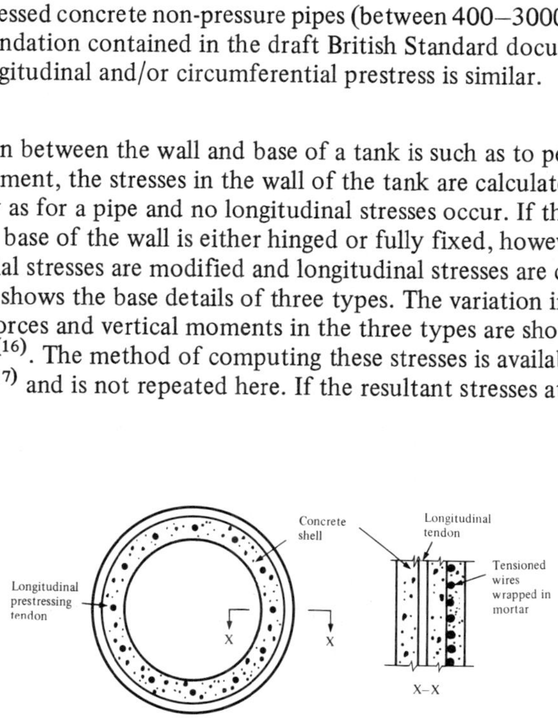

Figure 15.48 Prestressed concrete pipes non-cylinder type (BS 4625)

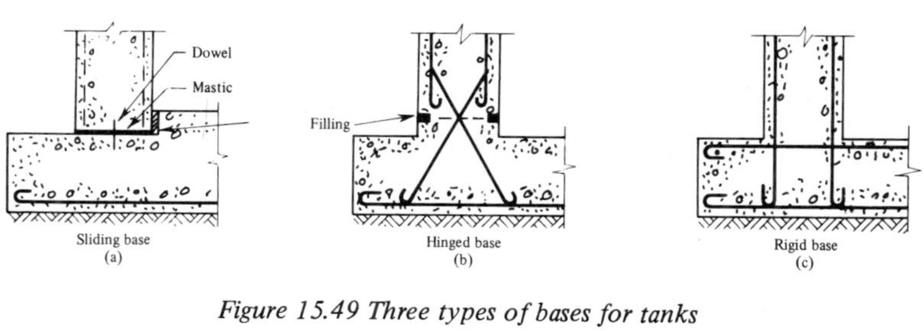

Figure 15.49 Three types of bases for tanks

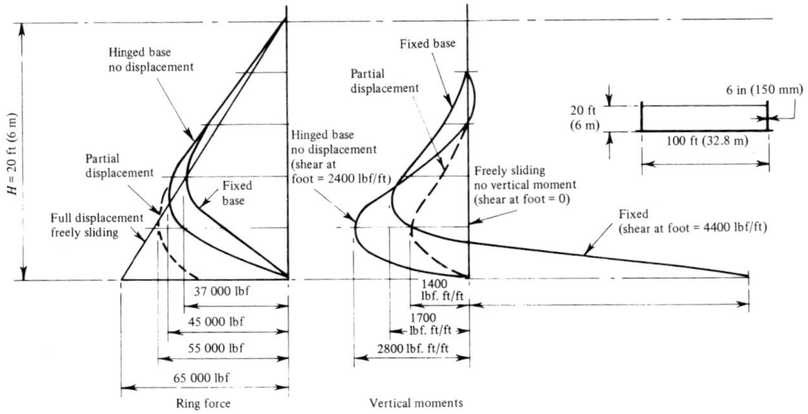

Figure 15.50 Comparison of ring forces and bending moment in a tank with three types of bases. (after Creasy)

load are required to be approximately zero, then the effective prestress can be computed, independently of the type of junction, from the expression

$$f_{pe} = \frac{whD}{2t}$$

in which w is the density of the contained liquid, h the depth at which the prestress is required, D the diameter, and t the thickness of the tank. As in the case of pipes, the prestressing force per unit length is therefore $P_t = \frac{whD}{R_0}$, and this will vary from a maximum at the bottom to zero at the top of the tank for free base. However, for hinged and rigid base the position of maximum prestressing force will be in accordance with Figure 15.50 – somewhere in the lower third of the tank. Care should be taken to ensure that the stresses at the junction of the wall and base which occur when the prestress is applied are not excessive.

Again, in calculating the value of R_0, it is desirable to assume that the full losses occur. In calculating R_0 due attention should be paid to 'friction loss' where applicable. In systems where tensioned wires are wrapped around the cylinder (e.g. Preload and BBRV) there is no friction loss.

Tendons are protected by guniting. This, however, was found to be unsatisfactory in corrosive atmosphere.

Alternatively, the tendons can be placed in ducts within the tank wall and pressure grouting of duct is carried after stressing. Here also, to minimize friction losses, the tendons are in segments and overlapped. For tensioning convenience, intermediate piers are introduced. Instead of duct, the 'unbonded' tendons, as described in Chapter 1 can be used with advantage. These unbonded tendons can be cast into the concrete and subsequently tensioned and, as already mentioned, no further grouting is necessary. This often results in a very economic structure. In assuming the ultimate moment capacity of the section, however, due consideration should be given to the unbonded condition of the tendons.

REFERENCES

1. MATTOCK, A. H. and KAAR, P. H. Contnuous precast prestressed concrete bridges. *Journal of the PCA Research and Development Laboratories,* Vol. 2, No. 1, January 1960. pp. 23–24.
2. ANDERSON, A. R. Stretched-out AASHO-PCI beams type III and IV for larger span highway bridges. Journal of *Precast Concrete Institute,* Vol. 18, No. 5. pp. 32–49.
3. BAXTER, J. W., BIRKETT, E. M. and GIFFORD, E. W. H. The narrows bridge, Perth, Western Australia. *Proceedings of the Institution of Civil Engineers,* September, 1961, Vol. 20, pp. 39–84.
4. GLANVILLE, W. H. *et al.* An investigation of the stresses in reinforced concrete piles during driving. London, H.M.S.O, 1938. pp. 111.
5. SMITH, E. A. Technical paper No. 20. Pile calculations by the wave equation. *Concrete and Constructional Engineering,* June, 1958.
6. STROBEL, G. C. and HEALD, J. Theoretical and practical discussion of the design, testing and use of pretensioned prestressed concrete piling. *Journal of the Prestressed Concrete Institute,* Vol. 5, September, 1961. pp. 22–33.
7. GERWICK, B. C. Torsion in concrete piles during driving. *Journal of the Prestressed Concrete Institute,* Vol. 4, June, 1959, pp. 58–63.
8. BERTERO, V. V. *et al.* A seismic design of prestressed concrete piling. Presented at the FIP Congress, New York, 25 May 1974. pp. 9.
9. TIMOSHENKO, S. P. *Theory of plates and shells.* New York, McGraw-Hill, 1940. pp. 492.
10. LUNDGREN, H. *Cylindrical shells.* Vol. 1. *Cylindrical roofs.* Copenhagen, Danish Technical Press, 1949. pp. 360.
11. HOSSDROF, H. Prefabricated north light shell construction. *Schweizerische Bauzeitung,* Vol. 80, No. 50. 13 December, 1962. pp. 838–843.
12. ABELES, P. W. and LUCAS, J. The plastic behaviour of a prestressed concrete shell with and without anti-twist adjustments (based on a model test). *Non-classical shell problems: Proceedings of IASS symposium held in Warsaw, 2–5 September 1963.* Amsterdam, North-Holland Publishing Co., 1964. pp. 1084–1100.
13. PORTLAND CEMENT ASSOCIATION. *Elementary analysis of hyperbolic paraboloid shells.* Chicago, 1960. pp. 20. 15085. 01D.
14. BOBROWSKI, J. and BARDHAN-ROY, B. K. Simplified design and construction technique of hyperbolic paraboloid shells. *Proceedings of IASS Congress on Large-Span Shells,* Leningrad, 6–9 September 1966. Moscow, TSINIS, 1968. Vol. 2. pp. 68–79.
15. DAVIES, J. D. Discontinuity in circumferentially prestressed cylinders. *Civil Engineering and Public Works Review,* Vol. 55, No. 643. February 1960, pp. 225–227.
16. CREASY, L. R. *Prestressed concrete cylindrical tanks.* London, Contractors Record, 1961, pp. 216.
17. GRAY, W. S. and MANNING, G. P. *Reinforced concrete reservoirs and tanks.* 4th edition, London, Concrete Publications Ltd, 1960, pp. 189.

CHAPTER 16

DESIGN FOR FIRE RESISTANCE

16.1 General considerations

The fire resistance indicates the ability of a construction to withstand fire for a specified period during which the structure shall not fail under a definite loading (the entire service load or part of it).

The requirements for fire resistance for a definite building construction are governed by the purpose for which the building is to be used, its floor area or cubic capacity and contents. The distance from the nearest fire station and/or water supply, the type of occupancy, the contents and the provision of sprinklers are of influence, particularly from the point of view of insurance premium. The U.S.A. model by-laws, for example, specify minimum resistance to fire for domestic purposes, of ½ h, if the building does not exceed 50 ft (15 m) in height or 2500 ft^2 (230 m^2) in floor area; and 1 h when these limits are exceeded for public buildings and warehouses which are not used predominantly for storage. Generally, three different periods are specified, namely ½ h, if the building does not exceed 50 ft (15 m) in height or 125 000 ft^3 (3500 m^3) in capacity; 1 h for a building between 50 ft and 70 ft (15 m and 23 m) in height or between 125 000 and 250 000 ft^3 (3500 and 7500 m^3) in capacity and not exceeding 7500 ft^2 (700 m^2) in area; and 2 h if these limits are exceeded. For warehouses which are used predominantly for storage, periods of resistance of ½ h, 1 h, 2 h and 4 h, are required according to the height and capacity of the building.

In Great Britain the fire resistance required varies from ½ h to 4 h depending on the purpose group of the building, floor area and height, as specified in the Building Regulations. For particulars see Table 16.1[(1)].

The fire resistance of an element is expressed in terms of time as established in standard fire tests (in accordance with BS 476; Part 8: 1972 in Great Britain) which are controlled to standard time–temperature curves as shown in Figure 16.1. In this figure the standard time–temperature curves as set out in the United Kingdom, USA and Germany are shown. It is seen that up to ½ h fire duration, corresponding to an increase in temperature up to 820°C (1500°F) the three standards are almost identical. Figure 16.2 illustrates the average temperatures in the case of a real fire based on British research[(2)]. In his diagram the individual figures: 60, 30, 15 and 7.5 indicate the material of wood in kilogrammes per square metre to be burnt, whereas the figures in brackets: (¼) and (½) represent the proportion of openings in the walls. Thus, an amount of 60 kg/m^2 wood for a case in which only a quarter of the wall area consists of

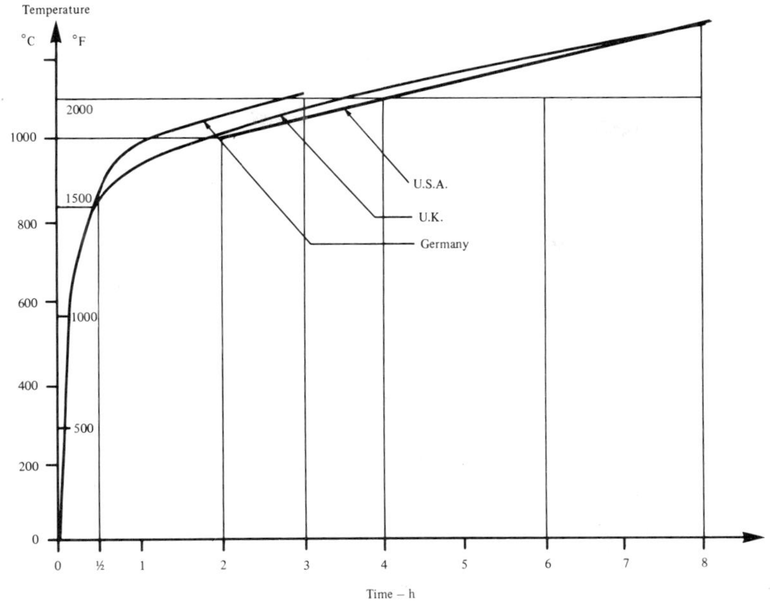

Figure 16.1 Standard time-temperature curve

openings represents the worst solution, whereas a case in which 7½ kg/m^2 wood and openings equal to half of the wall area represents the least dangerous fire loading.

It is seen that the required fire resistance rating for various buildings greatly depends on the amount of material which may burn and the type of construction available. Obviously, if the building is a warehouse containing a material which only burns slowly, taking longer than 4 h, it is hardly possible to ensure a satisfactory solution. It is often stated that a rather shorter time is sufficient, particularly when sprinklers are provided. However, this does not seem to be justified because there have been cases where the sprinklers did not work properly and it is therefore preferable to use constructions with higher fire resistance ratings.

Fire resistance of structural concrete is assumed to have reached its limit when:

(1) The service load cannot be supported any longer. (Limit state of stability)
(2) The structural member ceases to prevent the passage of flames and combustion gases used. (Limit state of integrity–flame barrier limit)
(3) An excessive transfer of heat to the unexposed side occurs. (Limit state of integrity–thermal limit)

The latter two criteria are applicable to floors only, which perform a separating function. Although they are the criteria for failure in fire, in a structural sense they could be compared with the serviceability limit since the structure can be repaired after the fire without undue cost, provided the deformations are not excessive.

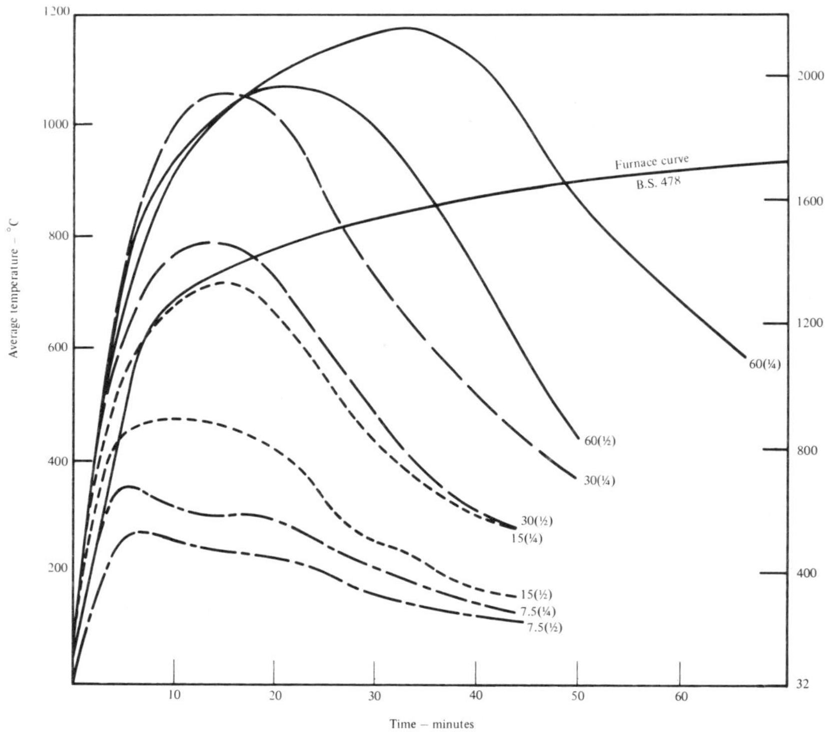

Figure 16.2 Average computed temperature as a function of time

Classification of fire resistance is derived from standard fire tests which are carried out above a furnace and fire endurance is defined by the time which elapses during which the construction can withstand the conditions based on the specified increasing temperature.

As already stated in Chapter 3 it has become evident that the conditions for simply supported beams are much more unfavourable than those for continuous beams or for beams which are built-in at the ends or capable of having thermal restraint. In various countries specific regulations have been made which do not quite agree with each other and are often of an arbitrary nature, except in the United States where more practical considerations apply from the point of view of fire insurance and the requirement of possible repair. There is a great difference between theoretical tests and practical experience in the case of actual fire. The first named author had an opportunity to visit a newly built bungalow in North Carolina at which a fire occurred following a collision involving an oil drum. Its duration was 1½ h during which temperatures of 1400°F (750°C) must have been reached since glass melted; the walls were badly damaged and had to be rebuilt, but the prestressed lightweight concrete floor and roof consisting of double tee-beams without special insulation retained their full carrying capacity. This was ascertained from a subsequent loading test[3].

Table 16.1 Fire rating requirement according to British Building Regulations

Purpose group	Dimensions, specifying maximum limits values otherwise indicated			Minimum period of fire resistance		
	Height (in feet)	Floor area (in thousand square feet)	Capacity (in thousand cubic feet)	Elements above ground (in hours)	Elements below ground (in hours)	
(1)	(2)	(3)	(4)	(5)	(6)	
I	N/L	N/L	N/L	½	1(*a*)	x
II	90	20	N/L	1	1½	
	over 90	20	N/L	1½	2	
III	25	5	N/L	½	1	x
	50	2.5	N/L	1(*b*)	1	
	90	30	300	1	1½	
	over 90	20	200	1½	2	
IV	25	2.5	N/L	0	1(*c*)	x
	25	5	N/L	½	1	
	50	N/L	125	1(*b*)	1	
	90	50	500	1	1½	
	N/L	N/L	N/L	1½	2	
V	25	1.5	N/L	0	1(*c*)	x
	25	5	N/L	½	1	
	50	N/L	125	1(*b*)	1	
	90	10	250	1	2	
	N/L	20	250	2	4	
VI	25	2.5	N/L	0	1(*c*)	x
	25	N/L	60	½	1	
	50	N/L	150	1(*b*)	1	
	90	N/L	300	1	2	
	90	N/L	1000	2	4	
	over 90	20	200	2	4	
VII	25	2.5	N/L	0	1(*c*)	x
	25	5	N/L	½	1	
	50	N/L	125	1(*b*)	1	
	90	50	500	1	1½	
	N/L	N/L	N/L	1½	2	
VIII	25	1.5	N/L	0	1(*c*)	x
	25	3	N/L	½	1	
	50	N/L	60	1(*b*)	1	
	50	N/L	125	1	2	
	90	N/L	250	2	4	
	90	N/L	750	4	4	
	over 90	10	N/L	4	4	

Notes

(*a*) The period is half an hour for elements forming part of a basement storey which has an area not exceeding 500 ft^2.

(*b*) This period is reduced to half an hour in respect of a floor which is not a compartment floor, except as to the beams which support the floor or any part of the floor which contributes to the structural support of the building as a whole.

(*c*) No fire resistance is required if the elements form part of a basement storey which has an area not exceeding 500 ft^2.

x The items thus marked are applicable only to buildings, not to compartments, except in relation to purpose group III (Other residential); see also regulations E7(2)(*a*) and E8(7)(*a*).

In the above Table the different purpose groups are defined as follows:

I	Small residential (private dwelling)
II	Institutional (e.g. hospital, school etc.)
III	Other residential (flats etc.)
IV	Office
V	Shops
VI	Factories
VII	Places of assembly other than those covered in I to VI
VIII	Storage

16.2 The different methods of dimensioning prestressed concrete members for a definite fire resistance rating

There are three possibilities of dimensioning prestressed concrete members to resist fire for a definite time, relating to the generally accepted time–temperature curve in Figure 16.1.

(1) Tests of full-sized members giving specific minimum dimensions of standard constructions.
(2) Tables showing minimum dimensions in which the cover of the steel (in fact the distance from the centroid of the steel) and the minimum dimensions are specified. These tables vary for different countries and are mostly based on conclusions made from tests, or interpolation (and extrapolation) of test results.
(3) Analytical method. When the individual conditions are known, or can be reasonably assumed, fire resistance may be established from computation. In this case, the effect of continuity and/or actual restraint can be taken into account, and thermal expansion and restraint if any may also be considered.

The diagram, Figure 16.3, obtained from Kordina[4] of Braunschweig Technical University, shows the strains of small concrete specimens, subjected to various compressive stresses ranging between 0–60 per cent of the concrete strength at 20°C, exposed to fire tests. It can be seen from Figure 16.3 that at zero stress there is only tensile strain whereas with compressive stresses up to 40 per cent of the strength, the strain is first tensile and becomes, after considerable increase in temperature, compressive, whilst with stresses 50 to 60

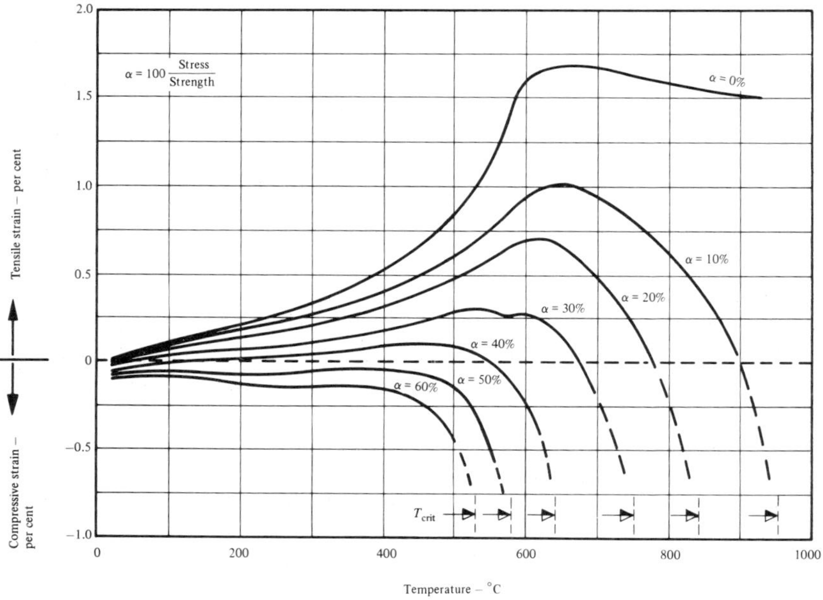

Figure 16.3 Strain in concrete specimens at varying temperatures according to Kordina and Schneider[4]

per cent of the strength it is compressive all the time. The diagram also illustrates the maximum compressive strain at the critical temperatures which occur for different concrete strengths.

In the following, methods 2 and 3 for dimensioning prestressed concrete members to resist fire are dealt with separately.

16.3 Design based on tests

Fire tests have already been discussed in Chapter 3. Whenever possible the results of such tests should be preferred in determining the fire resistance of similar constructions. Test results as well as observations of structures under actual fire have proved that detailing significantly influences the performance of the structure.

The excellent performance of the prestressed concrete construction in Avianca Building, Bogota, was, no doubt, largely due to good detailing of the individual precast members as well as the entire structure[(5)].

16.4 Dimensioning on the basis of tables

National Codes of Practice specify certain cover and/or minimum dimensions for structural members for a specific period of fire rating. These vary to a great extent in different countries. In the USA the P.C.I. Committee on Fire Resistance Rating adopted a Code provision in December 1971 which, however, has not been published on the consideration that some values are rather conservative. This proposal is reproduced in Table 16.2. P.C.I. also published a report[(6)] on the Fire Resistance of post-tensioned structures. The recommendations in the report with respect to cover etc. corresponding to different periods of fire rating are virtually identical to those shown in the table.

In the P.C.I. proposal a continuous beam or slab is defined as that which has been designed for continuity at least for superimposed dead load and full live load, and has at least 20 per cent of the maximum negative moment reinforcement made continuous throughout the span.

The American proposal also includes restrained beams and slabs which are framed into a building in such a manner that restraints to thermal expansion occur although structurally the support condition could be simply supported.

In the United Kingdom the Building Research Establishment has recently published a report[(7)] in which recommendations have been made with respect to the minimum dimensions (thickness, width and cover) of various structural members for different periods of fire resistance. These recommendations, in as much as they are applicable to concrete structures, have already been incorporated in the 'Interim Guidance' by a Joint Committee of the Institution of Structural Engineers and the Concrete Society entitled *Design and Detailing of Concrete Structures for Fire Resistance*[(8)] and are also expected to be included in the British Code of Practice. The recommendations relating to prestressed concrete members – dense as well as lightweight aggregate concrete – are reproduced here in Tables 16.3a and 16.3b.

The 'width' in these Tables (16.3a and b) refers to that at the level of the lowest reinforcement and the 'cover' to the average distance of all tensile steel from the heated surface. The structure is considered continuous if adequate end fixity is provided to develop restraining forces capable of resisting end

Table 16.2 Minimum dimensions for prestressed concrete members in dense concrete (American proposal)

Member	End conditions	Spacings	Minimum dimensions in mm for fire resistance of (Figures in brackets are in inches)					
				4 h	3 h	2 h	1½ h	1 h
Beams	Simply supported		Width	305 (12)	255 (10)	195 (7.70)	155 (6.10)	115 (4.30)
			Cover	95 (3.70)	80 (3.20)	65 (2.50)	50 (2.00)	30 (1.58)
	Continuous	1.2 m (4 ft) or less on centres	Width	110 (4.10)	98 (3.80)	79 (3.10)	65 (2.50)	65 (2.50)
			Cover	45 (1.70)	38 (1.50)	33 (1.30)	33 (1.30)	33 (1.30)
		Over 1.2 m (4 ft) on centres	Width	195 (7.70)	155 (6.10)	115 (4.50)	115 (4.50)	115 (4.50)
			Cover	65 (2.50)	50 (2.00)	45 (1.70)	45 (1.70)	45 (1.70)
Slabs (Floors with plain soffit)	Simply supported		Thickness	178 (7.00)	158 (6.20)	127 (5.00)	110 (4.30)	90 (3.50)
			Cover	71 (2.80)	50 (2.00)	46 (1.80)	38 (1.50)	28 (1.10)
	Continuous		Thickness	178 (7.00)	158 (6.20)	127 (5.00)	110 (4.30)	90 (3.50)
			Cover	25 (1.00)	25 (1.00)	20 (0.80)	20 (0.80)	20 (0.80)

moments present under normal loading conditions as well as those induced due to heating of the elements[7,8].

When the clear cover to the outermost steel is large (over 40 mm in case of dense concrete and over 50 mm in case of lightweight concrete) supplementary

Table 16.3a Minimum dimensions for prestressed concrete members in dense concrete (Density > 2300 kg/m³) (B.R.E. proposal)

Member	End conditions	Minimum dimensions in mm for fire resistance of (Figures in brackets are in inches)					
			4 h	3 h	2 h	1½ h	1 h
Beams	Simply supported	Width*	280 (11.0)	240 (9.5)	200 (7.9)	150 (5.9)	120 (4.7)
		Cover	90 (3.5)	80 (3.2)	70 (2.8)	55 (2.2)	40 (1.6)
	Continuous	Width*	240 (9.5)	200 (7.9)	150 (5.9)	120 (4.7)	100 (3.9)
		Cover	80 (3.2)	70 (2.8)	55 (2.2)	40 (1.6)	30 (1.2)
Slabs (Floors with plain soffit)	Simply supported	Thickness	170 (6.7)	150 (5.9)	125 (4.9)	110 (4.3)	95 (3.7)
		Cover	65 (2.6)	55 (2.2)	40 (1.6)	30 (1.2)	20 (0.8)
	Continuous	Thickness	170 (6.7)	150 (5.9)	125 (4.9)	110 (4.3)	95 (3.7)
		Cover	55 (2.2)	45 (1.8)	35 (1.4)	25 (1.0)	20 (0.8)
Floors with open soffit (ribbed)	Simply supported	Thickness	150 (5.9)	135 (5.3)	115 (4.5)	105 (4.1)	90 (3.5)
		Width	200 (7.9)	175 (6.9)	150 (5.9)	135 (5.3)	110 (4.3)
		Cover	75 (3.0)	65 (2.6)	55 (2.2)	45 (1.8)	35 (1.4)
	Continuous	Thickness	150 (5.9)	135 (5.3)	115 (4.5)	105 (4.1)	90 (3.5)
		Width	175 (6.9)	150 (5.9)	125 (4.9)	110 (4.3)	75 (3.0)
		Cover	65 (2.6)	55 (2.2)	45 (1.8)	35 (1.4)	25 (1.0)

**For* I *beams width of web (b_w) shall not be less than half of width shown here.*

reinforcements in the shape of metal lath or wire fabric or stirrups need be provided within 20 mm from the exposed face. The amount of fabric reinforcement shall not be less 0.5 kg/m^2 with wire spacing not exceeding 100 mm. Spacing of stirrups shall not exceed 200 mm centre to centre.

A comparison of the Tables 16.2 and 16.3a shows that while for simply

Table 16.3b Minimum dimensions for prestressed concrete members in lightweight aggregate concrete (Density < 2000 kg/m^3) (B.R.E. proposal)

Member	End conditions	Minimum dimensions in mm for fire resistance of (Figures in brackets are in inches)					
			4 h	3 h	2 h	1½ h	1 h
Beams	Simply supported	Width*	250 (9.8)	200 (7.9)	160 (6.3)	130 (5.1)	110 (4.3)
		Cover	75 (3.0)	65 (2.6)	55 (2.2)	45 (1.8)	30 (1.2)
	Continuous	Width*	200 (7.9)	150 (5.9)	125 (4.9)	100 (3.9)	90 (3.5)
		Cover	65 (2.6)	55 (2.2)	45 (1.8)	35 (1.4)	25 (1.0)
Slabs (Floors with plain soffit)	Simply supported	Thickness	150 (5.9)	135 (5.3)	115 (4.5)	105 (4.1)	90 (3.5)
		Cover	60 (2.4)	45 (1.8)	35 (1.4)	30 (1.2)	20 (0.8)
	Continuous	Thickness	150 (5.9)	135 (5.3)	115 (4.5)	105 (4.1)	90 (3.5)
		Cover	45 (1.8)	35 (1.4)	30 (1.2)	25 (1.0)	20 (0.8)
Floors with open soffit (ribbed)	Simply supported	Thickness	130 (5.1)	115 (4.5)	100 (3.9)	95 (3.7)	85 (3.4)
		Width	175 (6.9)	150 (5.9)	125 (4.9)	110 (4.3)	90 (3.5)
		Cover	65 (2.6)	55 (2.2)	45 (1.8)	35 (1.4)	30 (1.2)
	Continuous	Thickness	130 (5.1)	115 (4.5)	100 (3.9)	95 (3.7)	85 (3.4)
		Width	150 (5.9)	125 (4.9)	110 (4.3)	90 (3.5)	75 (3.0)
		Cover	55 (2.2)	45 (1.8)	35 (1.4)	30 (1.2)	25 (1.0)

**For* I *beams width of web* (b_w) *shall not be less than half the width shown here*

supported end conditions the American requirements are normally more severe, for continuous structures, on the other hand, these requirements are appreciably less stringent than in the British recommendations.

An interesting point in the American proposals is that for continuous beams spaced at close centres (4 ft i.e. 1.2 m or less) the requirements with regard to the cover and width have been significantly reduced. In other words the fire resistance of closely spaced continuous beams is considered to be appreciably higher than that of widely spaced continuous beams under similar conditions. Such a distinction is perhaps justifiable from the consideration that closely spaced beams tend to act like an anisotropic slab and thus ensure better transverse distribution of load. The capacity of load distribution to adjacent ribs improves fire performance since it may reduce the risk (generally present in primary beams) of sudden collapse due to spalling. The recognition of this fact is also evident in the BRE report where a separate table has been provided for ribbed slabs which require less stringent requirements with respect to dimensions, cover, etc. than beams, although not to the same extent as in the American proposals for beams spaced on 4 ft (1.2 m) or less centres.

Obviously there is no clear cut line of demarcation (beams having slightly more spacing than 4 ft may not act very differently); but the demarcation line apparently has been drawn in the American proposal on the basis of practical experience especially with the consideration of standard double tee units which have rib centres at 4 ft (1.2 m) usually.

16.5 Design for fire resistance: analytical approach

The practice of determining the fire resistance of a structure on the basis of tabulated data, although based on general tests, has some disadvantages. Firstly it fails to recognise the importance of detailing which, in fact, controls the mechanism of failure, and secondly the tests themselves are rather limited in scope to give a basis for general application to structures with widely varying conditions, surroundings and interactions.

A good deal of information is now available which enables a designer to properly understand the failure mode and pattern of a structure when exposed to a fire environment and thus make an analytical approach to determine its fire resistance from limit state of stability considerations. The design principles based on test performances and consistent with the limit state philosophy of design have also been formulated for some cases. The application of these principles is, however, still confined to flexural members (slabs, beams, etc.) only and where the failure is governed by yielding of tensile steel. It is not yet possible to reliably extend the analytical approach to structural members predominantly under compression such as columns and walls.

16.5.1 Heat transmissions and temperature distribution in concrete

The classical equation for heat flow in a concrete mass is given by

$$\frac{K_1}{C\gamma}\left\{\frac{\delta^2 T}{\delta^2 x}+\frac{\delta^2 T}{\delta^2 y}\right\}=\frac{\delta T}{\delta t}$$

where T = temperature
t = time
γ = density of concrete

C = specific heat of concrete
K_1 = thermal conductivity

The solution of this differential equation is rather involved even with some simplified assumptions.

Thanks to international research and experiments, the insulating properties of concrete, rate of penetration of heat and temperature distribution in concrete exposed to standard fire tests, are more reliably known now.

Temperature distribution in slabs and beams – in dense as well as lightweight aggregate concrete – due to exposure to fire, equivalent to a standard time/temperature curve, is shown in Figures 16.4 and 16.5 (a to f). These figures, reproduced from reference 8, are very handy and useful for design purposes.

It should be noted that slabs (also walls) have only one face which is exposed to fire, i.e., heat penetrates through only one surface. In beams, however, heat penetrates from three exposed faces (four in columns). This is why, in the case of beams, temperature distribution is greatly influenced by the width of the rib.

16.5.2 Material properties at elevated temperatures

Both steel and concrete lose a high proportion of strength at high temperature. E values, creep and other properties are also greatly affected by elevated temperature although their influence in the failure of the structure is negligible.

Figures 16.6 and 16.7 show the reduction of strength respectively of concrete and steel with temperature based on American tests[6]. On the basis of these as well as the British and European tests the guidelines prepared by the Joint Committee of the Institution of Structural Engineers and the Concrete Society[8] produced idealised strength reduction curves for both concrete

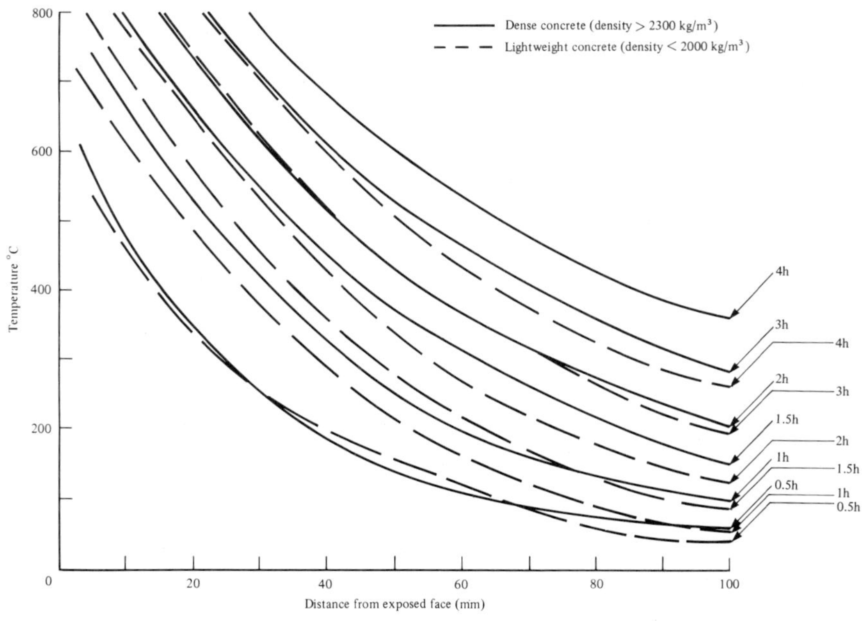

Figure 16.4 Temperature distribution in a concrete slab

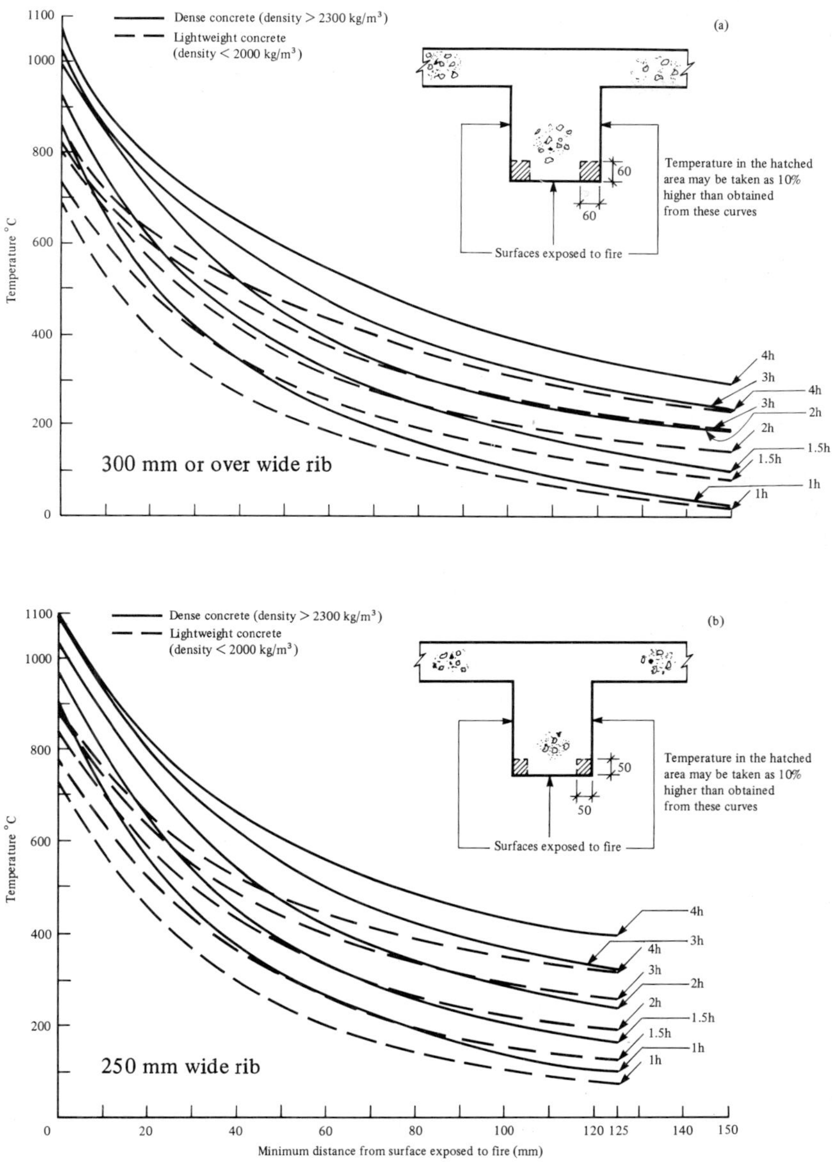

Figure 16.5 Temperature distribution in a concrete rib or beam

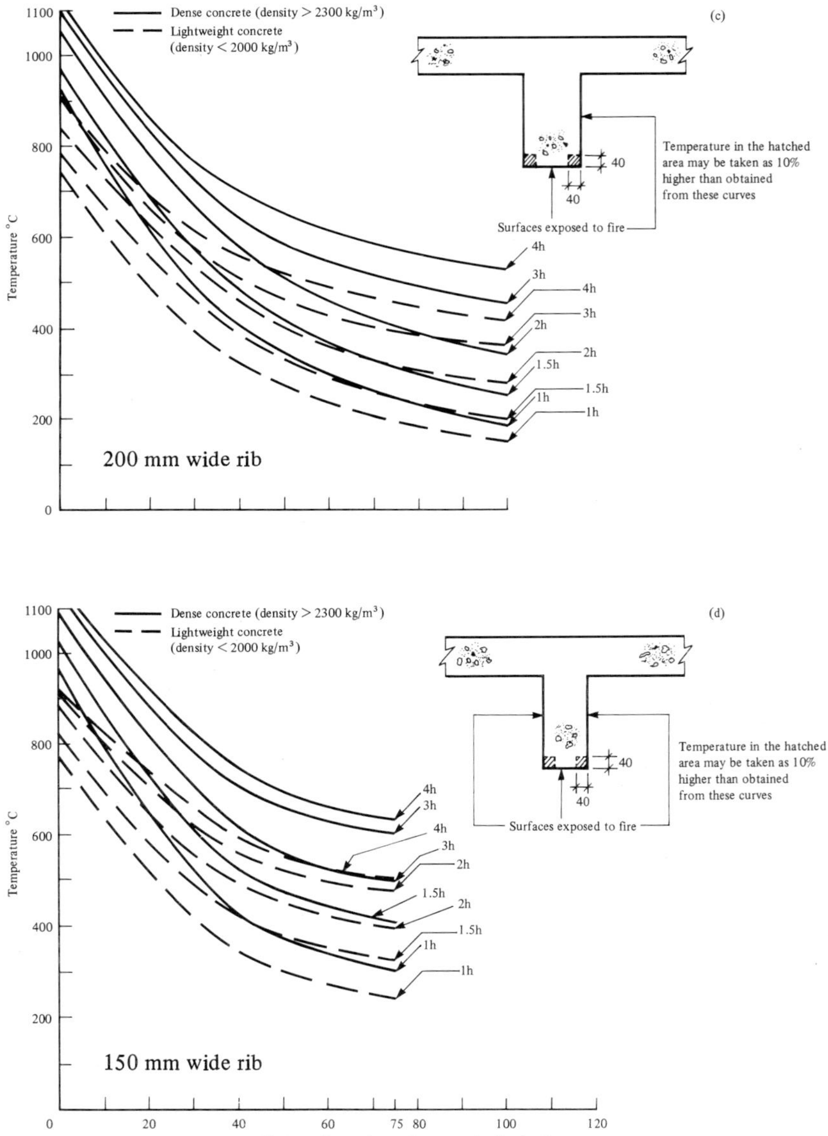

Figure 16.5 Temperature distribution in a concrete rib or beam

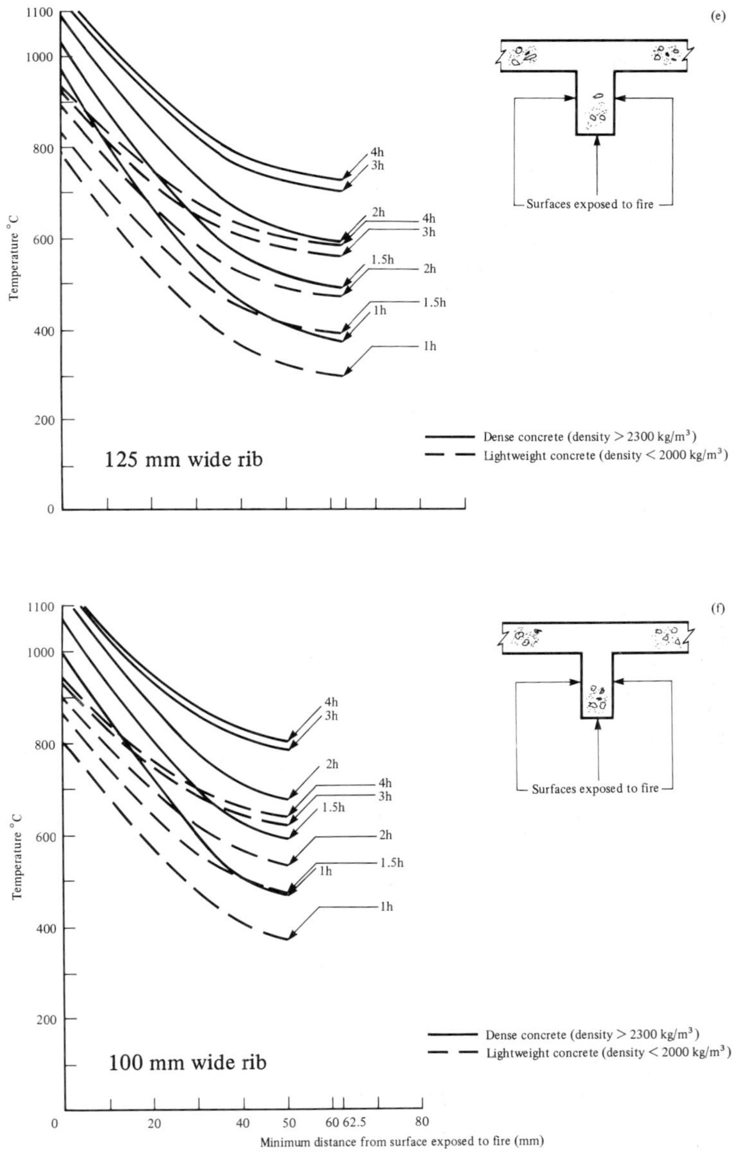

Figure 16.5 Temperature distribution in a concrete rib or beam

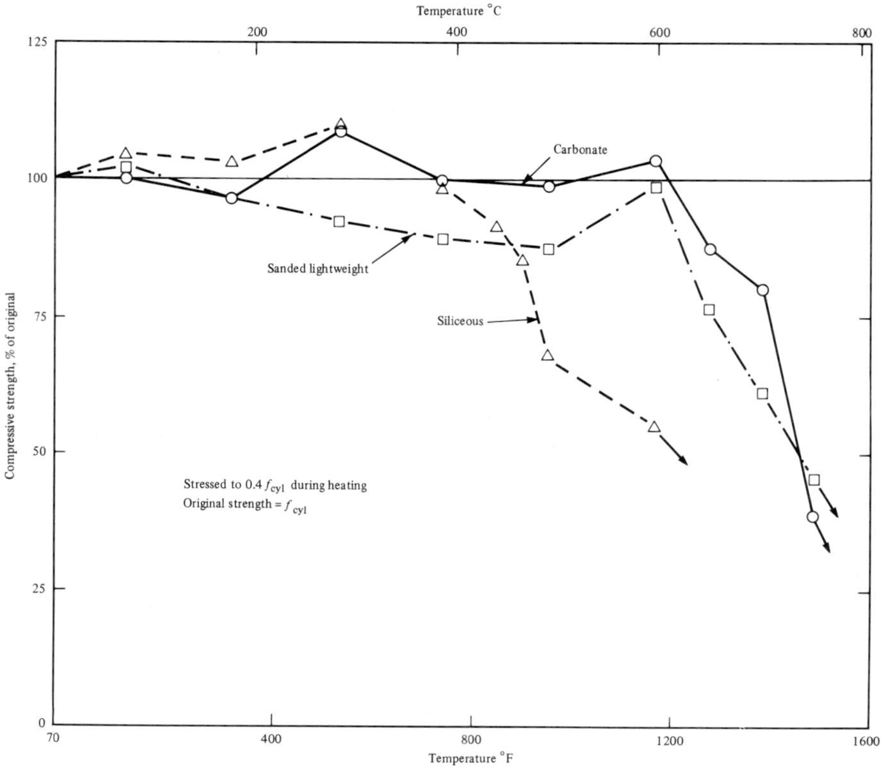

Figure 16.6 Compressive strength of concrete at high temperatures

and steel for design purposes. These curves are simple and easy to use and at the same time, being lower bounds, give results on the safe side. Figures 16.8 and 16.9 have been reproduced from the Guidelines – reference 8.

It can be seen that the lightweight aggregate concrete retains its original strength (i.e. the strength at 20°C) up to 500°C; thereafter the reduction takes place in such a way that only 40% of the original strength is retained at 800°C. For dense concrete reduction generally commences at 350°C and only 20% of the original strength is retained at 800°C.

For reinforcement bars, 50% of the original strength is lost when the temperature reaches 550°C. Prestressing tendons, on the other hand, lose half of their strength at about 400°C.

16.5.3 Design of flexural members

The member or element concerned is first designed for normal gravity and/or lateral loads as shown in previous chapters and then checked for appropriate period of fire resistance. The limit state of collapse condition needs only to be considered.

16.5.3.1 Partial safety factors

In the Guidelines[8] the following partial safety factors have been suggested for design of fire resistance:

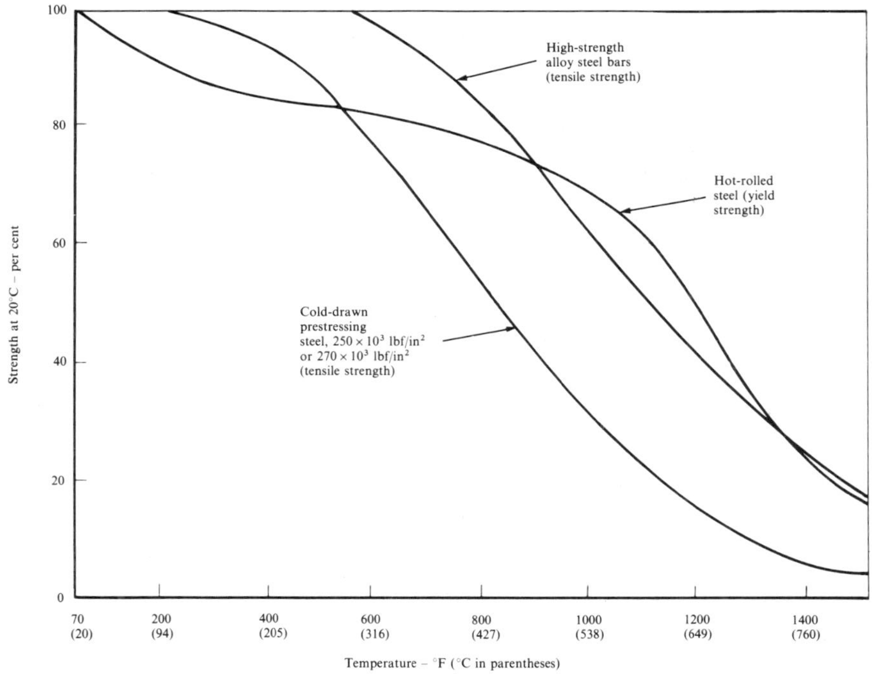

Figure 16.7 Temperature-strength relationship for hot-rolled and cold-drawn steel

1. Materials:
   Concrete (γ_{mc}) = 1.30
   Steel (γ_{ms}) = 1.00
2. Loads:
   Dead load (γ_g) = 1.05
   Live load (γ_q) = 1.00

As the requirements of fire resistance are applicable to completed buildings only it will be justified to use an age factor of at least 1.2 to concrete strength related to 28 day cube tests.

16.5.3.2 Design for pure flexure

The mechanics of the design are to determine by using design graphs in Figures 16.4 and 16.5 (or by calculation if possible) as appropriate, the temperature of concrete at the level of tensile reinforcement corresponding to the period of exposure to fire (controlled to the standard time/temperature curve) and from that the residual strength of steel by means of curves in Figure 16.9 or any other reliable way. As in the case of normal temperature design, for prestressing tendons the original strength will be related to f_{su} (or f_{sk}) whereas for reinforcement bars the original strength will be f_y or the yield strength.

The average temperature of concrete in the compression region is similarly determined and from that the residual strength of concrete in compression using the graph in Figure 16.8.

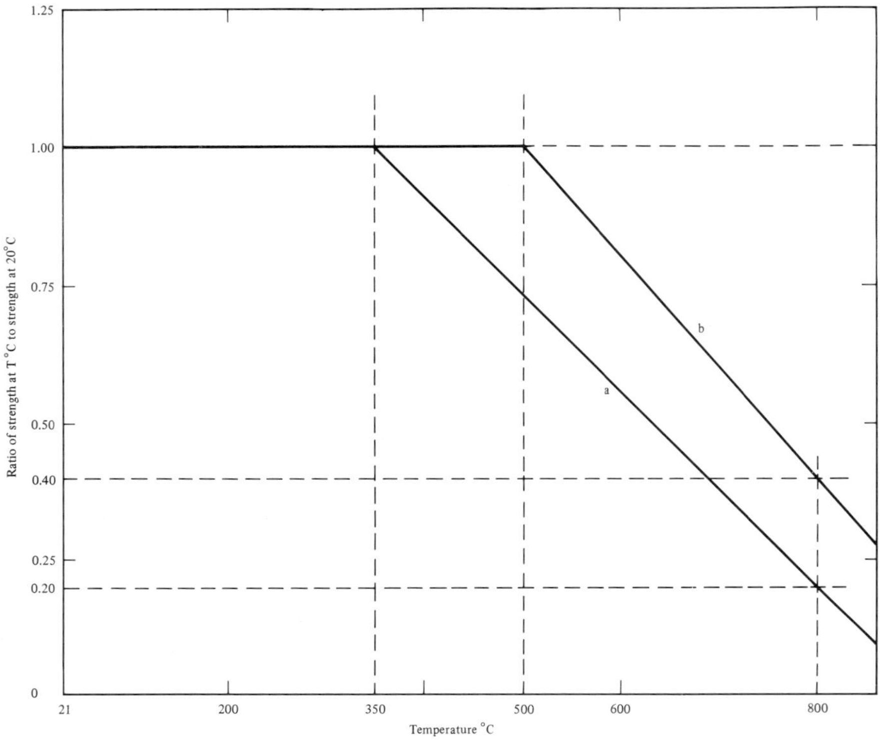

Figure 16.8 Design curves for variation of concrete strength with temperature
a. dense concrete
b. lightweight aggregate concrete

Having thus obtained the actual strength level of concrete and steel, the moment of resistance of the element is computed at the critical section broadly in accordance with the formulae given in Chapter 8, and compared with the moment due to applied load. If the simplified rectangular stress-block is used for calculation of moment of resistance, the uniform concrete stress is taken as $0.67\,f_{cT}o/\gamma_{mc}$ where $f_{cT}o$ is the concrete strength at temperature $_To$.

Alternatively a semi-parabolic stress-strain diagram similar to that for concrete in CP 110* may be used with the variations that the maximum strain may be taken as 0.0060 at temperature above 500°C, and $\gamma_{mc} = 1.3$.
Thus if $\Psi_{ST}o$ and $\Psi_{CT}o$ respectively represent the fraction of normal strength of steel and concrete (at 20°C) at temperature $_To$. Then for a rectangular beam:

$$F_{suT}o = b.d_n.0.67\,(1.2 \times f_{cu} \times \Psi_{CT}o)/\gamma_{mc}$$

$$\gamma_{mc} = 1.3$$

1.2 represents the age factor

*(see Figure 8.7)

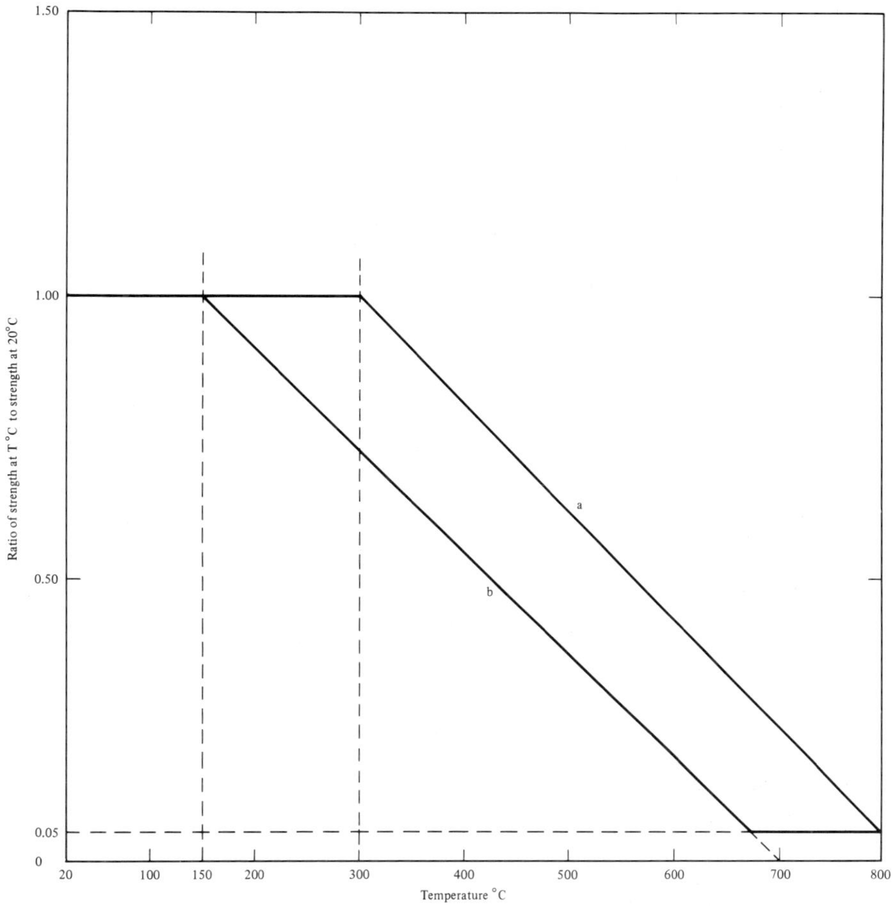

Figure 16.9 Design curves for variation of steel strength or yield stress with temperature
a. high-yield reinforcement bars, mild steel reinforcement bars, high strength alloy steel bars
b. prestressing wires or strands

As can be seen from Figure 16.2 the intensity of fire for a short period may be above the standard time/temperature curve. The effect of this (sometimes known as temperature shock) may be taken care of by ignoring 5 to 10 mm (1/4 in to 3/8 in) thickness of concrete near to exposed surfaces in the calculation.

$$F_{suTO} = A_s \,.\, (f_{su} \,.\, \Psi_{STO})/\gamma_{ms}$$

when $F_{suTO} = F_{cuTO}$

$$M_{RTO} = F_{suTO} \,.\, (d - d_n/2)$$

$$M_{RTO} \geqslant M_{aTO} \text{ due to } (1.05\, G_k + 1.00\, Q_k)$$

16.5.3.3 Shear

When the shear span/effective depth ratio is between 2 and 7 the combined effect of shear and bending may be critical. The equation 7.37a (or 7.37b or 7.37c depending on the unit used) may be applied making allowance for reduction of strength of concrete and steel at high temperature. The modified formula will be:

$$M_{T^O} = 0.875\,.\,d\,.\,l_s\,.\left(0.342\,b_1 + \frac{0.3\,M_{RT^O}}{d^2}\sqrt{\frac{z}{l_s}}\right) \times 4\sqrt{\frac{16.66}{\Phi\,.\,F_{su}\,\Psi_{ST^O}}} \not> M_{RT^O}$$

Notations are as explained in Chapter 7 (*7.12.4.2*)

If $M_{aT^O} > M_{T^O}$

then stirrups should be provided such that

$$\frac{A_{sv}}{s} \geqslant \frac{V_{aT^O} - \dfrac{M_{T^O}}{l_s}}{f_y\,\Psi_{ST^O}}$$

where V_{aT^O} is the shear at the section due to $(1.05\,G_k + 1.0\,Q_k)$

When the shear span/effective depth is 2 or less, the equation 7.34 as modified to take account of elevated temperature may be used.

Thus

$$V_{crT^O} = B\,.\,\frac{1.2 \times (0.7 f_{cu}\,\Psi_{CT^O})}{\gamma_{mc}}\,.\,\text{Sin}\;\theta$$

assuming $f_{cyl} = 0.7\,f_{cu}$

$\gamma_{mc} = 1.3$; (1.2 representing age factor of concrete)

V_{crT^O} = shear capacity at temperature $_{T^O}$

The other notations are as explained in Chapter 7 (*7.12.4.1*)

If $V_{crT^O} < V_{aT^O}$ then reinforcement should be provided in accordance with 7.35 and 7.36. In these equations, however, f_y should be replaced by $f_y\,\Psi_{ST^O}$ and γ_{ms} should be taken as 1.

16.5.3.4 Continuity and end restraint

The continuing reinforcements at support, even when provided just as anti-crack steel, considerably improves the fire resistance of a member or element provided the detailing is satisfactory. Being near the remote face from the fire, such reinforcement remains relatively unheated and thus retains a high proportion of normal temperature strength and consequently develops an appreciable moment of resistance which results in re-distribution of the applied moment[8]. This increase of moment shifts the point of inflexion further from the support. To prevent failure therefore the adequate amount of top reinforcement must be effectively continued well into the span to take care of such negative moment. The calculation procedure is as shown in the preceding section. The moment of resistance both at supports and in span should be calculated, and as long as the sum of these two resistance moments is greater than $M_{aT}o$, i.e $\frac{(1.05\,G_k + 1.00\,Q_k).\,l}{8}$, the structure will not collapse. Since at the support shear will be significant, the lesser value of $M_{RT}o$ or $M_{T}o$ should be taken as the support moment.

It is also recommended in the Guidelines that at the point of contraflexure of such structures hanger bars in the shape of stirrups should be provided in accordance with

$$A_{sv} = \frac{V_{aT}o}{f_y\,\Psi_{ST}o}$$ where $V_{aT}o$ = shear at point of contraflexure due to applied load.

It should be borne in mind that in the case of end spans such continuity moment may set up torsion in the supporting beam and the magnitude of the negative moment may then be governed by the torsional resistance of the supporting member. A worked out example of calculating fire resistance of a prestressed double tee floor with ends built into edge beams having adequate torsional restraint has been given in reference 8.

The supporting structure may also restrain the thermal expansion of the member exposed to fire and thus generate thrust which will act as an additional prestressing force and normally increase the moment capacity of the member. As a consequence the fire resistance of the member will be enhanced. Depending on the detailing at the bearing, the line of application of this thrust may also be above the longitudinal axis of the beam in which case the moment capacity will, in fact, be reduced.

A method to calculate fire resistance of a member on that principle has been suggested by Gustaferro(9) based on experimental results of the Portland Cement Association in conjunction with fire research specialists(10, 11).

16.5.3.5 Detailing

The importance of detailing cannot be emphasised too strongly. Whether in a fire condition or in a normal temperature situation the performance of a structure is greatly influenced by the detailing. The Joint Report on the Fire Resistance of Concrete Structures(12) clearly shows that almost anything that increases the load bearing capacity and overall stability of a structure under normal temperature, also invariably increases its resistance to fire damage. The reverse is also true. As under fire the whole structure is tested to its ultimate condition – a situation which otherwise very rarely occurs – any weakness in the detailing shows up very prominently in fire.

Essentially the detailing should be such that the assumptions made in the design remain valid. As indicated earlier all normal good detailing practices should be observed to ensure fire resistance of the structure. In addition special care should be given to anchorage of reinforcement as at elevated temperature the bond strength of concrete reduces appreciably. Attention is also required at laps and splicing especially where spalling may occur. The top bars should extend far into the span so that adequate effective steel is available for negative moment even when the point of contraflexure moves as a result of redistribution of the moment. Also at the point of contraflexure sufficient hanger bars in the shape of properly anchored stirrups are required.

16.6 Fire resistance and insurance premiums

In a publication of the Prestressed Concrete Institute a comparison is given of the American Insurance premium based on a research report prepared by Lin and Associates Inc.[13] It is apparent from the tables presented in the report that insurance premiums may be reduced by 40 per cent to 50 per cent if the prestressed roof or floor construction has a fire rating of 2 h, provided the supporting structure has a similar fire rating. The investigation relates to warehouse, office block, school building and supermarket design. Although it is not known whether the differences are as great in other countries it is obvious that sufficient attention needs to be paid to the fire resistance consideration at the early stages of the design.

REFERENCES

1. KENNARD, C. and DUFTON, J. *The Building Regulations.* London, Butterworth, 1966. p. 44-45.
2. BUTCHER, E. G., CHITTY, T. R. and ASHTON, L. A. *The temperature attained by steel in building fires.* London, HMSO, 1966. Fire Research Technical Paper 15.
3. ABELES, P. W. *Introduction to prestressed concrete.* London, Concrete Publications Ltd., Vol. 1, 1964. pp. 1–385.
4. KORDINA, K and SCHNEIDER, U. Zum mechanischen Verhatten von Normalbeton unter instationärer Wärmebeanspruchung (The mechanical behaviour of ordinary concrete under temporary instationery strain). *Beton*, Dusseldorf, 1975, No. 1.
5. JANNEY, W. and ELSTNER, R. Fire damage to the Avianca Building, Bogota. FIP Congress, 1974, New York.
6. GUSTAFERRO, A. H. Fire resistance of post-tensioned structures. *Journal of the Prestressed Concrete Institute,* Vol. 18, March–April 1973, No. 2. pp. 38–63.
7. READ, R. E. H., ADAMS, F. C. and COOKE, G. M. E. *Guidelines for the construction of fire resisting structural elements.* London, HMSO, 1980. Building Research Establishment Report.
8. *Design and detailing of concrete structures for fire resistance.* Interim guidance by a Joint Committee of the Institution of Structural Engineers and the Concrete Society, April 1978.
9. GUSTAFERRO, A. H. How to design prestressed concrete for a specific fire endurance. FIP Congress, 1974, New York.
10. SALVAGGIO, S. L. and CARLSON, C. C. Restraint in fire tests of concrete floors and roofs. Philadelphia, *American Society for Testing and Materials*, 1967. STP 422. pp. 21–39.
11. ISSEN, L. A., GUSTAFERRO, A. H. and CARLSON, C. C. Fire tests of concrete members: An improved method for estimating thermal restraint forces. Philadelphia, *American Society for Testing and Materials,* 1970. STP 464. pp. 153–180.
12. *Fire resistance of concrete structures.* Institution of Structural Engineers and the Concrete Society, Birmingham, 1975.
13. PRESTRESSED CONCRETE INSTITUTE. Prestressed concrete resists fire. Chicago, 1968. pp. 12 R202-68. (Containing research results by T. Y. Lin and Associates Inc., Chicago).

CHAPTER 17

SPECIFICATION

The following specimen clauses relate to prestressed concrete only. They should be considered as additional to those required for concrete and reinforced concrete which are given in other works, e.g. C. E. Reynolds, *Reinforced Concrete Designer's Handbook.* Concrete Publications Limited.

(1) Moulds for prestressed concrete work shall be of such a nature as to allow the concrete members to deform when the prestress is applied, that is, parts of the mould should be made removable. The bottom portions of the moulds should be prevented from adhering to the concrete and the sides should be designed so as to permit striking without effecting the stability of the remaining formwork.

(2) In order to obtain concrete of uniform strength, where members are cast in a long-line bed, the difference in strength at transfer between beams of the same type cast in the same line shall not exceed 15 per cent. The manufacturer shall satisfy the engineer by tests on special cubes cured under the same conditions as the members that this condition is complied with.

(3) Prestressed concrete work shall, unless otherwise specified, have a smooth finish on all surfaces exposed in the completed work. Surfaces which will be in contact with concrete cast in place to form a composite construction shall be purposely left very rough, in order to provide an adequate key. The valleys and ridges of any roughening system shall not be parallel with the direction of stressing.

(4) Prestressed concrete shall be moist cured until the prestress is applied.

(5) The prestressing force applied shall be that shown on the drawings.

(6) In order that the specified force is maintained after the tensioned steel has been anchored, an allowance must be made in assessing the value of the applied force for loss of stress due to the yield in the anchorages and for slip when the steel is wedged after tensioning.

(7) In long-line beds due consideration shall be given to the friction caused by the varying shape and number of diaphragms.

(8) With pre-tensioning, the specified force shall be maintained by the use of approved fixing devices at the ends of the tensioned steel during concreting and curing, until the concrete has attained the strength specified. The tensioned steel shall then be released gradually and uniformly.

(9) No post-tensioning of the steel shall take place until the concrete has attained the specified strength as ascertained from tests on cubes cured and hardened under the same conditions as the concrete of the member. Before stressing, the wires or strands in each cable shall be of equal length and tension.

(10) With post-tensioning, the required elongation and corresponding or appropriate prestressing force are as shown on the drawings. The elongation is based on assumed frictional losses; should the specified elongation not be obtained with the specified force, then the force shall be modified as directed by the engineer.

(11) Measurement of the force shall be by an accurate gauge indicating the force of the hydraulic jack and also by the elongation of the steel, after allowing for any initial slip in the wedges or jaws of the grips.

(12) The prestressing force should be applied from one or both ends as specified on the drawings.

(13) Each wire, pair of wires, cable, strand or bar shall be attached in such a manner as to ensure that the elongation is uniform throughout its length. Should any tensioned steel or anchorage break or be damaged, it should be replaced if so ordered.* Throughout all tensioning operations a spare jack capable of maintaining the designated load should be available.

(14) An approved system of post-tensioning shall comply with the following requirements.

(a) The method of tensioning and final anchorage shall be of such a nature that the controlled total force is imposed and maintained without slip relative to the concrete after the application of prestress.
(b) The tensioning apparatus shall efficiently transfer the force from the steel to the concrete by means of the anchorage without inducing dangerous secondary stresses in the steel, the anchorage, or the concrete.
(c) The anchorage system shall be secure against both maintained and fluctuating load, and against the effect of shock.
(d) The anchorage itself shall be adequately protected against corrosion following the completion of the final stressing operation.
(e) The cable, strand, or bar after being placed in the duct or sheath shall be so arranged as to allow an unrestricted flow of grout and ensure that the post-tensioned steel is completely embedded and the duct effectively filled, unless specific instructions are given on the drawing that the steel is to be left unbonded. The method used for achieving this shall be approved by the engineer.

(15) All tensioning shall be carried out in the order specified or approved. After tensioning, the wires, strands, or bars shall be secured in position and cut off as directed.

(16) The men employed shall be experienced in this class of work. Every precaution shall be taken to ensure the safety of the workmen during the tensioning period.

(17) At transfer of the prestress, the concrete shall have the cube strengths as shown on the drawings, or as specified. These requirements shall be ascertained from works tests on three cubes manufactured and cured under the same conditions as the concrete member.

(18) Sufficient cubes shall be made so that if the stipulated strength is not achieved, further cubes will be available for such additional testing as may be required.

*If the broken or damaged steel is a small proportion (say 2 per cent) of the total, replacement need not be insisted upon.

(19) At transfer of the prestress, the measured deflection of the beam shall not exceed the limits given on the drawings, or as specified.

(20) No cracks will generally be permitted on account of (a) shrinkage, settlement of formwork, or damage in handling before application of prestress, (b) local weakness of concrete at or immediately following prestressing, or (c) improper handling and transporting after application of prestress. However, if due to (c) hair-cracks occur in the compressive zone the engineer may, at his discretion, consider these to be harmless.

(21) Joints in the spaces between a series of precast concrete elements which are to be prestressed together shall be made precisely in accordance with the requirements on the drawings.

(22) Suitable extensions or connections of the ducts through the gaps shall be provided and steps taken to ensure that the jointing material does not enter the duct or press against the prestressing steel. The holes for the prestressing steel shall be accurately made and meet one another in true alignment at the ends and shall permit unobstructed passage of the grout.

(23) The stressing operation shall not take place until the filling material in the joint has attained a strength equal to at least 1½ times the stress at transfer.

(24) After post-tensioned wires, cables, strands, or bars have been tensioned and anchored they shall be grouted up solid throughout, suitable outlets with plugs being provided to prevent development of air locks unless otherwise specified. The method of mixing and injecting the grout shall be approved by the engineer before grouting is commenced. The pressure at which the grout is to be pumped into the duct shall be approved by the engineer and shall not normally exceed 80 lbf/in^2 (5.6 kgf/cm^2; 0.55 N/mm^2).

(25) The grout for cables and bars shall preferably be either a neat cement grout with a water/cement ratio not exceeding 0.4 or cement mortar (1:1¾ by weight). An approved expanding plasticising admixture should be used. All ducts and holes shall be thoroughly washed out with water and dried up with compressed air immediately before grouting, after the stressing of the members has been completed. Grouting is to be continuously carried out, the discharge from consecutive air holes being observed throughout. The mixer shall be thoroughly cleaned before grouting is commenced. A complete spare grouting plant for immediate use in cases of emergency or breakdown of the equipment in operation should be available.

(26) Where grooves are provided in which the prestressing steel is placed, they shall be cleaned before the mortar is inserted.

(27) The mortar shall consist of one part by weight of cement to three parts of sand with a water/cement ratio not exceeding 0.4 and shall be thoroughly compacted in the grooves, preferably by the use of vibrators, and subsequently cured.

(28) Alternatively, mortar comprising one part of cement to one part by weight of sand together with an approved expanding plasticising agent and less water may be used. External cables shall be embedded in mortar of these proportions, an upper cover being placed over the cables after vibration is completed to ensure that expansion occurs uniformly within the original profile.

(29) When performance tests are specified the loading should be as shown on the drawings. The test loadings may be such that (a) the calculated bending tensile stresses in the concrete will be 750 lbf/in^2 (52.5 kgf/cm^2; 5.2 N/mm^2)

for beams with pre-tensioned steel and 650 (45.5 kgf/cm^2; 4.48 N/mm^2) for beams with post-tensioned steel and having a cube strength at 28 days of about 7500 lbf/in^2 (525 kgf/cm^2; 52 N/mm^2) and (b) the calculated bending compressive stresses will not exceed 60 per cent of the cube strength, whichever is the less. No visible cracks should appear when the full test load is applied.

(30) Before performance tests are commenced, the specified cube strength at 28 days should have been obtained. Special cubes, cured under the same conditions as the members to be tested, are required for this purpose.

(31) The permissible test deflections, shown on the drawings, are based on the assumption that the beams are tested approximately one month after transfer of stress.

(Note. The value of E corresponding to a cube strength of 7500 lbf/in^2 (525 kgf/cm^2; 52 N/mm^2) may be expected to be 5.75 $\times$ 10^6 lbf/in^2 (4 $\times$ 10^5 kgf/cm^2; 3.96 $\times$ 10^4 N/mm^2), but the maximum permissible deflection may have been based on a lower value, such as 4 $\times$ 10^6 lbf/in^2 (2.8 $\times$ 10^5 kgf/cm^2; 2.76 $\times$ 10^4 N/mm^2) to allow for creep due to sustained loading.)

(32) If visible hair-cracks occur, or if the deflection at the centre at maximum load exceeds the given value, the prestressed member will be rejected. If the residual deflection after removal of loading exceeds 10 per cent (for pre-tensioning) or 15 per cent (for post-tensioning) of the measured maximum deflection, the beam may be rejected at the discretion of the engineer, together with the other beams produced on the same tensioning bed at the same time. If, however, the other beams are also tested and comply with the requirements they will be accepted.

(33) The loading for performance tests shall be applied by jacks calibrated to an accuracy of ±2% with provision for measurements of loads to the nearest ¼ ton and deflection to 0.001 in. (0.025 mm).

APPENDIX

DESIGN PHILOSOPHY

Some subjects which were discussed during the FIP Congress, New York 1974, have already been embodied in Chapter 15 and Chapter 16. Further points which seem to be of special interest are raised in this Appendix.

A.1 Ten principles

The first named author in his books *Introduction to prestressed concrete* Vol. 1 and 2, has put forward ten principles which are suitable for prestressed concrete and general life. They are briefly summarized here because they appear to be very useful for the designer also.

1. You cannot have everything. (Each solution has advantages and disadvantages which ought to be weighed up.)
2. You cannot have something for nothing. (One has mostly to pay in one way or another for something which is offered as a free gift into the bargain: not-withstanding the possibility of aiming at a solution offering the optimum advantages.)
3. It is never too late (e.g. to alter a design or to strengthen a structure before it collapses, or to adjust or even change principles previously employed in the light of increased knowledge and experience).
4. There is no progress without considered risk. (While it is important to ensure sufficient safety, over-conservatism can never lead to an understanding of novel structures.)
5. The proof of the pudding is in the eating. (This is in direct connection with the previous principle indicating the necessity of tests.)
6. Simplicity is always an advantage but beware of over-simplification. (This may, for example, lead to theoretical calculations which are not always correct, or to simplifications in detailing which do not cover all conditions.)
7. Do not generalize but rather qualify the specific circumstances. (Serious misunderstandings may be caused by unreserved generalizations.)
8. The important question is 'how good' not 'how cheap'. (A cheap price given by an inexperienced Contractor usually results in bad work, similarly the use of cheap, unproved appliances may result in later replacement.)
9. We live and learn. (It is always possible to increase one's own knowledge and experience although this may already be old knowledge to others.)
10. There is nothing completely new. (Nothing is achieved instantaneously but by a step-by-step development.)

A.2 Standardization – suitability – economy
In this time of standardization it is the task of the engineer to combine economy with repetition in prefabrication, but employing imagination in order to avoid unsuitable repetition, just for the sake of simplification. A good example is the development of inverted T-beams as illustrated in Figures A1, A2 and A3. Section A shows its use for a composite bridge beam deck as introduced by British Railways in 1948, the purpose being that the precast, prestressed concrete beam was as light as possible and sufficiently capable of carrying the composite section whereas the composite section carries the additional load (dead and live load). Thus, a minimum cross-section was used to reduce the cost (which was at that time based on the concrete quantity) and to allow easy transport. Jan Bobrowski and Partners developed the same basic idea to further different applications. Section B shows its use for the cantilever section of the grandstand, at Doncaster, built 1968–9. (Lightweight concrete was used in order to reduce the weight to suit the capacity of the crane available.) Section C illustrates the suspended gutter beam at the grandstand at Sandown Park, built 1972–3.

Bridge beams
British Railways
Eastern Region
1948

Figure A1

Doncaster
Racecourse

1968/9

Figure A2

Sandown Park
Racecourse

1972/3

Figure A3

A.3 Safety – simplicity – economy
The advantages of partial prestressing are now well recognised. The judicious application of partial prestressing not only leads to economic solution but also improves the overall performance of the structure. In order to ensure a safe solution it is essential, as already stated several times, that collapse should be avoided under any foreseeable circumstances. Under working load notwithstanding the three different classes of prestressed concrete, it does not seem a real progress to use Class (3) structures with permanently open cracks. In the same way it does not seem economical and necessary to employ Class (1) structures which are always in compression, particularly in view of the fact that 'always' is hardly ever entirely achieved, because of the possibility that increased losses may occur and temperature strains may cause cracking.

Although one type of solution is not likely to be the ideal or best for all conditions, the authors are of the opinion that the best value for money can be obtained in most cases if the type of partial prestressing embodied in type (ii) in Figure 1.3 (p. 9) is followed. This is virtually a combination of Class (1) [for permanent or frequently occurring service load] and Class (3) [for rarely occurring maximum service load] of CP 110 and therefore ensures that the structure remains crack-free under the permanent or frequently occurring service load. Cracks are allowed under the maximum service load but these close as soon as the temporary overload is withdrawn. Obviously to make such designs effective and economic a realistic assessment of the frequently occurring part of the service load needs to be made based on actual surveys and research. The CEB/FIP Model Code has made some assessments for certain types of structures; but these require further rationalization.

It should be emphasized that for the ultimate condition, the safety factor shall, however, be related to the maximum service load (i.e., inclusive of frequently occurring as well as rarely occurring part of the superload). The design then becomes very simple: the collapse condition governs the concrete compression zone and the ultimate steel force while the permanent or frequently occurring service load determines the effective prestressing force. The width of the cracks which may occur should not be of any concern as these will be temporary; under normal conditions there will be no cracks.

There are a number of formulae available for calculation of crack width of partially prestressed beams. Some are quite simple while others are highly complicated. These formulae give such widely different results that it is difficult to find the right one for a particular application.

Where the maximum service load is permanent, as in warehouses, the above solution may not be appropriate.

A.4 Philosophy of design

The designer should be in a position to adjust the design to the actual appropriate conditions. Thus it should be based on the properties of the concrete to be used, curing conditions, time of loading and so on. However, it is often not possible to know these circumstances at the time of design and often necessary to make unfavourable assumptions to cover any possible conditions. It is obviously much better if all the conditions are known, including the ability and experience of the contractor and/or concrete works.

Codes of Practice should therefore not limit a designer but should give him freedom to develop new possibilities. They should, however, set sufficient guidelines to enable the less experienced engineer to design safe constructions.

A.5 Engineer – Architect – Contractor

Co-operation between designer, architect and contractor (main contractor and/or precast contractor) is very important. The authors of this book favour the use of prefabrication while some engineers prefer concrete cast in place. In the latter case excellent results are possible only if the contractor is very reliable with regard to material and workmanship and has sufficient experience. Even in such a case rigorous supervision with regard to agreement between design specification and actual execution of work is essential. It often occurs that discrepancies between design specification and execution lead to unpleasant arguments which can only be avoided if close co-operation between designer and site engineer or production supervisor exists and the latter is fully aware of the reasons on which the design specification was based. Alternative proposals of adequate efficiency should be permitted.

As far as co-operation between engineer and architect is concerned, it often happens that the latter makes changes at a very late stage. This should be avoided since small alterations by the architect often may cause extensive changes in the design. It is therefore advisable that the designer should endeavour to explain to the architect the basic principle of the design as far as possible.

A.6 Prefabrication versus concrete in place

When prefabricated members are used, this may result in reduction of the actual building time: also supervision should be easier since repetition of similar members should be aimed at by the designer. If mistakes have occurred with precast concrete it is often possible to replace members before they are used. This is impossible with cast in place concrete.

For constructions using in-situ and prefabricated members knitted together, the question of progressive collapse has been introduced lately. Obviously the danger is the same in in-situ structures if extraordinary condtions apply (e.g. extensive explosions) and the detailing of the reinforcement is not excellent. However, the possibility of progressive collapse can be avoided with precast constructions by proper design and if connections are well detailed.

A.7 Statically-indeterminate structures

In this book the use of simply supported pre-tensioned members to carry the dead load and then to obtain statical indeterminacy for the super load with the help of post-tensioned or reinforced added concrete or welded connection has been preferred to statically-indeterminate structures made monolithically. The latter has the disadvantages of considerable secondary stresses and/or parasitic forces.

A.8 Limit-state design

The designer ought to know the ambient conditions at manufacture or building in order to obtain reliable assumptions regarding shrinkage and creep. With regard to these phenomena, age of the member and the season of time when loads are applied, or when e.g. added concrete is placed, should also be known. In this respect experience is important for the designer to assess the necessary assumptions which are needed for the basic design. Careful detailing greatly increases the resistance of a structure to all conditions to which it may be subjected including fire.

The following implications are connected with limit-state design:

(i) Prestressed concrete is particularly suitable for separate design for service load and collapse load because of the different stress conditions, as already stated before.

(ii) The progressive designer should base the design on the condition that for dead load and normal service load no tensile stresses occur even after maximum possible losses.

Any crack which may occur under abnormal service load of temporary duration will close when the excess load disappears.

(iii) The maximum temporary service load at which cracks may develop should be safely determined on the condition that no permanent deformation occurs under that load. This should preferably include resistance to fire for sufficient time so that the structure could be re-used with the minimum of repair work.

(iv) The serviceability condition should ensure satisfactory durability for a definite lifetime, to resist chemical and other influences.

(v) Collapse load should be of such magnitude that unforeseen, possible but not probable conditions are taken into account. It should be related to the maximum service load.

PART 2

DESIGN DATA

CONTENTS

NOTES ON DESIGN CHARTS

In preparing the charts given in the following pages, attention has been directed in general towards properties which are specifically related to prestressed concrete sections and components. This has led largely to the exclusion of more general data, which are already available in other publications; in particular, it will be found that the charts are often complementary to those included in *Reinforced Concrete Designer's Handbook* by C. E. Reynolds. Exceptions are Charts 6 and 7 relating to the shearing and torsional properties of sections, and Chart 1 in which is given a graphical method for determining principal stresses. The first two are included because calculated values of shearing and torsional stresses in prestressed concrete under working-load conditions are much more realistic than is the case with reinforced concrete; the determination of the principal tensile stress is needed more often in the design of prestressed concrete members than in the case of reinforced concrete or steelwork, and Chart 1 may serve as a reminder of the graphical method of calculating these stresses.

An unsymmetrical I-section is usually the most efficient for prestressed concrete, although the advantage over a symmetrical I-section in many cases is only marginal. The areas and properties of unsymmetrical I-sections are given in Charts 10 to 13; those of symmetrical I-sections and T-sections can be obtained by selecting suitable ratios of dimensions, and are given separately in Charts 8 and 9.

Details of prestressing steel are given in Chart 2, and the calculation of losses of the prestress is facilitated by means of Charts 3 and 4. It is also thought that comprehensive details of the many post-tensioning systems and anchors now available will be helpful to designers and detailers; these are given in Charts 16–49, which include all the major post-tensioning systems in use in Great Britain and the U.S.A. at the time of writing, together with some other systems which are used in Europe. A guide to the systems used in Great Britain is given in Chart 3 in which the approximate prestressing forces obtained with single anchors are listed in a convenient form.

CHART No. 1
DETERMINATION OF PRINCIPAL STRESSES

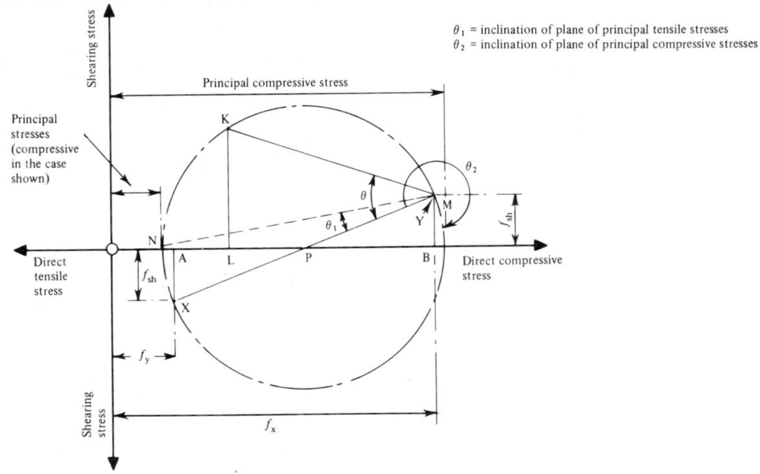

GENERAL CASE
f_x = longitudinal stress (due to prestress and bending)
f_y = vertical stress (due to vertical prestress, if any)
f_{sh} = shearing stress in homogeneous section
PROCEDURE

(1) Plot OA = f_y; OB = f_x; AX = BY = f_{sh}; join XY to cut OB at P.
(2) With centre P, radius PX (= PY), draw circle.
(3) To find stresses on any plane inclined at angle θ (clockwise) to vertical, plot $X\hat{Y}K = \theta$ and draw KL perpendicular to OM. Then OL = direct stress and KL = shearing stress.
(4) For planes of principal stress, shearing stress = 0. Hence principal stresses are OM (on plane at clockwise angle $X\hat{Y}M$) and ON (on plane at clockwise angle $X\hat{Y}N$). Since OP $= \frac{1}{2}(f_x + f_y)$ and AP $= \frac{1}{2}(f_x - f_y)$ it follows that PM $= \sqrt{\frac{1}{4}(f_x - f_y)^2 + f_x^2}$ (see Chapter 7).

Note: See also formulae on page 139.

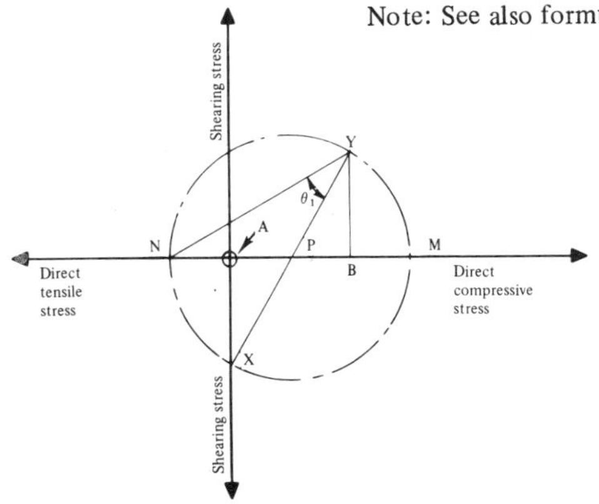

PARTICULAR CASE – PRINCIPAL TENSILE STRESS AT SUPPORT OF PRESTRESSED BEAM
f_x small; f_y = 0 (usually); f_{sh} large.
PROCEDURE

(1) Since f_y = 0, points A and O coincide. Plot OB = f_x, OX = BY = f_{sh}.
(2) Draw XPY and circle as previously.
(3) Principal tensile stress = ON, on plane inclined at angle $X\hat{Y}N = \theta_1$ to vertical.

Type	Nominal diameter		Characteristic tensile strength			Characteristic breaking load		
	in.	mm	lbf/in²	kgf/cm²	N/mm²	lbf	kgf	kN
Drawn strand	0.7	18	246 500	17 330	1700	85 400	38 700	380
Drawn strand	0.6	15.2	263 900	18 550	1820	67 400	30 580	300
Standard strand	0.6	15.2	237 800	16 720	1640	51 000	23 140	227
Drawn strand	0.5	12.7	271 100	19 060	1870	43 120	21 300	209
Standard strand	0.5	12.5	253 700	17 840	1750	34 050	16 820	165
Standard strand	0.366	9.3	259 600	18 250	1790	21 000	9530	93.5
Wire	0.276	7	227 600	16 000	1570	13 570	6160	60.4
	0.2	5	227 600	16 000	1570	6920	3140	30.8
	0.157	4	249 400	17 530	1720	4880	2210	21.7
	0.127	3.25	249 400	17 530	1720	3210	1460	14.3
Bar	1.57	40	145 000	10 190	1000	280 900	127 420	1250
	1.25	32	145 000	10 190	1000	179 800	81 550	800
	0.98	25	145 000	10 190	1000	112 370	50 970	500
	0.787	20	145 000	10 190	1000	73 030	33 100	325

CHART No. 3
REDUCTION FACTORS FOR PRE-TENSIONED STEEL

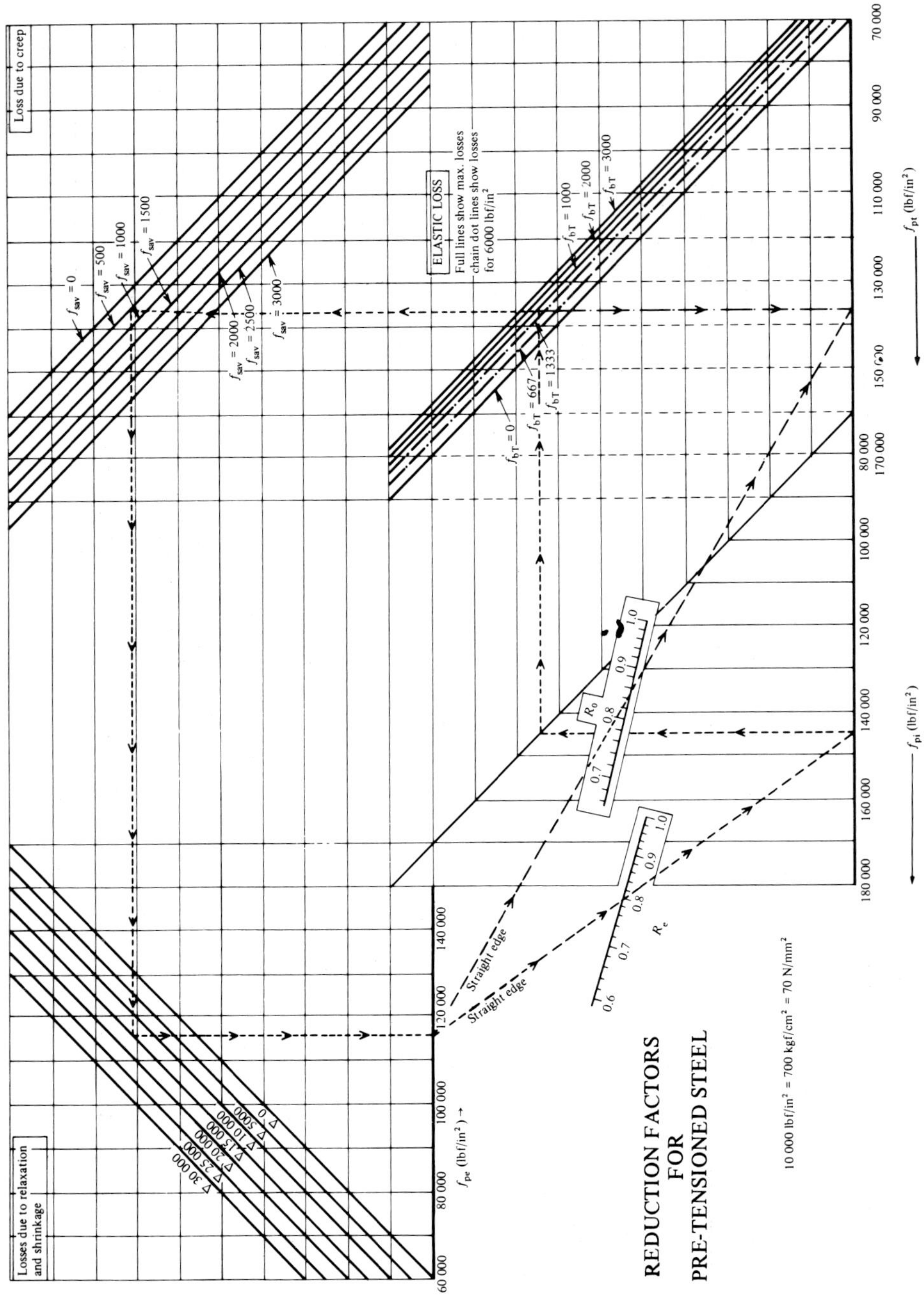

Note: For an example of the use of this chart see page 120.

CHART No. 4
REDUCTION FACTORS FOR POST-TENSIONED STEEL

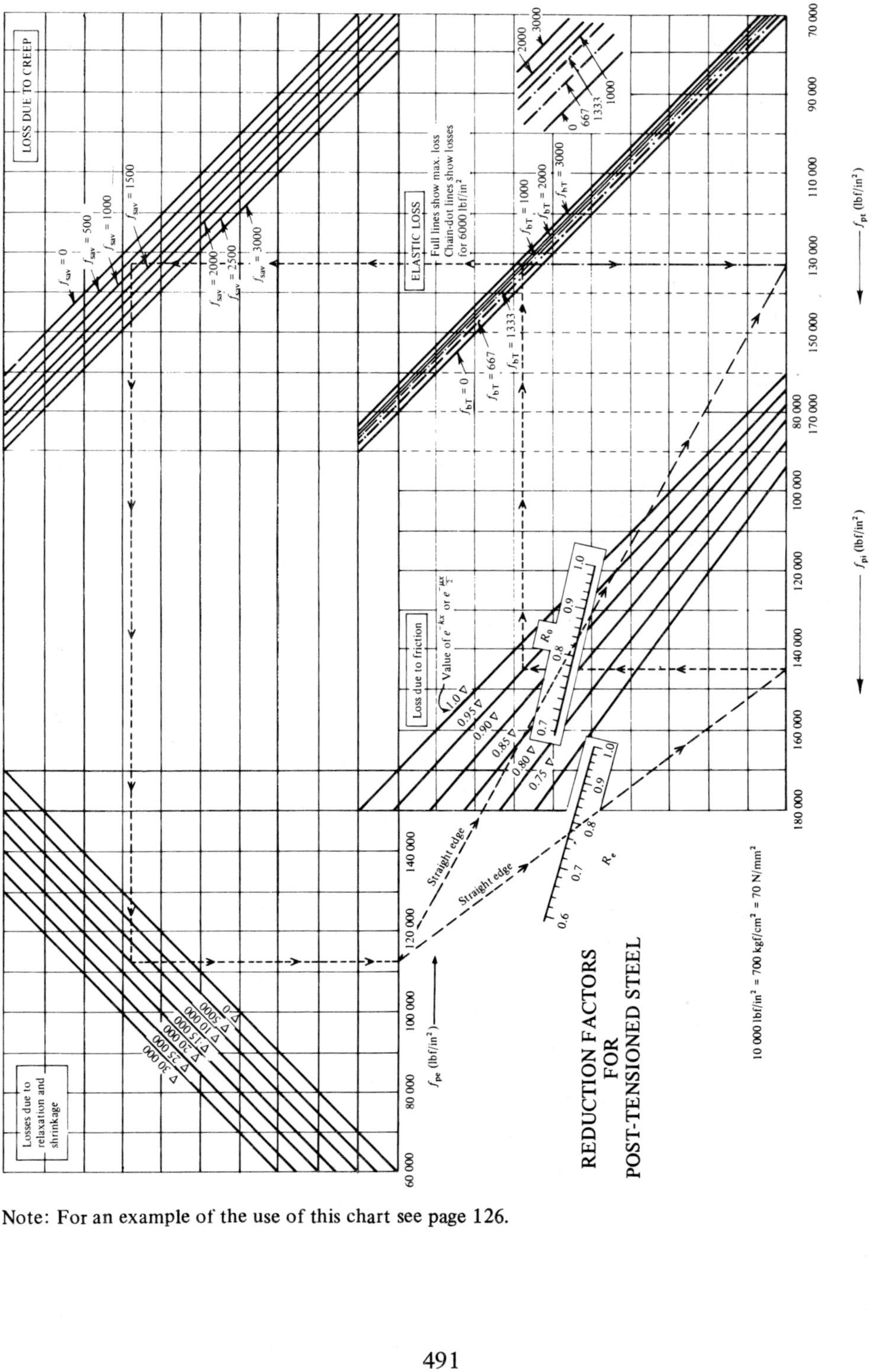

Note: For an example of the use of this chart see page 126.

CHART No. 5

SECTION AND STRESS DIAGRAM	PROPERTIES OF SECTION

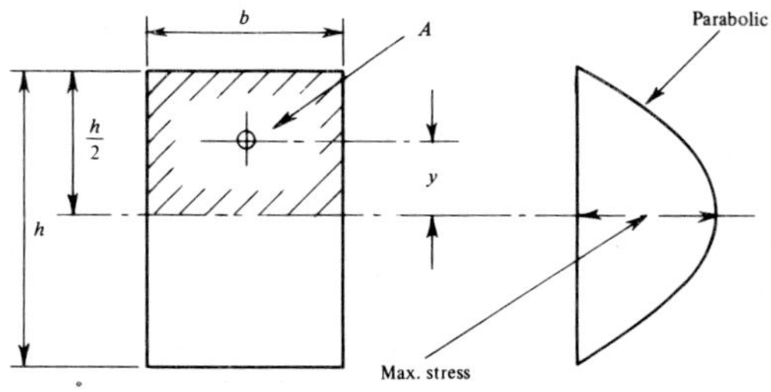

$A = \frac{bh}{2} \quad y = \frac{h}{4}$

$I = \frac{bh^3}{12}$

$\frac{Ay}{bI} = \frac{1.5}{bh}$

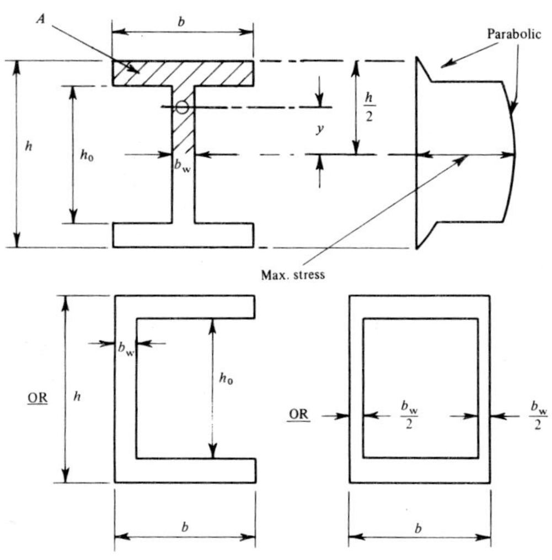

$Ay = \frac{b\,(h^2 - h_0^2)}{8} + \frac{b_w h_0^2}{8};$

$b_t = b_b = b$

$I = \frac{b\,(h^3 - h_0^3)}{12} + \frac{b_w h_0^3}{12}.$

$\frac{Ay}{b_w I} = \frac{1.5}{b_w}\left\{\frac{b(h^2 - h_0^2) + b_w h_0^2}{b\,(h^3 - h_0^3) + b_w h_0^3}\right\}$

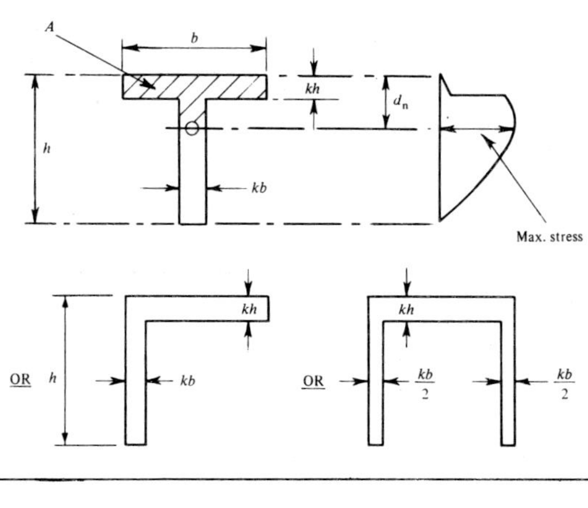

Centroid: $d_n = \frac{h\,(1 + k - k^2)}{2\,(2 - k)}$

$\frac{Ay}{b_w I} = \frac{3}{2kbh}$ approx.

SECTION AND STRESS DIAGRAM	PROPERTIES OF SECTION

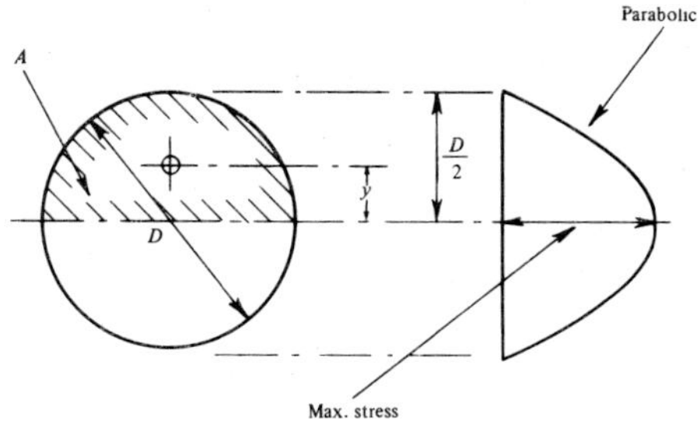

$$A = \frac{\pi D^2}{8}; y = \frac{2D}{3\pi}$$

$$b = D; I = \frac{\pi D^4}{64}$$

$$\frac{Ay}{bI} = \frac{16}{3\pi D^2}$$

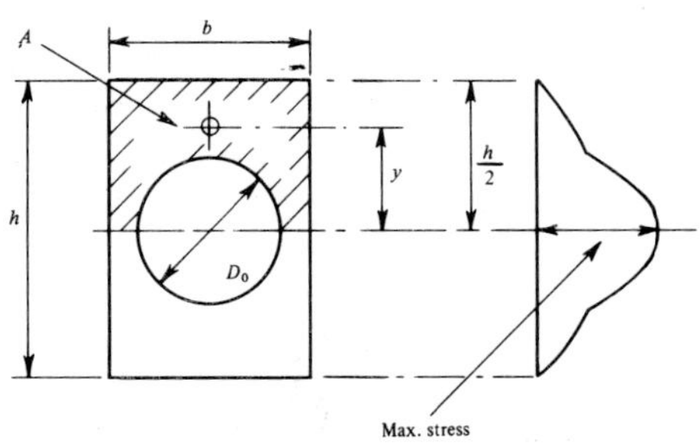

When $b = h$

$$A = \frac{bh}{2} - \frac{\pi D_0^2}{8}; Ay = \frac{bh^2}{8} - \frac{D_0^3}{12}$$

$$b_w = b - D_0; I = \frac{bh^3}{12} - \frac{\pi D_0^4}{64}$$

$$\frac{Ay}{b_w I} = \frac{1}{(b - D_0)}\left[\frac{3\,bh^2 - 2D_0^3}{2\,bh^3 - \frac{3}{8}\pi D_0^4}\right]$$

$$\frac{Ay}{b_w I} = \frac{1}{(h - D_0)}\left[\frac{3h^3 - 2D_0^3}{2h^4 - \frac{3}{8}\pi D_0^4}\right]$$

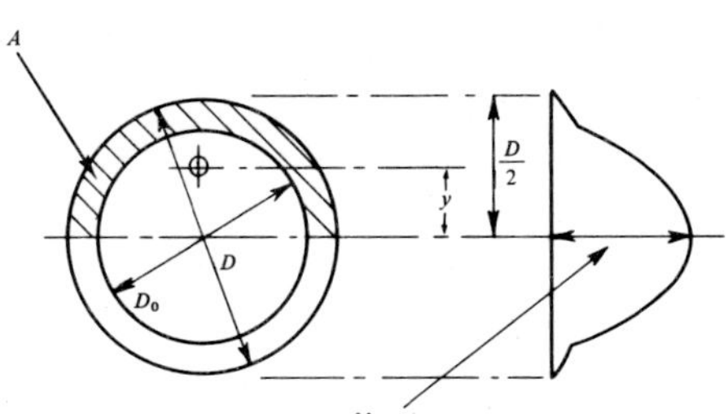

$$Ay = \frac{1}{12}(D^3 - D_0^3)$$

$$b = D - D_0; I = \frac{\pi(D^4 - D_0^4)}{64}$$

$$\frac{Ay}{bI} = \frac{16(D^2 + DD_0 + D_0^2)}{3\pi(D^4 - D_0^4)}$$

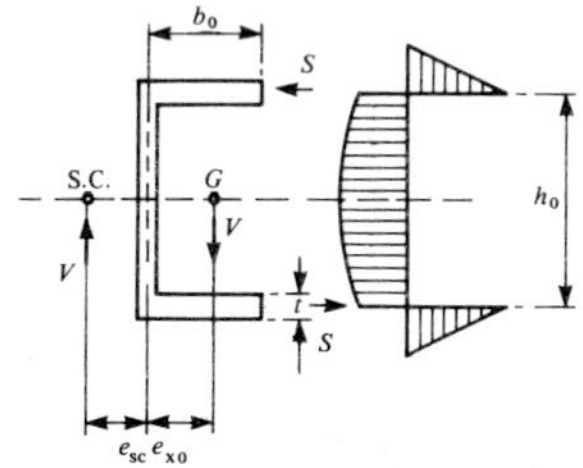

The shear centre S.C. is the point in an unsymmetrical cross-section through which the resultant forces originating from shear and bending pass if twisting is to be avoided. Let G be the gravity centre of a channel section (coordinates x and y and distance from web e_{x0}) and let S.C. be the shear centre the distance from G is x_{sc} and y_{sc} which is in this case zero). The two couples of forces are occuring in the two flanges. Sh_0 and the other in the web and through the shear centre $V(x_{sc} + e_{x0})$, must be in equilibrium. This results in

$$x_{sc} = e_{x0} + \frac{b_0^2 h_0 t}{4 I_x}$$

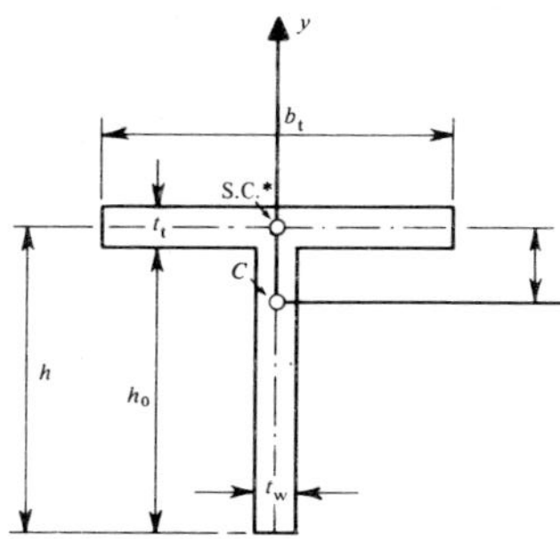

$y_{sc} = \frac{h_0 t_w h}{2(b_t t_t + h_0 t_w)}$

$X_{sc} = 0$

(1)

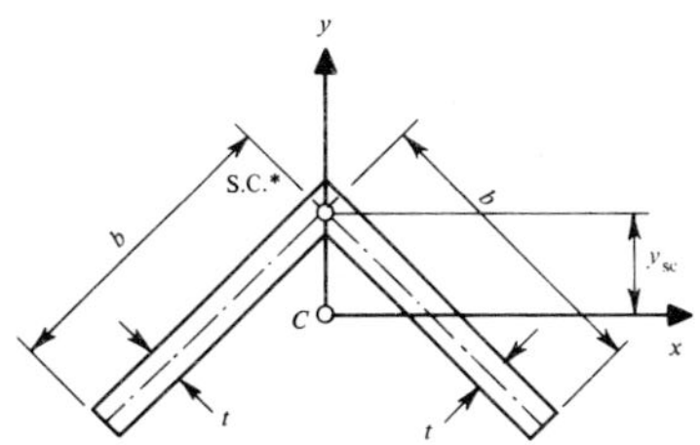

$y_{sc} = \frac{b}{4}\sqrt{2}$

$x_{sc} = 0$

(2)

Note: C. = Centroid
S.C. = Shear Centre
Since the flanges and webs are very thin the distances are taken from centre to centre

*Shear Centre of a section consisting of two rectangles only will always be at the intersection of the centroidal axis of the rectangles.

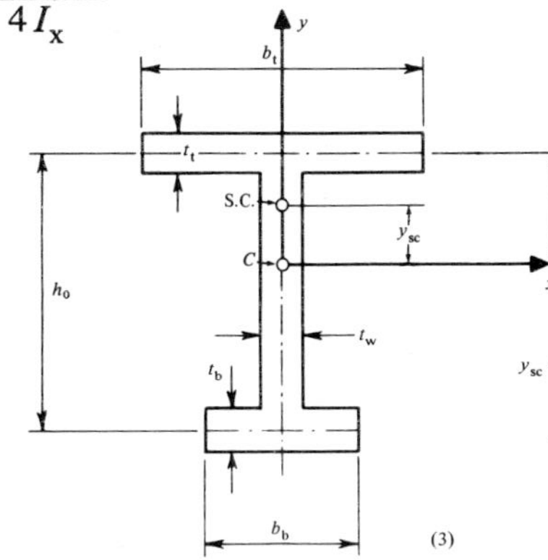

$y_{sc} = \left[e_t - \frac{h_0 t_b b_b^3}{t_b b_b^3 + t_t b_t^3}\right]$

$x_{sc} = 0$

(3)

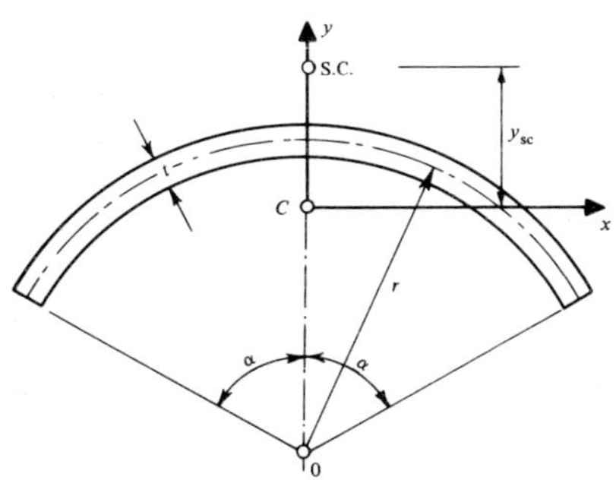

$y_{sc} = r\left[\frac{2 \sin\alpha - \alpha \cos\alpha}{\alpha - \sin\alpha \cos\alpha} - \frac{\sin\alpha}{\alpha}\right]$

$x_{sc} = 0$

(4)

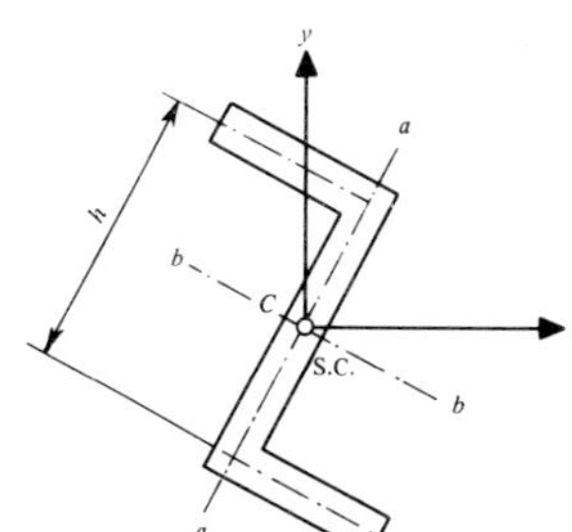

(5)

$x_{sc} = 0$
$y_{sc} = 0$

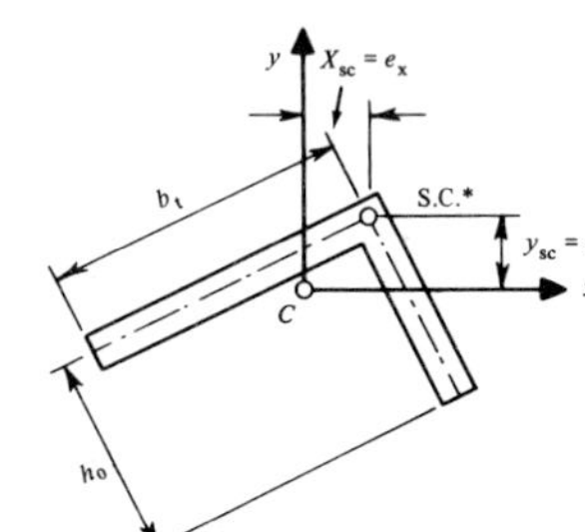

(6)

TORSIONAL PROPERTIES OF SECTIONS

SECTION	MAX. STRESS FOR UNIT TORQUE
Circle 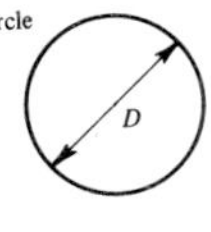	$\frac{16}{\pi D^3}$
Annulus 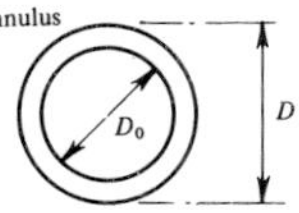	$\frac{16D}{\pi(D^4 - D_0^4)}$
Ellipse 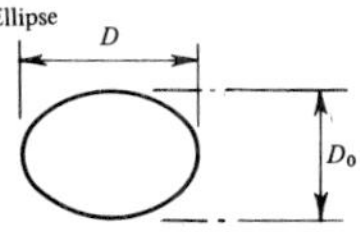	$\frac{16}{\pi D D_0^2}$
Equilateral Triangle 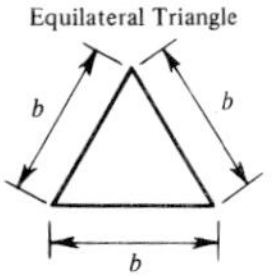	$\frac{20}{b^3}$
Regular Hexagon, Area = A 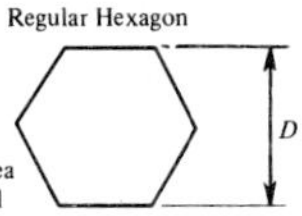	$\frac{4.6}{AD}$
Regular Octagon, Area = A 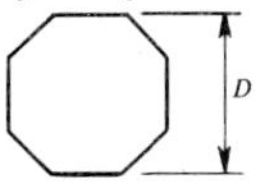	$\frac{4.5}{AD}$
Any section with no re-entrant angles	$\frac{40 I_p y}{A^4}$ in which I_p = Polar second moment of area; A = Area; y = Distance from centroid to furthest edge
Square, Area = A 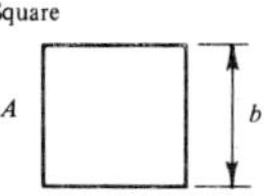	$\frac{4.8}{Ab}$
Rectangle, Area = A 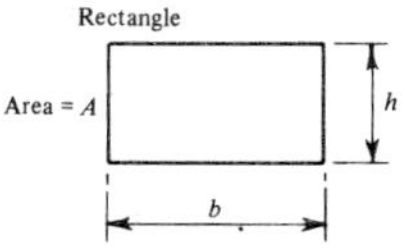	$\frac{3 + 1.8\frac{h}{b}}{Ah}$
Trapezium, Centroid 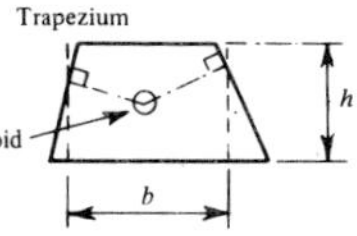	$\frac{3 + 1.8\frac{h}{b}}{bh^2}$
Narrow rectangles ($\frac{h}{b} \rightarrow 0$)	$\frac{3}{Ah}$
Sections composed of narrow rectangles (I, Z, and T sections)	$\frac{3b_w}{\Sigma bA}$ (Approx.) in which b_w = breadth of web; ΣbA = sum of products of Area × Thickness of component rectangles.

POSITION OF MAXIMUM TORSIONAL STRESS

Circle and Annulus:	At circumference
I, Z, and T sections:	At face of web adjacent to centroid
All other sections:	At points where faces are tangential to greatest inscribed circle.

NOTE.—To obtain the maximum torsional stress in a section, the appropriate stress given in the chart should be multiplied by the torsional moment. It is necessary to make sure that the units of the stress are consistent with those of the torsional moment.

CHART No. 8
SECTION MODULI AND MOMENTS OF INERTIA
SYMMETRICAL I-SECTIONS (PARALLEL FLANGES)

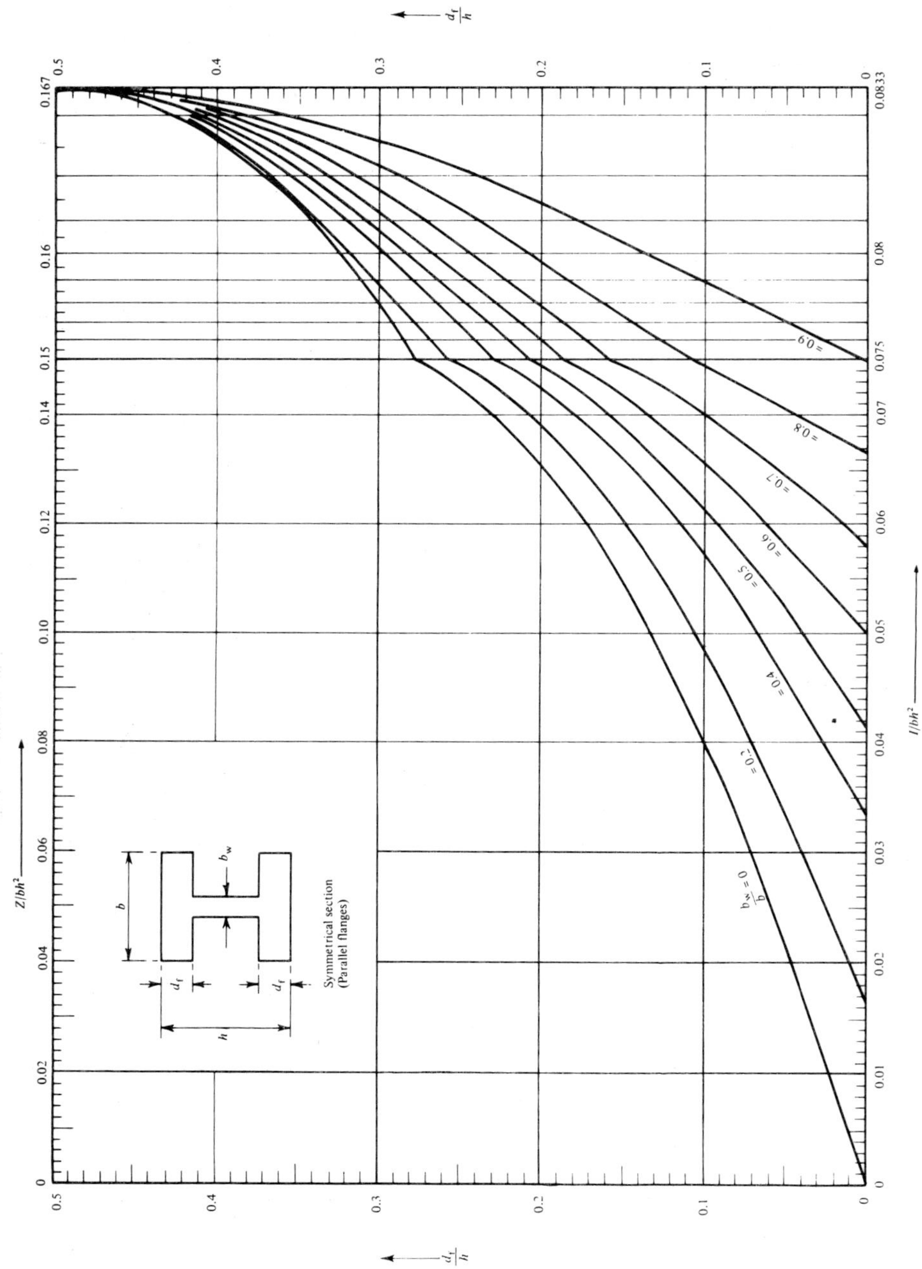

Example.–If $d_f = 0.2\,h$, $b_w = 0.2\,b$, then $I = 0.0725\,bh^3$ and $Z = 0.137\,bh^2$

CHART No. 9
SECTION MODULI AND MOMENTS OF INERTIA
SYMMETRICAL I-SECTIONS (TAPERED FLANGES)

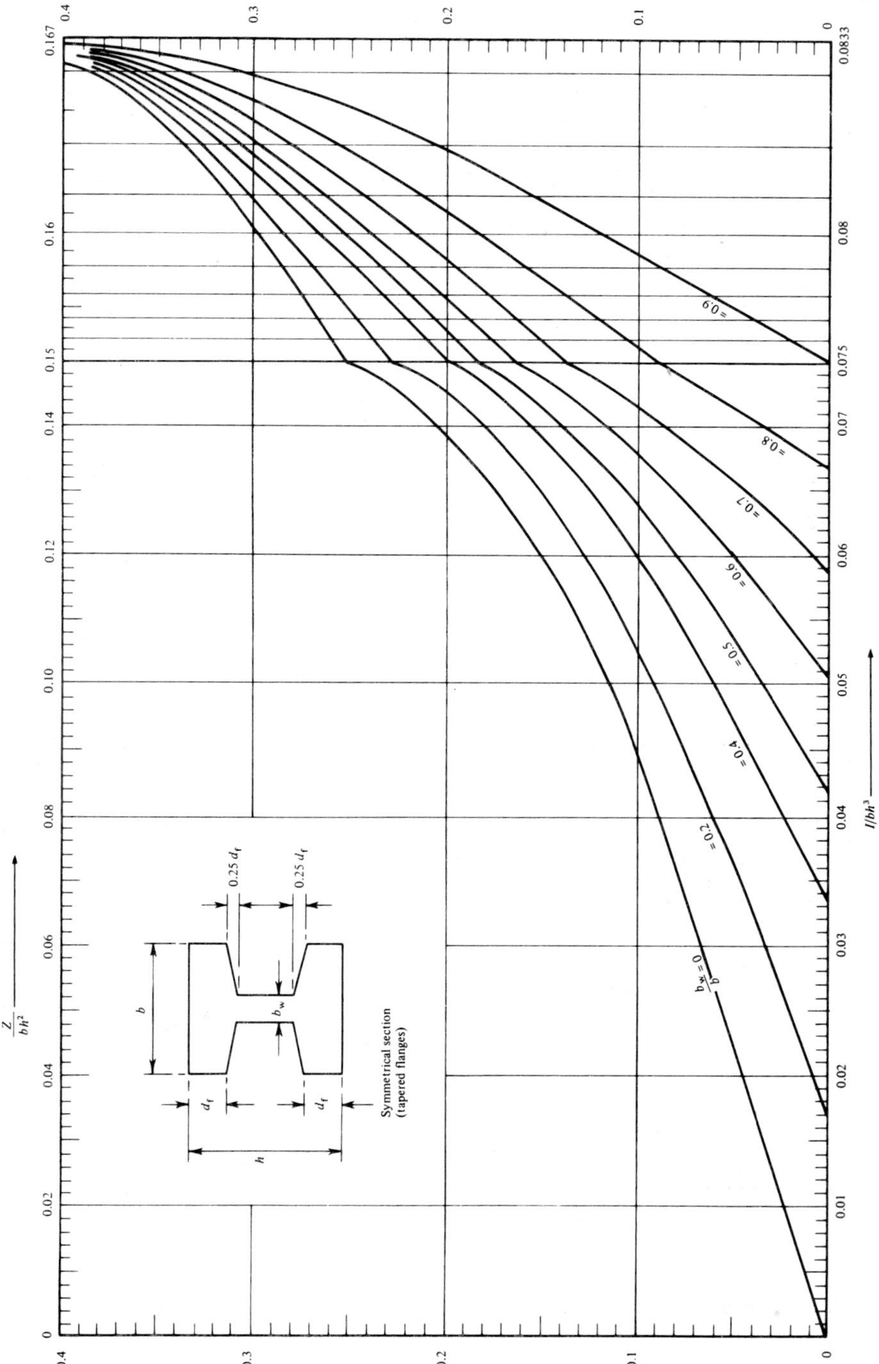

Example.—If $d_f = 0.2\,h$, $b_w = 0.2\,b$, then $I = 0.0725\,bh^3$ and $Z = 0.145\,bh^2$

CHART No. 10
AREAS OF I-SECTIONS (PARALLEL FLANGES)

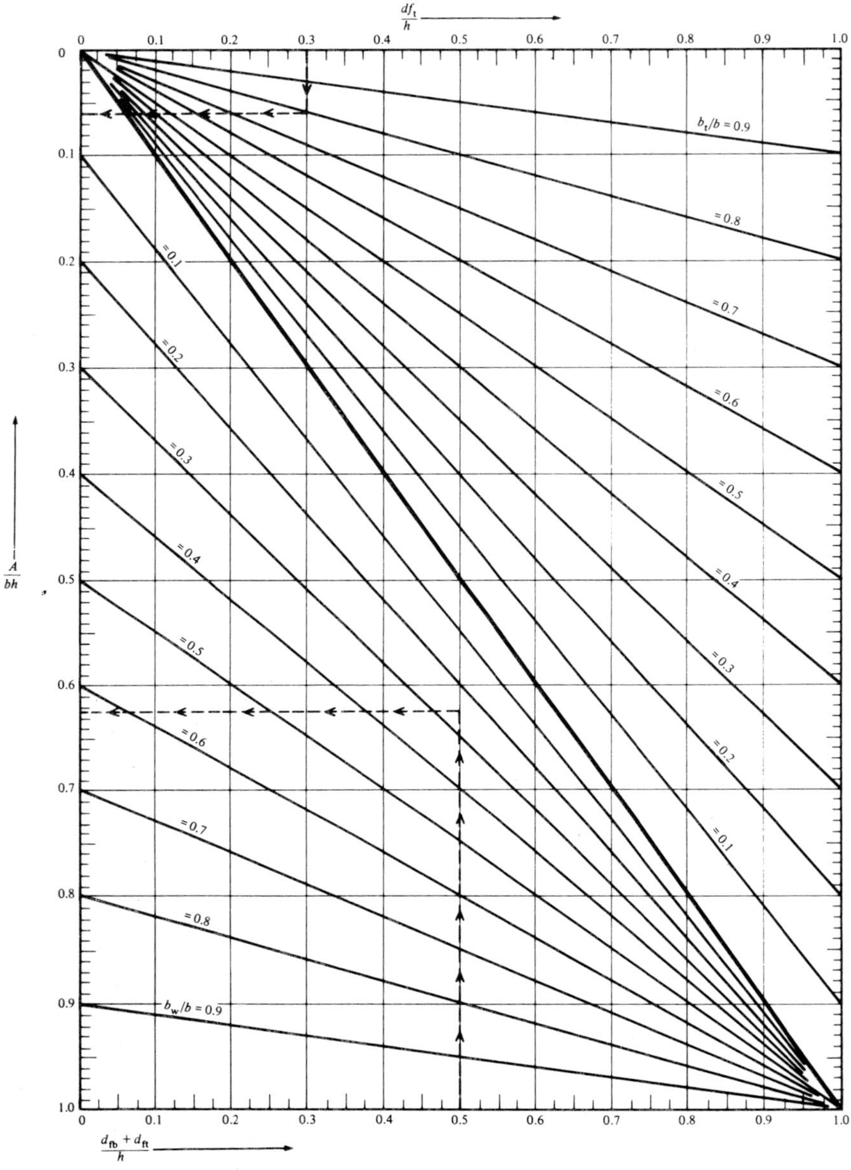

Note.–For notation and examples see page 500.

AREAS OF I-SECTIONS (TAPERED FLANGES)

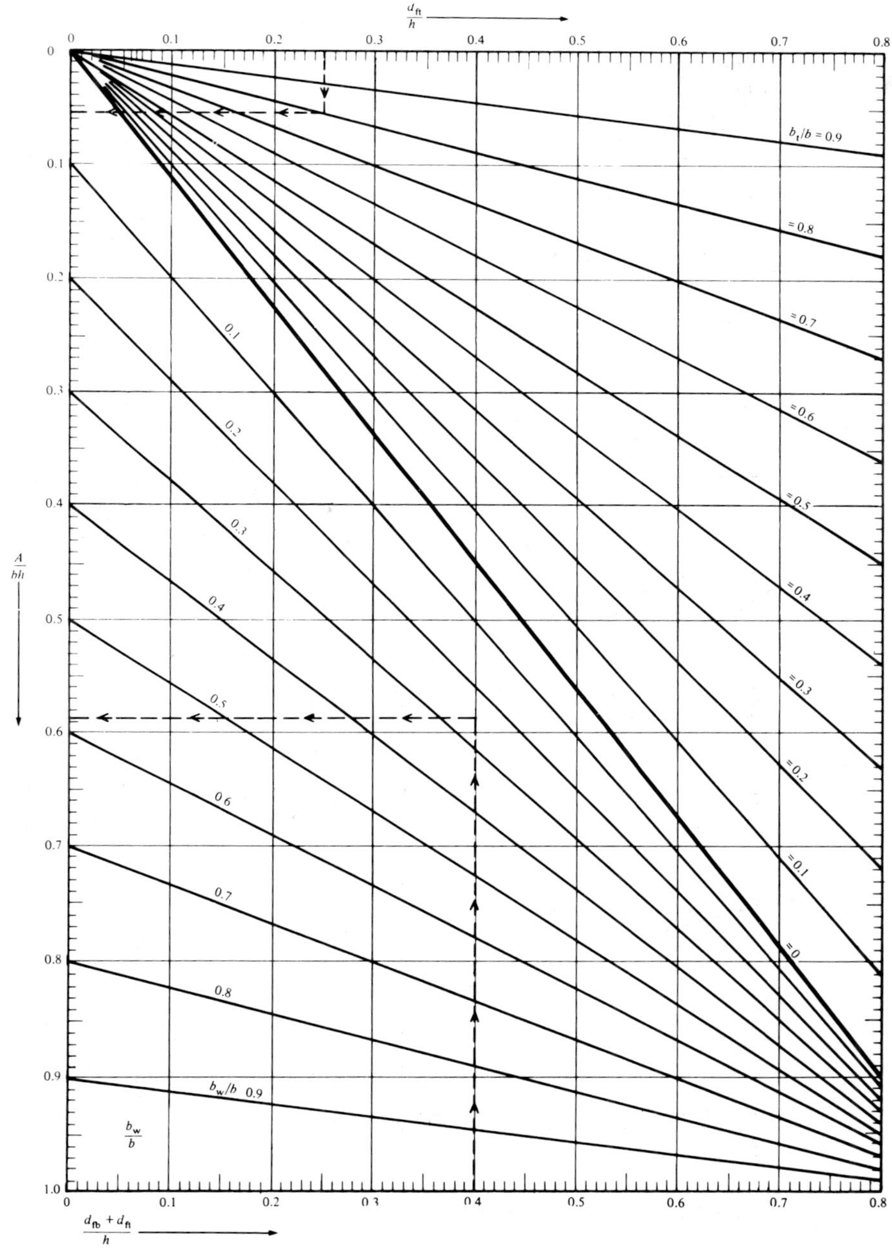

Note.—For notation and examples see page 500.

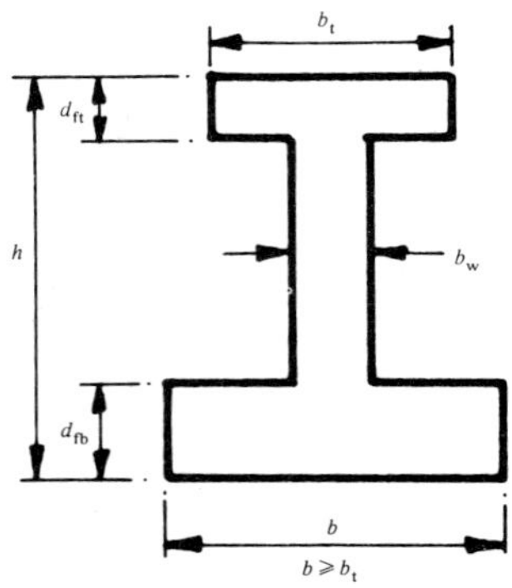

Profile and notation for Chart 10

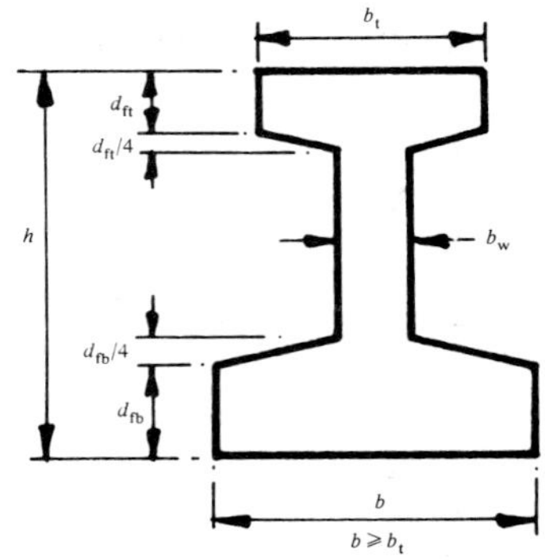

Profile and notation for Chart 11

For any required section, calculate $\frac{b_w}{b}; \frac{b_t}{b}; \frac{d_{fb}}{h}; \frac{d_{ft}}{h}$. Enter the chart at the bottom, with the appropriate value of $\frac{d_{fb} + d_{ft}}{h}$, and read the value at the left corresponding to $\frac{b_w}{b}$. For sections in which $b_t = b$ (that is, symmetrical sections) or in which $d_{ft} = 0$ (that is, T-sections), this is the correct answer.

When b_t is less than b but greater than b_w, enter the chart at the top with the appropriate value of $\frac{d_{ft}}{h}$ and read the value at the left corresponding to $\frac{b_t}{b}$. Subtract this result from the first value.

Examples

(1) Parallel flanges; $d_{fb} = 0.2h$; $d_{ft} = 0.3h$; $b_t = 0.8b$; $b_w = 0.25b$.
Area from Chart 10 (shown in dotted lines) = $0.625bh - 0.060bh = 0.565bh$.
(Calculated value = $0.565bh$.)

(2) Tapered flanges; $d_{fb} = 0.15h$; $d_{ft} = 0.25h$; $b_t = 0.8b$; $b_w = 0.25b$.
Area from Chart 11 (shown in dotted lines) = $0.588bh - 0.055bh = 0.533bh$.
(Calculated value = $0.532bh$.)

(3) T-section with tapered flanges; $d_{fb} = 0.2h$; $d_{ft} = 0$; $b_t = b_w = 0.25b$.
Area from Chart 11 (not marked on chart) = $0.42bh$.
(Calculated value = $0.419bh$.)

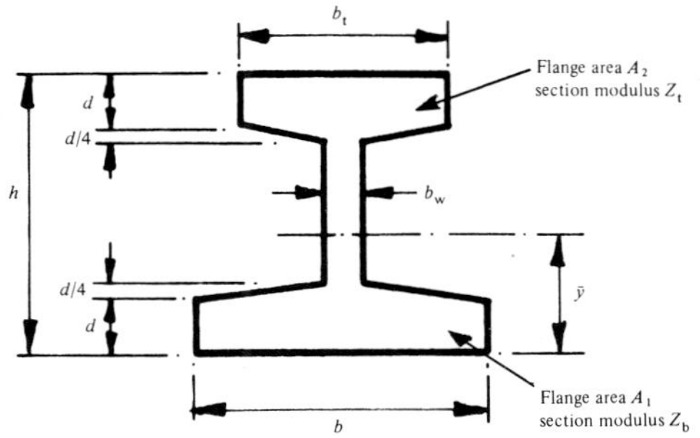

Profile and notation for Charts 12 and 13

Charts 12 and 13 are based on the following assumptions (see diagram above):

(1) $b_w = 0.1b$ (Chart 12) or $0.2b$ (Chart 13).
(2) Flanges have equal depths.
(3) Flanges are tapered by an amount equal to 0.25 × depth of flange.

Notes

(*a*) Although the charts are prepared for values of $b_w = 0.1b$ and $0.2b$, small variations from these values have little effect on the accuracy of the result.

(*b*) The application of these charts to the selection of sections to suit design data is described on page 191, to which reference should be made. The notes given on the charts indicate some further possible applications; they do not replace the descriptions given on page 191 and in connection with Examples 9.2.1, 9.2.2, and 9.2.3.

(*c*) Profiles other than unsymmetrical I-sections are obtained by using appropriate values for the variables. Particular examples are:

(1) Symmetrical I-sections: $b_t = b$.
(2) Rectangular sections: $d_f = 0$.
(3) Inverted T-sections: $b = b_w$.

For sections with parallel flanges, a value of d_f equal to 0.9 times the actual depth of flange of the section should be used. Since $d_{fb} = d_{ft}$, these symbols are abbreviated to d in Charts 12 and 13 only.

CHART No. 14
LOADING TERMS N FOR VARIOUS CABLE PROFILES

PRIMARY MOMENT	LOADING TERMS N	PRIMARY MOMENT	LOADING TERMS N
1 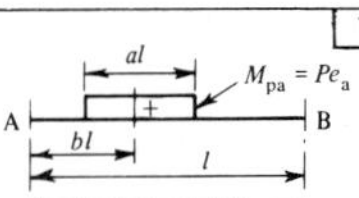 Constant eccentricity over length al	$N_{AB} = -6a(1-b)M_{pa}$ $N_{BA} = -6ab\,M_{pa}$	2 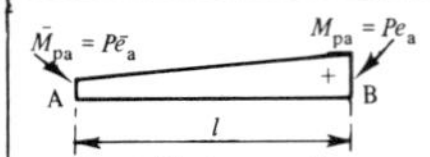 Eccentricity increases from $\bar{e}_a$ at A or e_a at B	$N_{AB} = -(2\bar{M}_{pa} + M_{pa})$ $N_{BA} = -(\bar{M}_{pa} + 2M_{pa})$
S 3 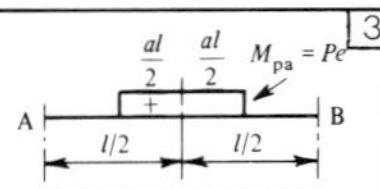 Constant eccentricity over central length al	$N_{AB} = N_{BA} = -3a\,M_{pa}$	S 4 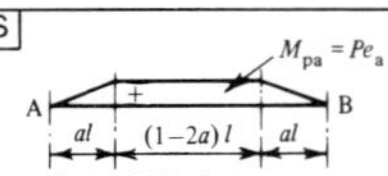 Trapezoidal primary moment diagram	$N_{AB} = N_{BA} =$ $-3(1-a)M_{pa}$
5 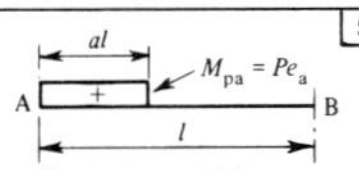 Constant eccentricity over length al from support A	$N_{AB} = -3a(2-a)M_{pa}$ $N_{BA} = -3a^2M_{pa}$	6 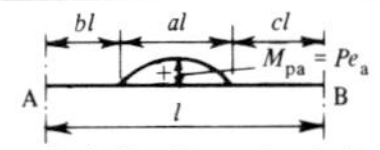 Parabolic cable over length al	$N_{AB} = -2a(a+2c)M_{pa}$ $N_{BA} = -2a(a+2b)M_{pa}$
S 7 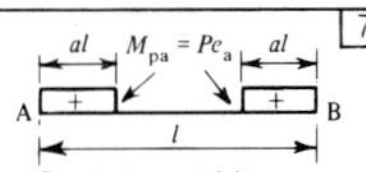 Constant eccentricity over length al from each support	$N_{AB} = N_{BA} = -6a\,M_{pa}$	8 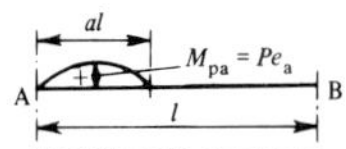 Parabolic cable over length al from support A	$N_{AB} = -2a(2-a)M_{pa}$ $N_{BA} = -2a^2M_{pa}$
S 9 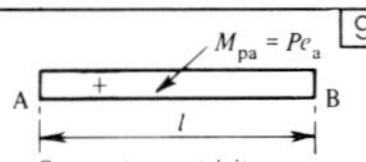 Constant eccentricity over entire length	$N_{AB} = N_{BA} = -3M_{pa}$	S 10 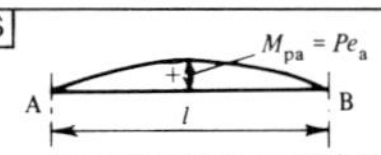 Parabolic cable over entire length	$N_{AB} = N_{BA} = -2M_{pa}$
11 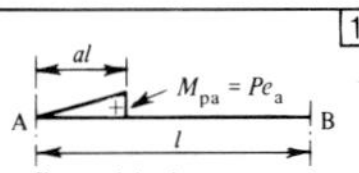 Eccentricity increases over length al from support A	$N_{AB} = -a(3-2a)M_{pa}$ $N_{BA} = -2a^2M_{pa}$	12 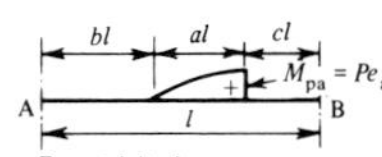 Eccentricity increases parabolically over length al	$N_{AB} = -a\left(4c + \frac{3a}{2}\right)M_{pa}$ $N_{BA} = -a\left(4b + \frac{5a}{2}\right)M_{pa}$
13 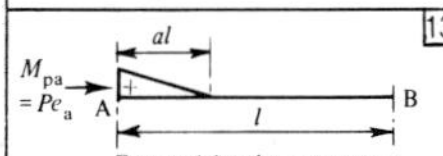 Eccentricity decreases over length al from support A	$N_{AB} = -a(3-a)M_{pa}$ $N_{BA} = -a^2M_{pa}$	14 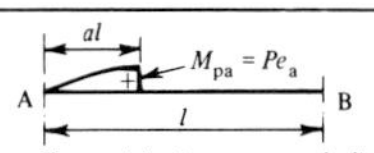 Eccentricity increases parabolically over entire length al from A	$N_{AB} = -4a\left(1 - \frac{5a}{8}\right)M_{pa}$ $N_{BA} = -\frac{5a^2}{2}M_{pa}$
15 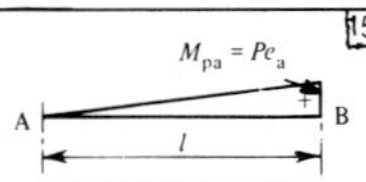 Eccentricity increases over entire length	$N_{AB} = -M_{pa}$ $N_{BA} = -2M_{pa}$	16 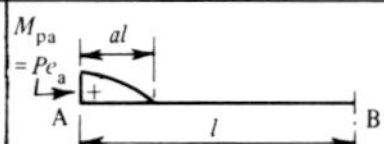 Eccentricity decreases parabolically over length al from A	$N_{AB} = -4a\left(1 - \frac{3a}{8}\right)M_{pa}$ $N_{BA} = -\frac{3a^2}{2}M_{pa}$
17 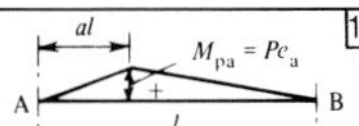 Eccentricity increases from both ends	$N_{AB} = -(2-a)M_{pa}$ $N_{BA} = -(1+a)M_{pa}$	18 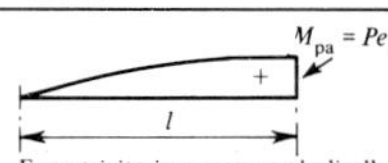 Eccentricity increases parabolically over entire length	$N_{AB} = -\frac{3}{2}M_{pa}$ $N_{BA} = -\frac{5}{2}M_{pa}$
S 19 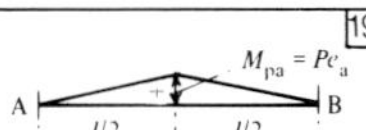 Eccentricity increases from both ends to centre	$N_{AB} = N_{BA} = -\frac{3}{2}M_{pa}$	S 20 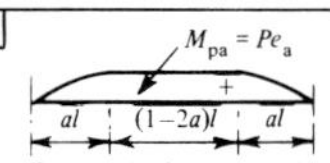 Eccentricity increases parabolically over length al from both ends	$N_{AB} = N_{BA} =$ $-(3-2a)M_{pa}$
21 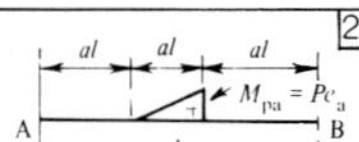 Eccentricity increases from zero over length al	$N_{AB} = -3a\left(c + \frac{a}{3}\right)M_{pa}$ $N_{BA} = -3a\left(b + \frac{2a}{3}\right)M_{pa}$	S Denotes symmetrical Diagram: $N_{BA} = N_{AB}$	

Note. Span ratios a and b in this chart correspond to ratios α in Chapter 14.

CHART No. 15
LOADING TERMS N FOR VARIOUS LOADINGS

LOADING	LOADING TERMS N_{AB}, N_{BA}	LOADING	LOADING TERMS N_{AB}, N_{BA}
1 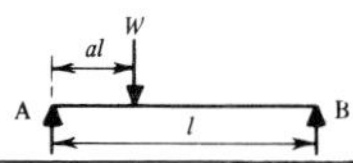	$N_{AB} = -a(1-a)(2+a)Wl$ $N_{BA} = -a(1-a)^2\,Wl$	S 2 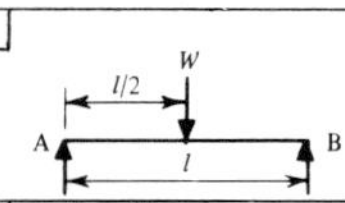	$N_{AB} = N_{BA} = -\frac{3}{8}\,Wl$
S 3 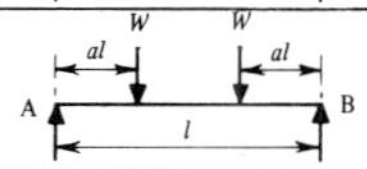	$N_{AB} = N_{BA} = -3a(1-a)\,Wl$	NS 4 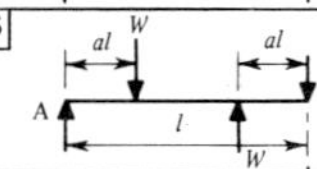	$N_{AB} = -N_{BA} = -a(1-a)(1-2a)\,Wl$
S 5 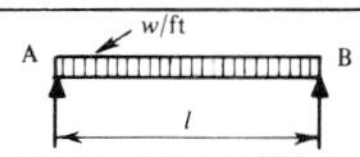	$N_{AB} = N_{BA} = -\frac{wl^2}{4}$	NS 6 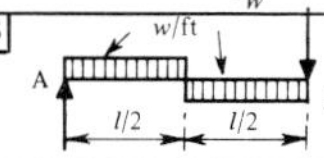	$N_{AB} = -N_{BA} = -\frac{1}{32}\,wl^2$
7 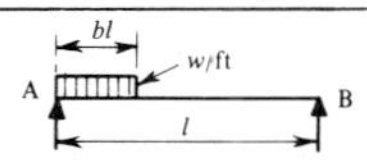	$N_{AB} = -\frac{b^2}{4}(2-b)^2 wl^2$ $N_{BA} = -\frac{b^2}{4}(2-b^2)wl^2$	8 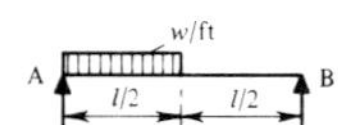	$N_{AB} = -\frac{9}{64}\,wl^2$ $N_{BA} = -\frac{7}{64}\,wl^2$
S 9 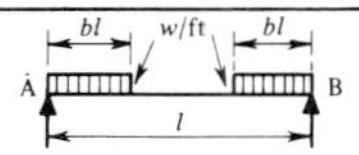	$N_{AB} = N_{BA} = -\frac{b^2}{2}(3-2b)wl^2$	NS 10 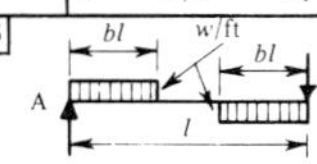	$N_{AB} = -N_{BA} = -\frac{b^2}{2}(1-2b+b^2)wl^2$
11 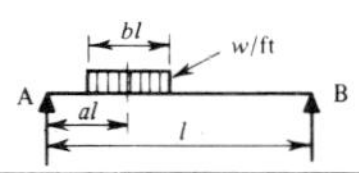	$N_{AB} = -\frac{b}{4}(1-a)(8a-4a^2-b^2)wl^2$ $N_{BA} = -\frac{ab}{4}(4-4a^2-b^2)\,wl^2$	S 12 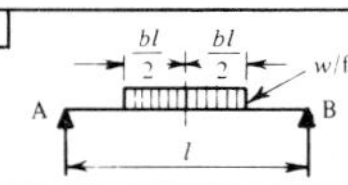	$N_{AB} = -N_{BA} = -\frac{b}{8}(3-b^2)wl^2$
S 13 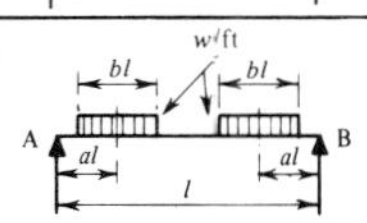	$N_{AB} = N_{BA} =$ $-\frac{b}{4}(12a-12a^2-b^2)wl^2$	NS 14 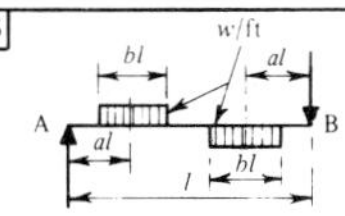	$N_{AB} = -N_{BA} =$ $-\frac{b}{4}(4a-12a^2-b^2+8a^3+2ab^2)wl^2$
15 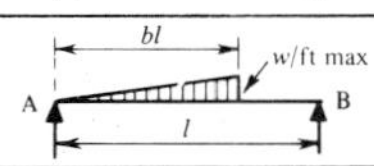	$N_{AB} = -\frac{b^2}{60}(40-45b+12b^2)wl^2$ $N_{BA} = -\frac{b^2}{15}(5-3b^2)wl^2$	16 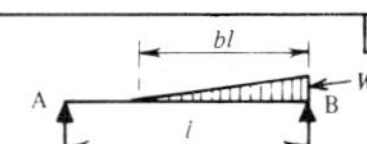	$N_{AB} = -\frac{b^2}{60}(10-3b^2)wl^2$ $N_{BA} = -\frac{b^2}{60}(20-15b+3b^2)wl^2$
17 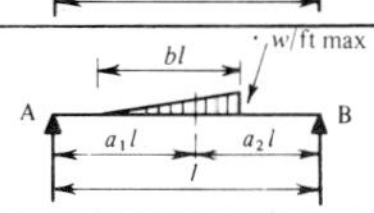	$N_{AB} = -\frac{b}{540}\{270a_2(1-a_1^2) - b^2(45a_2+2b)\} \times wl^2$ $N_{BA} = -\frac{b}{540}\{270a_1(1-a_1^2) - b^2(45a_1-2b)\} \times wl^2$	18 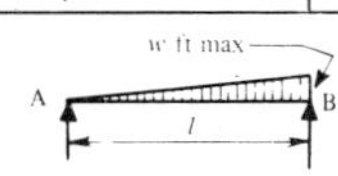	$N_{AB} = -\frac{7}{60}\,wl^2$ $N_{BA} = -\frac{2}{15}\,wl^2$
S 19 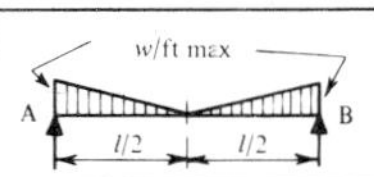	$N_{AB} = N_{BA} = -\frac{3}{32}\,wl^2$	NS 20 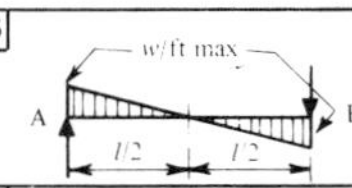	$N_{AB} = -N_{BA} = -\frac{1}{60}\,wl^2$
21 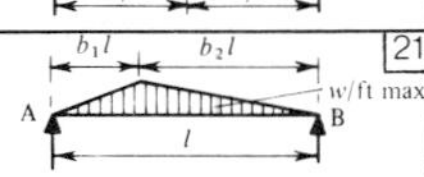	$N_{AB} = -(1+b_2)(7-3b_2^2)\frac{wl^2}{60}$ $N_{BA} = -(1+b_1)(7-3b_1^2)\frac{wl^2}{60}$	S 22 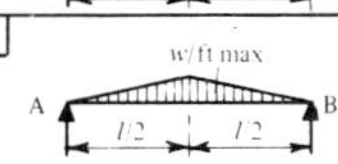	$N_{AB} = N_{BA} = -\frac{5}{32}\,wl^2$
S 23 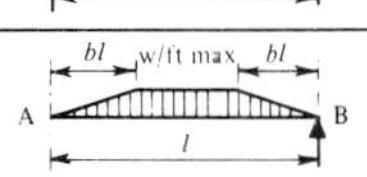	$N_{AB} = N_{BA} = -\frac{1}{4}(1-2b^2+b^3)\,wl^2$	S 24 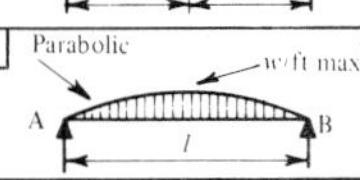	$N_{AB} = N_{BA} = -\frac{1}{5}\,wl^2$
25 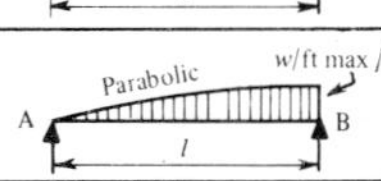	$N_{AB} = -\frac{1}{6}\,wl^2$ $N_{BA} = -\frac{11}{60}\,wl^2$	26 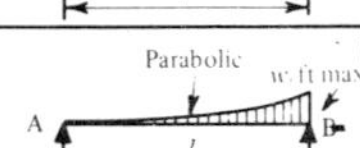	$N_{AB} = -\frac{1}{15}\,wl^2$ $N_{BA} = -\frac{1}{12}\,wl^2$
S 27 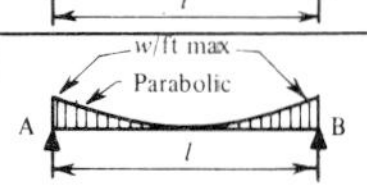	$N_{AB} = N_{BA} = -\frac{1}{20}\,wl^2$	28 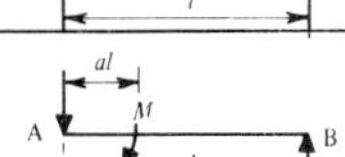	$N_{AB} = -(2-6a+3a^2)M$ $N_{BA} = -(1-3a^2)M$
S 29 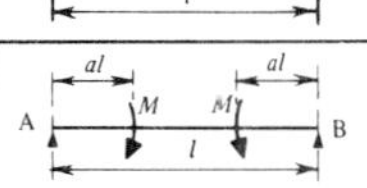	$N_{AB} = N_{BA} = -3(1-2a)M$	NS 30 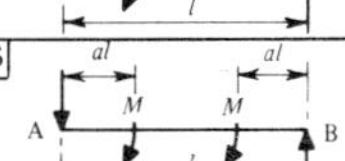	$N_{AB} = -N_{BA} = -(1-6a+6a^2)M$

S Denotes symmetrical loading: $N_{AB} = N_{BA}$

NS Denotes loading with negative symmetry: $N_{AB} = -N_{BA}$

Note. Span ratios a and b in this chart correspond to ratios α in Chapter 14.

POST-TENSIONING SYSTEMS

Charts 16 to 49 on pages 506 to 554 give dimensions and details of most of the 'Tendon System' now in use in Great Britain, United States and the Continent of Europe. They can be based on the development of 5 original systems, as described by the first author's book as summarised in the following initial sequence of their development.

(1) The Coyne System, usually used for anchorage of retaining walls and dams in bed-rocks. It consists of a bundle of straight wires, embedded in a bulkhead of concrete and held by bond, the other end being fixed to an anchor block. Tensioning is obtained in one operation by forcing the anchor head away from the end of the structure and inserting spacing members to maintain the separation.

(2) The Freyssinet System. Its essential feature is the conical shape of the anchorage, comprising a male and female cone. The former is embedded at the end of the structure and the male cone is inserted after tensioning by means of a double acting jack. A number of tendons being secured between the female and male cones.

(3) The Magnel System. In this case not all the wires are tensioned in one operation, but in pairs and secured by means of conical wedges, against a distribution plate.

(4) The Roebling Systems, consists of a stranded cable, having two anchor heads, the cable being made to length. The forces are transferred by anchor plates one at each end and the force is secured by means of a screw thread.

(5) The Bar Systems (Macalloy and Dywidag Systems). The ends of the bar have screw threads, the bar being anchored by a plate by a nut similar to the Roebling system.

The Magnel and Roebling systems are only of historical importance.

The charts 16–49 of different post-tensioning systems have been updated on the basis of information supplied by the manufacturers or obtained from their catalogues and brochures. No responsibility for the accuracy of the information is taken by the authors. Co-operation of the manufacturers is gratefully acknowledged. The systems are divided in two groups: I, relating to wires and strands, and II relating to bar systems mainly used in Great Britain.

CHART No. 16 12, 8 AND 4 WIRE SPIRAL ANCHORAGES

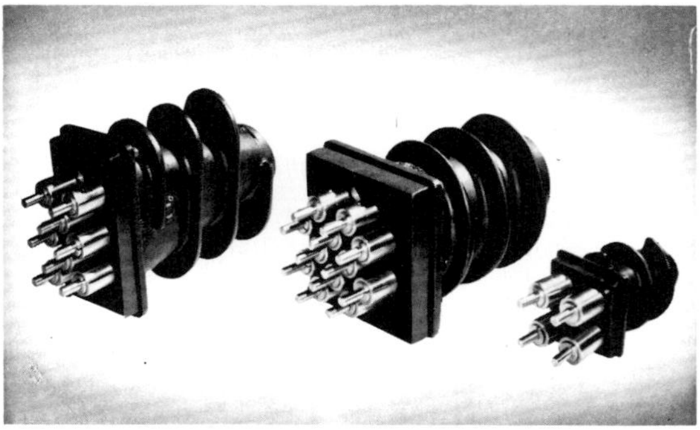

12, 8 & 4 WIRE SPIRAL ANCHORAGES
These anchorages have been in general use for 7mm (0.276 in.) wire for many years. They are still a highly competitive solution to many post tensioning applications, and are generally available only on special order.

RECOMMENDED CENTRES

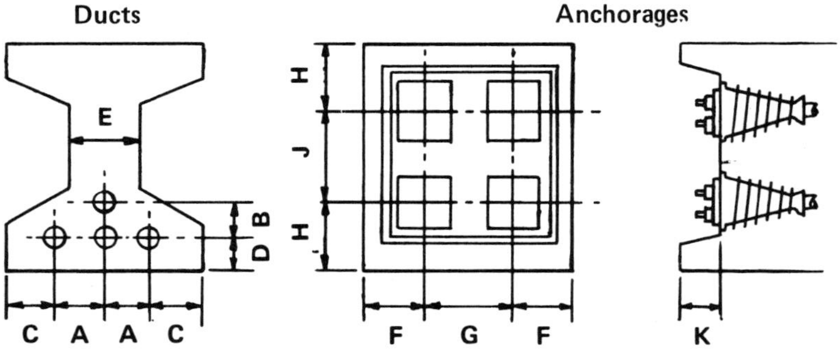

Spacers or Profilers

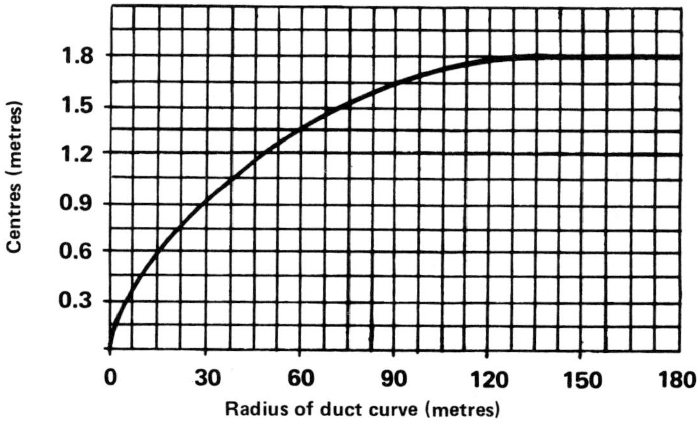

Wire systems

TENDON		P.S.FORCE (TONNES)		ANCHORAGE			DUCT DIA	A	B	C	D	E	F	G	H	J	K
INCH	MM	ULT.	70%	TYPE	FLANGE SIZE	LENGTH											
12/0.276	12ϕ7	72.9	51.0	S4	124 x 121	165	42	70	65	65	65	130	100	150	100	150	75
8/0.276	8ϕ7	48.5	33.9	S3	140 x 76	178	39	65	60	65	65	130	90	100	120	165	75
4/0.276	4ϕ7	24.25	16.95	S2	70 x 70	83	25	50	45	60	60	115	85	95	85	95	75
2/0.276	2ϕ7	12.13	8.45														

1 tonne force = 1000 kgf = 2200 lbf = 9810 Nf = 9.81 kNf

CCL SYSTEM (CONTINUED)
CHART No. 17 STRANDFORCE ANCHORAGE FOR 4, 5 or 10 STRAND SYSTEMS

STRANDFORCE ANCHORAGE
The anchorage is widely used and is available for tendons of 10/0.7 in. (18 mm) strands as shown, or for 5/0.7 in. (18 mm) strands or for 4/0.5 in. (13 mm) or 4/0.6 in. (15 mm) strands.

DUCTS

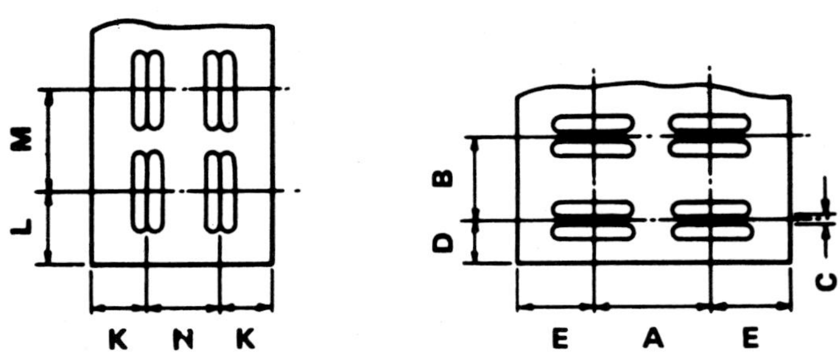

ANCHORAGES

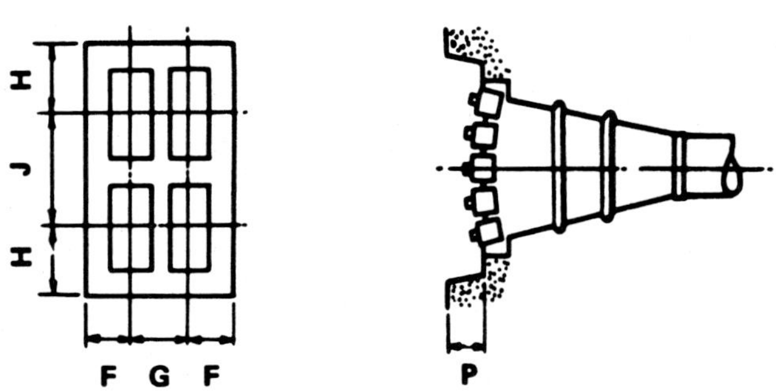

CHART No. 17 (CONTINUED) DATA FOR STRANDFORCE ANCHORAGES

	TENDON	P.S. FORCE kN		DUCT EXT. DIM.	ANCHORAGE DIMENSIONS		RECOMMENDED CENTRES														JACK REF
		ULT	70%		FLANGE SIZE	LENGTH L	A	B	C	D	E	F	G	H	J	K	L	M	N	P	
R1	4/13 STD	660	462	75 X 27	235 X 80	248	125		25		90	95	130	210	325	65	80	105	80	80	A,B
	4/13 SUP	736	515	75 X 27			125		25		90	95	130	210	325	65	80	105	80	80	A,B
	4/13 D	836	585	75 X 27			130		25		90	100	140	220	335	65	80	105	80	85	B
	4/15 STD	908	636	75 X 27			130		25		90	100	140	220	335	65	80	105	80	85	B,C D
	4/15 SUP	1000	700	75 X 27			130		25		90	100	140	220	335	65	80	105	80	85	B,C, D
R2	5/18 STD	1850	1295	112 X 30	362 X 89	394	165		25		110	100	140	280	460	65	95	140	80	90	C,D
	5/18 D	1900	1330	112 X 30			165		25		110	100	140	280	460	65	95	140	80	90	C,D
R3	10/18 STD	3700	2590	2 No. 112 X 30	362 X 171	394	165	110	25	80	110	190	230	280	460	75	95	140	110	90	C,D
	10/18 D	3800	2660	2 No. 112 X 30			165	110	25	80	110	190	230	280	460	75	95	140	110	90	C,D

CCL SYSTEM (CONTINUED)
CHART No. 18 CABCO STRAND ANCHORAGES

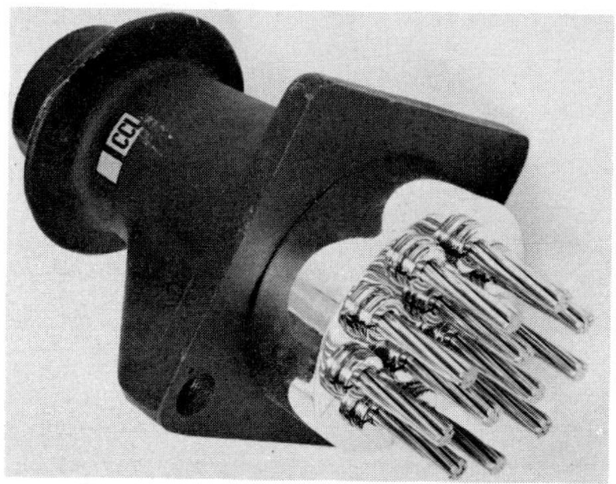

CCL Cabco strand anchorages are used for single strand stressing for use with tendons incorporating four to 12 strands, and strand diameters catered for are 0.5 in. (12.7 mm), 0.6 in. (15.2 mm) and 0.7 in. (18 mm) in either standard, super or Dyform/Compact form.

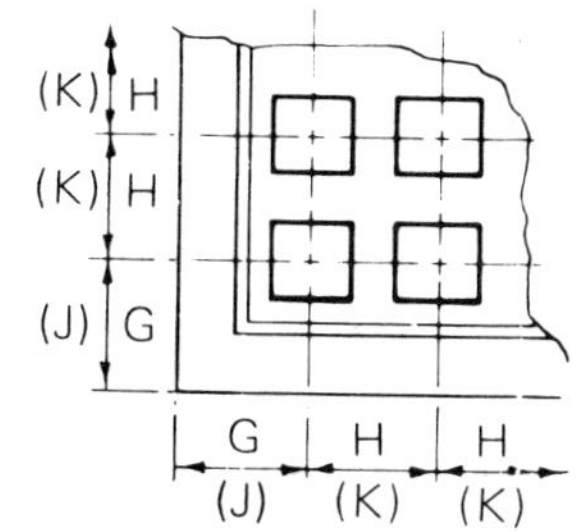

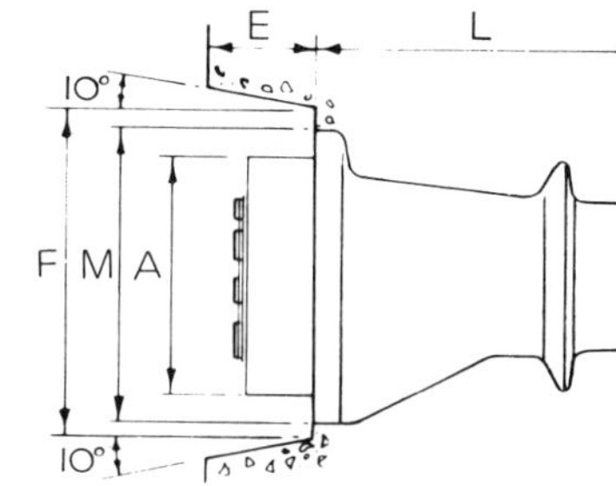

Data for CABCO anchorage

	TENDON	P.S. FORCE kN		DUCT I.D. mm	ANCHORAGE NOMINAL DIMENSIONS				SETTING OUT DIMENSIONS				TENDON MASS kg/m	JACK REF
		ULT	70%		Flange Size M	Length L	A	B	E	F	G	H		
S4	4/13STD	660	462	42	121 x 121	165	114 sq	120	120	150	110	145	2.9	A,B
	4/13SUP	736	515	42			114 sq	120	120	150	110	145	3.2	A,B
U2	4/15STD	908	636	51	175 x 175	230	130	140	105	200	140	190	4.4	B,C,D
	4/15SUP	1000	700	51			130	140	105	200	140	190	4.5	B,C,D
	4/15D	1200	840	51			130	140	105	200	140	195	5.2	C,D
	4/18STD	1480	1036	51			130	190	115	200	140	215	6.6	C,D
	4/18D	1520	1064	51			130	190	115	200	140	215	7.0	C,D
	7/13STD	1155	808	51			130	140	95	200	140	195	5.2	A,B
	7/13SUP	1288	902	51			130	140	95	200	140	195	5.6	A,B
	7/13D	1463	1024	51			130	140	95	200	140	215	6.2	B

Cabco anchorage

U3	7/15STD	1589	1112	63*	215 x 215	250	150	155	105	240	160	220	7.7	B,C,D
	7/15SUP	1750	1225	63*			150	155	105	240	160	220	7.9	B,C,D
	7/15D	2100	1470	63*			150	155	105	240	160	250	9.1	C,D
	12/13STD	1980	1386	75			150	145	100	240	160	250	8.9	A,B
	12/13SUP	2208	1546	75			150	145	100	240	160	250	9.6	A,B
U4	7/18STD	2590	1813	75*	245 x 245	330	180	190	115	270	180	255	11.5	C,D
	7/18D	2660	1862	75*			180	190	115	270	180	255	12.2	C,D
	12/13D	2508	1756	75*			180	150	105	270	180	255	10.6	B
	12/15STD	2724	1907	80			180	170	105	270	190	255	13.2	B,C,D
	12/15SUP	3000	2100	80			180	170	105	270	190	265	13.5	B,C,D
U5	12/15D	3600	2520	80	265 x 265	360	200	170	115	290	205	295	15.6	C,D

Looped anchorages

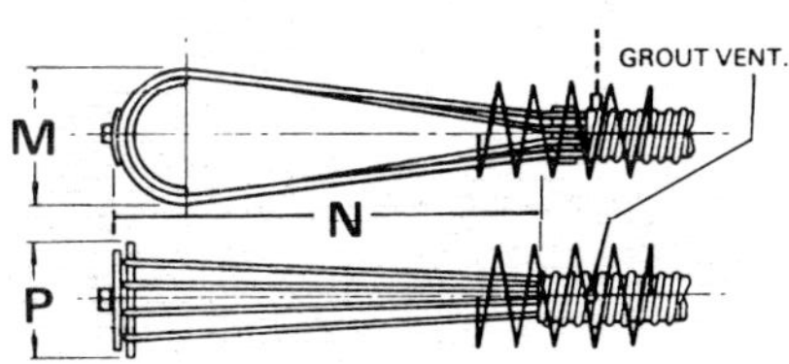

	S4	U2	U3	U4	U5	U6	U7	U8	R1	R2	R3
M	250	250	250	250	300	300	300	300	250	300	300
N	600	600	700	800	800	900	1000	1200	600	800	900
P	100	170	285	400	450	500	700	900	170	290	425

NOTES- 1. Looped dead end anchorages should only be used on short tnedons (not greater than 20m) and when it is practical to bend the strands on site.
2. Odd strands are looped round the anchorage clamped and returned into the sheathing for a length of 750 mm. 3. End block reinforcement is required at this anchorage (please consult data sheet overleaf for details).

CHART No. 19 MULTIFORCE ANCHORAGE

12 STRANDS 12.7 mm

28 STRANDS 18 mm

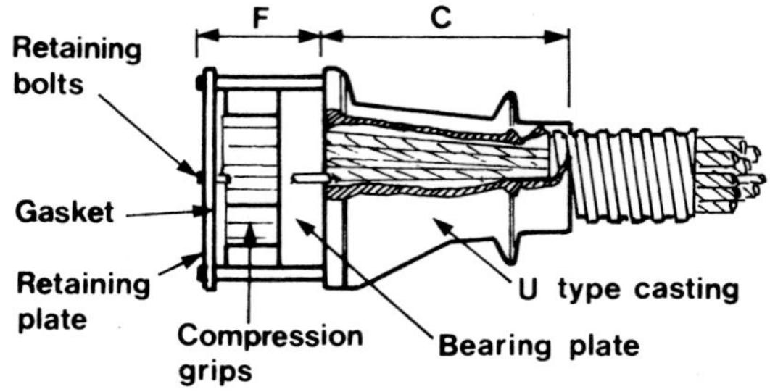

DEAD END ANCHORAGE

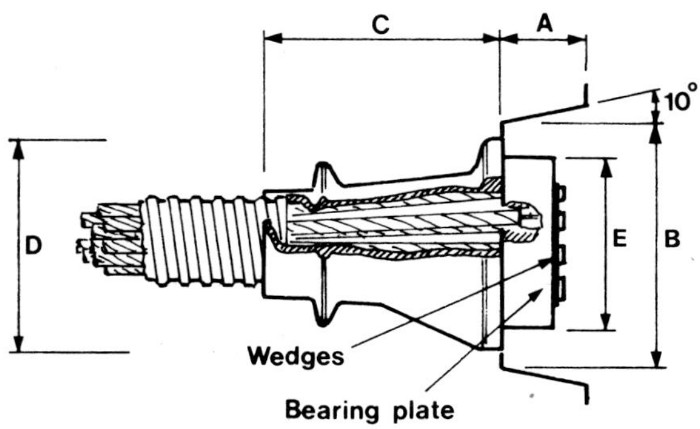

STRESSING ANCHORAGE

The Multiforce range of anchorages cater for simultaneous stressing of all the strands in tendons ranging from 12 strands of 0.5 in. diameter (12.7 mm) and 0.6 in. diameter (15.2 mm) up to 28/0.7 in. (18 mm). Standard, super and Dyform/Compact strands are catered for in the systems.

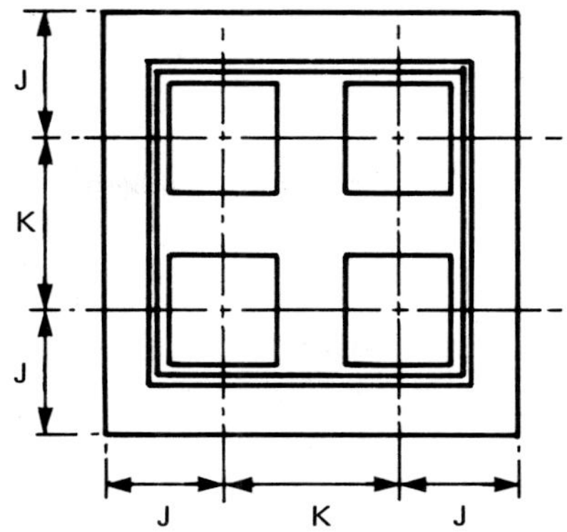

Anchorage data for U3 multiforce system

Tendon	Stressing Head	Sheathing mm		Anchorage Nominal Dimensions				Setting Out Dimensions			
		I.D.		C	D	E	F	A	B	J	K
12/13 STD	Multimatic 2000 M	75 / 63		250	215	150	135	100	240	160	250
12/13 SUP		75 / 63		250	215	150	135	100	240	160	250

Anchorage data for U5 multiforce system

Tendon	Stressing Head	Sheathing mm		Anchorage Nominal Dimensions				Setting Out Dimensions			
		I.D.		C	D	E	F	A	B	J	K
19/13 STD	Multimatic 3000 M	81		360	265	200	155	120	290	205	295
19/13 SUP		81		360	265	200	155	120	290	205	295
12/15 STD		80		360	265	200	155	115	290	200	275
12/15 SUP		80		360	265	200	155	115	290	200	275
12/15 DYF		80		360	265	200	155	115	290	205	295

Anchorage data for U6 multiforce system

Tendon	Stressing Head	Sheathing mm		Anchorage Nominal Dimensions				Setting Out Dimensions			
		I.D.		C	D	E	F	A	B	J	K
25/13 STD	Multimatic 5000 M	90		500	300	230	155	120	325	220	375
25/13 SUP		90		500	300	230	155	120	325	220	375
13/15 DYF		90/84		500	300	230	160	120	325	215	355

Anchorage data for U7 multiforce system

Tendon	Stressing Head	Sheathing mm		Anchorage Nominal Dimensions				Setting Out Dimensions			
		I.D.		C	D	E	F	A	B	J	K
31/13 STD	Multimatic 5000 M	100		520	335	250	170	135	360	242	432
31/13 SUP		100		520	335	250	170	135	360	242	432
19/15 STD		100		520	335	250	180	130	360	230	380
19/15 SUP		100		520	335	250	180	130	360	230	380
19/15 DYF		100		520	335	250	180	130	360	242	432

Anchorage data for U8 multiforce system

Tendon	Stressing Head	Sheathing mm		Anchorage Nominal Dimensions				Setting Out Dimensions			
		I.D.		C	D	E	F	A	B	J	K
19/18 DYF	Multimatic 7000 M	105		600	380	275	230	140	405	300	495

CHART No. 20 UNBONDED ANCHORAGES

(For 12.7 mm single strand and 15 mm single strand systems)

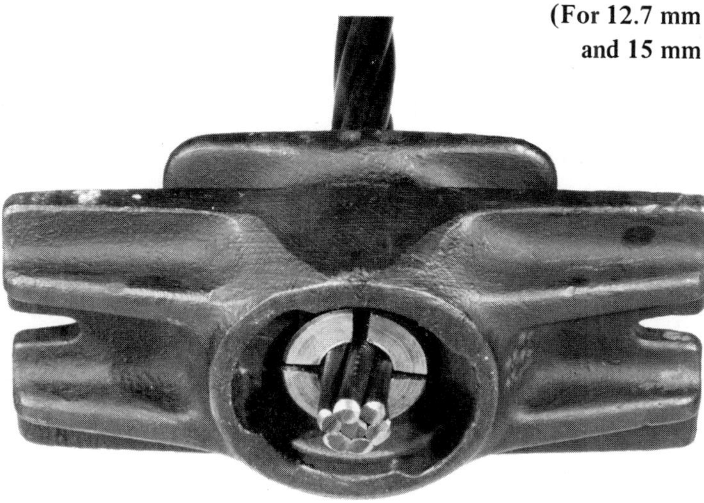

UNBONDED ANCHORAGES
These anchorages are now being used extensively with unbonded tendons in slabs.

PSC FREYSSINET SYSTEM
CHART No. 21 THE 'K' RANGE SYSTEMS

This range of compact efficient anchorages which include tendon forces up to 1000 tonnes, is stressed with central holes tensioning jacks throughout thus permitting a compact anchorage, blind end anchorage and coupler solution for tendon service loads.

DATA FOR 'K' RANGE SYSTEMS

13 mm 'K' MONOGROUP RANGE			
Anchorage type	No. of strands	Sheath diameter (internal)	System and jack type
1 K13	1	25 mm	Mono strand
7 K13	2 to 7	55 mm	K100
12 K13	8 to 12	65 mm	K200
19 K13	13 to 19	85 mm	K350
27 K13	20 to 27	95 mm	K500
37 K13	28 to 37	110 mm	K700
55 K13	38 to 55	130 mm	K1000
Characteristic strength of each strand = 186 kN			

15 mm 'K' MONOGROUP RANGE			
Anchorage type	No. of strands	Sheath diameter (internal)	System and jack type
1 K15	1	25 mm	Mono strand
7 K15	2 to 7	65 mm	K200
12 K15	8 to 12	75 mm	K350
15 K15	13 to 15	85 mm	K350
19 K15	16 to 19	95 mm	K500
27 K15	20 to 27	110 mm	K700
37 K15	28 to 37	130 mm	K1000
Characteristic strength of each strand = 265 kN			

'K' range systems for drawn/compact strands			
7 K15D	4 to 7 (15 mm)	65 mm	K200
12 K15D	8 to 12 (15 mm)	85 mm	K350
19 K15D	13 to 19 (15 mm)	95 mm	K500
19 K18D	13 to 19 (18 mm)	110 mm	K700
Characteristic strength of each 15 mm strand = 300 kN Characteristic strength of each 18 mm strand = 380 kN			

CHART No. 22 ANCHORAGE DIMENSIONS

Anchorages (castings)

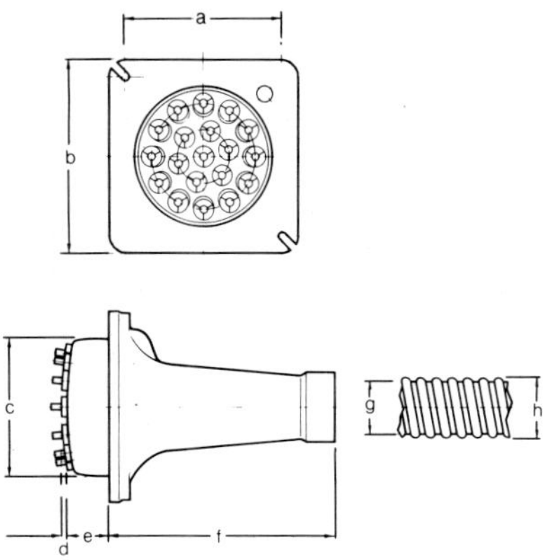

	1M13	1M15	7K13 4K15	12K13 7K15	12K15	19K13	27K13 19K15	37K13 27K15	55K13 37K15
a	100	103	130	170	200	225	250	300	370
b	125 x 70	130 x 85	160	210	245	270	315	365	450
c	—	—	115	140	160	190	220	260	320
d	10	10	10	10	10	10	10	10	10
e	—	—	50	60	60	60	65	80	95
f	50	50	103	145	190	290	385	435	465
g	—	—	55	65	75	85	95	110	130
h	16	20	61	71	81	91	101	118	140

Anchorages (fabricated)

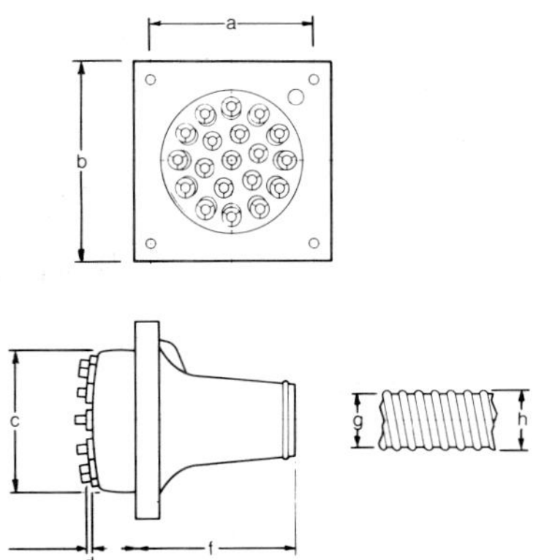

	1M13	1M15	7K13 4K15	12K13 7K15	12K15	19K13	27K13 19K15	37K13 27K15	55K13 37K15
a	90	90	130	170	200	210	250	300	420
b	110 x 65	115 x 85	170	225	260	285	340	400	495
c	40	50	115	140	160	190	220	260	320
d	10	10	10	10	10	10	10	10	10
e	40	45	50	60	60	60	65	80	95
f	15	15	100	155	175	355	412	400	480
g	—	—	55	65	75	85	95	110	130
h	16	20	61	71	81	91	101	118	140

Dimensions given in these tables relate to anchorages for use with normal strands up to the following maximum breaking load:

13 mm (0.5"): 18.7 tonnes, 185 kN, 41.3 kips

15 mm (0.6"): 26.5 tonnes, 260 kN, 58.6 kips

PSC FREYSSINET SYSTEM (CONTINUED)
CHART No. 23 DETAILING INFORMATION

Anchorage detailing

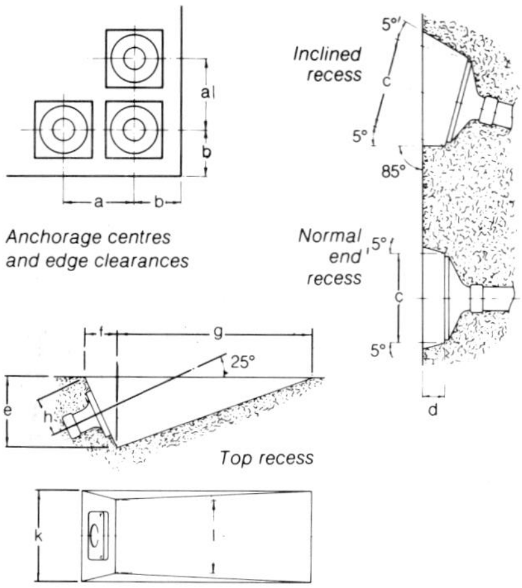

	1M13	1M15	7K13 4K15	12K13 7K15	12K15	19K13	27K13 19K15	37K13 27K15	55K13 37K15
a	200	200	200	270	300	325	375	450	525
b	90	90	135	175	200	225	250	300	375
c	90	90	180	250	300	350	500	550	650
d	90	90	100	120	125	125	140	150	160
e	—	—	225	270	360	380	450	550	635
f	—	—	105	125	170	180	210	255	300
g	—	—	700	875	1000	1000	1250	1500	1750
h	—	—	115	150	175	200	235	275	325
k	—	—	330	380	450	450	570	690	800
l	—	—	270	300	375	375	480	580	670

Couplers

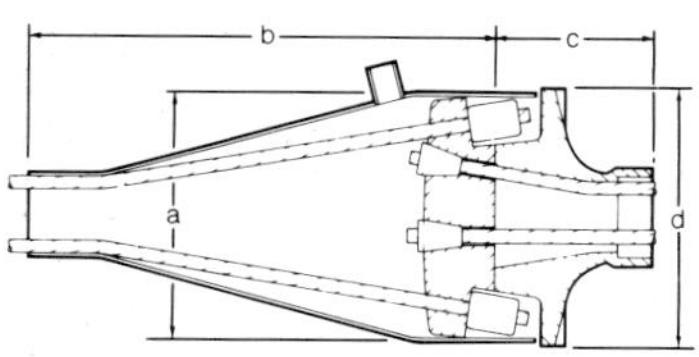

	1M13	1M15	7K13 4K15	12K13 7K15	12K15	19K13	27K13 19K15	37K13 27K15	55K13 37K15
a	For coupling use pretensioning couplers.		160	204	230	260	308	Dimensions for larger couplers are available on request.	
b			450	555	600	490	680		
c			120	166	120	280	362		
d			160	210	245	270	315		

The range of standard couplers is the 'flange' type in which the continuing cable is attached by pre-applied 'FU' type swaged grips.

Blind ends

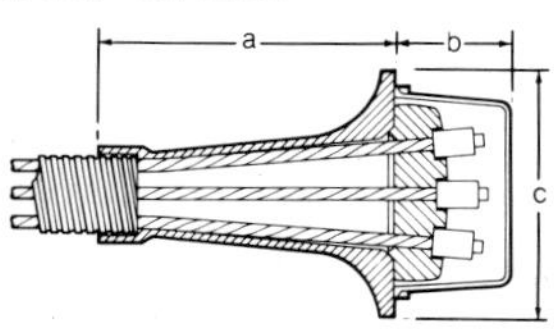

		7K13 4K15	12K13 7K15	12K15	19K13	27K13 19K15	37K13 27K15	55K13 37K15
a	Enclosed FU Swage type.	160	210	245	270	315	365	450
b		103	145	190	290	385	435	465
c		115	115	127	120	145	165	180

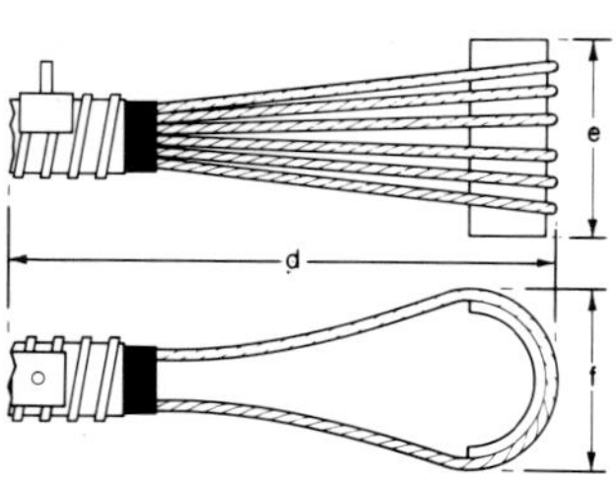

			6K13	12K13	12K15
d	Looped type	—	500	600	700
e	—	—	200	250	250
f	—	—	250	250	300

The standard blind end anchorage is similar to the normal anchorage but the jaws are replaced by 'FU' swaged grips, a permanently attached anchorage pre-applied to the strand with a swaging jack.

Looped anchorages are practical only for the smaller tendons and details are given for the 6K13, 12K13 and 12K15 systems.

A-type stressing anchorage

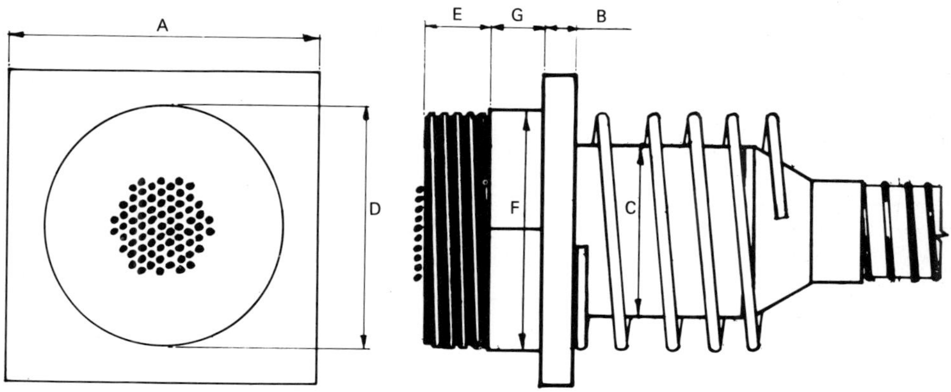

TENDON REFERENCE		A24	A34	A42	A55	A61	A73	A85	A97	A109
No 7 mm ϕ wires	n	24	34	42	55	61	73	85	97	109
Characteristic strength	kN	1542	2185	2699	3535	3920	4692	5463	6234	7005
Jacking force 70%	kN	1080	1530	1890	2474	2744	3284	3824	4364	4904
Jacking force 75%	kN	1157	1639	2024	2651	2940	3519	4097	4676	5254
Jacking force 80%	kN	1234	1748	2159	2828	3136	3753	4370	4987	5604
Bearing plate										
Side length	A	205	240	270	305	325	355	380	410	430
Thickness	B	25	30	35	40	45	55	60	65	75
Trumpet O/D	C	72	83	87	97	106	110	115	125	130
Anchor head										
Thread diameter	D	110	124	139	158	158	172	185	202	202
Standard length	E	44	50	56	64	64	69	74	81	81
Chocks ϕ	F	117	131	147	168	168	182	196	214	214
Chocks min thickness		TO BE SPECIFIED								
Sheathing I/D		50	60	65	75	80	85	95	100	105
Sheathing O/D		58	68	73	83	88	93	103	108	113

Dimensions in mm

CHART No. 25 B-TYPE STRESSING ANCHORAGE DETAILS

B-type stressing anchorage

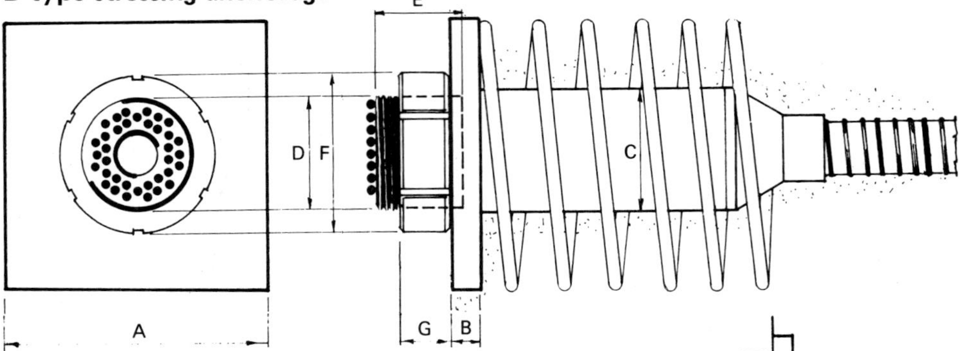

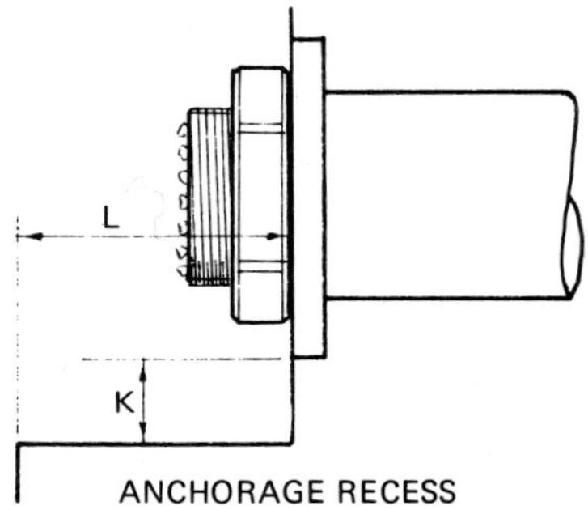

ANCHORAGE RECESS

TENDON REFERENCE		B8	B16	B24	B34
No 7 mm ϕ wires	n	8	16	24	34
Characteristic strength		514	1028	1542	2185
Jacking force 70%		354	720	1080	1395
Jacking force 75%		386	771	1157	1494
Jacking force 80%		411	823	1234	1594
Bearing plate					
Side length sq.	A	150	175	220	250
Thickness	B	14	15	20	25
Trumpet O/D	C	87	120	133	154
Anchor head					
Thread diameter	D	75	100	115	130
Standard length	E	40	60	80	90
Locknut					
Diameter	F	105	135	155	180
Thickness	G	22	30	40	50
Sheathing					
I/D		30	40	50	60
O/D		38	48	58	68
Recess					
Min. B.P. clearance	K	100	100	100	100
Min. depth	L	130	110	130	140
Combined B.P.					
Horizontal min.	M	133	171	197	235
Vertical min.	N	138	171	197	235
Edge distance	O	75	90	110	125
BUB coupler					
Trumpet length	P	275	335	405	455
Trumpet ϕ	Q	100	120	133	154
BSB coupler					
Trumpet length	R	225	260	315	350
Trumpet ϕ	S	87	120	133	154

Forces in kN/dimensions in mm

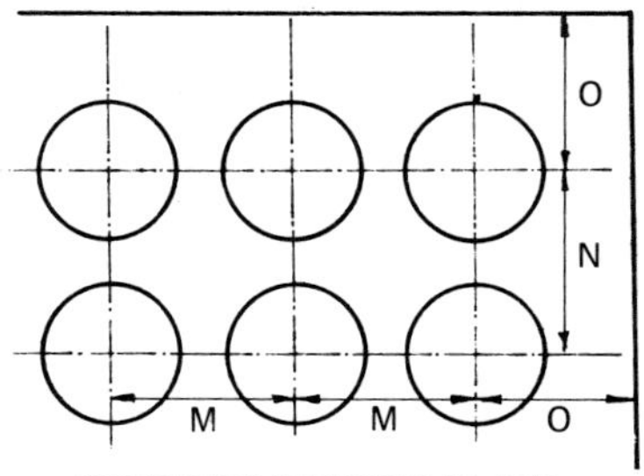

COMBINED BEARING PLATE

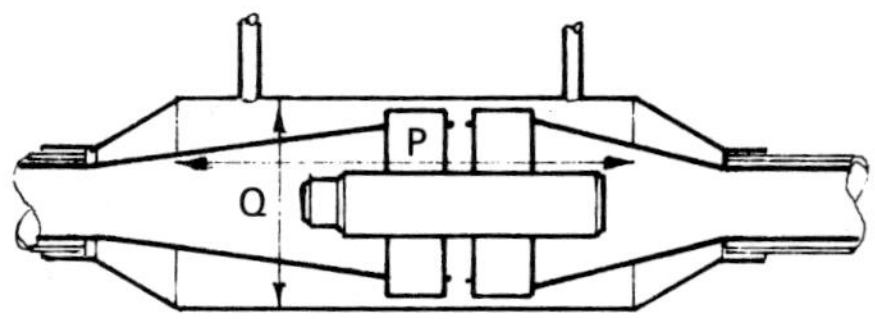

UNSTRESSED COUPLER

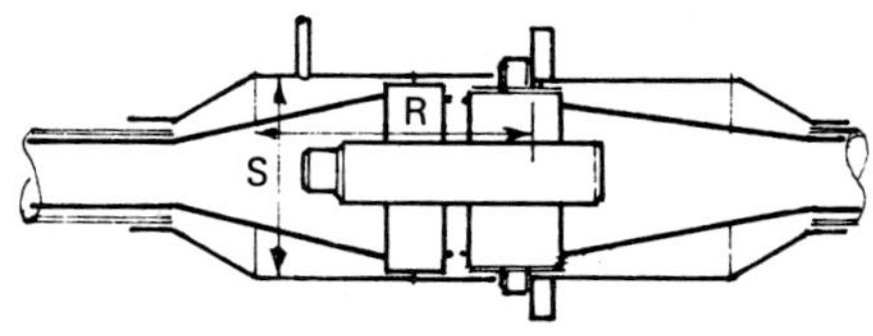

STRESSED COUPLER

C-type stressing anchorage

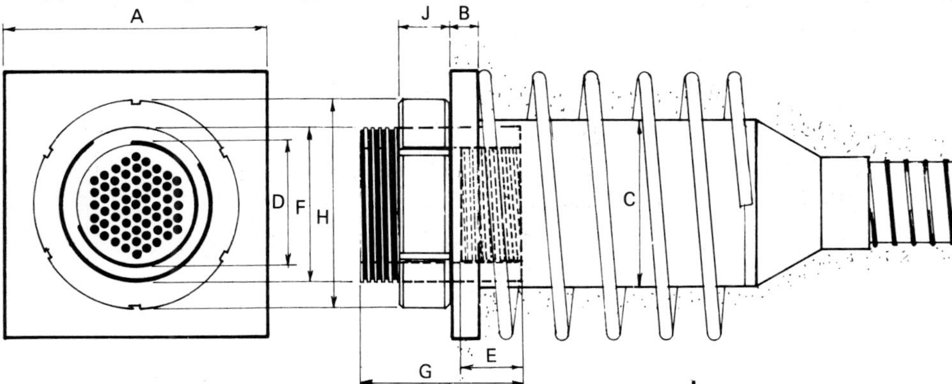

TENDON REFERENCE		C31	C42	C55
No 7 mm ϕ wires	n	31	42	55
Characteristic strength	kN	1992	2699	3535
Jacking force 70%	kN	1395	1890	2828
Jacking force 75%	kN	1494	2024	2651
Jacking force 80%	kN	1594	2159	2474
Bearing plate				
Side length	A	250	280	300
Thickness	B	25	35	35
Trumpet O/D	C	130	154	165
Anchor head				
Thread diameter	D	80	90	98
Standard length	E	46	53	63
Pull sleeve				
O/D	F	118	130	144
Standard length	G	104	118	138
Locknut				
Diameter	H	162	178	198
Thickness	J	39	45	53
Sheathing				
I/D		60	65	75
O/D		68	73	83
Recess				
Min. B.P. clearance	M	100	100	100
Min. depth	N	154	168	188
Combined B.P.				
Horizontal min.	O	235	241	267
Vertical min.	P	235	267	305
Edge distance	Q	125	140	150
CUC coupler				
Trumpet length	R	140	160	180
Trumpet O/D	S	145	154	165
Coupler length	T	120	140	160
Coupler O/D	U	120	130	144
CSC coupler				
Trumpet length	V	250	290	330
Trumpet O/D	W	130	154	165
Coupler length	X	140	160	180
Coupler O/D	Y	120	130	144

Dimensions in mm

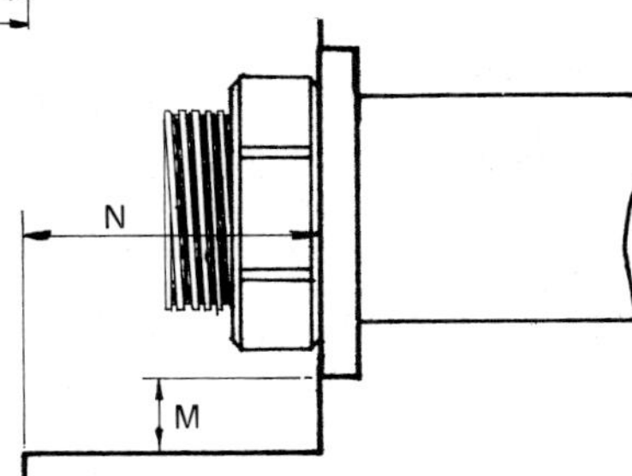

ANCHORAGE RECESS

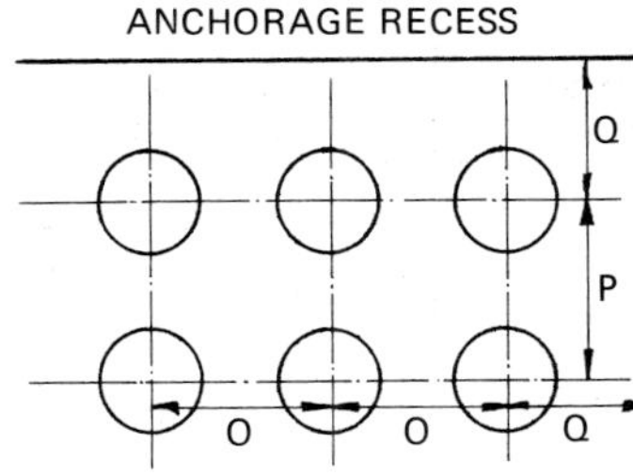

COMBINED BEARING PLATE

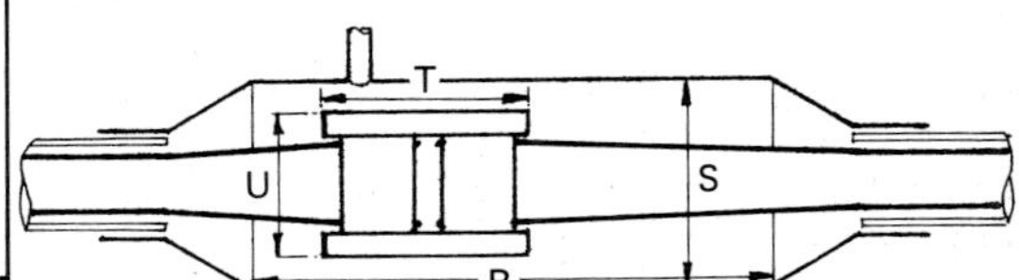

CUC COUPLER

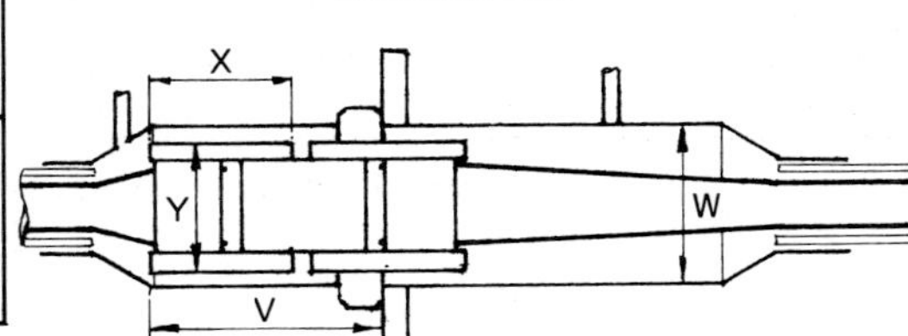

CSC COUPLER

CHART No. 27 'L'-TYPE STRESSING ANCHORAGE DETAILS

L-type stressing anchorage

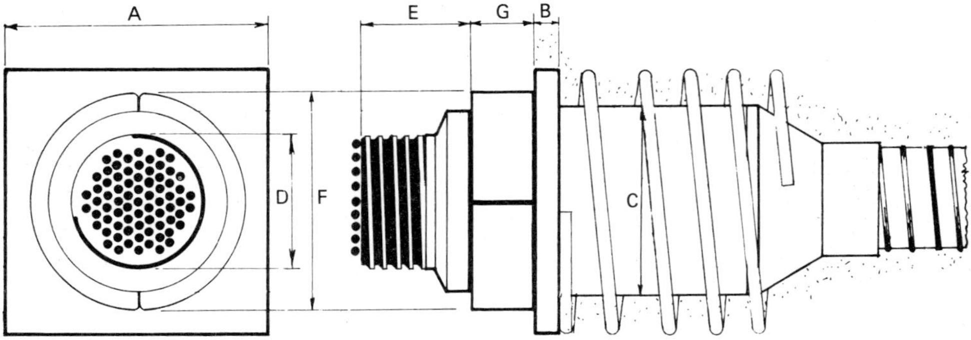

TENDON REFERENCE		L61	L73	L85	L97	L109	L121
No. 7 mm ϕ wires	n	61	73	85	97	109	121
Characteristic strength		3920	4692	5463	6234	7005	7777
Jacking force 70%		2744	3284	3824	4364	4904	5444
Jacking force 75%		2940	3519	4097	4676	5254	5832
Jacking force 80%		3136	3753	4370	4987	5604	6221
Bearing plate							
Side length	A	305	335	360	385	408	425
Thickness	B	40	45	50	50	55	60
Trumpet O/D	C	194	203	219	229	245	245
Anchor head							
Thread diameter	D	115	125	130	145	150	155
Standard length	E	107	110	120	126	134	146
Chocks ϕ	F	208	228	250	262	280	294
Chocks min. thick.	G	44	48	52	55	58	62
Sheathing							
I/D		80	85	95	100	105	110
O/D		88	93	103	108	113	118
Recess							
Min. B.P. clearance	K	100	100	100	100	100	100
Min. depth	L	205	210	225	235	245	260
Combined B.P.							
Horizontal min.	M	335	360	375	420	450	460
Vertical min.	N	280	310	345	355	375	425
Edge distance	O	155	170	180	195	205	215
LUL coupler							
Coupler length	P	180	195	210	215	235	250
Coupler O/D	Q	150	161	174	186	197	208
Trumpet length	R	200	210	220	235	245	255
Trumpet O/D	S	165	180	195	205	215	230
LSL coupler							
Coupler length	T	185	190	210	215	230	245
Coupler ϕ	U	155	168.5	175	195	198	210
Trumpet length	V	227	235	256	265	282	303
Trumpet O/D	W	165	180	195	205	215	230
Larger sizes up to 187/7 do exist							

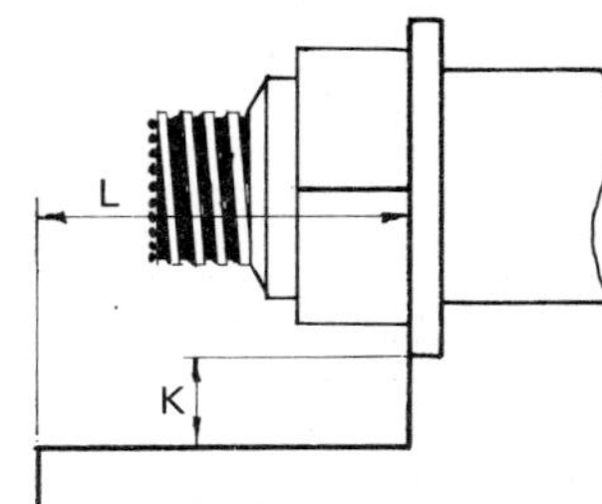

ANCHORAGE RECESS

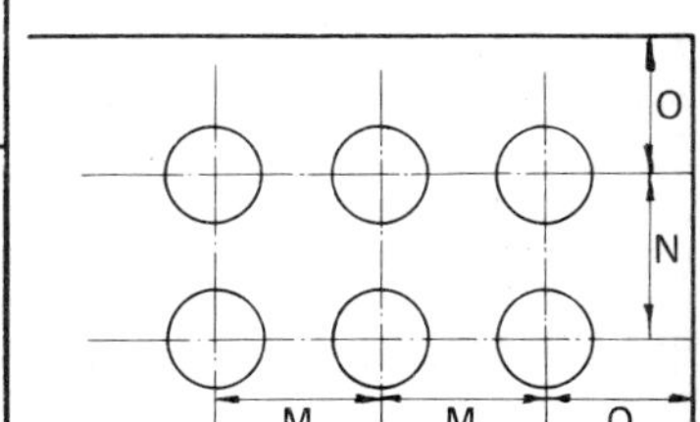

COMBINED BEARING PLATE

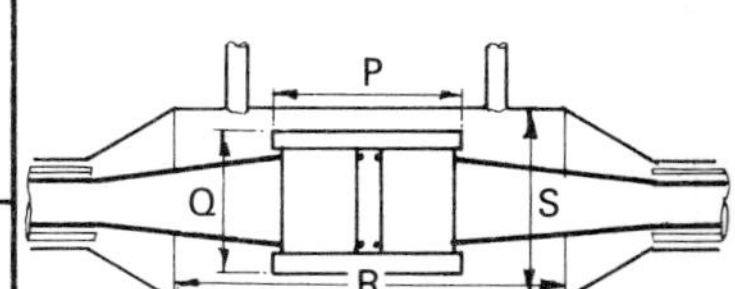

LUL COUPLER

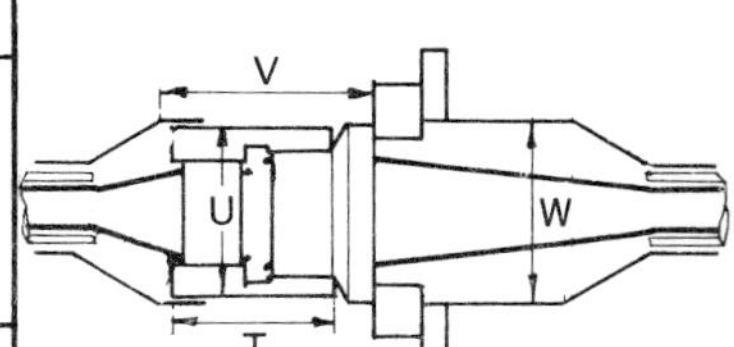

LSL COUPLER

Forces in kN/dimensions in mm

D-type fixed anchorage

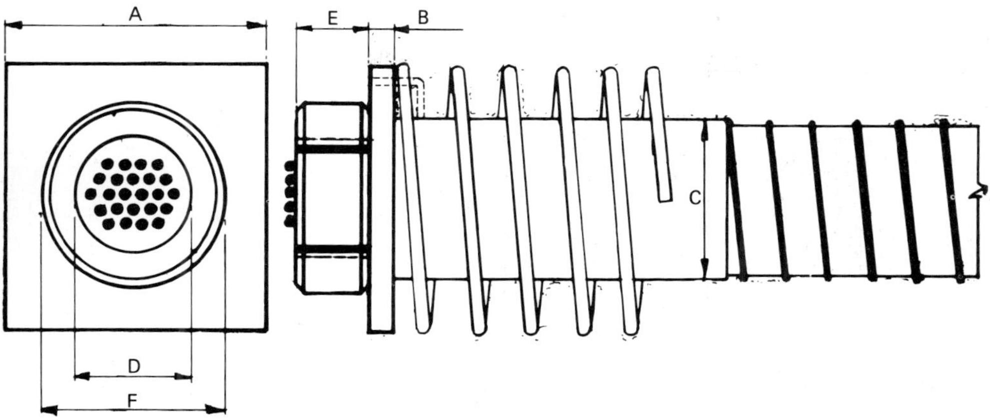

TENDON REFERENCE		D8	D16	D24	D42	D55	D73	D85	D97	D109	D121
No. 7 mm ϕ wires	n	8	16	24	42	55	73	85	97	109	121
Characteristic strength		1542	2185	2699	3535	3920	4692	5463	6234	7005	
Jacking force 70%		1080	1530	1890	2474	2744	3284	3824	4364	4904	
Jacking force 75%		1157	1639	2024	2651	2940	3519	4097	4676	5254	
Jacking force 80%		1234	1748	2159	2828	3136	3753	4370	4987	5604	
Bearing plate											
Side length	A	125	165	210	250	300	310	360	370	385	400
Thickness	B	15	20	25	35	40	50	50	50	55	55
Trumpet I/D	C	70	80	95	110	120	140	145	160	160	165
Anchor head											
Thread diameter	D	50	60	75	90	98	120	124	139	139	145
Standard length	E	35	45	50	53	63	80	90	92	102	110
Locknut											
Thickness	E	35	45	50	53	63	80	90	92	102	110
Diameter	F	90	105	125	145	168	195	205	225	225	245
Sheathing											
I/D		60	70	85	100	110	130	135	150	150	155
O/D		68	78	93	108	118	138	143	158	158	163

Forces in kN/dimensions in mm

CHART No. 29 E & F-TYPE FIXED ANCHORAGE DETAILS

F-type fixed anchorage

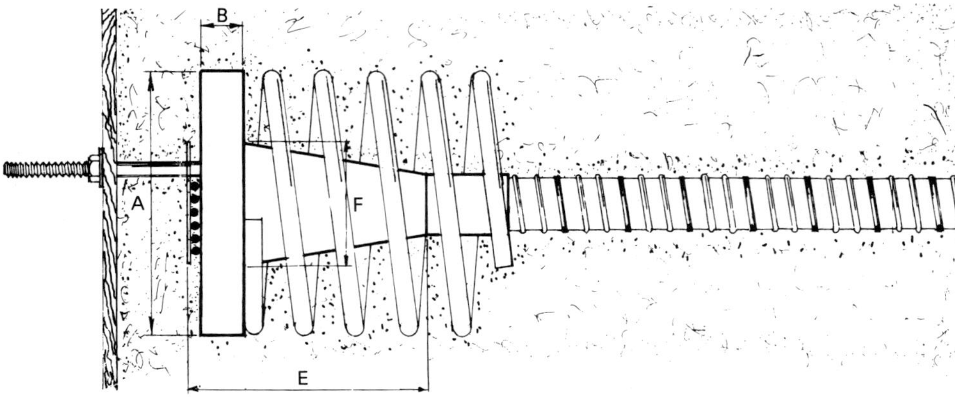

E-type fixed anchorage

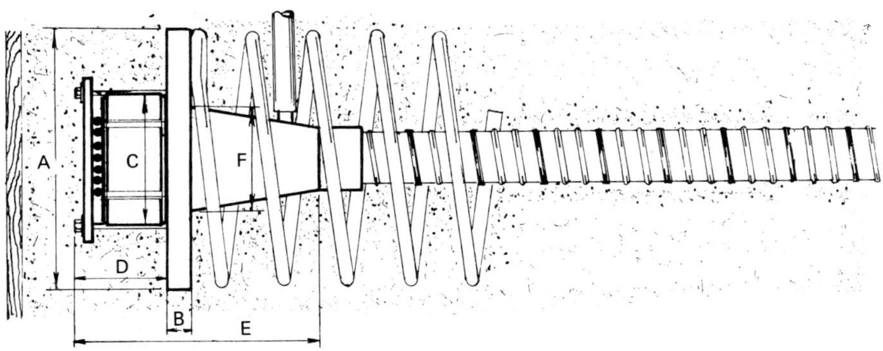

TENDON REFERENCE		F16	F24	F34	E42	E55	E73	E85	E97	E109	E121
No. 7 mm ϕ wires	n	16	24	34	42	55	73	85	97	109	121
Characteristic strength	kN	1028	1542	2185	2699	3535	4692	5463	6234	7005	7777
Jacking load 70%	kN	720	1080	1530	2159	2474	3284	3824	4364	4903	5443
Jacking load 75%	kN	771	1157	1639	2024	2651	3519	4097	4676	5254	5832
Jacking load 80%	kN	823	1233	1748	2159	2828	3753	4370	4987	5604	6221
Bearing plate											
Side length	A	160	230	250	280	300	335	360	385	405	425
Thickness	B	20	40	50	45	50	60	60	70	70	80
Anchorhead											
O/D	C	–	–	–	140	155	180	195	205	220	230
Thickness	D	–	–	–	55	65	85	90	95	100	105
Anchor length	E	270	335	410	243	278	328	348	378	393	423
Trumpet O/D	F	96	114	152	90	95	110	130	135	145	150
Sheathing I/D		40	50	60	65	75	85	95	100	105	110
Sheathing O/D		48	58	68	73	83	93	103	108	113	118

Dimensions in mm

S-type fixed anchorage

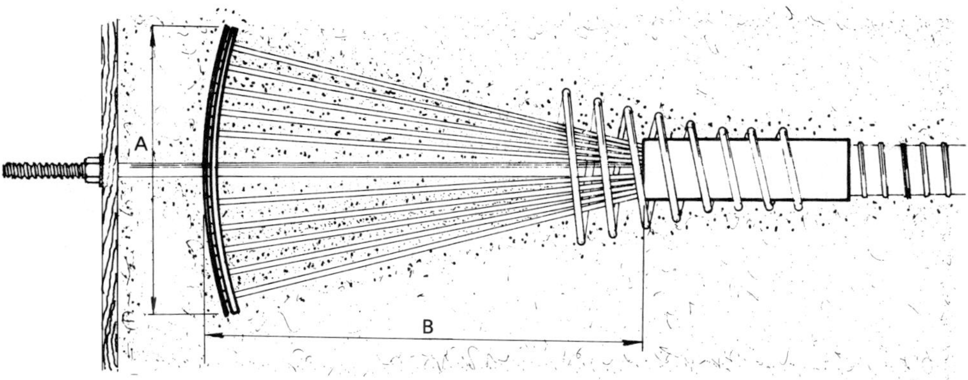

TENDON REFERENCE		S16	S24	S34	S42	S55	S73	S85	S97	S109	S121
No. 7 mm ϕ wires	n	16	24	34	42	55	73	85	97	109	121
Characteristic strength	kN	1028	1542	2185	2699	3535	4692	5463	6234	7005	7777
Jacking load 70%	kN	720	1080	1530	1890	2474	3284	3824	4364	4904	5444
Jacking load 75%	kN	771	1157	1639	2024	2651	3519	4097	4676	5254	5832
Jacking load 80%	kN	823	1234	1748	2159	2828	3753	4370	4987	5604	6221
Square type											
Plate side	A	180	220	260	300	340	400	425	450	–	–
Anchor length	B	450	550	650	750	850	975	1050	1100	–	–
Rectangular type											
Plate length	A	220	300	360	400	500	550	–	–	–	–
Plate width		150	160	180	200	220	250	–	–	–	–
Anchor length	B	450	550	650	750	850	975	–	–	–	–
Long type											
Plate length	A	400	560	560	650	700	–	–	–	–	–
Plate width		80	80	120	140	160	–	–	–	–	–
Anchor length	B	600	700	750	850	950	–	–	–	–	–
Sheathing											
I/D		40	50	60	65	75	85	95	100	105	110
O/D		48	58	68	73	83	93	103	108	113	118

BBRV SYSTEM (CONTINUED)
CHART No. 31 CURVATURE & FIXING DETAILS

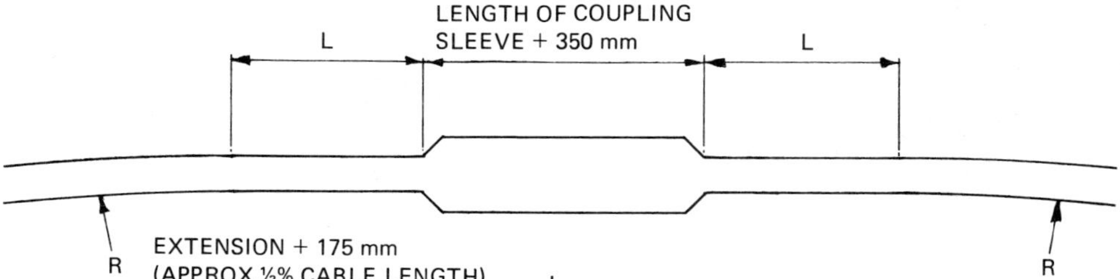

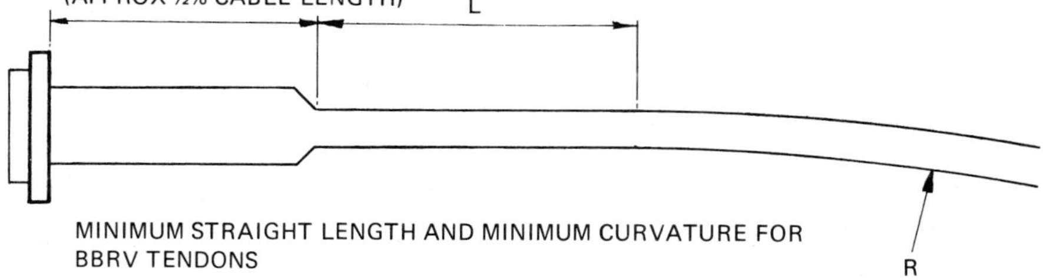

MINIMUM STRAIGHT LENGTH AND MINIMUM CURVATURE FOR BBRV TENDONS

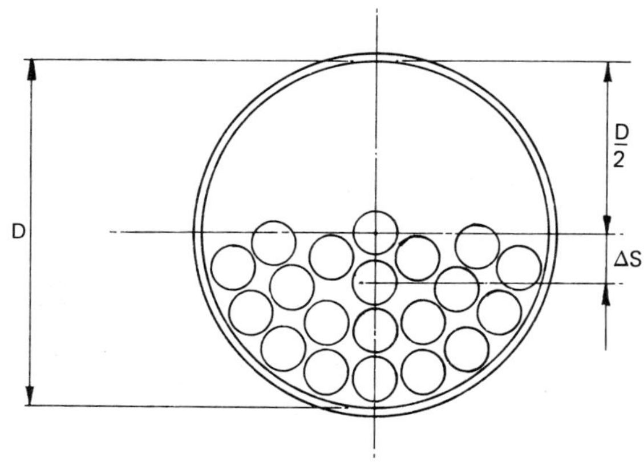

ECCENTRICITY OF TENDON IN DUCT

Number of wires (7ϕ)	R. min. mm	L. min. mm	D. min. mm	ΔS mm
8	2000	500	30	8
16	2000	700	40	8
24	2500	900	50	7
34	2500	1100	55	10
42	3000	1300	65	9
55	3000	1500	75	9
79	3500	1700	85	13
85	4000	1700	95	19
109	4500	1800	105	19
139	5000	1900	115	18
163	5500	2000	125	20
187	6000	2200	135	27

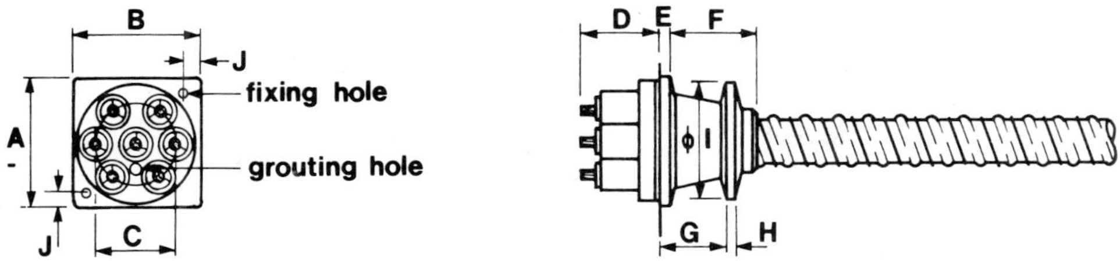
B
J
fixing hole
A
grouting hole
J
C
D
E
F
ϕ I
G
H

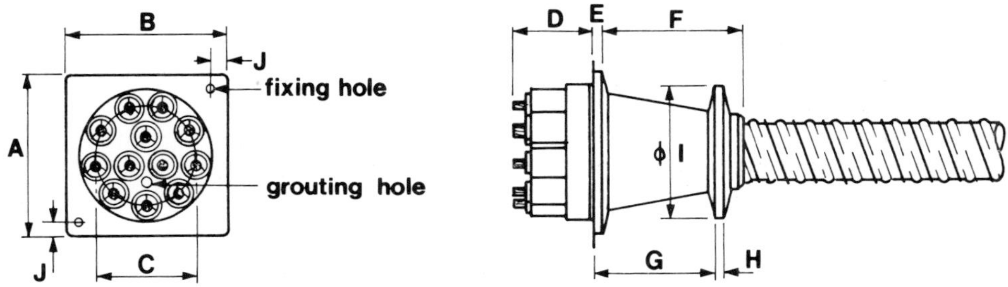
B
J
fixing hole
A
grouting hole
J
C
D
E
F
ϕ I
G
H

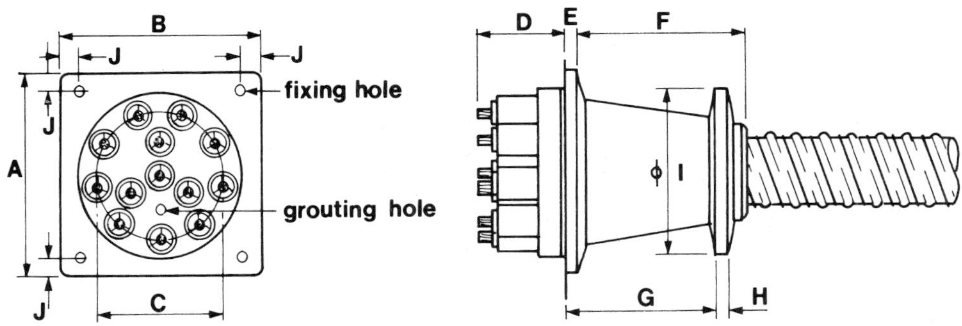
B
J
J
fixing hole
J
A
grouting hole
J
C
D
E
F
ϕ I
G
H

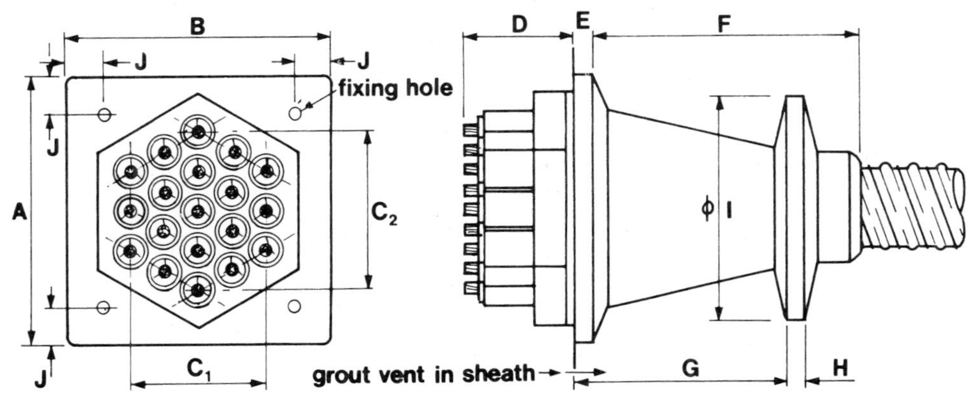
B
J
J
fixing hole
J
A
C2
J
C1
D
E
F
ϕ I
grout vent in sheath
G
H

7 Strand system

Strand ins mm	Dims	A	B	C	D	E	F	G	H	I	J	Sheath Int. Dia.
0.5	ins	6	6	$3\frac{3}{4}$	$3\frac{3}{4}$	$\frac{1}{2}$	$4\frac{1}{8}$	$3\frac{1}{8}$	$\frac{1}{2}$	$5\frac{1}{2}$	$\frac{5}{8}$	2
13	mm	152	152	95	95	13	105	80	13	140	16	51
	cm	15.2	15.2	9.5	9.5	1.3	10.5	8.0	1.3	14.0	1.6	5.1
0.6	ins	7	7	4	$4\frac{1}{2}$	$\frac{1}{2}$	5	4	$\frac{1}{2}$	6	$\frac{5}{8}$	$2\frac{1}{2}$
15	mm	178	178	102	114	13	127	102	13	152	16	64
	cm	17.8	17.8	10.2	11.4	1.3	12.7	10.2	1.3	15.2	1.6	6.4
0.7	ins	$8\frac{1}{2}$	$8\frac{1}{2}$	$4\frac{3}{4}$	$5\frac{1}{4}$	$\frac{1}{2}$	$7\frac{3}{8}$	$6\frac{3}{8}$	$\frac{1}{2}$	7	$\frac{3}{4}$	3
18	mm	216	216	121	133	13	187	162	13	178	19	76
	cm	21.6	21.6	12.1	13.3	1.3	18.7	16.2	1.3	17.8	1.9	7.6

12 Strand system

Strand ins mm	Dims	A	B	C	D	E	F	G	H	I	J	Sheath Int. Dia.
0.5	ins	$8\frac{1}{2}$	$8\frac{1}{2}$	$5\frac{1}{8}$	$4\frac{1}{4}$	$\frac{1}{2}$	$7\frac{3}{8}$	$6\frac{3}{8}$	$\frac{1}{2}$	7	$\frac{3}{4}$	3
13	mm	216	216	130	108	13	187	162	13	178	19	76
	cm	21.6	21.6	13.0	10.8	1.3	18.7	16.2	1.3	17.8	1.9	7.6
0.6	ins	10	10	$5\frac{5}{8}$	5	$\frac{5}{8}$	$7\frac{1}{2}$	$6\frac{1}{2}$	$\frac{5}{8}$	8	1	$3\frac{1}{2}$
15	mm	254	254	143	127		191	165	16	203	26	89
	cm	25.4	25.4	14.3	12.7	1.6	19.1	16.5	1.6	20.5	2.6	8.9
0.7	ins	11	11	$6\frac{1}{2}$	6	$\frac{3}{4}$	$9\frac{1}{4}$	$8\frac{1}{4}$	$\frac{3}{4}$	9	1	4
18	mm	280	280	165	152	19	235	209	19	228	26	102
	cm	28.0	28.0	16.5	15.2	1.9	23.5	20.9	1.9	22.8	2.6	10.2

13 Strand system

Strand ins mm	Dims	A	B	C	D	E	F	G	H	I	J	Sheath Int. Dia.
0.6	ins	11	11	$6\frac{1}{2}$	5	$\frac{3}{4}$	$9\frac{1}{4}$	$8\frac{1}{4}$	$\frac{3}{4}$	9	1	$3\frac{1}{2}$
15	mm	280	280	165	127	19	235	209	19	228	26	89
	cm	28.0	28.0	16.5	12.7	1.9	23.5	20.9	1.9	22.8	2.6	8.9

19 Strand system

Strand ins mm	Dims	A	B	C_1	C_2	D	E	F	G	H	I	J	Sheath Int. Dia.
0.6	ins	13	13	$6\frac{1}{2}$	$7\frac{1}{2}$	6	1	$12\frac{3}{4}$	$8\frac{3}{4}$	1	11	$1\frac{3}{4}$	$3\frac{3}{4}$
15	mm	330	330	165	191	152	26	324	222	26	279	44	95
	cm	33.0	33.0	16.5	19.1	15.2	2.6	32.4	22.2	2.6	27.9	4.4	9.5
0.7	ins	$14\frac{1}{2}$	$14\frac{1}{2}$	$7\frac{3}{8}$	$8\frac{5}{8}$	$6\frac{1}{2}$	1	$14\frac{5}{8}$	$10\frac{5}{8}$	1	12	2	$4\frac{1}{8}$
18	mm	368	368	187	219	165	26	371	27.0	26	305	51	105
	cm	36.8	36.8	18.7	21.9	16.5	2.6	37.1	27.0	2.6	30.5	5.1	10.5

imperial metric (S.I.) continental

CHART No. 33 SCD SYSTEMS FOR CIRCULAR STRUCTURES INTERNAL TENDONS

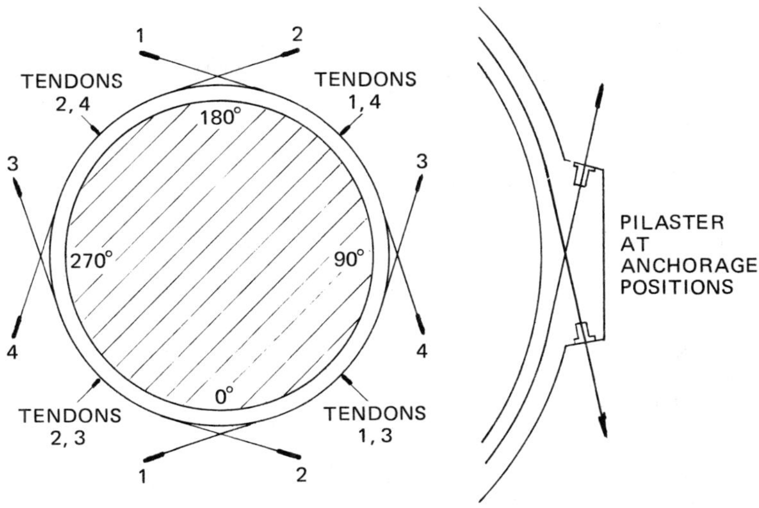

Strand dia. (mm)	Anchorage dimensions (mm)			Strand details					
				Characteristic strength (kN)			Tendon force at 70% (kN)		
	A	B	C	Normal	Super	Dyform	Normal	Super	Dyform
13	89	140	98	165	184	209	115	128	146
15	89	140	98	227	250	300	158	175	210
18	102	140	98	–	–	380	–	–	266

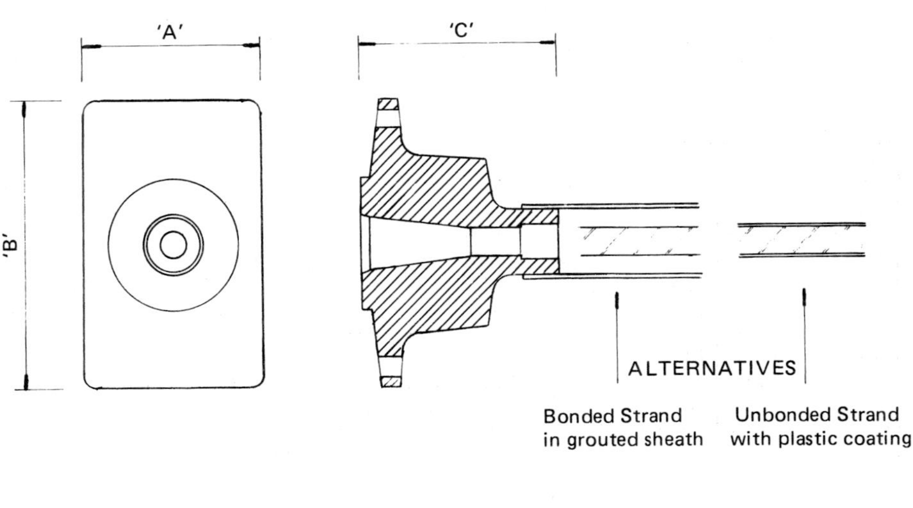

SCD SYSTEM (CONTINUED)
CHART No. 34 SCD SYSTEMS FOR CIRCULAR STRUCTURES
EXTERNAL TENDONS

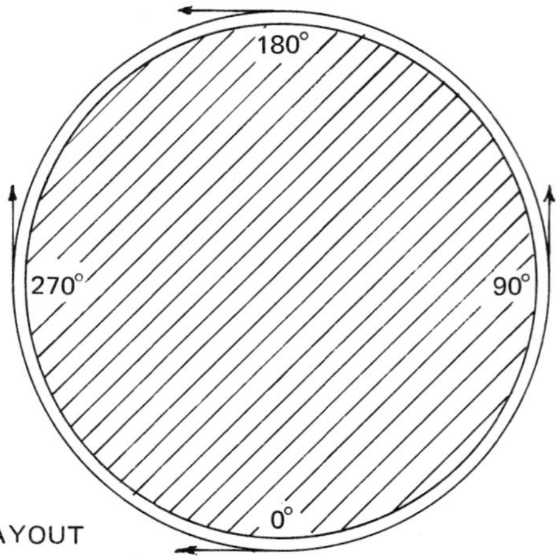

TYPICAL STRESSING LAYOUT

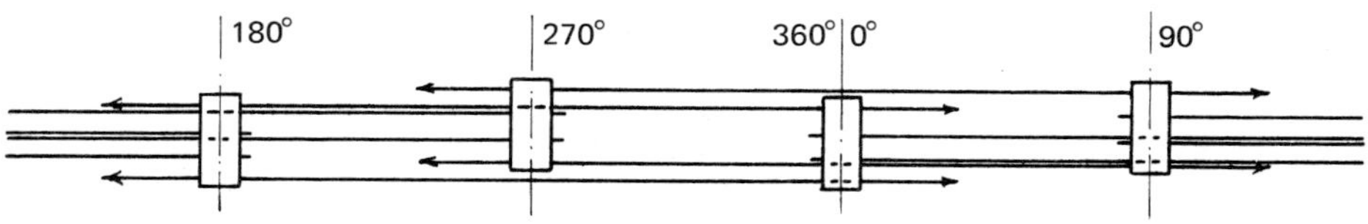

DEVELOPED TENDON

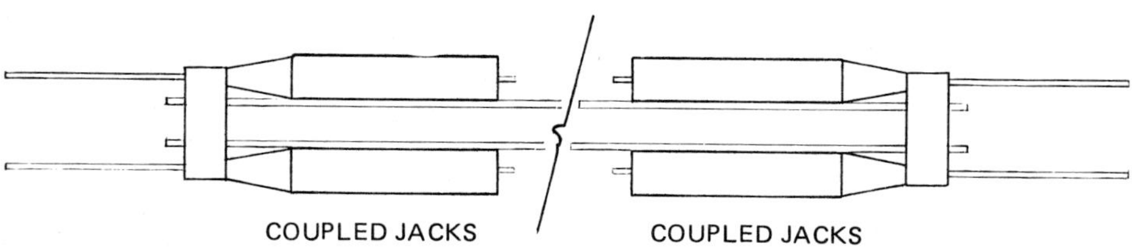

JACKING SYSTEM

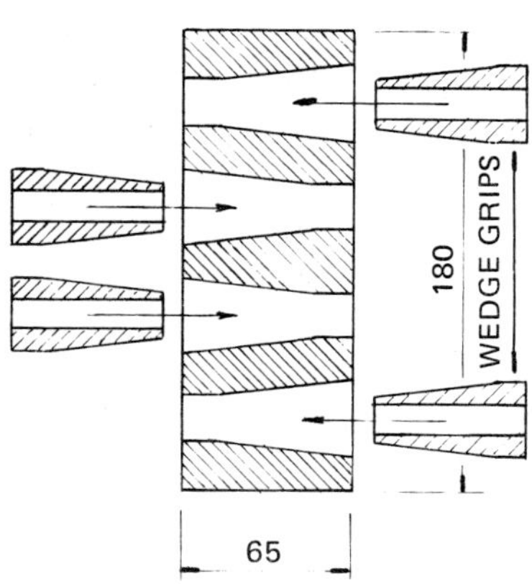

DOUBLE OPPOSED ANCHOR BLOCK

Tendon sizes	Characteristic strength (kN)			Tendon force @ 70%		
	N	S	D	N	S	D
2/13	330	368	418	231	257	292
4/13	660	736	836	462	515	562

Tendons are supported at designed positions by special hangers.
Skid bars fixed between tendon and wall to reduce friction.
Cable protection by gunite or concrete.

CHART No. 35 TENDON CHARACTERISTICS

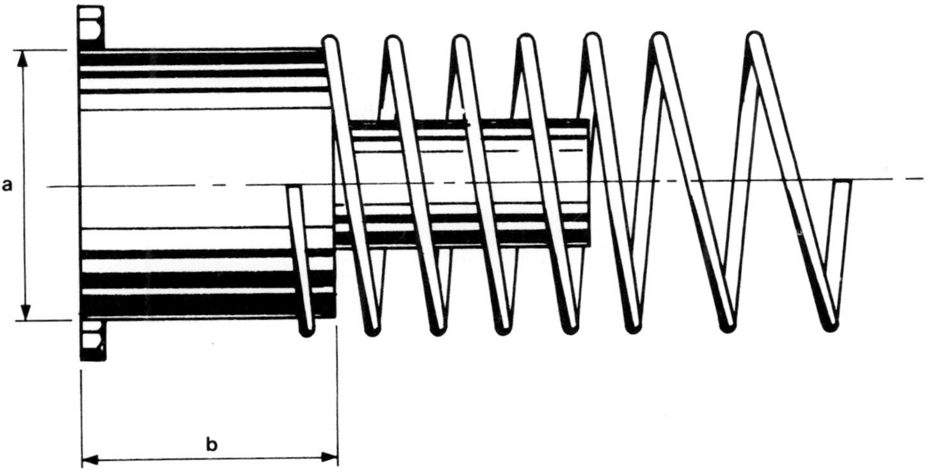

Strand dia,	No. of strand	A_s	Jacking force 270 kf strand	Socket size	Duct dia.	Jack 110 T	Jack 200 T
$\frac{3}{8}$	3	.25	55 kf	170 K	2 in.	•	
	6	.51	110 kf	170 K	2 in.	•	
	8	.68	147 kf	170 K	2 in.	•	
	10	.85	183 kf	170 K	2 in.	•	
	12	1.02	220 kf	170 K	2 in.	•	
$\frac{7}{16}$	3	.34	74 kf	170 K	2 in.	•	
	*6	.69	149 kf	170 K	2 in.	•	
	*8	.92	199 kf	170 K	2 in.	•	
	10	1.15	249 kf	230 K	$2\frac{1}{2}$ in.		•
	*12	1.38	298 kf	230 k	$2\frac{1}{2}$ in.		•
$\frac{1}{2}$	3	.46	99 kf	170 K	2 in.	•	
	*6	.92	198 kf	170 K	2 in.	•	
	*8	1.22	266 kf	230 K	$2\frac{1}{2}$ in.		•
	10	1.53	331 kf	300 K	$2\frac{1}{2}$ in.		•
	*12	1.84	397 kf	300 K	$2\frac{1}{2}$ in.		•

*Recommended as the most economical combinations.

kf = kip force

CHART No. 35 (CONTINUED)

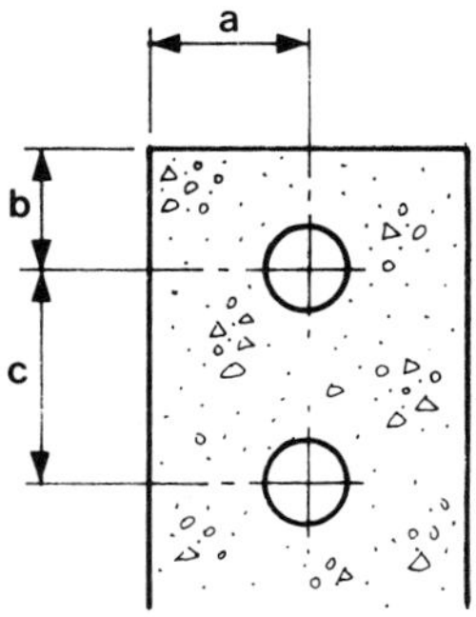

	170 kf	230 kf	300 kf
a	3″	4″	6″
b	4″	4″	6″
c	8″	8″	10″
d	8″	8″	10″

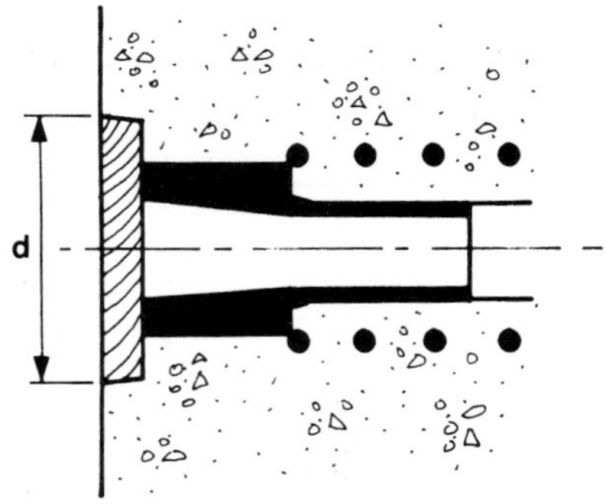

	170 kf	230 kf	300 kf
DIMENSIONS a/b	$3\frac{5}{8}$ in./4 in.	$5\frac{1}{4}$ in./4 in.	$6\frac{1}{4}$ in./5 in.
SPIRAL	12 turns of $\frac{1}{4}$ in. ϕ bar	12 turns of $\frac{1}{4}$ in. ϕ bar	12 turns of $\frac{3}{8}$ in. ϕ bar
WEIGHT	9.5 lbs.	24.5 lbs.	45 lbs.
MAX. DESIGN LOAD	170 kf	230 kf	300 kf

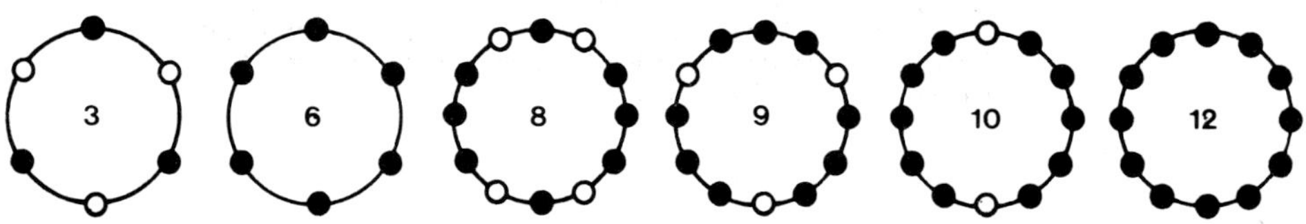

Arrangement of bearing plates

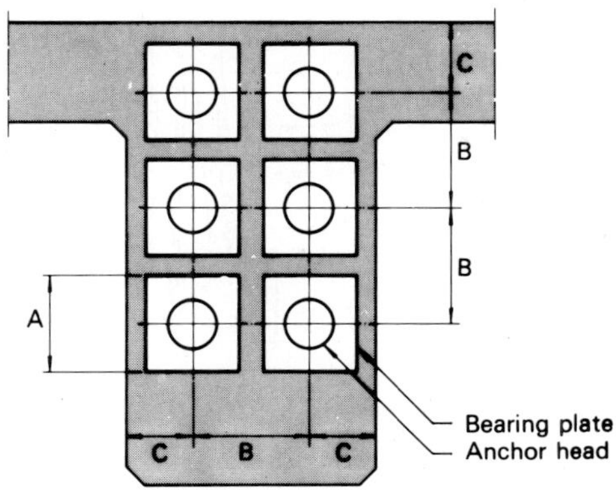

B: spiral diameter + 20 mm

C: ½ spiral diameter + cover according to the standards (min. 20 mm)

Note: With concrete of more than 35 N/mm^2 cube strength the dimensions of the bearing plates and spirals can be reduced. For detailed information please contact the local VSL licensee or Losinger Ltd., VSL International, Berne, Switzerland.

Block-outs and space requirements for jacks

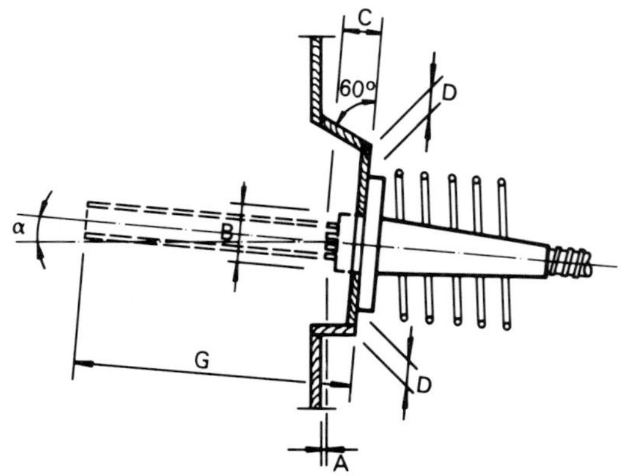

Jack type	Dmin	E	F	G	H
ZPE-20FJ	—	90	112	300	1200
ZPE-30	30	100	140	600	1100
ZPE-60	30	140	180	650	1100
ZPE-7A	30	200	280	800	1300
ZPE-12/St 2	50	200	310	670	1300
ZPE-19	50	250	390	850	1500
ZPE-460/31	60	300	485	700	1500
ZPE-500	80	330	550	1150	2000
ZPE-1000	80	450	790	1300	2200
ZPE-1400	100	500	850	1500	2400

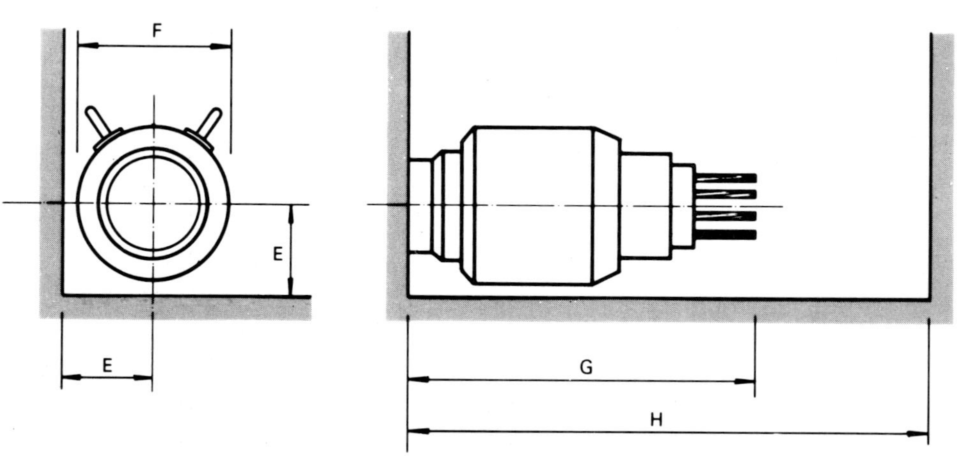

Type E

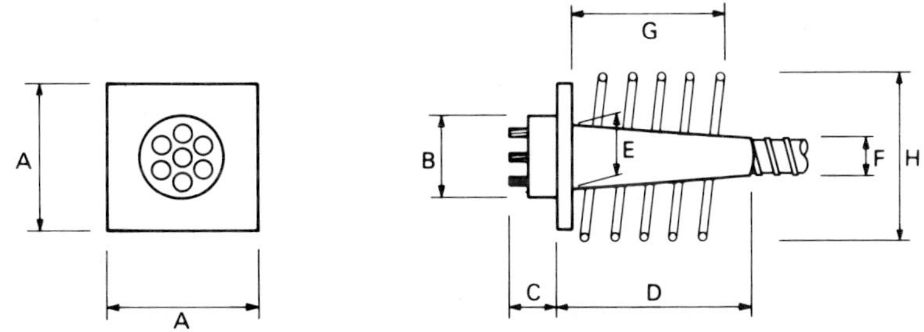

Type E	5-1		5-3		5-4		5-7		5-12		5-19		5-22		5-31		5-37		5-42		5-55	
Fptk ≤*	186	209	186	209	186	209	186	209	186	209	186	209	186	209	186	209	186	209	186	209	186	209
A	70	75	120	130	140	150	180	200	250	280	300	350	330	370	400	420	420	480	450	500	520	575
B ø	42	45	80	80	85	85	110	110	150	155	180	200	200	210	230	250	250	270	290	320	320	350
C	75	75	90	90	90	90	90	90	90	100	105	115	115	130	130	145	145	160	160	175	180	195
D	80	82	205	210	210	215	215	225	275	280	445	450	530	540	610	615	555	565	600	610	755	765
E ø	35	35	50	50	55	55	74	74	104	104	135	135	150	150	172	172	188	188	201	201	230	230
F ø	35	35	50	50	55	55	62	62	77	77	92	92	97	97	112	112	132	132	142	142	155	155
G	90	100	135	150	150	165	200	240	275	300	330	390	360	390	420	520	455	560	490	560	560	640
H ø	80	90	145	165	170	195	230	265	310	350	395	450	430	470	500	565	550	620	590	660	680	760

Type E	6-1		6-2		6-3		6-4		6-7		6-12		6-19		6-31		6-37		6-42		6-55	
Fptk ≤*	272	308	272	308	272	308	272	308	272	308	272	308	272	308	272	308	272	308	272	308	272	308
A	80	90	120	130	150	160	170	180	220	250	300	320	370	400	470	500	520	550	550	590	620	670
B ø	53	53	80	85	85	95	110	110	132	132	170	175	220	230	270	280	300	310	320	340	360	390
C	80	80	90	90	90	90	90	90	100	100	110	117	130	140	160	175	180	195	180	195	220	235
D	80	85	205	205	215	215	215	215	225	230	415	420	585	590	765	770	910	920	975	980	1050	1050
E ø	35	35	50	50	56	56	65	65	84	84	118	118	150	150	192	192	215	215	232	232	255	255
F ø	35	35	50	50	56	56	55	55	72	72	92	92	112	112	142	142	155	155	165	165	185	185
G	100	100	150	150	150	180	200	240	250	300	330	360	420	480	520	585	585	650	650	660	720	750
H ø	100	105	145	160	185	200	215	230	290	315	380	410	480	520	620	670	680	740	730	790	840	890

*F_{ptk}: breaking load of strand in kN

Type EC

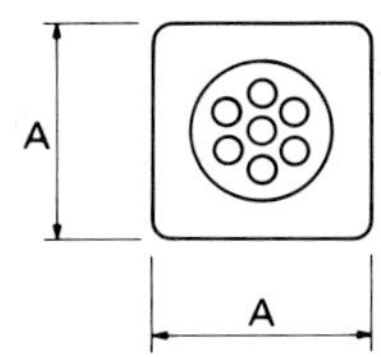

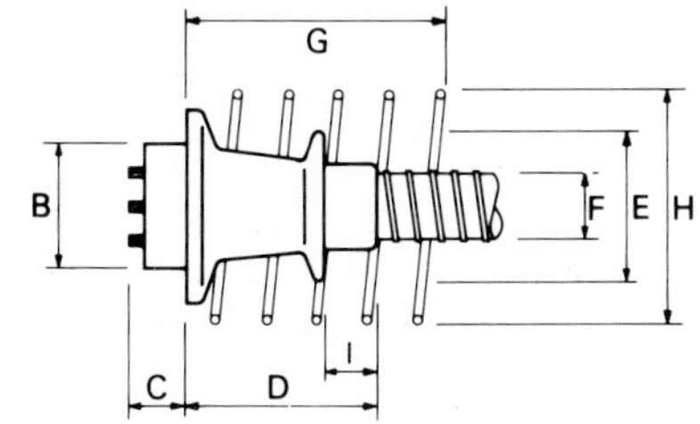

Type EC	5-4		5-7		5-12		5-19		5-22		5-31		5-37		5-42		5-55
Fptk ≦ *	186	209	186	209	186	209	186	209	186	209	186	209	186	209	186	209	209
A	135	135	165	175	215	225	270	280	290	305	340	355	370	385	395	415	460
D	125	125	155	155	215	215	285	285	335	335	365	365	360	360	380	380	460
E ø	100	100	125	130	160	170	200	210	220	230	255	265	275	285	295	310	350
F ø	48	48	60	60	75	75	90	90	95	95	110	110	130	130	140	140	153
G	145	210	210	230	285	345	345	380	380	440	440	500	475	540	540	605	650
H ø	160	160	200	220	270	290	350	370	370	400	440	470	480	510	510	550	600
I	50	50	55	55	55	55	55	55	60	60	65	65	75	75	75	75	80

Type EC	6-4		6-7		6-12		6-19		6-31		6-37	
Fptk ≦ *	272	308	272	308	272	308	272	308	272	308	272	308
A	150	155	190	205	250	265	310	330	390	420	430	460
D	155	155	170	170	245	245	305	305	350	350	450	450
E ø	115	120	145	155	190	200	235	250	295	320	320	350
F ø	53	53	70	70	90	90	110	110	140	140	153	153
G	210	210	260	285	345	345	440	440	540	605	605	650
H ø	200	200	250	270	330	350	400	440	510	550	550	600
I	60	60	55	55	60	60	65	65	70	70	80	80

*F_{ptk}: breaking load of strand in kN

B and C as for type E

CHART No. 38 DEAD END ANCHORAGE DETAILS

Type H

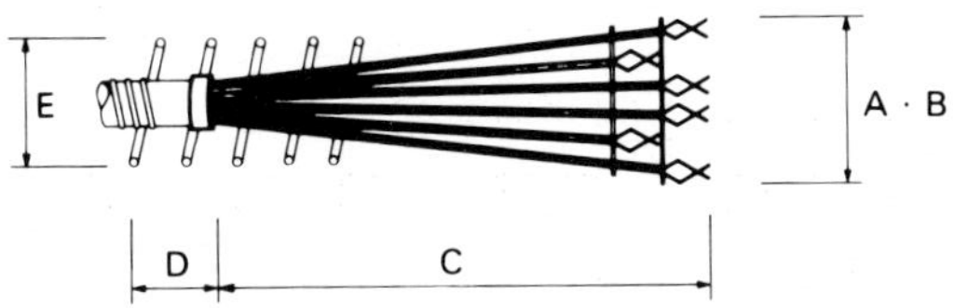

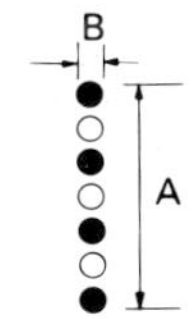

Type I

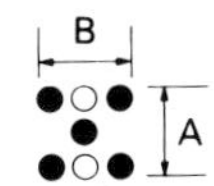

Type II

Type H	5-3	5-4		5-7		5-12		5-19		5-22		5-31		5-37		5-42		5-55	
Type	I	I	II	I	II	I	II	I	II	I	II	I	II	I	II	I	II	I	II
A	230	310	150	370	170	350	310	470	310	570	390	670	470	870	570	870	570	1170	670
B	70	70	170	70	190	190	270	190	390	190	390	220	430	310	430	350	550	350	550
C	930	930	930	1280	1280	1280	1130	1280	1280	1280	1280	1280	1280	1680	1680	1680	1680	1980	1980
D	145	145	145	155	155	155	155	155	155	155	155	155	155	165	165	165	165	185	185
E ø	—	—	—	180	180	200	200	230	230	300	300	350	350	350	350	400	400	400	400

Type H	6-3	6-4		6-7		6-12		6-19		6-31		6-37	
Type	I	I	II	I	II	I	II	I	II	I	II	I	II
A	290	390	190	450	210	430	390	570	390	810	570	1050	690
B	90	90	210	90	230	230	330	230	470	260	510	370	510
C	950	950	950	1300	1300	1300	1150	1300	1300	1700	1700	2000	2000
D	145	145	145	155	155	155	155	155	155	165	165	185	185
E ø	—	—	—	200	200	230	230	300	300	400	400	400	400

The arrangement of the bulbs (type I or II) is chosen according to the space available in the structure.

Type U

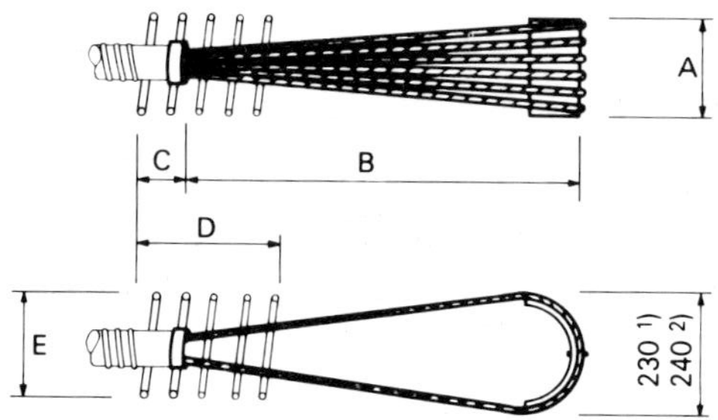

Type U	5-1	5-3	5-7	5-12	5-19	5-22	5-31
A	15	70	160	280	400	440	640
B	600	600	600	700	800	900	1300
C	—	78	112	114	126	126	126
D	—	188	262	264	316	316	316
E ø	—	130	190	220	260	260	260

Type U	6-1	6-2	6-3	6-4	6-7	6-12	6-19
A	35	75	145	145	255	400	690
B	600	600	600	600	700	900	1450
C	—	78	112	112	114	126	126
D	—	188	262	262	264	316	316
E ø	—	130	190	190	220	260	260

1) for units with 13 mm (0.5'') strands
2) for units with 15 mm (0.6'') strands

Type P

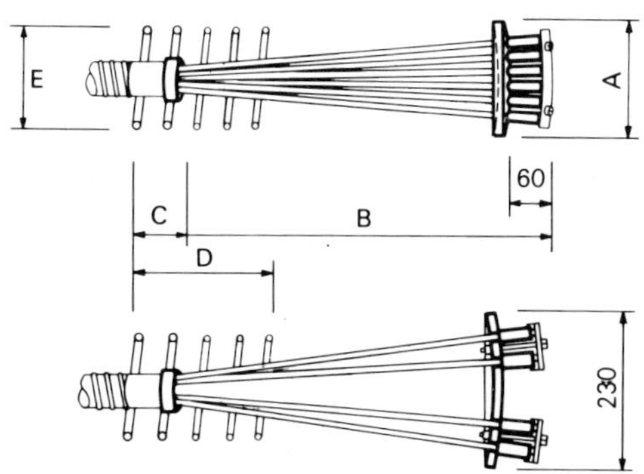

Type P	5-3	5-7	5-12	5-19	5-22	5-31
A	70	140	210	350	420	560
B	300	350	400	500	600	800
C	85	110	120	135	135	135
D	188	262	264	316	316	316
Eø	130	190	220	260	260	260

Type P	6-2	6-3	6-4	6-7	6-12	6-19
A	60	100	100	185	280	470
B	250	300	350	400	500	700
C	85	110	110	120	135	135
D	188	262	262	264	316	316
Eø	130	190	190	220	260	260

Couplers

Type K

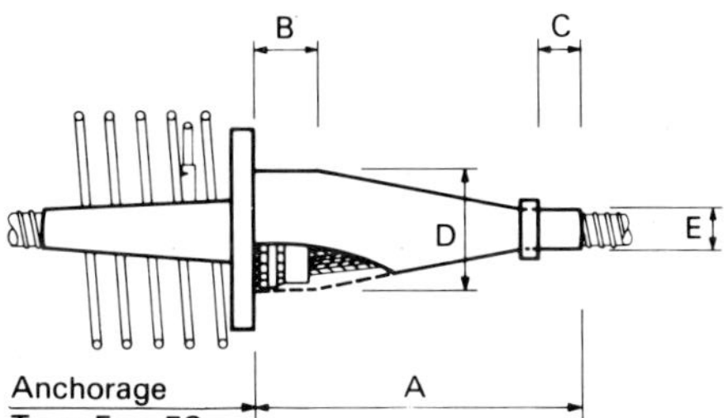

Type K	5-3	5-7	5-12	5-19	5-22	5-31	5-37	5-42
A	390	500	570	670	750	1100	1200	1180
B	140	140	140	140	140	140	180	180
C	30	60	60	80	90	90	120	130
D ø	130	170	200	240	260	350	390	395
E ø	45	62	77	92	97	112	132	142

Type K	6-2	6-3	6-4	6-7	6-12	6-19	6-31	6-37
A	350	470	490	560	660	750	1000	1280
B	150	160	160	160	160	160	180	200
C	30	60	60	70	80	90	130	130
D ø	140	150	160	190	240	280	360	430
E ø	47	50	55	72	92	112	142	155

Type V

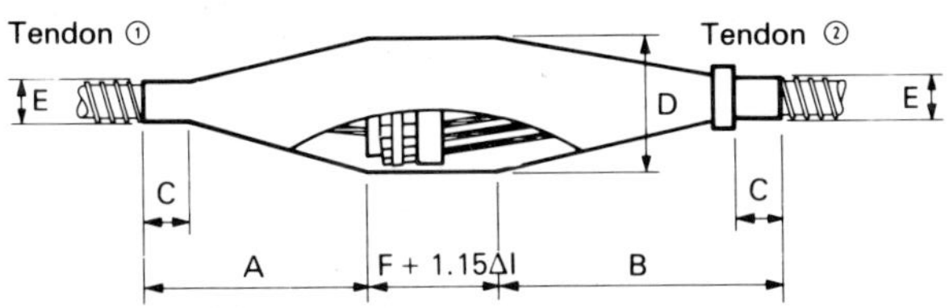

Δl = Tendon elongation

Type V	5-3	5-7	5-12	5-19	5-22	5-31
A	250	290	350	540	640	710
B	270	380	450	555	640	990
D	140	180	210	250	270	360
F	205	205	205	205	205	205

Type V	6-2	6-3	6-4	6-7	6-12	6-19
A	240	280	280	305	510	690
B	220	330	350	420	525	620
D	150	160	170	200	250	290
F	215	225	225	225	225	225

C and E as for type K

Centre stressing anchorages

Type Z

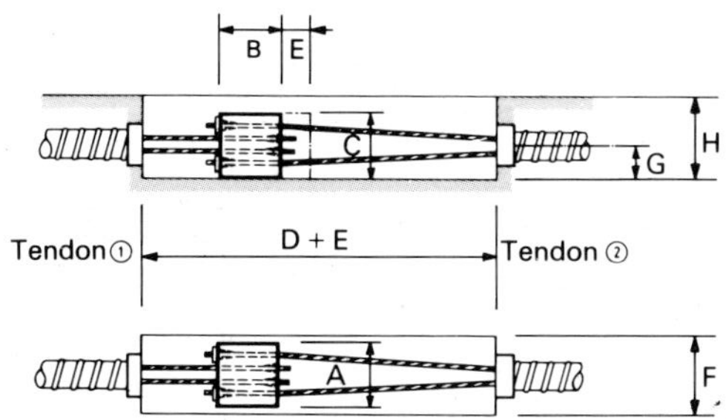

Type Z	5-2	5-4	5-6	5-12	6-2	6-4	6-6	6-12
A	130	160	200	280	140	170	210	300
B	60	70	90	140	70	80	100	160
C	80	90	130	140	90	100	140	160
D[1]	560	720	890	1440	620	1130	1320	1910
E	Elongation tendon 2				Elongation tendon 2			
F	170	200	240	320	180	210	250	340
G	60	65	85	90	65	70	90	100
H[1]	140	150	190	200	150	160	200	220

[1]) Dimensions for curved concrete surfaces upon request

Type ZU

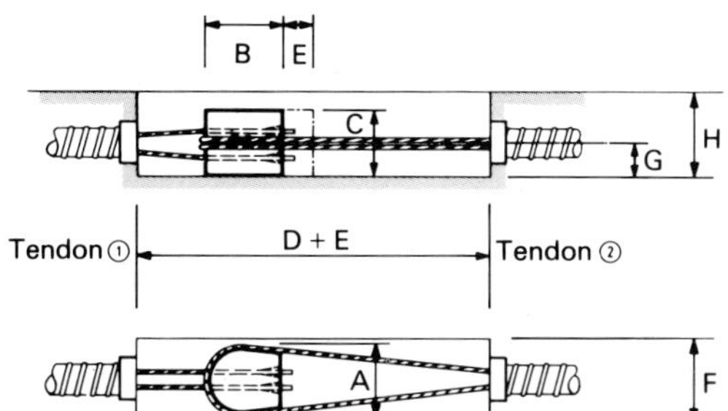

Type ZU	5-2	5-4	5-6	6-2	6-4	6-6
A	165	165	185	170	170	200
B	103	103	113	105	105	120
C	80	88	95	90	92	105
D[1]	655	835	925	700	815	930
E	Elongation tendon 2			Elongation tendon 2		
F	200	200	220	210	220	240
G	60	65	70	65	65	75
H[1]	140	150	155	145	150	170

[1]) Dimensions for curved concrete surfaces upon request

CHART No. 40 TYPES OF ANCHORAGES AND DETAILS OF TYPE A41

Type of anchorage	Number and diameter of wires	Allowable force (kN)
AK 10	1/12.2 mm diameter	101
AK 41	4/12.2 mm diameter	404
AK 124	12/12.2 mm diameter	1212
AK 163	12/14 mm diameter	1595

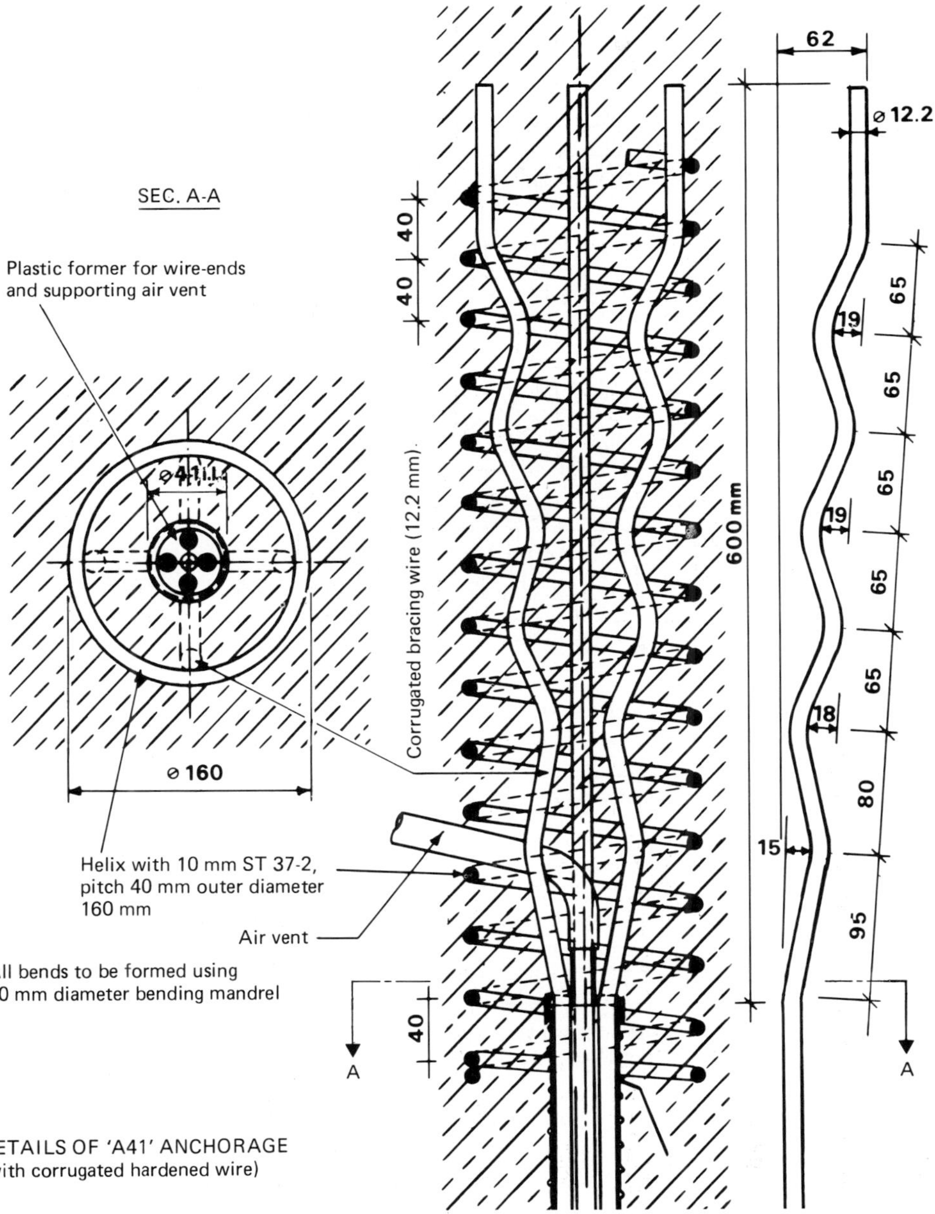

DETAILS OF 'A41' ANCHORAGE
(with corrugated hardened wire)

LEOBA SYSTEM (CONTINUED)
CHART No. 41 AK 124 ANCHORAGE (AK 163 SIMILAR)

A. DETAILS OF EXTERNALLY APPLIED AK 124 ANCHORAGE

ASSEMBLY FOR CASTING

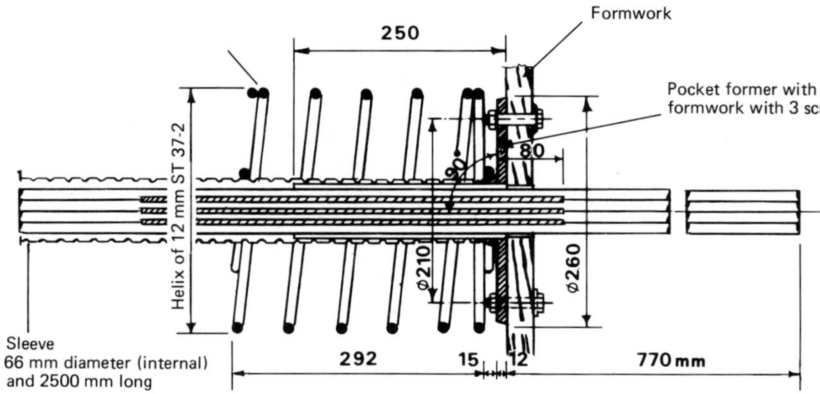

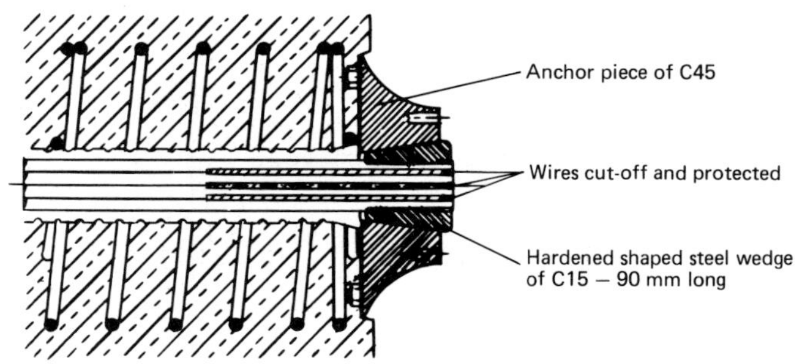

ANCHORAGE AFTER COMPLETION OF STRESSING

Length of wire (cm)	Length of stressing anchor end with one end stressing* in M.
10	for anchor end
35	0 – 25
50	25 – 60
65	60 – 95
80	95 – 130

*With stressing at both ends the lengths should be doubled

B. DETAILS OF CAST-IN AK 124 ANCHORAGE

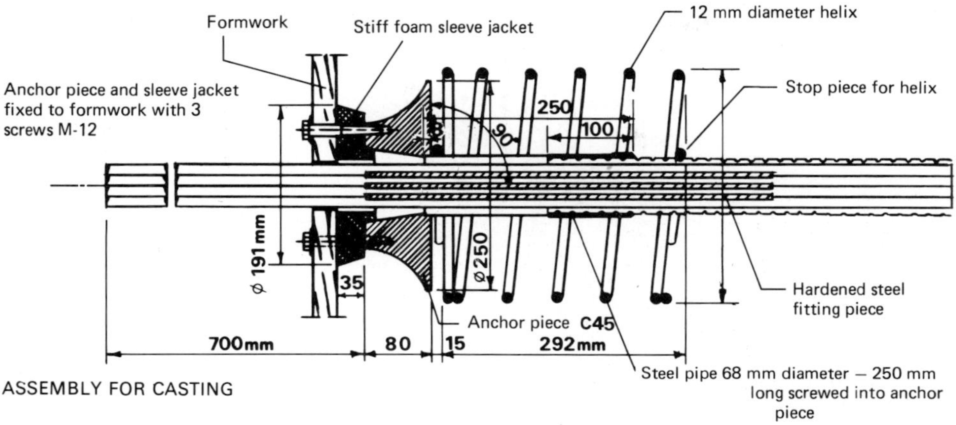

ASSEMBLY FOR CASTING

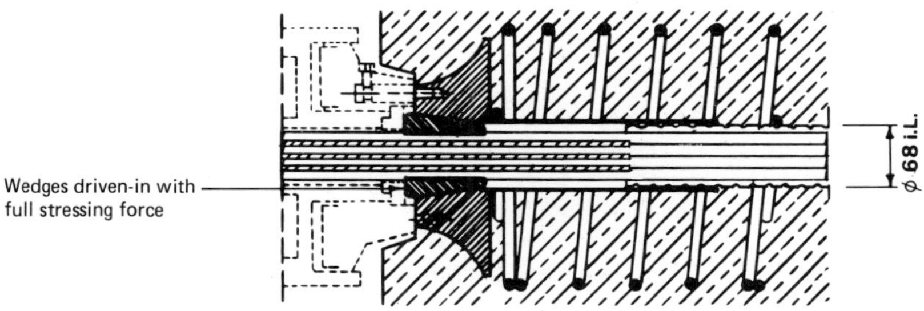

STRESSING OF TENDONS

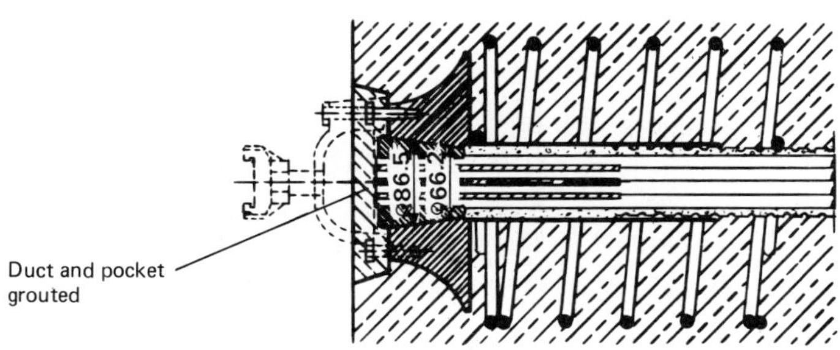

GROUTING AFTER STRESSING

PZ SYSTEM
CHART No. 42

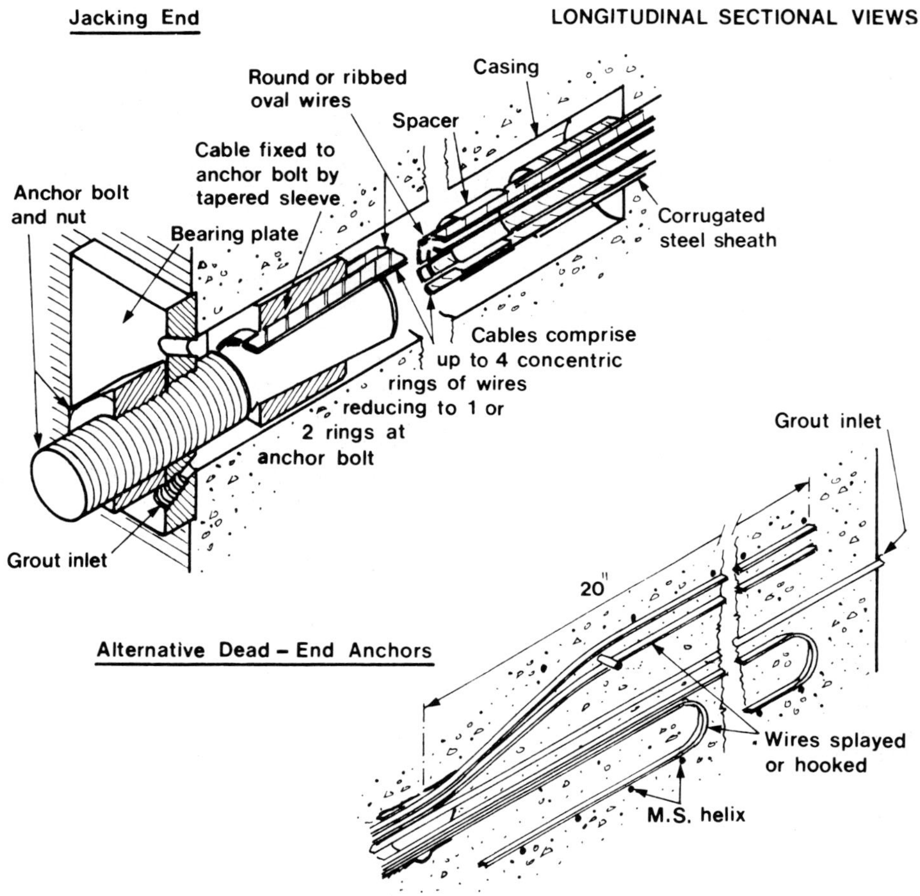

Details of Cables — Oval Wires

Dia. of Duct (inches):	0.91	1.20	1.66	2.17	2.37
Size of Wire (inches):	0.315 × 0.118		0.354 × 0.166		
No. of Wires: No. of Rings: Nos of Wires in Rings:	6 1 6	13 2 8, 5	16 2 10, 6	32 3 14, 11, 7	44 4 17, 13, 9, 5

Details of Cables — Circular Wires

Dia. of Duct (inches):	0.91	1.20		1.66		2.17			2.37	
Size of Wire (inches):	0.196		0.237	0.196	0.237	0.196	0.237	0.276	0.196	0.237
No. of Wires: No. of Rings: Nos of Wires in Rings:	6 1 6	13 1 13	9 1 9	24 2 15, 9	16 2 10, 6	50 2 31, 19	32 2 20, 12	24 2 15, 9	60 3 33, 18, 9	46 3 23, 15, 8

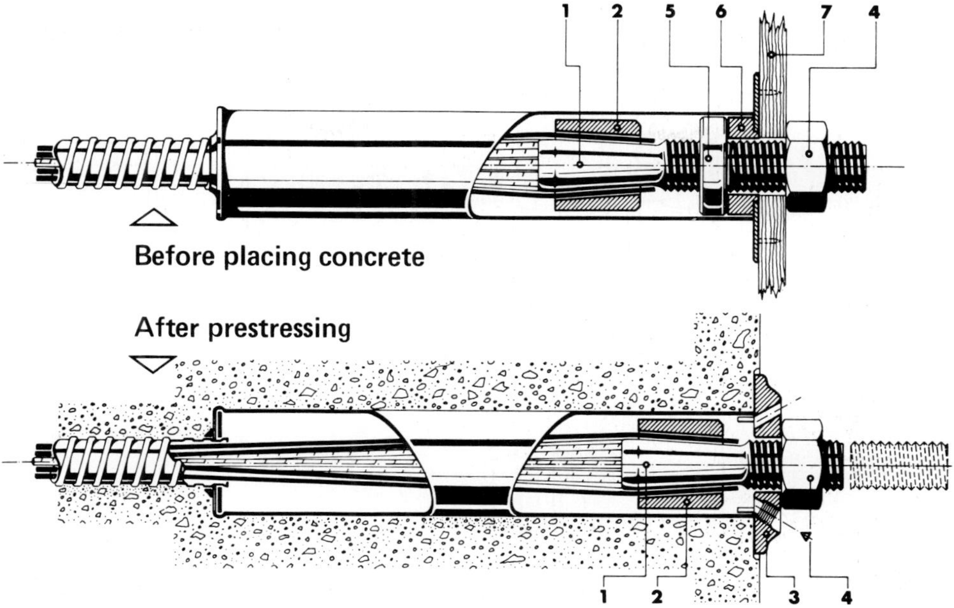

1 Anchor bolt with cone
2 Conical anchor jacket
3 Anchor plate
4 Anchor nut
5 Round centering nut
6 Temporary erection plate
7 Shuttering

Three to sixty prestressing wires are combined with a prestressing bar to anchorage

THE MACALLOY SYSTEM

CHART No. 43 DIAGRAMATIC VIEW OF MACALLOY THREADED ANCHORAGE

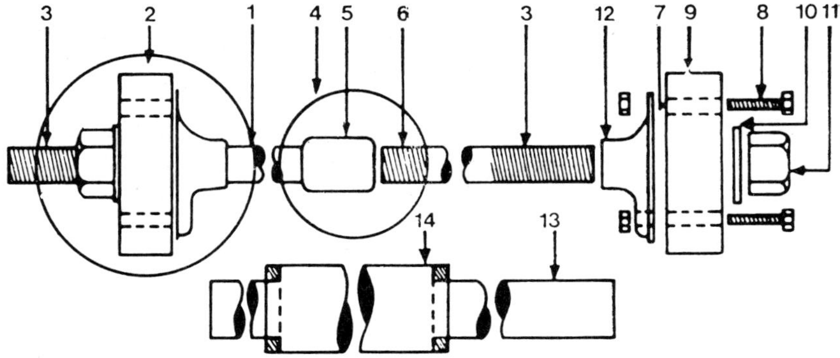

THREADED ANCHORAGE

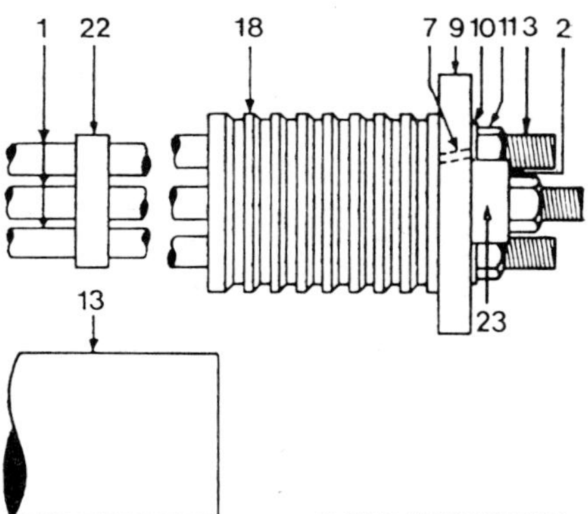

3-BAR ANCHORAGE

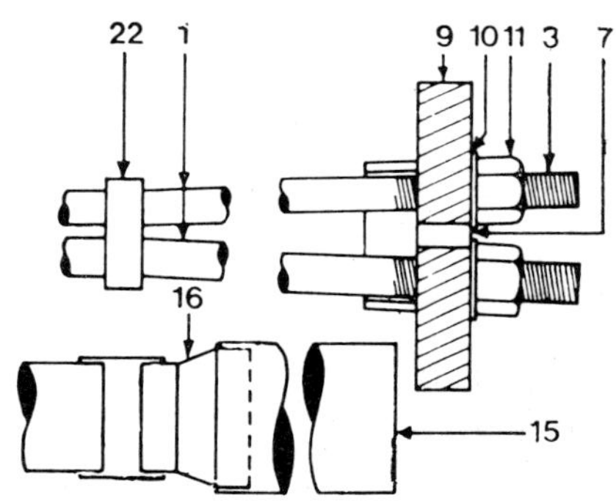

4-BAR ANCHORAGE

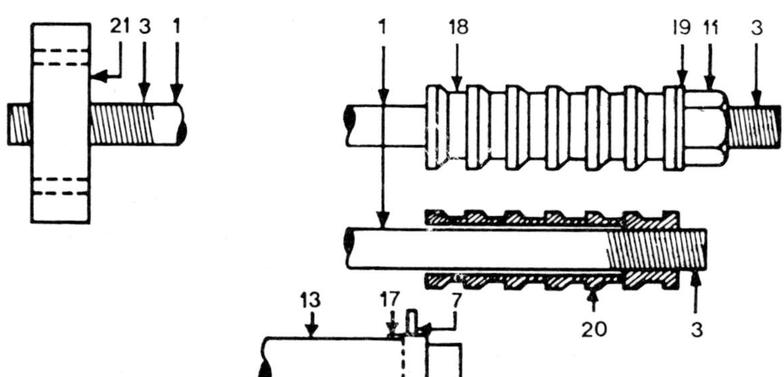

TUBULAR SLEEVE ANCHORAGES

1 Bar
2 Threaded anchorage
3 Jacking or anchor thread
4 Coupled joint
5 Coupler
6 Coupler thread
7 Grouting hole
8 Nuts and bolts

9 End plate
10 Washer
11 Nut
12 Grouting flange
13 Sheathing for bar
14 Sheathing at coupler
15 4-bar end sheathing
16 4-bar sheathing reducer

17 Sheath connection for sleeve
18 Sleeve anchor
19 Sleeve washer
20 Tapped sleeve anchor
21 Threaded end plate
22 3- or 4-bar spacer
23 Bridge for 3-bar anchorage

CHART No. 44 TABLE FOR BARS AND FITTINGS

	ITEM	UNIT	BAR DIAMETER mm 20	25	32	40	4×32	3×40	4×40
BARS	Sectional area	mm^2	314·2	490·9	804·3	1256·6	3217	3770	5026
	Weight per metre	kg	2·637	4·043	6·513	10·298	26·05	30·89	41·19
	Metre run of bar per tonne	m	379	247	154	97	38	32	24
	Characteristic load	kN	325	505	830	1300	3320	3900	5200
	Prestress at 70% characteristic	kN	228	354	581	910	2324	2730	3640
	Prestress after 15% losses	kN	193	301	494	774	1975	2320	3094
	Minimum centres of anchorages	mm	100	120	150	175	300	175	350
NUTS	Nut reference 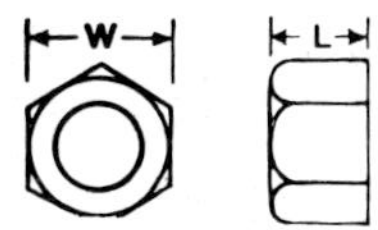		N20	N25	N32	N40	N32	N40	N40
	Length L	mm	25	32	40	50	40	50	50
	Width across flats W	mm	42	48	57	66	57	66	66
WASHERS	Washer reference 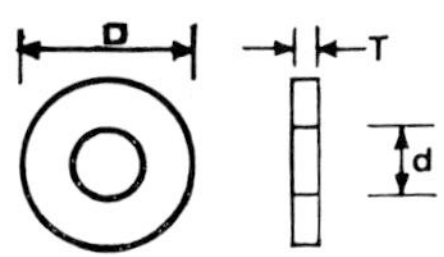		SW20	SW25	SW32	SW40	SW32	SW40	SW40
	Outside diameter D	mm	50	60	70	80	70	80	80
	Inside diameter d	mm	23	28	35	43	35	43	43
	Thickness T	mm	5	5	5	5	5	5	5
STANDARD COUPLERS	Coupler reference 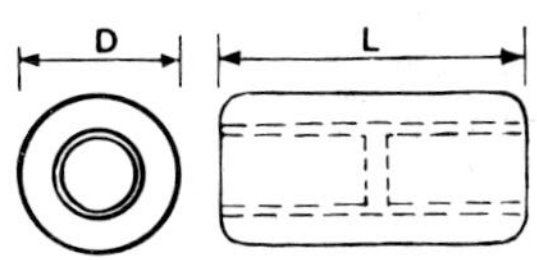		(E) C20	(E) C25	(E) C32	(E) C40	C32	C40	C40
	Length L	mm	55	75	100	125	100	125	125
	Outside diameter (standard) D	mm	35	42·5	50	60	50	60	60
	Outside diameter (economy) D	mm	40	50	60	75	—	—	—

CHART No. 44 (CONTINUED)

	ITEM	UNIT	BAR DIAMETER mm 20	25	32	40	4 × 32	3 × 40	4 × 40
COUPLERS	R & LH Round–diameter	mm	35	42·5	50	60	50	60	60
	R & LH Hexagon–across flats	mm	42	48	57	66	—	—	—
	R & LH Turnbuckle–across flats	mm	42	52	61	70	—	—	—
DUCTS	Bar duct or sheathing inside diameter	mm	40	40	42·5	50	110	160 × 65	140
	Sheathing i/d minimum	mm	27	32	39	47			
	Coupler or end sheathing i/d	mm	50	57·5	65	75	165	160 × 65	180
	Coupler sheath–minimum	mm	45	52·5	60	70			
GROUTING FLANGES	Flange reference 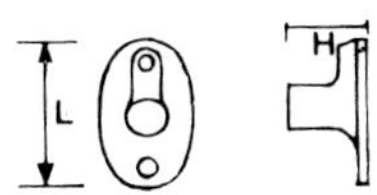		GF20	GF25	GF32	GF40			
	Length L	mm	125	125	125	140			
	Height H	mm	40	40	40	40			
THREADS	Pitch	mm	1·581	1·814	2·117	2·540	2·117	2·540	2·540
	Thread length–jacking end	mm	250	250	250	250	250	250	250
	Thread length–dead end	mm	100	100	100	100	100	100	100
	Thread length–coupler	mm	40	45	55	65	55	65	65
BAR LENGTHS	Bars jacked one end–length over plates +	mm	230	230	230	230	230	230/380	230
	Bars jacked both ends–length over plates +	mm	300	300	300	300	300	300/450	300
	Bars with tapped plate–length over plates +	mm	175	175	175	175	—	—	—

THE MACALLOY SYSTEM (CONTINUED)
CHART No. 45 TABLE FOR END PLATES

	Item	Unit	20	25	32	40	4 × 32	4 × 40	4 × 40
			Bar diameter mm						
STANDARD END PLATES	Plate reference		P20	P25	P32	P40	P432	P340	P440
	Length	mm	100	100	125	150	300	300	400
	Width	mm	100	100	125	150	300	175	350
	Thickness	mm	25	40	50	60	60	50	70
	Hole diameter	mm	26	31	38	46	4 × 35	3 × 43	4 × 43
EMBEDDED ANCHORAGES — TAPPED PLATE	Plate reference		TP20	TP25	TP32	TP40			
	Length	mm	100	100	125	150			
	Width	mm	100	100	125	150			
	Thickness	mm	25	40	50	60			
	Tapped hold diameter	mm	20	25	32	40			
EMBEDDED ANCHORAGES — SLEEVE	Sleeve reference		S20	S25	S32	S40		S340	
	Length	mm	160	200	200	220		270	
	Outside diameter	mm	50	70	70	85		211×86	
	Inside diameter	mm	30	40	40	50		175×50	
THREAD MUFFS	Muff reference		TM20	TM25	TM32	TM40	TM32	TM40	TM40
	Tread pitch	mm	1 587	1 814	2 117	2 540	2 117	2 540	2 540
	Outside diameter	mm	24.5	29.5	36	44	36	44	44
	Length	mm	50	50	50	50	50	50	50
BAR SPAC-INGS	Usual minimum spacing of bar centres at anchorages	mm	100	120	150	175	90	65	95
INTERIOR SPACERS	Spacer reference	mm					IS432	IS340	IS440
	Outside diameter	mm					110	156×63	140
	Length	mm					45	50	50

THE MACALLOY SYSTEM (CONTINUED)
CHART No. 46 TABLE FOR DETAILS OF MACALLOY JACKS

Jack Mark		7	8	10	12	13
Capacity kN		700	500	700	1000	400
Stroke mm		150	150	40	100	70
Bar Diameter mm		20 25 32	20 & 25	20 25 32	32 & 40	20 & 25
Jack Dimensions	A (max)	700	635	340	575	436
	B	100	100	100	115	60
	C	405	380	130	335	223
	D (min)	115	100	100	120	113
	E	180	170	180	145	102
	F	215	190	190	230	153
	G	930	860	460	800	580
Anchorage Recess Dimensions	H (min)	190	180	190	150	160
	J (min)	115	115	115	125	80
	K (max)	100	100	100	115	70
	L	380	360	380	175	320
	M	760	760	760	200	300
	N	95	90	95	125	80
	P	95	90	95	125	80

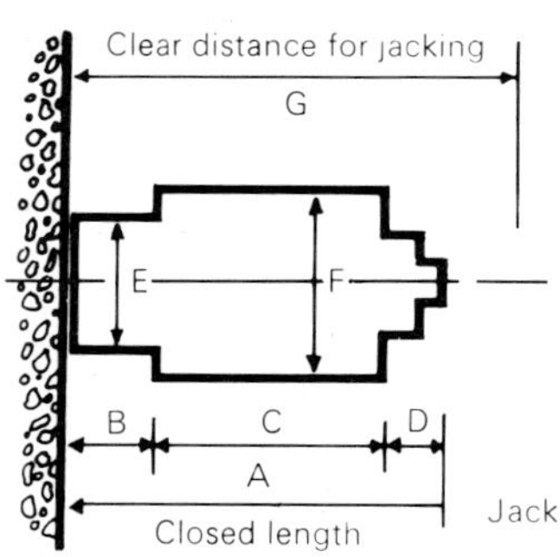

Jack

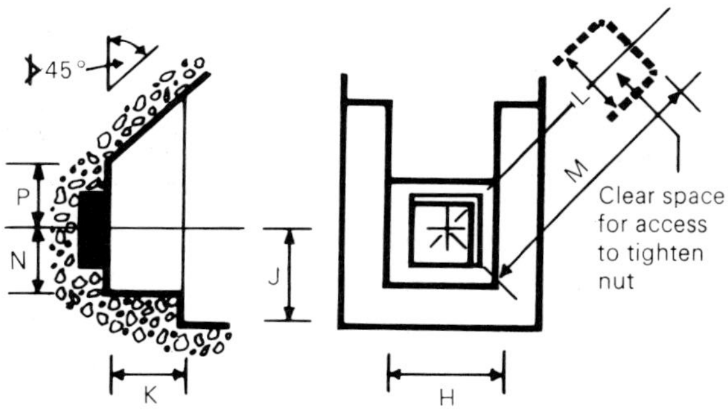

Anchorage recess

DESIGN PROPERTIES OF STRESSTEEL BARS

Nominal Bar Size ϕ in.	Ultimate Strength Guaranteed Minimum	
	Regular 145 ksi	Special 160 ksi
$\frac{3}{4}$	64	71
$\frac{7}{8}$	87	96
1	114	126
$1\frac{1}{8}$	144	159
$1\frac{1}{4}$	178	196
$1\frac{3}{8}$	215	238

† Design properties indicated are in accordance with ACI Building Code 318-63, Sections 2606 and 2607. Temporary jacking stresses up to 0.8f's are permitted to overcome losses due to tendon friction, anchorage seating and elastic shortening. Losses due to creep, shrinking and steel relaxation should be deducted from the recommended initial tensioning load to obtain actual final design load. Actual final design load, after losses are accounted for, may be less than 0.6f's.

1 kip force = 1000 lbf = 454 kgf = 4.45 kN

BAR ANCHORAGE DETAILING INFORMATION

Wedges and Nuts

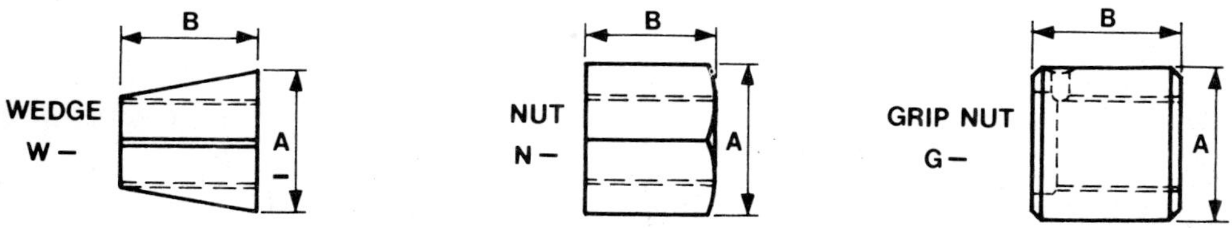

Bar ϕ in.	WEDGES Part No.	A	B	Wt. # Ea.	NUTS* Part No.	A	B	Wt. # Ea.	GRIP NUTS Part No.	A	B	Wt. # Ea.
$\frac{3}{4}$	W6	$1\frac{3}{8}$	$1\frac{1}{4}$	.2	N6	$1\frac{3}{8}$	$1\frac{1}{4}$	.5	G6	$1\frac{7}{8}$	$1\frac{3}{4}$	1.0
$\frac{7}{8}$	W7	$1\frac{3}{4}$	$1\frac{1}{2}$	.3	N7	$1\frac{5}{8}$	$1\frac{7}{16}$	.7	G7	$2\frac{1}{16}$	2	1.5
1	W8	$1\frac{3}{4}$	$1\frac{1}{2}$	.5	N8	$1\frac{7}{8}$	$1\frac{5}{8}$	1.0	G8	$2\frac{1}{4}$	$2\frac{1}{4}$	2.5
$1\frac{1}{8}$	W9	2	$1\frac{3}{4}$	.7	N9	$2\frac{1}{8}$	$1\frac{13}{16}$	1.5	G9	$2\frac{1}{2}$	$2\frac{1}{2}$	2.7
$1\frac{1}{4}$	W10	$2\frac{1}{4}$	2	.8	N10	$2\frac{1}{4}$	2	2.0	G10	$2\frac{3}{4}$	$2\frac{5}{8}$	3.2
$1\frac{3}{8}$	W11	$2\frac{3}{8}$	$2\frac{3}{16}$	1.1	N11	$2\frac{3}{8}$	2	2.0	G11	3	$2\frac{7}{8}$	4.2

All data subject to revision as new developments are made.
All dimensions in inches

*$\frac{3}{16}$ in. washer shipped with each nut.

DYWIDAG SYSTEM
CHART No. 48

THE DYWIDAG THREADBAR

The DYWIDAG Post-tensioning System uses for tendons hot rolled alloy steel bars, proofstretched up to the yield point. It is a special bar, a bar with rolled on rib deformations over its entire length, which act as threads. Thus the bar can be cut at any desired point and is ready either to screw a coupler or an anchorage on. The thread deformations also provide maximum bonding, important for ultimate load behaviour of a structural member or for anchoring through bond.

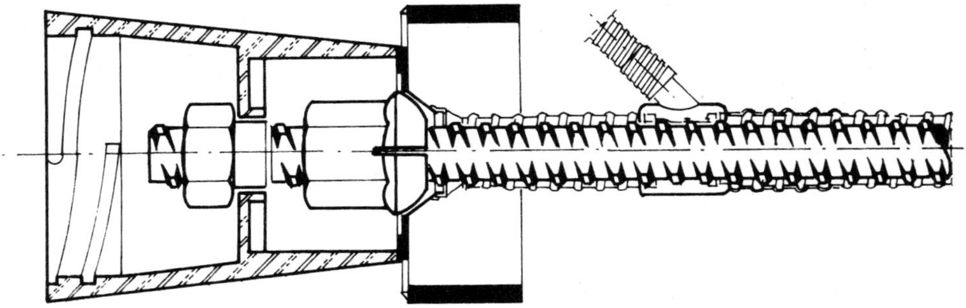

Bell Anchorage

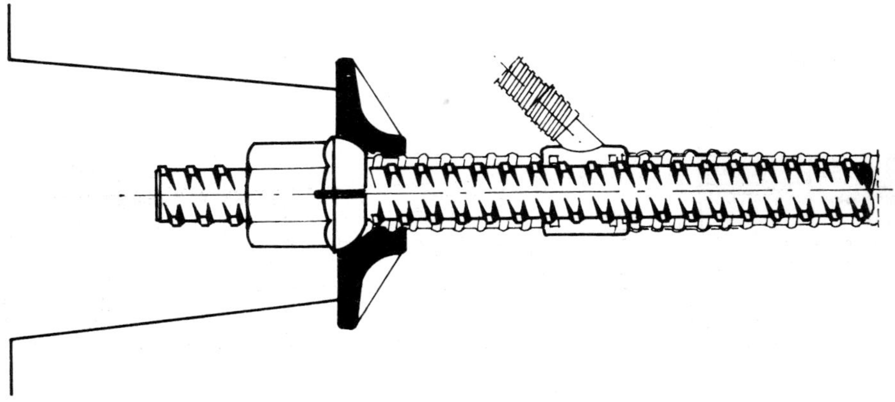

Anchorage with Ribbed Plate (square or rectangular)

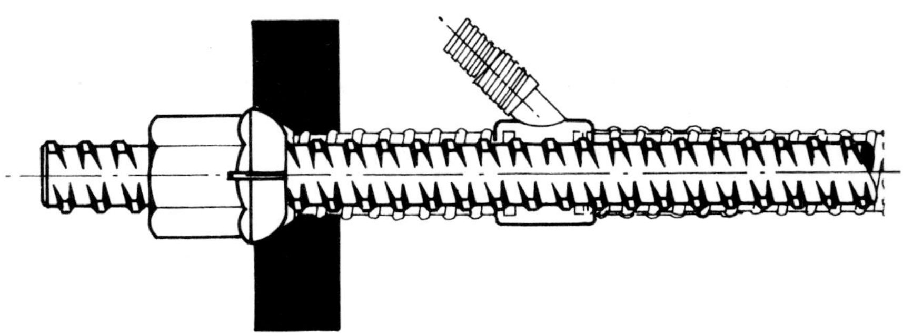

Anchorage with Solid Plate (square or rectangular)

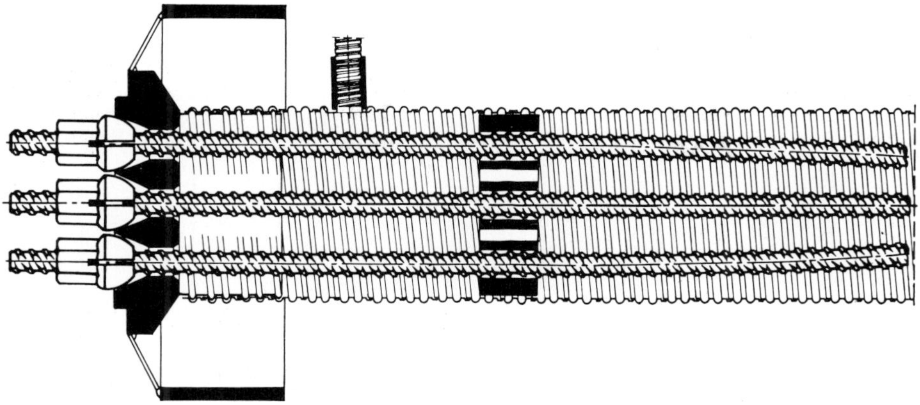

Bell Anchorage

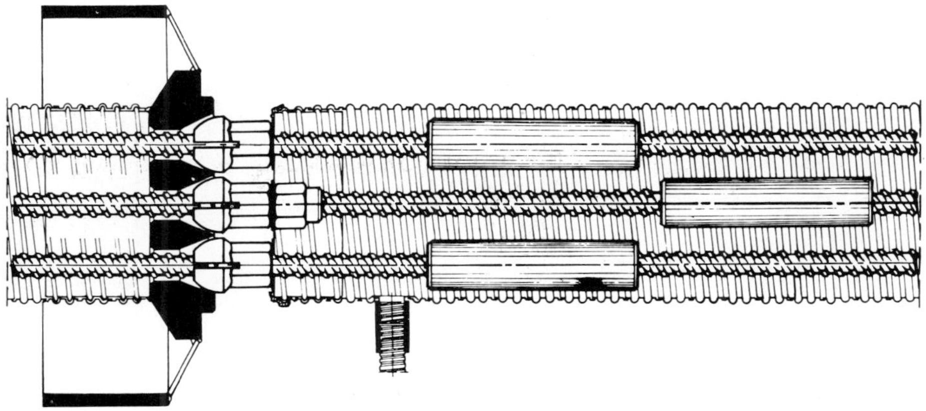

Tendon Splice

(normally only necessary to continue the tendon from a stressed anchorage)

Dywidag system of anchorages are available for diameters ranging from 5/8'' to 1-3/8'' with ultimate strength between from 155–160 kip/in² (1.07–1.1 kN/mm²)

DYWIDAG SYSTEM (CONTINUED)
CHART No. 49 DETAILS OF DYWIDAG SINGLE BAR TENDON THREAD BAR

Grade of steel			835/1030					1080/1230			
	Thread bar dia. mm →		15	16	26.5	32	36	26.5	32	36	
Bn 250	Distance between center of bells	cm	12	—	16	23	28	22	29	32	Bell anchorage
	Distance from center of bell to the edge of the concrete	cm	8	—	12	13.5	16	13	16.5	18	
Bn 350	Distance between center of bells	cm	12	—	16	22	27	21 (31)	26 (38)	29 (42)	
	Distance from center of bell to the edge of the concrete	cm	8	—	11	13	15.5	12.5 (18)	15 (21)	16.5 (23)	
Bn 450	Distance between center of bells	cm	12	14	16	21	25	20 (31)	24 (38)	27 (42)	
	Distance from center of bell to the edge of the concrete	cm	8	9	10	12.5	14.5	12 (18)	14 (21)	15.5 (23)	
Bn 450	Distance between center of plates	cm	—	—	—	—	—	y:14 x:26	y:15 x:34	y:17 x:40	QR plate anchorage (surface mountable)
	Distance from center of plate to the edge of the concrete	cm	—	—	—	—	—	y:9 x:15	y:9.5 x:19	y:10.5 x:22	
Bn 250	Distance between center of plates	cm	—	—	—	—	—	—	—	30	Plate anchorage ribbed square
	Distance from center of plate to the edge of the concrete	cm	—	—	—	—	—	—	—	17	
Bn 350	Distance between center of plates	cm	—	—	—	—	—	—	—	28	
	Distance from center of plate to the edge of the concrete	cm	—	—	—	—	—	—	—	16	
Bn 450	Distance between center of plates	cm	—	11.5	—	20	22	—	22	25	
	Distance from center of plate to the edge of the concrete	cm	—	8	—	12	13	—	13	15	

Bn 250	Distance between center of plates	cm					y:16 x:44	y:16 x:28	y:16 x:44	y:16 x:55	Plate anchorage ribbed rectangular
	Distance from center of plate to the edge of the concrete	cm	— —	— —	— —	— —	y:10(8) x:24	y:10(8) x:16	y:10(8) x:24	y:10(8) x:30	
Bn 350	Distance between center of plates	cm	— —	— —	— —	— —	y:16 x:36	y:16 x:23	y:16 x:36	y:16 x:44	
	Distance from center of plate to the edge of the concrete	cm	—	—	—	—	y:10(8) x:20	y:10(8) x:14	y:10(8) x:20	y:10(8) x:24	
Bn 450	Distance between center of plates	cm	— —	y:8 x:16	y:12 x:20	y:12 x:30	y:16 x:28	y:16 x:18	y:16 x:28	y:16 x:33	
	Distance from center of plate to the edge of the concrete	cm	— —	y:6(4) x:10	y:8(6) x:12	y:8(6) x:17	y:10(8) x:16	y:10(8) x:11	y:10(8) x:16	y:10(8) x:19	
Bn 250 LBn 250	Distance between center of plates	cm	—	—	—	24	—	—	—	—	Surface mountable plate anchorage solid square
	Distance from center of plate to the edge of the concrete	cm	—	—	—	14	—	—	—	—	
Bn 350 LBn 350	Distance between center of plates	cm	—	—	—	22	—	—	—	—	
	Distance from center of plate to the edge of the concrete	cm	—	—	—	13	—	—	—	—	
Bn 450 LBn 450	Distance between center of plates	cm	—	—	—	20	23	—	23	—	
	Distance from center of plate to the edge of the concrete	cm	—	—	—	12	13.5	—	13.5	—	
Bn 250 LBn 250	Distance between center of plates	cm	—	—	—	—	y:16 x:44	y:16 x:28	y:16 x:44	y:16 x:55	Surface mountable plate anchorage solid rectangular
	Distance from center of plate to the edge of the concrete	cm	—	—	—	—	y:10(8) x:24	y:10(8) x:16	y:10(8) x:24	y:10(8) x:30	
Bn 350 LBn 350	Distance between center of plates	cm	—	—	—	—	y:16 x:36	y:16 x:23	y:16 x:36	y:16 x:44	
	Distance from center of plate to the edge of the concrete	cm	—	—	—	—	y:10(8) x:20	y:10(8) x:14	y:10(8) x:20	y:10(8) x:24	
Bn 450 LBn 450	Distance between center of plates	cm	—	—	y:13 x:20	y:13 x:30	y:16 x:28	y:16 x:18	y:16 x:28	y:16 x:33	
	Distance from center of plate to the edge of the concrete	cm	—	—	y:8.5 x:12	y:8.5 x:17	y:10(8) x:16	y:10(8) x:11	y:10(8) x:16	y:10(8) x:19	

Nominal diameter	Yield load	Ultimate load	Anchorage available
15 mm	157 kN	191 kN	Bell anchorage
16 mm	266 kN	296 kN	Bell anchorage Plate anchorage Plate anchorage (can be surface mounted)
26.5 mm	460 kN	568 kN	Bell anchorage Plate anchorage (rect.) Plate anchorage (rect.) (can be surface mounted)
32 mm	671 kN	828 kN	Bell anchorage Plate anchorage Plate anchorage (can be surface mounted)
36 mm	850 kN	1049 kN	Bell anchorage Plate anchorage (square) Plate anchorage (rect.) (can be surface mounted)
	1099 kN	1252 kN	Bell anchorage Plate anchorage Plate anchorage (rect.) (can be surface mounted)